ARCTIC OCEAN
Barents Sea
Arctic Circle
SIBERIA
RUSSIA
URAL MTS.
See Europe inset map below
KAZAKHSTAN
MONGOLIA
GOBI (DESERT)
UZBEKISTAN
TIAN SHAN
KYRGYZSTAN
GEORGIA
ARMENIA
TURKMENISTAN
TAJIKISTAN
NORTH KOREA
SOUTH KOREA
JAPAN
TURKEY
AZER.
CHINA
SYRIA
IRAN
ATLAS MTS.
TUNISIA
LEBANON
IRAQ
AFGHANISTAN
HIMALAYAS
ISRAEL
JORDAN
BHUTAN
MOROCCO
KUWAIT
PACIFIC OCEAN
East China Sea
ALGERIA
LIBYA
BAHRAIN
PAKISTAN
NEPAL
WESTERN SAHARA (Morocco)
SAHARA
QATAR
EGYPT
Thar Desert
Arabian Peninsula
BANGLADESH
Tropic of Cancer
TAIWAN
SAHEL
UNITED ARAB EMIRATES
INDIA
South China Sea
MAURITANIA
SAUDI ARABIA
LAOS
MALI
NIGER
OMAN
Deccan Plateau
BURMA (MYANMAR)
SENEGAL
ERITREA
YEMEN
THAILAND
PHILIPPINES
THE GAMBIA
CHAD
SUDAN
DJIBOUTI
Arabian Sea
VIETNAM
BURKINA FASO
Bay of Bengal
CAMBODIA
GUINEA
TOGO
NIGERIA
Ethiopian Highlands
COTE D'IVOIRE
BENIN
CENTRAL AFRICAN REPUBLIC
SOUTH SUDAN
ETHIOPIA
SRI LANKA
BRUNEI
PALAU
MICRONESIA
GHANA
CAMEROON
MALDIVES
LIBERIA
SOMALIA
MALAYSIA
EQ. GUINEA
UGANDA
Equator
SAO TOME & PRINCIPE
GABON
RWANDA
KENYA
SINGAPORE
SIERRA LEONE
REPUBLIC OF THE CONGO
BURUNDI
INDONESIA
DEMOCRATIC REPUBLIC OF THE CONGO
PAPUA NEW GUINEA
GUINEA-BISSAU
TANZANIA
SEYCHELLES
COMOROS
ANGOLA
MALAWI
INDIAN OCEAN
TIMOR LESTE
SOLOMON ISLANDS
ZAMBIA
MADAGASCAR
NAMIBIA
ZIMBABWE
MAURITIUS
BOTSWANA
Tropic of Capricorn
MOZAMBIQUE
AUSTRALIA
Kalahari Desert
Great Victoria Desert
ATLANTIC OCEAN
SWAZILAND
SOUTH AFRICA
LESOTHO
140°E
0°
20°E
40°E
60°E
80°E
100°E
120°E
40°N
20°N
0°
20°S
40°S
ANTARCTICA
EUROPE
20°W
0°
20°E
40°E
60°E
ICELAND
Arctic Circle
SWEDEN
FINLAND
0 250 500 Miles
0 250 500 Kilometers
NORWAY
60°N
ESTONIA
RUSSIA
UNITED KINGDOM
DEN.
LATVIA
RUSS.
LITH.
IRELAND
NETH.
POLAND
BELARUS
GERMANY
BELG.
CZECH REP.
UKRAINE
LUX.
SLVK.
ATLANTIC OCEAN
LIECH.
AUS.
HUNGARY
MOLDOVA
SWITZ.
SLOVE.
FRANCE
ITALY
CRO.
SERB.
ROMANIA
BOS. & HERZ.
KOS.
BULGARIA
PORTUGAL
SPAIN
ANDORRA
MONT.
MAC.
ALB.
40°N
GREECE
TURKEY
MALTA
MOROCCO
ALGERIA
TUNISIA
CYPRUS

The Cultural Landscape

The Cultural Landscape

AN INTRODUCTION TO HUMAN GEOGRAPHY

ELEVENTH EDITION

JAMES M. RUBENSTEIN

MIAMI UNIVERSITY, OXFORD, OHIO

PEARSON

Boston Columbus Indianapolis New York San Francisco Upper Saddle River
Amsterdam Cape Town Dubai London Madrid Milan Munich Paris Montréal Toronto
Delhi Mexico City São Paulo Sydney Hong Kong Seoul Singapore Taipei Tokyo

Geography Editor: Christian Botting
Senior Marketing Manager: Maureen McLaughlin
Project Editor: Anton Yakovlev
Executive Development Editor: Jonathan Cheney
Assistant Editor: Kristen Sanchez
Editorial Assistant: Bethany Sexton
Senior Marketing Assistant: Nicola Houston
Director of Development: Jennifer Hart
Managing Editor, Geosciences and Chemistry: Gina M. Cheselka
Production Project Manager: Janice Stangel
Production Management: Kelly Keeler, Element LLC
Design Manager: Derek Bacchus
Interior and Cover Designer: Naomi Schiff
Illustrations and Cartography: Kevin Lear, International Mapping
Image Permissions Manager: Maya Melenchuk
Photo Researcher: Stefanie Ramsey
Text Permissions Manager: Joe Croscup
Text Permission Researcher: Stephen Barker
Media Producer: Ziki Dekel
Media Editor: Miriam Adrianowicz
Operations Specialist: Michael Penne
Cover Photo Credit: © Picture Contact BV / Alamy

Credits and acknowledgments borrowed from other sources and reproduced, with permission, in this textbook appear **on pages CR-1 and CR-2**.

Library of Congress Cataloging-in-Publication Data
Rubenstein, James M.
The cultural landscape / James M. Rubenstein. – 11th ed.
p. cm.
Includes index.
ISBN 978-0-321-83158-3
1. Human geography. 2. Human geography–Textbooks. 3. Human geography–Study and teaching. I. Title.
GF41.R82 2014
304.2–dc23 2012033067

2 3 4 5 6 7 8 9 10—RRD—15 14 13

www.pearsonhighered.com
ISBN-10: 0-321-83158-6; ISBN-13: 978-0-321-83158-3 (Student Edition)
ISBN-10: 0-321-86303-8; ISBN-13: 978-0-321-86303-4 (Instructor's Review Copy)

BRIEF CONTENTS

DEDICATION

This book is dedicated to Bernadette Unger, Dr. Rubenstein's wife, who has been by his side through many books, as well as to the memory of his father, Bernard W. Rubenstein. Dr. Rubenstein also gratefully thanks the rest of his family for their love and support.

CONTENTS

2 Population and Health 42

3 Migration 76

4 Folk and Popular Culture 106

KEY ISSUE 1

KEY ISSUE 2

KEY ISSUE 3

KEY ISSUE 4

5 Languages 140

KEY ISSUE 1

KEY ISSUE 2

KEY ISSUE 3

KEY ISSUE 4

6 Religions 180

7 Ethnicities 224

12 Services and Settlements 428

13 Urban Patterns 458

PREFACE

Geography is the study of where things are located on Earth's surface and the reasons for the location. The word *geography*, invented by the ancient Greek scholar Eratosthenes, is based on two Greek words. *Geo* means "Earth," and *graph* means "to write." Geographers ask two simple questions: where? and why? Where are people and activities located across Earth's surface? Why are they located in particular places? *The Cultural Landscape* seeks to answer these questions as they relate to our contemporary world. The book provides an accessible, in-depth, and up-to-date introduction to human geography for majors and non-majors alike.

New to the 11th Edition

This edition brings substantial changes in both organization and content.

New Organization

A long-time strength of this book has been its clear, easy-to-use organization and outline. Electronic versions of the books now coexist with traditional print format. Traditional textbooks must be formatted to facilitate reading on tablets and computers, while not compromising the pedagogic strengths of traditional print formats. Organizational features from previous editions have been retained and considerably strengthened for this electronic age through the addition of several new features:

- Each two-page spread is now self-contained. As a result, maps and photos appear next to where they are discussed in the text. No more rifling through the book to find a map that has been discussed on one page but doesn't actually appear until several pages later.
- Two-page spreads now begin with a Learning Outcome for the material on that spread. The Learning Outcome helps the reader focus on the most important point presented on each spread.
- Most two-page spreads now contain a *Pause and Reflect* feature to stimulate further thought on the material presented in the spread.
- Each chapter is still outlined around four *Key Issues*, as in previous editions. New to this edition is a *Check-In* feature at the end of each of the four *Key Issues*. The *Check-In* summarizes the principal points made regarding the *Key Issue* that was just concluded.
- The end-of-chapter spreads summarize all the *Key Issues* and *Learning Outcomes* and presents a *Thinking Geographically* essay/discussion question as well as a Google Earth activity specific to each *Key Issue*.

New Content

Issues of sustainability and resource management, depletion and misuse of Earth's resources, and prospects for a sustainable future are increasingly central to the understanding of the demographic, cultural, political, and economic patterns, problems, and policies that human geographers study. Material that in previous editions appeared in a separate chapter at the end of the book has been integrated into the discussion of other topics.

- Chapter 1 (Basic Concepts) includes a new *Key Issue* that introduces the concept of sustainability.
- Chapter 2 is reframed as Population and Health. As the rate of population growth declines from its peak during the second half of the twentieth century, population geography is increasingly concerned with the health of humans, not just their fertility and mortality. A new *Key Issue* has been added that addresses regional variations in medical conditions and practices.
- Chapter 3 (Migration) includes discussion of recent legal and political controversies over migration in the United States and Europe, including the border control legislation enacted by the state of Arizona that was upheld in part and voided in part by the U.S. Supreme Court.
- Chapter 4 (Folk and Popular Culture) includes a new key issue concerning sustainability challenges faced by folk and popular cultures, especially recycling of the material artifacts of popular culture.
- Chapter 5 (Languages) has been reorganized to help students understand the worldwide distribution of languages right off the bat.
- Chapter 6 (Religions) also now begins with an overview of global patterns, before important features of diversity among religions are introduced.
- Chapter 7 (Ethnicities) opens with a description of the complex ethnic heritage of a prominent American – President Obama. Most of the material related to nationalities transferred to Chapter 8.
- Chapter 8 (Political Geography) includes an expanded discussion of gerrymandering as a result of redistricting in accordance with the 2010 U.S. Census. The chapter also addresses the events of Arab Spring.
- Chapter 9 (Development) contains a new *Key Issue* that discusses the importance of energy in sustainable development. The chapter also discusses reasons underlying the severe global recession that began in 2008, as well as reasons poor economic conditions have lingered, especially in Europe.

- Chapter 10 (now called Food and Agriculture) includes a new *Key Issue* that focuses on regional variations of food preferences and needs.
- Chapter 11 (now called Industry and Manufacturing) has a new *Key Issue* that addresses the importance of reducing industrial pollution in promoting sustainable development.
- Chapter 13 (Urban Patterns) includes results from the 2010 U.S. Census.
- Current data and information are integrated into all text, tables, and maps from the 2010 U.S. Census, 2012 Population Reference Bureau, Population Data and other important sources.
- This 11th edition is now supported by **MasteringGeography™** with Pearson eText, the most widely used and effective online homework, tutorial, and assessment system for the sciences. Assignable media and activities include MapMaster™ interactive maps, *Encounter Human Geography* Google Earth explorations, geography videos, geoscience animations, Thinking Spatially and Data Analysis activities on the toughest topics in geography, end-of-chapter questions, reading quizzes, and Test Bank questions. See page XVIII for more detailed information.

Human Geography as a Social Science

The main purpose of this book is to introduce students to the study of human geography as a social science by emphasizing the relevance of geographic concepts to human problems. It is intended for use in college-level introductory human or cultural geography courses, as well as the equivalent advanced placement course in high school. At present, human geography is the fastest-growing course in the AP curriculum.

A central theme in this book is a tension between two important themes—globalization and cultural diversity. In many respects, we are living in a more unified world economically, culturally, and environmentally. The actions of a particular corporation or country affect people around the world. For example, geographers examine the prospects for an energy crisis by relating the distributions of energy production and consumption. Geographers find that the users of energy are located in places with different social, economic, and political institutions than are the producers of energy. The United States and Japan consume far more energy than they produce, whereas Russia and Saudi Arabia produce far more energy than they consume.

This book argues that after a period when globalization of the economy and culture has been a paramount concern in geographic analysis, local diversity now demands equal time. People are taking deliberate steps to retain distinctive cultural identities. They are preserving little-used languages, fighting fiercely to protect their religions, and carving out distinctive economic roles. Local diversity even extends to addressing issues, such as the energy crisis, that at first glance are considered global. For example, Israel is working with the French carmaker Renault and the Silicon Valley company Project Better Place to encourage electric vehicles by installing tens of thousands of recharging stations. Brazil has passed laws to require more use of biofuels, produced from crops grown in Brazil and processed in factories there. Meanwhile, the United Arab Emirates has invested in a subway system as an alternative to motor vehicles, even though the country is one of the world's leading producers of petroleum.

Divisions within Geography

Because geography is a broad subject, some specialization is inevitable. At the same time, one of geography's strengths is its diversity of approaches. Rather than being forced to adhere rigorously to established disciplinary laws, geographers can combine a variety of methods and approaches. This tradition stimulates innovative thinking, although students who are looking for a series of ironclad laws to memorize may be disappointed.

Human versus Physical Geography

Geography is both a physical science and a social science. When geography concentrates on the distribution of physical features, such as climate, soil, and vegetation, it is a physical science. When it studies cultural features, such as language, industries, and cities, geography is a social science. This division is reflected in some colleges, where physical geography courses may carry natural science credit while human and cultural geography courses carry social science credit.

While this book is concerned with geography from a social science perspective, one of the distinctive features of geography is its use of natural science concepts to help understand human behavior. The distinction between physical and human geography reflects differences in emphasis, not an absolute separation.

Topical versus Regional Approach

Geographers face a choice between a topical approach and a regional approach. The topical approach, which is used in this book, starts by identifying a set of important cultural issues to be studied, such as population growth, political disputes, and economic restructuring. Geographers using the topical approach examine the location of different aspects of the topic, the reasons for the observed pattern, and the significance of the distribution.

The alternative approach is regional. Regional geographers select a portion of Earth and study the environment, people, and activities within the area. The regional geography approach is used in courses on Europe, Africa, Asia, and other areas of the world. Although this book is organized by topics, geography students should be aware of the location of places in the world. A separate index section lists the book's maps by location. One indispensable aid in the study of regions is an atlas, which can also be used to find unfamiliar places that pop up in the news.

Descriptive versus Systematic Method

Whether using a topical or a regional approach, geographers can select either a descriptive or a systematic method. Again, the distinction is one of emphasis, not an absolute separation. The descriptive method emphasizes the collection of a variety of details about a particular location. This method has been used primarily by regional geographers to illustrate the uniqueness of a particular location on Earth's surface. The systematic method emphasizes the identification of several basic theories or techniques developed by geographers to explain the distribution of activities.

This book uses both the descriptive and systematic methods because total dependence on either approach is unsatisfactory. An entirely descriptive book would contain a large collection of individual examples not organized into a unified structure. A completely systematic approach suffers because some of the theories and techniques are so abstract that they lack meaning for the student. Geographers who depend only on the systematic approach may have difficulty explaining important contemporary issues.

Outline of Main Topics

The book discusses the following main topics:

- What basic concepts do geographers use? Chapter 1 provides an introduction to ways that geographers think about the world. Geographers employ several concepts to describe the distribution of people and activities across Earth, to explain reasons underlying the observed distribution, and to understand the significance of the arrangements.
- Where are people located in the world? Why do some places on Earth contain large numbers of people or attract newcomers while other places are sparsely inhabited? Chapters 2 and 3 examine the distribution and growth of the world's population, as well as the movement of people from one place to another.
- How are different cultural groups distributed? Chapters 4 through 8 analyze the distribution of different cultural traits and beliefs and the problems that result from those spatial patterns. Important cultural traits discussed in Chapter 4 include food, clothing, shelter, and leisure activities. Chapters 5 through 7 examine three main elements of cultural identity: languages, religions, and ethnicities. Chapter 8 looks at political problems that arise from cultural diversity. Geographers look for similarities and differences in the cultural features at different places, the reasons for their distribution, and the importance of these differences for world peace.
- How do people earn a living in different parts of the world? Human survival depends on acquiring an adequate food supply. One of the most significant distinctions in the world is whether people produce their food directly from the land or buy it with money earned by performing other types of work. Chapters 9 through 12 look at the three main ways of earning a living: agriculture, manufacturing, and services. Chapter 13 discusses cities, the centers for economic as well as cultural activities.

Suggestions for Use

This book can be used in an introductory human or cultural geography course that extends over one semester, one quarter, or two quarters. An instructor in a one-semester course could devote one week to each of the chapters, leaving time for examinations. In a one-quarter course, the instructor might need to omit some of the book's material. A course with more of a cultural orientation could use Chapters 1 through 8. If the course has more of an economic orientation, then the appropriate chapters would be 1 through 3 and 8 through 13. A two-quarter course could be organized around the culturally oriented Chapters 1 through 8 during the first quarter and the more economically oriented Chapters 9 through 13 during the second quarter. Topics of particular interest to the instructor or students could be discussed for more than one week.

Acknowledgments

For a book that has been through many editions to maintain its leadership position, stale and outdated material and methods must be cleared out to make way for the fresh and contemporary. It is all too easy for an author in the twenty-first century to rely on practices that brought success in the twentieth century. Strong proactive leadership is required from the publisher to push an already strong book to loftier aspirations. This leadership is especially critical during a period when the teaching and learning environment is changing much more rapidly than even in the late twentieth century.

A major reason for the long-term success of this book has been the quality of leadership in geography at Pearson Education. The key members of Pearson's hands-on revision team are:

- Christian Botting, geography editor at Pearson Education, who has now led the team through four of my book projects. Christian's skills have made him ideally positioned to proactively bring together scientific books with twenty-first century technology and pedagogy.
- Anton Yakovlev, geography project manager at Pearson Education, who has managed three book projects with me now. Anton not only keeps impeccable control of what has to be done when, he has been more proactive than any previous project manager in initiating many great ideas.
- Jonathan Cheney, executive development editor at Pearson Education, who has undertaken the detailed

editorial development of the manuscript. Instead of passively editing line-by-line, Jonathan is proactive in adjusting the outdated material and suggesting fresh directions.

Prior to Christian, two individuals served as geography editors for most of the past three decades. Paul F. Corey, who is now president of Science, Business and Technology at Pearson, guided development of the third, fourth, and fifth editions of this book. Dan Kaveney guided development of the sixth, seventh, eighth, and ninth editions.

Because Pearson is the dominant publisher of college geography textbooks, the person in charge of geography wields considerable influence in shaping what is taught in the nation's geography curriculum. I will always value the sound judgment, outstanding vision, and friendship of both Paul and Dan, and I am gratified that Christian has quickly and successfully assumed the leadership position.

Others at Pearson who have been especially helpful on this project include Bethany Sexton, geography editorial assistant; Gina Cheselka, geosciences production managing editor; Maureen McLaughlin, senior marketing manager; Kristen Sanchez, assistant editor; Ziki Dekel and Tim Hainley, media producers; and others.

In this age of outsourcing, Pearson works with many independent companies to create books. This edition has been the beneficiary of a top-notch team:

- Kelly Keeler, senior project manager for Higher Education at Element LLC, directed the flow of production work to the author.
- Kitty Wilson handled the copyediting work with sensitivity.
- Stefanie Ramsay found great photos.
- Kevin Lear, senior project manager at International Mapping, and his team, produced outstanding maps for this book. Back in the 1980s, when he was just getting started as a professional cartographer, Kevin produced GIS-generated full-color maps for the second edition of this book, the first time that either GIS or full color had been used in a geography text—and a major reason for launching this book's success.

I am grateful for the great work done on a variety of print and digital ancillaries by Craig S. Campbell, Youngstown State University; Matt Cartlidge, University of Nebraska–Lincoln; John Conley, Saddleback College; Stephen Davis, University of Illinois–Chicago; Sarah Goggin, Cyprus College; and Marc Healy, Elgin Community College.

I would also like to extend a special thanks to all of my colleagues who have, over the years, offered a good deal of feedback and constructive criticism. Colleagues who served as reviewers as we prepared the 10th edition are Patricia Boudinot, George Mason University; Henry Bullamore, Frostburg State University; Caitie Finlayson, Florida State University; Jeff Gordon; Bowling Green State University; Richard J. Grant, University of Miami; Marc Healy, Elgin Community College; Scott Hunt, Columbus State Community College; Jonathan Leib, Old Dominion University; Max Lu, Kansas State University; Debra Matthews, Boise State University; Lashale Pugh, Youngstown State University; Roger Seyla, University of Cincinnati; Suzanne Struve, Blinn College; Scott Therkalsen, Grossmont College; and David Wishart, University of Nebraska–Lincoln.

THE TEACHING AND LEARNING PACKAGE

In addition to producing the text itself, the authors and publisher have been pleased to work with a number of talented people to produce an excellent instructional package.

For Teachers and Students

MasteringGeography™ with Pearson eText

The **Mastering** platform is the most widely used and effective online homework, tutorial, and assessment system for the sciences. It delivers self-paced tutorials that provide individualized coaching, focus on your course objectives, and are responsive to each student's progress. The Mastering system helps teachers maximize class time with customizable, easy-to-assign, and automatically graded assessments that motivate students to learn outside class and arrive prepared for lecture.

MasteringGeography offers:

- **Assignable activities** that include MapMaster™ Interactive Map activities, *Encounter Human Geography* Google Earth Explorations, Video activities, Geoscience Animation activities, Map Projection activities, Thinking Spatially and Data Analysis activities on the toughest topics in geography, end-of-chapter questions and exercises, reading quizzes, and Test Bank questions.
- **Student Study Area** with MapMaster™ interactive maps, videos, Geoscience Animations, web links, videos, glossary flashcards, "In the News" RSS feeds, chapter quizzes, an optional Pearson eText that includes versions for iPad and Android devices, and more.

Pearson eText gives students access to the text whenever and wherever they can access the Internet. The eText pages look exactly like the printed text, and include powerful interactive and customization functions, and links to media.

- **Practicing Geography: Careers for Enhancing Society and the Environment by Association of American Geographers (0321811151)** This book examines career opportunities for geographers and geospatial professionals in business, government, nonprofit, and educational sectors. A diverse group of academic and industry professionals share insights on career planning, networking, transitioning between employment sectors, and balancing work and home life. The book illustrates the value of geographic expertise and technologies through engaging profiles and case studies of geographers at work.
- **Teaching College Geography: A Practical Guide for Graduate Students and Early Career Faculty by Association of American Geographers (0136054471)** This two-part resource provides a starting point for becoming an effective geography teacher from the very first day of class. Part One addresses "nuts-and-bolts" teaching issues. Part Two explores being an effective teacher in the field, supporting critical thinking with GIS and mapping technologies, engaging learners in large geography classes, and promoting awareness of international perspectives and geographic issues.
- **Aspiring Academics: A Resource Book for Graduate Students and Early Career Faculty by Association of American Geographers (0136048919)** Drawing on several years of research, this set of essays is designed to help graduate students and early career faculty start their careers in geography and related social and environmental sciences. *Aspiring Academics* stresses the interdependence of teaching, research, and service—and the importance of achieving a healthy balance of professional and personal life—while doing faculty work. Each chapter provides accessible, forward-looking advice on topics that often cause the most stress in the first years of a teaching appointment.
- **Television for the Environment *Earth Report* Geography Videos on DVD (0321662989)** This three-DVD set helps students visualize how human decisions and behavior have affected the environment and how individuals are taking steps toward recovery. With topics ranging from the poor land management promoting the devastation of river systems in Central America, to the struggles for electricity in China and Africa, these 13 videos from Television for the Environment's global *Earth Report* series recognize the efforts of individuals around the world to unite and protect the planet.
- **Television for the Environment *Life* World Regional Geography Videos on DVD (013159348X)** From the Television for the Environment's global *Life* series, this two-DVD set brings globalization and the developing world to the attention of any world regional geography course. These 10 full-length video programs highlight matters such as the growing number of homeless children in Russia, the lives of immigrants living in the United States trying to aid family still living in their native countries, and the European conflict between commercial interests and environmental concerns.
- **Television for the Environment *Life* Human Geography Videos on DVD (0132416565)** This three-DVD set is designed to enhance any human geography course. These DVDs include 14 full-length video programs from Television for the Environment's global *Life* series, covering a wide array of issues affecting people and places in the contemporary world, including the serious health risks of pregnant women in Bangladesh, the social inequalities of the "untouchables" in the Hindu caste system, and Ghana's struggle to compete in a global market.

For Teachers

- **Instructor Resource Manual Download (0321841158)** The *Instructor Resource Manual* written by John Conley of Saddleback College, follows the new organization of the main text. Each chapter of the *Instructor Resource Manual* opens with a specific introduction highlighting core learning objectives presented in the specific chapter. The *Instructor Resource Manual* includes Icebreakers to start classroom discussion, Challenges to Comprehension, Review/Reflection Questions, answers to the Pause and Reflect and Thinking Geographically questions found in the text, and Additional Resources to examine during classroom sessions or to assign to students.
- **TestGen/Test Bank (0321841166)** TestGen is a computerized test generator that lets teachers view and edit *Test Bank* questions, transfer questions to tests, and print the test in a variety of customized formats. Authored by Stephen Davis of the University of Illinois at Chicago, this *Test Bank* includes approximately 1,000 multiple-choice, true/false, and short-answer/essay questions. Questions are correlated against the revised U.S. National Geography Standards, chapter-specific learning outcomes, and Bloom's Taxonomy to help teachers better map the assessments against both broad and specific teaching and learning objectives. The *Test Bank* is also available in Microsoft Word® and can be imported into Blackboard.
- **Instructor Resource DVD (032184114X)** This DVD provides everything teachers need where they want it. The *Instructor Resource DVD* helps make teachers more effective by saving them time and effort. All digital resources can be found in one well-organized, easy-to-access place. This DVD includes:
  - All textbook images as JPEGs, PDFs, and PowerPoint™ presentations
  - Pre-authored Lecture Outline PowerPoint™ presentations, which outline the concepts of each chapter with embedded art and can be customized to fit teachers' lecture requirements
  - CRS "Clicker" Questions in PowerPoint™ format, which correlate to the U.S. National Geography Standards, chapter-specific learning outcomes, and Bloom's Taxonomy
  - The TestGen software, *Test Bank* questions, and answers for both Macs and PCs
  - Electronic files of the *Instructor Resource Manual* and *Test Bank*

This Instructor Resource Center content is also available completely online via the Instructor Resources section of MasteringGeography and www.pearsonhighered.com/irc.

For Students

- **Goode's World Atlas, 22nd Edition (0321652002)** *Goode's World Atlas* has been the world's premiere educational atlas since 1923—and for good reason. It features more than 250 pages of maps, from definitive physical and political maps to important thematic maps that illustrate the spatial aspects of many important topics. The 22nd edition includes 160 pages of new, digitally produced reference maps, as well as new thematic maps on global climate change, sea level rise, CO_2 emissions, polar ice fluctuations, deforestation, extreme weather events, infectious diseases, water resources, and energy production.
- **Dire Predictions: Understanding Global Warming by Michael Mann and Lee R. Kump (0136044352)** This text is for any science or social science course in need of a basic understanding of Intergovernmental Panel on Climate Change (IPCC) reports. Periodic reports from the IPCC evaluate the risk of climate change brought on by humans. But the sheer volume of scientific data remains inscrutable to the general public, particularly to those who may still question the validity of climate change. In just over 200 pages, this practical text presents and expands upon the essential findings in a visually stunning and undeniably powerful way to the lay reader. Scientific findings that provide validity to the implications of climate change are presented in clear-cut graphic elements, striking images, and understandable analogies.

Pearson's Encounter Series

Pearson's *Encounter* series provides rich, interactive explorations of geoscience concepts through Google Earth™ activities, exploring a range of topics in regional, human, and physical geography. For those who do not use MasteringGeography, all chapter explorations are available in print workbooks as well as in online quizzes, at www.mygeoscienceplace.com, accommodating different classroom needs. Each exploration consists of a worksheet, online quizzes, and a corresponding Google Earth™ KMZ file:

- ***Encounter Human Geography*** Workbook and Website by Jess C. Porter (0321682203)
- ***Encounter World Regional Geography*** Workbook and Website by Jess C. Porter (0321681754)
- ***Encounter Physical Geography*** Workbook and Website by Jess C. Porter and Stephen O'Connell (0321672526)
- ***Encounter Geosystems*** Workbook and Website by Charlie Thomsen (0321636996)
- ***Encounter Earth*** Workbook and Website by Steve Kluge (0321581296)

A proven path to learning

The text's consistent chapter structure and supporting pedagogy provides a learning path that identifies and reinforces important issues and outcomes.

Key Issues form a learning path

Key Issues highlight the four main points around which each chapter is organized. At the end of each Key Issue section, a Check-In summarizes the main focus of the section.

KEY ISSUE 4

Why Do Some Regions Face Health Threats?

- Epidemiologic Transition
- Infectious Diseases
- Health Care

Learning Outcome 2.4.1
Summarize the four stages of the epidemiologic transition.

NEW! Learning Outcomes in each Key Issue section identify the skills and knowledge students will gain from each section.

CHECK IN: KEY ISSUE 4

Why do some regions face health threats?

- ✓ The epidemiologic transition has four stages of distinctive diseases.
- ✓ A resurgence of infectious diseases may signal a possible stage 5 of the epidemiologic transition.
- ✓ The provision of health care varies sharply between developed and developing countries.

Check Ins at the end of each section summarize the main points.

KEY ISSUE 4

Why Do Regions Face Health Threats?

The epidemiologic transition is a change in a society's distinctive types of diseases. Health care is better in developed countries, but even they are threatened by infectious diseases diffused through modern means of transportation.

LEARNING OUTCOME 2.4.1: Summarize the four stages of the epidemiologic transition.

- Stage 1 was characterized by pestilence and famine, stage 2 by pandemics, and stages 3 and 4 by degenerative diseases.

LEARNING OUTCOME 2.4.2: Summarize the reasons for a stage 4 and possible stage 5 of the epidemiologic transition.

- Evolution, poverty, and increased connections may influence the resurgence of infectious diseases.

LEARNING OUTCOME 2.4.3: Describe the diffusion of AIDS.

LEARNING OUTCOME 2.4.4: Understand reasons for variations in health care between developed and developing countries.

- Health care varies widely around the world because developing countries generally lack resources to provide the same level of health care as developed countries.

LEARNING OUTCOME 2.4.5: Understand reasons for variations in health between developed and developing countries.

NEW! Reviews of Key Issues close out each chapter with a recap of Learning Outcomes that summarize and reinforce significant concepts.

Active learning reinforces Key Issues

NEW! Pause and Reflect Questions are integrated throughout the chapters, giving students a chance to stop and check their understanding of the reading.

Pause and Reflect 2.4.4
Why do men have lower life expectancies than women in most countries?

Pause and Reflect 2.4.5
Why might levels of hospital beds and physicians be lower in North America than in other developed countries?

End-of-chapter questions:

- **Thinking Geographically** are application-oriented sections that allow students to explore issues more intensively.
- **NEW!** Engaging end-of-chapter features include exercises that explore Key Issues using **Google Earth.**

THINKING GEOGRAPHICALLY 2.4: Health-care indicators for the United States do not always match those of other developed countries. What reasons might explain these differences?

GOOGLE EARTH 2.4: Several hundred thousand died, some from infectious diseases, after an earthquake hit Haiti January 12, 2010, the date this Google Earth image was taken. The roof of the cathedral in the capital Port au Prince collapsed. What other evidence of the earthquake can be seen in images from January 2010?

Explore human geography in a cultural landscape

How can teachers hold the attention of today's students? By using stories and examples that emphasize the relevance of geographic concepts tools, technologies, and to universal human concerns such as health, equality, and sustainability.

NEW! Sustainability and Inequality in Our Global Village features in each chapter discuss current social, economic, and environmental topics relevant to the chapter themes.

SUSTAINABILITY AND INEQUALITY IN OUR GLOBAL VILLAGE

Ethnic Cleansing and Drought

More than 2 million Somalis—one-fourth of the country's population—are classified as refugees or internally displaced persons. As elsewhere in sub-Saharan Africa, continued fighting among ethnic groups and the absence of a strong national government able to maintain order have contributed to the large number of refugees.

Adding to the woes of the Somali people, the worst drought in 60 years hit the country in 2010 and 2011, especially in the south (Figure 7-48). It is impossible to count the number of Somalis forced to migrate because of famine rather than civil war; both factors probably affect most Somalis. Because of the civil war, much of the food and water sent by international relief organizations could not get through to the people in need. Improved weather in 2012 resulted in a larger harvest, and more supplies were reaching people.

▲ FIGURE 7-48 **SOMALIA** Somali victims of fighting and famine line up for food and medical assistance in 2011.

International organizations distributed seeds and dug irrigation canals to help in the longer term, but a renewal of fighting or a bit less rainfall could push the country back into famine.

Contemporary Geographic Tools offer readers a wealth of representations and perspectives to better understand issues, using geographic methods and online tools and technologies such as geographic information systems, aerial photography, remote sensing, and Google Earth.

SUSTAINABILITY AND INEQUALITY IN OUR GLOBAL VILLAGE

Climate Change in the South Pacific

One consequence of global warming is a rise in the level of the oceans. The large percentage of the world's population—including one-half of Americans—who live near the sea face increased threat of flooding. The threat is especially severe for island countries in the Pacific Ocean; they could be wiped off the map entirely.

Kiribati is a collection of approximately 32 small islands, one of the world's most isolated countries (Figure 11-33). Despite its extreme isolation, global forces threaten Kiribati's existence. Rising sea levels due to global warming threaten Kiribati because the entire country is within a few meters of sea level. Two of Kiribati's islands—Tebua Tarawa and Abanuea—have already disappeared.

Kiribati and other Pacific island microstates are atolls—that is, islands made of coral reefs. A coral is a small sedentary marine animal that has a horny or calcareous skeleton. Corals form colonies, and the skeletons build up to form coral reefs. Coral is very fragile. Humans are attracted to coral for its beauty and the diversity of species it supports, but handling coral can kill it. The threat of global warming to coral is especially severe: Coral stays alive in only a narrow range of ocean temperatures, between 23°C and 25°C (between 73°F and 77°F), so global warming threatens the ecology of Kiribati, even if it remains above sea level.

Kiribati has an emergency response to rising sea levels. The government has negotiated with Fiji to purchase 2,000 hectares (5,000 acres) of land on the island of Vanua Levu to relocate people from Kiribati someday.

◄ FIGURE 11-33 **KIRIBATI** Global warming may cause the oceans to rise, submerging small island countries such as Kiribati.

CONTEMPORARY GEOGRAPHIC TOOLS

Claiming Ellis Island

Twelve million immigrants to the United States between 1892 and 1954 were processed at Ellis Island, situated in New York Harbor (Figure 3-33). Incorporated as part of the Statue of Liberty National Monument in 1965, Ellis Island was restored and reopened in 1990 as a museum of immigration. Before building the immigration center, the U.S. government used Ellis Island as a fort and powder magazine beginning in 1808.

An 1834 agreement approved by the U.S. Congress gave Ellis Island to New York State and gave the submerged lands surrounding the island to New Jersey. When the agreement was signed, Ellis Island was only 1.1 hectares (2.75 acres), but beginning in the 1890s, the U.S. government enlarged the island, eventually to 10.6 hectares (27.5 acres).

New Jersey state officials claimed that the 10.6-hectare Ellis Island was part of their state, not New York. The claim was partly a matter of pride on the part of New Jersey officials to stand up to their more glamorous neighbor. After all, Ellis Island was only 400 meters (1,300 feet) from the New Jersey shoreline, yet tourists—like immigrants a century ago—are transported by ferry to Lower Manhattan more than a mile away. More practically, the sales tax collected by the Ellis Island museum gift shop was going to New York rather than to New Jersey.

After decades of dispute, New Jersey took the case to the U.S. Supreme Court. In 1998, the Supreme Court ruled 6–3 that New York owned the original island but that New Jersey owned the rest. New York's jurisdiction was set as the low waterline of the original island. Critical evidence in the decision was a series of maps prepared by New Jersey Department of Environmental Protecti officials using Geographi tion System (GIS). NJD scanned into an image f U.S. coast map that was to be the most reliable ma era. The image file of th was brought into ArcView the low waterline shown map was edited and dep a series of dots. The perin current island was map global positioning syst surveying.

After ruling in favo Jersey's claim, the Suprem ected the NJDEP to deline cise boundary between th again using GIS. Overlayin low waterline onto the c identified New York's te the rest of the current isla ermined to belong to New

▲ FIGURE 3-33 **ELLIS ISLAND** Ellis Island is in the foreground, Jersey City, New Jersey, is to the left, and Manhattan, New York is to the rear.

Modular design: The ultimate in flexibility and effectiveness

Exceptionally clear organization and a modular approach mean ease of use for both students and teachers.

NEW! Modular organization simplifies lesson planning for teachers, studying for students. The Eleventh Edition is more tightly organized into modules that work as a unit, providing flexibility for students and teachers.

30 THE CULTURAL LANDSCAPE

KEY ISSUE 4

Why Are Some Human Actions Not Sustainable?

- Sustainability and Resources
- Sustainability and Human–Environment Relationships

Learning Outcome 1.4.1
Describe the three pillars of sustainability.

Geography is distinctive because it encompasses both social science (human geography) and natural science (physical geography). This book focuses on human geography but doesn't forget that humans are interrelated with Earth's atmosphere, land, water, and vegetation, as well as with its other living creatures.

From the perspective of human geography, nature offers a large menu of resources available for people to use. A **resource** is a substance in the environment that is useful to people, economically and technologically feasible to access, and socially acceptable to use. A substance is merely part of nature until a society has a use for it. Food, water, minerals, soil, plants, and animals are examples of resources.

Sustainability and Resources

Earth's resources are divided between those that are renewable and those that are not:

- A **renewable resource** is produced in nature more rapidly than it is consumed by humans.
- A **nonrenewable resource** is produced in nature more slowly than it is consumed by humans.

Geographers observe two major misuses of resources:

- Humans deplete nonrenewable resources, such as petroleum, natural gas, and coal.
- Humans destroy otherwise renewable resources through pollution of air, water, and soil.

The use of Earth's renewable and nonrenewable natural resources in ways that ensure resource availability in the future is **sustainability.** Efforts to recycle metals, paper, and plastic, develop less polluting industrial processes, and protect farmland from suburban sprawl are all examples of practices that contribute to a more sustainable future.

THREE PILLARS OF SUSTAINABILITY

According to the United Nations, sustainability rests on three pillars: environment, economy, and society. The UN report *Our Common Future* is a landmark work in recognizing sustainability as a combination of natural and human elements. The report, released in 1987, is frequently called the Brundtland Report, named for the chair of the World Commission on Environment and Development, Gro Harlem Brundtland, former prime minister of Norway.

Sustainability requires curtailing the use of nonrenewable resources and limiting the use of renewable resources to the level at which the environment can continue to supply them indefinitely. To be sustainable, the amount of timber cut down in a forest, for example, or the number of fish removed from a body of water must remain at a level that does not reduce future supplies.

The Brundtland Report argues that sustainability can be achieved only by bringing together environmental protection, economic growth, and social equity (Figure 1-38). The report is optimistic about the possibility of promoting environmental protection at the same time as economic growth and social equity.

THE ENVIRONMENT PILLAR. The sustainable use and management of Earth's natural resources to meet human needs such as food, medicine, and recreation is **conservation.** Renewable resources such as trees and wildlife are conserved if they are consumed at a less rapid rate than they can be replaced. Nonrenewable resources such as petroleum and coal are conserved if we use less today in order to maintain more for future generations (Figure 1-38, left).

Conservation differs from **preservation,** which is the maintenance of resources in their present condition, with as little human impact as possible. Preservation takes the view that the value of nature does not derive from human needs and interests but from the fact that every plant and animal living on Earth has a right to exist and should be preserved, regardless of the cost. Preservation does not regard nature as a resource for human use. In contrast, conservation is compatible with development but only

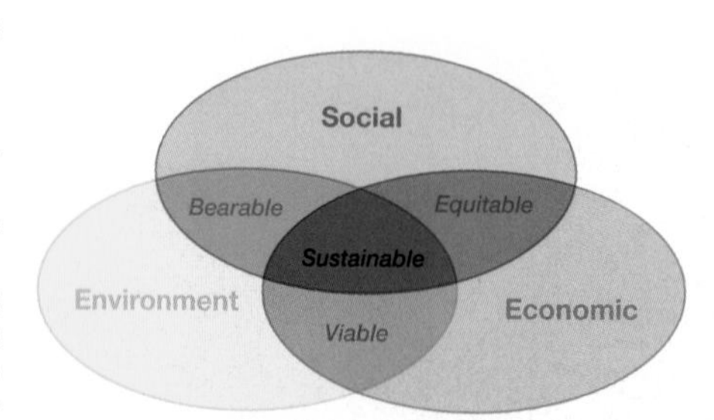

▲ **FIGURE 1-37 THREE PILLARS OF SUSTAINABILITY** The UN's Brundtland Report considers sustainability to be a combination of environmental protection, economic development, and social equity.

Updated coverage and recent data on the most current human geography issues includes:

- Expanded emphasis on resource issues and sustainability integrated throughout
- Dedicated coverage of medical and health geography and the challenges and threats of access
- Revised discussion of food agriculture incorporating critical issues such as scarcity of food and water resources
- Gender and women's issues
- Political geography coverage capturing the results of the 2012 U.S. elections as well as a number of recent Supreme Court decisions (e.g., redistricting/gerrymandering, and migration) and the implications of other world events
- Integrated discussion of development and inequality reflecting the state of the world economy and the widening class gap
- New applications of cultural phenomena, from professional sports and music to social networking

Contemporary photos and maps bring human geography to life

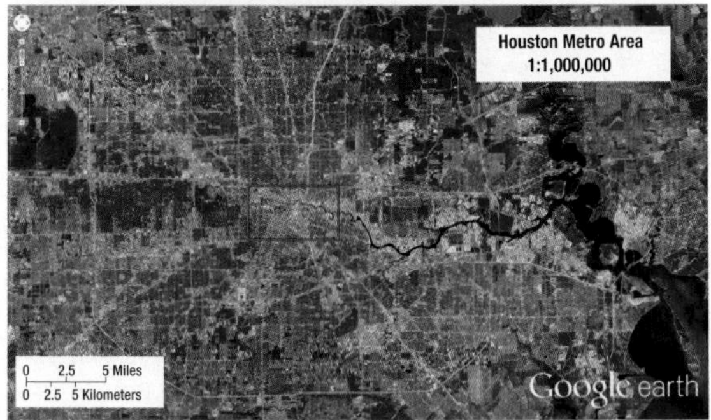

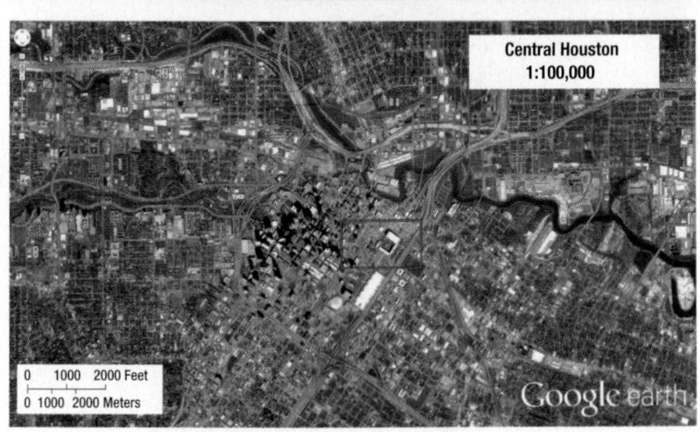

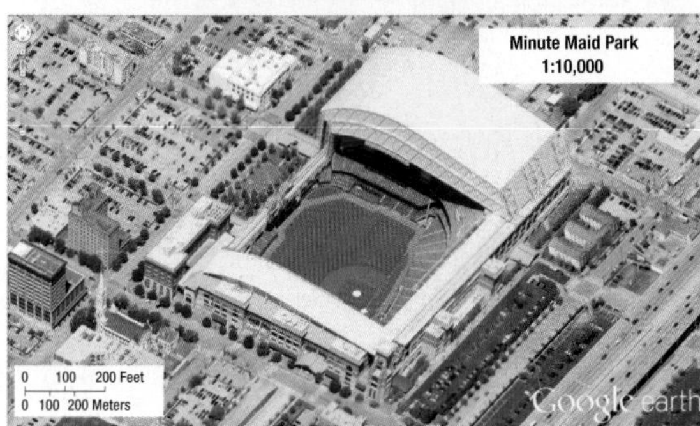

▲ FIGURE 1-8 **MAP SCALE** The four images show southeast Texas (first), the city of Houston (second), downtown Houston (third), and Minute Maid Park (fourth). The map of southeastern Texas has a fractional scale of 1:10,000,000. Expressed as a written statement, 1 inch on the map represents 10 million inches (about 158 miles) on the ground. Look what happens to the scale on the other three maps. As the area covered gets smaller, the maps get more detailed, and 1 inch on the map represents smaller distances.

Restyled and modernized maps use the latest census and population data to enhance the program's overall effectiveness. Key maps within MapMaster™ in MasteringGeography™ connect the text directly to online media and assessment.

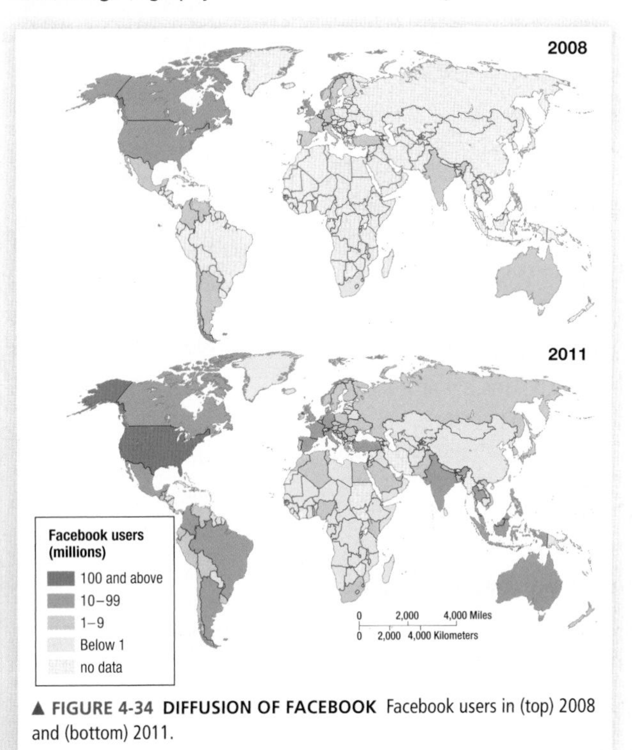

▲ FIGURE 4-34 **DIFFUSION OF FACEBOOK** Facebook users in (top) 2008 and (bottom) 2011.

Updated with more than 40% new photos

▲ FIGURE 4-38 **PROTESTORS SHARING INFORMATION DURING ARAB SPRING** Two Egyptian protesters took photographs with their mobile phones when Egyptian riot police fired tear gas during an Arab Spring protest in 2011.

MasteringGeography™ www.masteringgeography.com

The Mastering online homework, tutorial, and assessment system helps teachers focus on their course objectives by delivering self-paced tutorials that provide students with individualized coaching and respond to each student's progress.

Tools for improving geographic literacy and exploring Earth's dynamic landscape

MapMaster™ is a powerful interactive map tool that presents assignable layered thematic and place name interactive maps at world and regional scales for students to test their geographic literacy and spatial reasoning skills, and explore the modern geographer's tools.

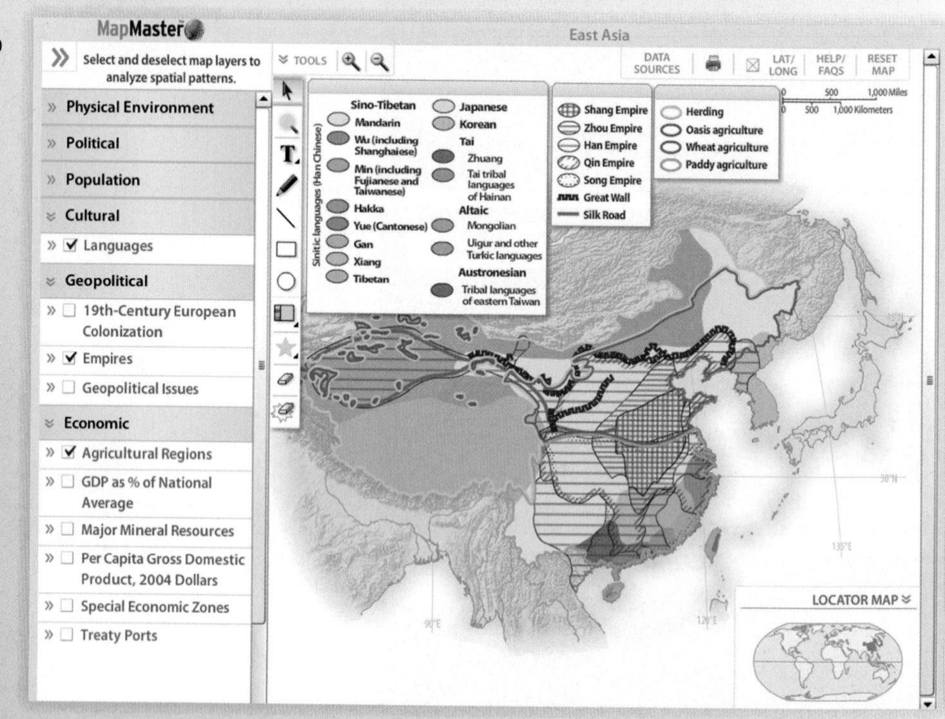

MapMaster Layered Thematic Interactive Map Activities act as a mini-GIS tool, allowing students to layer various thematic maps to analyze spatial patterns and data at regional and global scales and answer multiple-choice and short-answer questions organized by region and theme.

NEW! Layered Thematic Map Features

- 90 new map layers
- Zoom and annotation functionalities
- All maps updated with data from the 2010 U.S. Census, as well as current data from the United Nations, and the Population Reference Bureau

MapMaster Place Name Interactive Map Activities have students identify place names of political and physical features at regional and global scales, explore select recent country data from the CIA World Factbook, and answer associated assessment questions.

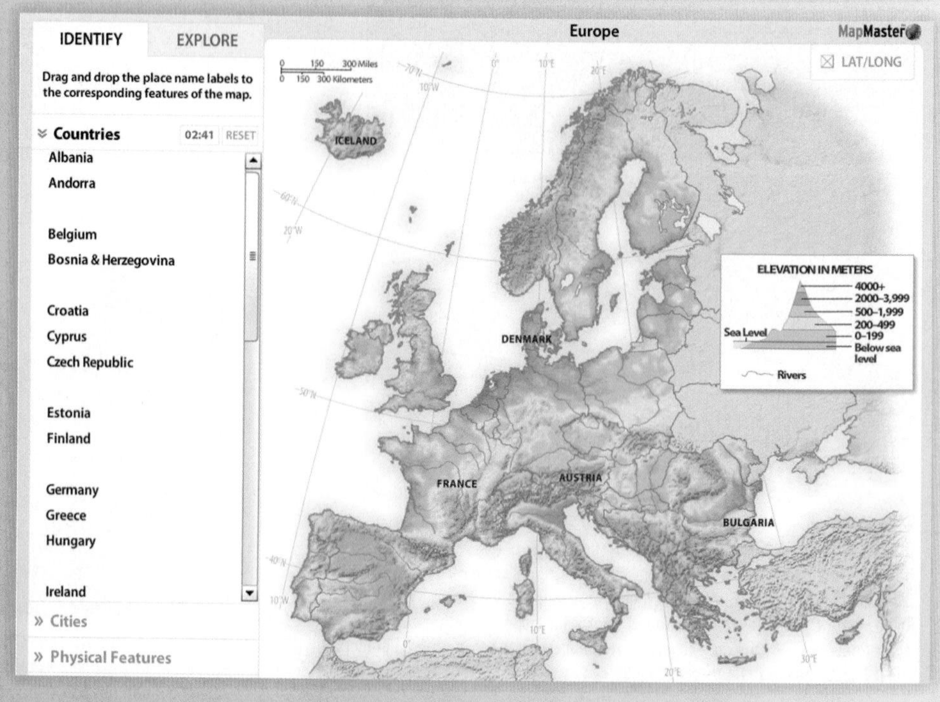

Help students develop a sense of place and spatial reasoning skills

Encounter Activities provide rich, interactive explorations of geography concepts using the dynamic features of **Google Earth™** to visualize and explore Earth's landscape. Dynamic assessment includes questions related to core human geography concepts. All explorations include corresponding Google Earth KMZ media files, and questions include hints and specific wrong-answer feedback to help coach students towards mastery of the concepts.

Geography videos provide students a sense of place and allow them to explore a range of locations and topics. Covering issues of economy, development, globalization, climate and climate change, culture, etc., there are 10 multiple choice questions for each video. These video activities allow teachers to test students' understanding and application of concepts, and offer hints and wrong-answer feedback.

Thinking Spatially and Data Analysis Activities help students master the toughest concepts to develop spatial reasoning and critical thinking skills by identifying and labeling features from maps, illustrations, graphs, and charts. Students then examine related data sets, answering multiple-choice and increasingly higher order conceptual questions, which include hints and specific wrong-answer feedback.

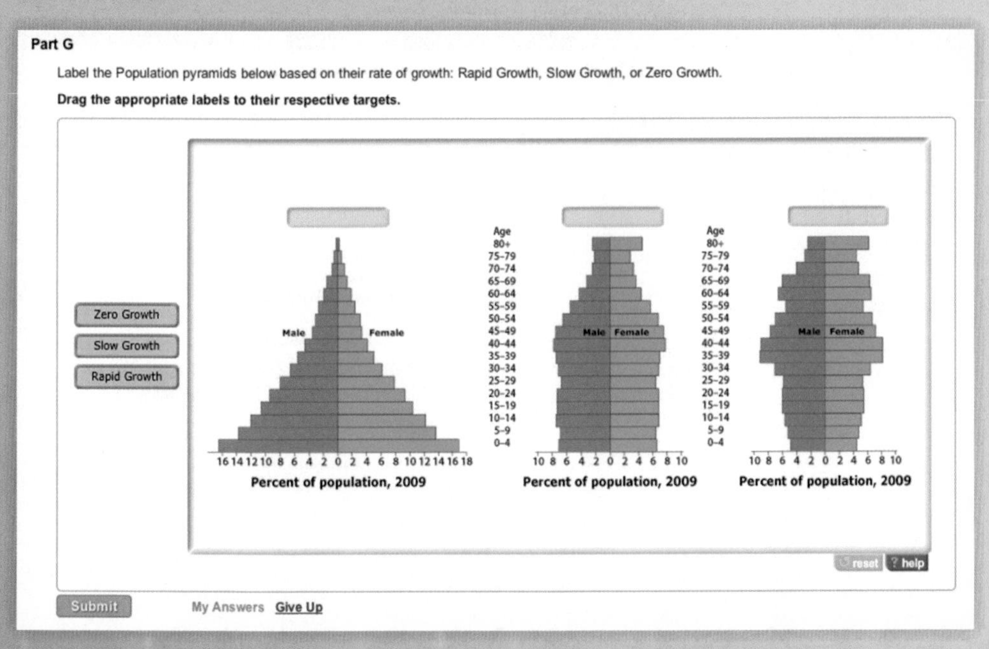

Student Resources in MasteringGeography

- MapMaster™ interactive maps
- Practice chapter quizzes
- Geography videos
- "In the News" RSS feeds
- Glossary flashcards
- Optional Pearson eText and more

Callouts to MasteringGeography appear at the end of each chapter to direct students to extend their learning beyond the textbook.

MasteringGeography™ www.masteringgeography.com

With the Mastering gradebook and diagnostics, you'll be better informed about your students' progress than ever before. Mastering captures the step-by-step work of every student—including wrong answers submitted, hints requested, and time taken at every step of every problem—all providing unique insight into the most common misconceptions of your class.

Quickly monitor and display student results

The **Gradebook** records all scores for automatically graded assignments. Shades of red highlight struggling students and challenging assignments.

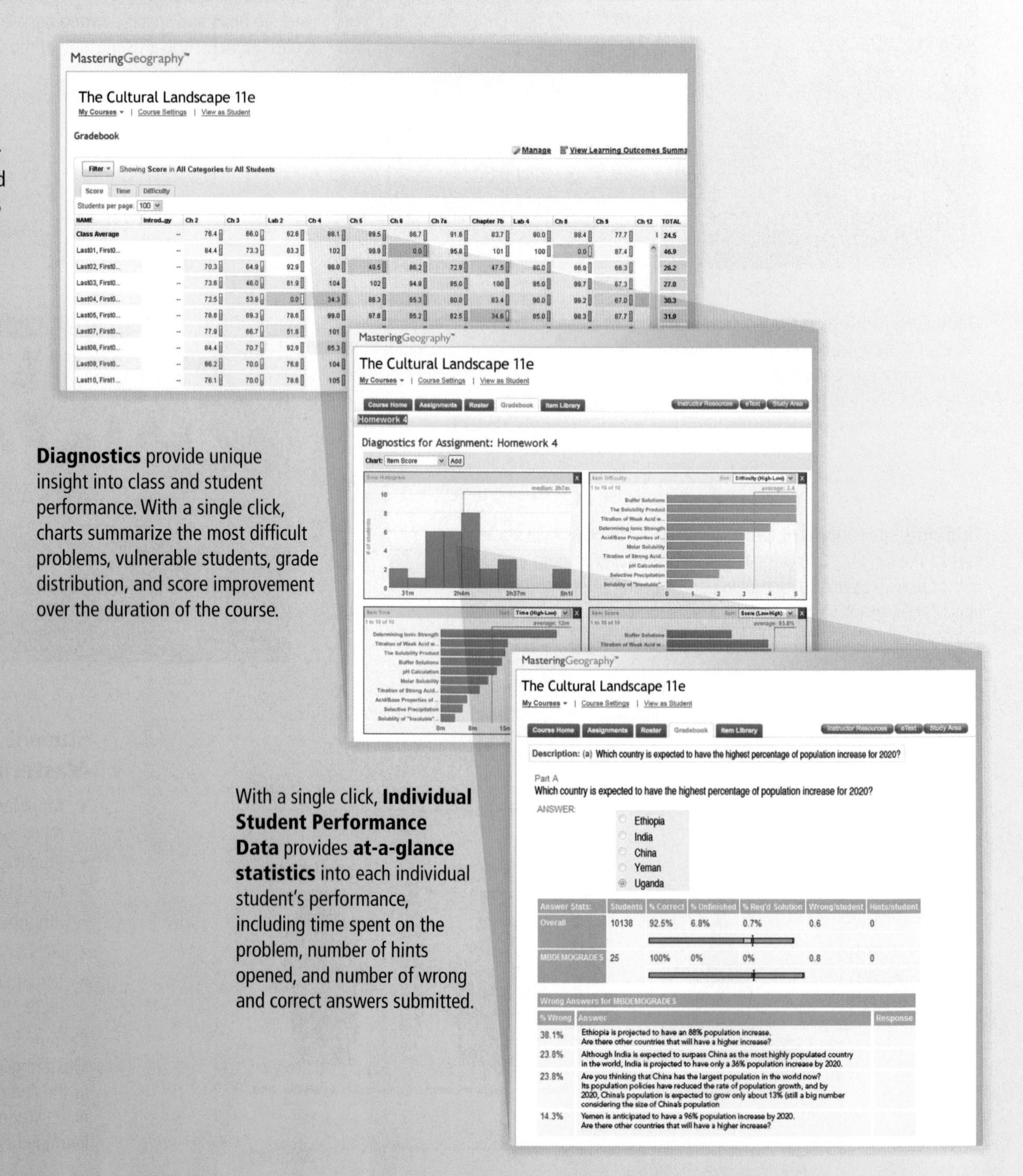

Diagnostics provide unique insight into class and student performance. With a single click, charts summarize the most difficult problems, vulnerable students, grade distribution, and score improvement over the duration of the course.

With a single click, **Individual Student Performance Data** provides **at-a-glance statistics** into each individual student's performance, including time spent on the problem, number of hints opened, and number of wrong and correct answers submitted.

Quickly measure student performance against learning outcomes

Learning Outcomes

MasteringGeography provides quick and easy access to information on student performance against your learning outcomes and makes it easy to share those results.

- Quickly add your own learning outcomes, or use publisher-provided ones, to track student performance and report it to your administration.
- View class and individual student performance against specific learning outcomes.
- Effortlessly export results to a spreadsheet that you can further customize and/or share with your chair, dean, administrator, and/or accreditation board.

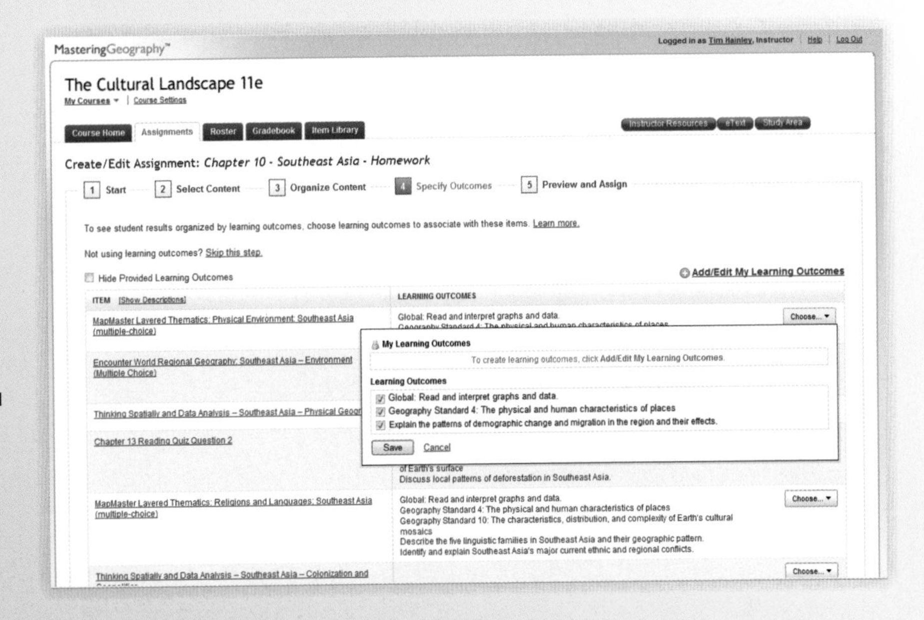

Easy to make your own

Customize publisher-provided problems or quickly add your own. MasteringGeography makes it easy to edit any questions or answers, import your own questions, and quickly add images, links, and files to further enhance the student experience.

Upload your own video and audio files from your hard drive to share with students, as well as record video from your computer's webcam directly into MasteringGeography—no plug-ins required. Students can download video and audio files to their local computer or launch them in Mastering to view the content.

NEW! The Pearson eText app is a great companion to Pearson's eText browser-based book reader. It allows existing subscribers who view their Pearson eText titles on a Mac or PC to additionally access their titles in a bookshelf on the iPad or Android Tablets either online or via download.

Pearson eText gives students access to ***The Cultural Landscape: An Introduction to Human Geography,*** **11th Edition** whenever and wherever they can access the Internet. The eText pages look exactly like the printed text, and include powerful interactive and customization functions. Users can create notes, highlight text in different colors, create bookmarks, zoom, click hyperlinked words and phrases to view definitions, and view as a single page or as two pages. Pearson eText also links students to associated media files, enabling them to view an animation as they read the text, and offers a full-text search and the ability to save and export notes. The Pearson eText also includes embedded URLs in the chapter text with active links to the Internet.

The Cultural Landscape

Chapter

1 Basic Concepts

Why are these people driving around your neighborhood with a camera on their roof? Page 12

Where were your North Face clothes made? Page 21

KEY ISSUE 1

How Do Geographers Describe Where Things Are?

Mapping then and now p. 5

Mapmaking has come a long way, from sticks and shells to satellite mashups. Why are maps so important to geographers?

KEY ISSUE 2

Why Is Each Point On Earth Unique?

My place in the world p. 14

Where am I? The tiny spot on Earth that each of us inhabits is a special place to us—and for good reason.

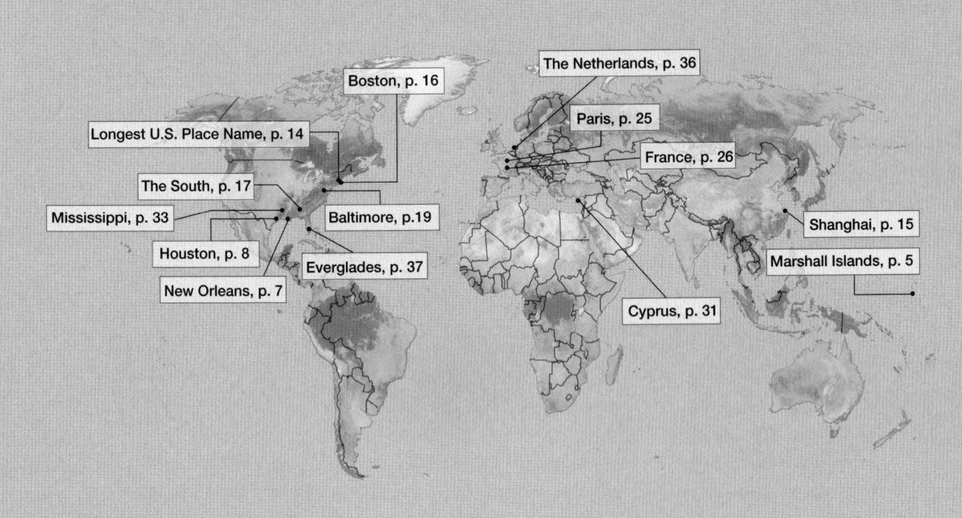

▲ Geographers see people everywhere, including this Muslim clergyman in Afghanistan, Twittering on his smart phone, being pulled in opposite directions by two factors—globalization and local diversity. Modern communications and technology foster globalization, pulling people into greater cultural and economic interaction with others. At the same time, people are searching for more ways to express their unique cultural traditions and economic practices.

KEY ISSUE 3

Why Are Different Places Similar?

A world of similarities and differences p. 21

We are bound together with the rest of the world—whether we like it or not. How do we fit into a global economy and society?

KEY ISSUE 4

Why Are Some Human Actions Not Sustainable?

Caring for Earth p. 31

Earth has been entrusted to us. Will we leave it in better shape than we inherited it—or in worse shape?

Introducing

Basic Concepts

What do you expect from this geography course? You may think that geography involves memorizing lists of countries and capitals. Perhaps you associate geography with photographic essays of exotic places in popular magazines. Contemporary geography is the scientific study of where people and activities are found across Earth's surface and the reasons why they are found there.

In his framework of all scientific knowledge, the German philosopher Immanuel Kant (1724–1804) compared geography and history:

Geographers . . .	Historians . . .
identify the location of important places and explain why human activities are located beside one another.	identify the dates of important events and explain why human activities follow one another chronologically.
ask where and why.	ask when and why.
organize material spatially.	organize material chronologically.
recognize that an action at one point on Earth can result from something happening at another point, which can consequently affect conditions elsewhere.	recognize that an action at one point in time can result from past actions that can in turn affect future ones.

History and geography differ in one especially important manner: A historian cannot enter a time machine to study other eras firsthand; however, a geographer can enter an automobile or airplane to study Earth's surface. This ability to reach other places lends excitement to the discipline of geography—and geographic training raises the understanding of other spaces to a level above that of casual sightseeing.

To introduce human geography, we concentrate on two main features of human behavior—culture and economy. The first half of the book explains why the most important cultural features, such as major languages, religions, and ethnicities, are arranged as they are across Earth. The second half of the book looks at the locations of the most important economic activities, including agriculture, manufacturing, and services.

This chapter introduces basic concepts that geographers employ to address their "where" and "why" questions. Many of these concepts are words commonly employed in English but given particular meaning by geographers:

- **KEY ISSUE 1** looks at geography's most important tool—mapping. Accurate maps are constructed from satellite imagery, such as Figures 1-1 and 1-2.
- **KEY ISSUE 2** addresses the first of two principal "why" questions. Geographers want to know why each point on Earth is in some ways unique. For example, why does Figure 1-2 have some bright points and some dark areas?
- **KEY ISSUE 3** looks at why different places on Earth have similar features. For example, what common features distinguish the bright areas in Figure 1-2?
- **KEY ISSUE 4** discusses sustainability. Distinctive to geography is the importance given to relationships between human activities and the physical environment. For example, what are the relationships between the tan areas in Figure 1-1 and the dark areas in Figure 1-2? This book focuses on human geography, but it doesn't forget that we also need to understand how humans interact with Earth's atmosphere, water, vegetation, and other living creatures.

▶ **FIGURE 1-1 SATELLITE IMAGE: DAYTIME** The composite image was assembled by the Geosphere Project of Santa Monica, California. Thousands of images were recorded over a 10-month period by satellites of the National Oceanographic and Atmospheric Administration. The images were then electronically assembled, much like a jigsaw puzzle.

KEY ISSUE 1

How Do Geographers Describe Where Things Are?

- Maps
- Contemporary Tools

The word *geography*, invented by the ancient Greek scholar Eratosthenes, is based on two Greek words. *Geo* means "Earth," and *graphy* means "to write." Geography is the study of where things are found on Earth's surface and the reasons for the locations. Human geographers ask two simple questions: Where are people and activities found on Earth? Why are they found there?

Thinking geographically is one of the oldest human activities (Figure 1-3). Perhaps the first geographer was a prehistoric human who crossed a river or climbed a hill, observed what was on the other side, returned home to tell about it, and scratched the route in the dirt. Perhaps the second geographer was a friend or relative who followed the dirt drawing to reach the other side.

Maps

Geography's most important tool for thinking spatially about the distribution of features across Earth is a map. A **map** is a two-dimensional or flat-scale model of Earth's surface, or a portion of it. A map is a scale model of the real world, made small enough to work with on a desk or computer. It can be a hasty here's-how-to-get-to-the-party sketch, an elaborate work of art, or a precise computer-generated product. For centuries, geographers have worked to perfect the science of mapmaking, called **cartography.**

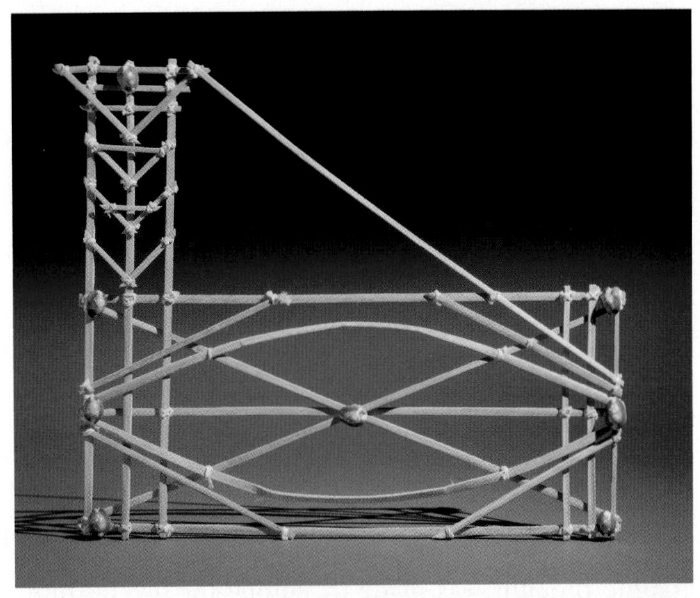

▲ FIGURE 1-3 **POLYNESIAN "STICK CHART"** A "stick chart" is a type of ancient map created by people living in the present-day Marshall Islands in the South Pacific Ocean. Islands were shown with shells, and patterns of swelling of waves were shown with palm strips.

Contemporary cartographers are assisted by computers and satellite imagery.

Geography is immediately distinguished from other disciplines by its reliance on maps to display and analyze information. A map serves two purposes:

- **As a reference tool.** A map helps us to find the shortest route between two places and to avoid getting lost along the way. We consult maps to learn where in the world something is found, especially in relationship to a place we know, such as a town, body of water, or highway. The maps in an atlas or a road map are especially useful for this purpose.
- **As a communications tool.** A map is often the best means for depicting the distribution of human activities or physical features, as well as for thinking about reasons underlying a distribution.

▶ FIGURE 1-2 **SATELLITE IMAGE: NIGHTTIME** The portion of Earth illuminated at night reflects the distribution of electricity. The dark areas are either sparsely inhabited areas, such as deserts and mountains, or areas where people are too poor to have electricity.

EARLY MAPMAKING

Learning Outcome 1.1.1
Explain differences between early maps and contemporary maps.

The earliest maps were reference tools—simple navigation devices designed to show a traveler how to get from Point A to Point B. Eratosthenes (276?–194? B.C.), the first person of record to use the word *geography*, prepared one of the earliest maps of the known world (Figure 1-4). Ptolemy (A.D. 100?–170?) produced maps that were not improved upon for more than 1,000 years, based on information collected by merchants and soldiers who traveled through the Roman Empire.

After Ptolemy, little progress in mapmaking or geographic thought was made in Europe for several hundred years. Maps became less mathematical and more fanciful, showing Earth as a flat disk surrounded by fierce animals and monsters. Geographic inquiry continued, though, outside Europe. Pei Xiu, the "father of Chinese cartography," produced an elaborate map of China in A.D. 267. Building on Ptolemy's long-neglected work, Muhammad al-Idrisi (1100–1165?), a Muslim geographer, prepared a world map and geography text in 1154 (Figure 1-5).

Mapmaking as a reference tool revived during the Age of Exploration and Discovery. Columbus, Magellan, and other explorers who sailed across the oceans in search of trade routes and resources in the fifteenth and sixteenth centuries required accurate maps to reach desired destinations without wrecking their ships. In turn, cartographers took information collected by the explorers to create more accurate maps. German cartographer Martin Waldseemuller (1470?–1520) produced the first map with the label "America"; he wrote on the map (translated from Latin) "from Amerigo the discoverer . . . as if it were the land of Americus, thus 'America'". Abraham Ortelius (1527–1598), a Flemish cartographer, created the first modern atlas (Figure 1-6).

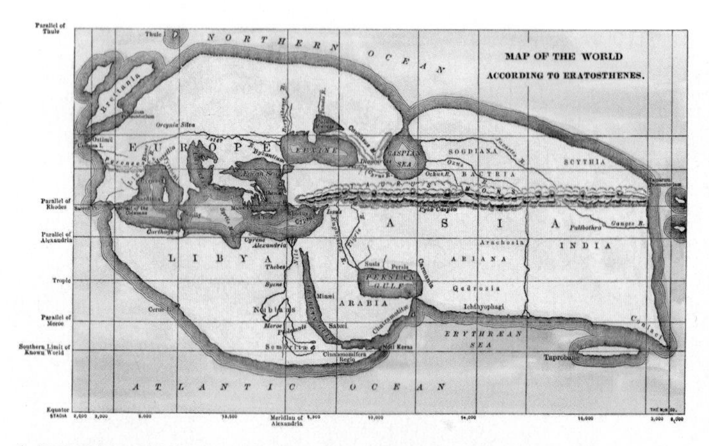

▲ **FIGURE 1-4 WORLD MAP BY ERATOSTHENES, 194? B.C.** This is a nineteenth-century reconstruction of the map produced by Eratosthenes.

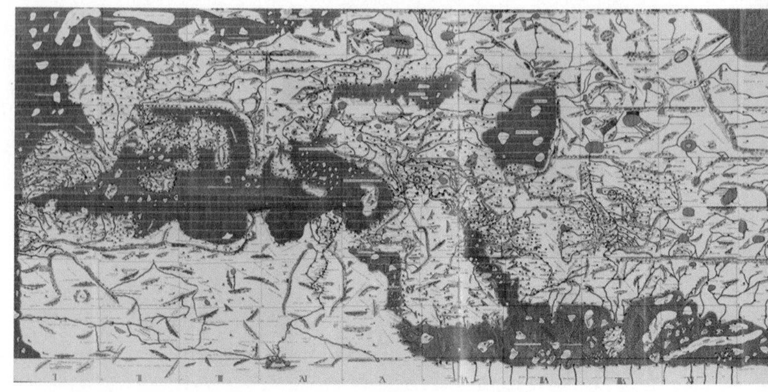

▲ **FIGURE 1-5 WORLD MAP BY AL-IDRISI, 1154** Al-Idrisi built on Ptolemy's map, which had been neglected for nearly a millennium.

By the seventeenth century, maps accurately displayed the outline of most continents and the positions of oceans. Bernhardus Varenius (1622–1650) produced *Geographia Generalis*, which stood for more than a century as the standard treatise on systematic geography.

Pause and Reflect 1.1.1
What is one main difference between Eratosthenes's world map (Figure 1-4) and the word map of Ortelius (Figure 1-6)?

CONTEMPORARY MAPPING

Contemporary maps are still created as tools of reference, but human geographers now make use of maps primarily as tools of communication. Maps are geographers' most essential tool for displaying geographic information and for offering geographic explanation. The feature on page 7 includes a contemporary use of maps to demonstrate issues of sustainability and inequality in New Orleans.

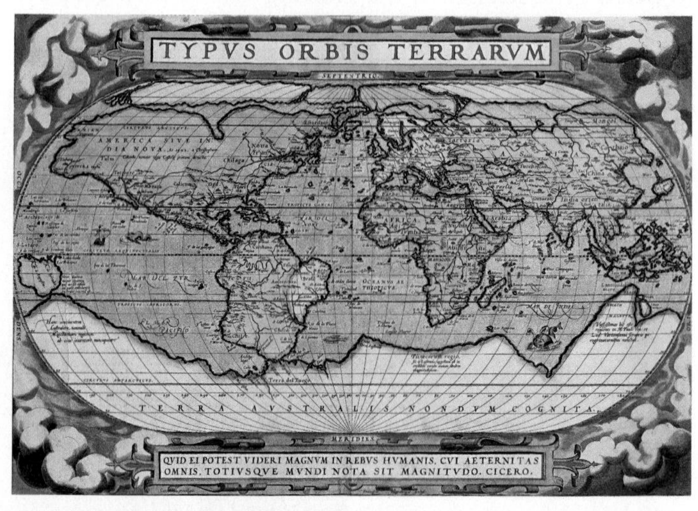

▲ **FIGURE 1-6 WORLD MAP BY ORTELIUS, 1571** This map was one of the first to show the extent of the Western Hemisphere, as well as Antarctica.

SUSTAINABILITY AND INEQUALITY IN OUR GLOBAL VILLAGE

Mapping a Disaster: Hurricane Katrina

Hurricane Katrina, one of the strongest hurricanes ever to hit the United States, struck in 2005. It killed 1,836 people and was the costliest natural disaster in U.S. history, measured in the dollar value of the destruction. The aftermath of Katrina provides a useful introduction to geographic perspectives on contemporary global issues of sustainability and inequality. Is a city like New Orleans—below sea level and protected by aging levees—sustainable in an era of rising sea levels and stronger hurricanes? Why did Katrina affect residents of New Orleans so unequally, with lower-income people much more likely to die or become homeless than more wealthy people?

Hurricanes such as Katrina form in the Atlantic Ocean during the late summer and autumn and gather strength over the warm waters of the Gulf of Mexico. When a hurricane passes over land, it can generate a powerful storm surge that floods low-lying areas. New Orleans was especially vulnerable because the site of the city is below sea level. To protect it and other low-lying cities from flooding, government agencies had constructed a complex system of levees, dikes, seawalls, canals, and pumps (Figure 1-7, left). Two days after the hurricane hit, the flood-protection levees in New Orleans broke, flooding 80 percent of the city (Figure 1-7, bottom).

Human geographers are especially concerned with the inequality of the destruction. Katrina's victims were primarily poor, African American, and older individuals (Figure 1-7, center). They lived in the lowest-lying areas, most vulnerable to flooding, and many lacked transportation, money, and information that would have enabled them to evacuate in advance of the storm. In contrast, the wealthy portions of New Orleans, such as tourist attractions like the Vieux Carré (French Quarter), were spared the worst because they were located on slightly higher ground. The slow and incompetent response to the destruction by local, state, and federal emergency teams was attributed by many analysts to the victims' lack of a voice in the political, economic, and social life of New Orleans and other impacted communities.

Inequalities persist several years after the hurricane (Figure 1-7, right). Five years after Katrina, according to the 2010 census, a large percentage of African Americans had still not returned to New Orleans. According to the census, the population of New Orleans declined from 484,674 in 2000 to 343,829 in 2010. African Americans accounted for 84 percent of the decline because most of the houses that remained damaged from the hurricane were in predominantly African American neighborhoods. The percentage of African Americans in New Orleans declined from 67 percent in 2000 to 60 percent in 2010.

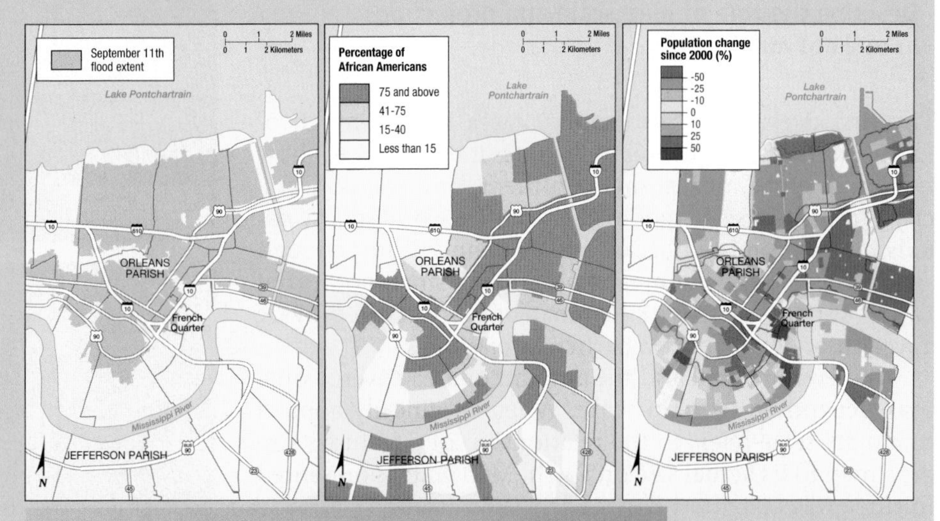

▲ FIGURE 1-7 **SUSTAINABILITY AND INEQUALITY IN NEW ORLEANS** (left) Extent of flooding in New Orleans from storm surge after Katrina. (middle) Two-thirds of the population of New Orleans was African American, but the area spared the flooding was less than one-fourth African American. (right) The percentage of homes that have been fixed up and reoccupied since Katrina is lower in the areas that had relatively large African American populations than in other areas. (bottom) Flooded neighborhood in New Orleans nine days after Katrina.

MAP SCALE

Learning Outcome 1.1.2
Describe the role of map scale and projections in making maps.

The first decision a cartographer faces is how much of Earth's surface to depict on the map. Is it necessary to show the entire globe, or just one continent, or a country, or a city? To make a scale model of the entire world, many details must be omitted because there simply is not enough space. Conversely, if a map shows only a small portion of Earth's surface, such as a street map of a city, it can provide a wealth of detail about a particular place.

The level of detail and the amount of area covered on a map depend on its **map scale.** When specifically applied to a map, scale refers to the relationship of a feature's size on a map to its actual size on Earth. Map scale is presented in three ways (Figure 1-8).

- A *ratio or fraction* shows the numerical ratio between distances on the map and Earth's surface. A scale of 1:24,000 or 1/24,000 means that 1 unit (for example, inch, centimeter, foot, finger length) on the map represents 24,000 of the same unit (for example, inch, centimeter, foot, finger length) on the ground. The unit chosen for distance can be anything, as long as the units of measure on both the map and the ground are the same. The 1 on the left side of the ratio always refers to a unit of distance *on the map,* and the number on the right always refers to the *same unit* of distance *on Earth's surface.*
- A *written scale* describes the relationship between map and Earth distances in words. For example, the statement "1 inch equals 1 mile" on a map means that 1 inch on the map represents 1 mile on Earth's surface. Again, the first number always refers to map distance and the second to distance on Earth's surface.
- A *graphic scale* usually consists of a bar line marked to show distance on Earth's surface. To use a bar line, first determine with a ruler the distance on the map in inches or centimeters. Then hold the ruler against the bar line and read the number on the bar line opposite the map distance on the ruler. The number on the bar line is the equivalent distance on Earth's surface.

Maps often display scale in more than one of these three ways.

The appropriate scale for a map depends on the information being portrayed. A map of a downtown area, such as Figure 1-8, bottom, may have a scale of 1:10,000, whereas a map of southeast Texas (Figure 1-8, top) may have a scale of 1:10,000,000. One inch represents about 1/6 mile on the downtown Houston map and about 170 miles on the southeast Texas map.

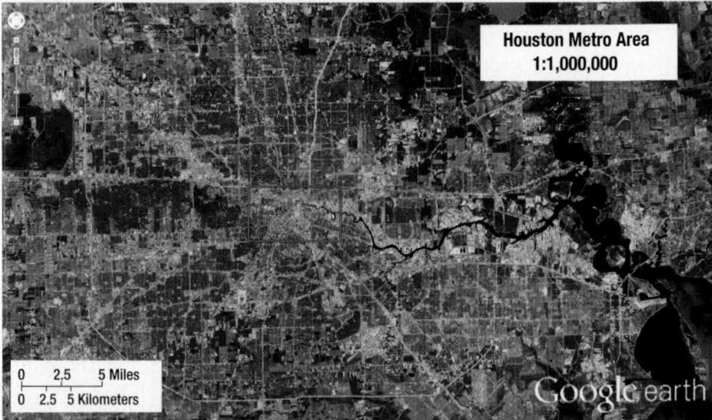

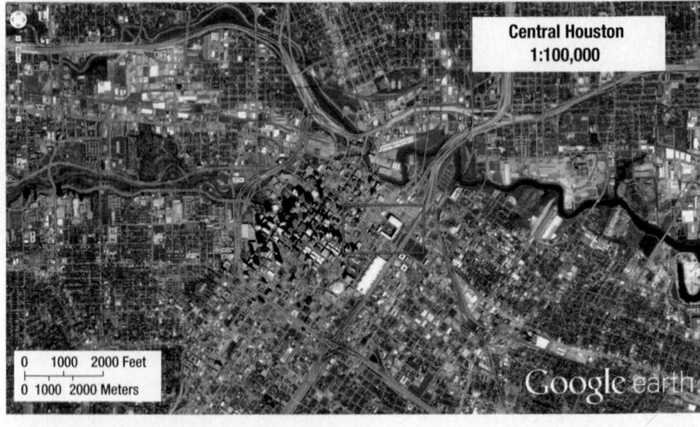

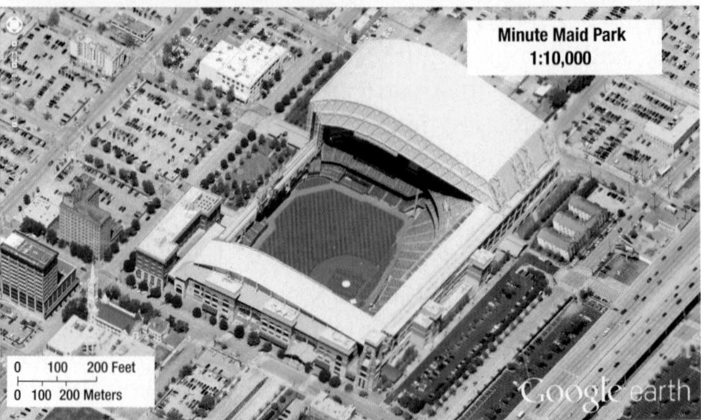

▲ **FIGURE 1-8 MAP SCALE** The four images show (top) southeast Texas (second), the city of Houston (third), downtown Houston, and (bottom) Minute Maid Park. The map of southeastern Texas has a fractional scale of 1:10,000,000. Expressed as a written statement, 1 inch on the map represents 10 million inches (about 158 miles) on the ground. Look what happens to the scale on the other three maps. As the area covered gets smaller, the maps get more detailed, and 1 inch on the map represents smaller distances.

At the scale of a small portion of Earth's surface, such as a downtown area, a map provides a wealth of details about the place. At the scale of the entire globe, a map must omit many details because of lack of space, but it can effectively communicate processes and trends that affect everyone.

PROJECTION

Earth is very nearly a sphere and is therefore accurately represented with a globe. However, a globe is an extremely limited tool with which to communicate information about Earth's surface. A small globe does not have enough space to display detailed information, whereas a large globe is too bulky and cumbersome to use. And a globe is difficult to write on, photocopy, display on a computer screen, or carry in the glove box of a car. Consequently, most maps—including those in this book—are flat. Three-dimensional maps can be made but are expensive and difficult to reproduce.

Earth's spherical shape poses a challenge for cartographers because drawing Earth on a flat piece of paper unavoidably produces some distortion. Cartographers have invented hundreds of clever methods of producing flat maps, but none has produced perfect results. The scientific method of transferring locations on Earth's surface to a flat map is called **projection** (Figure 1-9).

The problem of distortion is especially severe for maps depicting the entire world. Four types of distortion can result:

1. The *shape* of an area can be distorted, so that it appears more elongated or squat than in reality.
2. The *distance* between two points may become increased or decreased.
3. The *relative size* of different areas may be altered, so that one area may appear larger than another on a map but is in reality smaller.
4. The *direction* from one place to another can be distorted.

Most of the world maps in this book, such as Figure 1-9 center, are *equal area projections*. The primary benefit of this type of projection is that the relative sizes of the landmasses on the map are the same as in reality. The projection minimizes distortion in the shapes of most landmasses. Areas toward the North and South poles—such as Greenland and Australia—become more distorted, but they are sparsely inhabited, so distorting their shapes usually is not important.

To largely preserve the size and shape of landmasses, however, the projection in Figure 1-9 center forces other distortions:

- The Eastern and Western hemispheres are separated into two pieces, a characteristic known as interruption.
- The meridians (the vertical lines), which in reality converge at the North and South poles, do not converge at all on the map. Also, they do not form right angles with the parallels (the horizontal lines).
- The Robinson projection, in Figure 1-9 right, is useful for displaying information across the oceans. Its major disadvantage is that by allocating space to the oceans, the land areas are much smaller than on interrupted maps of the same size.
- The Mercator projection, in Figure 1-9 left, has several advantages: Shape is distorted very little, direction is consistent, and the map is rectangular. Its greatest disadvantage is that relative size is grossly distorted toward the poles, making high-latitude places look much larger than they actually are.

Pause and Reflect 1.1.2
What type of projection would be best for a world map of population density? Why?

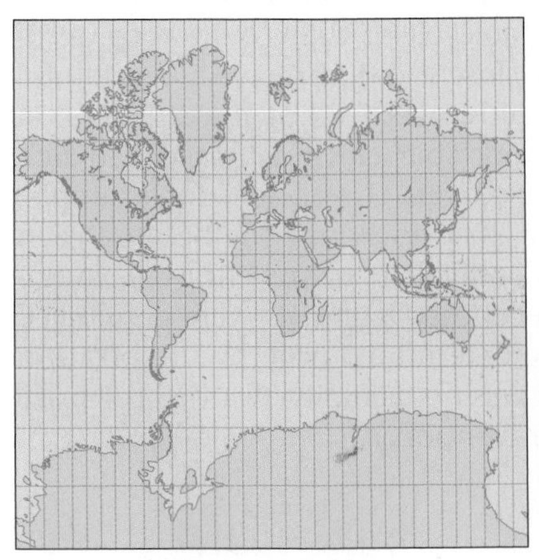

Mercator Projection

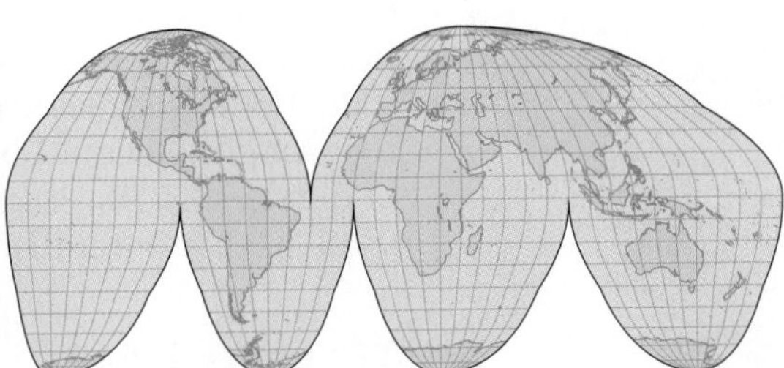

Goode Homolosine Projection

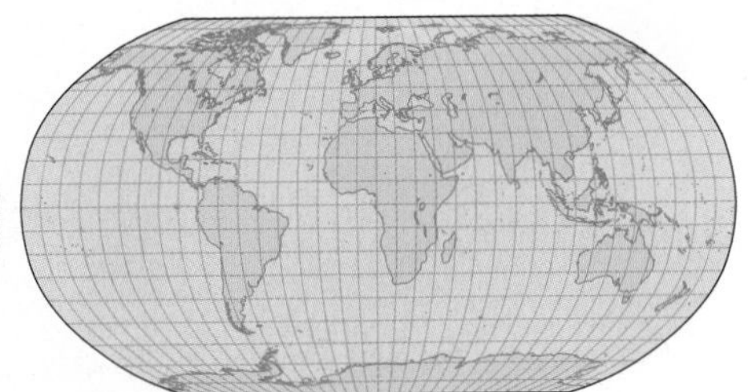

Robinson Projection

▲ **FIGURE 1-9 PROJECTION**
(left) Mercator projection, (center) equal area projection, (right) Robinson projection. Compare the sizes of Greenland and South America on these maps. Which of the two landmasses is actually larger?

GEOGRAPHIC GRID

Learning Outcome 1.1.3

Explain how latitude and longitude are used to locate points on Earth's surface.

The geographic grid is a system of imaginary arcs drawn in a grid pattern on Earth's surface. The location of any place on Earth's surface can be described precisely by meridians and parallels, two sets of imaginary arcs drawn in a grid pattern on Earth's surface (Figure 1-10). The geographic grid plays an important role in telling time:

- A **meridian** is an arc drawn between the North and South poles. The location of each meridian is identified on Earth's surface according to a numbering system known as **longitude.**

The meridian that passes through the Royal Observatory at Greenwich, England, is 0° longitude, also called the **prime meridian.** The meridian on the opposite side of the globe from the prime meridian is 180° longitude. All other meridians have numbers between 0° and 180° east or west, depending on whether they are east or west of the prime meridian. For example, Belo Horizonte, Brazil, is located at 44° west longitude and Baghdad, Iraq, at 44° east longitude.

- A **parallel** is a circle drawn around the globe parallel to the equator and at right angles to the meridians. The numbering system to indicate the location of a parallel is called **latitude.**

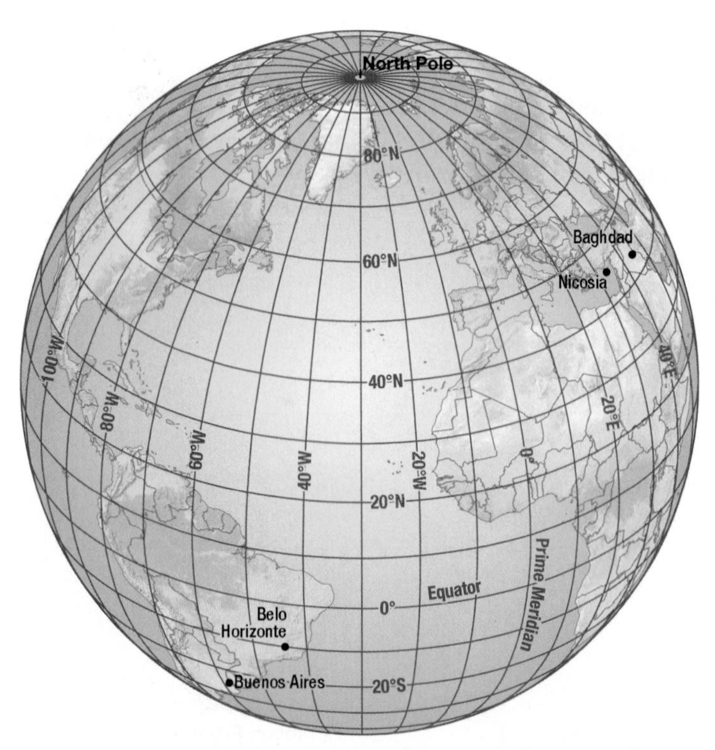

▲ **FIGURE 1-10 GEOGRAPHIC GRID** Meridians are arcs that connect the North and South poles. The meridian through Greenwich, England, is the prime meridian, or 0° longitude. Parallels are circles drawn around the globe parallel to the equator. The equator is 0° latitude, and the North Pole is 90° north latitude.

The equator is 0° latitude, the North Pole 90° north latitude, and the South Pole 90° south latitude. Nicosia, Cyprus, is located at 35° north latitude and Buenos Aires, Argentina, at 35° south latitude.

Latitude and longitude are used together to identify locations. For example, Denver, Colorado, is located at 40° north latitude and 105° west longitude.

The mathematical location of a place can be designated more precisely by dividing each degree into 60 minutes (′) and each minute into 60 seconds (″). For example, the official mathematical location of Denver, Colorado, is 39°44′ north latitude and 104°59′ west longitude. The state capitol building in Denver is located at 39°42′2″ north latitude and 104°59′04″ west longitude. GPS systems typically divide degrees into decimal fractions rather than minutes and seconds. The Colorado state capitol, for example, is located at 39.714444° north latitude and 84.984444° west longitude.

Measuring latitude and longitude is a good example of how geography is partly a natural science and partly a study of human behavior. Latitudes are scientifically derived by Earth's shape and its rotation around the Sun. The equator (0° latitude) is the parallel with the largest circumference and is the place where every day has 12 hours of daylight. Even in ancient times, latitude could be accurately measured by the length of daylight and the position of the Sun and stars.

On the other hand, 0° longitude is a human creation. Any meridian could have been selected as 0° longitude because all meridians have the same length and all run between the poles. The 0° longitude runs through Greenwich because England was the world's most powerful country when longitude was first accurately measured and the international agreement was made.

Inability to measure longitude was the greatest obstacle to exploration and discovery for many centuries. Ships ran aground or were lost at sea because no one on board could pinpoint longitude. In 1714, the British Parliament enacted the Longitude Act, which offered a prize equivalent to several million in today's dollars to the person who could first measure longitude accurately.

English clockmaker John Harrison won the prize by inventing the first portable clock that could keep accurate time on a ship—because it did not have a pendulum. When the Sun was directly overhead of the ship—noon local time—Harrison's portable clock set to Greenwich time could say it was 2 P.M. in Greenwich, for example, so the ship would be at 30° west longitude because each hour of difference was equivalent to traveling 15° longitude. (Most eighteenth-century scientists were convinced that longitude could be determined only by the position of the stars, so Harrison was not actually awarded the prize until 40 years after his invention.)

TELLING TIME

Longitude plays an important role in calculating time. Earth as a sphere is divided into 360° of longitude (the degrees from 0° to 180° west longitude plus the degrees from 0° to 180° east longitude).

As Earth rotates daily, these 360 imaginary lines of longitude pass beneath the cascading sunshine. If we let every fifteenth degree of longitude represent one time zone, and divide the 360° by 15°, we get 24 time zones, or one for each hour of the day. By international agreement, **Greenwich Mean Time (GMT)**, or Universal Time (UT), which is the time at the prime meridian (0° longitude), is the master reference time for all points on Earth.

Each 15° band of longitude is assigned to a standard time zone (Figure 1-11). The eastern United States, which is near 75° west longitude, is therefore 5 hours earlier than GMT (the 75° difference between the prime meridian and 75° west longitude, divided by 15° per hour, equals 5 hours). Thus when the time in New York City in the winter is 1:32 P.M. (or 13:32 hours, using a 24-hour clock), it is 6:32 P.M. (or 18:32 hours) GMT. During the summer, many places in the world, including most of North America, move the clocks ahead one hour; so in the summer when it is 6:32 P.M. GMT, the time in New York City is 2:32 P.M.

When you cross the **International Date Line,** which, for the most part, follows 180° longitude, you move the clock back 24 hours, or one entire day, if you are heading eastward toward America. You turn the clock ahead 24 hours if you are heading westward toward Asia. To see the need for the International Date Line, try counting the hours around the world from the time zone in which you live. As you go from west to east, you add 1 hour for each time zone. When you return to your starting point, you will reach the absurd conclusion that it is 24 hours later in your locality than it really is. Therefore—if it is 6:32 A.M. *Monday* in Auckland, when you get to Honolulu, it will be 8:32 A.M. *Sunday* because the International Date Line lies between Auckland and Honolulu.

The International Date Line for the most part follows 180° longitude. However, several islands in the Pacific Ocean belonging to the countries of Kiribati and Samoa, as well as to New Zealand's Tokelau territory, moved the International Date Line several thousand kilometers to the east. Samoa and Tokelau moved it in 2011 so that they could be on the same day as Australia and New Zealand, their major trading partners. Kiribati moved it in 1997 so that it would be the first country to see each day's sunrise. Kiribati hoped that this feature would attract tourists to celebrate the start of the new millennium on January 1, 2000 (or January 1, 2001, when sticklers pointed out the new millennium really began). But it did not.

Pause and Reflect 1.1.3

Compare the stick chart in Figure 1-3 with the geographic grid in Figure 1-10. What are their similarities and differences?

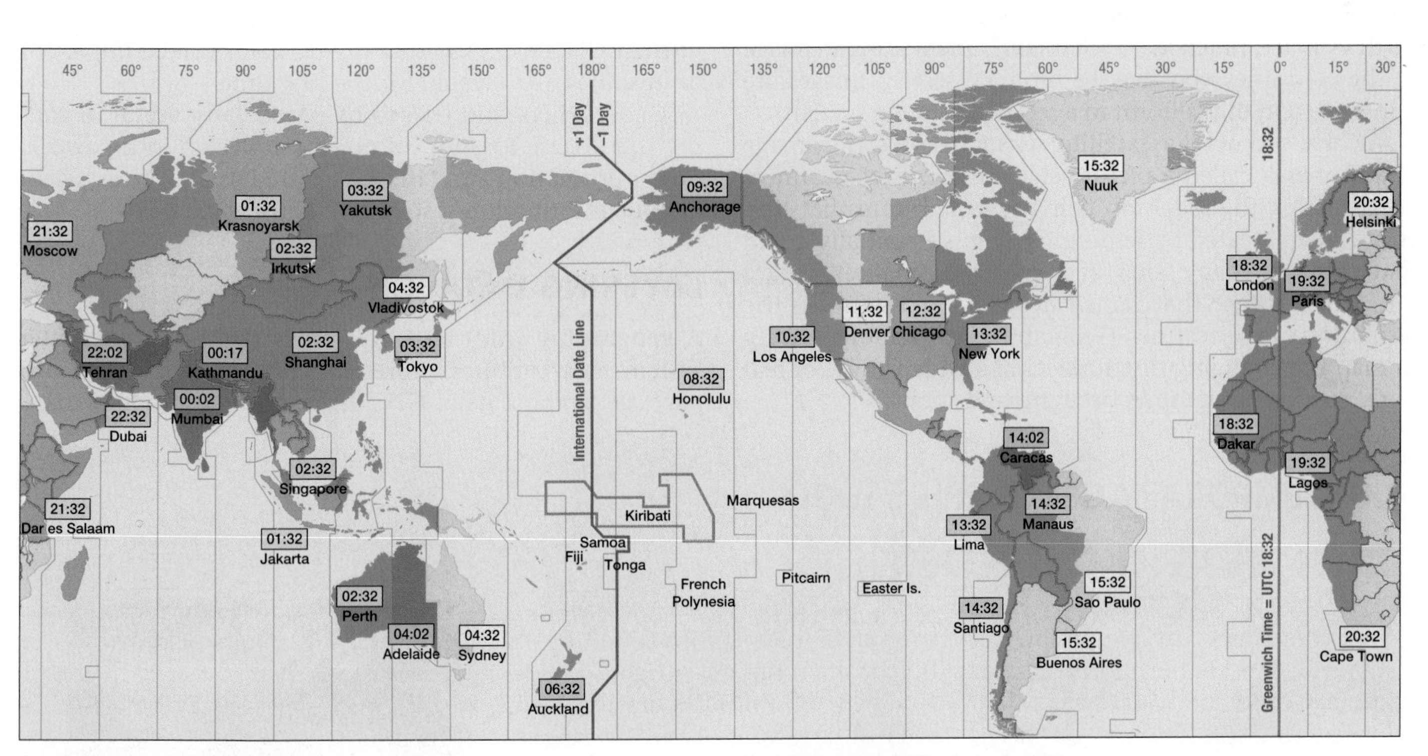

▲ **FIGURE 1-11 TIME ZONES**

The United States and Canada share four standard time zones:

- Eastern, near 75° west, is 5 hours earlier than GMT.
- Central, near 90° west, is 6 hours earlier than GMT.
- Mountain, near 105° west, is 7 hours earlier than GMT.
- Pacific, near 120° west, is 8 hours earlier than GMT.

The United States has two additional standard time zones:

- Alaska, near 135° west, is 9 hours earlier than GMT.
- Hawaii-Aleutian, near 150° west, is 10 hours earlier than GMT.

Canada has two additional standard time zones:

- Atlantic, near 60° west, is 4 hours earlier than GMT.
- Newfoundland is 3½ hours earlier than GMT; the residents of Newfoundland assert that their island, which lies between 53° and 59° west longitude, would face dark winter afternoons if it were in the Atlantic Time Zone and dark winter mornings if it were 3 hours earlier than GMT.

Contemporary Tools

Learning Outcome 1.1.4

Identify contemporary analytic tools, including remote sensing, GPS, and GIS.

Having largely completed the formidable task of accurately mapping Earth's surface, geographers have turned to **geographic information science (GIScience)**, which involves the development and analysis of data about Earth acquired through satellite and other electronic information technologies. GIScience helps geographers to create more accurate and complex maps and to measure changes over time in the characteristics of places.

GIScience is made possible by satellites in orbit above Earth sending information to electronic devices on Earth to record and interpret information. Satellite-based information allows us to know the precise location of something on Earth and data about that place.

COLLECTING DATA: REMOTE SENSING

The acquisition of data about Earth's surface from a satellite orbiting Earth or from other long-distance methods is known as **remote sensing.** Remote-sensing satellites scan Earth's surface, much like a television camera scans an image in the thin lines you can see on a TV screen. Images are transmitted in digital form to a receiving station on Earth.

At any moment a satellite sensor records the image of a tiny area called a picture element, or pixel. Scanners are detecting the radiation being reflected from that tiny area. A map created by remote sensing is essentially a grid that contains many rows of pixels. The smallest feature on Earth's surface that can be detected by a sensor is the resolution of the scanner. Geographers use remote sensing to map the changing distribution of a wide variety of features, such as agriculture, drought, and sprawl.

PINPOINTING LOCATIONS: GPS

The system that accurately determines the precise position of something on Earth is the **Global Positioning System (GPS).** The GPS in the United States includes three elements:

- Satellites placed in predetermined orbits by the U.S. military (24 in operation and 3 in reserve).
- Tracking stations to monitor and control the satellites.
- A receiver that can locate at least 4 satellites, figure out the distance to each, and use this information to pinpoint its own location.

GPS is most commonly used for navigation, as discussed in the Contemporary Geographic Tools box. Pilots of aircraft and ships stay on course with GPS. On land, GPS detects a vehicle's current position, the motorist programs the desired destination into a GPS device, and the device provides instructions on how to reach the destination. GPS can also be used to find the precise location of a vehicle, enabling a motorist to summon help in an emergency or monitoring the progress of a delivery truck or position of a city bus. Cell phones equipped with GPS allow individuals to share their whereabouts with others.

GPS devices enable private individuals to contribute to the production of accurate digital maps, through web sites such as Google's OpenStreetMap.org. Travelers can enter information about streets, buildings, and bodies of water in their GPS devices so that digital maps can be improved or in some cases created for the first time.

Geographers find GPS to be particularly useful in coding the precise location of objects collected in fieldwork. That information can later be entered as a layer in a geographic information system (GIS), discussed next.

LAYERING DATA: GIS

A **geographic information system (GIS)** is a computer system that captures, stores, queries, analyzes, and displays geographic data. GIS can be used to produce maps

CONTEMPORARY GEOGRAPHIC TOOLS
Electronic Navigation

Two companies are responsible for supplying most of the information fed into navigation devices: Navteq, short for Navigation Technologies, and Tele Atlas, originally known as Etak. Tele Atlas, based in the Netherlands was founded in 1984, and Navteq, based in the United States, was founded a year later. Navteq and Tele Atlas get their information from what they call "ground truthing." Hundreds of field researchers drive around, building the database. One person drives, while the other feeds information into a notebook computer (Figure 1-12). Hundreds of attributes are recorded, such as crosswalks, turn restrictions, and name changes. Thus, electronic navigation systems ultimately depend on human observation.

A reflection of the growing importance of navigation technology, Navteq and Tele Atlas were both acquired in 2008 by larger communications companies (Nokia and Tom Tom, respectively).

▲ **FIGURE 1-12 GPS** Navteq researchers at work in Florida.

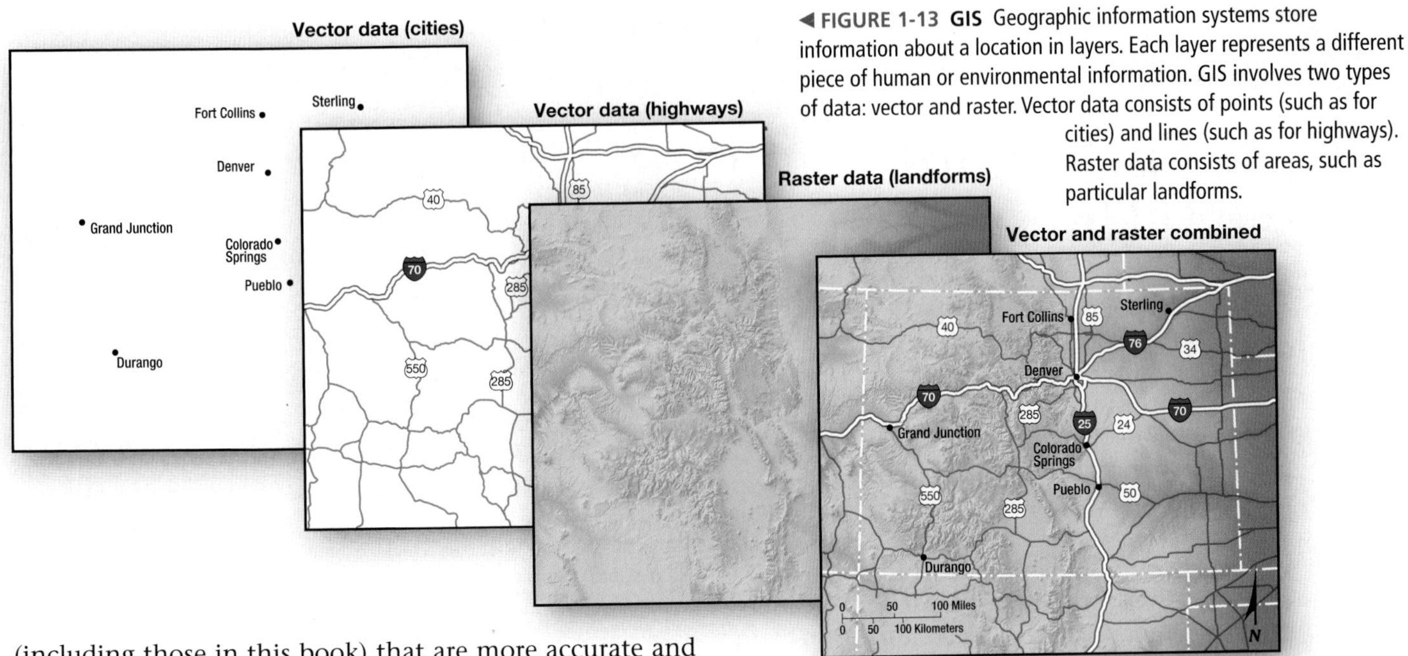

◀ **FIGURE 1-13 GIS** Geographic information systems store information about a location in layers. Each layer represents a different piece of human or environmental information. GIS involves two types of data: vector and raster. Vector data consists of points (such as for cities) and lines (such as for highways). Raster data consists of areas, such as particular landforms.

(including those in this book) that are more accurate and attractive than those drawn by hand.

The position of any object on Earth can be measured and recorded with mathematical precision and then stored in a computer. A map can be created by asking the computer to retrieve a number of stored objects and combine them to form an image. In the past, when cartographers drew maps with pen and paper, a careless moment could result in an object being placed in the wrong location, and a slip of the hand could ruin hours of work. GIS is more efficient than pen and ink for making a map: Objects can be added or removed, colors brightened or toned down, and mistakes corrected (as long as humans find them!) without having to tear up the paper and start from scratch.

Each type of information can be stored in a layer. For example, separate layers could be created for boundaries of countries, bodies of water, roads, and names of places. A simple map might display only a single layer by itself, but most maps combine several layers (Figure 1-13), and GIS permits construction of much more complex maps than can be drawn by hand.

Layers can be compared to show relationships among different kinds of information. For example, to protect hillsides from development, a geographer may wish to compare a layer of recently built houses with a layer of steep slopes. GIS enables geographers to calculate whether relationships between objects on a map are significant or merely coincidental. For example, maps showing where cancer rates are relatively high and low (such as those in Figure 1-25) can be combined with layers showing the location of people with various incomes and ethnicities, the location of different types of factories, and the location of mountains and valleys.

MIXING DATA: MASHUPS

Computer users have the ability to do their own GIS because mapping services provide access to the application programming interface (API), which is the language that links a database such as an address list with software such as mapping. The API for mapping software, available at such sites as www.google.com/apis/maps, enables a computer programmer to create a mashup that places data on a map.

The term *mashup* refers to the practice of overlaying data from one source on top of one of the mapping services; the term comes from the hip-hop practice of mixing two or more songs. A mashup map can show the locations of businesses and activities near a particular street or within a neighborhood in a city. The requested information could be all restaurants within 1 kilometer (0.6 mile) of an address or, to be even more specific, all pizza parlors. Mapping software can show the precise locations of commercial airplanes currently in the air, the gas stations with the lowest prices, and current traffic tie-ups on highways and bridges.

Pause and Reflect 1.1.4

State a question you have about the area where you live. Now describe a mashup that you could create using GIS that would answer your question.

CHECK-IN: KEY ISSUE 1

How Do Geographers Describe Where Things Are?

- ✓ **Maps are tools of reference and increasingly tools of communication. Reading a map requires recognizing its scale and projection.**
- ✓ **Contemporary mapping utilizes electronic technologies, such as remote sensing, GPS, and GIS.**

KEY ISSUE 2

Why Is Each Point on Earth Unique?

- Place: A Unique Location
- Region: A Unique Area

Learning Outcome 1.2.1
Identify geographic characteristics of places, including toponym, site, and situation.

A **place** is a specific point on Earth distinguished by a particular characteristic. Every place occupies a unique location, or position, on Earth's surface. Although each place on Earth is in some respects unique, in other respects it is similar to other places. The interplay between the uniqueness of each place and the similarities among places lies at the heart of geographic inquiry into why things are found where they are.

Place: A Unique Location

Humans possess a strong sense of place—that is, a feeling for the features that contribute to the distinctiveness of a particular spot on Earth—perhaps a hometown, vacation destination, or part of a country. Describing the features of a place is an essential building block for geographers to explain similarities, differences, and changes across Earth. Geographers think about where particular places are located and the combination of features that make each place on Earth distinct.

Geographers describe a feature's place on Earth by identifying its **location**, the position that something occupies on Earth's surface. In doing so, they consider three ways to identify location: place name, site, and situation.

PLACE NAMES

Because all inhabited places on Earth's surface—and many uninhabited places—have been named, the most straightforward way to describe a particular location is often by referring to its place name. A **toponym** is the name given to a place on Earth.

A place may be named for a person, perhaps its founder or a famous person with no connection to the community, such as George Washington. Some settlers select place names associated with religion, such as St. Louis and St. Paul, whereas other names derive from ancient history, such as Athens, Attica, and Rome, or from earlier occupants of the place (Figure 1-14).

A place name may also indicate the origin of its settlers. Place names commonly have British origins in North America and Australia, Portuguese origins in Brazil, Spanish origins elsewhere in Latin America, and Dutch origins in South Africa. Some place names derive from features of the physical environment. Trees, valleys, bodies of water, and other natural features appear in the place names of most languages.

▲ **FIGURE 1-14 LONGEST U.S. PLACE NAME** The longest place name in the United States may be Lake Chargoggagoggmanchauggagoggchaubunagungamaugg, Massachusetts. One hypothesis is that the name is Algonquian language for "fishing place at the boundaries—neutral meeting grounds." Others believe that the original meaning is unknown, and the current meaning and spelling are recent inventions.

The Board of Geographical Names, operated by the U.S. Geological Survey, was established in the late nineteenth century to be the final arbiter of names on U.S. maps. In recent years the board has been especially concerned with removing offensive place names, such as those with racial or ethnic connotations.

SITE

The second way that geographers describe the location of a place is by **site**, which is the physical character of a place. Important site characteristics include climate, water sources, topography, soil, vegetation, latitude, and elevation. The combination of physical features gives each place a distinctive character.

Site factors have always been essential in selecting locations for settlements, although people have disagreed on the attributes of a good site, depending on cultural values. Some have preferred a hilltop site for easy defense from attack. Others have located settlements near convenient river-crossing points to facilitate communication with people in other places.

Humans have the ability to modify the characteristics of a site. Central Boston is more than twice as large today as it was during colonial times (Figure 1-15). Colonial Boston was a peninsula connected to the mainland by a very narrow neck. During the nineteenth century, a dozen major projects filled in most of the bays, coves, and marshes. A major twentieth-century landfill project created Logan Airport. Several landfill projects continue into the twenty-first century. The central areas of New York and Tokyo have also been expanded through centuries of landfilling in nearby bodies of water, substantially changing these sites.

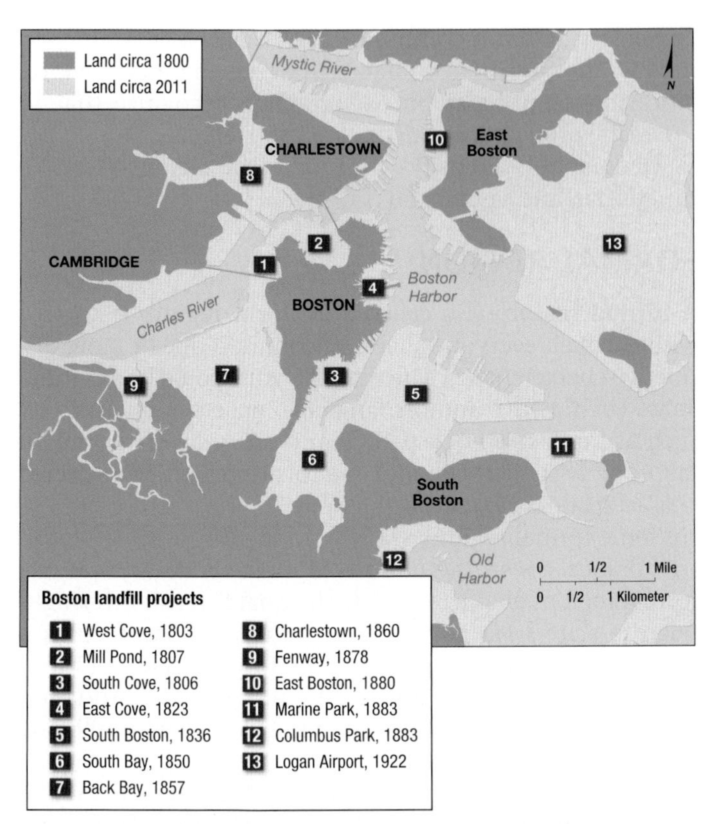

▲ **FIGURE 1-15 CHANGING SITE OF BOSTON** The site of Boston has been altered by filling in much of Boston Harbor, primarily during the nineteenth century.

SITUATION

Situation is the location of a place relative to other places. Situation is a valuable way to indicate location, for two reasons—finding an unfamiliar place and understanding its importance.

First, situation helps us find an unfamiliar place by comparing its location with a familiar one. We give directions to people by referring to the situation of a place: "It's down past the courthouse, on Locust Street, after the third traffic light, beside the yellow-brick bank." We identify important buildings, streets, and other landmarks to direct people to the desired location.

Second, situation helps us understand the importance of a location. Many locations are important because they are accessible to other places. For example, because of its situation, Shanghai has become a center for the trading and distribution of goods across Asia and the Pacific Ocean (Figure 1-16). Shanghai is situated near the confluence of the Yangtze River and the East China Sea. The port of Shanghai has become the world's largest.

Pause and Reflect 1.2.1

How would you describe the site and situation of the place where you live? (Use online maps or an atlas to help analyze the characteristics of your location.)

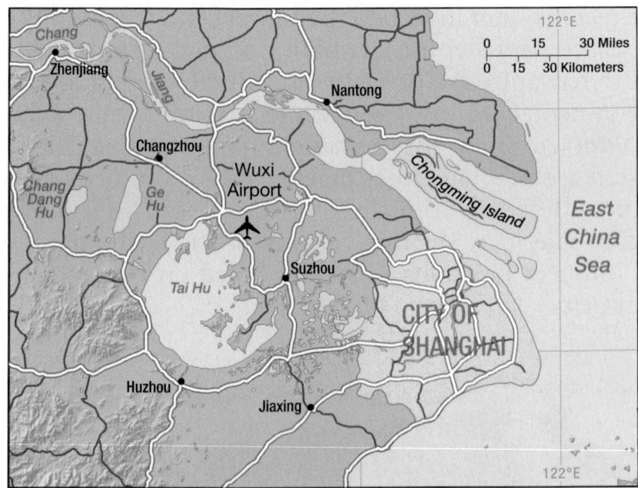

▲ **FIGURE 1-16 SITE AND SITUATION OF SHANGHAI** The site of the city of Shanghai is along the south bank of Yangtze River. The situation of Shanghai, near the mouth of the Yangtze, where it flows into the East China Sea, is critical in making the city the world's largest port.

Region: A Unique Area

Learning Outcome 1.2.2
Identify the three types of regions.

The "sense of place" that humans possess may apply to a larger area of Earth rather than to a specific point. An area of Earth defined by one or more distinctive characteristics is a **region.** A particular place can be included in more than one region, depending on how the region is defined.

The designation *region* can be applied to any area larger than a point and smaller than the entire planet. Geographers most often apply the concept at one of two scales:

- Several neighboring countries that share important features, such as those in Latin America.
- Many localities within a country, such as those in southern California.

A region derives its unified character through the **cultural landscape**—a combination of cultural features such as language and religion, economic features such as agriculture and industry, and physical features such as climate and vegetation. The southern California region can be distinguished from the northern California region, for example.

The contemporary **cultural landscape approach** in geography—sometimes called the **regional studies** approach—was initiated in France by Paul Vidal de la Blache (1845–1918) and Jean Brunhes (1869–1930). It was later adopted by several American geographers, including Carl Sauer (1889–1975) and Robert Platt (1880–1950). Sauer defined cultural landscape as an area fashioned from nature by a cultural group. "Culture is the agent, the natural area the medium, the cultural landscape is the result."

People, activities, and environment display similarities and regularities within a region and differ in some way from those of other regions. A region gains uniqueness from possessing not a single human or environmental characteristic but a combination of them. Not content to merely identify these characteristics, geographers seek relationships among them. Geographers recognize that in the real world, characteristics are integrated.

Geographers identify three types of regions—formal, functional, and vernacular.

FORMAL REGION

A **formal region**, also called a **uniform region**, is an area within which everyone shares in common one or more distinctive characteristics. The shared feature could be a cultural value such as a common language, an economic activity such as production of a particular crop, or an environmental property such as climate. In a formal region, the selected characteristic is present throughout.

Some formal regions are easy to identify, such as countries or local government units. Montana is an example of a formal region, characterized with equal intensity throughout the state by a government that passes laws, collects taxes, and issues license plates. The formal region of Montana has clearly drawn and legally recognized boundaries, and everyone living within them shares the status of being subject to a common set of laws.

In other kinds of formal regions, a characteristic may be predominant rather than universal. For example, we can distinguish formal regions within the United States characterized by a predominant voting for Republican candidates, although Republicans do not get 100 percent of the votes in these regions—nor in fact do they always win (Figure 1-17).

A cautionary step in identifying formal regions is the need to recognize the diversity of cultural, economic, and environmental factors, even while making a generalization. Problems may arise because a minority of people in a region speak a language, practice a religion, or possess resources different from those of the majority. People in a region may play distinctive roles in the economy and hold different positions in society based on their gender or ethnicity.

▲ **FIGURE 1-17 FORMAL REGIONS** The three maps show the winner by region in the (left) 2004, (center) 2008, and (right) 2012 presidential elections. The extensive areas of support for Democrats (blue) and Republicans (red) are examples of formal regions. (left) In 2004, Democrat John Kerry won most of the states in the Northeast, Upper Midwest, and Pacific Coast regions, while Republican George W. Bush won the remaining regions. (center) In 2008, Democrat Barack Obama won the election by capturing some states in regions that had been won entirely by the Republican four years earlier. (right) In 2012, Democrat Obama won reelection because he carried nearly the same states as four years earlier.

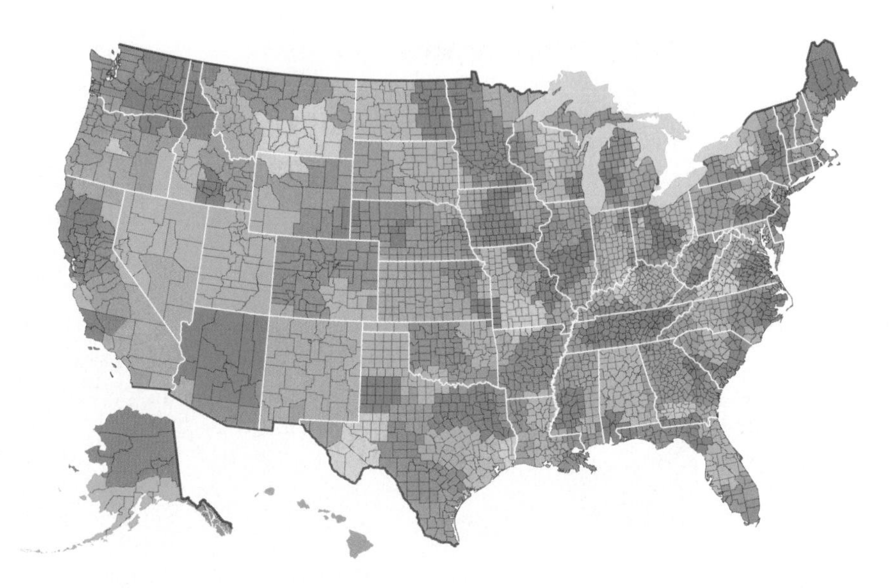

◀ **FIGURE 1-18 FUNCTIONAL REGIONS** The United States is divided into functional regions based on television markets, which are groups of counties served by a collection of TV stations. Many of these TV market functional regions cross state lines.

FUNCTIONAL REGION

A **functional region**, also called a **nodal region**, is an area organized around a node or focal point. The characteristic chosen to define a functional region dominates at a central focus or node and diminishes in importance outward. The region is tied to the central point by transportation or communications systems or by economic or functional associations.

Geographers often use functional regions to display information about economic areas. A region's node may be a shop or service, with the boundaries of the region marking the limits of the trading area of the activity. People and activities may be attracted to the node, and information may flow from the node to the surrounding area.

An example of a functional region is the reception area of a TV station. A TV station's signal is strongest at the center of its service area (Figure 1-18). At some distance from the center, more people are watching a station originating in another city. That place is the boundary between the nodal regions of the two TV market areas. Similarly, a department store attracts fewer customers from the edge of a trading area, and beyond that edge, customers will most likely choose to shop elsewhere.

New technology is breaking down traditional functional regions. TV stations are broadcast to distant places by cable, satellite, or Internet and through the Internet customers can shop at distant stores.

VERNACULAR REGION

A **vernacular region**, or **perceptual region**, is an area that people believe exists as part of their cultural identity. Such regions emerge from people's informal sense of place rather than from scientific models developed through geographic thought.

A useful way to identify a perceptual region is to get someone to draw a **mental map**, which is an internal representation of a portion of Earth's surface. A mental map depicts what an individual knows about a place, containing personal impressions of what is in the place and where the place is located. On a college campus, a senior is likely to have a more detailed and "accurate" map than a first-year student.

As an example of a vernacular region, Americans frequently refer to the South as a place with environmental, cultural, and economic features perceived to be quite distinct from those of the rest of the United States (Figure 1-19). Many of these features can be measured. Economically, the South is a region of high cotton production and low high school graduation rates. Culturally, the South includes the states that joined the Confederacy during the Civil War and where Baptist is the most prevalent religious denomination. Environmentally, the South is a region where the last winter frost occurs in March, and rainfall is more plentiful in winter than in summer. Southerners and other Americans alike share a strong sense of the American South as a distinctive place that transcends geographic measurement. The perceptual region known as the South is a source of pride to many Americans—and for others it is a place to avoid.

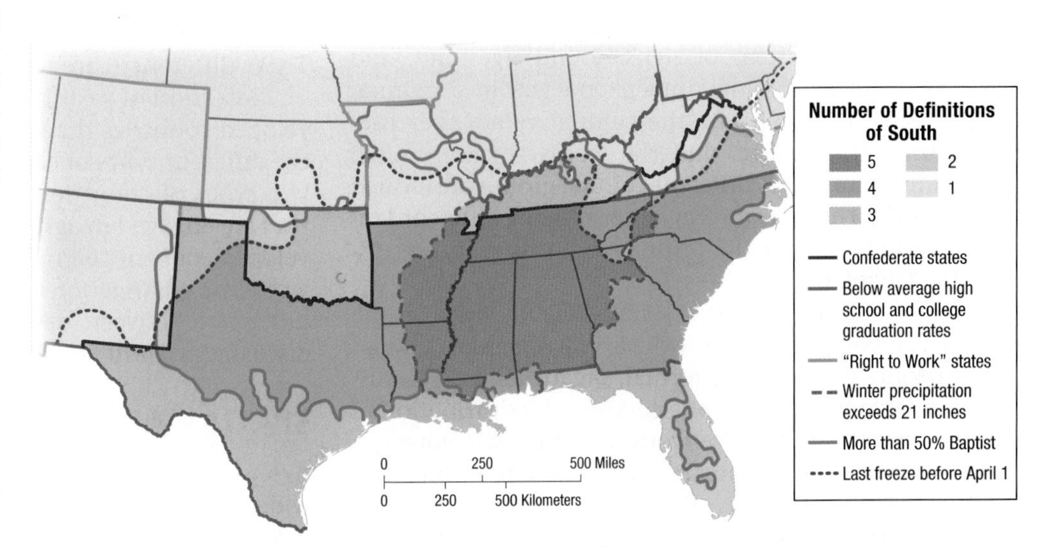

▶ **FIGURE 1-19 VERNACULAR REGIONS** The South is popularly distinguished as a distinct vernacular region within the United States, according to a number of factors, such as mild climate, propensity for growing cotton, and importance of the Baptist Church.

REGIONS OF CULTURE

Learning Outcome 1.2.3
Describe two geographic definitions of culture.

In thinking about *why* each region on Earth is distinctive, geographers refer to **culture**, which is the body of customary beliefs, material traits, and social forms that together constitute the distinct tradition of a group of people. Geographers distinguish groups of people according to important cultural characteristics, describe where particular cultural groups are distributed, and offer reasons to explain the observed distribution.

In everyday language, we think of *culture* as the collection of novels, paintings, symphonies, and other works produced by talented individuals. A person with a taste for these intellectual outputs is said to be "cultured." Intellectually challenging culture is often distinguished from *popular* culture, such as TV. *Culture* also refers to small living organisms, such as those found under a microscope or in yogurt. *Agriculture* is a term for the growing of living material at a much larger scale than in a test tube.

The origin of the word *culture* is the Latin *cultus*, which means "to care for." Culture is a complex concept because "to care for" something has two very different meanings:

- To care *about*—to adore or worship something, as in the modern word *cult*
- To take care *of*—to nurse or look after something, as in the modern word *cultivate*

Geography looks at both of these facets of the concept of culture to see why each region in the world is unique.

CULTURE: WHAT PEOPLE CARE ABOUT. Geographers study why the customary ideas, beliefs, and values of a people produce a distinctive culture in a particular place. Especially important cultural values derive from a group's language, religion, and ethnicity. These three cultural traits are both an excellent way of identifying the location of a culture and the principal means by which cultural values become distributed around the world.

Language is a system of signs, sounds, gestures, and marks that have meanings understood within a cultural group. People communicate the cultural values they care about through language, and the words themselves tell something about where different cultural groups are located (Figure 1-20). The distribution of speakers of different languages and reasons for the distinctive distribution are discussed in Chapter 5.

Religion is an important cultural value because it is the principal system of attitudes, beliefs, and practices through which people worship in a formal, organized way. As discussed in Chapter 6, geographers look at the distribution of religious groups around the world and the different ways that the various groups interact with their environment.

Ethnicity encompasses a group's language, religion, and other cultural values, as well as its physical traits. A group possesses these cultural and physical characteristics as a product of its common traditions and heredity. As addressed in Chapter 7, geographers find that problems of conflict and inequality tend to occur in places where more than one ethnic group inhabits and seeks to organize the same territory.

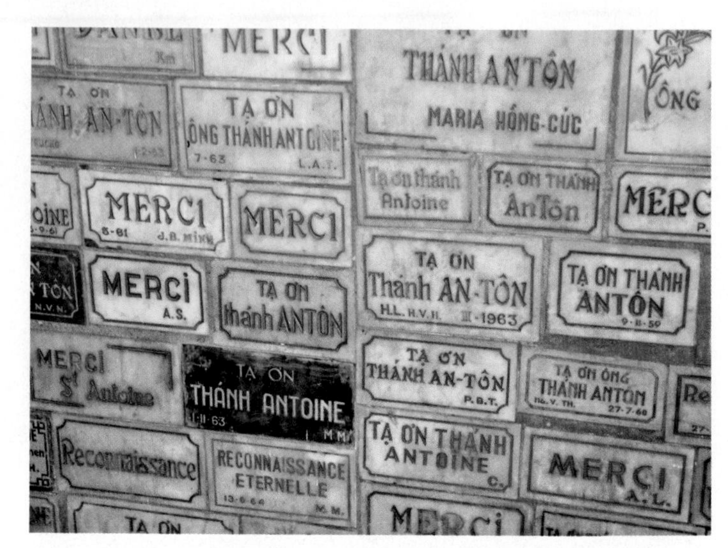

▲ **FIGURE 1-20 CULTURE: WHAT PEOPLE CARE ABOUT** Language and religion are important elements of culture that people care about. These tiles in French and Vietnamese are in the Basilica of Our Lady of the Immaculate Conception (Notre-Dame Basilica) in Ho Chi Minh City, Vietnam. When the basilica was constructed in the late nineteenth century, France was the colonial ruler of Vietnam. "Ta ón Thánh Antôn" is Vietnamese for "Thanks to Saint Anthony" (the patron saint of lost and stolen items).

CULTURE: WHAT PEOPLE TAKE CARE OF. The second element of culture of interest to geographers is production of material wealth—the food, clothing, and shelter that humans need in order to survive and thrive. All people consume food, wear clothing, build shelter, and create art, but different cultural groups obtain their wealth in different ways.

Geographers divide the world into regions of developed countries and regions of developing countries. Various shared characteristics—such as per capita income, literacy rates, TVs per capita, and hospital beds per capita—distinguish developed regions and developing regions. These differences are reviewed in Chapter 9.

Possession of wealth and material goods is higher in developed countries than in developing countries because of the different types of economic activities carried out in the two types of countries. Most people in developing countries are engaged in agriculture, whereas most people in developed countries earn their living through performing services in exchange for wages. This fundamental economic difference between developed and developing regions is discussed in more detail in Chapters 10 through 13.

SPATIAL ASSOCIATION

A region can be constructed to encompass an area of widely varying scale, from a very small portion of Earth to a very large portion. Different conclusions may be reached concerning a region's characteristics, depending on its scale. Consider the percentage of Americans who die each year from cancer. Death rates vary widely among scales within the United States (Figure 1-21):

- At the scale of the United States, the Great Lakes and South regions have higher levels of cancer than the West.
- At the scale of the state of Maryland, the eastern region has a higher level of cancer than the western region.
- At the scale of the city of Baltimore, Maryland, lower levels of cancer are found in the northern region.

To explain why regions possess distinctive features, such as a high cancer rate, geographers try to identify cultural, economic, and environmental factors that display similar spatial distributions. By integrating other spatial information about people, activities, and environments, we can begin to see factors that may be associated with regional differences in cancer:

- At the national scale, the Great Lakes region may have higher cancer rates in part because the distribution of cancer is spatially associated with the distribution of factories.
- At the state scale, Baltimore City may have higher cancer rates because of a concentration of people with lower levels of income and education. People living in the rural Eastern Shore region may be exposed to run-off of chemicals from farms into the nearby Chesapeake Bay, as well as discharges carried by prevailing winds from factories further west.
- At the urban scale, neighborhoods on the north side of Baltimore City contain a higher percentage of people with high incomes and are further from the city's factories and port facilities.

Pause and Reflect 1.2.3

For each map in Figure 1-21, write a question that you could ask about the data on the map at that scale. How do your questions change as the map's scale changes?

CHECK-IN: KEY ISSUE 2

Why Is Each Point On Earth Unique?

✓ **Location is identified through name, site, and situation.**

✓ **Regions can be formal, functional, or vernacular.**

✓ **Culture encompasses what people care about and what people take care of.**

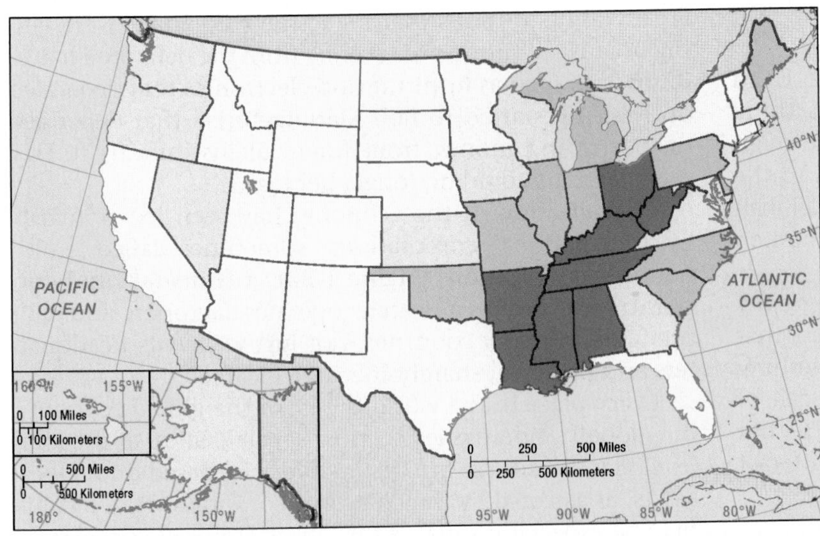

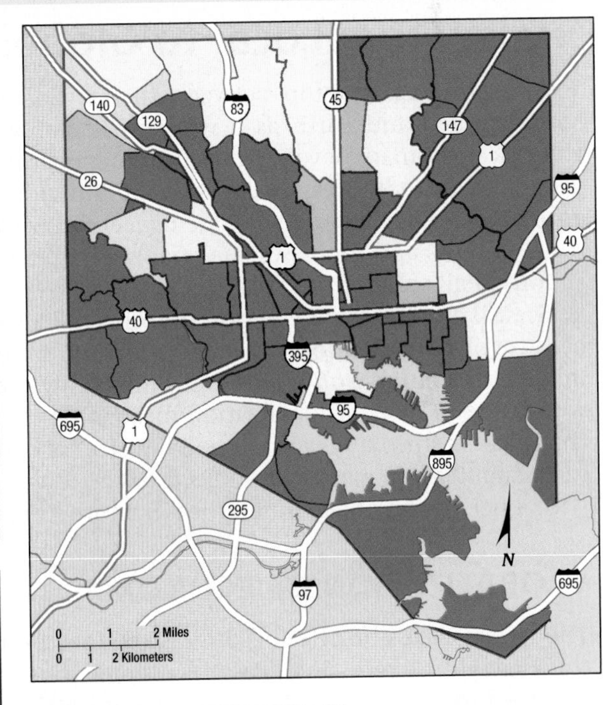

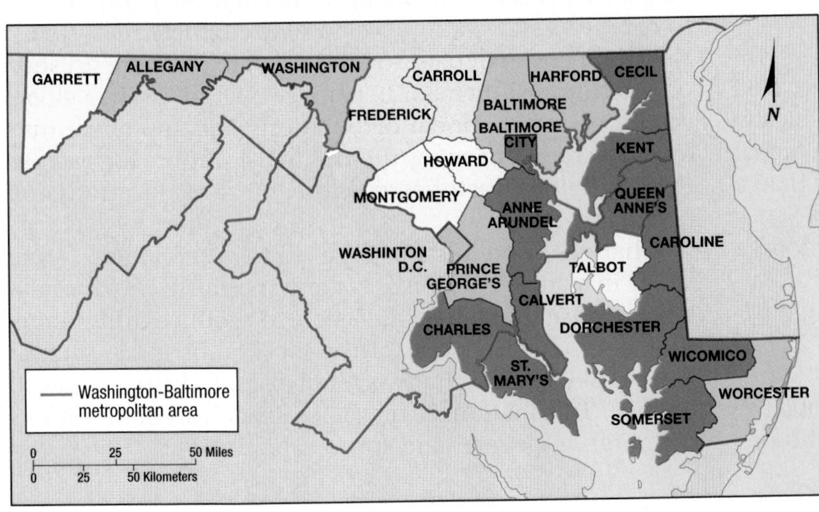

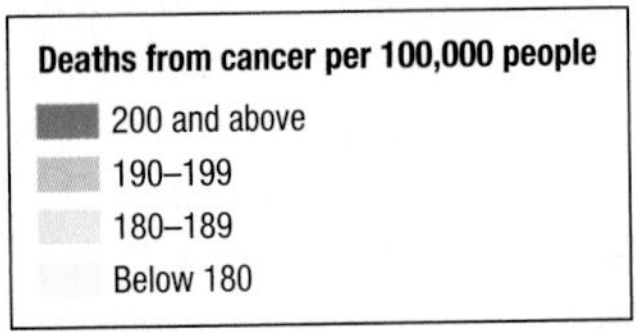

▲ **FIGURE 1-21 SPATIAL ASSOCIATION** On the national scale, the Great Lakes and South regions have higher cancer rates than the West. On the scale of the state of Maryland, the eastern region has a higher cancer rate than the western region. On the urban scale, southern and western neighborhoods of Baltimore City have higher cancer rates than northwestern ones. Geographers try to understand the reason for such variations.

KEY ISSUE 3

Why Are Different Places Similar?

- **Scale: From Local to Global**
- **Space: Distribution of Features**
- **Connections between Places**

Learning Outcome 1.3.1
Give examples of changes in economy and culture occurring at global and local scales.

Although accepting that each place or region on Earth may be unique, geographers recognize that human activities are rarely confined to one location. Three basic concepts—scale, space, and connections—help geographers explain why similarities among places and regions do not result from coincidence.

Scale: From Local to Global

Scale is the relationship between the portion of Earth being studied and Earth as a whole. Geographers think about scale at many levels, from local to global. Although geographers study every scale from the individual to the entire Earth, increasingly they are concerned with global-scale patterns and processes. Geographers explain human actions at all scales, from local to global.

Scale is an increasingly important concept in geography because of **globalization**, which is a force or process that involves the entire world and results in making something worldwide in scope. Globalization means that the scale of the world is shrinking—not literally in size, of course, but in the ability of a person, an object, or an idea to interact with a person, an object, or an idea in another place.

GLOBALIZATION OF ECONOMY

The severe recession that began in 2008 has been called the first global recession. Past recessions were typically confined to one country or region. In contrast, the global economy declined in 2009 for the first time in more than a half-century. The fate of a home buyer in the United States was tied to the fate of a banker in the United Kingdom, a sales clerk in Japan, a clothing maker in China, and a construction worker in Nigeria. All were caught in a global-scale web of falling demand and lack of credit.

The global financial crisis began in the United States and Europe with the bursting of the housing bubble. A **housing bubble** is a rapid increase in the value of houses followed by a sharp decline in their value. Housing prices had risen very rapidly for a number of years, primarily because very low interest rates made it possible for more people to borrow more money to buy more houses:

- Poorer people bought houses for the first time because financial institutions were willing to lend them money even though they were at a high risk of not being able to repay the debt.
- Wealthy people bought second and third homes as investments, taking advantage of the low rates for borrowing money. They were betting that prices would continue to escalate, enabling them to resell the houses at a profit. The wealthy also invested money in funds that directly or indirectly provided the loans to high-risk people.
- The government encouraged low-income families to buy houses even though they were at risk of not repaying the loans. Less government regulation and oversight of the financial industry made it easier for abusive practices to occur.

Declining demand for housing led to falling prices. Many people owed more on their houses than the houses were now worth if they tried to sell them. Ultimately, many defaulted on their loans and walked away from the houses, leaving them vacant and derelict.

The crisis spread from housing through the economy. Financial institutions that made the risky loans were failing because of the loss of revenue from the defaulted loans. Businesses such as furniture and electronics that depended on housing started to fail. Manufacturers that depended on borrowing money from financial institutions to buy raw materials could no longer get loans.

Globalization of the economy has been led primarily by transnational corporations, sometimes called multinational corporations (Figure 1-22). A **transnational corporation** conducts research, operates factories, and sells products in many countries, not just where its headquarters and principal shareholders are located.

Every place in the world is part of the global economy, but globalization has led to more specialization at the local level. Each place plays a distinctive role, based on its local assets, as assessed by transnational corporations. A locality may be especially suitable for a transnational corporation to conduct research, to develop new engineering systems, to extract raw materials, to produce parts, to store finished products, to sell them, or to manage operations. In a global economy, transnational corporations remain competitive by correctly identifying the optimal location for each of these activities. Factories are closed in some locations and opened in others.

Changes in production have led to a spatial division of labor, in which a region's workers specialize in particular tasks. Transnationals decide where to produce things in response to characteristics of the local labor force, such as level of skills, prevailing wage rates, and attitudes toward unions. Transnationals may close factories in locations with high wage rates and strong labor unions.

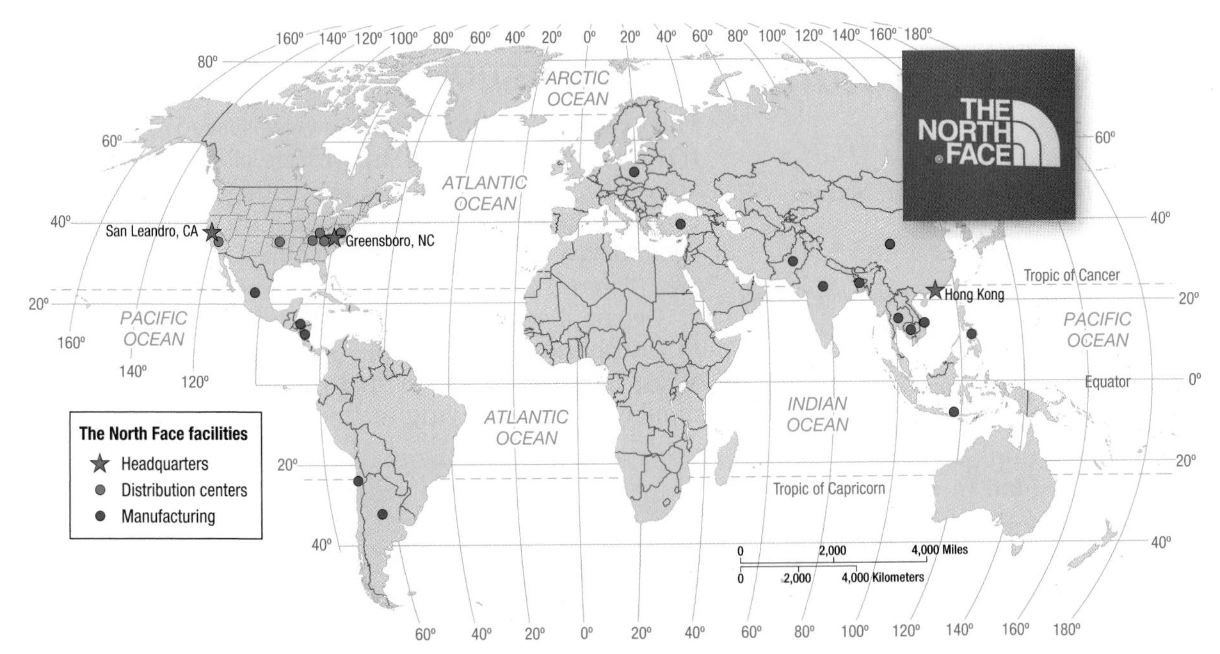

▲ **FIGURE 1-22 GLOBALIZATION OF ECONOMY** Most North Face clothing is manufactured in Latin America and Asia. The company's headquarters is in San Leandro, California, the headquarters of its parent VF Corporation is in Greensboro, North Carolina, and manufacturing is managed from its Hong Kong office.

GLOBALIZATION OF CULTURE

Geographers observe that increasingly uniform cultural preferences produce uniform "global" landscapes of material artifacts and of cultural values. Fast-food restaurants, service stations, and retail chains deliberately create a visual appearance that varies among locations as little as possible (Figure 1-23). That way, customers know what to expect, regardless of where in the world they happen to be.

Underlying the uniform cultural landscape is globalization of cultural beliefs and forms, especially religion and language. Africans, in particular, have moved away from traditional religions and have adopted Christianity or Islam, religions shared with hundreds of millions of people throughout the world. Globalization requires a form of common communication, and the English language is increasingly playing that role.

As more people become aware of elements of global culture and aspire to possess them, local cultural beliefs, forms, and traits are threatened with extinction. The survival of a local culture's distinctive beliefs, forms, and traits may be threatened by interaction with such social customs as wearing jeans and Nike shoes, consuming Coca-Cola and McDonald's hamburgers, and communicating using cell phones and computers.

Yet despite globalization, cultural differences among places not only persist but actually flourish in many places. Global standardization of products does not mean that everyone wants the same cultural products. The communications revolution that promotes globalization of culture also permits preservation of cultural diversity. TV, for example, was once limited to a handful of channels displaying one set of cultural values. With the distribution of programming through cable, satellite, and Internet, people now can choose from hundreds of programs in many languages.

With the globalization of communications, people in two distant places can watch the same TV program. At the same time, with the fragmentation of the broadcasting market, two people in the same house can watch different programs. Groups of people on every continent may aspire to wear jeans, but they might live with someone who prefers skirts. In a global culture, companies can target groups of consumers with similar tastes in different parts of the world.

Pause and Reflect 1.3.1
Give examples from your own community of (a) a cultural element that is local and (b) a cultural element that reflects the globalization of culture.

◀ **FIGURE 1-23 GLOBALIZATION OF CULTURE** McDonald's has more than 32,000 restaurants in 117 countries. To promote global uniformity of its restaurants, the company erects signs around the world that include two golden arches.

Space: Distribution of Features

Learning Outcome 1.3.2
Identify the three properties of distribution across space.

Space refers to the physical gap or interval between two objects. Geographers observe that many objects are distributed across space in a regular manner, for discernible reasons. Spatial thinking is the most fundamental skill that geographers possess to understand the arrangement of objects across Earth. Geographers think about the arrangement of people and activities found in space and try to understand why those people and activities are distributed across space as they are.

Look around the space you currently occupy—perhaps a classroom or a bedroom. Tables and chairs are arranged regularly, perhaps in a row in a classroom or against a wall at home. The room is located in a building that occupies an organized space—along a street or a side of a quadrangle. Similarly, the community containing the campus or house is part of a system of communities arranged across the country and around the world.

Geographers explain how features such as buildings and communities are arranged across Earth. On Earth as a whole, or within an area of Earth, features may be numerous or scarce, close together or far apart. The arrangement of a feature in space is known as its **distribution.** Geographers identify three main properties of distribution across Earth—density, concentration, and pattern.

DISTRIBUTION PROPERTIES: DENSITY

Density is the frequency with which something occurs in space. The feature being measured could be people, houses, cars, trees, or anything else. The area could be measured in square kilometers, square miles, hectares, acres, or any other unit of area.

Remember that a large number of a feature does not necessarily lead to a high density. Density involves two measures—the number of a feature and the land area. China is the country with the largest number of people—approximately 1.4 billion—but it does not have the world's highest density. The Netherlands, for example, has only 17 million people, but its density of 400 persons per square kilometer is much higher than China's 140 persons per square kilometer. The reason is that the land area of China is 9.6 million square kilometers, compared to only 37,000 square kilometers for the Netherlands.

High population density is also unrelated to poverty. The Netherlands is one of the world's wealthiest countries, and Mali one of the world's poorest. Yet the Netherlands' density of 400 persons per square kilometer is much larger than Mali's density of 12 persons per square kilometer (see Chapter 2 for more about density).

DISTRIBUTION PROPERTIES: CONCENTRATION

The extent of a feature's spread over space is its **concentration.** If the objects in an area are close together, they are *clustered*; if relatively far apart, they are *dispersed.* To compare the level of concentration most clearly, two areas need to have the same number of objects and the same size area (Figure 1-24).

Geographers use concentration to describe changes in distribution. For example, the distribution of people across the United States is increasingly dispersed. The total number of people living in the United States is growing

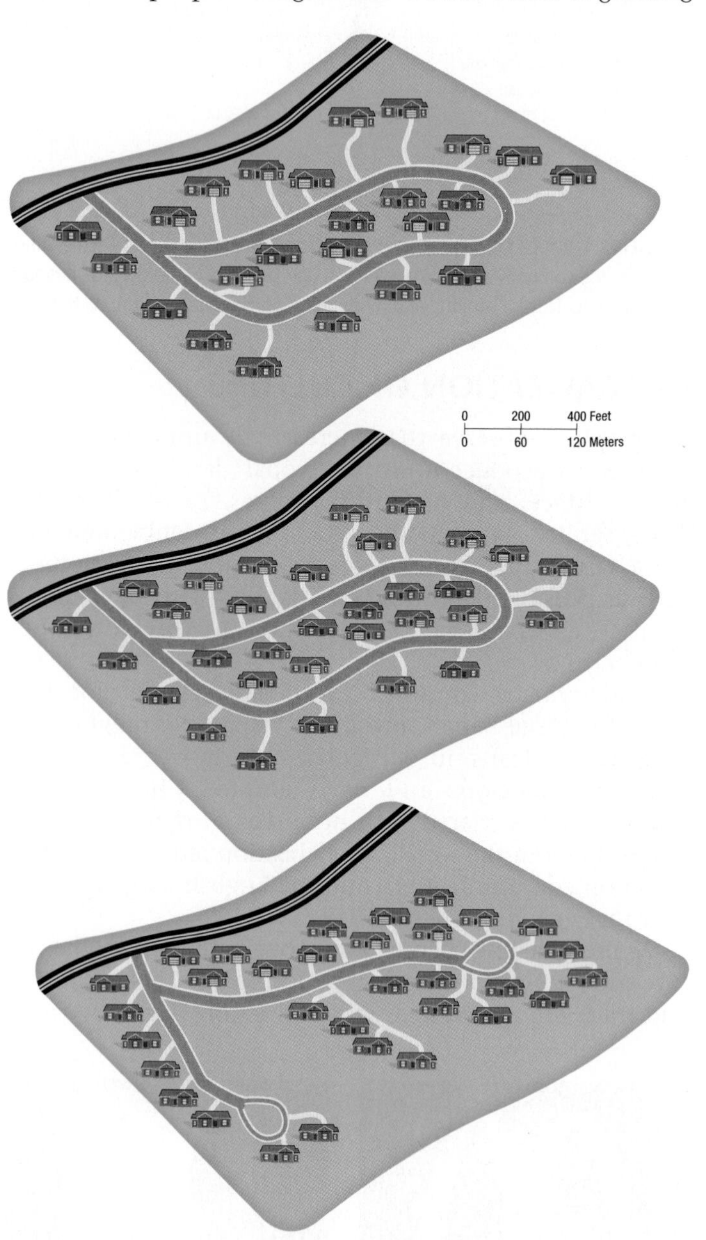

▲ **FIGURE 1-24 DISTRIBUTION OF HOUSES** The top plan for a residential area has a lower density than the middle plan (24 houses compared to 32 houses on the same 82-acre piece of land), but both have dispersed concentrations. The middle and lower plans have the same density (32 houses on 82 acres), but the distribution of houses is more clustered in the lower plan. The lower plan has shared open space, whereas the middle plan provides a larger, private yard surrounding each house.

▲ **FIGURE 1-25 DISTRIBUTION OF BASEBALL TEAMS** The changing distribution of North American baseball teams illustrates the difference between density and concentration.

slowly—less than 1 percent per year—and the land area is essentially unchanged. But the population distribution is changing from relatively clustered in the Northeast to more evenly dispersed across the country.

Concentration is not the same as density. Two neighborhoods could have the same density of housing but different concentrations. In a dispersed neighborhood, each house has a large private yard, whereas in a clustered neighborhood, the houses are close together and the open space is shared as a community park.

The distribution of major-league baseball teams illustrates the difference between density and concentration (Figure 1-25). After remaining unchanged during the first half of the twentieth century, the distribution of major-league baseball teams changed during the second half of the twentieth century. The major leagues expanded from 16 to 30 teams in North America between 1960 and 1998, thus increasing the density. At the same time, 6 of the 16 original teams moved to other locations. In 1952, every team was clustered in the Northeast United States, but the moves dispersed several teams to the West Coast and Southeast. These moves, as well as the spaces occupied by the expansion teams, resulted in a more dispersed distribution.

DISTRIBUTION PROPERTIES: PATTERN

The third property of distribution is **pattern**, which is the geometric arrangement of objects in space. Some features are organized in a geometric pattern, whereas others are distributed irregularly. Geographers observe that many objects form a linear distribution, such as the arrangement of houses along a street or stations along a subway line.

Objects are frequently arranged in a square or rectangular pattern. Many American cities contain a regular pattern of streets, known as a grid pattern, which intersect at right angles at uniform intervals to form square or rectangular blocks. The system of townships, ranges, and sections established by the Land Ordinance of 1785 is another example of a square or grid pattern (Figure 1-26).

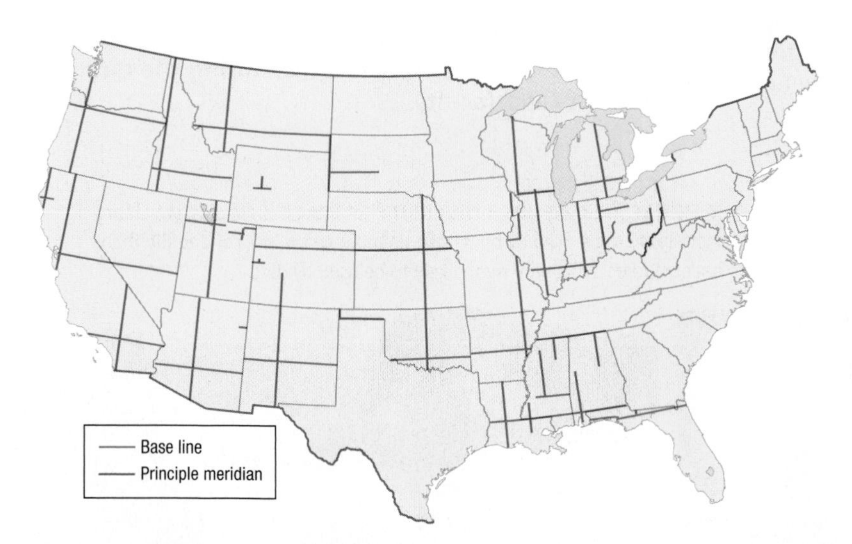

N

T24N R1W

T24N R1E

T23N R1W

6	5	4	3	2	1
7	8	9	10	11	12
18	17	16	15	14	13
19	20	21	22	23	24
30	29	28	27	26	25
31	32	33	34	35	36

T22N R1W

T22N R1E

▲ **FIGURE 1-26 PATTERN: TOWNSHIP AND RANGE** (left) To facilitate the numbering of townships, the U.S. Land Ordinance of 1785 designated several north–south lines as principal meridians and several east–west lines as base lines. (right) As territory farther west was settled, additional lines were delineated. Townships are typically 6 miles by 6 miles.

CULTURAL IDENTITY IN SPACE

Learning Outcome 1.3.3
Describe different ways in which geographers approach aspects of cultural identity such as gender, ethnicity, and sexuality.

Patterns in space vary according to gender, ethnicity, and sexuality. Geographers study these cultural traits because they are important in explaining why people sort themselves out in space and move across the landscape in distinctive ways. Critical geographers are especially concerned with the way in which the movement across space and the resulting distribution of activities perpetuate traditional roles of gender, ethnicity, and sexuality.

DISTRIBUTION ACROSS SPACE. The importance of space is learned as a child. Which child—the boy or girl—went to Little League and which went to ballet lessons? To which activity is substantially more land allocated in a city—ballfields or dance studios? (See Figure 1-27.) Space may be designed to appeal to a particular cultural group—or repel that group. A bar that appeals to whites may be uncomfortable for persons of color. A park that attracts African Americans may be uncomfortable for whites. One person's haven may generate fear for safety in others. *Behavioral geography* is a branch of human geography that emphasizes the importance of understanding the psychological basis for individual human actions.

Openly homosexual men and lesbian women may be attracted to some locations to reinforce spatial interaction with other gays. Some communities have relatively high concentrations of same-sex couples (Figure 1-28). Similarly, within communities that attract a concentration of gays, businesses that appeal primarily to gays may not be distributed uniformly (Figure 1-29). These communities and neighborhoods may be seen as offering a sympathetic haven for homosexuals and lesbians through inclusive policies and business practices. *Humanisitic geography* is a branch of human geography that emphasizes the different ways that individuals perceive their surrounding environment.

MOVEMENT ACROSS SPACE. Traditional roles and relationships influence how people move across space. For example, consider the spatial patterns typical of a household that consists of a husband and wife:

- **Movement by gender: Husband.** He gets in his car in the morning and drives from home to work, where he parks the car and spends the day. In the late afternoon, he collects the car and drives home. The location of the home was selected primarily to ease his daily commute to work.
- **Movement by gender: Wife.** She drives the children to school in the morning, walks the dog, drives to the supermarket, and visits her mother. In the afternoon, she drives the children from school to Little League or ballet lessons. Most American women are now employed at work outside the home, adding a substantial complication to an already complex pattern of moving across urban space. Where is her job located? The family house was already selected largely for access to her husband's place of employment, so she may need to travel across town. Who leaves work early to drive a child to a doctor's office? Who takes a day off work when a child is home sick?
- **Movement by ethnicity.** Movement across space varies by ethnicity because in many neighborhoods the residents are virtually all white people or virtually all persons of color. For example, most African Americans in Dayton, Ohio, live on the west side, whereas the east side is home to a virtually all-white population. As a result, when office workers are heading home from downtown Dayton, persons of color are driving or waiting for buses on the westbound streets, whereas whites are moving on the eastbound streets.

Pause and Reflect 1.3.3
Using your own campus as the example, describe how movement across space varies during the day for students and faculty.

▼ **FIGURE 1-27 GENDER DIVERSITY IN SPACE** Ballfields, which are more likely to be used by boys, take up more space in a community than ballet studios, which are more likely to be used by girls.

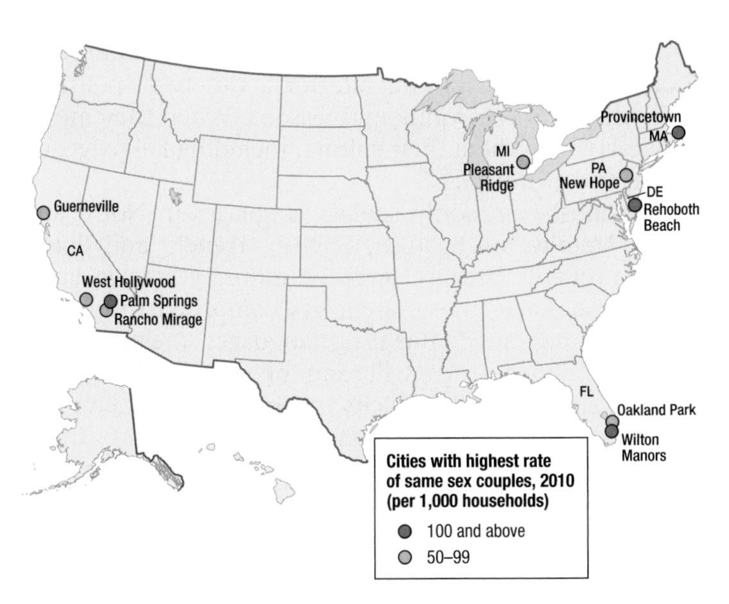

▲ **FIGURE 1-28 CONCENTRATION OF GAYS** These 10 cities have the highest percentages of same-sex couples living together.

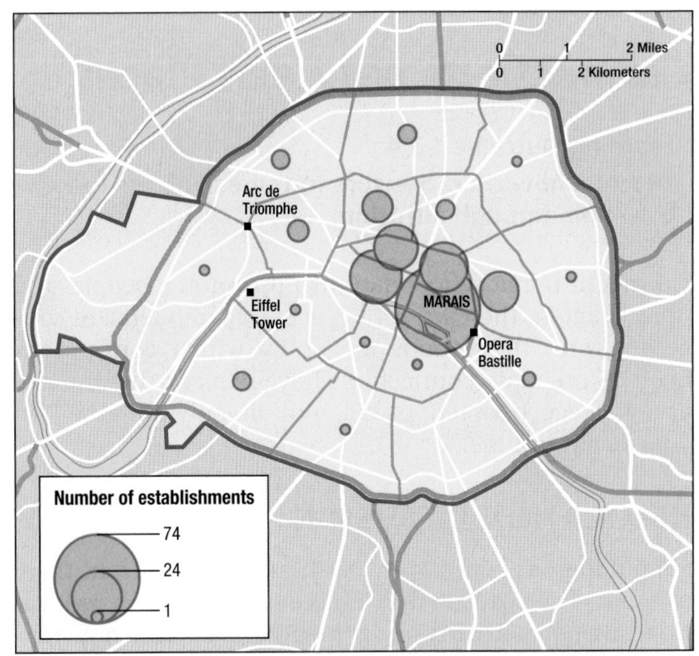

▲ **FIGURE 1-29 GAY-ORIENTED BUSINESSES IN PARIS** In Paris, 140 businesses appealing primarily to gays were identified through four 2004 guidebooks for gay travelers and residents. Gay-oriented businesses were found to be highly clustered in the Marais district of central Paris.

CULTURAL IDENTITY IN CONTEMPORARY GEOGRAPHY THOUGHT

Poststructuralist geography emphasizes the need to understand multiple perspectives regarding space. The experiences of women differ from those of men, blacks from whites, and gays from straights. It is important to listen to and to record what different groups have to say about their environment.

Cultural groups compete to organize space. Poststructuralist geographers are especially concerned with cultural groups that are dominated in space, especially women, ethnic minorities, and gays, as well as confrontations that result from the domination. Distinctive spatial patterns by gender, race, and sexual orientation are constructed by the attitudes and actions of others.

Critical geographers use their studies to focus on the needs and interests of the underprivileged. Although it is illegal to discriminate against people of color, spatial segregation persists. In many places in the world, it is legal to discriminate against gays (Figure 1-30).

All academic disciplines and workplaces have proclaimed sensitivity to issues of cultural diversity. For geographers, concern for cultural diversity is not merely a politically correct expediency; it lies at the heart of geography's spatial tradition. Nor is geographers' deep respect for the dignity of all cultural groups merely a matter of political correctness; it lies at the heart of geography's understanding of space.

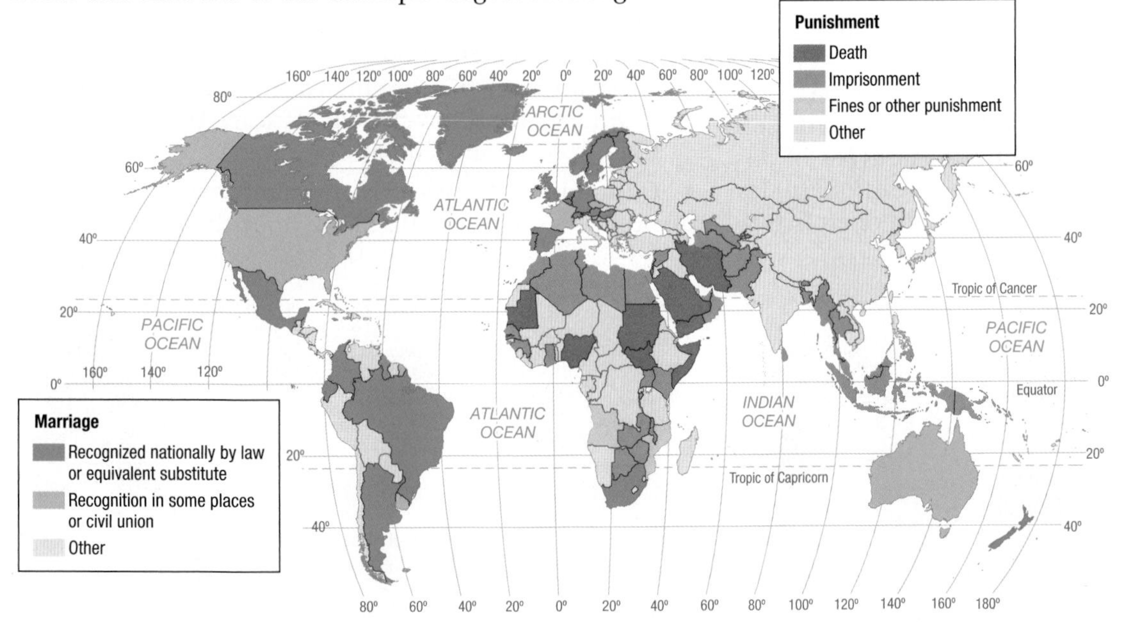

◀ **FIGURE 1-30 SEXUAL DIVERSITY IN SPACE** The International Lesbian, Gay, Bisexual, Trans and Intersex Association maps the distribution of laws that discriminate on the basis of gender. The harshest laws against male–male or female–female relationships are found in sub-Saharan Africa and Southwest Asia and North Africa. Laws supporting male–male or female–female marriage or equivalent substitute are found primarily in Europe and Latin America.

Connections between Places

Learning Outcome 1.3.4

Describe how characteristics can spread across space over time through diffusion.

Connection refers to relationships among people and objects across the barrier of space. Geographers are concerned with the various means by which connections occur. More rapid connections have reduced the distance across space between places, not literally in miles, of course, but in time.

RELOCATION DIFFUSION

Something originates at a hearth and diffuses from there to other places. A **hearth** is a place from which an innovation originates. **Diffusion** is the process by which a characteristic spreads across space from one place to another over time. Geographers document the location of nodes and the processes by which diffusion carries things elsewhere over time.

How does a hearth emerge? A cultural group must be willing to try something new and must be able to allocate resources to nurture the innovation. To develop a hearth, a group of people must also have the technical ability to achieve the desired idea and the economic structures, such as financial institutions, to facilitate implementation of the innovation.

As discussed in subsequent chapters, geographers can trace the dominant cultural, political, and economic features of the contemporary United States and Canada primarily to hearths in Europe and the Middle East. Other regions of the world also contain important hearths. In some cases an idea, such as an agricultural practice, may originate independently in more than one hearth. In other cases, hearths may emerge in two regions because two cultural groups modify a shared concept in two different ways.

For a person, an object, or an idea to have interaction with persons, objects, or ideas in other regions, diffusion must occur. Geographers observe two basic types of diffusion—relocation and expansion. The spread of an idea through physical movement of people from one place to another is termed **relocation diffusion**. We shall see in Chapter 3 that people migrate for a variety of political, economic, and environmental reasons. When they move, they carry with them their culture, including language, religion, and ethnicity.

The most commonly spoken languages in North and South America are Spanish, English, French, and Portuguese, primarily because several hundred years ago Europeans who spoke those languages comprised the largest number of migrants. Thus these languages spread through relocation diffusion. We will examine the diffusion of languages, religions, and ethnicity in Chapters 5 through 7.

Introduction of a common currency, the euro, in 12 European countries in 2002 gave scientists an unusual opportunity to measure relocation diffusion from hearths (Figure 1-31). Although a single set of paper money was issued, each of the 12 countries minted its own coins in proportion to its share of the region's economy. A country's coins were initially distributed only inside its borders, although the coins could also be used in the other 11 countries. Scientists in France took month-to-month samples to monitor the proportion of coins from each of the other 11 countries. The percentage of coins from a particular country is a measure of the level of relocation diffusion to and from France.

EXPANSION DIFFUSION

The spread of a feature from one place to another in an additive process is **expansion diffusion.** This expansion may result from one of three processes:

- **Hierarchical diffusion** is the spread of an idea from persons or nodes of authority or power to other persons or places (Figure 1-32). Hierarchical diffusion may result from the spread of ideas from political leaders, socially elite people, or other important persons to others in the community. Innovations may also originate in a particular node or core region of power, such as a large urban center, and diffuse later to isolated rural areas on the periphery. Hip-hop or rap music is an example of an innovation that originated in urban areas, though it diffused from low-income African Americans rather than from socially elite people.
- **Contagious diffusion** is the rapid, widespread diffusion of a characteristic throughout the population. As

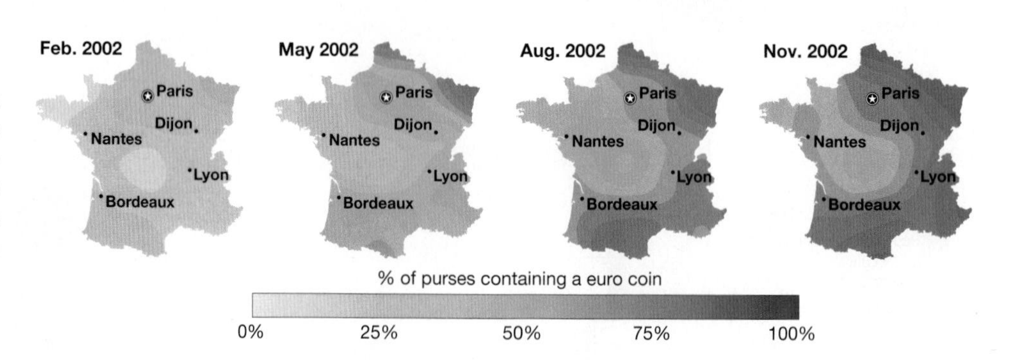

▶ **FIGURE 1-31 RELOCATION DIFFUSION: EURO COINS**
Introduction of a common currency, the euro, in 12 European countries on January 1, 2002, gave scientists an unusual opportunity to measure relocation diffusion. The percentage of euro coins circulating in France but minted in other countries is a measure of the level of relocation diffusion into France.

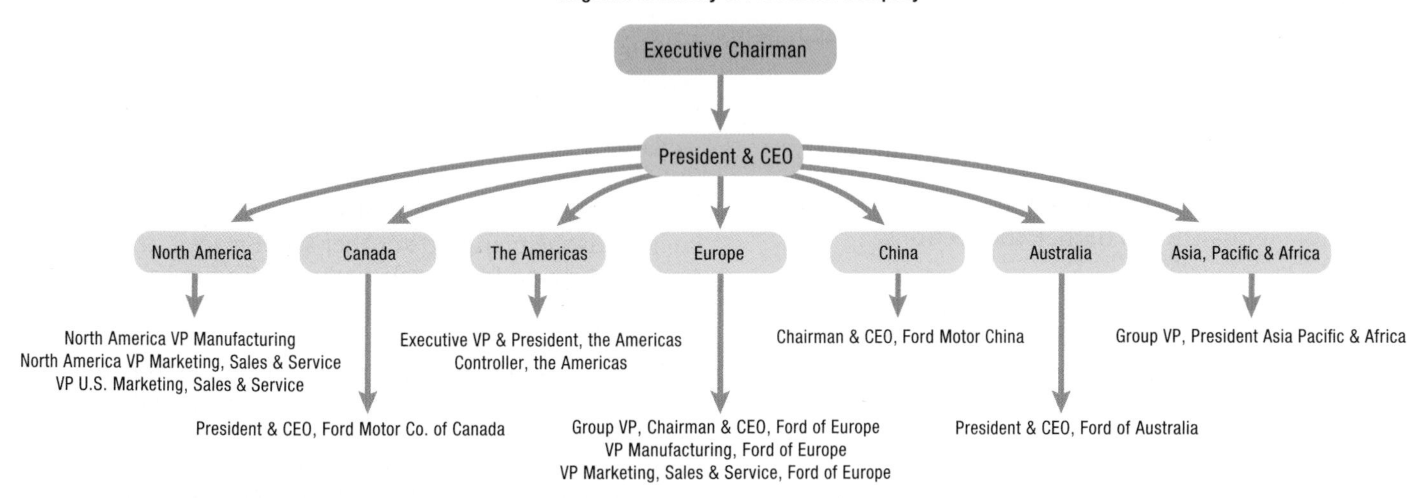

▲ **FIGURE 1-32 HIERARCHICAL DIFFUSION: FORD LEADERSHIP**
Ford Motor Company's top executives are organized by world regions, according to where the company sells most of its vehicles.

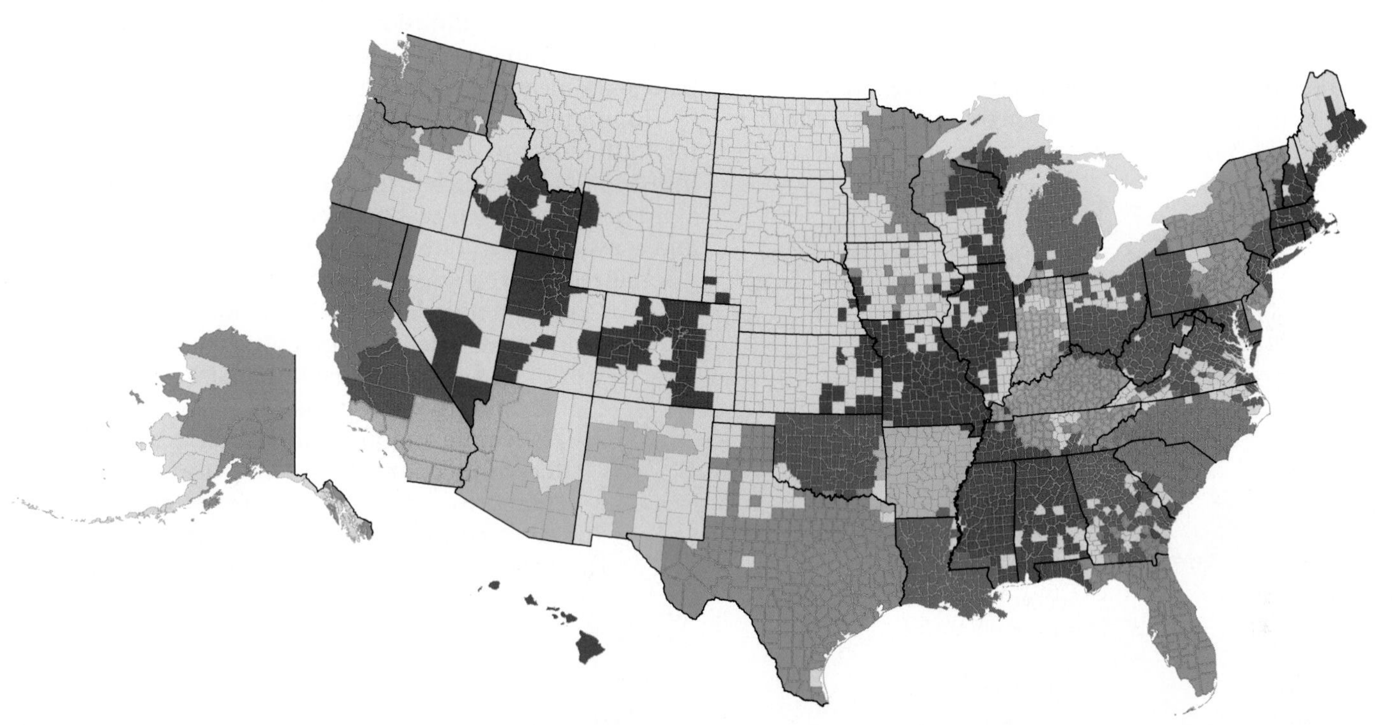

▲ **FIGURE 1-33 CONTAGIOUS DIFFUSION: TEXT MESSAGING**
Cities are hearths of diffusion for many features. AT&T, with the help of Massachusetts Institute of Technology and IBM, mapped the origin and destination of all SMS messages for a month. Each region on the map shows a clustering of senders and recipients of messages. Areas in gray had too few senders or recipients to map.

the term implies, this form of diffusion is analogous to the spread of a contagious disease, such as influenza. Contagious diffusion spreads like a wave among fans in a stadium, without regard for hierarchy and without requiring permanent relocation of people. The rapid adoption throughout the United States of AIDS prevention methods and new medicines is an example of contagious diffusion. An idea placed on the World Wide Web spreads through contagious diffusion ("goes viral") because web surfers throughout the world have access to the same material simultaneously—and quickly (Figure 1-33).

- **Stimulus diffusion** is the spread of an underlying principle even though a characteristic itself apparently fails to diffuse. For example, innovative features of Apple's iPhone and iPad have been adopted by competitors.

Expansion diffusion occurs much more rapidly in the contemporary world than it did in the past. Hierarchical diffusion is encouraged by modern methods of communication, such as computers, texting, blogging, Twittering, and e-mail. Contagious diffusion is encouraged by use of the Internet, especially the World Wide Web. Stimulus diffusion is encouraged by all of the new technologies.

SPATIAL INTERACTION

Learning Outcome 1.3.5

Explain how places are connected through networks and how inequality can hinder connections.

In the past, most connections among cultural groups required the physical movement of settlers, explorers, and plunderers from one location to another. As recently as A.D. 1800, people traveled in the same ways and at about the same speeds as in 1800 B.C.—they were carried by an animal, took a sailboat, or walked.

The farther away someone is from another, the less likely the two are to interact. Contact diminishes with increasing distance and eventually disappears. This trailing-off phenomenon is called **distance decay.** In the contemporary world, distance decay is much less severe because connection between places takes much less time. Geographers apply the term **space–time compression** to describe the reduction in the time it takes for something to reach another place (Figure 1-34).

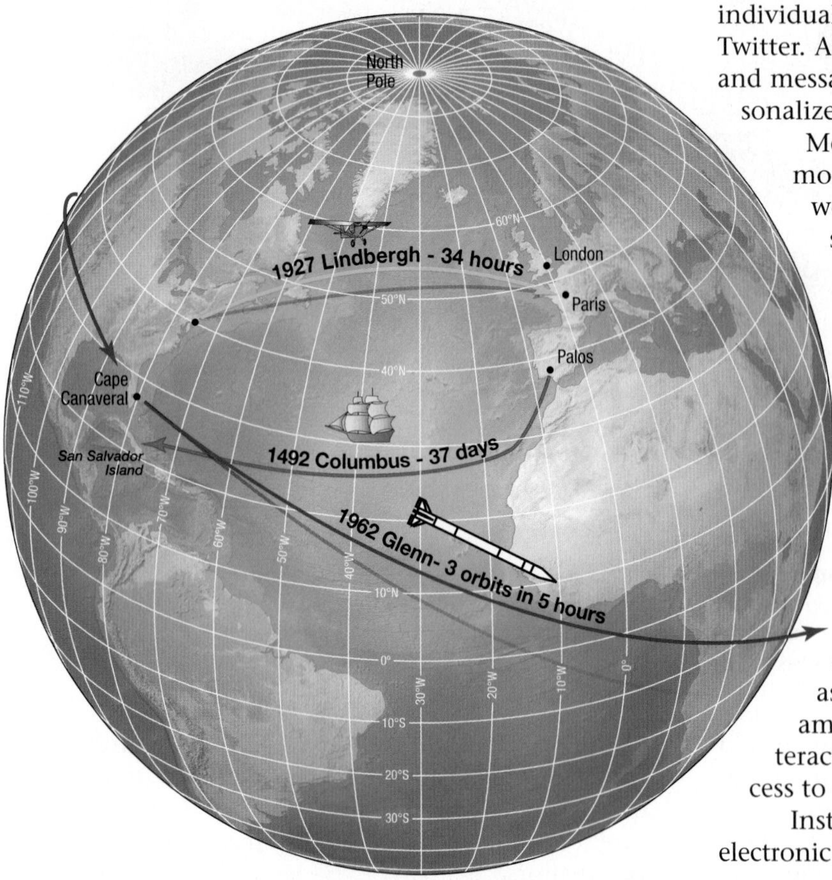

▲ FIGURE 1-34 **SPACE–TIME COMPRESSION** Transportation improvements have shrunk the world. In 1492, Christopher Columbus took 37 days (nearly 900 hours) to sail across the Atlantic Ocean from the Canary Islands to San Salvador Island. In 1912, the *Titanic* was scheduled to sail from Queenstown (now Cobh), Ireland, to New York in about 5 days, although two-thirds of the way across, after 80 hours at sea, it hit an iceberg and sank. In 1927, Charles Lindbergh was the first person to fly nonstop across the Atlantic, taking 33.5 hours to go from New York to Paris. In 1962, John Glenn, the first American to orbit in space, crossed above the Atlantic in about a half-hour and circled the globe three times in 5 hours.

Interaction takes place through a **network,** which is a chain of communication that connects places. Some airlines, for example, have networks known as "hub-and-spokes". With a hub-and-spokes network, an airline flies planes from a large number of places into one hub airport within a short period of time and then a short time later sends the planes to another set of places. In principle, travelers originating in relatively small towns can reach a wide variety of destinations by changing planes at the hub airport.

To be connected with another place in the modern world, we do not need to travel at all. Ideas that originate in a hearth are now able to diffuse rapidly to other areas through communications networks. One example is the TV network (for example, BBC in the United Kingdom, CBC in Canada, NBC in the United States), which comprises a chain of stations simultaneously broadcasting to distant places the same program, such as a football game. Through a communications network, diffusion from one place to another is instantaneous in time, even if the physical distance between places—as measured in kilometers or miles—is large.

Computers, tablets, and smart phones make it possible for individuals to set up their own connections through individually constructed networks such as Facebook and Twitter. At the touch of a button, we can transmit images and messages from one part of the world to our own personalized network around the world.

Modern networks make it possible for us to know more about what is happening elsewhere in the world, and space–time compression makes it possible for us to know it sooner. Distant places seem less remote and more accessible to us. With better connections between places, we are exposed to a constant barrage of cultural traits and economic initiatives from people in other regions, and perhaps we may adopt some of these cultural and economic elements.

UNEQUAL ACCESS

Electronic communications have played an especially important role in removing barriers to interaction between people who are physically far from each other. Physical barriers, such as oceans and deserts, can still retard interaction among people. In the modern world, barriers to interaction are more likely to derive from unequal access to electronics.

Instantaneous expansion diffusion, made possible by electronic communications, was once viewed as the "death" of geography because the ease of communications between distant places removed barriers to interaction. In reality, because of unequal access, geography matters even more than before.

People have unequal access to interaction in part because the quality of electronic service varies among places. Internet access depends on availability of electricity to power the computer and a service provider. Seconds count. Broadband service requires proximity to a digital subscriber line (DSL), a cable line, or other services. Most importantly, a person must be able to afford to pay for the communications equipment and service.

Global culture and economy are increasingly centered on the three core, or hearth, regions of North America, Europe, and Japan. These three regions have a large percentage of the world's advanced technology, capital to invest in new activities, and wealth to purchase goods and services. From "command centers" in the three major world cities of New York, London, and Tokyo, key decision makers employ modern telecommunications to send orders to factories, shops, and research centers around the world—an example of hierarchical diffusion. Meanwhile, "nonessential" employees of the companies can be relocated to lower-cost offices outside the major financial centers. For example, Fila maintains headquarters in Italy but has moved 90 percent of its production of sportswear to Asian countries. Mitsubishi's corporate offices are in Japan, but its electronics products are made in other Asian countries.

Countries in Africa, Asia, and Latin America contain three-fourths of the world's population and nearly all of its population growth. However, these countries find themselves on a periphery, or outer edge, with respect to the wealthier core regions of North America, Europe, and Japan. Global investment arrives from the core through hierarchical diffusion of decisions made by transnational corporations. People in peripheral regions, who once toiled in isolated farm fields to produce food for their families, now produce crops for sale in core regions or have given up farm life altogether and migrated to cities in search of jobs in factories and offices. As a result, the global economy has produced greater disparities than in the past between the levels of wealth and well-being enjoyed by people in the core and in the periphery. The increasing gap in economic conditions between regions in the core and periphery that results from the globalization of the economy is known as **uneven development** (Figure 1-35).

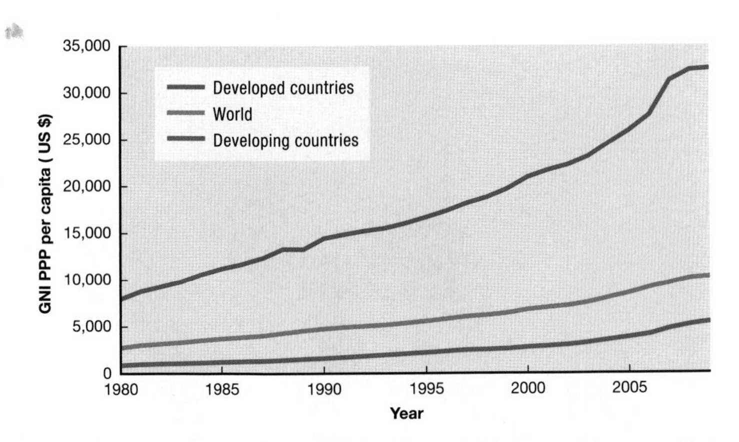

▲ **FIGURE 1-35 INCOME GAP BETWEEN RICH AND POOR COUNTRIES** Income has increased much more rapidly in developed countries than in developing ones.

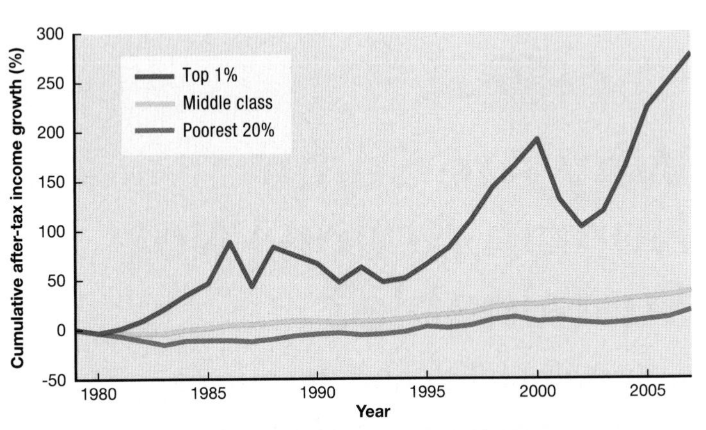

▲ **FIGURE 1-36 INCOME GROWTH OF THE WEALTHIEST 1 PERCENT** Between 1979 and 2007, the income of the wealthiest 1 percent in the United States grew by 278 percent, compared to an increase of approximately 34 percent for everyone else.

Economic inequality has also increased within countries. In the United States, the share of the national income held by the wealthiest 1 percent increased from 7 percent to 17 percent between 1979 and 2007, according to the Congressional Budget Office. The income of the wealthiest 1 percent increased by 278 percent, whereas the income of the poorest 20 percent increased by 18 percent, and the income of those in the middle increased by 38 percent (Figure 1-36).

In a global culture and economy, every area of the world plays some role intertwined with the roles played by other regions. Workers and cultural groups that in the past were largely unaffected by events elsewhere in the world now share a single economic and cultural world with other workers and cultural groups. The fate of an autoworker in Detroit is tied to investment decisions made in Mexico City, Seoul, Stuttgart, and Tokyo.

Pause and Reflect 1.3.5

What are the main differences between countries in the core regions and those in the periphery?

CHECK-IN: KEY ISSUE 3

Why Are Different Places Similar?

- ✓ **Geographers examine at all scales, though they are increasingly concerned with the global scale.**
- ✓ **Distribution has three properties—density, concentration, and pattern—and different cultural groups display different distributions in space.**
- ✓ **Places are connected through networks, and phenomena spread through relocation and expansion diffusion.**
- ✓ **In spite of space–time compression, peripheral regions in the global economy often have unequal access to the goods and services available in core regions.**

KEY ISSUE 4

Why Are Some Human Actions Not Sustainable?

- **Sustainability and Resources**
- **Sustainability and Human–Environment Relationships**

Learning Outcome 1.4.1
Describe the three pillars of sustainability.

Geography is distinctive because it encompasses both social science (human geography) and natural science (physical geography). This book focuses on human geography but doesn't forget that humans are interrelated with Earth's atmosphere, land, water, and vegetation, as well as with its other living creatures.

From the perspective of human geography, nature offers a large menu of resources available for people to use. A **resource** is a substance in the environment that is useful to people, economically and technologically feasible to access, and socially acceptable to use. A substance is merely part of nature until a society has a use for it. Food, water, minerals, soil, plants, and animals are examples of resources.

Sustainability and Resources

Earth's resources are divided between those that are renewable and those that are not:

- A **renewable resource** is produced in nature more rapidly than it is consumed by humans.
- A **nonrenewable resource** is produced in nature more slowly than it is consumed by humans.

Geographers observe two major misuses of resources:

- Humans deplete nonrenewable resources, such as petroleum, natural gas, and coal.
- Humans destroy otherwise renewable resources through pollution of air, water, and soil.

The use of Earth's renewable and nonrenewable natural resources in ways that ensure resource availability in the future is **sustainability.** Efforts to recycle metals, paper, and plastic, develop less polluting industrial processes, and protect farmland from suburban sprawl are all examples of practices that contribute to a more sustainable future.

THREE PILLARS OF SUSTAINABILITY

According to the United Nations, sustainability rests on three pillars: environment, economy, and society. The UN report *Our Common Future* is a landmark work in recognizing sustainability as a combination of natural and human elements. The report, released in 1987, is frequently called the Brundtland Report, named for the chair of the World Commission on Environment and Development, Gro Harlem Brundtland, former prime minister of Norway.

Sustainability requires curtailing the use of nonrenewable resources and limiting the use of renewable resources to the level at which the environment can continue to supply them indefinitely. To be sustainable, the amount of timber cut down in a forest, for example, or the number of fish removed from a body of water must remain at a level that does not reduce future supplies.

The Brundtland Report argues that sustainability can be achieved only by bringing together environmental protection, economic growth, and social equity (Figure 1-37). The report is optimistic about the possibility of promoting environmental protection at the same time as economic growth and social equity.

THE ENVIRONMENT PILLAR. The sustainable use and management of Earth's natural resources to meet human needs such as food, medicine, and recreation is **conservation.** Renewable resources such as trees and wildlife are conserved if they are consumed at a less rapid rate than they can be replaced. Nonrenewable resources such as petroleum and coal are conserved if we use less today in order to maintain more for future generations (Figure 1-38, left).

Conservation differs from **preservation**, which is the maintenance of resources in their present condition, with as little human impact as possible. Preservation takes the view that the value of nature does not derive from human needs and interests but from the fact that every plant and animal living on Earth has a right to exist and should be preserved, regardless of the cost. Preservation does not regard nature as a resource for human use. In contrast, conservation is compatible with development but only

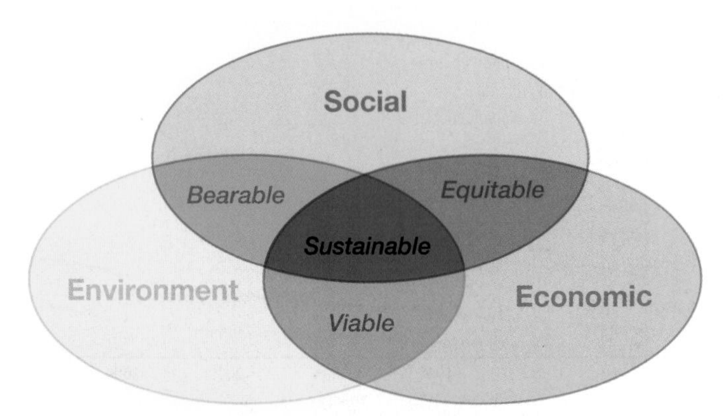

▲ FIGURE 1-37 **THREE PILLARS OF SUSTAINABILITY** The UN's Brundtland Report considers sustainability to be a combination of environmental protection, economic development, and social equity.

▲ **FIGURE 1-38 THREE PILLARS OF SUSTAINABILITY IN THE TROODOS MOUNTAINS OF CYPRUS**
Conservation of wildlife in the Troodos Mountains, Cyprus. (left) The environment pillar. The area is known for its outstanding rock formations. Much of the area is protected as national forests and UN World Heritage sites. (center) The economy pillar: Tourism is a major economic activity. (right) The social equity pillar. Local residents watch the tourists pass by. Some of the money generated by relatively wealthy tourists helps make life more bearable for residents living in a rugged environment.

if natural resources are utilized in a careful rather than a wasteful manner.

THE ECONOMY PILLAR. Natural resources acquire a monetary value through exchange in a marketplace (Figure 1-38 (center)). In a market economy, supply and demand are the principal factors determining affordability. The greater the supply, the lower the price; the greater the demand, the higher the price. Consumers will pay more for a commodity if they strongly desire it than if they have only a moderate desire. However, geographers observe that some goods do not reflect their actual environmental costs. For example, motorists sitting in a traffic jam do not have to pay a fee for the relatively high level of pollution their vehicles are emitting into the atmosphere.

The price of a resource depends on a society's technological ability to obtain it and to adapt it to that society's purposes. Earth has many substances that we do not use today because we lack the means to extract them or the knowledge of how to use them. Things that might become resources in the near future are potential resources.

THE SOCIETY PILLAR. Humans need shelter, food, and clothing to survive, so they make use of resources to meet these needs. Homes can be built of grass, wood, mud, stone, or brick. Food can be consumed by harvesting grains, fruit, and vegetables or by eating the flesh of fish, cattle, and pigs. Clothing can be made from harvesting cotton, removing skins from animals, or turning petroleum into polyester.

Consumer choices can support sustainability when people embrace it as a value. For example, a consumer might prefer clothing made of natural or recycled materials to clothing made directly from petroleum products. Society's values are the basis for choosing which resources to use (Figure 1-38 (right)).

SUSTAINABILITY'S CRITICS

Some environmentally oriented critics have argued that it is too late to discuss sustainability. The World Wildlife Fund (WWF), for example, claims that the world surpassed its sustainable level around 1980. The WWF Living Planet Report reaches its pessimistic conclusion by comparing the amount of land that humans are currently using with the amount of "biologically productive" land on Earth. "Biologically productive land" is defined as the amount of land required to produce the resources currently consumed and handle the wastes currently generated by the world's 7 billion people at current levels of technology.

The WWF calculates that humans are currently using about 13 billion hectares of Earth's land area, including 3 billion hectares for cropland, 2 billion for forest, 7 billion for energy, and 1 billion for fishing, grazing, and built-up areas. However, according to the WWF, Earth has only 11.4 billion hectares of biologically productive land, so humans are already using all of the productive land and none is left for future growth.

Others criticize sustainability from the opposite perspective: Human activities have not exceeded Earth's capacity, they argue, because resource availability has no maximum, and Earth's resources have no absolute limit because the definition of resources changes drastically and unpredictably over time. Environmental improvements can be achieved through careful assessment of the outer limits of Earth's capacity.

Critics and defenders of sustainable development agree that one important recommendation of the UN report has not been implemented—increased international cooperation to reduce the gap between more developed and less developed countries. Only if resources are distributed in a more equitable manner can poorer countries reduce the gap with richer countries.

EARTH'S PHYSICAL SYSTEMS

Learning Outcome 1.4.2
Describe the three abiotic physical systems.

Geographers classify natural resources as part of four interrelated systems. These four physical systems are classified as either biotic or abiotic. A **biotic** system is composed of living organisms. An **abiotic** system is composed of nonliving or inorganic matter. Three of Earth's four systems are abiotic:

- The **atmosphere:** a thin layer of gases surrounding Earth.
- The **hydrosphere:** all of the water on and near Earth's surface.
- The **lithosphere:** Earth's crust and a portion of upper mantle directly below the crust.

One of the four systems is biotic:

- The **biosphere:** all living organisms on Earth, including plants and animals, as well as microorganisms.

The names of the four spheres are derived from the Greek words for "stone" (*litho*), "air" (*atmo*), "water" (*hydro*), and "life" (*bio*).

ATMOSPHERE. A thin layer of gases surrounds Earth at an altitude up to 480 kilometers (300 miles). Pure dry air in the lower atmosphere contains approximately 78 percent nitrogen, 21 percent oxygen, 0.9 percent argon, 0.036 percent carbon dioxide, and 0.064 percent other gases (measured by volume). As atmospheric gases are held to Earth by gravity, pressure is created. Variations in air pressure from one location to another are responsible for producing such weather features as wind blowing, storms brewing, and rain falling.

The long-term average weather condition at a particular location is **climate.** Geographers frequently classify climates according to a system developed by German climatologist Vladimir Köppen. The modified Köppen system divides the world into five main climate regions that are identified by the letters A through E, as well as by names:

- A: Tropical Climates
- B: Dry Climates
- C: Warm Mid-Latitude Climates
- D: Cold Mid-Latitude Climates
- E: Polar Climates

The modified Köppen system divides the five main climate regions into several subtypes (Figure 1-39). For all but the B climate, the basis for the subdivision is the amount of precipitation and the season in which it falls. For the B climate, subdivision is made on the basis of temperature and precipitation.

Humans have a limited tolerance for extreme temperature and precipitation levels and thus avoid living in places that are too hot, too cold, too wet, or too dry. Compare the map of global climate to the distribution of population (see Figure 2-3). Relatively few people live in the Dry (B) and Polar (E) climate regions.

HYDROSPHERE. Water exists in liquid form in the oceans, lakes, and rivers, as well as groundwater in soil and rock. It can also exist as water vapor in the atmosphere, and as ice in glaciers. Over 97 percent of the world's water is in the oceans. The oceans supply the atmosphere with water vapor, which returns to Earth's surface as precipitation, the most

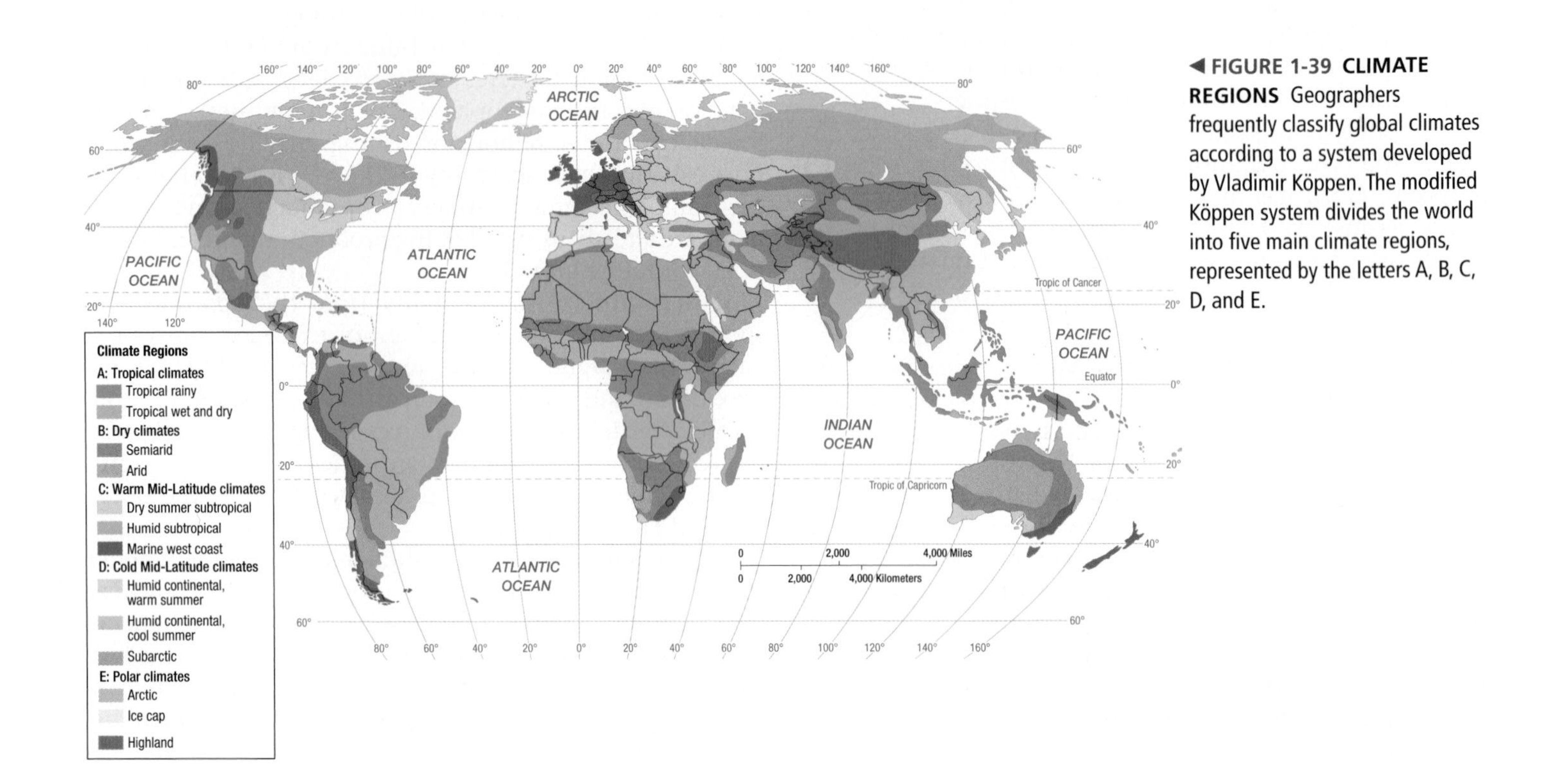

◀ **FIGURE 1-39 CLIMATE REGIONS** Geographers frequently classify global climates according to a system developed by Vladimir Köppen. The modified Köppen system divides the world into five main climate regions, represented by the letters A, B, C, D, and E.

▲ **FIGURE 1-40 MONSOON IN INDIA** People are working in a rice field during the rainy season.

important source of freshwater. Consumption of water is essential for the survival of plants and animals, and a large quantity and variety of plants and animals live in it. Because water gains and loses heat relatively slowly, it also moderates seasonal temperature extremes over much of Earth's surface.

The climate of a particular location influences human activities, especially production of the food needed to survive. People in parts of the A climate region, especially southwestern India, Bangladesh, and the Myanmar (Burma) coast, anxiously await the annual monsoon rain, which is essential for successful agriculture and provides nearly 90 percent of India's water supply (Figure 1-40). For most of the year, the region receives dry, somewhat cool air from the northeast. In June, the wind direction suddenly shifts, bringing moist, warm, southwesterly air, known as the *monsoon*, from the Indian Ocean. The monsoon rain lasts until September. In years when the monsoon rain is delayed or fails to arrive—in recent decades, at least one-fourth of the time—agricultural output falls and famine threatens in the countries of South Asia, where nearly 20 percent of the world's people live. The monsoon rain is so important in India that the words for "year," "rain," and "rainy season" are identical in many local languages.

LITHOSPHERE. Earth is composed of concentric spheres. The core is a dense, metallic sphere about 3,500 kilometers (2,200 miles) in radius. Surrounding the core is a mantle about 2,900 kilometers (1,800 miles) thick. The crust is a thin, brittle outer shell 8 to 40 kilometers (5 to 25 miles) thick. The lithosphere encompasses the crust, a portion of the mantle extending down to about 70 kilometers (45 miles). Powerful forces deep within Earth bend and break the crust to form mountain chains and shape the crust to form continents and ocean basins.

Earth's surface features, or landforms, vary from relatively flat to mountainous. Geographers find that the study of Earth's landforms—a science known as geomorphology—helps to explain the distribution of people and the choice of economic activities at different locations. People prefer living on flatter land, which generally is better suited for agriculture. Great concentrations of people and activities in hilly areas may require extensive effort to modify the landscape.

Topographic maps, published for the United States by the U.S. Geological Survey (USGS), show details of physical features, such as bodies of water, forests, mountains, valleys, and wetlands. They also show cultural features, such as buildings, roads, parks, farms, and dams. "Topos" are used by engineers, hikers, hunters, people seeking home sites, and anyone who really wants to see the lay of the land (Figure 1-41). The brown lines on the map are contour lines that show the elevation of any location. Lines are further apart in flatter areas and closer together in hilly areas.

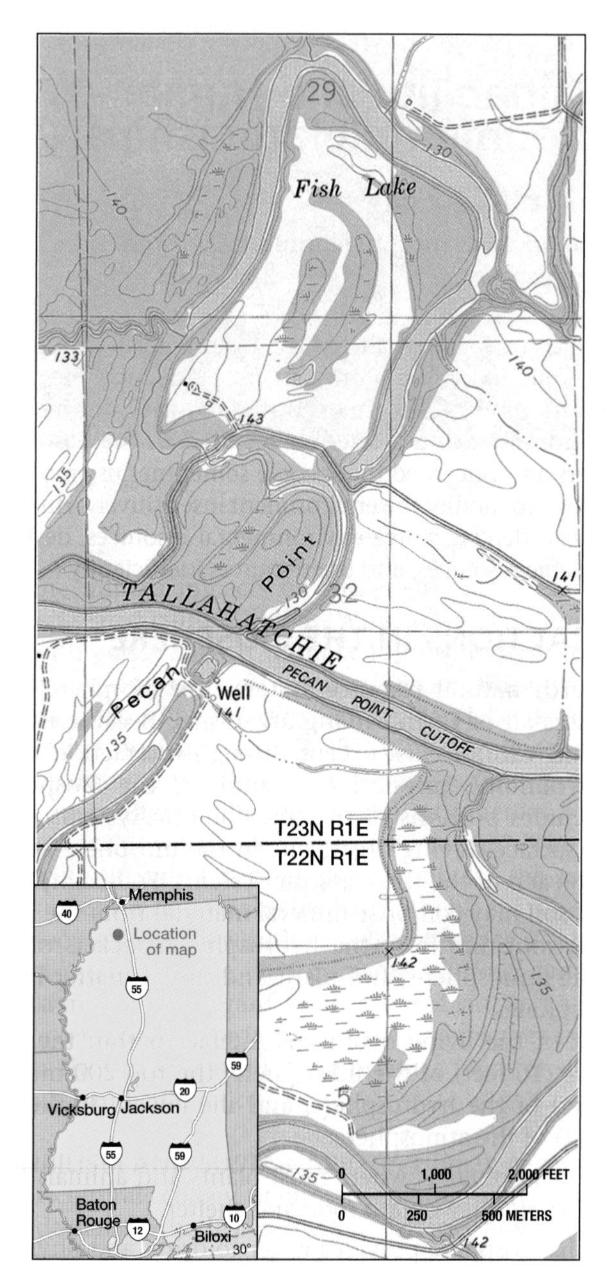

▲ **FIGURE 1-41 TOPOGRAPHIC MAP** A portion of a topographic map published by the U.S. Geological Survey shows physical features in northwestern Mississippi. The brown lines are contour lines that show the elevation of any location. The portion of the topo map shown here is part of sections 29 and 32 on the township and range map (Figure 1-30).

Pause and Reflect 1.4.2

Why would maps of Earth's hydrosphere, lithosphere, and biosphere be important in the quest for sustainability?

MODIFYING THE ENVIRONMENT

Learning Outcome 1.4.4
Compare ecosystems in the Netherlands and southern Louisiana.

Few ecosystems have been as thoroughly modified by humans as the Netherlands and Florida's Everglades. Because more than half of the Netherlands lies below sea level, most of the country today would be under water if it were not for massive projects to modify the environment by holding back the sea. Meanwhile, the fragile landscape of south Florida has been altered in insensitive ways.

THE NETHERLANDS: SUSTAINABLE ECOSYSTEM. The Dutch have a saying that "God made Earth, but the Dutch made the Netherlands." The Dutch have modified their environment with two distinctive types of construction projects—polders and dikes.

A **polder** is a piece of land that is created by draining water from an area. Polders, first created in the thirteenth century, were constructed primarily by private developers in the sixteenth and seventeenth centuries and by the government during the past 200 years. All together, the Netherlands has 6,500 square kilometers (2,600 square miles) of polders, comprising 16 percent of the country's land area (Figure 1-44). The Dutch government has reserved most of the polders for agriculture to reduce the country's dependence on imported food. Some of the polders are used for housing, and one contains Schiphol, one of Europe's busiest airports.

The second distinctive modification of the landscape in the Netherlands is the construction of massive dikes to prevent the North Sea, an arm of the Atlantic Ocean, from flooding much of the country. The Dutch have built dikes in two major locations—the Zuider Zee project in the north and the Delta Plan project in the southwest.

The Zuider Zee, an arm of the North Sea, once threatened the heart of the Netherlands with flooding. A dike completed in 1932 caused the Zuider Zee to be converted from a saltwater sea to a freshwater lake called Lake IJssel. Some of the lake has been drained to create several polders.

A second ambitious project in the Netherlands is the Delta Plan. Several rivers that flow through the Netherlands to the North Sea split into many branches and form a low-lying delta that is vulnerable to flooding. After a devastating flood in January 1953 killed nearly 2,000 people, the Delta Plan called for the construction of several dams to close off most of the waterways.

Once these two massive projects were finished, attitudes toward modifying the environment changed in the Netherlands. The Dutch scrapped plans to build additional polders in the IJsselmeer in order to preserve the lake's value for recreation.

The Dutch are deliberately breaking some of the dikes to flood fields. A plan adopted in 1990 called for returning 263,000 hectares (650,000 acres) of farms to wetlands or forests. Widespread use of insecticides and fertilizers on Dutch farms has contributed to contaminated drinking water, acid rain, and other environmental problems.

Global warming could threaten the Netherlands by raising the level of the sea around the country by between 20 and 58 centimeters (8 and 23 inches) within the next 100 years. Rather than build new dikes and polders, the Dutch have become world leaders in reducing the causes of global warming by acting to reduce industrial pollution and increase solar and wind power use, among other actions.

▲ **FIGURE 1-44 SUSTAINABLE ECOSYSTEM: THE NETHERLANDS** (left) The Dutch people have considerably altered the site of the Netherlands through creation of polders and dikes. (right) A polder in North Holland has been created by pumping the water from the site into the canal.

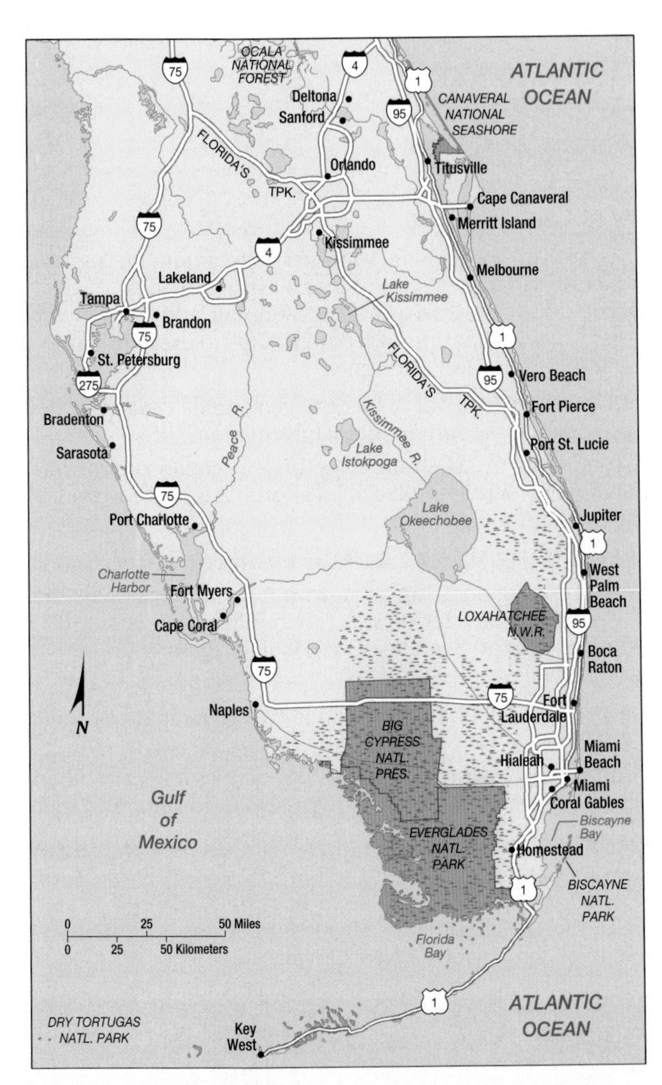

▲ **FIGURE 1-45 UNSUSTAINABLE ECOSYSTEM: SOUTH FLORIDA** To control flooding in central Florida, the U.S. Army Corps of Engineers straightened the course of the Kissimmee River, which had meandered for 160 kilometers (98 miles) from near Orlando to Lake Okeechobee. The water was rechanneled into a canal 90 meters wide (300 feet) and 9 meters deep (30 feet), running in a straight line for 84 kilometers (52 miles).

SOUTH FLORIDA: UNSUSTAINABLE ECOSYSTEM. Sensitive environmental areas in South Florida include barrier islands along the Atlantic and Gulf coasts, the wetlands between Lake Okeechobee and the Everglades National Park, and the Kissimmee River between Lake Kissimmee and Lake Okeechobee (Figure 1-45). These lowlands have been modified less sensitively than those in the Netherlands.

The Everglades was once a very wide and shallow freshwater river 80 kilometers (50 miles) wide and 15 centimeters (6 inches) deep, slowly flowing south from Lake Okeechobee to the Gulf of Mexico. A sensitive ecosystem of plants and animals once thrived in this distinctive landscape, but much of it has been destroyed by human actions.

The U.S. Army Corps of Engineers built a levee around Lake Okeechobee during the 1930s, drained the northern one-third of the Everglades during the 1940s, diverted the Kissimmee River into canals during the 1950s, and constructed dikes and levees near Miami and Fort Lauderdale during the 1960s. The southern portion of the Everglades became a National Park. These modifications opened up hundreds of thousands of hectares of land for growing sugarcane and protected farmland as well as the land occupied by the growing South Florida population from flooding. But they had unintended consequences for South Florida's environment.

Polluted water, mainly from cattle grazing along the banks of the canals, flowed into Lake Okeechobee, which is the source of freshwater for half of Florida's population. Fish in the lake began to die from the high levels of mercury, phosphorous, and other contaminants. The polluted water then continued to flow south into the National Park, threatening native vegetation such as sawgrass and endangering rare birds and other animals.

A 2000 plan called for restoring the historic flow of water through South Florida while improving flood control and water quality. A 2008 plan called for the state to acquire hundreds of thousands of acres of land from sugarcane growers. But to date, few elements of the plans to restore the Everglades have been implemented. One-half of the Everglades has been lost to development. In an ironic reminder of the Dutch saying quoted earlier, Floridians say, "God made the world in six days, and the Army Corps of Engineers has been tinkering with it ever since."

A generation ago, people concerned with environmental quality proclaimed, "Think global, act local." The phrase meant that the environment was being harmed by processes such as global warming that were global in scale, but it could be improved by actions, such as consuming less gasoline, that were local in scale. Contemporary geographers offer a different version of the phrase: "Think and act both global and local." All scales from local to global are important in geography—the appropriate scale depends on the specific subject.

Pause and Reflect 1.4.4

Both the Netherlands and the Florida Everglades face threats to their sustainability. Which is better positioned to face future challenges? Explain your answer.

CHECK-IN: KEY ISSUE 4

Why Are Some Human Actions Not Sustainable?

- ✓ **Sustainability combines environment, economy, and society.**
- ✓ **The interaction of humans and other living organisms with other physical systems results in ecosystems that may or may not be sustainable.**

Summary and Review

KEY ISSUE 1

How Do Geographers Describe Where Things Are?

Geography is most fundamentally a spatial science. Geographers use maps to display the location of objects and to extract information about places. Early geographers drew maps of Earth's surface based on exploration and observation. Today contemporary tools, such as remote sensing, GPS, and GIS, assist geographers in understanding reasons for observed regularities across Earth.

LEARNING OUTCOME 1.1.1: Explain differences between early maps and contemporary maps.

- Some of the earliest maps were used for navigation. Maps have had many other uses as tools of reference and communication.

LEARNING OUTCOME 1.1.2: Describe the role of map scale and projections in making maps.

- Contemporary maps indicate scale in three ways. Four types of distortion can occur in the transfer of Earth's round surface to a flat map.

LEARNING OUTCOME 1.1.3: Explain how latitude and longitude are used to locate points on Earth's surface.

- Latitude indicates position north or south of the equator, and longitude indicates position east or west of the prime meridian.

LEARNING OUTCOME 1.1.4: Identify contemporary analytic tools, including remote sensing, GPS, and GIS.

- Geographers today use the tools of Geographic Information Science (GIScience). Data gathered by remote sensing and GPS to measure changes over time and the characteristics of places can be combined and analyzed using geographic information systems (GIS).

THINKING GEOGRAPHICALLY 1.1: Mapping is partially a science, but it also involves a lot of human judgment. Provide examples of human judgment in mapping, such as in the creation of the geographic grid and in contemporary tools.

GOOGLE EARTH 1.1: What are the precise latitude and longitude of the U.S. Capitol building?

KEY ISSUE 2

Why Is Each Point on Earth Unique?

Every place on Earth is in some respects unique. Geographers also identify unique regions as areas distinguished by distinctive combinations of cultural as well as economic and environmental features. The distribution of features helps to explain why every place and every region is unique.

LEARNING OUTCOME 1.2.1: Identify geographic characteristics of places, including toponym, site, and situation.

- Location is the position something occupies on Earth. Geographers identify a place's location using place names, site, and situation.

LEARNING OUTCOME 1.2.2: Identify the three types of regions.

- A formal region is an area within which everyone shares distinctive characteristics. A functional region is an area organized around a node. A vernacular region is an area that people believe exists.

LEARNING OUTCOME 1.2.3: Describe two geographic definitions of culture.

- Culture can refer to cultural values such as language and religion, or to material culture such as food, clothing, and shelter.

THINKING GEOGRAPHICALLY 1.2: Describe the site and situation of your hometown.

GOOGLE EARTH 1.2: What characteristics of site and situation are visible in an aerial view of New Orleans?

Key Terms

Abiotic (p. 32) Composed of nonliving or inorganic matter.

Atmosphere (p. 32) The thin layer of gases surrounding Earth.

Biosphere (p. 32) All living organisms on Earth, including plants and animals, as well as microorganisms.

Biotic (p. 32) Composed of living organisms.

Cartography (p. 5) The science of making maps.

Climate (p. 32) The long-term average weather condition at a particular location.

Concentration (p. 22) The spread of something over a given area.

Connection (p. 26) Relationships among people and objects across the barrier of space.

Conservation (p. 30) The sustainable management of a natural resource.

Contagious diffusion (p. 26) The rapid, widespread diffusion of a feature or trend throughout a population.

Cultural ecology (p. 34) A geographic approach that emphasizes human–environment relationships.

Cultural landscape (p. 16) The fashioning of a natural landscape by a cultural group.

Culture (p. 18) The body of customary beliefs, social forms, and material traits that together constitute a group's distinct tradition.

Density (p. 22) The frequency with which something exists within a given unit of area.

KEY ISSUE 3

Why Are Different Places Similar?

Geographers work at all scales, from local to global. The global scale is increasingly important because few places in the contemporary world are totally isolated. Because places are connected to each other, they display similarities. Geographers study the interactions of groups of people and human activities across space, and they identify processes by which people and ideas diffuse from one location to another over time.

LEARNING OUTCOME 1.3.1: Give examples of changes in economy and culture occurring at global and local scales.

- Globalization means that the scale of the world is shrinking in terms of economy and culture.

LEARNING OUTCOME 1.3.2: Identify the three properties of distribution across space.

- Density is the frequency with which something occurs, concentration is the extent of spread, and pattern is the geometric arrangement.

LEARNING OUTCOME 1.3.3: Describe different ways in which geographers approach aspects of cultural identity such as gender, ethnicity, and sexuality.

- Males and females, whites and minorities, heterosexuals and homosexuals occupy different places and move across space differently.
- Critical geographers have developed different approaches to studying how different cultural groups perceive, experience, organize, and move through space.

LEARNING OUTCOME 1.3.4: Describe how characteristics can spread across space over time through diffusion.

- Something originates at a hearth and diffuses through either relocation diffusion (physical movement) or expansion diffusion (additive processes).

LEARNING OUTCOME 1.3.5: Explain how places are connected through networks and how inequality can hinder connections.

- Electronic communications have removed many physical barriers to interaction for those with access to them.

THINKING GEOGRAPHY 1.3: Imagine that a transportation device (perhaps like *Harry Potter's* floo powder) would enable all humans to travel instantaneously to any location on Earth. What might be the impact on the distribution of people and activities across Earth?

GOOGLE EARTH 1.3: How have the properties of distribution of Spring Valley, Nevada, changed over time?

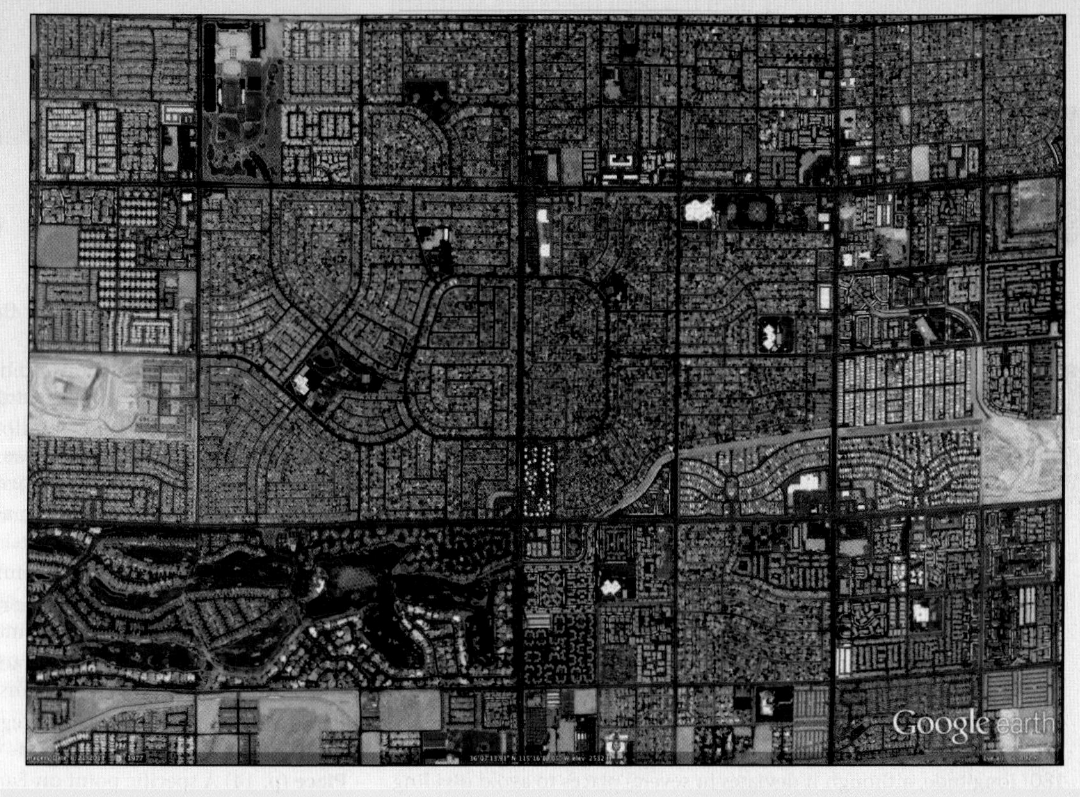

Diffusion (p. 26) The process of spread of a feature or trend from one place to another over time.

Distance decay (p. 28) The diminishing in importance and eventual disappearance of a phenomenon with increasing distance from its origin.

Distribution (p. 22) The arrangement of something across Earth's surface.

Ecology (p. 34) The scientific study of ecosystems.

Ecosystem (p. 34) A group of living organisms and the abiotic spheres with which they interact.

Environmental determinism (p. 34) A nineteenth- and early twentieth-century approach to the study of geography which argued that the general laws sought by human geographers could be found in the physical sciences. Geography was therefore the study of how the physical environment caused human activities.

Expansion diffusion (p. 26) The spread of a feature or trend among people from one area to another in an additive process.

Formal region (or uniform or homogeneous region) (p. 16) An area in which everyone shares in common one or more distinctive characteristics.

Functional region (or nodal region) (p. 17) An area organized around a node or focal point.

Geographic information science (GIScience) (p. 12) The development and analysis of data about Earth acquired through satellite and other electronic information technologies.

Chapter

2 Population and Health

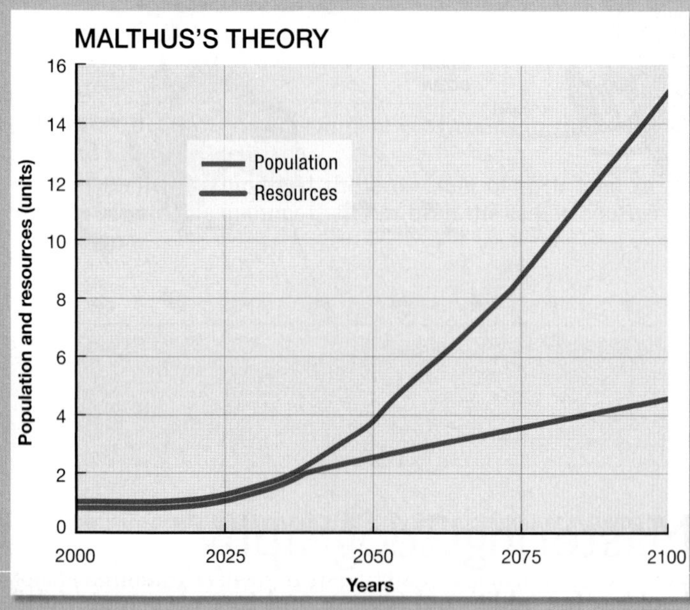

Why did Thomas Malthus think that this chart doomed humanity? Why do some people still think so? Page 60.

How did this water pump in London lead to the birth of GIS? Page 65.

KEY ISSUE 1

Where Is The World's Population Distributed?

A Crowded Planet—In Places p. 45

Manhattan's teeming throngs and Montana's wide open spaces—why aren't people spread out more evenly across Earth?

KEY ISSUE 2

Why Is Global Population Increasing?

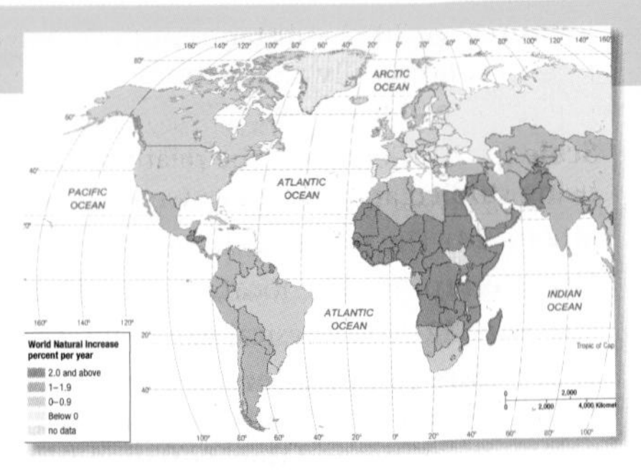

More and More of Us p. 50

All humans are born, and all must pass away—but their years on Earth vary, depending on where they live.

▲ This family lives in Jaipur, India. When these girls are older, how many children will they have? The answer matters for the future sustainability of the world because India adds more people to its population every year than any other country. Will these people have enough food? Will they have jobs?

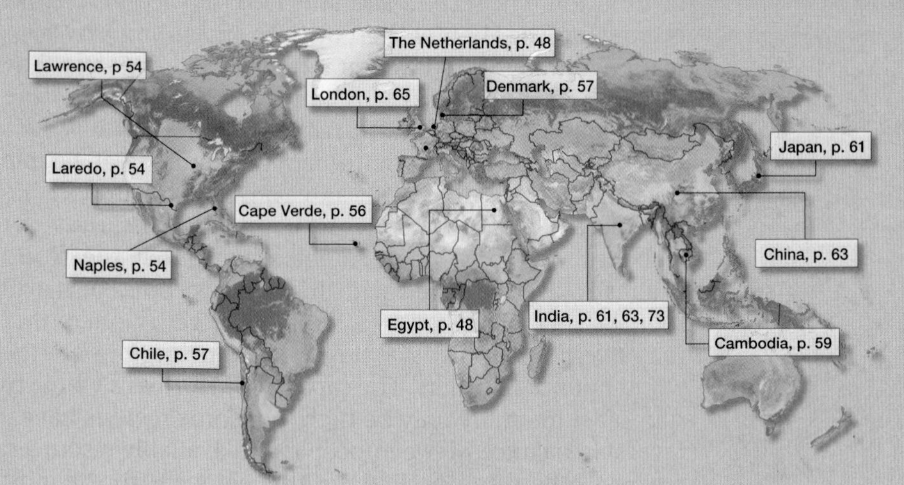

KEY ISSUE 3

Why Does Population Growth Vary Among Regions?

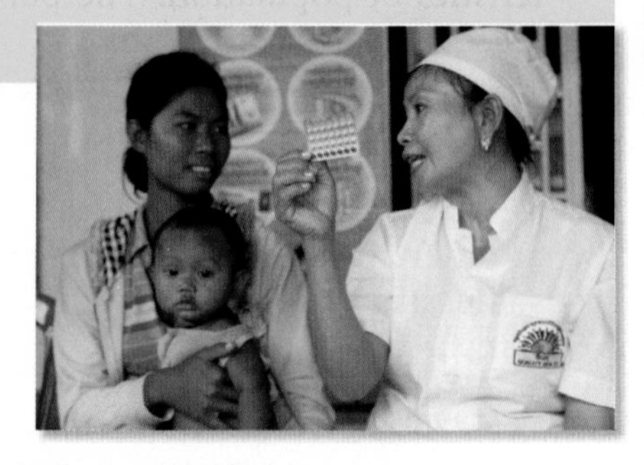

For Richer and for Poorer p. 56

It's probably not a surprise that poorer countries have higher birth rates. But why are the death rates higher in richer countries?

KEY ISSUE 4

Why Do Some Regions Face Health Threats?

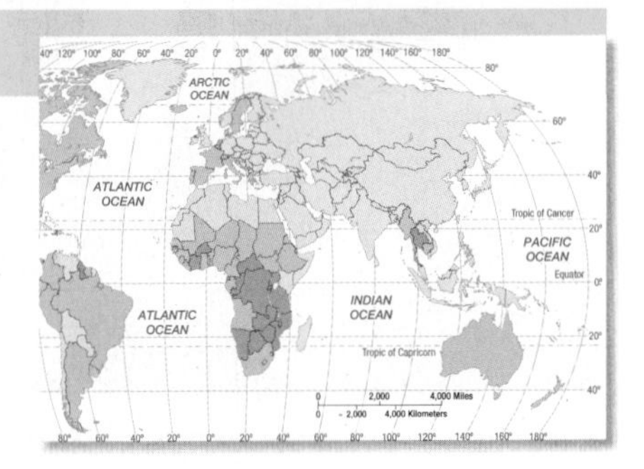

In Sickness and in Health p. 64

Humans are safer today from many once-frightening diseases, yet they are more vulnerable than ever before to new diseases.

Population Concentrations

Learning Outcome 2.1.1

Describe regions where population is clustered and where it is sparse.

Two-thirds of the world's inhabitants are clustered in four regions (Figure 2-3). The four population clusters occupy generally low-lying areas, with fertile soil and temperate climate. Most live near the ocean or near a river with easy access to an ocean, rather than in the interior of major landmasses.

CLUSTERS

The four major population clusters—East Asia, South Asia, Europe, and Southeast Asia—display differences in the pattern of occupancy of the land.

EAST ASIA. Nearly one-fourth of the world's people live in East Asia. The region, bordering the Pacific Ocean, includes eastern China, the islands of Japan, the Korean peninsula, and the island of Taiwan. The People's Republic of China is the world's most populous country and the fourth-largest country in land area. The Chinese population is clustered near the Pacific Coast and in several fertile river valleys that extend inland, though much of China's interior is sparsely inhabited mountains and deserts. More than one-half of the people live in rural areas where they work as farmers. In sharp contrast to China, more than three-fourths of all Japanese and Koreans are clustered in urban areas and work at industrial or service jobs.

SOUTH ASIA. Nearly one-fourth of the world's people live in South Asia, which includes India, Pakistan, Bangladesh, and the island of Sri Lanka. The largest concentration of people within South Asia lives along a 1,500-kilometer (900-mile) corridor from Lahore, Pakistan, through India and Bangladesh to the Bay of Bengal. Much of this area's population is concentrated along the plains of the Indus and Ganges rivers. Population is also heavily concentrated near India's two long coastlines—the Arabian Sea to the west and the Bay of Bengal to the east. Like the Chinese, most people in South Asia are farmers living in rural areas.

EUROPE. Europe includes four dozen countries, ranging from Monaco, with 1 square kilometer (0.7 square miles) and a population of 33,000, to Russia, the world's largest country in land area when its Asian part is included. In contrast to the three Asian concentrations, three-fourths of Europe's inhabitants live in cities, and fewer than 10 percent are farmers. The highest population concentrations in Europe are near the major rivers and coalfields of Germany and Belgium, as well as historic capital cities such as London and Paris.

SOUTHEAST ASIA. Around 600 million people live in Southeast Asia, mostly on a series of islands that lie between the Indian and Pacific oceans. Indonesia, which consists of 13,677 islands, is the world's fourth-most-populous country. The largest population concentration is on the island of Java, inhabited by more than 100 million people.

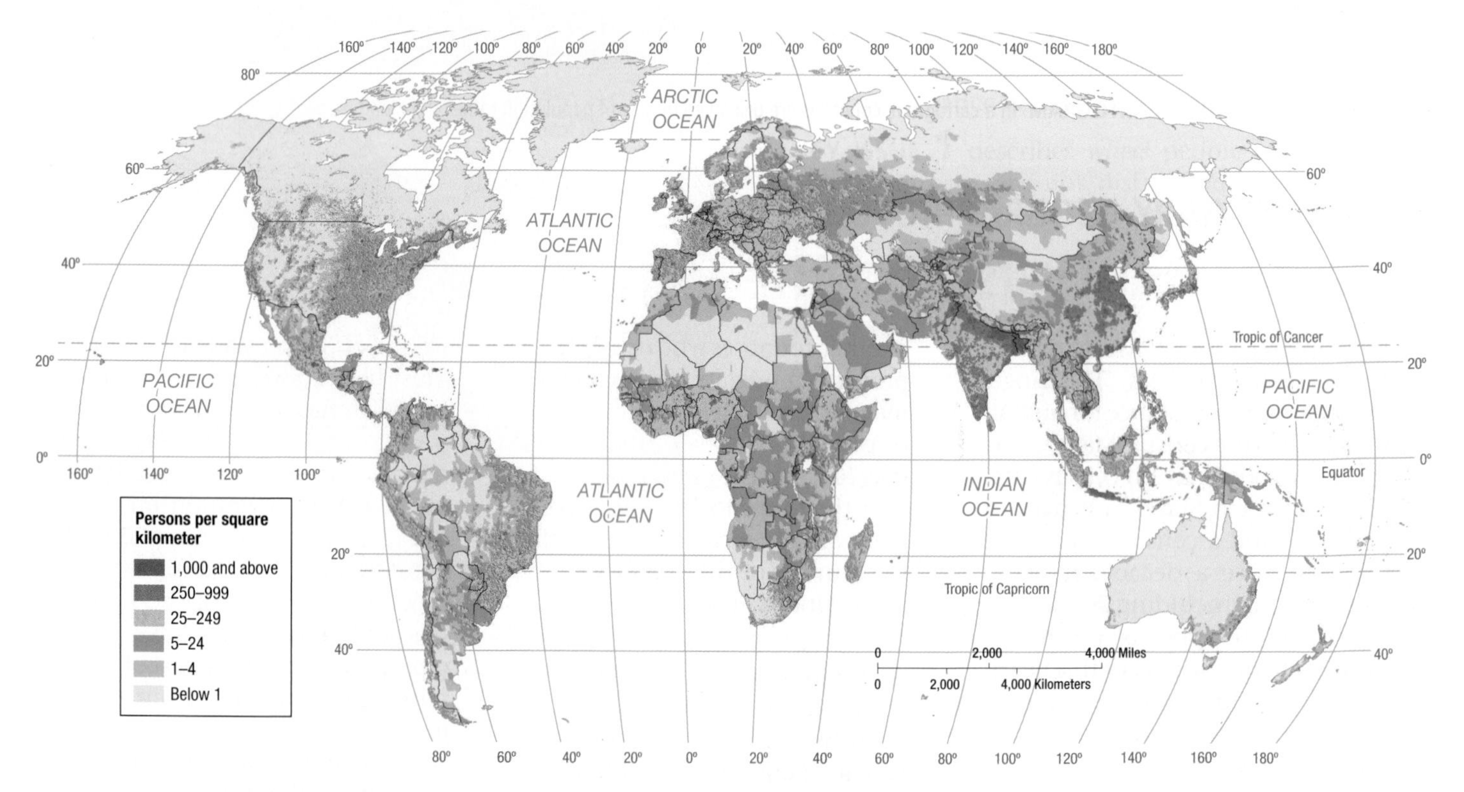

▲ **FIGURE 2-3 POPULATION DISTRIBUTION** People are not distributed uniformly across Earth's surface.

Several islands that belong to the Philippines contain high population concentrations, and population is also clustered along several river valleys and deltas at the southeastern tip of the Asian mainland, known as Indochina. Like China and South Asia, the Southeast Asia concentration is characterized by a high percentage of people working as farmers in rural areas.

OTHER CLUSTERS. The largest population concentration in the Western Hemisphere is in the northeastern United States and southeastern Canada. This cluster extends along the Atlantic Coast from Boston to Newport News, Virginia, and westward along the Great Lakes to Chicago. The largest cluster in Africa is along the Atlantic coast, especially the portion facing south. Nigeria is the most populous country in Africa. As in the three Asian concentrations, most West Africans work in agriculture.

Pause and Reflect 2.1.1
Why isn't North America one of the four major population clusters?

SPARSELY POPULATED REGIONS

Human beings avoid clustering in certain physical environments. Relatively few people live in regions that are too dry, too wet, too cold, or too mountainous for activities such as agriculture. The areas of Earth that humans consider too harsh for occupancy have diminished over time, whereas the portion of Earth's surface occupied by permanent human settlement—called the **ecumene**—has increased (Figure 2-4).

DRY LANDS. Areas too dry for farming cover approximately 20 percent of Earth's land surface. Deserts generally lack sufficient water to grow crops that could feed a large population, although some people survive there by raising animals, such as camels, that are adapted to the climate. Dry lands contain natural resources useful to people—notably, much of the world's oil reserves.

WET LANDS. Lands that receive very high levels of precipitation, located primarily near the equator, may also be inhospitable for human occupation. The combination of rain and heat rapidly depletes nutrients from the soil and thus hinders agriculture.

COLD LANDS. Much of the land near the North and South poles is perpetually covered with ice or the ground is permanently frozen (permafrost). The polar regions are unsuitable for planting crops, few animals can survive the extreme cold, and few humans live there.

HIGH LANDS. The highest mountains in the world are steep, snow covered, and sparsely settled. However, some high-altitude plateaus and mountain regions are more densely populated, especially at low latitudes (near the equator) where agriculture is possible at high elevations.

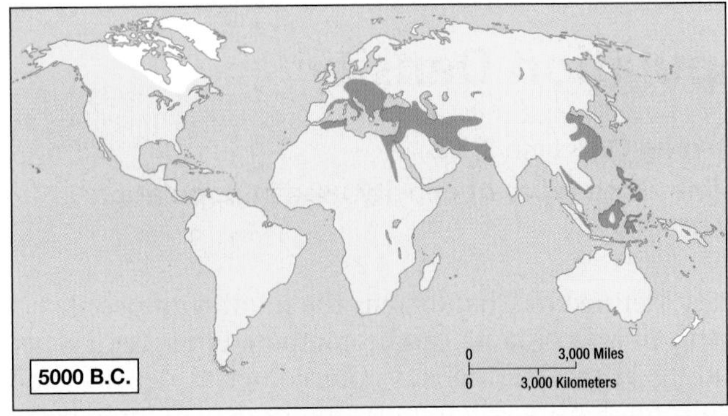

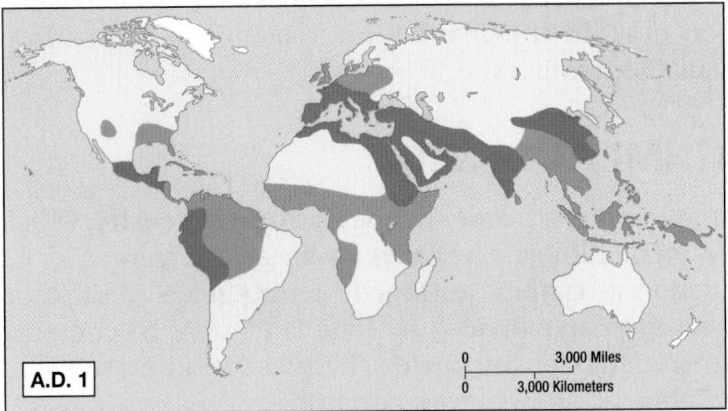

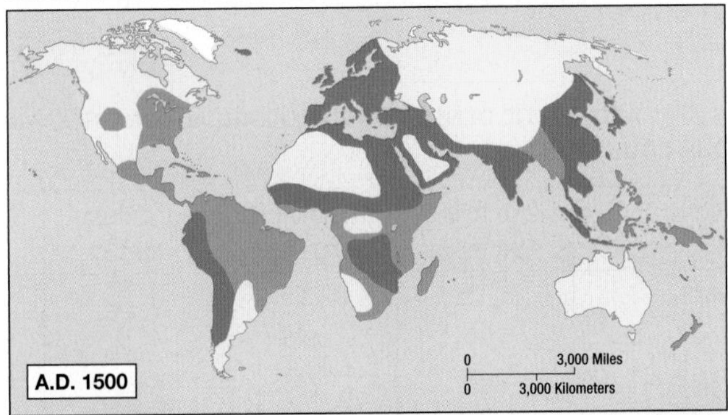

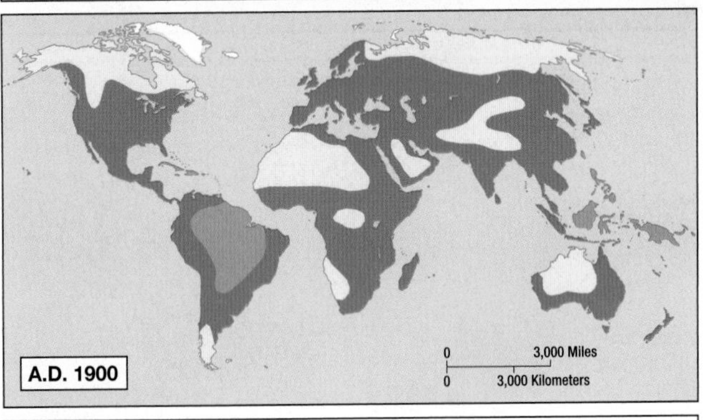

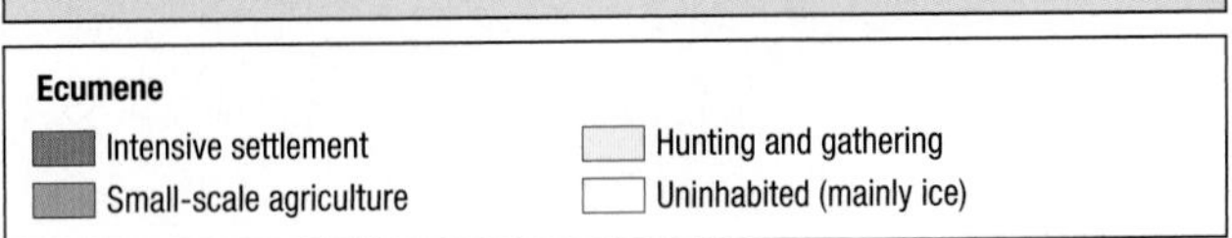

▲ **FIGURE 2-4 ECUMENE** Seven thousand years ago humans occupied only a small percentage of Earth's land area, primarily in Southwest Asia, Eastern Europe, and East Asia. Even 500 years ago much of North America and Asia lay outside the ecumene. Still, approximately three-fourths of the world's population live on only 5 percent of Earth's surface. The balance of Earth's surface consists of oceans (about 71 percent) and less intensively inhabited land.

Population Density

Learning Outcome 2.1.2
Define three types of density used in population geography.

Density, defined in Chapter 1 as the number of people occupying an area of land, can be computed in several ways, including arithmetic density, physiological density, and agricultural density. These measures of density help geographers describe the distribution of people in comparison to available resources.

ARITHMETIC DENSITY

Geographers most frequently use **arithmetic density**, which is the total number of objects in an area (Figure 2-5). In population geography, arithmetic density refers to the total number of people divided by total land area. To compute the arithmetic density, divide the population by the land area. Table 2-1 shows several examples.

Arithmetic density enables geographers to compare the number of people trying to live on a given piece of land in different regions of the world. Thus, arithmetic density answers the "where" question. However, to explain why people are not uniformly distributed across Earth's surface, other density measures are more useful.

PHYSIOLOGICAL DENSITY

Looking at the number of people per area of a certain type of land in a region provides a more meaningful population measure than arithmetic density. Land suited for agriculture is called arable land. In a region, the number of people supported by a unit area of arable land is called the **physiological density** (Figure 2-6).

Comparing physiological and arithmetic densities helps geographers understand the capacity of the land to yield enough food for the needs of the people. In Egypt, for example, the large difference between the physiological density and arithmetic density indicates that most of the country's land is unsuitable for intensive agriculture. In fact, all but 5 percent of Egyptians live in the Nile River valley and delta because it is the only area in the country that receives enough moisture (by irrigation from the river) to allow intensive cultivation of crops.

TABLE 2–1 ARITHMETIC DENSITIES, PHYSIOLOGICAL DENSITIES, AND AGRICULTURAL DENSITIES OF FOUR COUNTRIES

Country	Arithmetic Density	Physiological Density	Agricultural Density	Percentage Farmers	Percent Arable Land
Canada	3	65	1	2	0.5
United States	32	175	2	2	1.7
The Netherlands	400	1,748	23	3	0.01
Egypt	80	2,296	251	31	0.03

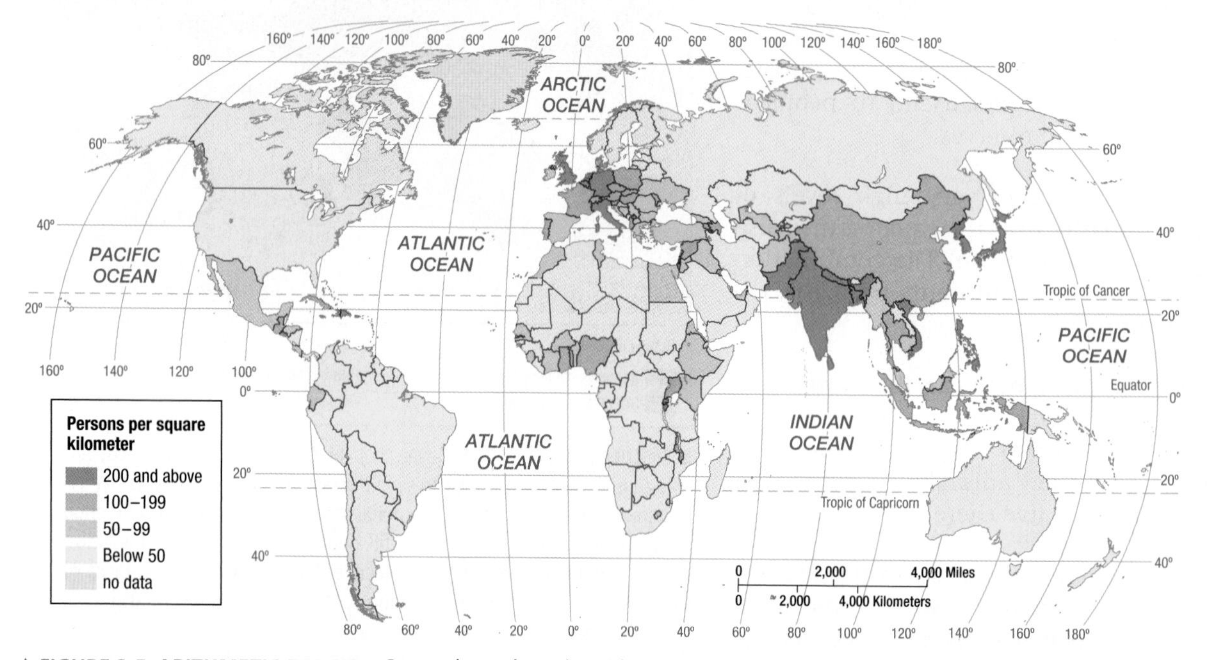

▲ **FIGURE 2-5 ARITHMETIC DENSITY** Geographers rely on the arithmetic density to compare conditions in different countries because the two pieces of information–total population and total land area–are easy to obtain. The highest arithmetic densities are found in Asia, Europe, and Central America. The lowest are in North and South America and South Pacific.

To understand relationships between population and resources in a country, geographers examine a country's physiological and agricultural densities together. For example, the physiological densities of both Egypt and the Netherlands are high, but the Dutch have a much lower agricultural density than the Egyptians. Geographers conclude that both the Dutch and Egyptians put heavy pressure on the land to produce food, but the more efficient Dutch agricultural system requires fewer farmers than does the Egyptian system.

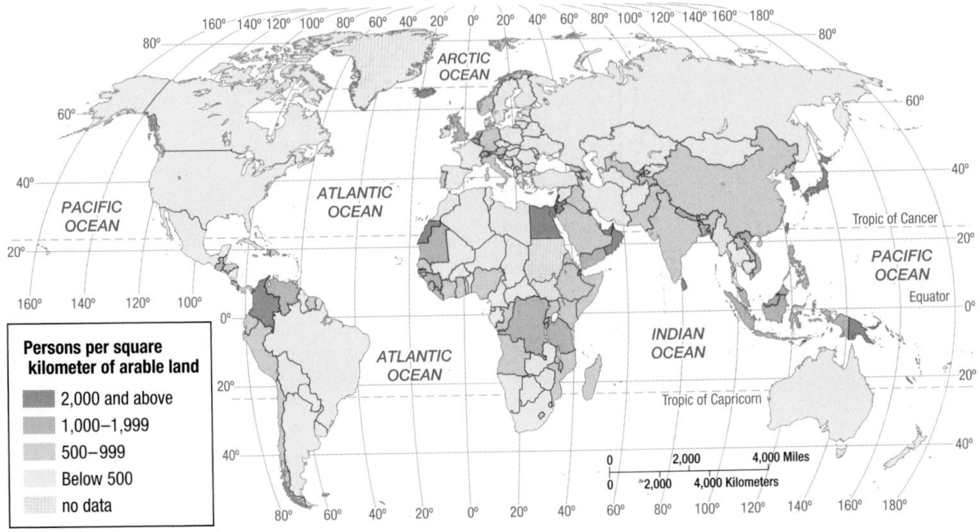

▲ **FIGURE 2-6 PHYSIOLOGICAL DENSITY** Physiological density provides insights into the relationship between the size of a population and the availability of resources in a region. The relatively large physiological densities of Egypt and the Netherlands demonstrates that crops grown on a hectare of land in these two countries must feed far more people than in the United States or Canada, which have much lower physiological densities. The highest physiological densities are found in Asia, sub-Saharan Africa, and South America. The lowest are in North America, Europe, and South Pacific.

Pause and Reflect 2.1.2

Name a country other than Egypt that has high physiological and agricultural densities.

AGRICULTURAL DENSITY

Two countries can have similar physiological densities but produce significantly different amounts of food because of different economic conditions. **Agricultural density** is the ratio of the number of farmers to the amount of arable land (Figure 2-7). Table 2-1 shows several examples.

Measuring agricultural density helps account for economic differences. Developed countries have lower agricultural densities because technology and finance allow a few people to farm extensive land areas and feed many people.

CHECK-IN: KEY ISSUE 1

Where Is The World's Population Distributed?

- ✓ **Most of the world's population is highly clustered in four regions.**
- ✓ **Arithmetic, physiological, and agricultural densities are different approaches to describing the distribution of people.**

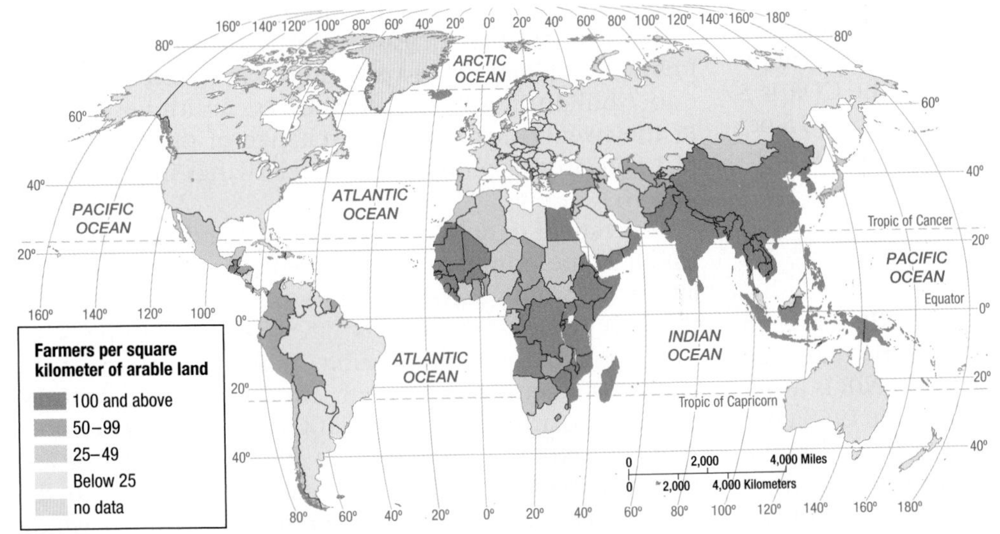

▲ **FIGURE 2-7 AGRICULTURAL DENSITY** The highest agricultural densities are found in Asia and sub-Saharan Africa. The lowest are in North America, Europe, and South Pacific.

KEY ISSUE 2

Why Is Global Population Increasing?

- **Components of Population Growth**
- **Population Structure**

Learning Outcome 2.2.1
Understand how to measure population growth through the natural increase rate.

Population increases rapidly in places where many more people are born than die, and it declines in places where deaths outnumber births. The population of a place also increases when people move in and decreases when people move out. This element of population change—migration—is discussed in Chapter 3.

COMPONENTS OF POPULATION GROWTH

Geographers most frequently measure population change in a country or the world as a whole by using three measures:

- **Crude birth rate (CBR)** is the total number of live births in a year for every 1,000 people alive in the society. A CBR of 20 means that for every 1,000 people in a country, 20 babies are born over a one-year period.
- **Crude death rate (CDR)** is the total number of deaths in a year for every 1,000 people alive in the society. Comparable to the CBR, the CDR is expressed as the annual number of deaths per 1,000 population.
- **Natural increase rate (NIR)** is the percentage by which a population grows in a year. It is computed by subtracting CDR from CBR, after first converting the two measures from numbers per 1,000 to percentages (numbers per 100). Thus if the CBR is 20 and the CDR is 5 (both per 1,000), then the NIR is 1.5 percent, or 15 per 1,000. The term *natural* means that a country's growth rate excludes migration.

NATURAL INCREASE

During the twenty-first century, the world NIR has been 1.2, meaning that the population of the world had been growing each year by 1.2 percent. The world NIR is lower today than its all-time peak of 2.2 percent in 1963, and it has declined sharply since the 1990s. However, the NIR during the second half of the twentieth century was high by historical standards. Most of humanity's

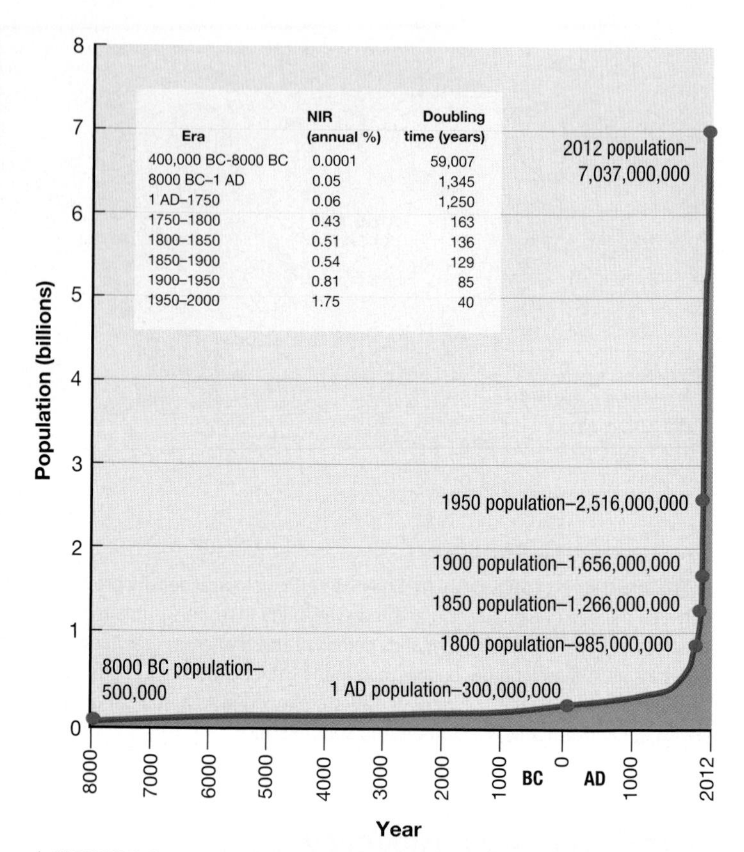

Era	NIR (annual %)	Doubling time (years)
400,000 BC-8000 BC	0.0001	59,007
8000 BC–1 AD	0.05	1,345
1 AD–1750	0.06	1,250
1750–1800	0.43	163
1800–1850	0.51	136
1850–1900	0.54	129
1900–1950	0.81	85
1950–2000	1.75	40

▲ **FIGURE 2-8 WORLD POPULATION THROUGH HISTORY** Through most of human history population growth was virtually nil. Population increased rapidly beginning in the eighteenth century.

several-hundred-thousand-year occupancy of Earth was characterized by an NIR of essentially zero, and Earth's population was unchanged, at perhaps a half-million (Figure 2-8).

About 82 million people are being added to the population of the world annually (Figure 2-9). This number represents a decline from the historic high of 87 million in 1990. The number of people added each year has dropped much more slowly than the NIR because the population base is much higher now than in the past.

World population increased from 3 to 4 billion in 14 years, from 4 to 5 billion in 13 years, and from 5 to 6 billion and 6 to 7 billion in 12 years. As the base continues to grow in the twenty-first century, a change of only one-tenth of 1 percent can produce very large swings in population growth.

The rate of natural increase affects the **doubling time,** which is the number of years needed to double a population, assuming a constant rate of natural increase. At the early twenty-first-century rate of 1.2 percent per year, world population would double in about 54 years. If the same NIR continued through the twenty-first century, global population in the year 2100 would reach 24 billion. When the NIR was 2.2 percent in 1963, doubling time was 35 years. Had the 2.2 percent rate continued into the twenty-first century, Earth's population in 2010 would

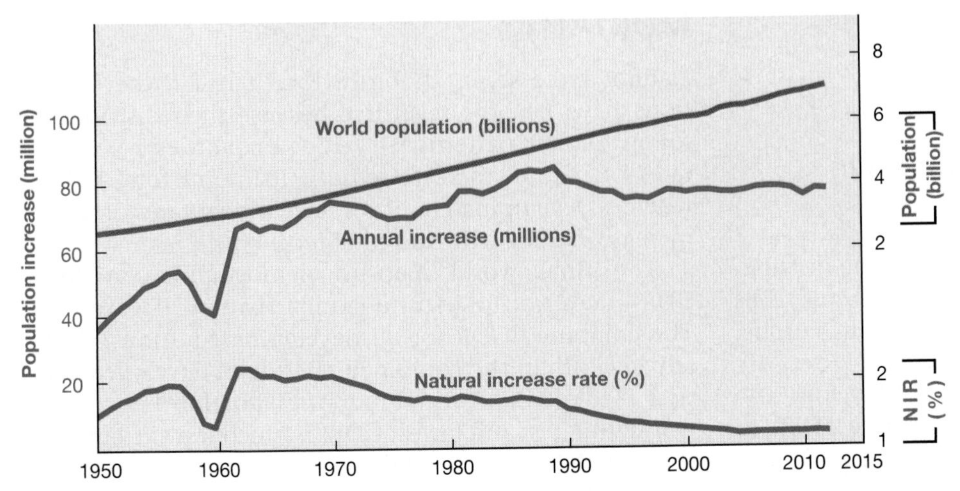

◀ **FIGURE 2-9 WORLD POPULATION GROWTH, 1950–2011** The NIR declined from its historic peak in the 1960s, but the number of people added each year has not declined very much because with world population increasing from 2.5 billion to more than 7 billion people during the period, the percentage has been applied to an ever larger base.

have been nearly 10 billion instead of nearly 7 billion. A 2.2 percent NIR through the twenty-first century would produce a total population of more than 50 billion in 2100.

More than 95 percent of the natural increase is clustered in developing countries (Figure 2-10). The NIR exceeds 2.0 percent in most countries of sub-Saharan Africa, whereas it is negative in Europe, meaning that in the absence of immigrants, population actually is declining. About one-third of the world's population growth during the past decade has been in South Asia, one-fourth in sub-Saharan Africa, and the remainder divided about equally among East Asia, Southeast Asia, Latin America, and Southwest Asia & North Africa.

Regional differences in NIRs mean that most of the world's additional people live in the countries that are least able to maintain them. To explain these variations in growth rates, geographers point to regional differences in fertility and mortality rates.

Pause and Reflect 2.2.1

The United States has an NIR of 0.6. Does that mean the doubling time is more than 54 years or less?

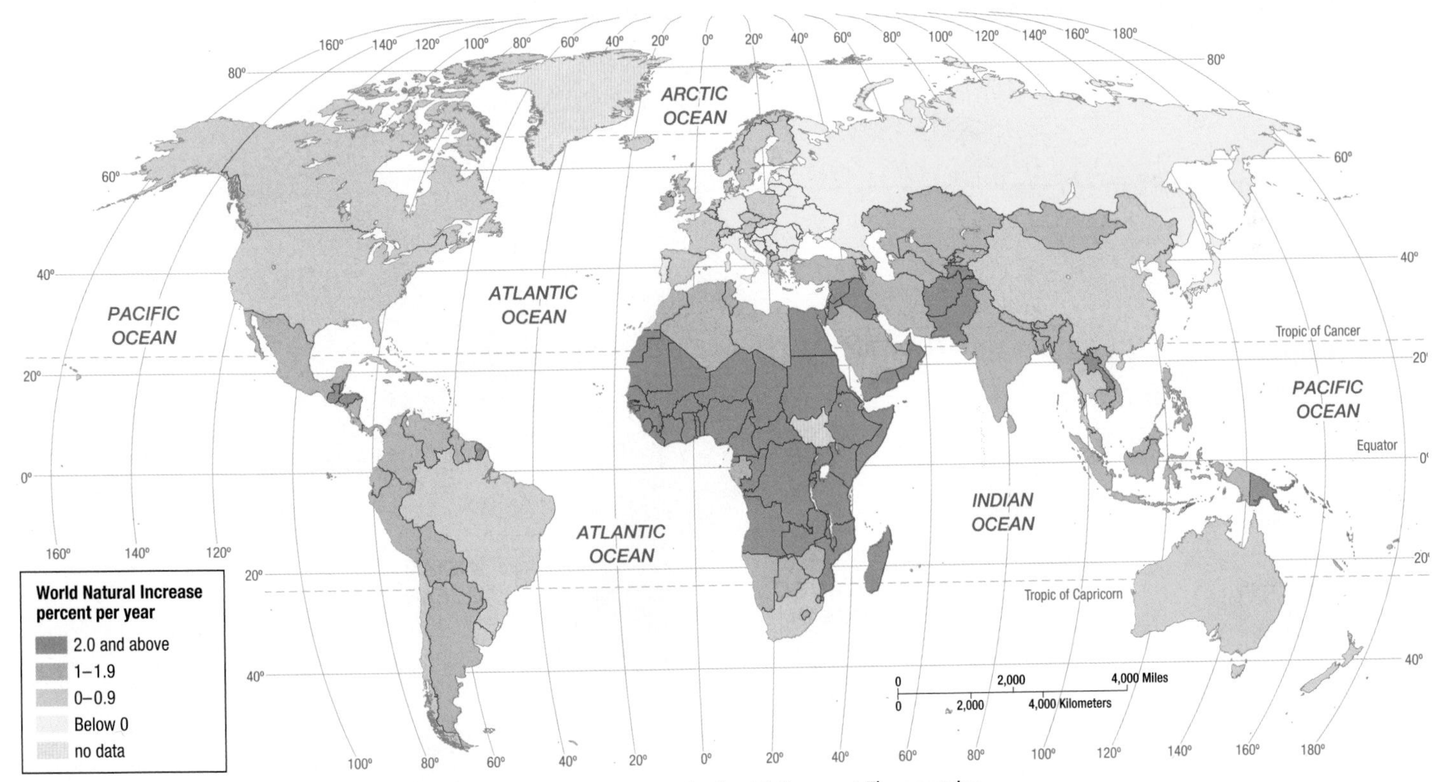

▲ **FIGURE 2-10 NATURAL INCREASE RATE** The world average is currently about 1.2 percent. The countries with the highest NIRs are concentrated in Africa and Southwest Asia.

FERTILITY

Learning Outcome 2.2.2

Understand how to measure births and deaths through CBR and CDR.

The world map of CBR (Figure 2-11) mirrors the distribution of NIR. As was the case with NIRs, the highest CBRs are in sub-Saharan Africa, and the lowest are in Europe. Many sub-Saharan African countries have a CBR over 40, whereas many European countries have a CBR below 10.

Geographers also use the **total fertility rate (TFR)** to measure the number of births in a society (Figure 2-12). The TFR is the average number of children a woman will have throughout her childbearing years (roughly ages 15 through 49). To compute the TFR, demographers assume that a woman reaching a particular age in the future will be just as likely to have a child as are women of that age today. Thus, the CBR provides a picture of a society as a whole in a given year, whereas the TFR attempts to predict the future behavior of individual women in a world of rapid cultural change.

The TFR for the world as a whole is 2.5, and, again, the figures vary between developed and developing countries. The TFR exceeds 5.0 in sub-Saharan Africa, compared to 2 or less in nearly all European countries.

Pause and Reflect 2.2.2

How does the TFR in your family compare to the overall figure for North America?

MORTALITY

Natural increase, crude birth, total fertility, the descriptions have become repetitive because their distributions follow similar patterns. Developed countries have lower rates of natural increase, crude birth, and total fertility, whereas developing countries have higher rates of natural increase, crude birth, and total fertility.

The final world map of demographic variables—CDR—does not follow the familiar pattern (Figure 2-13). The combined CDR for all developing countries is actually lower than the combined rate for all developed countries (Table 2-2). Furthermore, the variation between the world's highest and lowest CDRs is much less extreme than the variation in CBRs. The highest CDR in the world is 17 per 1,000, and the lowest is 1—a difference of 16—whereas

TABLE 2–2 COMPARING DEMOGRAPHIC FACTORS IN DEVELOPED AND DEVELOPING COUNTRIES

	Developed Countries	Developing Countries
Natural increase rate	0.2	1.4
Crude birth rate	11	22
Total fertility rate	1.7	2.6
Infant mortality rate	5	48
Life expectancy (years)	78	68
Crude death rate	10	8
Under age 15 (percent)	16	29
Age 65 (percent) and above	16	6

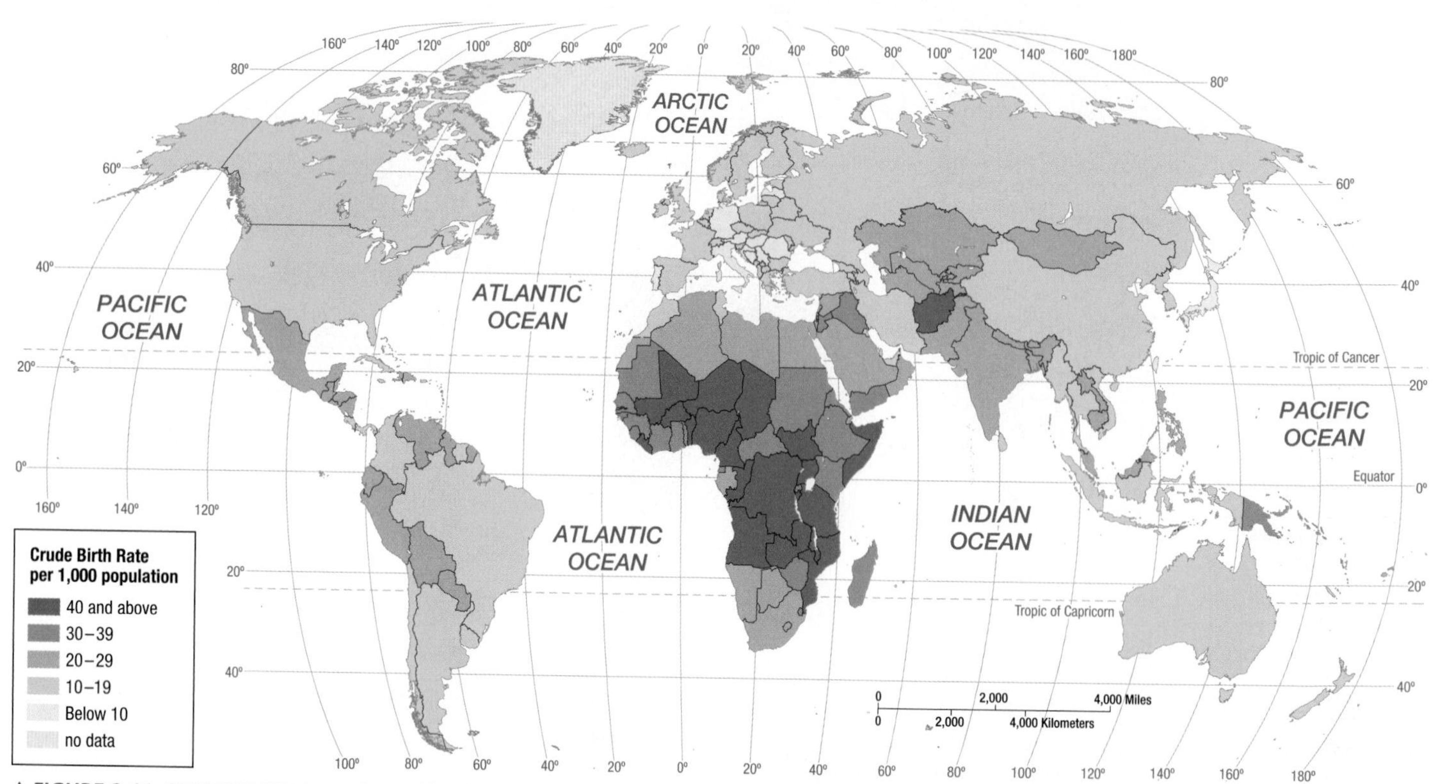

▲ **FIGURE 2-11 CRUDE BIRTH RATE (CBR)** The global distribution of CBRs parallels that of NIRs. The countries with the highest CBRs are concentrated in Africa and Southwest Asia.

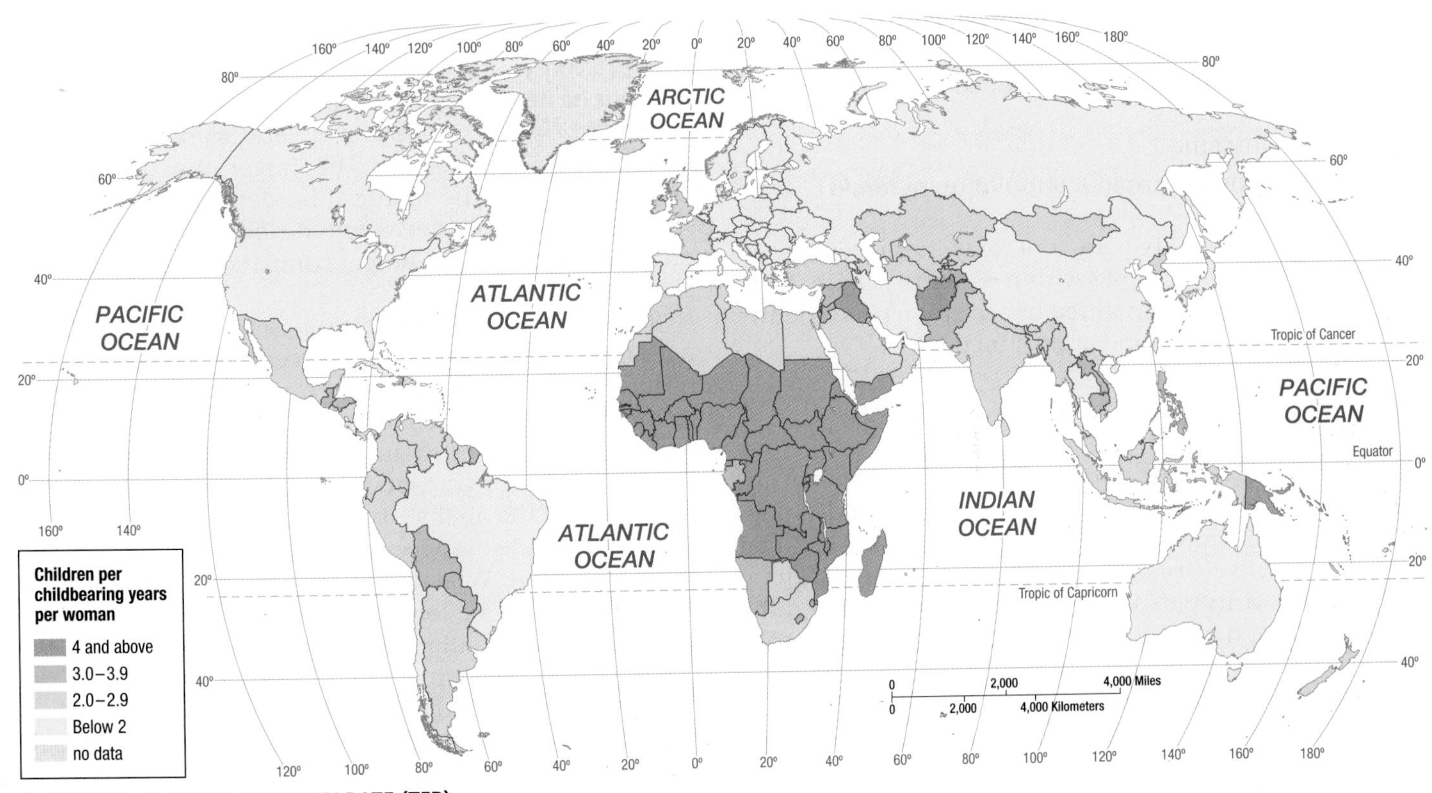

▲ **FIGURE 2-12 TOTAL FERTILITY RATE (TFR)**
As with NIRs and CBRs, the countries with the highest TFRs are concentrated in Africa and Southwest Asia.

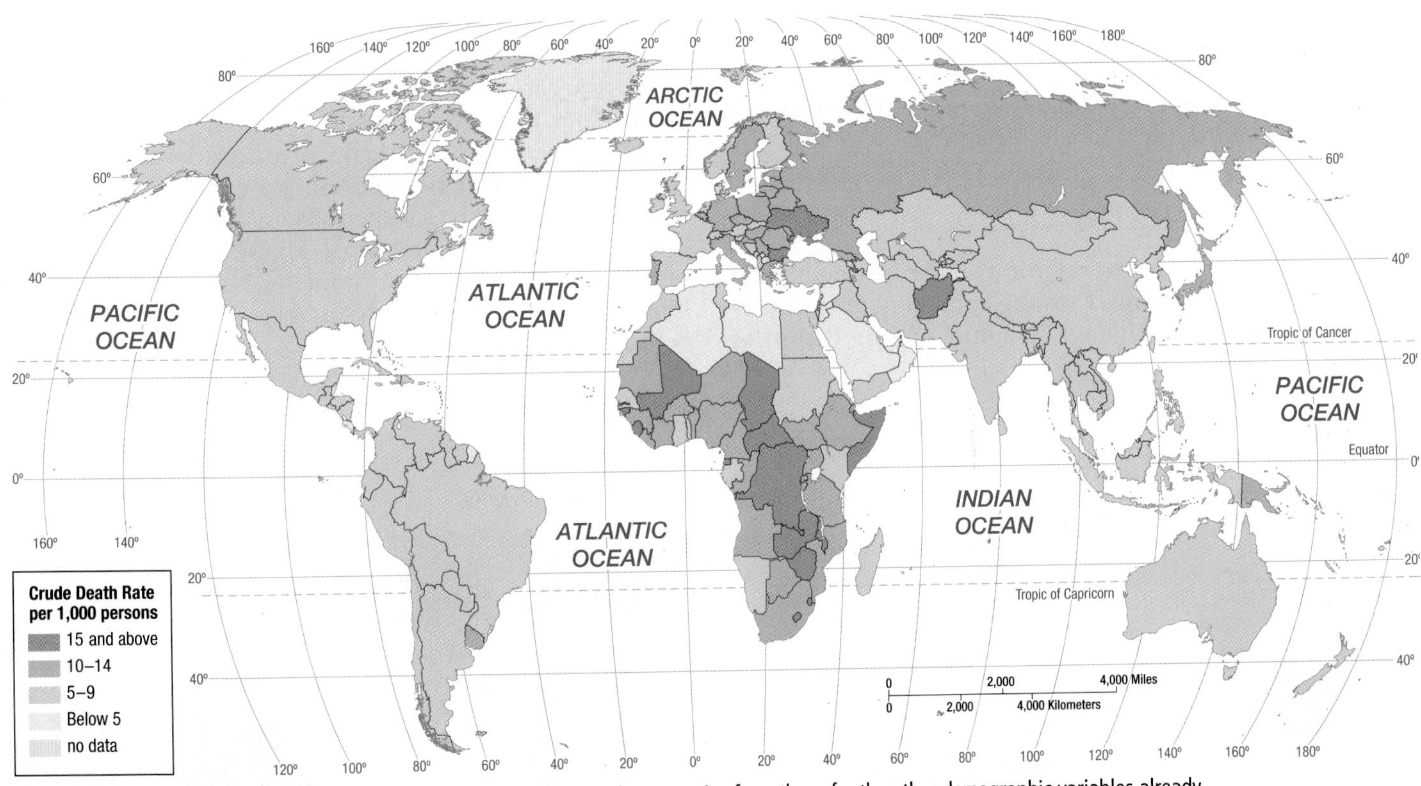

▲ **FIGURE 2-13 CRUDE DEATH RATE (CDR)** The global pattern of CDRs varies from those for the other demographic variables already mapped in this chapter. The demographic transition helps to explain the distinctive distribution of CDRs.

CBRs for individual countries range from 7 per 1,000 to 52, a spread of 45.

Why does Denmark, one of the world's wealthiest countries, have a higher CDR than Cape Verde, one of the poorest? Why does the United States, with its extensive system of hospitals and physicians, have a higher CDR than Mexico and nearly every country in Latin America? The answer is that the populations of different countries are at various stages in an important process known as the demographic transition, discussed later in this chapter.

Population Structure

Learning Outcome 2.2.3:
Understand how to read a population pyramid.

Fertility and mortality vary not only from country to country but also over time within a country. As a result, the number of people in different age groups in a country forms a pattern—the population structure.

POPULATION PYRAMIDS

A country's distinctive population structure can be displayed on a bar graph called a **population pyramid.** A population pyramid normally shows the percentage of the total population in five-year age groups, with the youngest group (zero to four years old) at the base of the pyramid and the oldest group at the top. The length of the bar represents the percentage of the total population contained in that group. By convention, males are usually shown on the left side of the pyramid and females on the right (Figure 2-14).

Population pyramids vary widely within the United States. For example, Laredo, Texas, which has a large Hispanic population, has a relatively broad pyramid, indicating a large percentage of children, whereas Naples, Florida, has a "reverse" pyramid, indicating a large percentage of elderly people. College towns, such as Lawrence, Kansas, have unusually shaped pyramids because of the exceptionally high percentage of people in their 20s.

DEPENDENCY RATIO

The age structure of a population helps to understand similarities and differences among countries. One important way to compare the age structure among countries is the **dependency ratio**, which is the number of people who are too young or too old to work, compared to the number of people in their productive years. People who are 0 to 14 years of age and 65 and over are normally classified as dependents. The larger the dependency ratio, the greater the financial burden on those who are working to support those who do not. The dependency ratio is 47 percent in Europe, compared to 85 percent in sub-Saharan Africa.

The high dependency ratio in sub-Saharan Africa derives from having a very high percentage of young people (Figure 2-15). Young dependents outnumber elderly ones by more than 14:1 in sub-Saharan Africa, whereas the numbers of people under 15 and over 65 are roughly equal in Europe. The large percentage of children in sub-Saharan Africa strains the ability of these relatively poor countries to provide needed services such as schools, hospitals, and day-care centers. When children reach the age of leaving school, jobs must be found for them, but the government must continue to allocate scarce resources to meet the needs of the still growing number of young people. On the other hand, the "graying" of the population places a burden on developed countries to meet their needs for income and medical care after they retire from jobs.

SEX RATIO

The number of males per 100 females in the population is the **sex ratio.** Developed countries have more females than males because on average women live seven years longer than men. Most Asian countries have more men than women, primarily because male babies greatly outnumber female babies, especially in the two most populous countries, China and India (Figure 2-16). The shortage of female babies in these countries has raised the possibility that a relatively large number of female fetuses are being aborted.

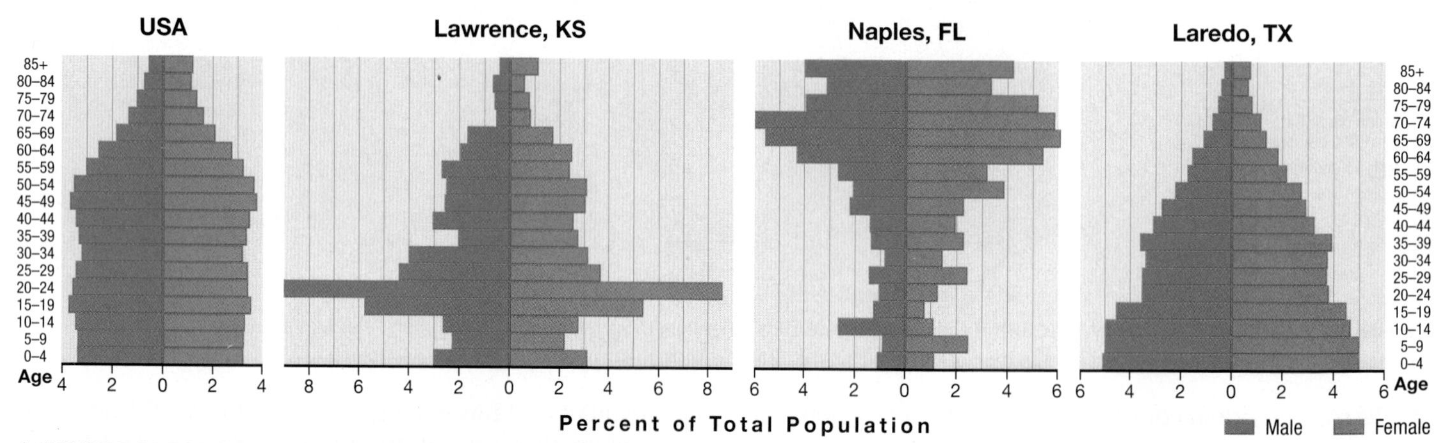

▲ **FIGURE 2-14 POPULATION PYRAMIDS FOR THE UNITED STATES AND SELECTED U.S. COMMUNITIES** Laredo has a broad pyramid, indicating higher percentages of young people and fertility rates. Lawrence has a high percentage of people in their twenties because it is the home of the University of Kansas. Naples has a high percentage of elderly people, especially women, so its pyramid is upside down.

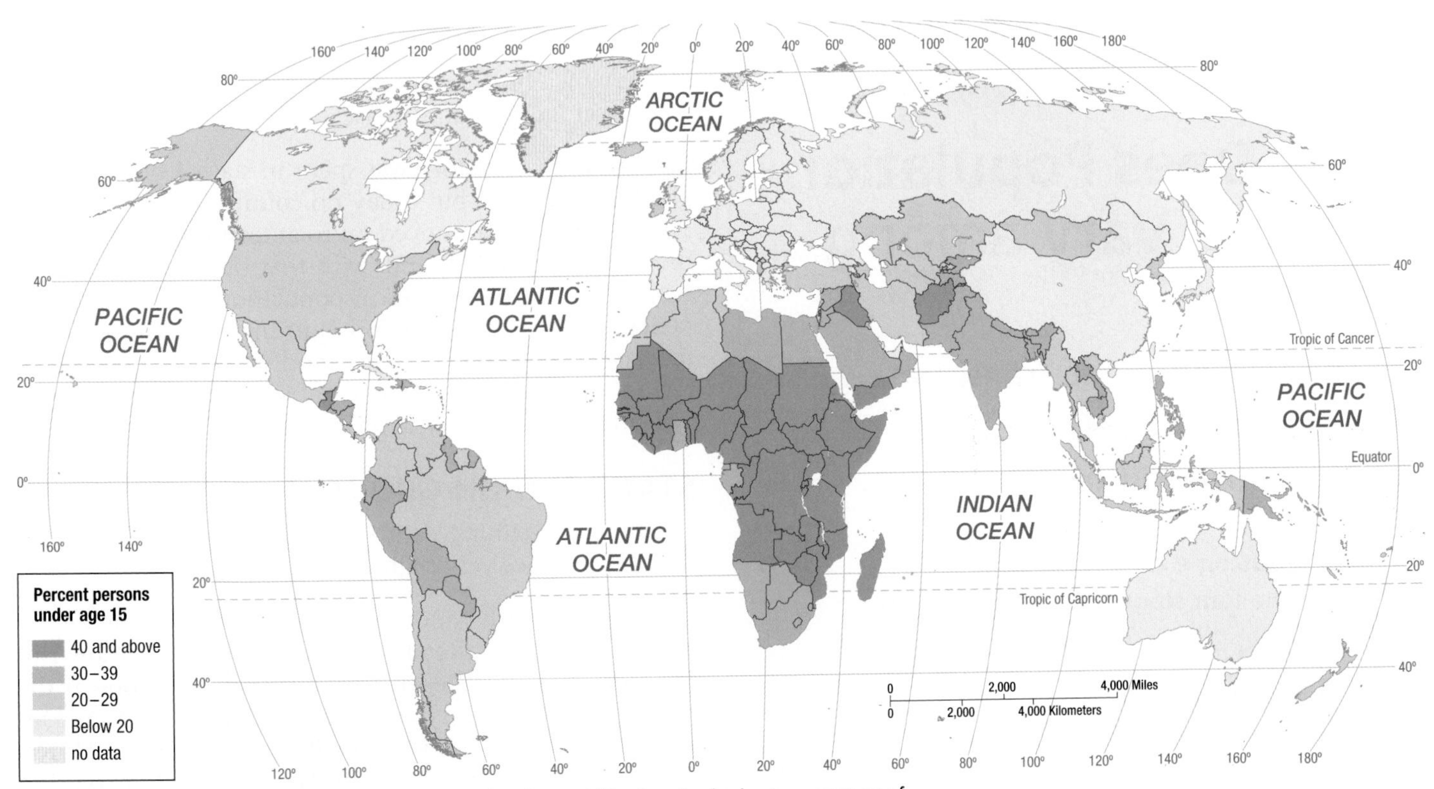

▲ **FIGURE 2-15 POPULATION UNDER AGE 15** Sub-Saharan Africa has the highest percentage of persons under age 15.

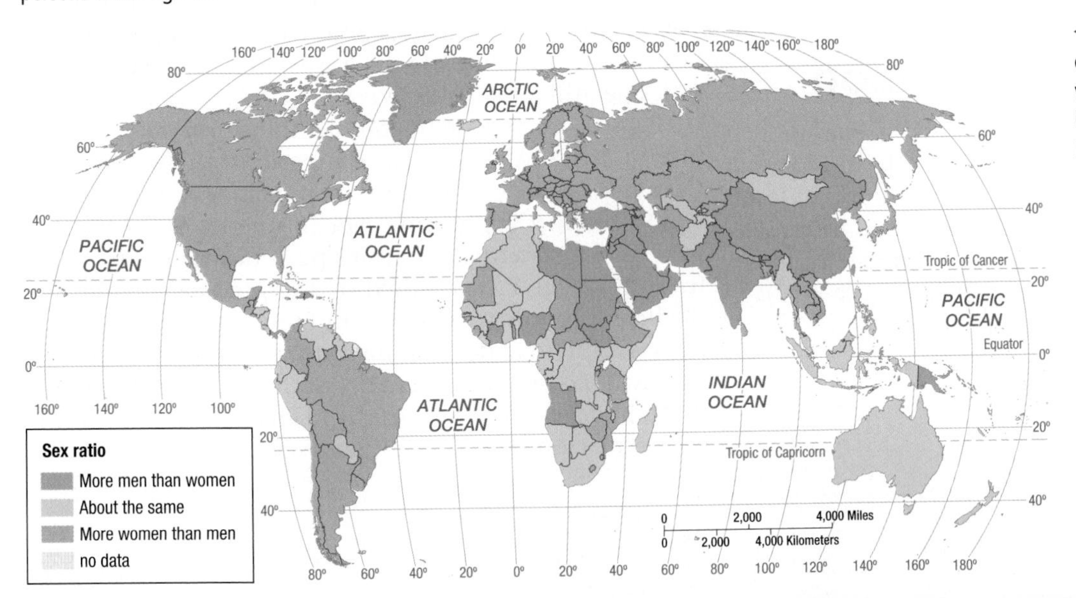

◀ **FIGURE 2-16 SEX RATIO** A map of the percentage of people over age 65 would show a reverse pattern, with the highest percentages in Europe and the lowest in Africa and Southwest Asia.

Pause and Reflect 2.2.3

Name a type of community that might have a lot more males than females.

CHECK-IN: KEY ISSUE 2

Why Is Global Population Increasing?

- ✓ **The NIR measures population growth as the difference between births and deaths.**
- ✓ **Births and deaths are measured using several indicators.**
- ✓ **A community's distinctive distribution by age and gender can be displayed in a population pyramid.**

KEY ISSUE 3

Why Does Population Growth Vary among Regions?

- The Demographic Transition
- Malthus on Overpopulation
- Population Futures

Learning Outcome 2.3.1
Describe the four stages of the demographic transition.

All countries have experienced some changes in NIR, CBR, and CDR, but at different times and at different rates. Why does global growth matter? In view of the current size of Earth's population and the NIR, will there soon be too many of us?

The Demographic Transition

The **demographic transition** is a process of change in a society's population from high crude birth and death rates and low rate of natural increase to a condition of low crude birth and death rates, low rate of natural increase, and higher total population. The process consists of four stages, and every country is in one of them (Figure 2-17).

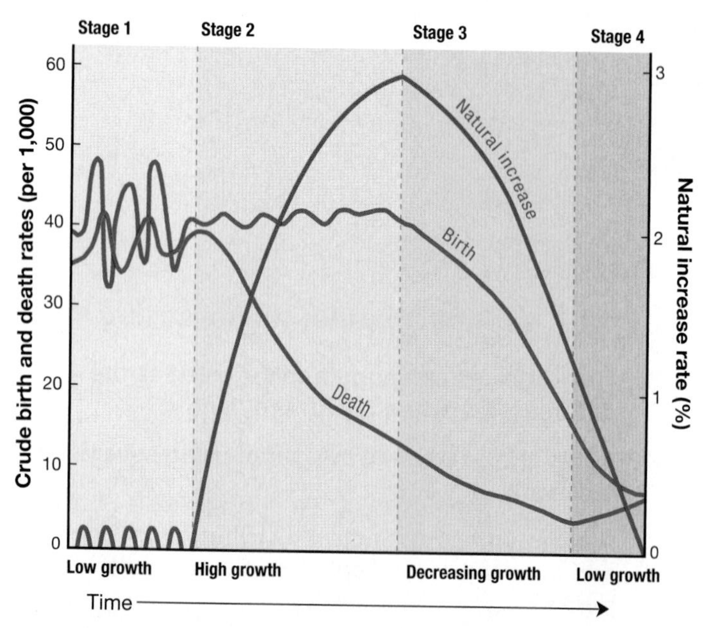

▲ FIGURE 2-17 **DEMOGRAPHIC TRANSITION MODEL** The demographic transition model consists of four stages.

STAGE 1: LOW GROWTH

Very high birth and death rates produce virtually no long-term natural increase.

Most of human history was spent in stage 1 of the demographic transition, but today no country remains in stage 1. Every nation has moved on to at least stage 2 of the demographic transition, and, with that transition, has experienced profound changes in population. For most of this period, people depended on hunting and gathering for food (see Chapter 10). When food was easily obtained, a region's population increased, but it declined when people were unable to locate enough animals or vegetation nearby.

STAGE 2: HIGH GROWTH

Rapidly declining death rates and very high birth rates produce very high natural increase.

Europe and North America entered stage 2 of the demographic transition after 1750, as a result of the **Industrial Revolution,** which began in the United Kingdom in the late eighteenth century and diffused to the European continent and North America (including the United States) during the nineteenth century. The Industrial Revolution was a conjunction of major improvements in manufacturing goods and delivering them to market (see Chapter 11). The result of this transformation was an unprecedented level of wealth, some of which was used to make communities healthier places to live.

Stage 2 of the demographic transition did not diffuse to Africa, Asia, and Latin America until around 1950 (Figure 2-18), and it made that transition for a different reason than in Europe and North America 200 years earlier. The late-twentieth-century push of developing countries into stage 2 was caused by the **medical revolution.** Medical technology invented in Europe and North America has diffused to developing countries. Improved medical practices have eliminated many of the traditional causes of death in developing countries and enabled more people to experience longer and healthier lives.

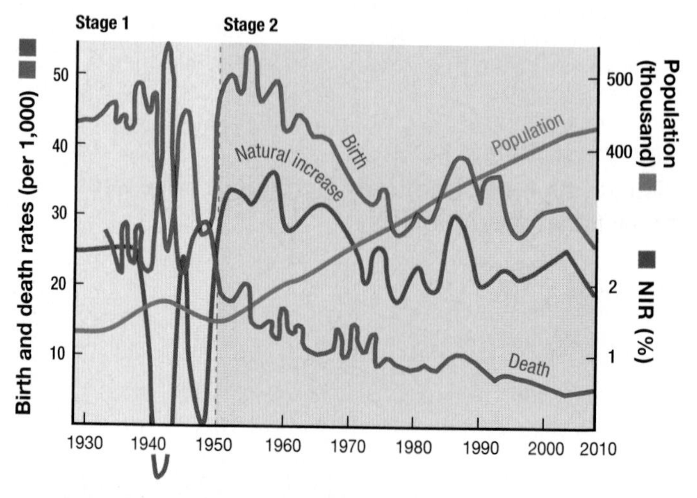

▲ FIGURE 2-18 **STAGE 2: CAPE VERDE** Cape Verde entered stage 2 of the demographic transition in approximately 1950, as indicated by the large gap between birth and death rates since then.

STAGE 3: DECREASING GROWTH

Birth rates rapidly decline, death rates continue to decline, and natural increase rates begin to moderate.

A country moves from stage 2 to stage 3 of the demographic transition when the CBR begins to drop sharply. The CDR continues to fall in stage 3 but at a much slower rate than in stage 2. The population continues to grow because the CBR is still greater than the CDR. But the rate of natural increase is more modest in countries in stage 3 than in those in stage 2 because the gap between the CBR and the CDR narrows.

A society enters stage 3 when people have fewer children. The decision to have fewer children is partly a delayed reaction to a decline in mortality,

Economic changes in stage 3 societies also induce people to have fewer offspring. People in stage 3 societies are more likely to live in cities than in the countryside and to work in offices, shops, or factories rather than on farms. Farmers often consider a large family to be an asset because children can do some of the chores. Urban homes are relatively small and may not have space to accommodate large families.

Most countries in Europe and North America (including the United States) moved from stage 2 to stage 3 of the demographic transition during the first half of the twentieth century. The movement took place during the second half of the twentieth century in many countries of Asia and Latin America, including Chile (Figure 2-19).

STAGE 4: LOW GROWTH

Very low birth and death rates produce virtually no long-term natural increase and possibly a decrease.

A country reaches stage 4 of the demographic transition when the CBR declines to the point where it equals the CDR and the NIR approaches zero. This condition is called **zero population growth (ZPG)**, a term often applied to stage 4 countries.

ZPG may occur when the CBR is still slightly higher than the CDR because some females die before reaching childbearing years, and the number of females in their childbearing years can vary. To account for these discrepancies, demographers more precisely define ZPG as the TFR that results in a lack of change in the total population over a long term. A TFR of approximately 2.1 produces ZPG.

Social customs again explain the movement to stage 4. Increasingly, women in stage 4 societies enter the labor force rather than remain at home as full-time homemakers. People who have access to a wider variety of birth-control methods are more likely to use some of them.

Denmark, like most other European countries, has reached stage 4 of the demographic transition (Figure 2-20). Denmark's population pyramid shows the impact of the demographic transition. Instead of a classic pyramid shape, Denmark has a column, demonstrating that the percentages of young and elderly people are nearly the same (Figure 2-21).

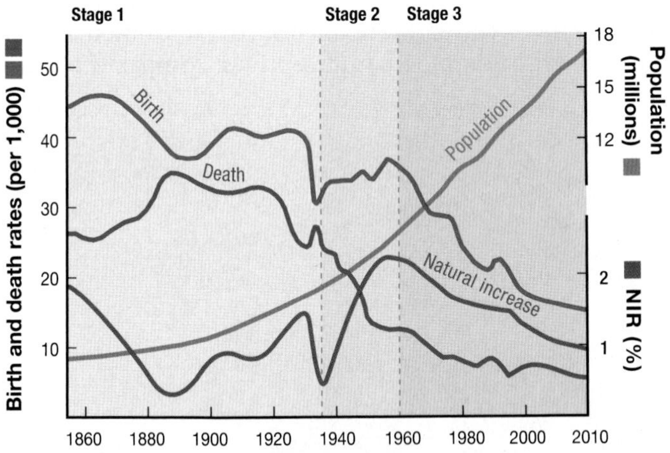

▲ **FIGURE 2-19 STAGE 3: CHILE** Chile entered stage 2 of the demographic transition in the 1930s, when death rates declined sharply, and stage 3 in the 1960s, when birth rates declined sharply.

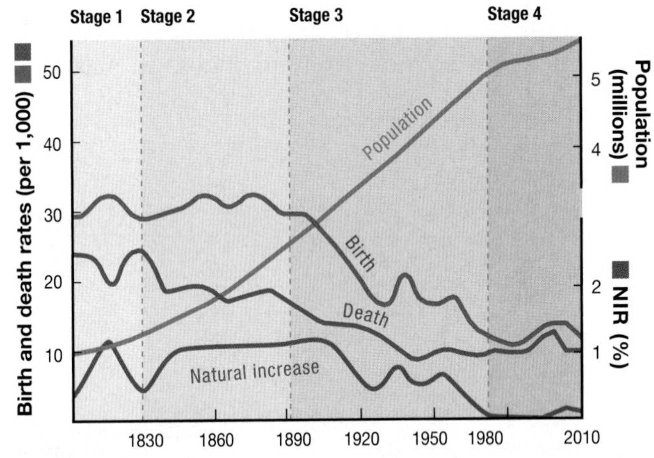

▲ **FIGURE 2-20 STAGE 4: DENMARK** Denmark has been in stage 4 of the demographic transition and has experienced virtually no change in total population since the 1970s.

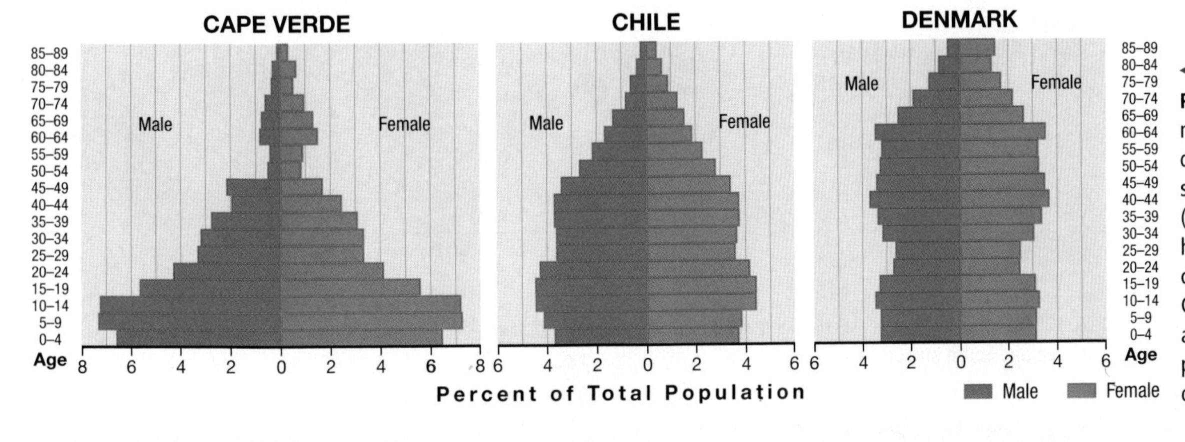

◀ **FIGURE 2-21 POPULATION PYRAMIDS** As a country moves through the demographic transition, the shape of the pyramid flattens. (left) Cape Verde's pyramid has a broad base, as is typical of a stage 2 country. (center) Chile's graph still resembles a pyramid. (right) Denmark's pyramid is flat, an indication of the aging of the population.

DECLINING BIRTH RATES

Learning Outcome 2.3.2
Summarize two approaches to reducing birth rates.

The CBR has declined rapidly since 1990, from 27 to 20 in the world as a whole and from 31 to 22 in developing countries (Figure 2-22). Two strategies have been successful in reducing birth rates:

1. **Lowering birth rates through education and health care.** One approach to lowering birth rates emphasizes the importance of improving local economic conditions (Figure 2-23). A wealthier community has more money to spend on education and health-care programs that promote lower birth rates. According to this approach:
   - With more women able to attend school and to remain in school longer, they would be more likely to learn employment skills and gain more economic control over their lives.
   - With better education, women would better understand their reproductive rights, make more informed reproductive choices, and select more effective methods of contraception.
   - With improved health-care programs, IMRs would decline through such programs as improved prenatal care, counseling about sexually transmitted diseases, and child immunization.
   - With the survival of more infants ensured, women would be more likely to choose to make more effective use of contraceptives to limit the number of children.
2. **Lowering birth rates through contraception.** The other approach to lowering birth rates emphasizes the importance of rapidly diffusing modern contraceptive methods (Figure 2-24). Economic development may promote lower birth rates in the long run, but the world cannot wait around for that alternative to take effect. Putting resources into family-planning programs can reduce birth rates much more rapidly. In developing countries, demand for contraceptive devices is greater than the available supply. Therefore, the most effective way to increase their use is to distribute more of them cheaply and quickly. According to this approach, contraceptives are the best method for lowering the birth rate.

Bangladesh is an example of a country that has had little improvement in the wealth and literacy of its people, but 56 percent of the women in the country used contraceptives in 2011 compared to 6 percent three decades earlier. Similar growth in the use of contraceptives has occurred in other developing countries, including Colombia, Morocco, and Thailand. Rapid growth in the acceptance of family planning is evidence that in the modern world, ideas can diffuse rapidly, even to places where people have limited access to education and modern communications.

The percentage of women using contraceptives is especially low in sub-Saharan Africa, so the alternative of distributing contraceptives could have an especially strong impact there. Fewer than one-fourth of women in sub-Saharan Africa employ contraceptives, compared to more

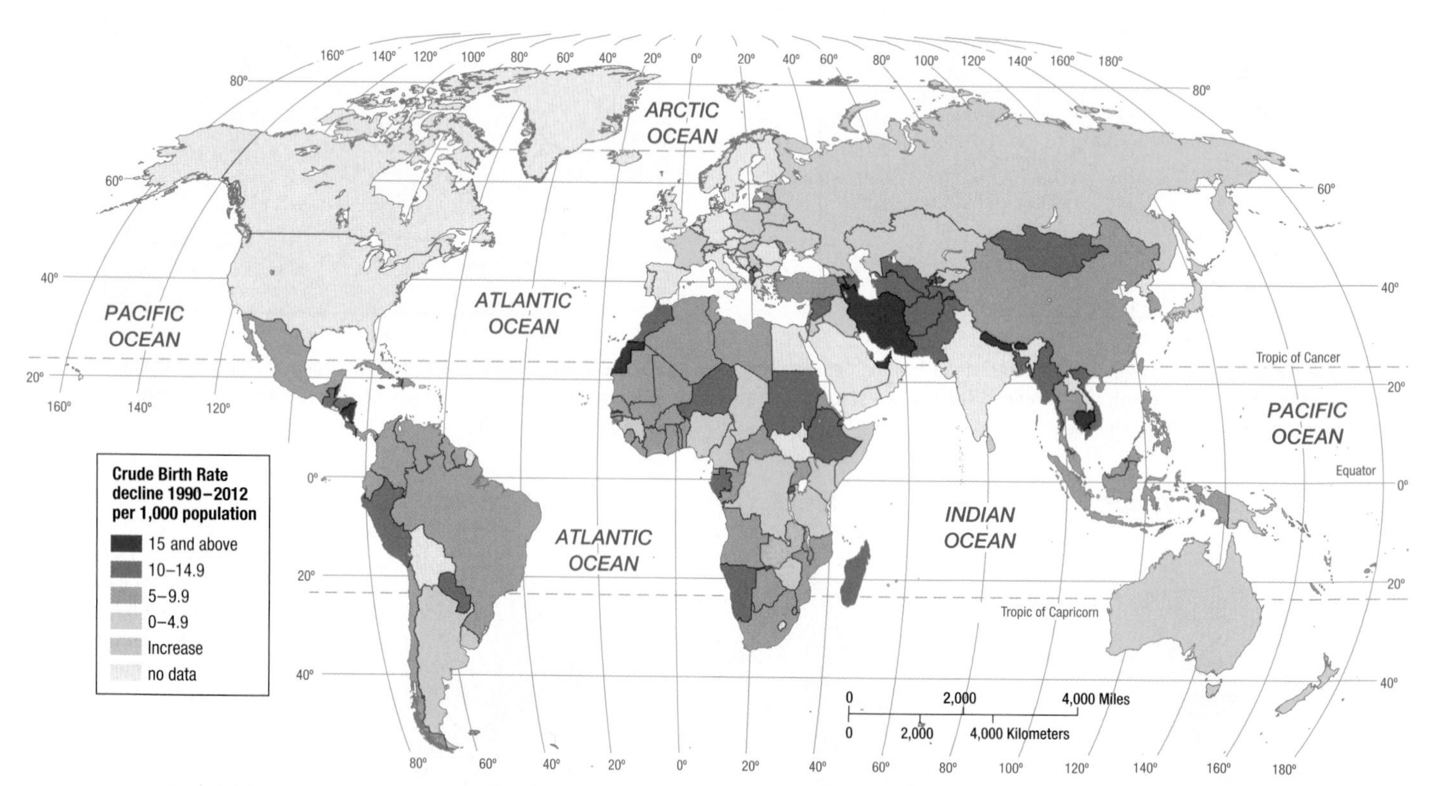

▲ **FIGURE 2-22 CBR CHANGE 1980–2012** The crude birth rate has declined in all but a handful of countries. Declines have been most rapid in Latin America and South and Southwest Asia.

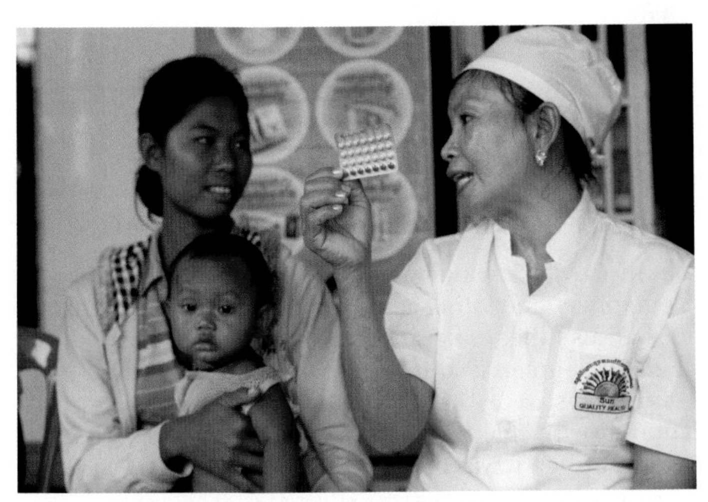

▲ **FIGURE 2-23 PROMOTING FEWER CHILDREN** Women talk about birth control at a health clinic in Kampong Cham, Cambodia.

than two-thirds in Asia and three-fourths in Latin America (Figure 2-25).

Regardless of which alternative is more successful, many oppose birth-control programs for religious and political reasons. Adherents of several religions, including Roman Catholics, fundamentalist Protestants, Muslims, and Hindus, have religious convictions that prevent them from using some or all birth-control methods. In the United States opposition is strong to terminating pregnancy by abortion, and the U.S. government has at times withheld aid to countries and family-planning organizations that advise abortion, even when such advice is only a small part of the overall aid program.

Pause and Reflect 2.3.2

Why have countries in Northern Europe had little if any decline in CBR since 1990?

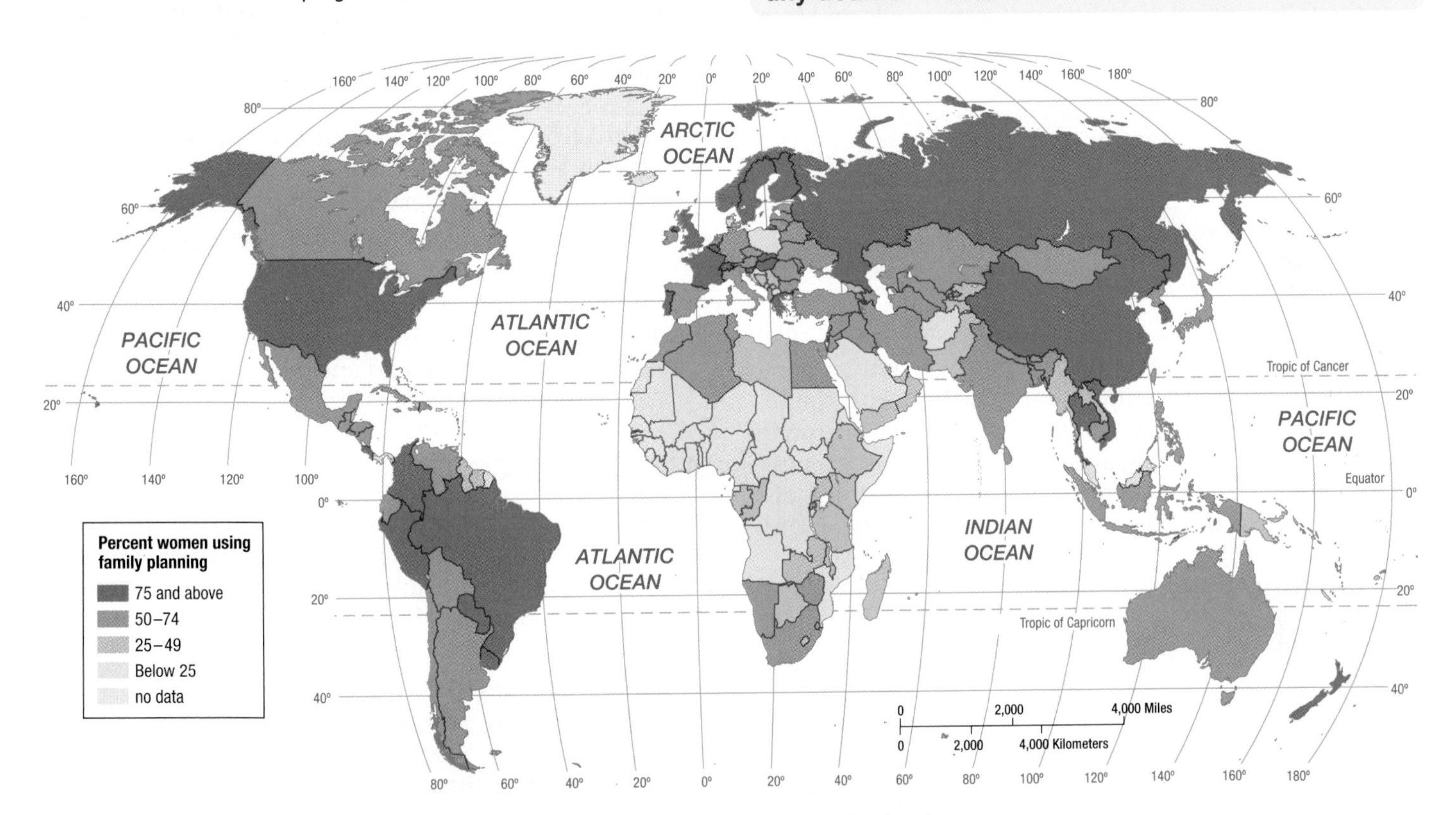

▲ **FIGURE 2-24 WOMEN USING FAMILY PLANNING** More than two-thirds of couples in developed countries use a family-planning method. Family-planning varies widely in developing countries. China reports the world's highest rate of family planning; the lowest rates are in sub-Saharan Africa.

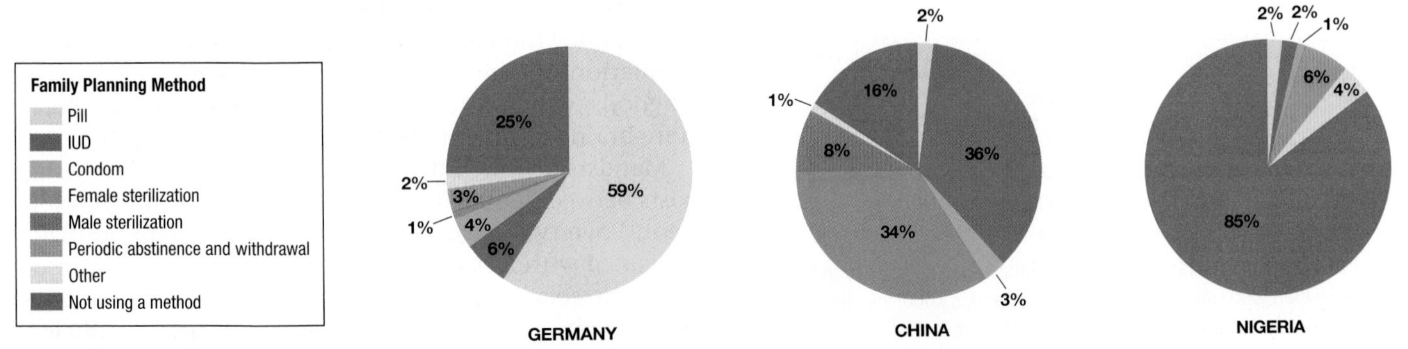

▲ **FIGURE 2-25 FAMILY PLANNING METHODS** The principal family-planning methods in developed countries like Germany are condoms and birth-control pills. The principal methods in China are intrauterine devices (IUDs) and female sterilization. People in sub-Saharan African countries such as Nigeria make minimal use of family-planning.

Malthus on Overpopulation

Learning Outcome 2.3.3

Summarize Malthus's argument about the relationship between population and resources.

English economist Thomas Malthus (1766–1834) was one of the first to argue that the world's rate of population increase was far outrunning the development of food supplies. In *An Essay on the Principle of Population*, published in 1798, Malthus claimed that the population was growing much more rapidly than Earth's food supply because population increased geometrically, whereas food supply increased arithmetically (Figure 2-26).

According to Malthus, these growth rates would produce the following relationships between people and food in the future:

Today:	1 person, 1 unit of food
25 years from now:	2 persons, 2 units of food
50 years from now:	4 persons, 3 units of food
75 years from now:	8 persons, 4 units of food
100 years from now:	16 persons, 5 units of food

Malthus stated made these conclusions several decades after England had become the first country to enter stage 2 of the demographic transition, in association with the Industrial Revolution. He concluded that population growth would press against available resources in every country unless "moral restraint" produced lower CBRs or unless disease, famine, war, or other disasters produced higher CDRs.

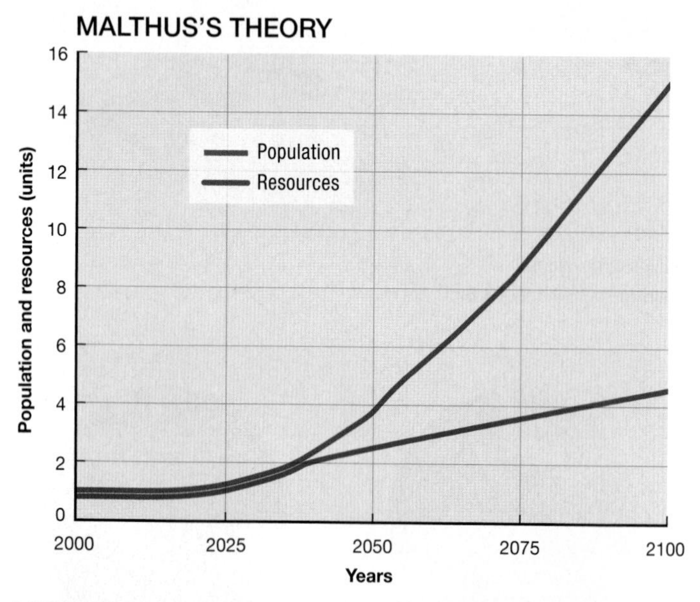

▲ **FIGURE 2-26 MALTHUS'S THEORY** Malthus expected population to grow more rapidly than food supply. The graph shows that if in 2000, the population of a place were 1 unit (such as 1 billion people) and the amount of resources were 1 unit (such as 1 billion tons of grain), then according to Malthus's theory, in 2100 the place would have around 15 billion people and 5 billion tons of grain).

Pause and Reflect 2.3.3

Calculate the units of population and food that Malthus predicted would exist in 200 years.

CONTEMPORARY NEO-MALTHUSIANS

Malthus's views remain influential today. Contemporary geographers and other analysts are taking another look at Malthus's theory because of Earth's unprecedented rate of natural increase during the twentieth century. Neo-Malthusians argue that two characteristics of recent population growth make Malthus's thesis even more frightening than when it was first written more than 200 years ago:

- In Malthus's time only a few relatively wealthy countries had entered stage 2 of the demographic transition, characterized by rapid population increase. Malthus failed to anticipate that relatively poor countries would have the most rapid population growth because of transfer of medical technology (but not wealth) from developed countries. As a result, the gap between population growth and resources is wider in some countries than even Malthus anticipated.
- World population growth is outstripping a wide variety of resources, not just food production. Neo-Malthusians paint a frightening picture of a world in which billions of people are engaged in a desperate search for food, water, and energy.

MALTHUS'S CRITICS

Malthus's theory has been severely criticized from a variety of perspectives. Criticism has been leveled at both the population growth and resource depletion sides of Malthus's equation.

RESOURCE DEPLETION. Many geographers consider Malthusian beliefs unrealistically pessimistic because they are based on a belief that the world's supply of resources is fixed rather than expanding.

POPULATION GROWTH. Critics disagree with Malthus's theory that population growth is a problem. To the contrary, a larger population could stimulate economic growth and, therefore, production of more food. A large population of consumers can generate a greater demand for goods, which results in more jobs. More people mean more brains to invent good ideas for improving life.

Marxists maintain that no cause-and-effect relationship exists between population growth and economic development. Poverty, hunger, and other social welfare problems associated with lack of economic development are a result of unjust social and economic institutions, not population growth. They argue that the world possesses sufficient resources to eliminate global hunger and poverty, if only these resources are shared equally.

MALTHUS'S THEORY AND REALITY

On a global scale, conditions during the past half-century have not supported Malthus's theory. Even though the human population has grown at its most rapid rate ever, world food production has consistently grown at a faster rate than the NIR since 1950. Malthus was fairly close to the mark on food production but much too pessimistic on population growth.

Overall food production has increased during the last half-century somewhat more rapidly than Malthus predicted. In India, for example, rice production has followed Malthus's expectations fairly closely, but wheat production has increased twice as fast as Malthus expected (Figure 2-27). Better growing techniques, higher-yielding seeds, and cultivation of more land have contributed to the expansion in the food supply (see Chapter 10). Many people in the world cannot afford to buy food or do not have access to sources of food, but these are problems of distribution of wealth rather than insufficient global production of food, as Malthus theorized.

It is on the population side of the equation that Malthus has proved to be inaccurate. His model expected population to quadruple during a half-century, but even in India—a country known for relatively rapid growth (see the next section)—population has increased more slowly than food supply.

However, neo-Malthusians point out that production of both wheat and rice has slowed in India in recent years, as shown in Figure 2-27. Without new breakthroughs in food production, India will not be able to keep food supply ahead of population growth.

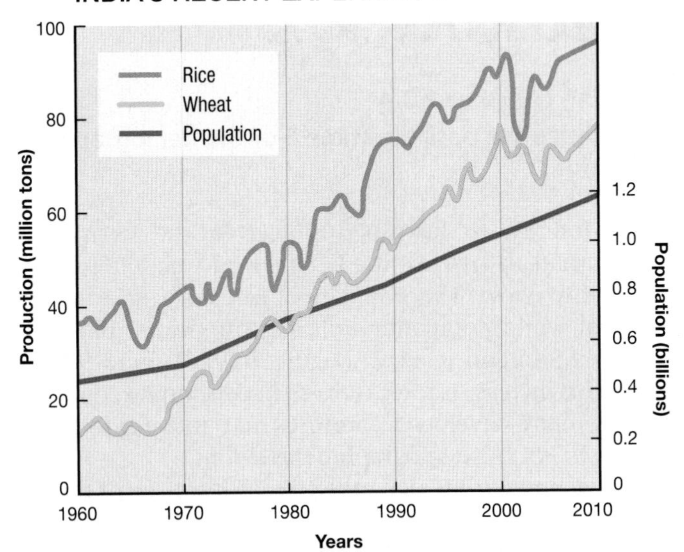

▲ **FIGURE 2-27 POPULATION AND FOOD PRODUCTION IN INDIA** Production of wheat and rice has increased more rapidly than has population.

JAPAN'S DECLINING POPULATION

Japan is an example of a country that faces the prospect of population decline in future, from 127 million in 2010 to 95 million in 2050, according to the Japanese government. With population decline will come a dramatic shift in the country's population structure (Figure 2-28). By 2050, the Japanese pyramid is expected to be reversed from that of 1950. Instead of a very high percentage of children, Japan will have a very high percentage of elderly people.

In the United States, the population is expected to continue to grow through immigration rather than through natural increase (see Chapter 3), but Japan discourages immigration. Japanese society, having placed a high value on social conformity for thousands of years, does not welcome outsiders from other cultural traditions.

With few immigrants, Japan faces a severe shortage of workers. Japan is addressing the labor force shortage primarily by encouraging more Japanese people to work, especially older people and women. Programs make it more attractive for older people to continue working, to receive more health-care services at home instead of in hospitals, and to borrow against the value of their homes to pay for health care. In the long run, more women in the labor force may translate into an even lower birth rate and therefore an even lower NIR in the future. Rather than combine work with child rearing, Japanese women are expected to make a stark choice: either marry and raise children or remain single and work.

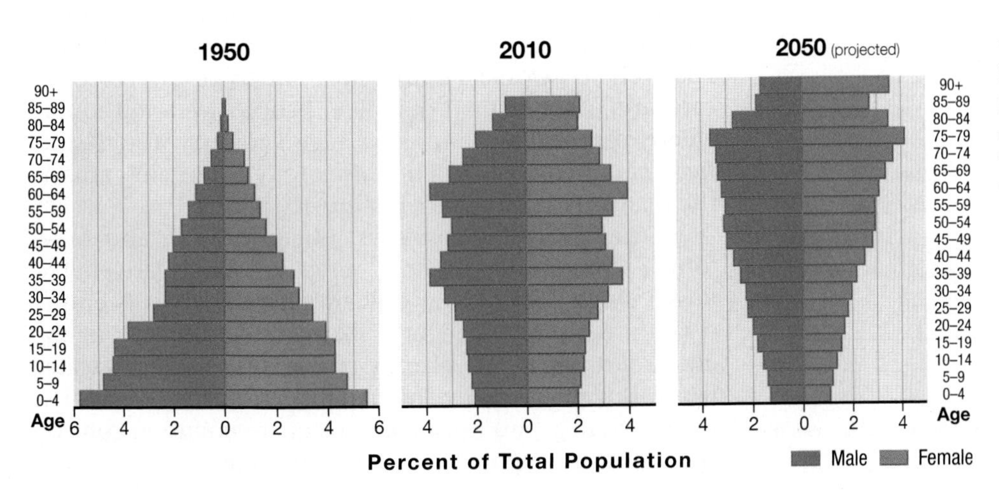

◀ **FIGURE 2-28 JAPAN'S CHANGING POPULATION PYRAMIDS** Japan's population pyramid has shifted from a broad base in 1950 to a rectangular shape. In the future, the bottom of the pyramid is expected to contract and the top to expand.

Population Futures

Learning Outcome 2.3.4
Summarize the possible stage 5 of the demographic transition.

NIR is forecast to be much slower in the twenty-first century than in the twentieth, but world population will continue to grow. The Population Reference Bureau forecasts that world population will increase from 7 billion in 2011 to 9.5 billion in 2050 (Figure 2-29). Around 97 percent of this increase is forecast to be in developing countries, whereas many developed countries may move into a possible stage 5 of the demographic transition.

Future population depends primarily on fertility. The United Nations forecasts that if the current TFR of 2.5 remains unchanged, world population would reach 12 billion in 2050. On the other hand, if TFR declines in the next few years to 1.5, world population would actually decline to 8 billion in 2050.

DEMOGRAPHIC TRANSITION POSSIBLE STAGE 5: DECLINE

A country that has passed through the four stages of the demographic transition has in some ways completed a cycle—from little or no natural increase in stage 1 to little or no natural increase in stage 4. Two crucial demographic differences underlie this process:

- The total population of the country is much higher in stage 4 than in stage 1.
- At the beginning of the demographic transition, the CBRs and CDRs are high—35 to 40 per 1,000—whereas at the end of the process the rates are very low, approximately 10 per 1,000.

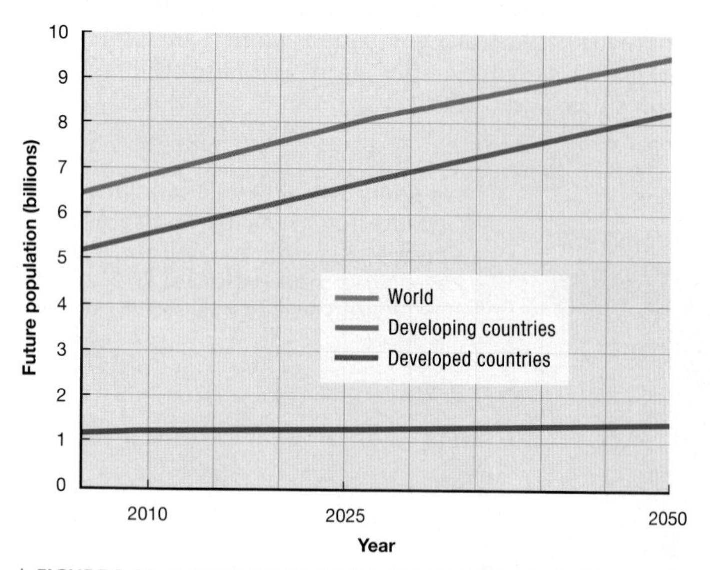

▲ **FIGURE 2-29 FUTURE POPULATION GROWTH** Nearly all of the world's population growth is forecast to be in developing countries.

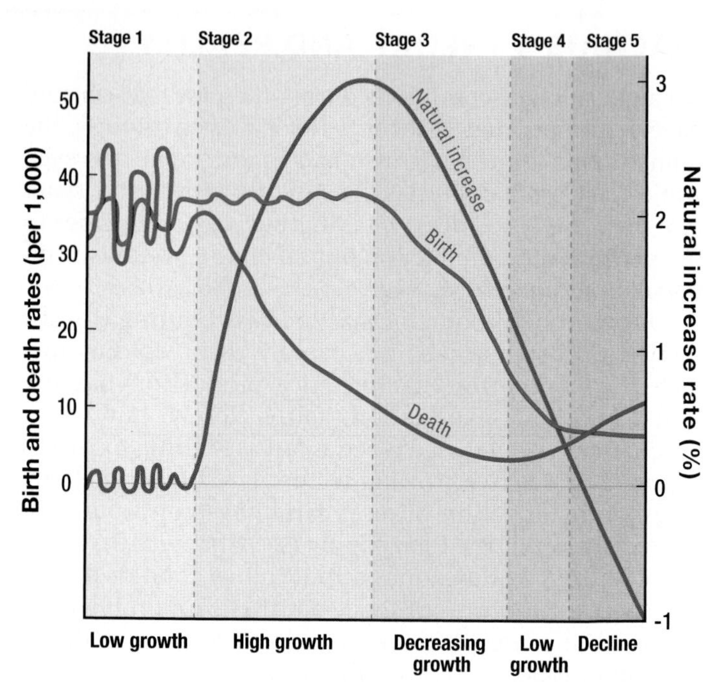

▲ **FIGURE 2-30 POSSIBLE DEMOGRAPHIC TRANSITION STAGE 5** Stage 5 of the demographic transition would be characterized by a negative NIR, because the CDR would be greater than the CBR.

The four-stage demographic transition is characterized by two big breaks with the past. The first break—the sudden drop in the death rate that comes from technological innovation—has been accomplished everywhere. The second break—the sudden drop in the birth rate that comes from changing social customs—has yet to be achieved in many countries.

Meanwhile, a possible stage 5 of the demographic transition is predicted by demographers for some developed countries. Stage 5 would be characterized by a very low CBR, an increasing CDR, and therefore a negative NIR (Figure 2-31). After several decades of very low birth rates, a stage 5 country would have relatively few young women aging into child-bearing years. As the smaller pool of women each chooses to have fewer children, birth rates would continue to fall even more than in stage 4.

The world's future population will definitely be older. The elderly support ratio is the number of working-age people (ages 15 to 64) divided by the number of persons 65 and older (Figure 2-30). A small number means that relatively few workers must contribute to pensions, health care, and other support that older people need. With more elderly people than children, a stage 5 country would experience an increased CDR because of high mortality among the relatively large number of elderly people.

Several European countries, notably Russia and other former Communist countries, already have negative NIRs. Russia's high CDR and low CBR are a legacy of a half-century of Communist rule. The low CBR may stem from a long tradition of strong family-planning programs and a deep-seated pessimism about having children in an uncertain world. The high CDR may be a legacy of inadequate pollution controls and inaccurate reporting by the Communists.

CHINA AND INDIA

The world's two most populous countries, China and India, will heavily influence future prospects for global overpopulation. These two countries—together encompassing more than one-third of the world's population—have adopted different family-planning programs. As a result of less effective policies, India adds 12 million more people each year than does China. Current projections show that India could surpass China as the world's most populous country around 2030.

INDIA'S POPULATION POLICIES. India, like most countries in Africa, Asia, and Latin America, remained in stage 1 of the demographic transition until the late 1940s. During the first half of the twentieth century, the Indian population increased modestly—less than 1 percent per year—and even decreased in some years because of malaria, famines, plagues, and cholera epidemics.

Immediately after gaining independence from England in 1947, India saw a sharp decline in death rate (to 20 per 1,000 in 1951), whereas the CBR remained relatively high (about 40). Consequently, the NIR jumped to 2 percent per year. In response to this rapid growth, India became the first country to embark on a national family-planning program, in 1952. The government has established clinics and has provided information about alternative methods of birth control. Birth-control devices have been distributed free or at subsidized prices. Abortions, legalized in 1972, have been performed at a rate of several million per year. All together, the government spends several hundred million dollars annually on various family-planning programs.

India's most controversial family-planning program was the establishment of camps in 1971 to perform sterilizations—surgical procedures in which people were made incapable of reproduction. A sterilized person was entitled to a payment, which has been adjusted several times but generally has been equivalent to the average monthly income in India. At the height of the program, in 1976, 8.3 million sterilizations were performed during a 6-month period, mostly on women.

The birth-control drive declined in India after 1976. Widespread opposition to the sterilization program grew in the country because people feared that they would be forcibly sterilized. The prime minister, Indira Gandhi, was defeated in 1977, and the new government emphasized the voluntary nature of birth-control programs. The term *family planning*, which the Indian people associated with the forced sterilization policy, was replaced by the term *family welfare* to indicate that compulsory birth-control programs had been terminated. Although Mrs. Gandhi served again as prime minister from 1980 until she was assassinated in 1984, she did not emphasize family planning during that time because of the opposition during her previous administration.

In the past several decades, government-sponsored family-planning programs in India have emphasized education, including advertisements on national radio and television networks and information distributed through local health centers. Given the cultural diversity of the Indian people, the national campaign has had only limited success. The dominant form of birth control continues to be sterilization of women, in many cases after the women have already borne several children.

CHINA'S POPULATION POLICIES. In contrast to India, China has made substantial progress in reducing its rate of growth. Since 2000, China has actually had a lower CBR than the United States.

The core of the Chinese government's family-planning program has been the One Child Policy, adopted in 1980. Under the One Child Policy, a couple needs a permit to have a child. Couples receive financial subsidies, a long maternity leave, better housing, and (in rural areas) more land if they agree to have just one child. The government prohibits marriage for men until they are age 22 and women until they are 20. To further discourage births, people receive free contraceptives, abortions, and sterilizations. Rules are enforced by a government agency, the State Family Planning Commission.

As China moves toward a market economy in the twenty-first century and as Chinese families become wealthier, the harsh rules in the One Child Policy have been relaxed, especially in urban areas. Clinics provide counseling on a wider range of family-planning options. Instead of fines, Chinese couples wishing a second child pay a "family-planning fee" to cover the cost to the government of supporting the additional person. Fears that relaxing the One Child Policy would produce a large increase in the birth rate have been unfounded. After a quarter-century of intensive educational programs, as well as coercion, the Chinese people have accepted the benefits of family planning.

Pause and Reflect 2.3.4
Why might China's One Child Policy result in many more male than female children?

CHECK-IN: KEY ISSUE 3

Why Does Population Growth Vary Among Regions?

- ✓ **The demographic transition has four stages characterized by varying rates of births, deaths, and natural increase.**
- ✓ **The CBR has declined since 1990 in all but a handful of countries.**
- ✓ **Malthus believed that population would outstrip resources, but critics argue that that hasn't been the case in the world as a whole.**

KEY ISSUE 4

Why Do Some Regions Face Health Threats?

- **Epidemiologic Transition**
- **Infectious Diseases**
- **Health Care**

Learning Outcome 2.4.1
Summarize the four stages of the epidemiologic transition.

As world NIR slows and the threat of overpopulation recedes, at least at a worldwide scale, geographers increasingly turn their attention to the health of the record number of people who are alive. Medical researchers have identified an **epidemiologic transition** that focuses on distinctive health threats in each stage of the demographic transition. Epidemiologists rely heavily on geographic concepts such as scale and connection because measures to control and prevent an epidemic derive from understanding its distinctive distribution and method of diffusion.

Epidemiologic Transition

The term *epidemiologic transition* comes from **epidemiology**, which is the branch of medical science concerned with the incidence, distribution, and control of diseases that are prevalent among a population at a special time and are produced by some special causes not generally present in the affected locality. The concept was originally formulated by epidemiologist Abdel Omran in 1971.

STAGE 1: PESTILENCE AND FAMINE (HIGH CDR)

In stage 1 of the epidemiologic transition, infectious and parasitic diseases were principal causes of human deaths, along with accidents and attacks by animals and other humans. Malthus called these causes of deaths "natural checks" on the growth of the human population in stage 1 of the demographic transition.

History's most violent stage 1 epidemic was the Black Plague (bubonic plague), which was probably transmitted to humans by fleas from migrating infected rats:

- The Black Plague originated among Tatars in present-day Kyrgyzstan.
- It diffused to present-day Ukraine when the Tatar army attacked an Italian trading post on the Black Sea.
- Italians fleeing the trading post carried the infected rats on ships west to the major coastal cities of Southeastern Europe in 1347.
- The plague diffused from the coast to inland towns and then to rural areas.
- It reached Western Europe in 1348 and Northern Europe in 1349.

About 25 million Europeans— at least one-half of the continent's population—died between 1347 and 1350. Five other epidemics in the late fourteenth century added to the toll in Europe. In China, 13 million died from the plague in 1380.

The plague wiped out entire villages and families, leaving farms with no workers and estates with no heirs. Churches were left without priests and parishioners, schools without teachers and students. Ships drifted aimlessly at sea after entire crews succumbed to the plague.

STAGE 2: RECEDING PANDEMICS (RAPIDLY DECLINING CDR)

Stage 2 of the epidemiologic transition has been called the *stage of receding pandemics*. A **pandemic** is disease that occurs over a wide geographic area and affects a very high proportion of the population. Improved sanitation, nutrition, and medicine during the Industrial Revolution reduced the spread of infectious diseases. Death rates did not decline immediately and universally during the early years of the Industrial Revolution. Poor people crowded into rapidly growing industrial cities had especially high death rates. Cholera—uncommon in rural areas—became an especially virulent epidemic in urban areas during the Industrial Revolution.

Construction of water and sewer systems had eradicated cholera by the late nineteenth century. However, cholera persists in several developing regions in stage 2 of the demographic transition, especially sub-Saharan Africa and South and Southeast Asia, where many people lack access to clean drinking water (Figure 2-31). Cholera has also been found on Hispaniola, the island shared by Haiti and the Dominican Republic, especially in the wake of an earthquake in 2010 that killed 200,000 and displaced 1 million.

A computer-based Geographic Information System was invented in the twentieth century, but the idea of overlaying maps to understand human and natural patterns is much older. A century before the invention of computers, GIS helped to explain and battle stage 2 pandemics.

Dr. John Snow (1813–1858) was a British physician, not a geographer. To fight one of the worst nineteenth century pandemics, cholera, Snow created a hand-made GIS in 1854. On a map of London's Soho neighborhood, Snow overlaid two other maps, one showing the addresses of cholera victims and the other the location of water pumps—which for the poor residents of Soho were the principal source of water for drinking, cleaning, and cooking (Figure 2-30).

The overlay maps showed that cholera victims were not distributed uniformly through Soho. Dr. Snow showed that

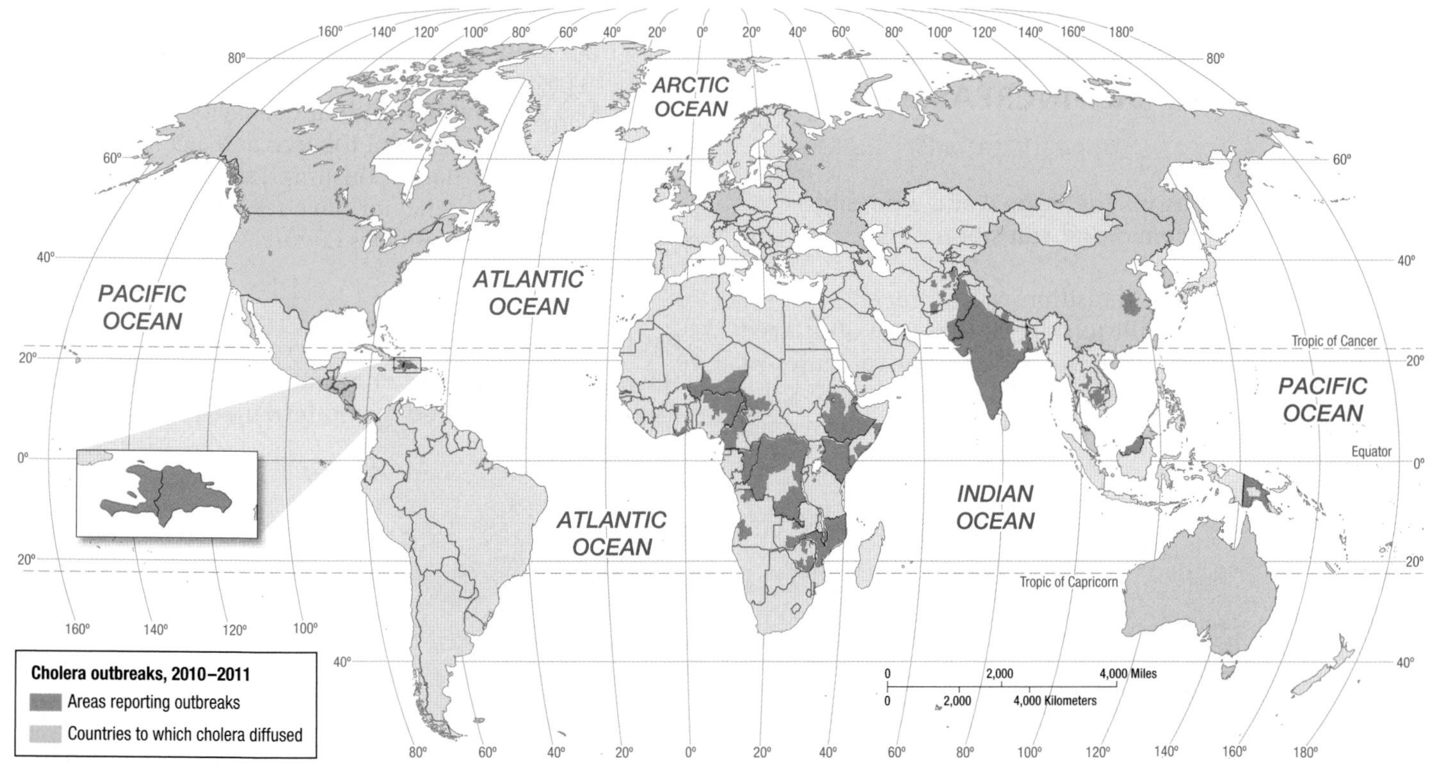

▲ **FIGURE 2-31 CHOLERA** Countries reporting cholera in recent years are found primarily in sub-Saharan Africa and South Asia.

a large percentage of cholera victims were clustered around one pump, on Broad Street. Tests at the Broad Street pump subsequently proved that the water there was contaminated. Further investigation revealed that contaminated sewage was getting into the water supply near the pump. Although no longer operative, the contaminated pump still stands in London and can be seen in the photo on page 42.

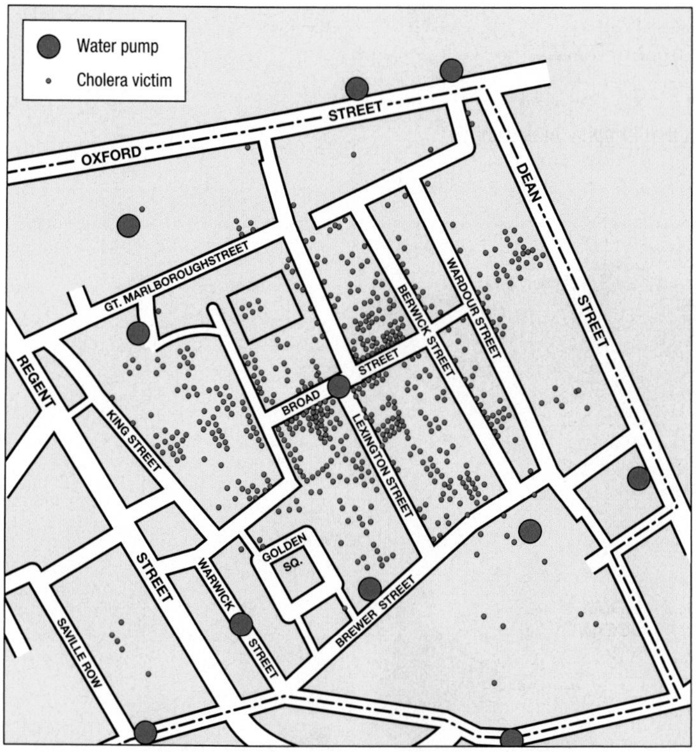

Before Dr. Snow's geographic analysis, many believed that epidemic victims were being punished for sinful behavior and that most victims were poor because poverty was considered a sin. Now we understand that cholera affects the poor because they are more likely to have to use contaminated water.

▲ **FIGURE 2-32 SIR JOHN SNOW'S CHOLERA MAP** In 1854, Dr. John Snow mapped the distribution of cholera victims and water pumps to prove that the cause of the infection was contamination of the pump near the corner of Broad and Lexington streets.

STAGE 3: DEGENERATIVE DISEASES (MODERATELY DECLINING CDR)

Stage 3 of the epidemiologic transition, the stage of degenerative and human-created diseases, is characterized by a decrease in deaths from infectious diseases and an increase in chronic disorders associated with aging. The two especially important chronic disorders in stage 3 are cardiovascular diseases, such as heart attacks, and various forms of cancer. The global pattern of cancer is the opposite of that for stage 2 diseases; sub-Saharan Africa and South Asia have the lowest incidence of cancer, primarily because of the relatively low life expectancy in those regions.

Pause and Reflect 2.4.1

In what climate zone are most of the countries that have experienced cholera recently?

STAGE 4: DELAYED DEGENERATIVE DISEASES (LOW BUT INCREASING CDR)

Learning Outcome 2.4.2
Summarize the reasons for Stage 4 and a possible stage 5 of the epidemiologic transition.

Omran's epidemiologic transition was extended by S. Jay Olshansky and Brian Ault to stage 4, the stage of delayed degenerative diseases. The major degenerative causes of death—cardiovascular diseases and cancers (Figure 2-33)—linger, but the life expectancy of older people is extended through medical advances. Through medicine, cancers spread more slowly or are removed altogether. Operations such as bypasses repair deficiencies in the cardiovascular system. Also improving health are behavior changes such as better diet, reduced use of tobacco and alcohol, and exercise. On the other hand, consumption of non-nutritious food and sedentary behavior have resulted in an increase in obesity in stage 4 countries (Figure 2-34).

Pause and Reflect 2.4.2
Have you had a parent or grandparent whose lifespan was extended by modern medical advances?

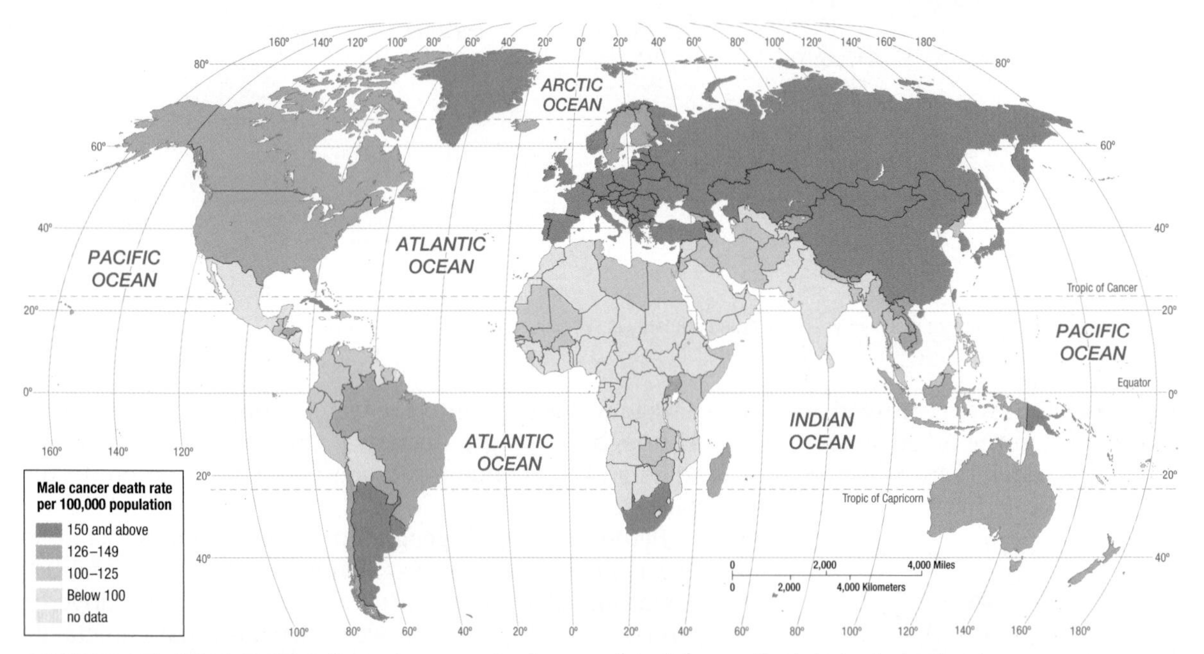

▲ **FIGURE 2-33 MALE CANCER** Cancer is an example of a cause of death for men that is higher in developed countries than in developing ones.

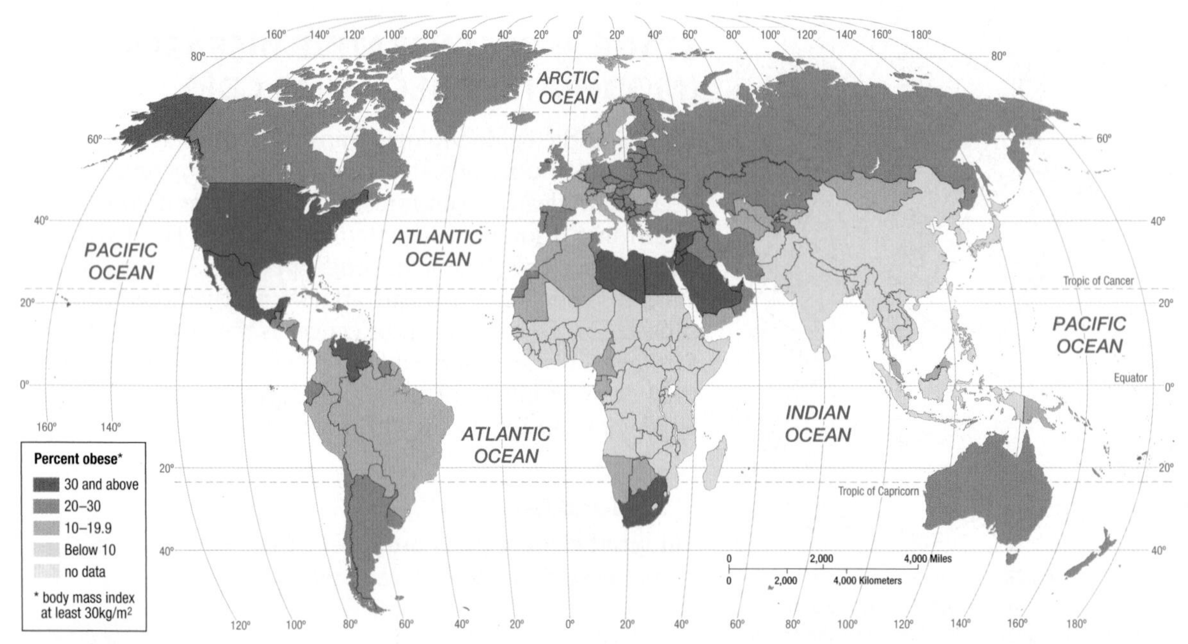

▲ **FIGURE 2-34 OBESITY** Obesity is a health problem in the United States and in Southwest Asia.

Infectious Diseases

Recall that in the possible stage 5 of the demographic transition, CDR rises because more of the population is elderly. Some medical analysts argue that the world is moving into stage 5 of the epidemiologic transition, brought about by a reemergence of infectious and parasitic diseases. Infectious diseases thought to have been eradicated or controlled have returned, and new ones have emerged. A consequence of stage 5 would be higher CDRs. Other epidemiologists dismiss recent trends as a temporary setback in a long process of controlling infectious diseases.

In a possible stage 5, infectious diseases thought to have been eradicated or controlled return, and new ones emerge. Three reasons help to explain the possible emergence of a stage 5 of the epidemiologic transition: evolution, poverty, and increased connections.

REASON FOR POSSIBLE STAGE 5: EVOLUTION

Infectious disease microbes have continuously evolved and changed in response to environmental pressures by developing resistance to drugs and insecticides. Antibiotics and genetic engineering contribute to the emergence of new strains of viruses and bacteria.

Malaria was nearly eradicated in the mid-twentieth century by spraying DDT in areas infested with the mosquito that carried the parasite. For example, new malaria cases in Sri Lanka fell from 1 million in 1955 to 18 in 1963. The disease returned after 1963, however, and now causes more than 1 million deaths worldwide annually. A major reason was the evolution of DDT-resistant mosquitoes.

REASON FOR POSSIBLE STAGE 5: POVERTY

Infectious diseases are more prevalent in poor areas than other places because unsanitary conditions may persist, and most people can't afford the drugs needed for treatment. Tuberculosis (TB) is an example of an infectious disease that has been largely controlled in developed countries but remains a major cause of death in developing countries (Figure 2-35). An airborne disease that is often called "consumption" and that damages the lungs, TB spreads principally through coughing and sneezing. TB was one of the principal causes of death among the urban poor in the nineteenth century during the Industrial Revolution.

The death rate from TB declined in the United States from 200 per 100,000 in 1900 to 60 in 1940 and 4 today. However, in developing countries, the TB rate is more than 10 times higher than in developed countries, and nearly 2 million people worldwide die from it annually. TB is more prevalent in poor areas because the long, expensive treatment poses a significant economic burden. Patients stop taking the drugs before the treatment cycle is completed.

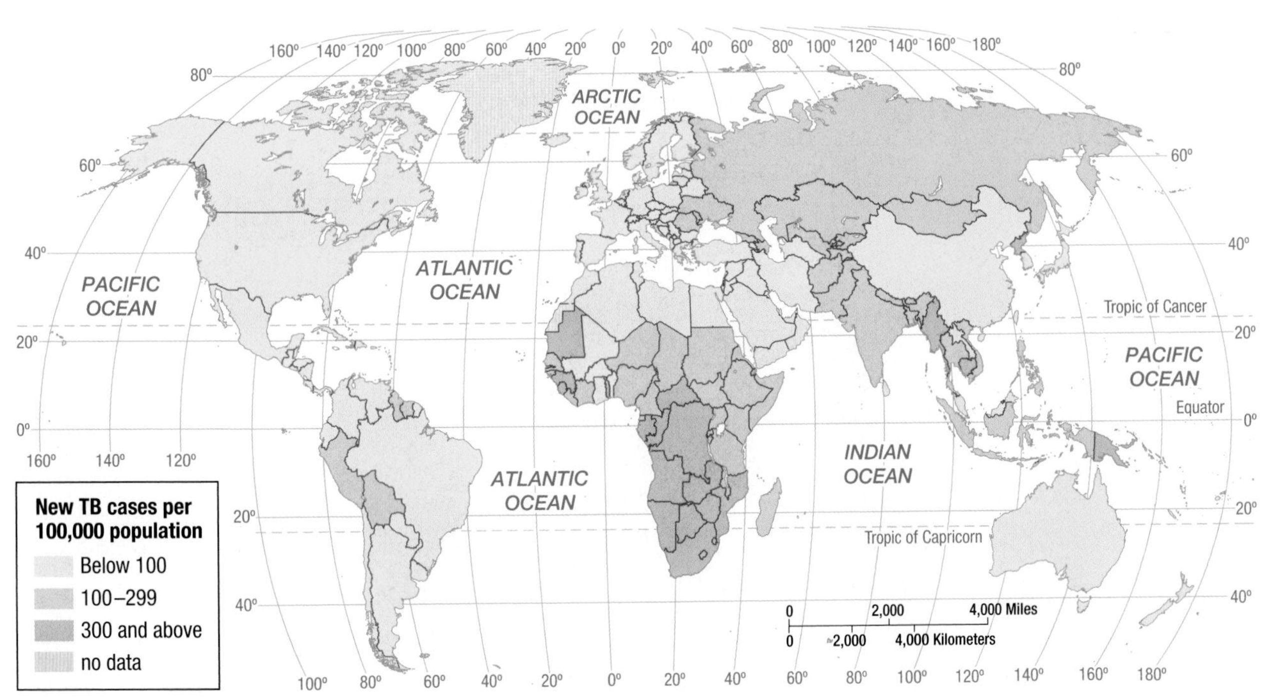

▲ **FIGURE 2-35 TUBERCULOSIS (TB) CASES** Death from tuberculosis is a good indicator of a country's ability to invest in health care, because treating the disease is expensive.

REASON FOR POSSIBLE STAGE 5: INCREASED CONNECTIONS

Learning Outcome 2.4.3
Describe the diffusion of AIDS.

Several dozen "new" pandemics, such as H1N1 (swine) flu and severe acute respiratory syndrome (SARS), have emerged over the past three decades and have spread through the process of relocation diffusion, discussed in Chapter 1. Motor vehicles allow rural residents to have greater connections with urban areas and for urban residents to easily reach rural areas. Airplanes allow residents of one country to easily connect with people in other countries. As they travel, people carry diseases with them and are exposed to the diseases of others.

The most lethal pandemic in recent years has been AIDS (acquired immunodeficiency syndrome). Worldwide, 30 million people died of AIDS from the beginning of the epidemic through 2010, and 34 million were living with HIV (human immunodeficiency virus, the cause of AIDS). The impact of AIDS has been felt most strongly in sub-Saharan Africa, home to 23 million of the world's 34 million HIV-positive people (Figure 2-36).

AIDS diffused from sub-Saharan Africa through relocation diffusion, both by Africans and by visitors to Africa returning to their home countries. AIDS entered the United States during the early 1980s through New York, California, and Florida (Figure 2-37). Not by coincidence, the three leading U.S. airports for international arrivals are in these three states (Figure 2-38). Though AIDS diffused to every state during the 1980s, these three states, plus Texas (a major port of entry by motor vehicle), accounted for half of the country's new AIDS cases in the peak year of 1993.

The number of new AIDS cases dropped rapidly in the United States during the 1990s and in sub-Saharan Africa in the 2000s. The decline resulted from the rapid diffusion of preventive methods and medicines such as AZT. The rapid spread of these innovations is an example of expansion diffusion rather than relocation diffusion.

Pause and Reflect 2.4.3
Have other pandemic diseases diffused rapidly in recent years?

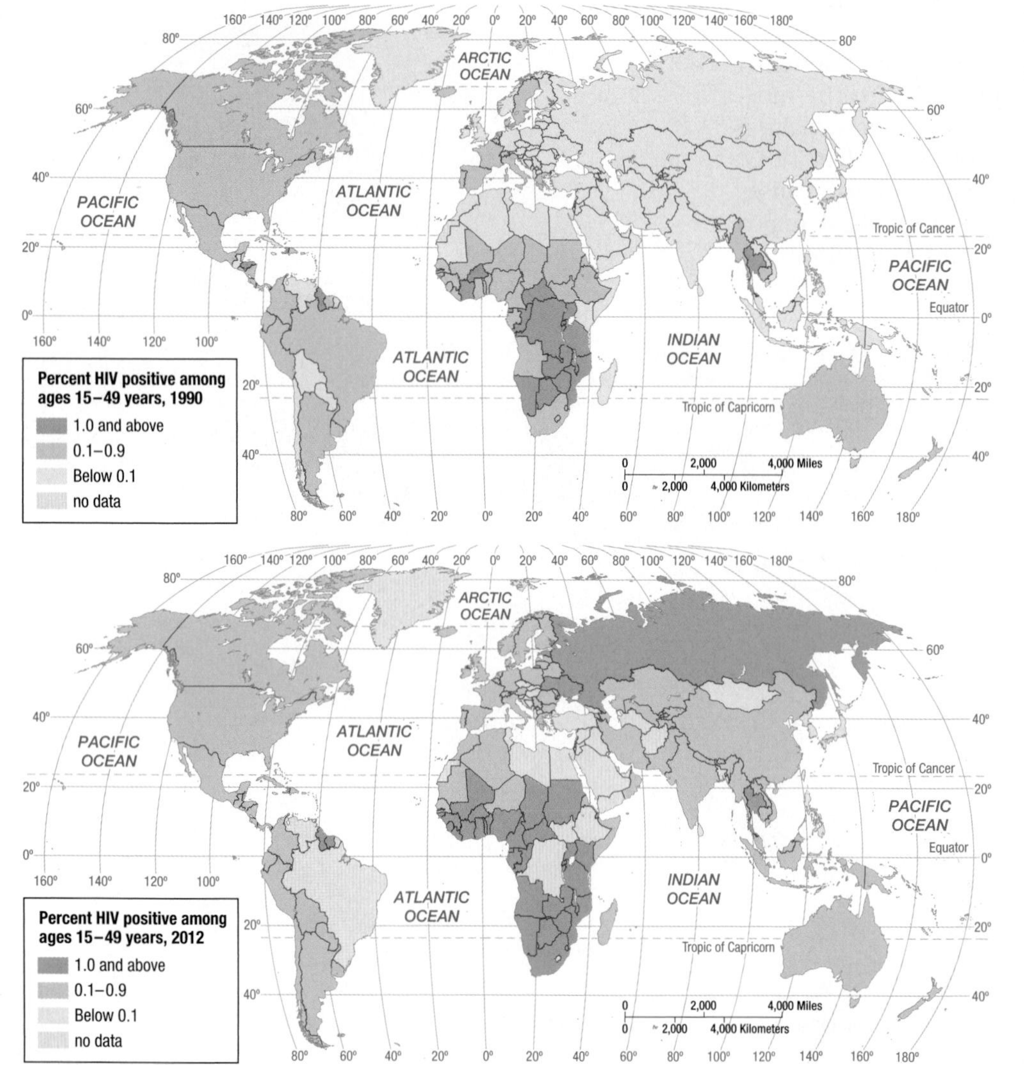

◀ **FIGURE 2-36 DIFFUSION OF HIV/AIDS** The highest rates of HIV infection are in sub-Saharan Africa and Russia.

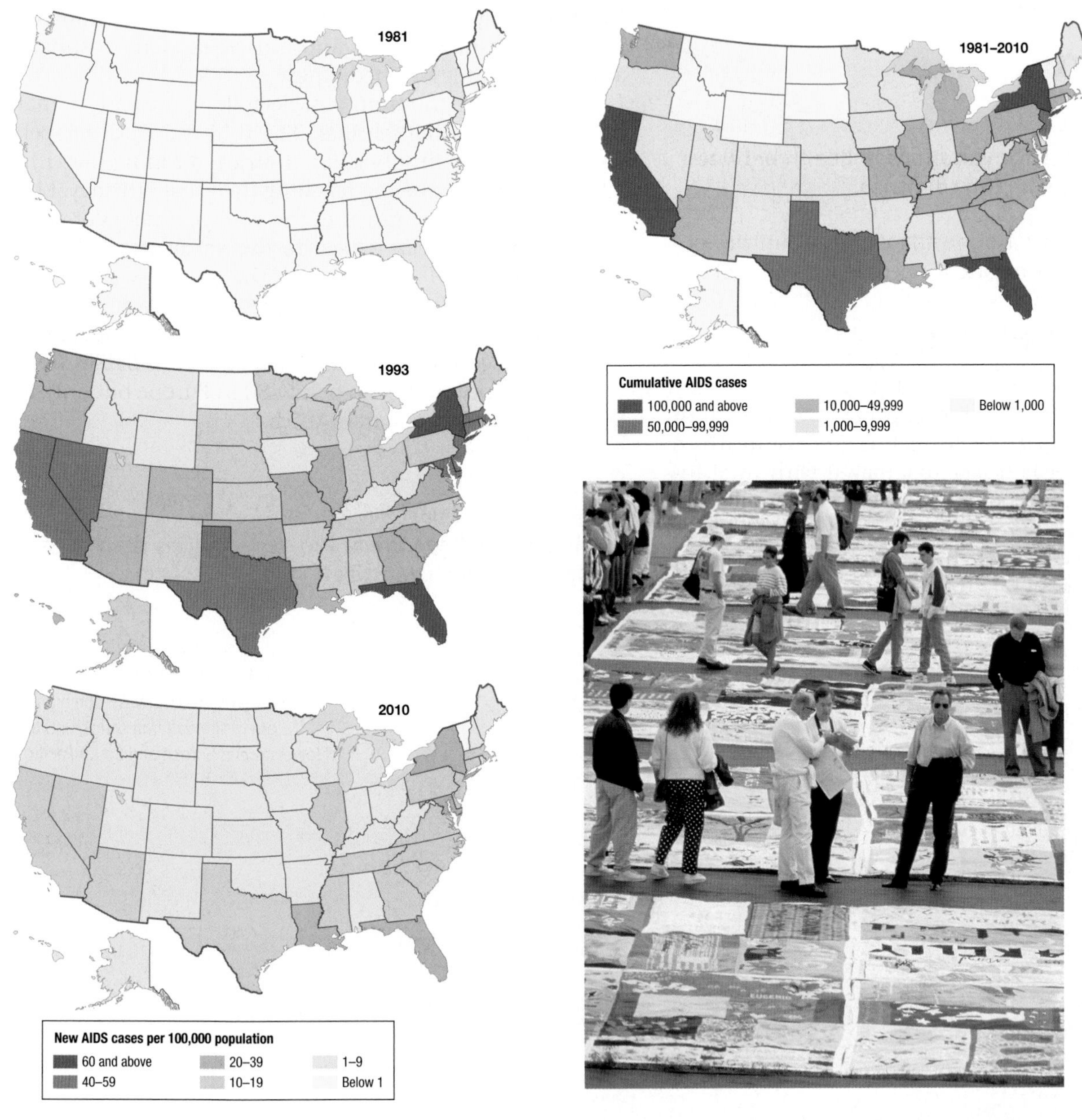

▲ FIGURE 2-37 **DIFFUSION OF HIV/AIDS IN THE UNITED STATES** AIDS diffused from states with relatively high immigration rates, such as California, Florida, and New York. The AIDS Memorial Quilt was assembled as a memorial to people who have died of AIDS.

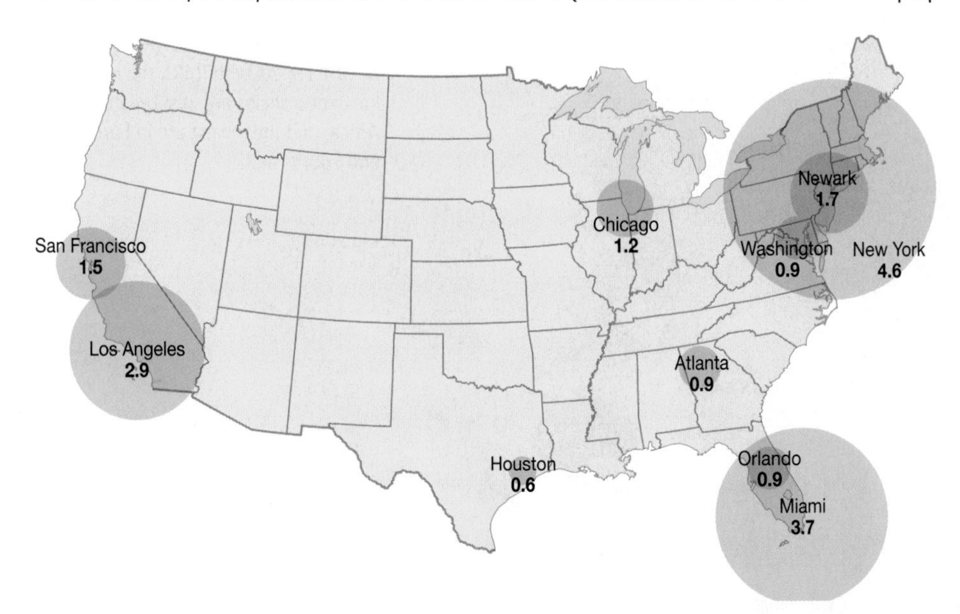

◀ FIGURE 2-38 **INTERNATIONAL PASSENGER ARRIVALS AT U.S. AIRPORTS 2011** Because AIDS arrived in the United States primarily through air travelers, the pattern of diffusion of AIDS in Figure 2-37 closely matches the distribution of international air passenger arrivals.

Health Care

Learning Outcome 2.4.4

Understand reasons for variations in health between developed and developing countries.

Health conditions vary around the world. Countries possess different resources to care for people who are sick.

INDICATORS OF HEALTH

Two important indicators of health in a country are the infant mortality rate and life expectancy. The **infant mortality rate (IMR)** is the annual number of deaths of infants under one year of age, compared with total live births (Figure 2-39). As is the case with the CBR and CDR, the IMR is usually expressed as the number of deaths among infants per 1,000 births rather than as a percentage (per 100). In general, the IMR reflects a country's health-care system. Lower IMRs are found in countries with well-trained doctors and nurses, modern hospitals, and large supplies of medicine.

The global distribution of IMRs follows the pattern that by now has become familiar. The IMR is 5 in developed countries and 80 in sub-Saharan Africa, meaning that 1 in 12 babies die there before reaching their first birthday. **Life expectancy** at birth measures the average number of years a newborn infant can expect to live at current mortality levels (Figure 2-40). Like most of the mortality and fertility rates discussed thus far, life expectancy is most favorable in the wealthy countries of Europe and least favorable in the poor countries of sub-Saharan Africa. Babies born today can expect to live to nearly 80 in Europe but only to less than 60 in sub-Saharan Africa.

Pause and Reflect 2.4.4

Why do men have lower life expectancies than women in most countries?

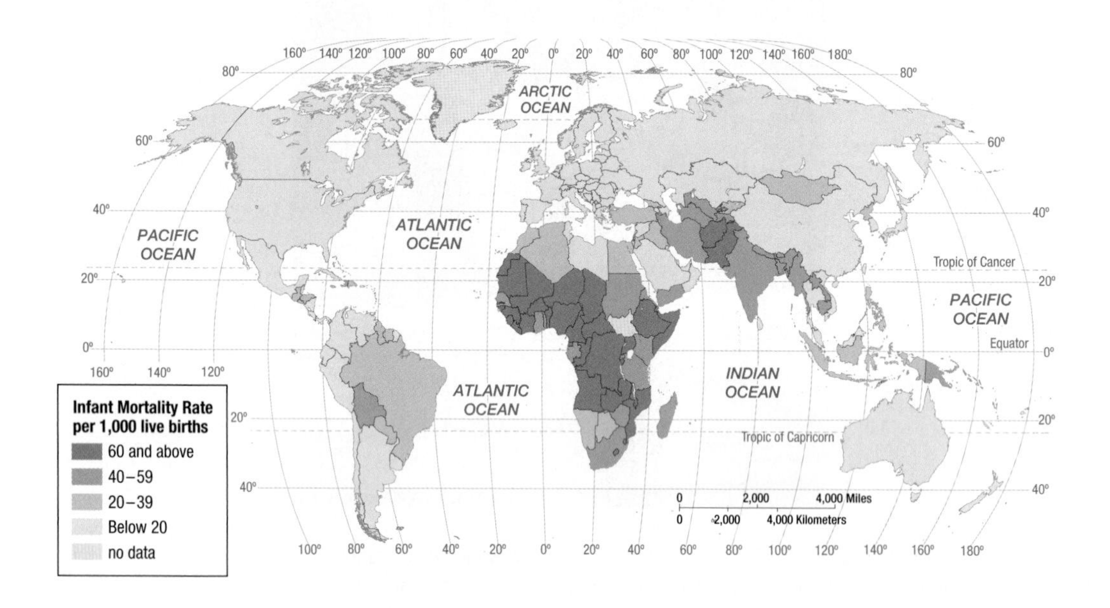

◀ **FIGURE 2-39 INFANT MORTALITY RATE (IMR)** The highest IMRs are in sub-Saharan Africa, and the lowest are in Europe and South Pacific.

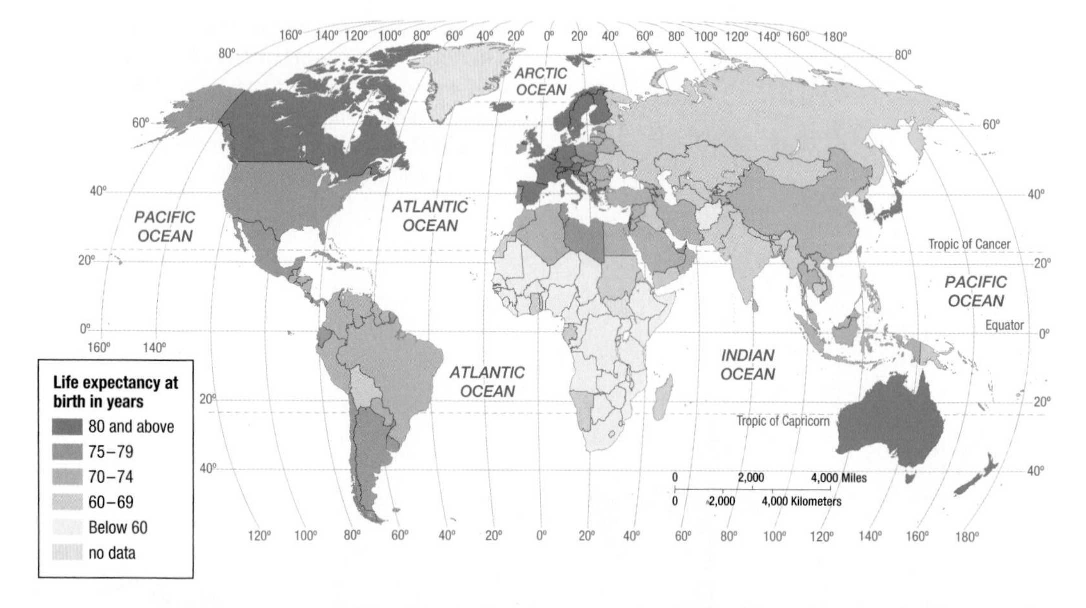

◀ **FIGURE 2-40 LIFE EXPECTANCY AT BIRTH** As with IMRs, the highest life expectancies are in sub-Saharan Africa, and the lowest are in Europe and South Pacific.

PROVISION OF HEALTH CARE

Even if they survive infancy, children remain at risk in developing countries. For example, 17 percent of children in developing countries are not immunized against measles, compared to 7 percent in developed countries. More than one-fourth of children lack measles immunization in South Asia and sub-Saharan Africa (Figure 2-41).

Developed countries use part of their wealth to protect people who, for various reasons, are unable to work. In these countries, some public assistance is offered to those who are sick, elderly, poor, disabled, orphaned, veterans of wars, widows, unemployed, or single parents. Annual per capita expenditure on health care exceeds $1,000 in Europe and $5,000 in the United States, compared to less than $100 in sub-Saharan Africa and South Asia (Figure 2-42).

Expenditures on health care exceed 15 percent of total government expenditures in Europe and North America compared to less than 5 percent in sub-Saharan Africa and South Asia (Figure 2-43). Countries in Northern Europe, including Denmark, Norway, and Sweden, typically provide the highest level of public-assistance payments. So not only do developed countries spend more on health care, they spend a higher percentage of their wealth on health care.

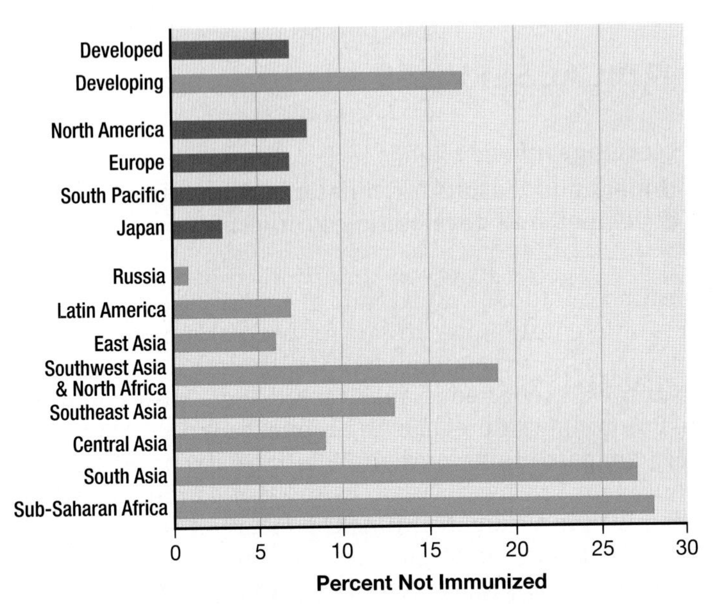

▲ **FIGURE 2-41 CHILDREN LACKING MEASLES IMMUNIZATION** The lowest rates of immunization are in sub-Saharan Africa and South Asia.

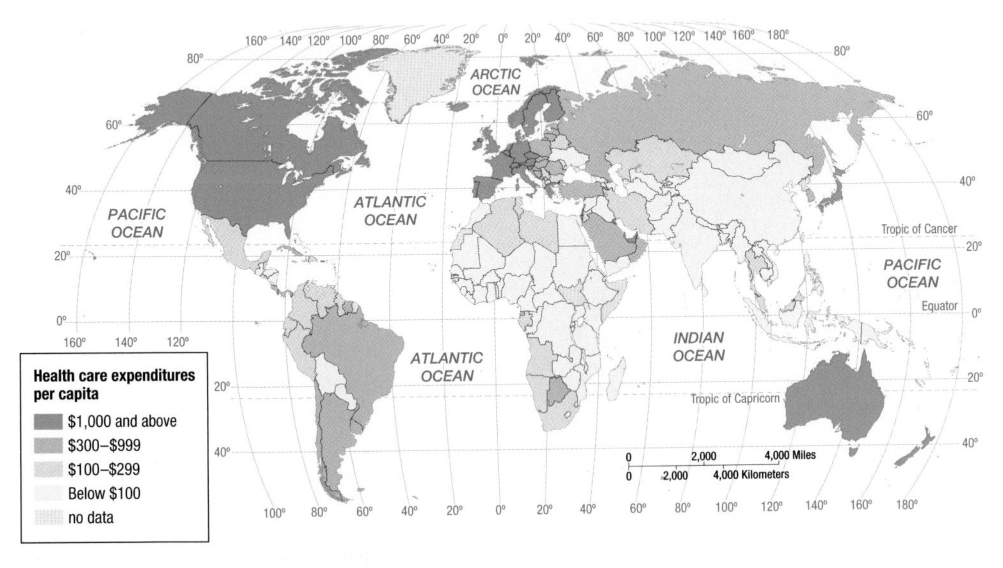

◀ **FIGURE 2-42 HEALTH CARE EXPENDITURES** The lowest levels of per capita health care expenditure are in sub-Saharan Africa and South Asia.

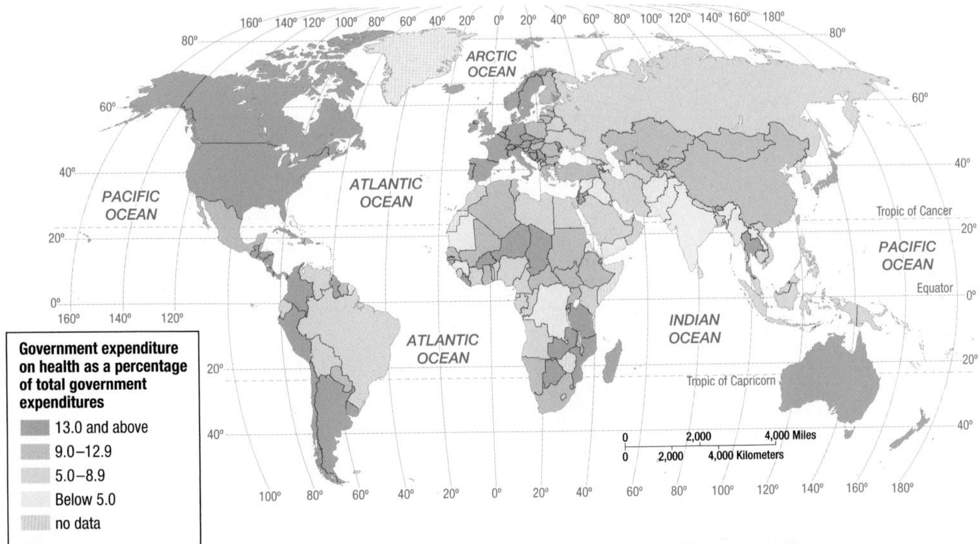

◀ **FIGURE 2-43 GOVERNMENT EXPENDITURES ON HEALTH CARE** The lowest levels of government expenditures are in Africa and Asia.

MEDICAL SERVICES

Learning Outcome 2.4.5
Understand reasons for variations in health between developed and developing countries.

Health conditions vary around the world. Countries possess different resources to care for people who are sick.

The high expenditure on health care in developed countries is reflected in medical facilities. Most countries in Europe have more than 50 hospital beds per 10,000 people, compared to fewer than 20 in sub-Saharan Africa and South and Southwest Asia (Figure 2-44). Europe has more than 30 physicians per 10,000 population, compared to fewer than 5 in sub-Saharan Africa (Figure 2-45).

In most developed countries, health care is a public service that is available at little or no cost. Government programs pay more than 70 percent of health-care costs in most European countries, and private individuals pay less than 30 percent. In developing countries, private individuals must pay more than half of the cost of health care (Figure 2-46). An exception to this pattern is the United States, a developed country where private individuals are required to pay an average of 55 percent of health care, more closely resembling the pattern in developing countries.

Developed countries are hard-pressed to maintain their current levels of public assistance. In the past, rapid economic growth permitted these states to finance generous programs with little difficulty. But in recent years economic growth has slowed, while the percentage of people needing public assistance has increased. Governments have faced a choice between reducing benefits and increasing taxes to pay for them. In some of the poorest countries, threats to health and sustainability are not so much financial as environmental. For a case in point, read the following Sustainability and Inequality in Our Global Village feature.

Pause and Reflect 2.4.5
Why might levels of hospital beds and physicians be lower in North America than in other developed countries?

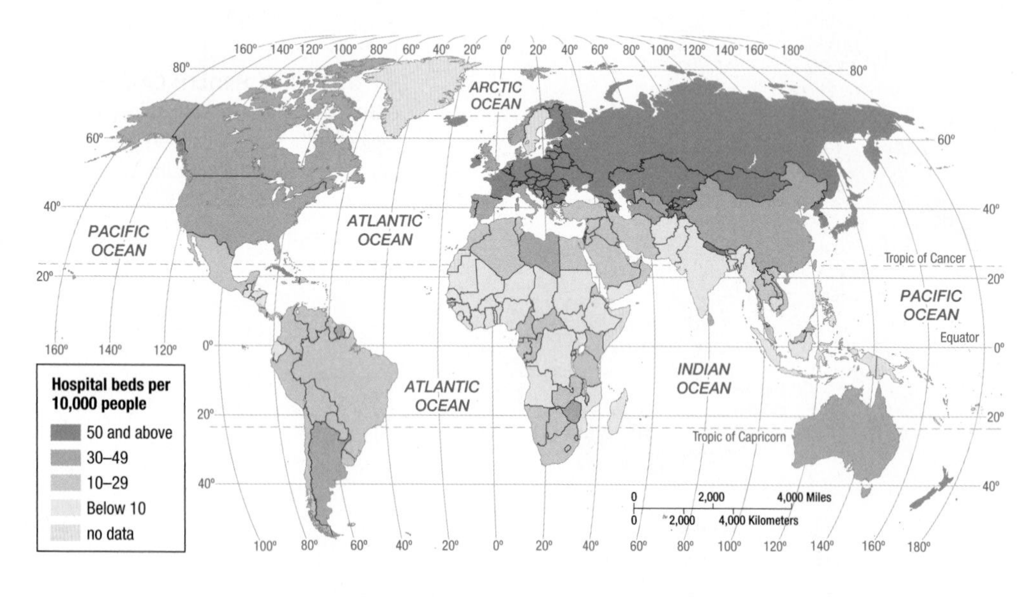

◀ **FIGURE 2-44 HOSPITAL BEDS PER 10,000 PEOPLE** The lowest rates are in sub-Saharan Africa and South Asia.

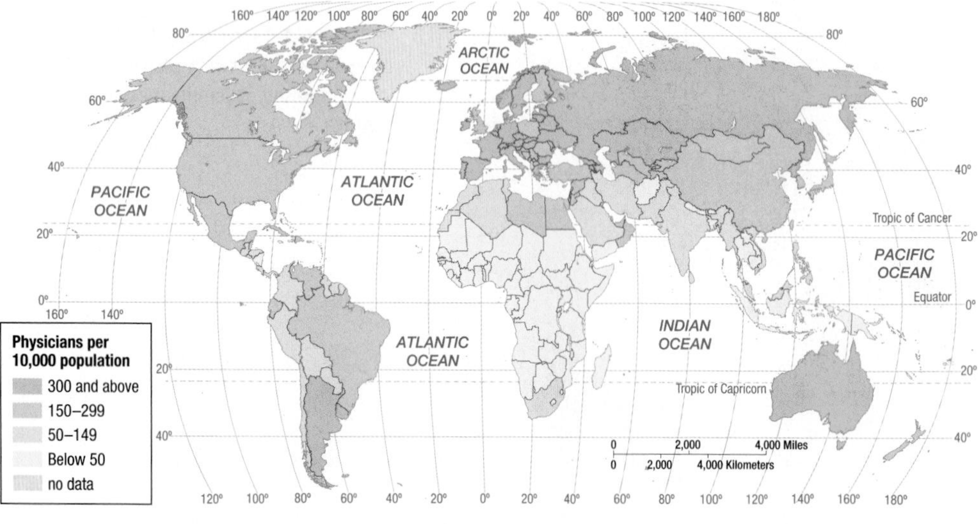

◀ **FIGURE 2-45 PHYSICIANS PER 10,000 PEOPLE** The lowest rates are in sub-Saharan Africa.

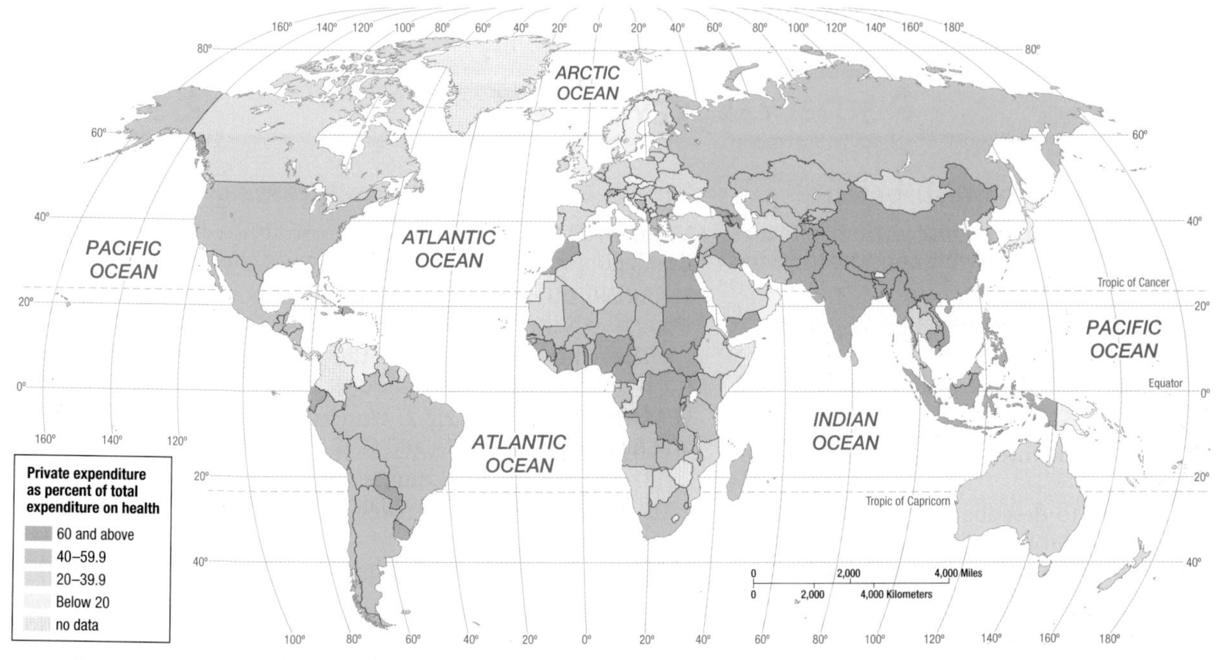

▲ FIGURE 2-46 **PUBLIC EXPENDITURES ON HEALTH CARE AS A SHARE OF TOTAL HEALTH CARE EXPENDITURES** The highest percentages are in Europe.

SUSTAINABILITY AND INEQUALITY IN OUR GLOBAL VILLAGE

Overpopulation in Sub-Saharan Africa

Overpopulation—too many people for the available resources—does not appear to be an immediate threat to the world, even in India, where the seven billionth human was said to have been born on October 31, 2011 (Figure 2-47). However, it does threaten areas within sub-Saharan Africa.

Sub-Saharan Africa was not classified in Key Issue 1 as one of the world's population concentrations. Geographers caution that the size, density, or clustering of population in a region is not an indication of overpopulation. Instead, overpopulation is a relationship between population and a region's level of resources. The capacity of the land to sustain life derives partly from characteristics of the natural environment and partly from human actions to modify the environment through agriculture, industry, and exploitation of raw materials. See for example, the image of Mali on page 44.

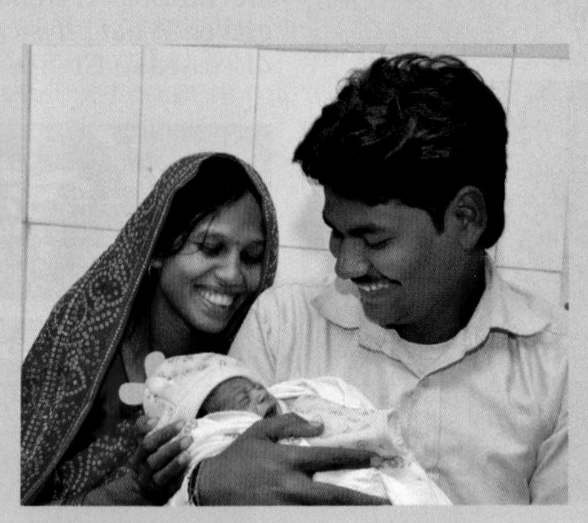

▲ FIGURE 2-47 **THE WORLD'S SEVEN BILLIONTH HUMAN**
Nargis Kumar, born October 31, 2011, to Vinita and Ajay Kumar, of Lucknow, India, was declared the world's seven billionth person by Plan International, a nongovernmental organization for children's' welfare.

The track toward overpopulation may already be irreversible in Africa. Rapid population growth has led to the inability of the land to sustain life in parts of the region. As the land declines in quality, more effort is needed to yield the same amount of crops. This extends the working day of women, who have the primary responsibility for growing food for their families. Women then regard having another child as a means of securing additional help in growing food.

CHECK IN: KEY ISSUE 4

Why Do Some Regions Face Health Threats?

- ✓ **The epidemiologic transition has four stages of distinctive diseases.**
- ✓ **A resurgence of infectious diseases may signal a possible stage 5 of the epidemiologic transition.**
- ✓ **The provision of health care varies sharply between developed and developing countries.**

Chapter 3 Migration

Why are these immigrants lining up in Dubai? Page 95

Why are these people watching a parade in New York? Page 100

KEY ISSUE 1

Where Are Migrants Distributed?

A World of Migrants 79

People are on the move around the world. Where are they heading, and where are they coming from?

KEY ISSUE 2

Where Do People Migrate Within a Country?

Moving Across Town or Across Country 84

Some people are moving into cities, while others are moving out of them.

▲ This tiny overcrowded boat is nearing Lampedusa, Italy, having sailed across the Mediterranean Sea from Tunisia. Geographers are interested in why people would risk such a dangerous journey. What pushed these people to set sail from Tunisia? What lured them to Italy, even though when they got there the authorities sent them right back to Tunisia?

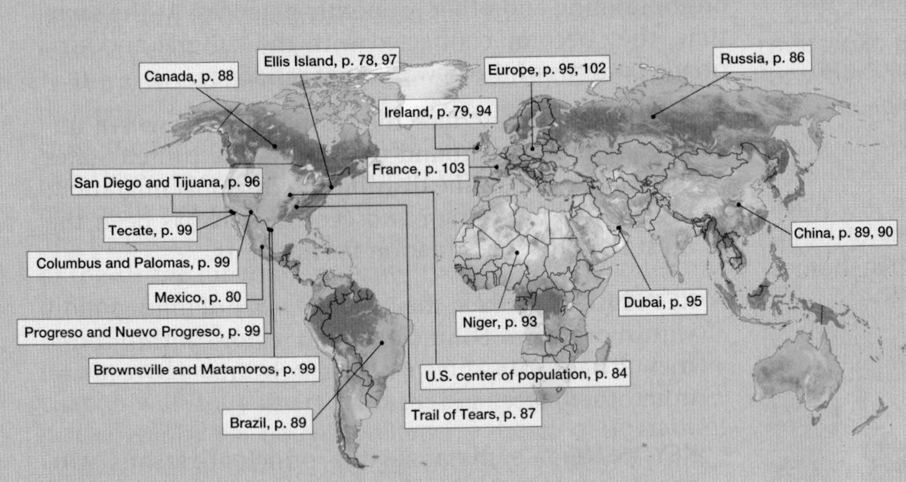

KEY ISSUE 3

Why Do People Migrate?

Pushing and Pulling 92

It takes a lot of motivation to pick up and move to a new home.

KEY ISSUE 4

Why Do Migrants Face Obstacles?

Where's the Welcome Mat? 96

Some immigrants are welcomed to their new homes, but others are told to leave.

Distance of Migration

Learning Outcome 3.1.1:
Describe the difference between international and internal migration.

Ravenstein's laws for the distance that migrants travel to their new homes:

- Most migrants relocate a short distance and remain within the same country.
- Long-distance migrants to other countries head for major centers of economic activity.

INTERNATIONAL AND INTERNAL MIGRATION

Migration can be divided into international migration and internal migration (Figure 3-4):

INTERNATIONAL MIGRATION. A permanent move from one country to another is **international migration**. International migration is further divided into two types:

- **Voluntary migration** implies that the migrant has chosen to move, especially for economic improvement (Figure 3-5).
- **Forced migration** means that the migrant has been compelled to move, especially by political or environmental factors.

The distinction between forced and voluntary migration is not clear-cut. Those who are migrating for economic reasons may feel forced by pressure inside themselves to migrate, such as to search for food or jobs, but they have not been explicitly compelled to migrate by the violent actions of other people.

INTERNAL MIGRATION. A permanent move within the same country is **internal migration**. Consistent with the distance-decay principle presented in Chapter 1, the farther away a place is located, the less likely that people will migrate to it. Thus, internal migrants are much more numerous than international migrants.

Internal migration can be divided into two types:

- **Interregional migration** is movement from one region of a country to another. Historically, the main type of interregional migration has been from rural to urban areas in search of jobs. In recent years, some developed countries have seen migration from urban to environmentally attractive rural areas.
- **Intraregional migration** is movement within one region. The main type of intraregional migration has been within urban areas, from older cities to newer suburbs.

Most people find migration within a country less traumatic than international migration because they find

▼FIGURE 3.4 INTERNATIONAL AND INTERNAL MIGRATION Mexico has international migration into the country from Central America and out of the country to the United States. Mexico also has internal migration, especially interregional migration to states near the U.S. border and intraregional migration into Mexico City.

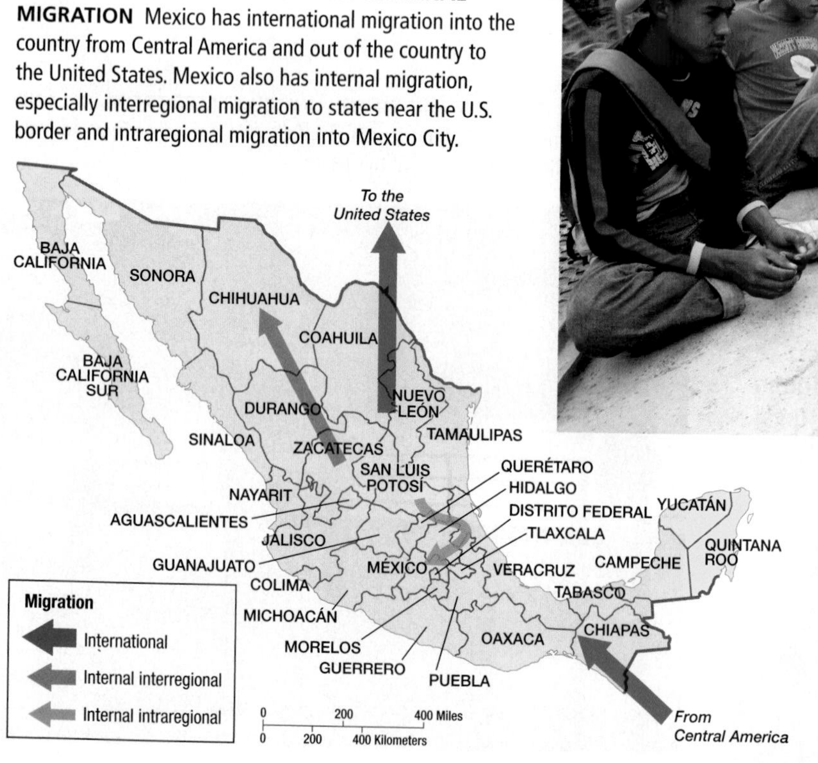

▲ FIGURE 3-5 INTERNATIONAL MIGRATION INTO MEXICO These immigrants from Honduras are traveling across Mexico on top of the train because they don't have enough money to pay for their travel.

familiar language, foods, broadcasts, literature, music, and other social customs after they move. Moves within a country also generally involve much shorter distances than those in international migration. However, internal migration can involve long-distance moves in large countries, such as in the United States and Russia.

Pause and Reflect 3.1.1
How many times have you moved? How many of these moves were international?

INTERNATIONAL MIGRATION PATTERNS

About 9 percent of the world's people are international migrants—that is, they currently live in countries other than the ones in which they were born. On a global scale, the three largest flows of migrants are:

- From Asia to Europe
- From Asia to North America
- From Latin America to North America

The global pattern reflects the importance of migration from developing countries to developed countries. Asia, Latin America, and Africa have net out-migration, and North America, Europe, and Oceania have net in-migration. Migrants from countries with relatively low incomes and high natural increase rates head for relatively wealthy countries, where job prospects are brighter.

The United States has more foreign-born residents than any other country: approximately 43 million as of 2010, and growing by around 1 million annually. Other developed countries have higher rates of net in-migration, including Australia and Canada, which are much less populous than the United States (Figure 3-6). The highest rates can be found in petroleum-exporting countries of Southwest Asia, which attract immigrants primarily from poorer countries in Asia to perform many of the dirty and dangerous functions in the oil fields.

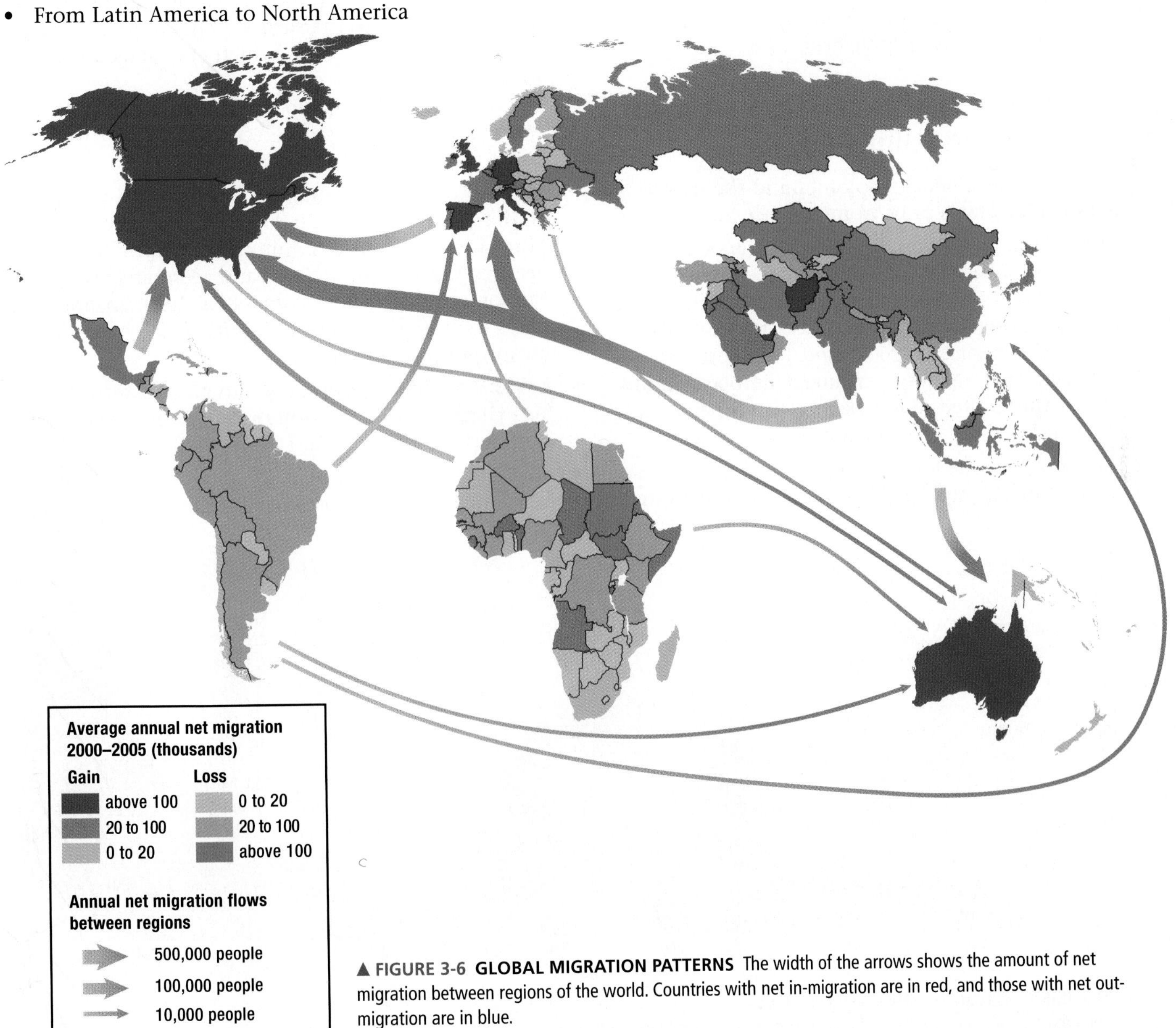

▲ **FIGURE 3-6 GLOBAL MIGRATION PATTERNS** The width of the arrows shows the amount of net migration between regions of the world. Countries with net in-migration are in red, and those with net out-migration are in blue.

U.S. Immigration Patterns

Learning Outcome 3.1.2
Identify the principal sources of immigrants during the three main eras of U.S. immigration.

The United States plays a special role in the study of international migration. The world's third–most-populous country is inhabited overwhelmingly by direct descendants of immigrants. About 75 million people migrated to the United States between 1820 and 2010, including 43 million who were alive in 2010.

The United States has had three main eras of immigration:

- Colonial settlement in the seventeenth and eighteenth centuries
- Mass European immigration in the late nineteenth and early twentieth centuries
- Asian and Latin American immigration in the late twentieth and early twenty-first centuries

U.S. IMMIGRATION: SEVENTEENTH AND EIGHTEENTH CENTURIES

Immigration to the American colonies and the newly independent United States came from two principal regions:

- **Europe.** About 2 million Europeans migrated to the American colonies and the newly independent United States prior to 1820. Permanent English colonies were established along the Atlantic Coast, beginning with Jamestown, Virginia, in 1607, and Plymouth, Massachusetts, in 1620. Ninety percent of European immigrants to the United States during this period came from Great Britain.
- **Sub-Saharan Africa.** Most African Americans are descended from Africans forced to migrate to the Western Hemisphere as slaves. During the eighteenth century, about 400,000 Africans were shipped as slaves to the 13 colonies that later formed the United States, primarily by the British. The importation of Africans as slaves was made illegal in 1808, but another 250,000 Africans were brought to the United States during the next half-century (see Chapter 7).

Most of the Africans were forced to migrate to the United States as slaves, whereas most Europeans were voluntary migrants—although harsh economic conditions and persecution in Europe blurred the distinction between forced and voluntary migration for many Europeans.

U.S. IMMIGRATION: MID-NINETEENTH TO EARLY TWENTIETH CENTURY

Between 1820 and 1920 approximately 32 million people immigrated to the United States. Nearly 90 percent emigrated from Europe. For European migrants, the United States offered a great opportunity for economic success. Early migrants extolled the virtues of the country to friends and relatives back in Europe, which encouraged still others to come.

Migration from Europe to the United States peaked at several points during the nineteenth and early twentieth centuries (Figure 3-7):

- **1840s and 1850s: Ireland and Germany.** Annual immigration jumped from 20,000 to more than 200,000. Three-fourths of all U.S. immigrants during those two decades came from Ireland and Germany. Desperate economic push factors compelled the Irish and Germans to cross the Atlantic. Germans also emigrated to escape political unrest.
- **1870s: Ireland and Germany.** Emigration from Ireland and Germany resumed following a temporary decline during the U.S. Civil War (1861–1865).
- **1880s: Scandinavia.** Immigration increased to 500,000 per year. Increasing numbers of Scandinavians, especially

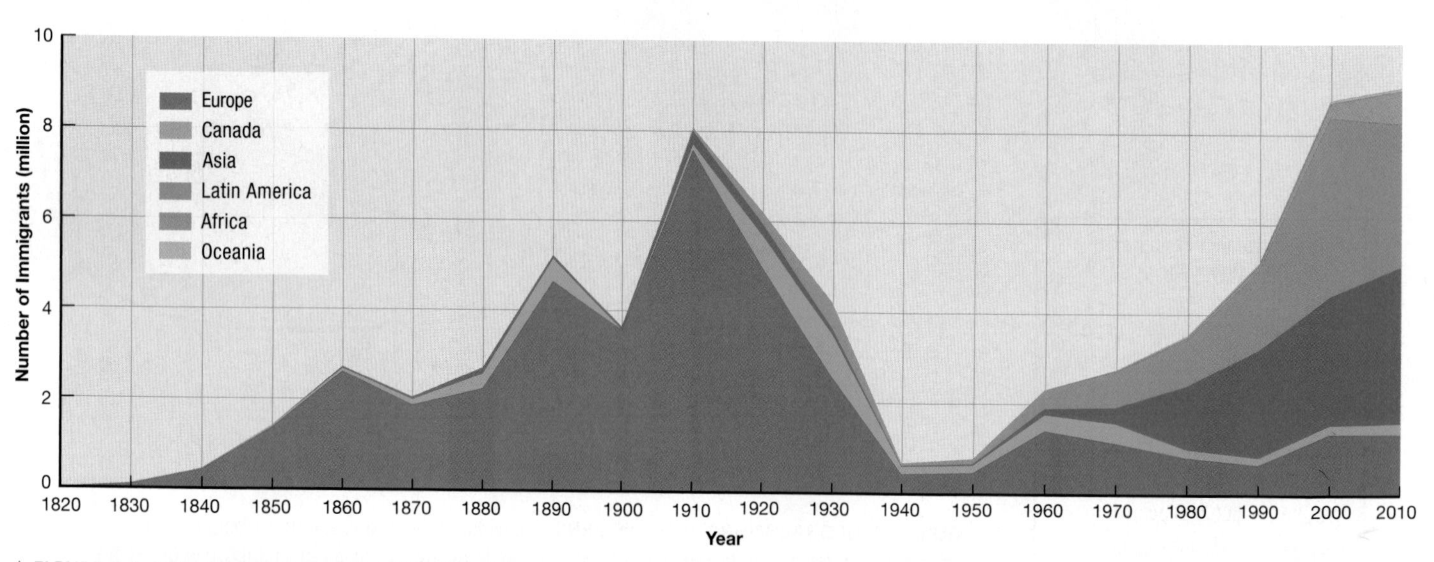

▲ **FIGURE 3-7 IMMIGRATION TO THE UNITED STATES**
Europeans comprised more than 90 percent of immigrants to the United States during the nineteenth century. Since the 1980s, Latin American and Asia have been the dominant sources of immigrants.

Swedes and Norwegians, joined Germans and Irish in migrating to the United States. The Industrial Revolution had diffused to Scandinavia, triggering a rapid population increase.

- **1905–1914: Southern and Eastern Europe.** Immigration to the United States reached 1 million. Two-thirds of all immigrants during this period came from Southern and Eastern Europe, especially Italy, Russia, and Austria-Hungary. The shift in the primary source of immigrants coincided with the diffusion of the Industrial Revolution to Southern and Eastern Europe, along with rapid population growth.

Among European countries, Germany has sent the largest number of immigrants to the United States, 7.2 million. Other major European sources include Italy, 5.4 million; the United Kingdom, 5.3 million; Ireland, 4.8 million; and Russia and the former Soviet Union, 4.1 million. About one-fourth of Americans trace their ancestry to German immigrants and one-eighth each to Irish and English immigrants.

Note that frequent boundary changes in Europe make precise national counts impossible. For example, most Poles migrated to the United States at a time when Poland did not exist as an independent country. Therefore, most were counted as immigrants from Germany, Russia, or Austria.

Pause and Reflect 3.1.2

In what stage of the demographic transition were European countries when they sent the most immigrants to the United States?

U.S. IMMIGRATION: LATE TWENTIETH TO EARLY TWENTY-FIRST CENTURY

Immigration to the United States dropped sharply in the 1930s and 1940s, during the Great Depression and World War II. The number of immigrants steadily increased beginning in the 1950s and then surged to historically high levels during the first decade of the twenty-first century.

More than three-fourths of the recent U.S. immigrants have emigrated from two regions:

- **Asia.** The leading sources of U.S. immigrants from Asia are China, the Philippines, India, and Vietnam.
- **Latin America.** Nearly one-half million emigrate to the United States annually from Latin America, more than twice as many as during the entire nineteenth century.

Recent immigrants are not distributed uniformly throughout the United States. More than one-half head for California, Florida, New York, or Texas (Figure 3-8).

Officially, Mexico passed Germany in 2006 as the country that has sent to the United States the most immigrants ever. Unofficially, because of the large number of unauthorized immigrants, Mexico probably became the leading source during the 1980s. In the early 1990s, an unusually large number of immigrants came from Mexico and other Latin American countries as a result of the 1986 Immigration Reform and Control Act, which issued visas to several hundred thousand people who had entered the United States in previous years without legal documents.

Although the pattern of immigration to the United States has changed from predominantly European to Asian and Latin American, the reason for immigration remains the same. Rapid population growth has limited prospects for economic advancement at home. Europeans left when their countries entered stage 2 of the demographic transition in the nineteenth century, and Latin Americans and Asians began to leave in large numbers in recent years after their countries entered stage 2. With poor conditions at home, immigrants were lured by economic opportunity and social advancement in the United States.

The motives for immigrating to the country may be similar, but the United States has changed over time. The United States is no longer a sparsely settled, economically booming country with a large supply of unclaimed land. In 1912, New Mexico and Arizona were admitted as the forty-seventh and forty-eighth states. Thus, for the first time in its history, all the contiguous territory of the country was a "united" state (other than the District of Columbia). This symbolic closing of the frontier coincided with the end of the peak period of emigration from Europe.

CHECK-IN: KEY ISSUE 1

Where Are Migrants Distributed?

- ✓ **Migration can be international (voluntary or forced) or internal (interregional or intraregional).**
- ✓ **Migration to the United States has occurred in three principal eras, with emigrants from different combinations of countries and regions predominating during each era.**

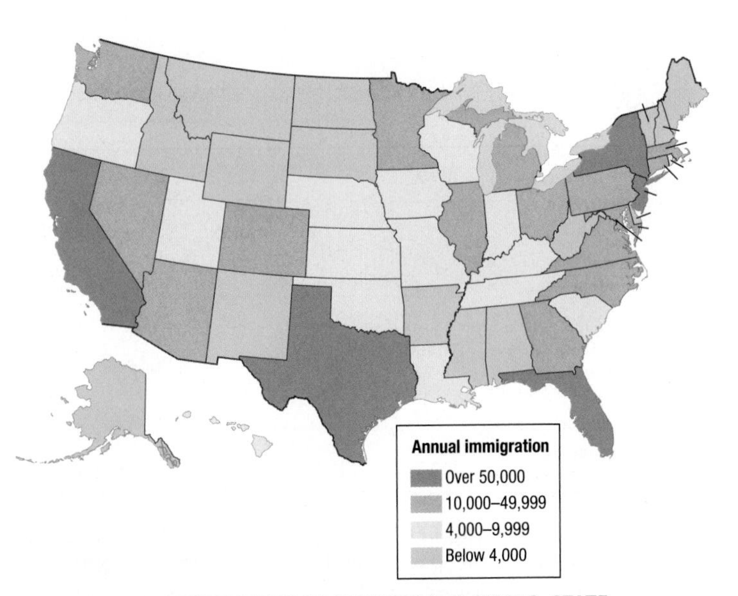

▲ **FIGURE 3-8 DESTINATION OF IMMIGRANTS BY U.S. STATE**
California, New York, Florida, and Texas are the leading destinations for immigrants.

KEY ISSUE 2

Where Do People Migrate within a Country?

- Interregional Migration
- Intraregional Migration

Learning Outcome 3.2.1
Describe the history of interregional migration in the United States.

Internal migration for most people is less disruptive than international migration. Two main types of internal migration are interregional (between regions of a country) and intraregional (within a region).

Interregional Migration

In the past, people migrated from one region of a country to another in search of better farmland. Lack of farmland pushed many people from the more densely settled regions of the country and lured them to the frontier, where land was abundant. Today, the principal type of interregional migration is from rural areas to urban areas. Most jobs, especially in services, are clustered in urban areas (see Chapter 12).

MIGRATION BETWEEN REGIONS OF THE UNITED STATES

An especially prominent example of large-scale internal migration is the opening of the American West. At the time of independence, the United States consisted of long-established settlements concentrated on the Atlantic Coast and a scattering of newer settlements in the territories west of the Appalachian Mountains. Through mass interregional migration, the interior of the continent was settled and developed.

CHANGING CENTER OF POPULATION. The U.S. Census Bureau computes the country's population center at the time of each census. The population center is the average location of everyone in the country, the "center of population gravity." If the United States were a flat plane placed on top of a pin, and each individual weighed the same, the population center would be the point where the population distribution causes the flat plane to balance on the pin.

The changing location of the population center graphically demonstrates the march of the American people across the North American continent over the past 200 years (Figure 3-9). The center has consistently shifted westward, although the rate of movement has varied in different eras:

- **1790: Hugging the coast.** This location reflects the fact that virtually all colonial-era settlements were near the Atlantic Coast. Few colonists ventured far from coastal locations because they depended on shipping links with Europe to receive products and to export raw materials. The Appalachian Mountains also blocked western development because of their steep slopes, thick

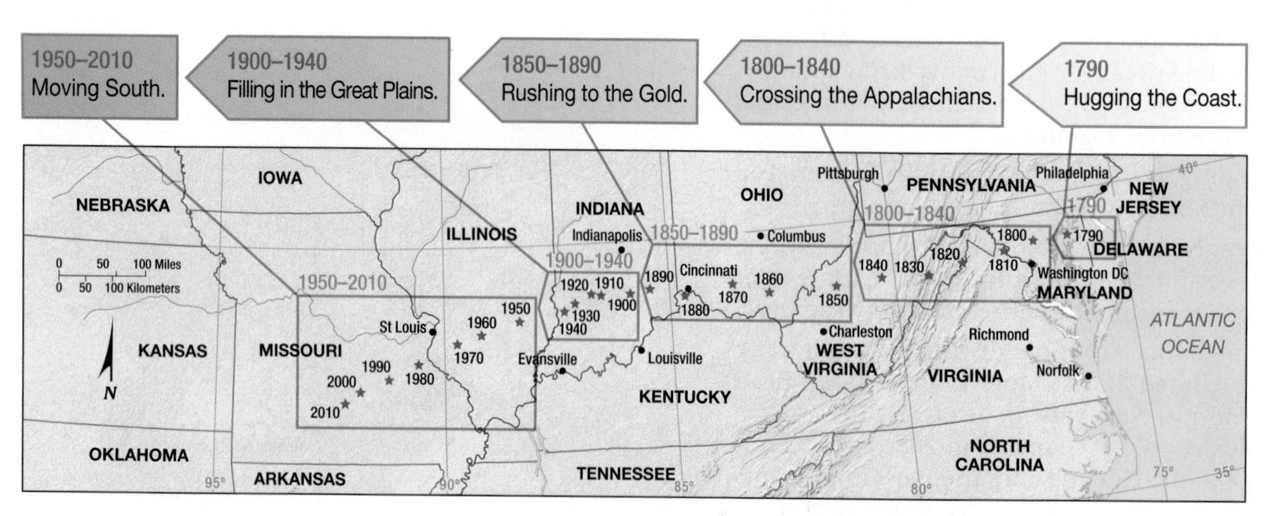

▲ **FIGURE 3-9 CHANGING CENTER OF U.S. POPULATION** The population center is the average location of everyone in the country, the "center of population gravity." If the United States were a flat plane placed on top of a pin, and each individual weighed the same, the population center would be the point where the population distribution causes the flat plane to balance on the head of a pin.

forests, and few gaps that allowed easy passage. The indigenous residents, commonly called "Indians," still occupied large areas and sometimes resisted the expansion of settlement.

- **1800–1840: Crossing the Appalachians.** Transportation improvements, especially the building of canals, helped to open the interior. Most important was the Erie Canal, which enabled people to travel inexpensively by boat between New York City and the Great Lakes. In 1840, the United States had 5,352 kilometers (3,326 miles) of canals. Encouraged by the opportunity to obtain a large amount of land at a low price, people moved into forested river valleys between the Appalachians and the Mississippi River. They cut down the trees and used the wood to build homes, barns, and fences.
- **1850–1890: Rushing to the gold.** The population center shifted westward more rapidly during this period. Rather than continuing to expand agriculture into the next available westward land, mid-nineteenth-century pioneers kept going all the way to California. The principal pull to California was the Gold Rush, beginning in the late 1840s. Pioneers during this period also passed over the Great Plains because of the physical environment. The region's dry climate, lack of trees, and tough grassland sod convinced early explorers such as Zebulon Pike that the region was unfit for farming, and maps at the time labeled the Great Plains as the Great American Desert.
- **1900–1940: Filling in the Great Plains.** The westward movement of the U.S. population center slowed during this period because emigration from Europe to the East Coast offset most of the emigration from the East Coast to the U.S. West. Also, immigrants began to fill in the Great Plains that earlier generations had bypassed. Advances in agricultural technology enabled people to cultivate the area. Farmers used barbed wire to reduce dependence on wood fencing, the steel plow to cut the thick sod, and windmills and well-drilling equipment to pump more water. The expansion of the railroads encouraged settlement of the Great Plains. The federal government gave large land grants to the railroad companies, which financed construction of their lines by selling portions to farmers. The extensive rail network then permitted settlers to transport their products to the large concentrations of customers in East Coast cities.
- **1950–2010: Moving south.** The population center resumed a more vigorous westward migration. It also moved southward, as Americans migrated to the South for job opportunities and warmer climate. The rapid growth of population and employment in the South has aggravated interregional antagonism. Some people in the Northeast and Midwest believe that southern states have stolen industries from them. In reality, some industries have relocated from the Northeast and Midwest, but most of the South's industrial growth comes from newly established companies.

Interregional migration has slowed considerably in the United States into the twenty-first century; net migration between each pair of regions is now close to zero. Regional differences in employment prospects have become less dramatic (Figure 3-10). The severe recession that began in 2008 discouraged people from migrating because of limited job prospects in all regions.

Pause and Reflect 3.2.1
What means of transportation were available to migrants crossing the United States during the different eras?

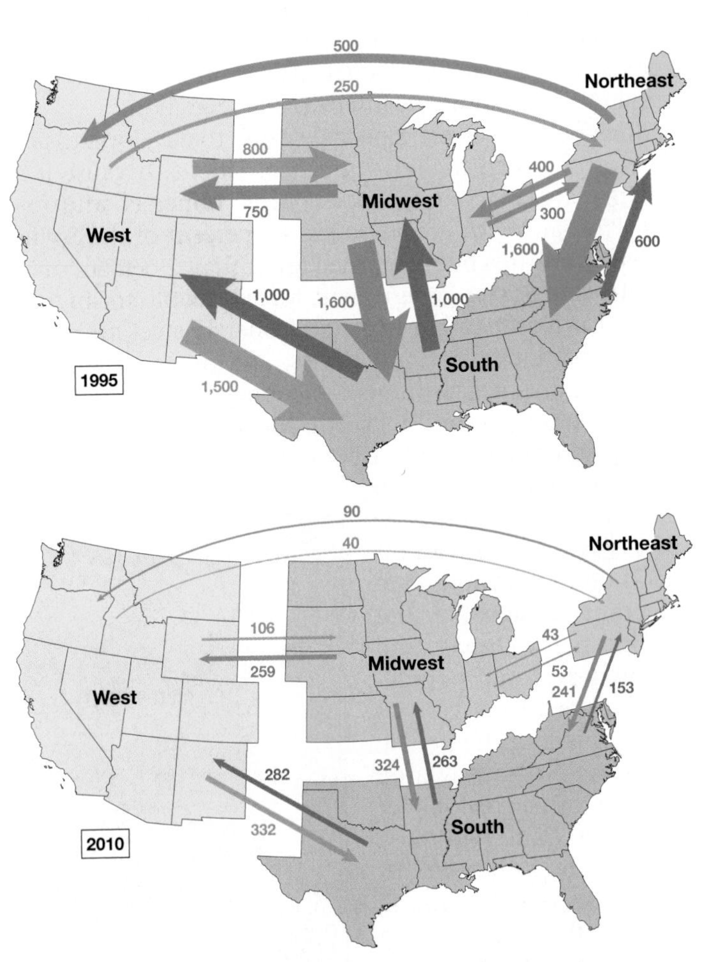

▲ **FIGURE 3-10 RECENT INTERREGIONAL MIGRATION IN THE UNITED STATES** Figures show average annual migration (in thousands) in 1995 (top) and 2010 (bottom).

MIGRATION BETWEEN REGIONS IN THE WORLD'S LARGEST COUNTRY

Learning Outcome 3.2.2
Describe interregional migration in Russia.

Long-distance interregional migration has been an important means of opening new regions for economic development in Russia. The population of Russia, the world's largest country in land area, is highly clustered in the western, or European, portion of the country. Much of the country, especially east of the Ural Mountains, is extremely sparsely inhabited (Figure 3-11). To open up the sparsely inhabited Asian portion of Russia, interregional migration was important in the former Soviet Union. Soviet policy encouraged factory construction near raw materials rather than near existing population concentrations (see Chapter 11). Not enough workers lived nearby to fill all the jobs at the mines, factories, and construction sites established in these remote, resource-rich regions. To build up an adequate labor force, the Soviet government had to stimulate interregional migration.

Soviet officials were especially eager to develop Russia's Far North, which included much of Siberia, because it is rich in natural resources—fossil fuels, minerals, and forests. The Far North encompassed 45 percent of the Soviet Union's land area but contained fewer than 2 percent of its people. The Soviet government forced people to migrate to the Far North to construct and operate steel mills, hydroelectric power stations, mines, and other enterprises. In later years, the government encouraged, instead, voluntary migration to the Far North, including higher wages, more paid holidays, and earlier retirement.

The incentives failed to pull as many migrants to the Far North as Soviet officials desired. People were reluctant because of the region's harsh climate and remoteness from population clusters. Each year, as many as half of the people in the Far North migrated back to other regions of the country and had to be replaced by other immigrants, especially young males willing to work in the region for a short period. One method the Soviet government used was to send a brigade of young volunteers, known as Komsomol, during school vacations to help construct projects, such as railroads (Figure 3-12).

▲ **FIGURE 3-12 INTERREGIONAL MIGRATION: RUSSIA** When Russia was still constituted as the principal components of the Soviet Union, workers migrated to Siberia to build rail lines, including these in 1930.

The collapse of the Soviet Union ended policies that encouraged interregional migration. In the transition to a market-based economy, Russian government officials no longer dictate "optimal" locations for factories.

Pause and Reflect 3.2.2
With Russia more closely linked economically to Europe, which region of the country is most likely to attract interregional migration?

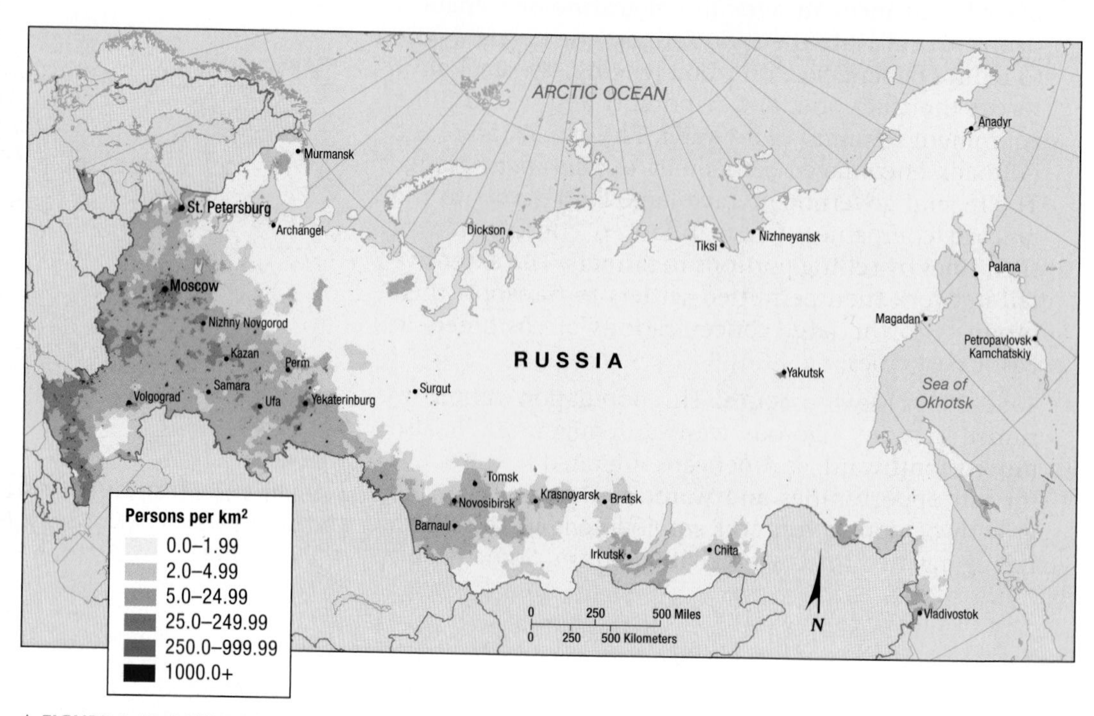

▲ **FIGURE 3-11 POPULATION DISTRIBUTION: RUSSIA** Russia's population is clustered in the west of the country, nearest to Europe.

SUSTAINABILITY AND INEQUALITY IN OUR GLOBAL VILLAGE

Trail of Tears

Like many other people, Native Americans also migrated west in the nineteenth century. But their migration was forced rather than voluntary. This inequality was written in law, when the Indian Removal Act of 1830 authorized the U.S. Army to remove five Indian tribes from their land in the southeastern United States and move them to Indian Territory (now the state of Oklahoma). The Choctaw were forced to emigrate from Mississippi in 1831, the Seminole from Florida in 1832, the Creek from Alabama in 1834, the Chickasaw from Mississippi in 1837, and the Cherokee from Georgia in 1838 (Figure 3-13). The five removals opened up 100,000 square kilometers (25 million acres) of land for whites to settle and relocated the tribes to land that was too dry to sustain their traditional ways of obtaining food. Approximately 46,000 Native Americans were estimated to have been uprooted, many of whom died in the long trek to the west. The route became known as the Trail of Tears; parts of it are preserved as a National Historic Trail (Figure 3-14).

▲ **FIGURE 3-13 TRAIL OF TEARS SCULPTURE** This sculpture in Chattanooga, Tennessee, commemorates the start of the path that the Cherokee were forced to take to relocate to Indian Territory (Oklahoma) in 1838.

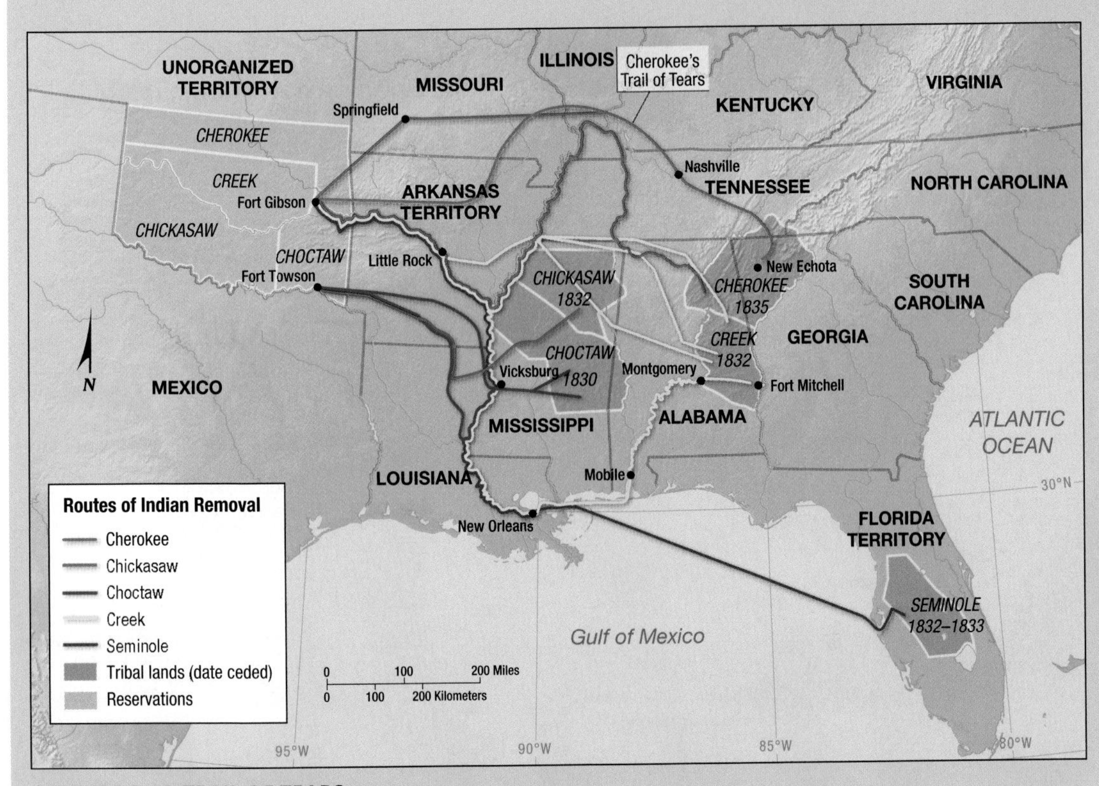

▲ **FIGURE 3-14 TRAIL OF TEARS**
These are the routes that the Cherokee, Chickasaw, Choctaw, Creek, and Seminole tribes took when they were forced to migrate in the early nineteenth century.

MIGRATION BETWEEN REGIONS IN OTHER LARGE COUNTRIES

Learning Outcome 3.2.3
Describe interregional migration in Canada, China, and Brazil.

The world's largest countries in land area other than Russia, and the United States are Canada, China, and Brazil. Government policies encourage interregional migration in Brazil and discourage it in China.

CANADA. As in the United States, Canada has had significant interregional migration from east to west for more than a century (Figure 3-15). The three westernmost provinces—Alberta, British Columbia, and Saskatchewan—are the destinations for most interregional migrants within Canada. Net out-migration is being recorded in provinces from Manitoba eastward (Figure 3-16).

▲ **FIGURE 3-15 INTERREGIONAL MIGRATION: CANADA IN 1900**
Dawson City, in Canada's Yukon Territory, was the destination in 1900 for many immigrants looking for gold.

CHINA. An estimated 100 million people have emigrated from rural areas in the interior of the country (Figure 3-17). They are headed for the large urban areas along the east coast, where jobs are most plentiful, especially in factories. The government once severely limited the ability of Chinese people to make interregional moves, but restrictions have been lifted in recent years.

BRAZIL. Most Brazilians live in a string of large cities near the Atlantic Coast. São Paulo and Rio de Janeiro have become two of the world's largest cities. In contrast, Brazil's tropical interior is very sparsely inhabited. To increase the attractiveness of the interior, the government moved its capital in 1960 from Rio to a newly built city called Brasília, situated 1,000 kilometers (600 miles) from the Atlantic Coast. Development of Brazil's interior has altered historic migration patterns. The coastal areas now have net out-migration, whereas the interior areas have net in-migration (Figure 3-18).

From above, Brasília's design resembles an airplane, with government buildings located at the center of the city and housing arranged along the "wings" (Figure 3-19). Thousands of people have migrated to Brasília in search of jobs. In a country with rapid population growth, many people will migrate where they think they can find employment. Many of these workers could not afford housing in Brasília and were living instead in hastily erected shacks on the outskirts of the city.

Pause and Reflect 3.2.3
In what ways are interregional migration in China and Brazil similar?

▼ **FIGURE 3-16 INTERREGIONAL MIGRATION: CANADA**
Population has been increasing more rapidly in the west of Canada, especially Alberta and Saskatchewan.

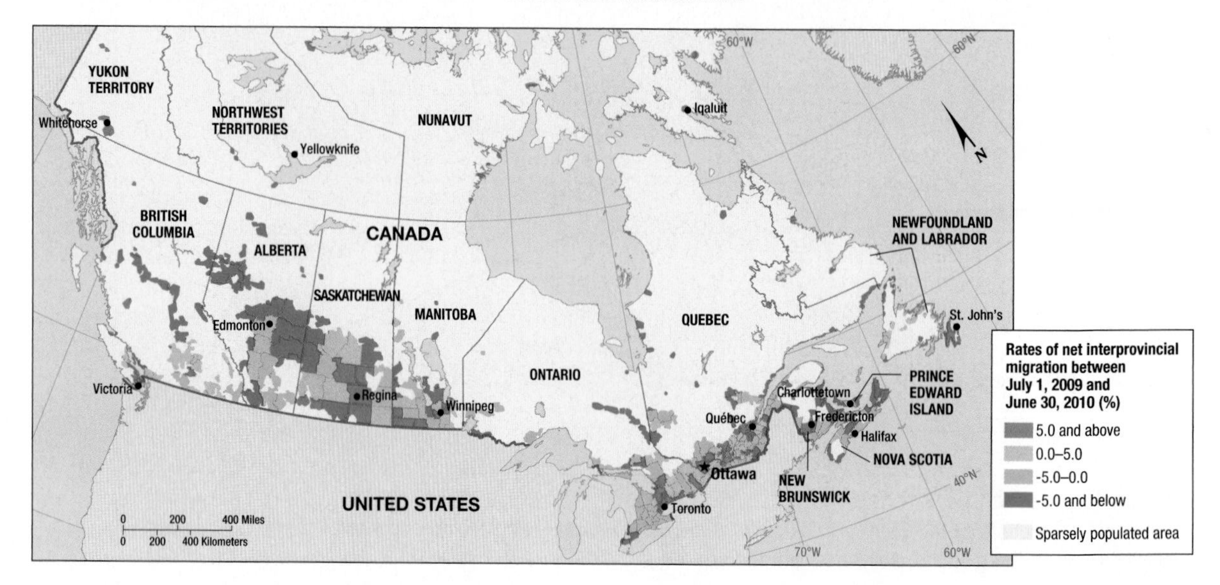

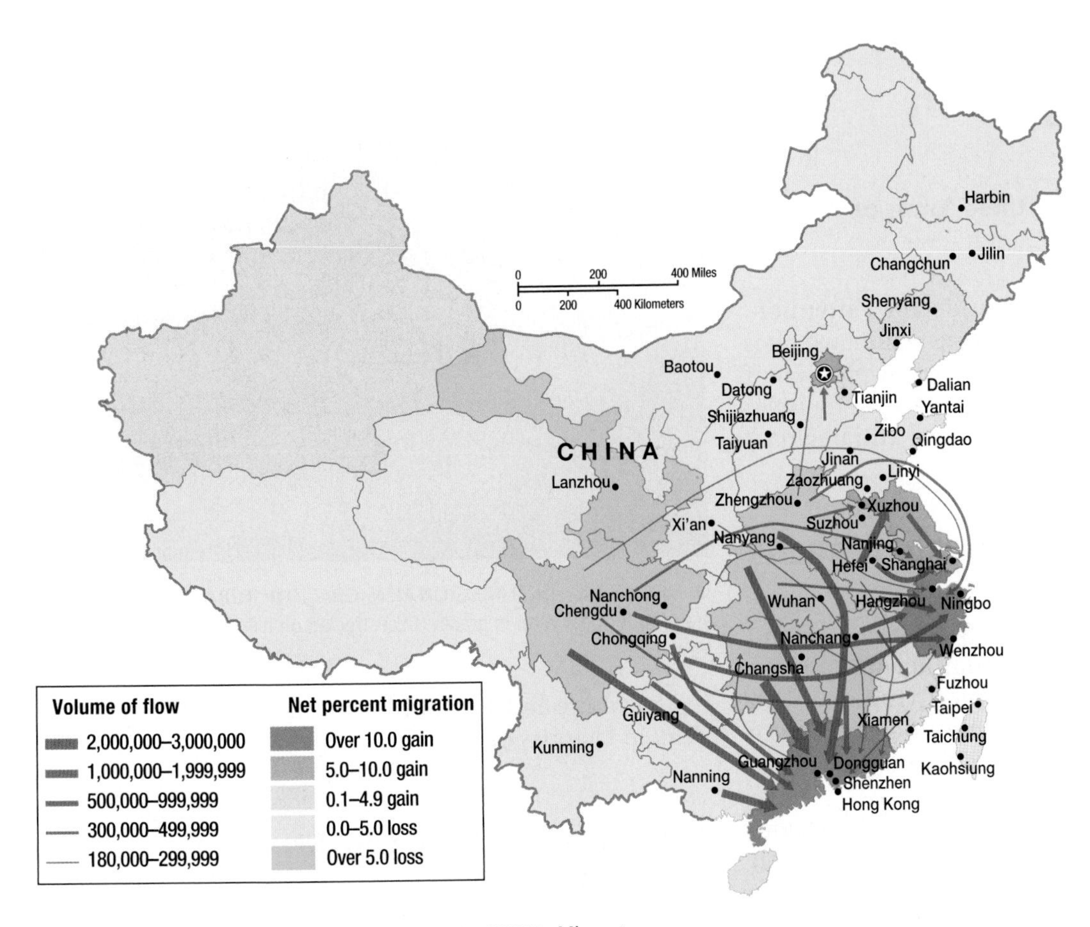

▲ **FIGURE 3-17 INTERREGIONAL MIGRATION: CHINA** Migrants are heading eastward towards the major cities.

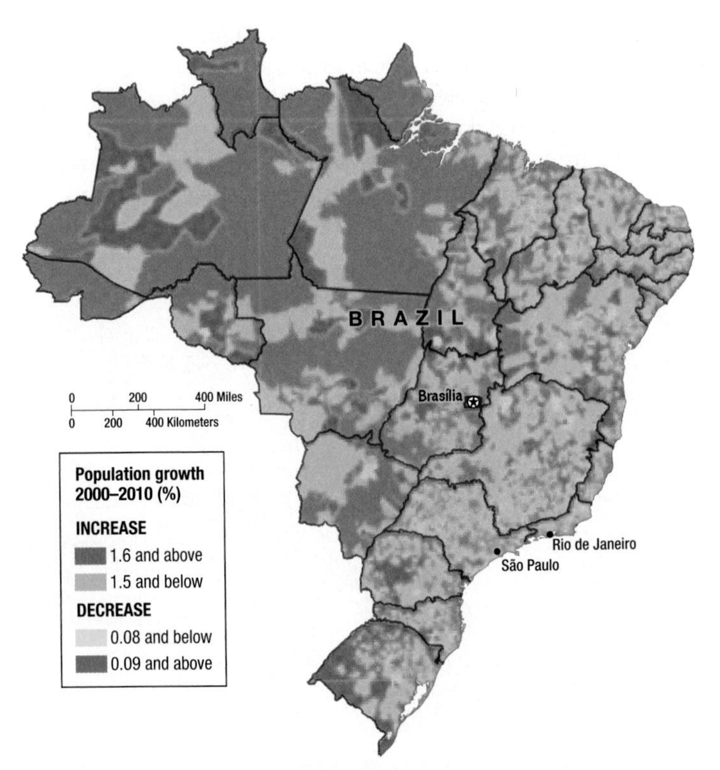

▲ **FIGURE 3-18 INTERREGIONAL MIGRATION: BRAZIL** Population is growing more rapidly in the interior of the country.

▲ **FIGURE 3-19 INTERREGIONAL MIGRATION: BRAZIL'S CAPITAL** From the air, Brasília looks like an airplane. The city was built beginning in 1960 to lure migrants from the country's large coastal cities.

Intraregional Migration

Learning Outcome 3.2.4

Explain differences among the three forms of intraregional migration.

Interregional migration attracts considerable attention, but far more people move within the same region, which is intraregional migration. Worldwide, the most prominent type of intraregional migration is from rural areas to urban areas. In the United States, the principal intraregional migration is from cities to suburbs.

MIGRATION FROM RURAL TO URBAN AREAS

Migration from rural (or nonmetropolitan) areas to urban (or metropolitan) areas began in the 1800s in Europe and North America as part of the Industrial Revolution (see Chapter 11). The percentage of people living in urban areas in the United States, for example, increased from 5 percent in 1800 to 50 percent in 1920 and 80 percent in 2010.

In recent years, urbanization has diffused to developing countries of Asia, Latin America, and Africa. Between 1950 and 2010, the percentage living in urban areas increased from 40 percent to 80 percent in Latin America, from 15 percent to 45 percent in Asia, and from 10 percent to 40 percent in sub-Saharan Africa. Worldwide, more than 20 million people are estimated to migrate each year from rural to urban areas (Figure 3-20).

As with interregional migrants, most people who move from rural to urban areas seek economic advancement. They are pushed from rural areas by declining opportunities in agriculture and are pulled to the cities by the prospect of work in factories or in service industries (Figure 3-21).

▲ **FIGURE 3-20 INTRAREGIONAL MIGRATION: CHINA** These migrants from rural to urban areas within China are waiting for trains to take them back to the countryside during holidays.

▲ **FIGURE 3-21 INTRAREGIONAL MIGRATION: RURAL TO URBAN** Housing for poor rural migrants is constructed on hills in the suburbs of Lima, Peru.

MIGRATION FROM URBAN TO SUBURBAN AREAS

Most intraregional migration in developed countries is from cities out to surrounding suburbs. The population of most cities in developed countries has declined since the mid-twentieth century, while suburbs have grown rapidly. Nearly twice as many Americans migrate from cities to suburbs each year as migrate from suburbs to cities (Figure 3-22). Comparable patterns are found in Canada and Europe.

The major reason for the large-scale migration to the suburbs is not related to employment, as is the case with other forms of migration. For most people, migration to suburbs does not coincide with changing jobs. Instead, people are pulled by a suburban lifestyle. Suburbs offer the opportunity

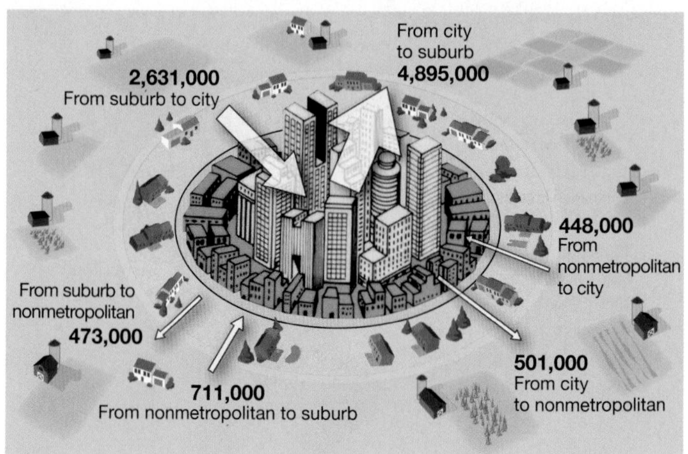

▲ **FIGURE 3-22 INTRAREGIONAL MIGRATION: UNITED STATES** This figure shows migration between cities, suburbs, and nonmetropolitan areas in 2010.

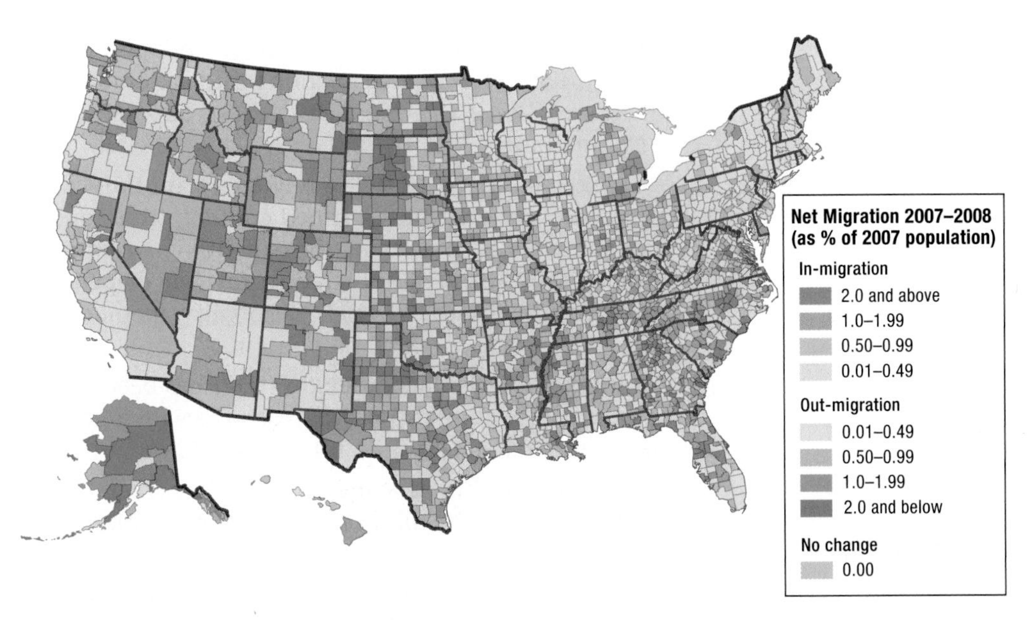

▲ **FIGURE 3-23 NET MIGRATION BY COUNTY** Rural counties experienced net in-migration in Rocky Mountain and southern states and net out-migration in Great Plains states.

to live in a detached house rather than an apartment, surrounded by a private yard where children can play safely. A garage or driveway on the property guarantees space to park cars at no extra charge. Suburban schools tend to be more modern, better equipped, and safer than those in cities. Cars and trains enable people to live in suburbs yet have access to jobs, shops, and recreational facilities throughout the urban area (see Chapter 13).

As a result of suburbanization, the territory occupied by urban areas has rapidly expanded. To accommodate suburban growth, farms on the periphery of urban areas are converted to housing and commercial developments, where new roads, sewers, and other services must be built.

MIGRATION FROM URBAN TO RURAL AREAS

Developed countries witnessed a new migration trend during the late twentieth century. For the first time, more people immigrated into rural areas than emigrated out of them. Net migration from urban to rural areas is called **counterurbanization**.

The boundary where suburbs end and the countryside begins cannot be precisely defined. Counterurbanization results in part from very rapid expansion of suburbs. But most counterurbanization represents genuine migration from cities and suburbs to small towns and rural communities.

As with suburbanization, people move from urban to rural areas for lifestyle reasons. Some are lured to rural areas by the prospect of swapping the frantic pace of urban life for the opportunity to live on a farm, where they can own horses or grow vegetables. Others move to farms but do not earn their living from agriculture; instead, they work in nearby factories, small-town shops, or other services. In the United States, evidence of counterurbanization can be seen primarily in the Rocky Mountain states. Rural counties in states such as Colorado, Idaho, Utah, and Wyoming have experienced net in-migration (Figure 3-23).

With modern communications and transportation systems, no location in a developed country is truly isolated, either economically or socially. Computers, tablets, and smart phones enable us to work anywhere and still have access to an international network. We can buy most products online and have them delivered within a few days. We can follow the fortunes of our favorite teams on television anywhere in the country, thanks to satellite dishes and webcasts.

Intraregional migration has slowed during the early twenty-first century as a result of the severe recession (Figure 3-24). Intraregional migrants, who move primarily for lifestyle reasons rather than for jobs, found that they couldn't get loans to buy new homes and couldn't find buyers for their old homes.

Pause and Reflect 3.2.3
What changes in communications and transportation might make counterurbanization easier or harder?

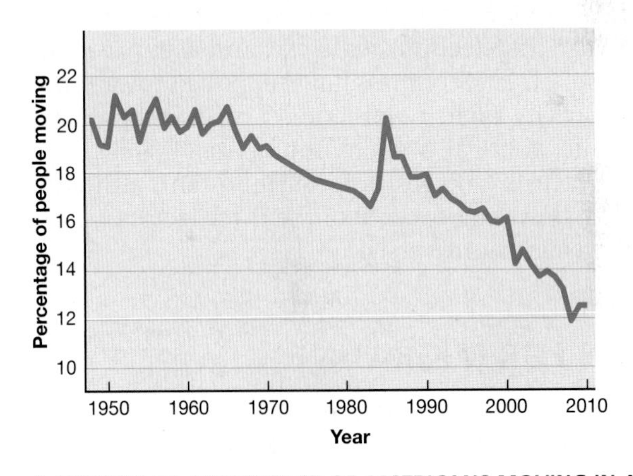

▲ **FIGURE 3-24 PERCENTAGE OF AMERICANS MOVING IN A YEAR** The percentage has declined from 20 percent in the 1980s to 12 percent in the 2010s.

CHECK IN: KEY ISSUE 2

Where Do People Migrate Within a Country?

- ✓ **Migration between regions is important within the United States, as well as within other large countries.**
- ✓ **Migration within countries takes several forms, including rural to urban, urban to suburban, and urban to rural.**

KEY ISSUE 3

Why Do People Migrate?

- **Reasons for Migrating**
- **Migrating to Find Work**
- **Characteristics of Migrants**

Learning Outcome 3.3.1
Provide examples of political, environmental, and economic push and pull factors.

People decide to migrate because of push factors and pull factors:

- A **push factor** induces people to move out of their present location.
- A **pull factor** induces people to move into a new location.

As migration for most people is a major step not taken lightly, both push and pull factors typically play a role. To migrate, people view their current place of residence so negatively that they feel pushed away, and they view another place so attractively that they feel pulled toward it.

We can identify three major kinds of push and pull factors: economic, political, and environmental. Usually, one of the three factors emerges as most important, although ranking the relative importance of the three factors can be difficult and even controversial.

REASONS FOR MIGRATING

Ravenstein's laws help geographers make generalizations about where and how far people migrate. The laws also sum up the reasons why people migrate:

- Most people migrate for economic reasons.
- Political and environmental factors also induce migration, although not as frequently as economic factors.

POLITICAL PUSH AND PULL FACTORS

Political factors can be especially compelling push factors, forcing people to emigrate from a country. Slavery was once an important political push factor. Millions of people were shipped to other countries as slaves or as prisoners, especially from sub-Saharan Africa to North America and Latin America, during the eighteenth and early nineteenth centuries (see Chapter 7).

Forced political migration now occurs because of political conflict. The United Nations High Commission for Refugees (UNHCR) recognizes three groups of forced political migrants:

- A **refugee** has been forced to migrate to another country to avoid the effects of armed conflict, situations of generalized violence, violations of human rights, or other disasters and cannot return for fear of persecution because of race, religion, nationality, membership in a social group, or political opinion.
- An **internally displaced person (IDP)** has been forced to migrate for similar political reasons as a refugee but has not migrated across an international border.
- An **asylum seeker** is someone who has migrated to another country in the hope of being recognized as a refugee.

The United Nations counted 10.6 million refugees, 14.7 million IDPs, and 838,000 asylum seekers in 2010 (Figure 3-25). The UNHCR also found that 198,000 refugees and 2.9 million IDPs had returned to their homes in 2010.

The largest number of refugees in 2010 was forced to migrate from Afghanistan and Iraq because of the continuing wars there. Countries bordering Afghanistan and Iraq, including Pakistan, Iran, and Syria, received the most refugees.

ENVIRONMENTAL PUSH AND PULL FACTORS

People sometimes migrate for environmental reasons, pulled toward physically attractive regions and pushed from hazardous ones. In this age of improved communications and transportation systems, people can live in environmentally attractive areas that are relatively remote and still not feel too isolated from employment, shopping, and entertainment opportunities.

Attractive environments for migrants include mountains, seasides, and warm climates. Proximity to the Rocky Mountains lures Americans to the state of Colorado, and the Alps pull French people to eastern France. Some migrants are shocked to find polluted air and congestion in these areas. The southern coast of England, the Mediterranean coast of France, and the coasts of Florida attract migrants, especially retirees, who enjoy swimming and lying on the beach. Of all elderly people who migrate from one U.S. state to another, one-third select Florida as their destination. Regions with warm winters, such as southern Spain and the southwestern United States, attract migrants from harsher climates.

Migrants are also pushed from their homes by adverse physical conditions. Water—either too much or too little—poses the most common environmental threat. Many people are forced to move by water-related disasters because they live in a vulnerable area, such as a floodplain (Figure 3-26). The **floodplain** of a river is the area subject to flooding during a specific number of years, based

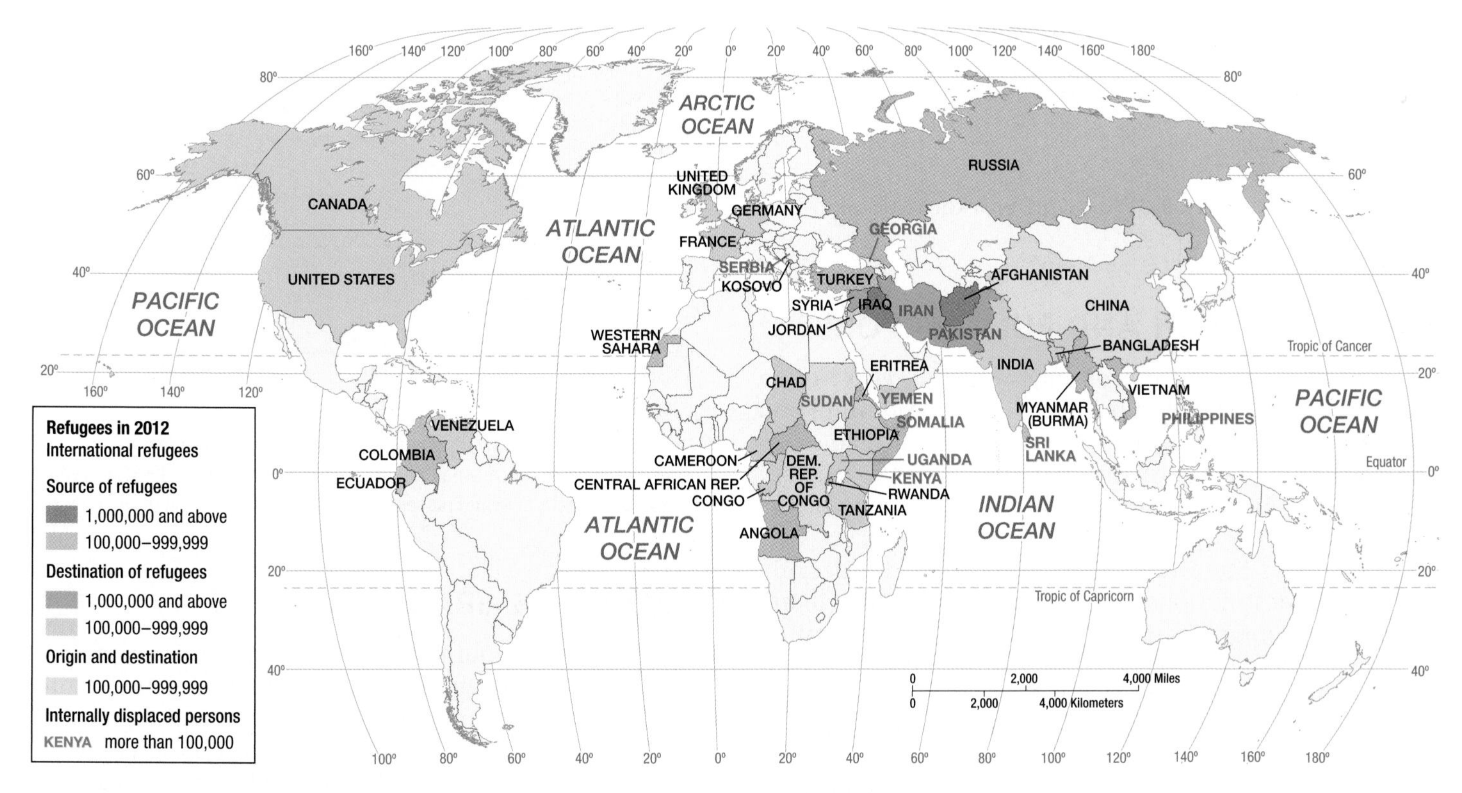

▲ **FIGURE 3-25 POLITICAL FACTORS: REFUGEES AND IDPS** The largest numbers of refugees originated in Southwest Asia and sub-Saharan Africa.

on historical trends. People living in the "100-year floodplain," for example, can expect flooding on average once every century. Many people are unaware that they live in a floodplain, and even people who do know often choose to live there anyway.

A lack of water pushes others from their land (Figure 3-27). Hundreds of thousands have been forced to move from the Sahel region of northern Africa because of drought conditions. The people of the Sahel have traditionally been pastoral nomads, a form of agriculture adapted to dry lands but effective only at low population densities (see Chapter 10).

The capacity of the Sahel to sustain human life—never very high—has declined recently because of population growth and several years of unusually low rainfall. Consequently, many of these nomads have been forced to move into cities and rural camps, where they survive on food donated by the government and international relief organizations.

▲ **FIGURE 3-26 FLOODING** Flooding of the Mississippi River in 2011 inundated farms in the floodplain.

▲ **FIGURE 3-27 DROUGHT** This man in Abala, Niger, is explaining that his animals have died because of drought.

Migrating to Find Work

Learning Outcome 3.3.2
Summarize the flows of migrant workers in Europe and Asia.

ECONOMIC PUSH AND PULL FACTORS

Most people migrate for economic reasons. People often emigrate from places that have few job opportunities and immigrate to places where jobs seem to be available. Because of economic restructuring, job prospects often vary from one country to another and within regions of the same country.

The United States and Canada have been especially prominent destinations for economic migrants. Many European immigrants to North America in the nineteenth century truly expected to find streets paved with gold. While not literally so gilded, the United States and Canada did offer Europeans prospects for economic advancement. This same perception of economic plenty now lures people to the United States and Canada from Latin America and Asia.

The relative attractiveness of a region can shift with economic change. Ireland was a place of net out-migration through most of the nineteenth and twentieth centuries. Dire economic conditions produced net out-migration in excess of 200,000 a year during the 1850s. The pattern reversed during the 1990s, as economic prosperity made Ireland a destination for immigrants, especially from Eastern Europe. However, the collapse of Ireland's economy as part of the severe global recession starting in 2008 brought a net out-migration to Europe (Figure 3-28).

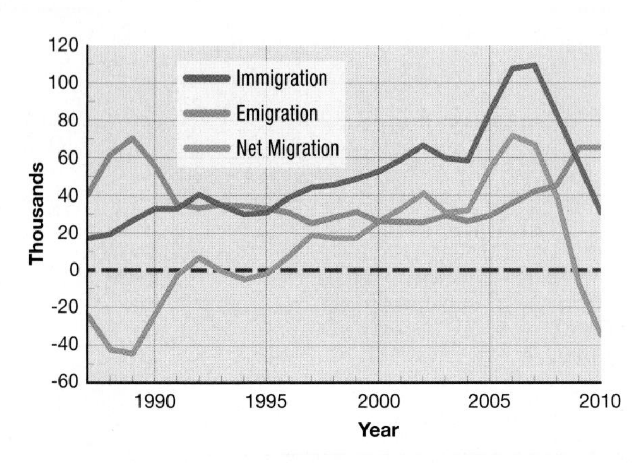

▲ **FIGURE 3-28 ECONOMIC MIGRATION: IRELAND** With few job prospects, Ireland historically had net out-migration until the 1990s. The severe recession of the early twenty-first century has brought net out-migration back to Ireland.

Pause and Reflect 3.3.1
What would it take for Ireland to once again have net in-migration?

It is sometimes difficult to distinguish between migrants seeking economic opportunities and refugees fleeing from government persecution. The distinction between economic migrants and refugees is important because the United States, Canada, and European countries treat the two groups differently. Economic migrants are generally not admitted unless they possess special skills or have a close relative already there, and even then they must compete with similar applicants from other countries. However, refugees receive special priority in admission to other countries.

People unable to migrate permanently to a new country for employment opportunities may be allowed to migrate temporarily. Prominent forms of temporary work are found in Europe and Asia.

EUROPE'S MIGRANT WORKERS

Of the world's 16 countries with the highest per capita income, 14 are in Northern and Western Europe. As a result, the region attracts immigrants from poorer regions located the south and east. These immigrants serve a useful role in Europe, taking low-status and low-skill jobs that local residents won't accept. In cities such as Berlin, Brussels, Paris, and Zurich, immigrants provide essential services, such as driving buses, collecting garbage, repairing streets, and washing dishes (Figure 3-29).

Although relatively low paid by European standards, immigrants earn far more than they would at home. By letting their people work elsewhere, poorer countries reduce their own unemployment problems. Immigrants also help their native countries by sending a large percentage of their earnings back home to their families. The injection of foreign currency then stimulates the local economy.

▲ **FIGURE 3-29 IMMIGRANTS IN EUROPE** Immigrant from North Africa cleans the streets in Paris.

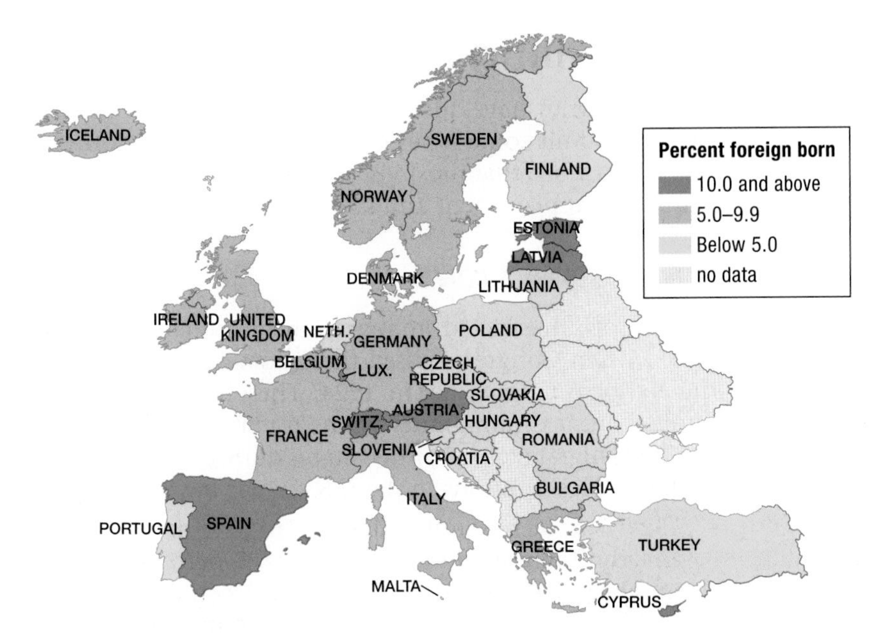

◀ **FIGURE 3-30 PERCENTAGE OF IMMIGRANTS IN EUROPE** Luxembourg and Switzerland have the highest percentages of immigrants.

Germany and other wealthy European countries operated a **guest worker** program mainly during the 1960s and 1970s. Immigrants from poorer countries were allowed to immigrate temporarily to obtain jobs. They were protected by minimum-wage laws, labor union contracts, and other support programs. The guest worker program was intended to be temporary. After a few years, the guest workers were expected to return home.

The first guest worker programs involved emigration from Southern European countries such as Italy, Portugal, and Spain. Northern European countries were then much wealthier and more economically developed and offered many more job opportunities. Turkey and North Africa replaced Southern Europe as the leading sources. Today, most immigrants in search of work in Europe come from Eastern Europe, such as Poland and Romania.

The term "guest worker" is no longer used in Europe, and the government programs no longer exist. Many immigrants who arrived originally under the guest worker program have remained permanently. They, along with their children and grandchildren, have become citizens of the host country. The foreign-born population exceeds 40 percent in Luxembourg and 20 percent in Switzerland. Among the most populous European countries, Spain has the highest share of foreign-born population (Figure 3-30). In Europe as a whole, though, the percentage of foreign-born residents is only one-half that of North America.

ASIA'S MIGRANT WORKERS

Asia is both a major source and a major destination for migrants in search of work:

- **China.** Approximately 40 million Chinese currently live in other countries, including 30 million in Southeast Asia, 5 million in North America, and 2 million in Europe. Chinese comprise three-fourths of the population in Singapore and one-fourth in Malaysia. Most migrants were from southeastern China. China's booming economy is now attracting immigrants from neighboring countries, especially Vietnamese, who are willing to work in China's rapidly expanding factories. Immigration from abroad pales in comparison to internal migration within China.
- **Southwest Asia.** The wealthy oil-producing countries of Southwest Asia have been major destinations for people from poorer countries in the region, such as Egypt and Yemen. During the late twentieth century, most immigrants arrived from South and Southeast Asia, including India, Pakistan, the Philippines, and Thailand (Figure 3-31). Working conditions for immigrants have been considered poor in some of these countries. The Philippine government determined in 2011 that only two countries in Southwest Asia—Israel and Oman—were "safe" for their Filipino migrants, and the others lacked adequate protection for workers' rights. For their part, oil-producing countries fear that the increasing numbers of guest workers will spark political unrest and abandonment of traditional Islamic customs.

Pause and Reflect 3.3.2
Why are street cleaning and construction jobs attractive for immigrants to Europe and Southwest Asia?

CHECK-IN: KEY ISSUE 3

Why Do People Migrate?

✓ **People migrate for a combination of political, environmental, and economic push and pull factors.**

✓ **Most people migrate in search of work.**

▲ **FIGURE 3-31 IMMIGRANTS IN SOUTHWEST ASIA** These immigrants in Dubai have lined up to get construction jobs.

KEY ISSUE 4

Why Do Migrants Face Obstacles?

- Controlling Migration
- Unauthorized Immigration
- Attitudes toward Immigrants

Learning Outcome 3.4.1
Identify the types of immigrants who are given preference to enter the United States.

An environmental or political feature that hinders migration is an **intervening obstacle**. The principal obstacle traditionally faced by migrants to other countries was environmental: the long, arduous, and expensive passage over land or by sea. Think of the cramped and unsanitary conditions endured by eighteenth- and nineteenth-century immigrants to the United States who had to sail across the Atlantic or Pacific Ocean in tiny ships. Or the mountains and deserts that European pioneers and displaced Native Americans were forced to cross in their westward migration across the North American continent.

Transportation improvements that have promoted globalization, such as motor vehicles and airplanes, have diminished the importance of environmental features as intervening obstacles. Today, the major obstacles faced by most immigrants are political. A migrant needs a passport to legally emigrate from a country and a visa to legally immigrate to a new country (Figure 3-32).

▲ **FIGURE 3-32 PASSPORT CONTROL** Backup at the border from Tijuana, Mexico (right) into the United States at San Diego.

Controlling Migration

Most countries have adopted selective immigration policies that admit some types of immigrants but not others. The two reasons that most visas are granted are for specific employment placement and family reunification.

U.S. QUOTA LAWS

The era of unrestricted immigration to the United States ended when Congress passed the Quota Act in 1921 and the National Origins Act in 1924. These laws established **quotas**, or maximum limits on the number of people who could immigrate to the United States during a one-year period. Key modifications in the U.S. quotas have included:

- **1924:** For each country that had native-born persons already living in the United States, 2 percent of their number (based on the 1910 census) could immigrate each year. This ensured that most immigrants would come from Europe.
- **1965:** Quotas for individual countries were replaced with hemisphere quotas (170,000 from the Eastern Hemisphere and 120,000 from the Western Hemisphere).
- **1978:** A global quota of 290,000 was set, including a maximum of 20,000 per country.
- **1990:** The global quota was raised to 700,000.

Because the number of applicants for admission to the United States far exceeds the quotas, Congress has set preferences:

- **Family reunification.** Approximately three-fourths of immigrants are admitted to reunify families, primarily spouses or unmarried children of people already living in the United States. The typical wait for a spouse to gain entry is currently about five years.
- **Skilled workers.** Exceptionally talented professionals receive most of the remainder of the quota.
- **Diversity.** A few immigrants are admitted by lottery under a diversity category for people from countries that historically sent few people to the United States.

The quota does not apply to refugees, who are admitted if they are judged genuine refugees. Also admitted without limit are spouses, children, and parents of U.S. citizens. The number of immigrants can vary sharply from year to year, primarily because numbers in these two groups are unpredictable.

Other countries charge that by giving preference to skilled workers, immigration policies in the United States and Europe contribute to a **brain drain**, which is a large-scale emigration by talented people. Scientists, researchers, doctors, and other professionals migrate to countries where they can make better use of their abilities.

Asians have made especially good use of the priorities set by the U.S. quota laws. Many well-educated Asians enter the United States under the preference for skilled workers. Once admitted, they can bring in relatives under the family-reunification provisions of the quota. Eventually, these immigrants can bring in a wider range of other relatives from Asia, through a process of **chain migration**, which is the migration of people to a specific location because relatives or members of the same nationality previously migrated there.

CONTEMPORARY GEOGRAPHIC TOOLS

Claiming Ellis Island

Twelve million immigrants to the United States between 1892 and 1954 were processed at Ellis Island, situated in New York Harbor (Figure 3-33). Incorporated as part of the Statue of Liberty National Monument in 1965, Ellis Island was restored and reopened in 1990 as a museum of immigration. Before building the immigration center, the U.S. government used Ellis Island as a fort and powder magazine beginning in 1808.

An 1834 agreement approved by the U.S. Congress gave Ellis Island to New York State and gave the submerged lands surrounding the island to New Jersey. When the agreement was signed, Ellis Island was only 1.1 hectares (2.75 acres), but beginning in the 1890s, the U.S. government enlarged the island, eventually to 10.6 hectares (27.5 acres).

New Jersey state officials claimed that the 10.6-hectare Ellis Island was part of their state, not New York. The claim was partly a matter of pride on the part of New Jersey officials to stand up to their more glamorous neighbor. After all, Ellis Island was only 400 meters (1,300 feet) from the New Jersey shoreline, yet tourists—like immigrants a century ago—are transported by ferry to Lower Manhattan more than a mile away. More practically, the sales tax collected by the Ellis Island museum gift shop was going to New York rather than to New Jersey.

After decades of dispute, New Jersey took the case to the U.S. Supreme Court. In 1998, the Supreme Court ruled 6–3 that New York owned the original island but that New Jersey owned the rest. New York's jurisdiction was set as the low waterline of the original island. Critical evidence in the decision was a series of maps prepared by New Jersey Department of Environmental Protection (NJDEP) officials using Geographic Information System (GIS). NJDEP officials scanned into an image file an 1857 U.S. coast map that was considered to be the most reliable map from that era. The image file of the old map was brought into ArcView, and then the low waterline shown on the 1857 map was edited and depicted using a series of dots. The perimeter of the current island was mapped using global positioning system (GPS) surveying.

After ruling in favor of New Jersey's claim, the Supreme Court directed the NJDEP to delineate the precise boundary between the two states, again using GIS. Overlaying the 1857 low waterline onto the current map identified New York's territory, and the rest of the current island was determined to belong to New Jersey.

▲ **FIGURE 3-33 ELLIS ISLAND** Ellis Island is in the foreground, Jersey City, New Jersey, is to the left, and Manhattan, New York is to the rear.

Attitudes toward Immigrants

Learning Outcome 3.4.3
Describe characteristics of immigrants to the United States.

Americans and Europeans share mixed views about immigration. They recognize that immigrants play an important economic role in their countries, but key features of immigration trouble Americans and Europeans. In the United States, the principal concern relates to unauthorized immigration. In Europe, the principal concern relates to cultural diversity.

CHARACTERISTICS OF MIGRANTS

Ravenstein noted distinctive gender and family-status patterns in his migration theories:

- Most long-distance migrants are male.
- Most long-distance migrants are adult individuals rather than families with children.

GENDER OF MIGRANTS

Ravenstein theorized that males were more likely than females to migrate long distances to other countries because searching for work was the main reason for international migration, and males were much more likely than females to be employed (Figure 3-40). This held true for U.S. immigrants during the nineteenth and much of the twentieth centuries, when about 55 percent were male. But the gender pattern reversed in the 1990s, and in the twenty-first century women constitute about 55 percent of U.S. immigrants.

Mexicans who come to the United States without authorized immigration documents—currently the largest group of U.S. immigrants—show similar gender changes. As recently as the late 1980s, males constituted 85 percent of the Mexican migrants arriving in the United States without proper documents, according to U.S. census and immigration service estimates. But since the 1990s, women have accounted for about half of the unauthorized immigrants from Mexico.

The increased female migration to the United States partly reflects the changing role of women in Mexican society. In the past, rural Mexican women were obliged to marry at a young age and to remain in the village to care for children. Now some Mexican women are migrating to the United States to join husbands or brothers already in the United States, but most are seeking jobs. At the same time, women feel increased pressure to get jobs in the United States because of poor economic conditions in Mexico.

AGE AND EDUCATION OF MIGRANTS

Ravenstein also believed that most long-distance migrants were young adults seeking work rather than children or elderly people. For the most part, this pattern continues for the United States:

- About 40 percent of immigrants are young adults between the ages of 25 and 39, compared to about 23 percent of the entire U.S. population.
- Immigrants are less likely to be elderly people; only 5 percent of immigrants are over age 65, compared to 12 percent of the entire U.S. population.
- Children under 15 comprise 16 percent of immigrants, compared to 21 percent for the total U.S. population. With the increase in women migrating to the United States, more children are coming with their mothers (Figure 3-41).
- Recent immigrants to the United States have attended school for fewer years and are less likely to have high school diplomas than are U.S. citizens. The typical

▲ **FIGURE 3-40 EMIGRANTS FROM MEXICO: MALE DAY LABORERS** Day laborers from Mexico have crossed the border into San Diego, California, to look for temporary work.

▲ **FIGURE 3-41 EMIGRANTS FROM MEXICO: WOMEN AND CHILDREN AT A PARADE IN NEW YORK** Children who have emigrated from Mexico display the Mexican flag at a Latinos Unidos parade in Brooklyn, New York.

unauthorized Mexican immigrant has attended school for four years, less than the average American but a year more than the average Mexican.

IMMIGRATION CONCERNS IN THE UNITED STATES

Americans are divided concerning whether unauthorized migration helps or hurts the country (Figure 3-42). This ambivalence extends to specific elements of immigration law:

- **Border patrols.** Americans would like more effective border patrols so that fewer unauthorized immigrants can get into the country, but they don't want to see money spent to build more fences along the border. The U.S. Department of Homeland Security has stepped up enforcement, including deportation of a record 390,000 unauthorized immigrants in 2010.
- **Workplace.** Most Americans recognize that unauthorized immigrants take jobs that no one else wants, so they support some type of work-related program to make them legal, and they oppose raids on workplaces in attempts to round up unauthorized immigrants. Most Americans support a path to U.S. citizenship for unauthorized immigrants.
- **Civil rights.** Americans favor letting law enforcement officials stop and verify the legal status of anyone they suspect of being an unauthorized immigrant. On the other hand, they fear that enforcement efforts that identify and deport unauthorized immigrants could violate the civil rights of U.S. citizens.
- **Local initiatives.** Polls show that most Americans believe that enforcement of unauthorized immigration is a federal government responsibility and do not support the use of local law enforcement officials to find unauthorized immigrants. On the other hand, residents of some states along the Mexican border favor stronger enforcement of authorized immigration.

The strongest state initiative has been Arizona's 2010 law that obligated local law enforcement officials, when practicable, to determine a person's immigration status. Under the Arizona law, foreigners are required to carry at all times documents proving they are in the country legally and to produce those documents upon request of a local law enforcement official. In 2012, the U.S. Supreme Court struck down several provisions of the law.

Although it does not share a border with Mexico, Alabama enacted a similar measure in 2011. The Alabama law also prohibited or restricted unauthorized immigrants from attending public schools and colleges. On the other hand, Texas, which has the longest border with Mexico, has not enacted harsh anti-immigrant laws, and more than 100 localities across the country have passed resolutions supporting more rights for unauthorized immigrants—a movement known as "Sanctuary City."

Controversy even extends to what to call the group of immigrants:

- *Unauthorized immigrant* is the term preferred by academic observers, including the authoritative Pew Hispanic Center, as a neutral term.
- *Undocumented immigrant* is the term preferred by groups that advocate for more rights for these individuals.
- *Illegal alien* is the term preferred by groups who favor tougher restrictions and enforcement of immigration laws.

Opposition to immigration into the United States predates the current era of most immigrants coming from Latin America and Asia. Hostility intensified when Italians, Russians, Poles, and other Southern and Eastern Europeans poured into the United States beginning in the late nineteenth century. Earlier European immigrants, mostly from Northern and Western Europe, had converted the forests and prairies of the vast North American interior into productive farms and had helped to extend U.S. control across the continent. By the early twentieth century, most Americans saw the frontier as closed and thought that therefore entry into the country should be closed as well.

A government study in 1911 reflected popular attitudes when it concluded that immigrants from Southern and Eastern Europe were racially inferior, "inclined toward violent crime," resisted assimilation, and "drove old-stock citizens out of some lines of work." A century later, many Americans have similar reactions to the arrival of large numbers of immigrants from Latin America and Asia.

Pause and Reflect 3.4.3
In what ways are reactions to immigrants today similar to those of a century ago?

▼ FIGURE 3-42 IMMIGRATION CONTROVERY IN THE UNITED STATES Demonstrations supporting (left) tighter immigration controls and (right) more rights for immigrants.

IMMIGRATION CONCERNS IN EUROPE

Learning Outcome 3.4.4
Compare American and European attitudes toward immigrants.

Attitudes toward immigration are also ambivalent in Europe. Europeans have more rights than ever before to migrate elsewhere within Europe, whereas non-Europeans face more restrictions than in the past.

SOURCES OF EUROPEAN IMMIGRATION. Agreements among European countries, especially the 1985 Schengen Treaty, give a citizen of one European country the right to hold a job, live permanently, and own property elsewhere. The removal of migration restrictions for Europeans has set off large-scale migration flows within the region. The principal flows are from the poorer countries of Europe to the richer ones, where job opportunities have been greater.

In recent years the largest flows within Europe have included (Figure 3-43):

- From Southeastern Europe, especially Romania, as well as Bulgaria, Albania, and Serbia, especially to Italy and Spain.
- From Eastern Europe, especially Poland, as well as Russia and Ukraine, especially to Germany, the United Kingdom, and Ireland.
- From Northern Europe, especially the United Kingdom and Germany, to attractive climates in Southern Europe, especially in Spain.

During the twentieth century, the largest flows within Europe were south to north, especially from Greece, Italy, Portugal, and Spain to France and Germany.

While migration within Europe has become easier and more common, it has become more difficult for non-Europeans to immigrate to a European country. During the twentieth century, large numbers of Turks and North Africans migrated to Europe. Germany's Turkish population remains the largest group of non-Europeans in Europe.

OPPONENTS OF IMMIGRATION. Most European countries are now in stage 4 of the demographic transition (very low or negative NIR) and have economies capable of meeting the needs of their people. The safety valve of emigration is no longer needed. To the contrary, population growth in Europe is fueled by immigration from other regions, a trend that many Europeans dislike.

Hostility to immigrants has become a central plank in the platform of political parties in many European countries. These parties blame immigrants for crime, unemployment, and high welfare costs. Above all, the anti-immigration parties fear that long-standing cultural traditions of the host country are threatened by immigrants who adhere to different religions, speak different languages, and practice different food and other cultural habits. From the standpoint of these parties, immigrants represent a threat

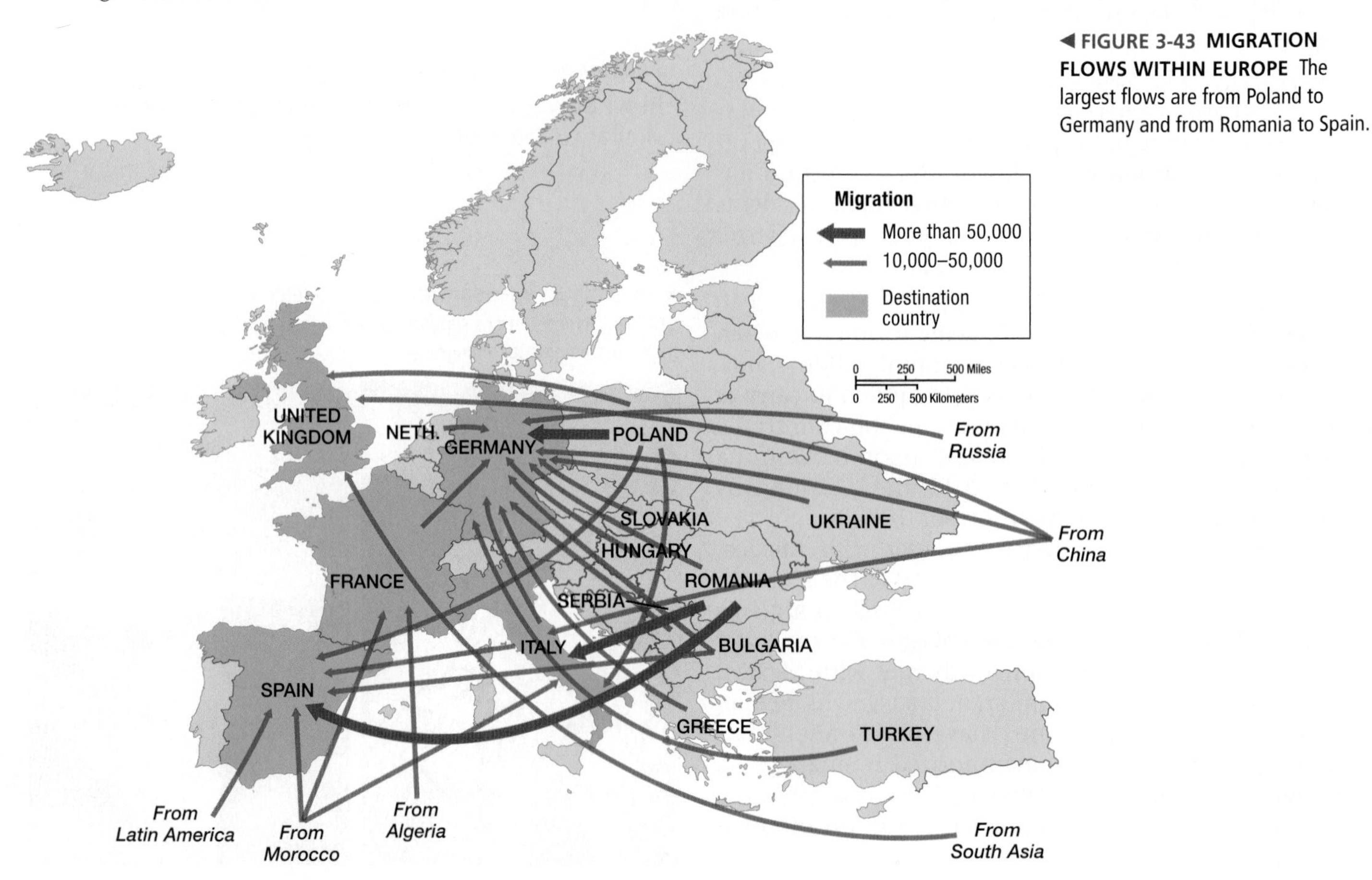

◀ **FIGURE 3-43 MIGRATION FLOWS WITHIN EUROPE** The largest flows are from Poland to Germany and from Romania to Spain.

to the centuries-old cultural traditions of the host country (Figure 3-44).

The severe global recession of the early twenty-first century has reduced the number of immigrants to the United States and Europe. With high unemployment and limited job opportunities in the principal destination countries, potential migrants have much less incentive to risk the uncertainties and expenses of international migration. Countries such as Ireland, Portugal, and Spain that had become destinations during the late twentieth century once again have net out-migration.

Pause and Reflect 3.4.4
How are attitudes towards immigrants similar in the United States and Europe?

EUROPEANS AS EMIGRANTS. The inhospitable climate for immigrants in Europe is especially ironic because Europe was the source of most of the world's emigrants, especially during the nineteenth century. Application of new technologies spawned by the Industrial Revolution—in areas such as public health, medicine, and food—produced a rapid decline in the CDR and pushed much of Europe into stage 2 of the demographic transition (high NIR). As the population increased, many Europeans found limited opportunities for economic advancement. Migration to the United States, Canada, Australia, and other regions of the world served as a safety valve, draining off some of that increase.

The emigration of 65 million Europeans has profoundly changed world culture. As do all migrants, Europeans brought their cultural heritage to their new homes. Because of migration, Indo-European languages are now spoken by half of the world's people (as discussed in Chapter 5), and Europe's most prevalent religion, Christianity, has the world's largest number of adherents (see Chapter 6). European art, music, literature, philosophy, and ethics have also diffused throughout the world.

Regions that were sparsely inhabited prior to European immigration, such as North America and Australia, have become closely integrated into Europe's cultural traditions. Distinctive European political structures and economic systems have also diffused to these regions. Europeans also planted the seeds of conflict by migrating to regions with large indigenous populations, especially in Africa and Asia. They frequently imposed political domination on existing populations and injected their cultural values with little regard for local traditions. Economies in Africa and Asia became based on raising crops and extracting resources for export to Europe rather than on growing crops for local consumption and using resources to build local industry. Many of today's conflicts in former European colonies result from past practices by European immigrants, such as drawing arbitrary boundary lines and discriminating among different local ethnic groups.

▲ **FIGURE 3-44 IMMIGRATION CONTROVERY IN FRANCE** Demonstrations supporting (top) more rights for immigrants and (bottom) tighter immigration controls.

CHECK-IN: KEY ISSUE 4

Why Do Migrants Face Obstacles?

- ✓ **Immigration is tightly controlled by most countries.**
- ✓ **The United States has more than 11 million unauthorized immigrants, mostly from Mexico.**
- ✓ **Americans and Europeans are divided on how to regard immigrants, especially unauthorized ones.**

Summary and Review

KEY ISSUE 1

Where Are Migrants Distributed?

On a global scale, the largest flows of migrants are from Asia to Europe and from Asia and Latin America to the United States. The United States receives by far the largest number of migrants.

LEARNING OUTCOME 3.1.1: Describe the difference between international and internal migration.

- Migration can be international (between countries, either voluntary or forced) or internal (within a country, either interregional or intraregional).

LEARNING OUTCOME 3.1.2: Identify the principal sources of immigrants during the three main eras of U.S. immigration.

- The United States has had three main eras of immigration. The principal source of immigrants has shifted from Europe during the first two eras to Latin America and Asia during the third (current) era.

THINKING GEOGRAPHICALLY 3.1: What is the impact of emigration on the place from which migrants depart?

GOOGLE EARTH 3.1: The Lower East Side Tenement Museum, at 103 Orchard St. in New York City, shows what life was like for European immigrants. Which of the three principal eras of immigration to the United States does the Tenement Museum describe?

KEY ISSUE 2

Where Do People Migrate Within a Country?

Historically, interregional migration was especially important in settling the frontiers of large countries such as Russia, Canada, the United States, China, and Brazil. The most important intraregional migration trends are from rural to urban areas within developing countries and from cities to suburbs within developed countries.

LEARNING OUTCOME 3.2.1: Describe the history of interregional migration in the United States.

- Migration within the United States has primarily occurred from east to west, though at varying rates. Recently, interregional migration has also occurred from north to south.

LEARNING OUTCOME 3.2.2: Describe interregional migration in Russia.

- The world's largest country has a distinctive pattern of interregional migration, a legacy of the era of Communist rule.

LEARNING OUTCOME 3.2.3: Describe interregional migration in Canada, China, and Brazil.

- Canada, China, and Brazil also have unequal population distributions. Canadians have been migrating from east to west, Chinese have been migrating from the rural interior to the large coastal cities, and Brazilians from the large coastal cities to the interior.

LEARNING OUTCOME 3.2.4: Explain differences among the three forms of intraregional migration.

- Three intraregional migration patterns are from rural to urban areas, from urban to suburban areas, and from urban to rural areas.

THINKING GEOGRAPHICALLY 3.2: In recent years, has your community seen net in-migration or net out-migration? What factors explain your community's net migration?

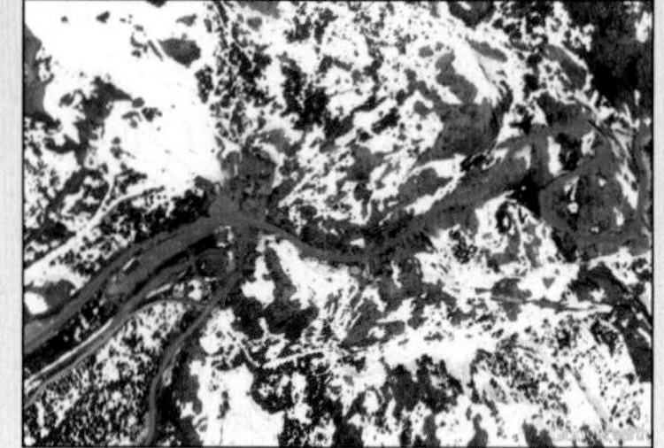

GOOGLE EARTH 3.2: The Donner Pass, through the Sierra Nevada Mountains, was one of the most difficult obstacles for early immigrants to the west. What is now the principal route through the area?

Key Terms

Asylum seeker (p. 92) Someone who has migrated to another country in the hope of being recognized as a refugee.

Brain drain (p. 96) Large-scale emigration by talented people.

Chain migration (p. 97) Migration of people to a specific location because relatives or members of the same nationality previously migrated there.

Circulation (p. 78) Short-term, repetitive, or cyclical movements that recur on a regular basis.

Counterurbanization (p. 91) Net migration from urban to rural areas in more developed countries.

Emigration (p. 78) Migration from a location.

Floodplain (p. 92) The area subject to flooding during a given number of years, according to historical trends.

Forced migration (p. 80) Permanent movement, usually compelled by cultural factors.

Guest worker (p. 95) A term once used for a worker who migrated to the developed countries of Northern and Western Europe, usually from Southern and Eastern Europe or from North Africa, in search of a higher-paying job.

Immigration (p. 78) Migration to a new location.

Internal migration (p. 80) Permanent movement within a particular country.

Internally displaced person (IDP) (p. 92) Someone who has been forced to migrate for similar political reasons as a refugee but has not migrated across an international border.

International migration (p. 80) Permanent movement from one country to another.

Interregional migration (p. 80) Permanent movement from one region of a country to another.

Intervening obstacle (p. 96) An environmental or cultural feature of the landscape that hinders migration.

Intraregional migration (p. 80) Permanent movement within one region of a country.

Migration (p. 78) A form of relocation diffusion involving a permanent move to a new location.

KEY ISSUE 3

Why Do People Migrate?

Migration is induced by a combination of push and pull factors. People feel compelled (pushed) to emigrate from a location for political, environmental, and economic reasons. Similarly, people are induced (pulled) to immigrate because of the political, environmental, and economic attractiveness of a new location.

LEARNING OUTCOME 3.3.1: Provide examples of political, environmental, and economic push and pull factors.

- People migrate because of a combination of push and pull factors. These factors may be political, environmental, and economic. Most people migrate for economic push and pull reasons.

LEARNING OUTCOME 3.3.2: Summarize the flows of migrant workers in Europe and Asia.

- People migrate for temporary work, especially from developing countries to developed countries, where they take jobs that are not desired by local residents.

THINKING GEOGRAPHICALLY 3.3: What factors motivated your family or your ancestors to migrate?

GOOGLE EARTH 3.3: This is Gulfport, Mississippi, on August 29, 2005, just after Hurricane Katrina hit. Set the time slider for July 11, 2005. What evidence of flood damage can be seen in the August 29 image?

KEY ISSUE 4

Why Do Migrants Face Obstacles?

Migrants have difficulty getting permission to enter other countries, and they face hostility from local citizens once they arrive. Immigration laws restrict the number of immigrants who can legally enter the United States. In Europe and Southwest Asia, temporary workers migrate to perform menial jobs.

LEARNING OUTCOME 3.4.1: Identify the types of immigrants who are given preference to enter the United States.

- Immigration is tightly controlled by most countries. The United States gives preference to immigrants with family members already in the country and to those who have special job skills.

LEARNING OUTCOME 3.4.2: Describe the population characteristics of unauthorized immigrants to the United States.

- The United States has more than 11 million unauthorized immigrants, who are in the country without proper documents. Most have emigrated from Mexico.

LEARNING OUTCOME 3.4.3: Describe characteristics of immigrants to the United States.

- In the past, most immigrants were males, but now an increasing share of immigrants to the United States are women and children.

LEARNING OUTCOME 3.4.4: Compare American and European attitudes toward immigrants.

- Americans and Europeans have divided and ambivalent attitudes toward the large number of immigrants, especially those arriving without proper documentation.

THINKING GEOGRAPHICALLY 3.4: Should the United States admit more or fewer immigrants for family reunification, or for job skills, or by random lottery?

GOOGLE EARTH 3.4: Gerrard Street is the center of Chinatown in London, England. Why do you think the street has been closed to vehicular traffic?

Migration transition (p. 79) A change in the migration pattern in a society that results from industrialization, population growth, and other social and economic changes that also produce the demographic transition.

Mobility (p. 78) All types of movement between location.

Net migration (p. 78) The difference between the level of immigration and the level of emigration.

Pull factor (p. 92) A factor that induces people to move to a new location.

Push factor (p. 92) A factor that induces people to leave old residences.

Quotas (p. 96) In reference to migration, laws that place maximum limits on the number of people who can immigrate to a country each year.

Refugees (p. 92) People who are forced to migrate from their home country and cannot return for fear of persecution because of their race, religion, nationality, membership in a social group, or political opinion.

Unauthorized immigrants (p. 98) People who enter a country without proper documents to do so.

Voluntary migration (p. 80) Permanent movement undertaken by choice.

Chapter

4 Folk and Popular Culture

Why is the man in the middle wearing a sweater to work? Page 109

Why are these people posing for the camera? Page 117

KEY ISSUE 1

Where Are Folk and Popular Leisure Activities Distributed?

What We Do For Fun p. 109

Music and sports can be folk or popular. The differences between them involve geography.

KEY ISSUE 2

Where Are Folk and Popular Material Culture Distributed?

Coke or Pepsi? p. 116

We all need food, clothing, and shelter. How we provide for these needs says a lot about our society's culture.

▲ These dancers from India are performing a Kathakali dance drama on the streets of New York as part of a dance festival. The drama depicts stories of Hindu gods Rama and Krishna. Audiences in New York can admire the costumes and the technical skills of the performers, but popular culture in the United States does not relate easily to the meaning of folk culture like this.

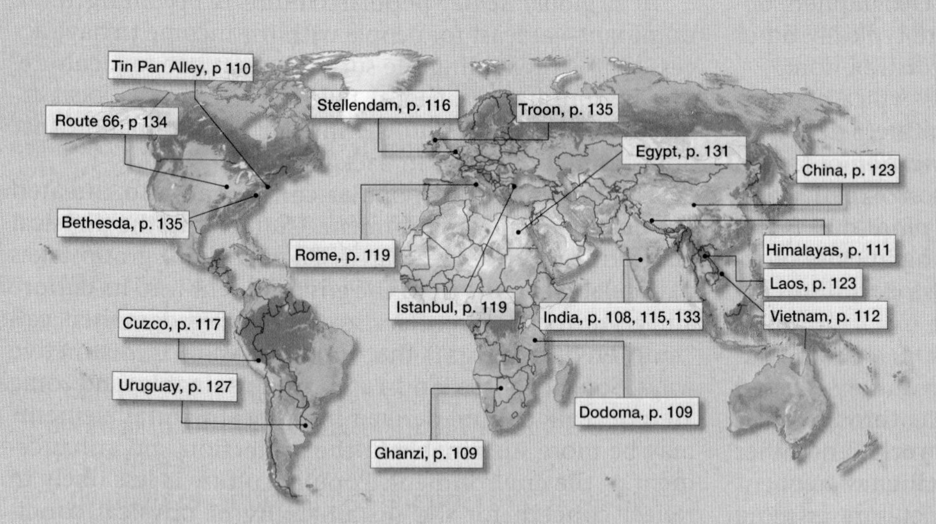

KEY ISSUE 3

Why Is Access to Folk and Popular Culture Unequal?

Accessing the World, if You Can p. 126

Watching TV is the world's most popular leisure activity. What about people who can't afford a TV? Or those who prefer Facebook, Twitter, and YouTube?

KEY ISSUE 4

Why Do Folk and Popular Culture Face Sustainability Challenges?

Don't Throw Away That Culture p. 132

The relentless push of popular culture can threaten the survival of folk culture—and the well-being of the entire planet.

Characteristics of Folk and Popular Culture

Learning Outcome 4.1.1
Compare the origin, diffusion, and distribution of folk and popular culture.

Each social custom has a unique spatial distribution, but in general, distribution is more extensive for popular culture than for folk culture. Two basic factors help explain the different spatial distributions of popular and folk cultures—the process of origin and the pattern of diffusion.

ORIGIN

Culture originates at a *hearth*, a center of innovation. Folk culture often has anonymous hearths, originating from anonymous sources, at unknown dates, through unidentified originators. It may also have multiple hearths, originating independently in isolated locations.

In contrast, popular culture is most often a product of developed countries, especially in North America and Europe. Popular culture is typically traceable to a specific person or corporation in a particular place, whereas folk culture typically has an unknown point of origin. For example, popular music as we know it today originated around 1900. At that time, the main popular musical entertainment in North America and Europe was the variety show, called the music hall in the United Kingdom and vaudeville in the United States. To provide songs for music halls and vaudeville, a music industry was developed in a district of New York that became known as Tin Pan Alley (Figure 4-6).

Popular music and other elements of popular culture, such as food and clothing, arise from a combination of advances in industrial technology and increased leisure time. Industrial technology permits the uniform reproduction of objects in large quantities (iPods, T-shirts, pizzas). Many of these objects help people enjoy leisure time, which has increased as a result of the widespread change in the labor force from predominantly agricultural work to predominantly service and manufacturing jobs.

▼ **FIGURE 4-6 ORIGIN OF POPULAR MUSIC** Shops such as this one in New York City's Tin Pan Alley sold popular music in records and sheet music in the early twentieth century.

DIFFUSION

Compared to popular culture, folk culture is transmitted from one location to another more slowly and on a smaller scale, primarily through relocation diffusion (migration). The spread of popular culture, such as popular music, typically follows the process of hierarchical diffusion, diffusing rapidly and extensively from hearths or nodes of innovation with the help of modern communications.

The diffusion of American popular music worldwide began in earnest during the 1940s, when the Armed Forces Radio Network broadcast music to American soldiers and to citizens of countries where American forces were stationed or fighting during World War II. In the late twentieth century, western dance music diffused rapidly from the United States to Europe, especially Detroit's techno music and Chicago's house music (Figure 4-7). Techno music was heavily influenced by soul, gospel, and ultimately African folk music. House music was heavily influenced by hip-hop that emerged in New York and other urban areas, which in turn diffused from funk, jazz, and again ultimately African folk music.

DISTRIBUTION

Popular culture is distributed widely across many countries, with little regard for physical factors. The distribution is influenced by the ability of people to access the material. The principal obstacle to access is lack of income to purchase the material.

A combination of local physical and cultural factors influences the distinctive distributions of folk culture. For example, in a study of artistic customs in the Himalaya Mountains, geographers P. Karan and Cotton Mather revealed that distinctive views of the physical environment emerge among neighboring cultural groups that are isolated. The study area, a narrow corridor of 2,500 kilometers (1,500 miles) in the Himalaya Mountains of Bhutan, Nepal, northern India, and southern Tibet (China), contains four religious groups: Tibetan Buddhists in the north, Hindus in the south, Muslims in the west, and Southeast Asian animists in the east (Figure 4-8). Despite their spatial proximity, limited interaction among these groups produces distinctive folk customs.

Through their choices of subjects of paintings, each group reveals how its culture mirrors the religions and individual views of the group's environment:

- **Buddhists.** In the northern region Buddhists paint idealized divine figures, such as monks and saints. Some of these figures are depicted as bizarre or terrifying, perhaps reflecting the inhospitable environment.

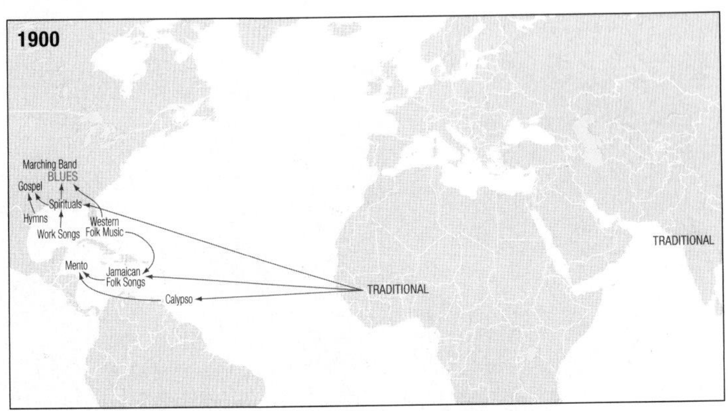

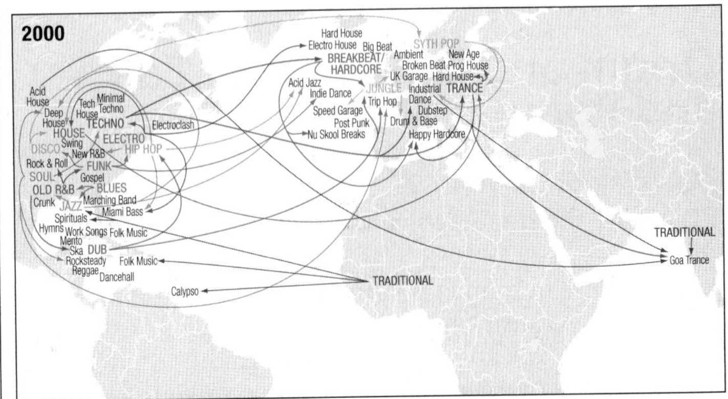

▲ **FIGURE 4-7 DIFFUSION OF WESTERN DANCE MUSIC** Popular dance music originated in the Western Hemisphere and diffused to Europe and Asia during the 1980s.

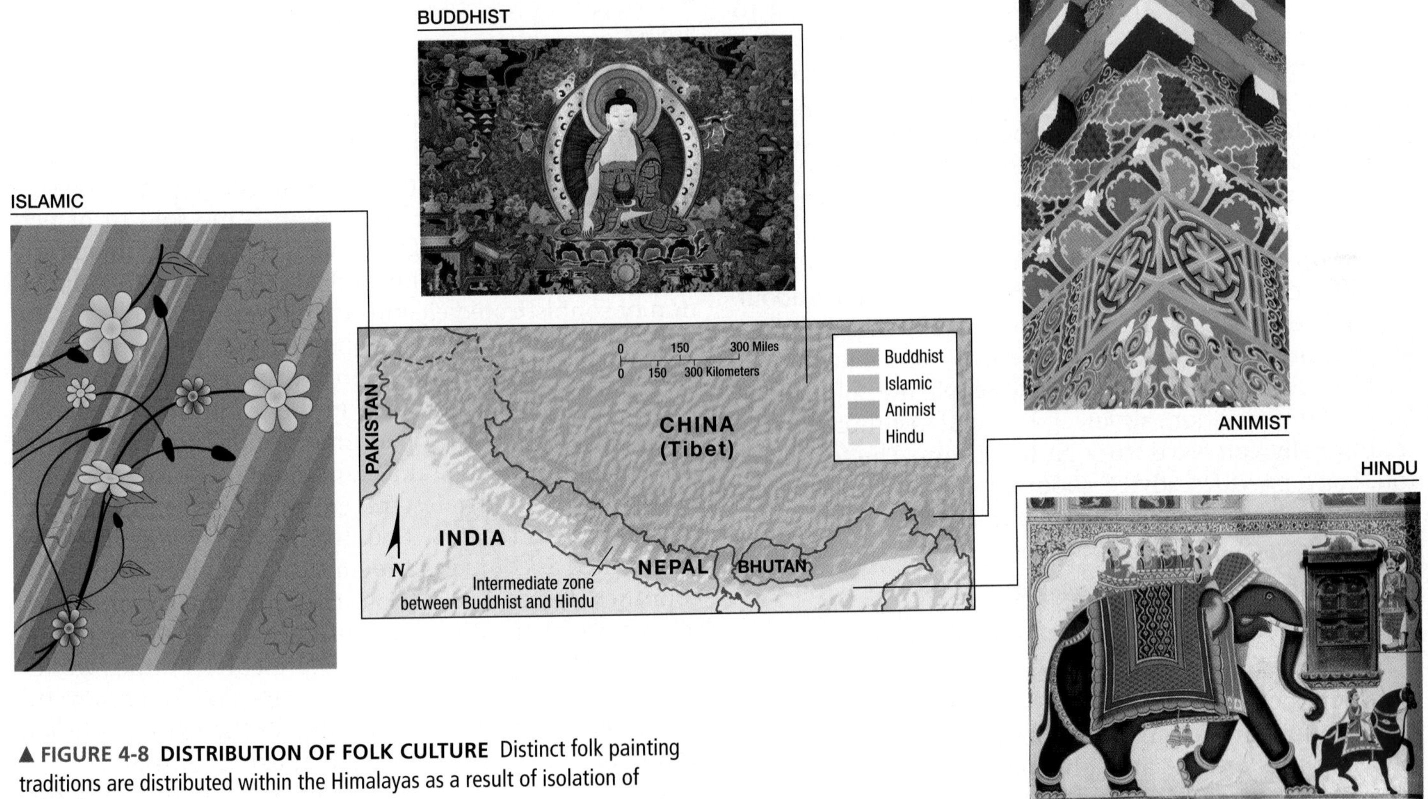

▲ **FIGURE 4-8 DISTRIBUTION OF FOLK CULTURE** Distinct folk painting traditions are distributed within the Himalayas as a result of isolation of cultural groups.

- **Hindus.** In the southern region Hindus create scenes from everyday life and familiar local scenes. Their paintings sometimes portray a deity in a domestic scene and frequently represent the region's violent and extreme climatic conditions.
- **Muslims.** In the western portion Muslims show the region's beautiful plants and flowers because the Muslim faith prohibits displaying animate objects in art. In contrast with the paintings from the Buddhist and Hindu regions, these paintings do not depict harsh climatic conditions.
- **Animists.** Animists from Myanmar (Burma) and elsewhere in Southeast Asia, who have migrated to the eastern region of the study area, paint symbols and designs that derive from their religion rather than from the local environment.

The distribution of artistic subjects in the Himalayas shows how folk customs are influenced by cultural institutions such as religion and by environmental processes such as climate, landforms, and vegetation. These groups display similar uniqueness in their dance, music, architecture, and crafts.

Pause and Reflect 4.1.1

What geographic factors account for the diversity of cultures in the Himalayas?

Origin and Diffusion of Folk and Popular Music

Learning Outcome 4.1.2
Compare the characteristics of folk and popular music.

Every culture in human history has had some tradition of music, argues music researcher Daniel Levitan. As music is a part of both folk and popular culture traditions, it can be used to illustrate the differences in the origin, diffusion, and distribution of folk and popular culture.

FOLK MUSIC

According to a Chinese legend, music was invented in 2697 B.C., when the Emperor Huang Ti sent Ling Lun to cut bamboo poles that would produce a sound matching the call of the phoenix bird. In reality, folk songs usually originate anonymously and are transmitted orally. A song may be modified from one generation to the next as conditions change, but the content is most often derived from events in daily life that are familiar to the majority of the people. As people migrate, folk music travels with them as part of the diffusion of folk culture.

Folk songs may tell a story or convey information about life-cycle events, such as birth, death, and marriage, or environmental features, such as agriculture and climate. For example, in Vietnam, where most people are subsistence farmers, information about agricultural technology was traditionally conveyed through folk songs. The following folk song provides advice about the difference between seeds planted in summer and seeds planted in winter:

> *Ma chiêm ba tháng không già*
> *Ma mùa tháng ruỗi ắt la'không non*[1]

This song can be translated as follows:

> While seedlings for the summer crop are not old when they are three months of age,
> Seedlings for the winter crop are certainly not young when they are one-and-a-half months old.

The song hardly sounds lyrical to a Western ear. But when English-language folk songs appear in cold print, similar themes emerge, even if the specific information conveyed about the environment differs.

Festivals throughout Vietnam feature music in locally meaningful environmental settings, such as hillsides or on water. Singers in traditional clothes sing about elements of daily life in the local village, such as the trees, flowers, and water source (Figure 4-9).

Pause and Reflect 4.1.2
What type of music do you like? Where does it fit in the popular music maps?

[1]From John Blacking and Joann W. Kealiinohomoku, eds., *The Performing Arts: Music and Dance* (The Hague: Mouton, 1979), 144. Reprinted by permission of the publisher.

▲ FIGURE 4-9 **VIETNAMESE FOLK MUSIC** Singers perform Quan Ho folk songs as part of the annual Lim Festival, which is held annually on the 13th to the 15th day of the first lunar month. Quan Ho folk music dates back more than 500 years and is recognized by UNESCO as part of humanity's intangible heritage.

POPULAR MUSIC

In contrast to folk music, popular music is written by specific individuals for the purpose of being sold to or performed in front of a large number of people. It frequently displays a high degree of technical skill through manipulation of sophisticated electronic equipment.

As with other elements of popular culture, popular musicians have more connections with performers of similar styles, regardless of where in the world they happen to live, than they do with performers of different styles who happen to live in the same community. The Landscape of Music project illustrates this point by depicting popular music as a world map, with different types of popular music represented as countries and musicians as places within the countries (Figure 4-10). "Countries" that are closer to each other have relatively similar musical styles. The most important musicians in each style are represented on the "world" map; the larger the size of the type, the more important the musician. Zooming in on a portion of the map reveals less important musicians within individual "countries."

In the past, according to Richard Florida, Charlotta Mellander, and Kevin Stolarick, musicians clustered in particular communities according to shared interest in specific styles, such as Tin Pan Alley in New York, Dixieland jazz in New Orleans, country in Nashville, and Motown in Detroit. Now with the globalization of popular music, musicians are less tied to the culture of particular places and instead increasingly cluster in communities where other creative artists reside, regardless of the particular style. In the United States, New York and Los Angeles attract the largest number of musicians so they can be near sources of employment and cultural activities that attract a wide variety of artists, not just performers of a specific type of music. Nashville is also a leading center for musicians, especially those performing country and gospel; it has the largest concentrations of musicians, when the number of musicians there is compared to a much smaller population than in New York and Los Angeles.

FIGURE 4-10 THE LANDSCAPE OF MUSIC The map shows relations among musicians. Musicians whose styles are closely related are depicted as inhabiting the same region. Musicians whose styles are unrelated are shown as inhabiting regions that are not adjacent to each other. The map was created by Yifan Hu, a researcher at AT&T Labs.

Popular musicians are also increasingly attracted to a handful of large clusters in order to have better access to agencies that book live performances, which have become increasingly important compared to recordings. Nearly all of the music festivals that attract the highest attendance are in Europe and the United States.

Connections between popular musicians are depicted in a transit map. Subway "lines" represent styles of popular music, and "interchanges" represent individuals who cross over between two styles. For example, Kanye West is placed at the interchange between hip-hop and soul, and Jimi Hendrix at the interchange between rock and blues and country (Figure 4-11).

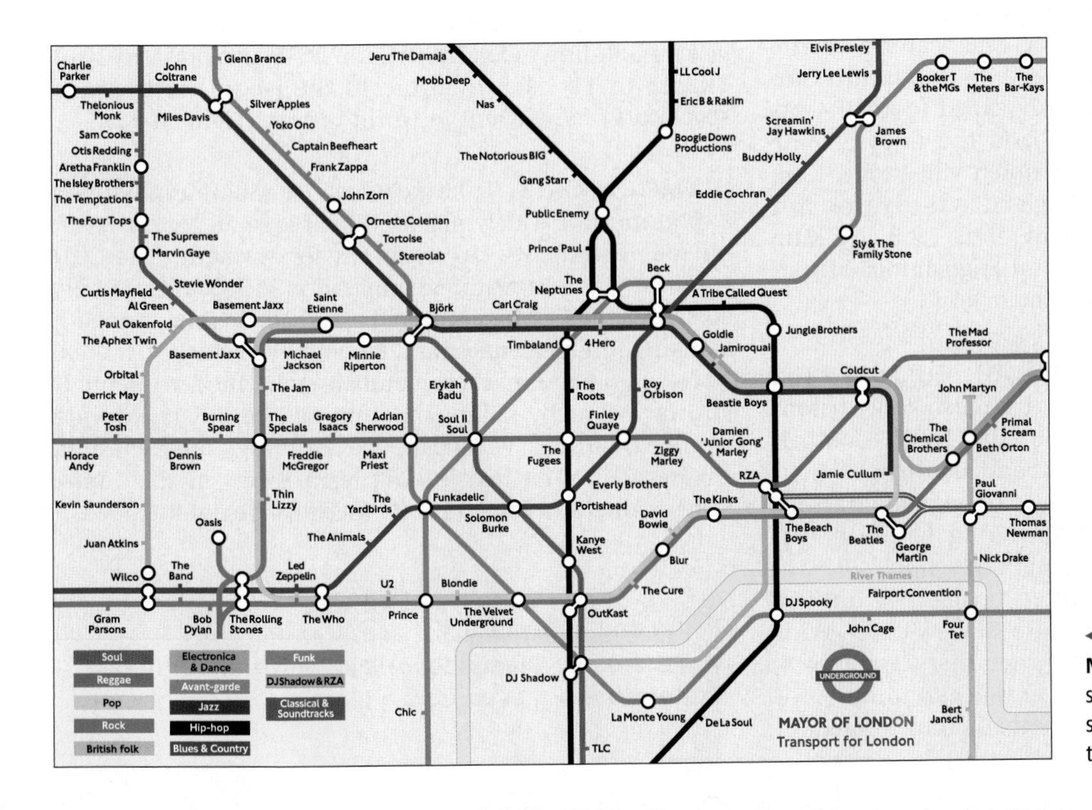

FIGURE 4-11 LONDON SUBWAY MAP OF POPULAR MUSIC This map showing relationships among musical styles is designed to look like the map of the London Underground (subway) system.

Origin and Diffusion of Folk and Popular Sports

Learning Outcome 4.1.3
Describe how sports have been transformed from folk to popular culture.

Many sports originated as isolated folk customs and were diffused like other folk culture, through the migration of individuals. The contemporary diffusion of organized sports, however, displays the characteristics of popular culture.

ORIGIN AND DIFFUSION OF POPULAR SPORTS

Soccer, the world's most popular sport—known in most of the world as football—originated as a folk custom in England during the eleventh century. It was transformed into a part of global popular culture beginning in the nineteenth century.

ORIGIN OF SOCCER: FOLK CULTURE. As with other folk customs, soccer's origin is obscure. The earliest documented contest took place in England in the eleventh century. According to football historians, after the Danish invasion of England between 1018 and 1042, workers excavating a building site encountered a Danish soldier's head, which they began to kick. "Kick the Dane's head" was imitated by boys, one of whom got the idea of using an inflated cow bladder.

Early football games resembled mob scenes. A large number of people from two villages would gather to kick the ball. The winning side was the one that kicked the ball into the center of the rival village. In the twelfth century, the game—by then commonly called football—was confined to smaller vacant areas, and the rules became standardized. Because football disrupted village life, King Henry II banned the game from England in the late twelfth century. It was not legalized again until 1603, by King James I. At this point, football was an English folk custom rather than a global popular custom.

DIFFUSION OF SOCCER: POPULAR CULTURE. The transformation of football from an English folk custom to global popular culture began in the 1800s. Football and other recreation clubs were founded in the United Kingdom, frequently by churches, to provide factory workers with organized recreation during leisure hours. Sport became a subject that was taught in school.

Increasing leisure time permitted people not only to participate in sporting events but also to view them. With higher incomes, spectators paid to see first-class events. To meet public demand, football clubs began to hire professional players. Several British football clubs formed an association in 1863 to standardize the rules and to organize professional leagues. Organization of the sport into a formal structure in the United Kingdom marks the transition of football from folk to popular culture.

The word *soccer* originated after 1863, when supporters of the game formed the Football Association. Association was shortened to *assoc*, which ultimately became twisted around into the word *soccer*. The terms *soccer* and *association football* also helped to distinguish the game from rugby football, which permits both kicking and carrying of the ball. Rugby originated in 1823, when a football player at Rugby School (in Rugby, England) picked up the ball and ran with it.

Beginning in the late 1800s, the British exported association football around the world, first to continental Europe and then to other countries. For example:

- **The Netherlands.** Dutch students returning from studies in the United Kingdom were the first to play football in continental Europe in the late 1870s.
- **Spain.** Miners in Bilbao adopted the sport in 1893, after seeing it played by English engineers working there.
- **Russia.** The English manager of a textile factory near Moscow organized a team at the factory in 1887 and advertised in London for workers who could play football. After the Russian Revolution in 1917, both the factory and its football team were absorbed into the Soviet Electric Trade Union. The team, renamed the Moscow Dynamo, became the country's most famous football team.

British citizens further diffused the game throughout the worldwide British Empire. In the twentieth century, soccer, like other sports, was further diffused by new communication systems, especially radio and TV.

The global popularity of soccer is seen in the World Cup, in which national soccer teams compete every four years, including in South Africa in 2010 and Brazil in 2014. Thanks to TV, each final breaks the record for the most spectators of any event in world history (Figure 4-12).

OLYMPIC SPORTS. To be included in the Summer Olympics, a sport must be widely practiced in at least 75 countries and on four continents (50 countries for women). The 2016 Summer Olympics features competition in 28 sports: archery, aquatics, athletics, badminton, basketball, boxing, canoeing/kayaking, cycling, equestrian, fencing, field hockey, football (soccer), golf, gymnastics, handball, judo, modern pentathlon, rowing, rugby, sailing, shooting, table tennis, taekwondo, tennis, triathlon, volleyball, weightlifting, and wrestling (Figure 4-13). The two leading team sports in the United States—American football and baseball—are not included.

Pause and Reflect 4.3.1
Are there any Olympic sports in which the United States does not even field a team?

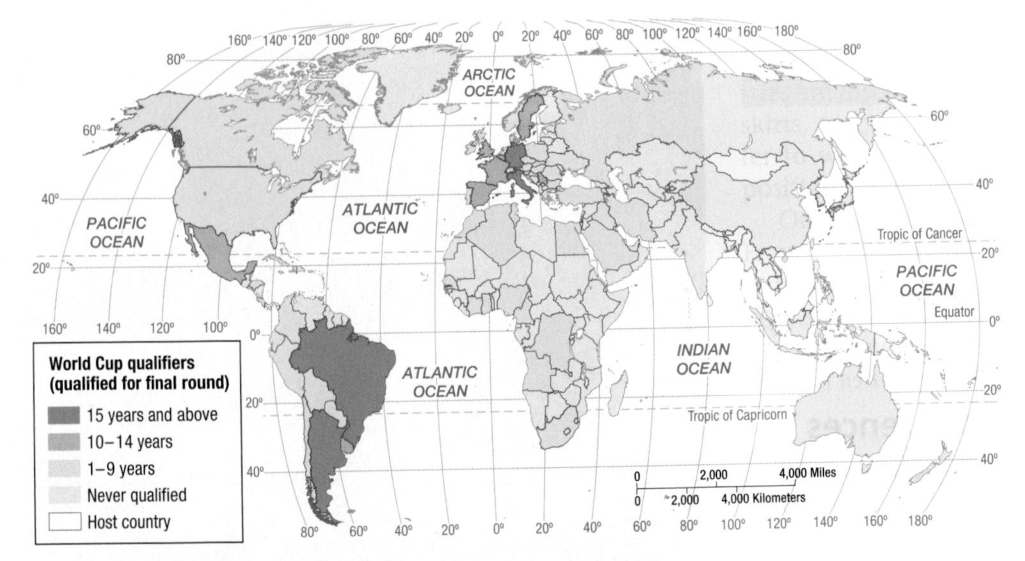

▲ **FIGURE 4-12 GLOBAL SPORTS: WORLD CUP** Most countries in Europe and Latin America have qualified for the World Cup finals. As soccer diffuses worldwide, qualification has increased in other regions.

- Australia rules football is a sport distinct from soccer and the football played in North America. Distinctive forms of football developed in Australia, as well as the United States and Canada, as a result of lack of interaction among sporting nations during the nineteenth century.

Despite the diversity in distribution of sports across Earth's surface and the anonymous origin of some games, organized spectator sports today are part of popular culture. The common element in professional sports is the willingness of people throughout the world to pay for the privilege of viewing, in person or on TV, events played by professional athletes.

▲ **FIGURE 4-13 GLOBAL SPORTS: OLYMPICS** Athletes from Turkey (in blue gear) and the United States (in red gear) compete in women's taekwondo at the 2012 Summer Olympic Games in London, United Kingdom.

CHECK-IN: KEY ISSUE 1

Where Are Folk and Popular Leisure Activities Distributed?

✓ **Folk culture and popular culture have distinctive patterns of origin, diffusion, and distribution.**

✓ **Folk leisure activities typically have anonymous origins, diffuse through relocation diffusion, and have limited distribution.**

✓ **Popular leisure activities typically originate with identifiable individuals or corporations, diffuse rapidly through hierarchical diffusion, and have widespread distribution.**

SURVIVING FOLK SPORTS

Most other sports have diffused less than soccer. Cultural groups still have their own preferred sports, which are often unintelligible to people elsewhere. Consider the following:

- Cricket is popular primarily in the United Kingdom and former British colonies, especially in South Asia, the South Pacific, and Caribbean islands (Figure 4-14).
- Wushu, martial arts that combine forms such as kicking and jumping with combat such as striking and wrestling, are China's most popular sports.
- Baseball, once confined to North America, became popular in Japan after it was introduced by American soldiers who occupied the country after World War II.

▲ **FIGURE 4-14 CRICKET** Boys play cricket in Katni, India.

Folk and Popular Food Preferences

Learning Outcome 4.2.2

Understand reasons for folk food preferences and taboos.

According to the nineteenth-century cultural geographer Vidal de la Blache, "Among the connections that tie [people] to a certain environment, one of the most tenacious is food supply; clothing and weapons are more subject to modification than the dietary regime, which experience has shown to be best suited to human needs in a given climate."

Food preferences are inevitably affected by the availability of products, but people do not simply eat what is available in their particular environment. Food preferences are strongly influenced by cultural traditions. What is eaten establishes one's social, religious, and ethnic memberships. The surest good way to identify a family's ethnic origins is to look in its kitchen.

FOLK FOOD CUSTOMS AND THE ENVIRONMENT

Folk food habits are embedded especially strongly in the environment. Humans eat mostly plants and animals—living things that spring from the soil and water of a region. Inhabitants of a region must consider the soil, climate, terrain, vegetation, and other characteristics of the environment in deciding to produce particular foods.

FOOD AND PLACE: THE CONCEPT OF TERROIR. The contribution of a location's distinctive physical features to the way food tastes is known by the French term **terroir**. The word comes from the same root as *terre* (the French word for "land" or "earth"), but terroir does not translate precisely into English; it has a similar meaning to the English expressions "grounded" and "sense of place." Terroir is the sum of the effects on a particular food item of soil, climate, and other features of the local environment.

FOODS TO CRAVE OR AVOID. In folk cultures, certain foods are eaten because their natural properties are perceived to enhance qualities considered desirable by the society. Here are some examples:

- The Abipone people in Paraguay eat jaguars, stags, and bulls to make them strong, brave, and swift. The Abipone believe that consuming hens or tortoises will make them cowardly.
- The Ainu people in Japan avoid eating otters because they are believed to be forgetful animals, and consuming them could cause loss of memory.
- The Mbum Kpau women in Chad do not eat chicken or goat before becoming pregnant. Abstaining from consumption of these animals is thought to help escape pain in childbirth and to prevent birth of a child with abnormalities. During pregnancy, the Mbum Kpau avoid meat from antelopes with twisted horns, which could cause them to bear offspring with deformities.

FOODS FROM A GARDEN: THE BOSTANS OF ISTANBUL. Bostans, which are small gardens inside Istanbul, Turkey, have been supplying the city with fresh produce for hundreds of years (Figure 4-18). According to geographer Paul Kaldjian, Istanbul has around 1,000 bostans, run primarily by immigrants from Cide, a rural village in Turkey's Kastamonu province. Bostan farmers are able to maximize yields from their small plots of land (typically 1 hectare) through what Kaldjian calls clever and efficient manipulation of space, season, and resources. In a bostan, 15 to 20 different types of vegetables are planted at different times of the year, and the choice is varied from year to year, in order to reduce the risk of damage from poor weather. Most of the work is done by older men, who prepare beds for planting, sow, irrigate, and operate motorized equipment, according to Kaldjian. Women weed, and both men and women harvest.

FOODS AND THE ENVIRONMENT. People adapt their food preferences to conditions in the environment. In Asia, rice is grown in milder, moister regions; wheat thrives in colder, drier regions. In Europe, traditional preferences for quick-frying foods in Italy resulted in part from fuel shortages. In Northern Europe, an abundant wood supply encouraged the slow stewing and roasting of foods over fires, which also provided home heat in the colder climate.

Soybeans, an excellent source of protein, are widely grown in Asia. In the raw state they are toxic and indigestible. Lengthy cooking renders them edible, but fuel is scarce in Asia. Asians have adapted to this environmental challenge by deriving from soybeans foods that do not require extensive cooking. These include bean sprouts (germinated seeds), soy sauce (fermented soybeans), and bean curd (steamed soybeans).

FOOD TABOOS

According to many folk customs, everything in nature carries a signature, or distinctive characteristic, based on its appearance and natural properties. Consequently, people may desire or avoid certain foods in response to perceived beneficial or harmful natural traits.

People refuse to eat particular plants or animals that are thought to embody negative forces in the environment. Such a restriction on behavior imposed by social custom is a **taboo**. Other social customs, such as sexual practices, carry prohibitions, but taboos are especially strong in the area of food. Some folk cultures may establish food taboos because of concern for the natural environment. These taboos may help to protect endangered animals or to conserve scarce natural resources. To preserve scarce animal species, only a

▲ **FIGURE 4-18 ISTANBUL VEGETABLE GARDEN** Geographer Paul Kaldjian sketched a typical bostan, a traditional vegetable garden in the center of Istanbul, Turkey. Bostans provide residents of the large city of Istanbul with a source of fresh vegetables.

few high-ranking people in some tropical regions are permitted to hunt, and the majority cultivate crops.

Relatively well-known taboos against consumption of certain foods can be found in the Bible. The ancient Hebrews were prohibited from eating a wide variety of foods, including animals that do not chew their cud or that have cloven feet and fish lacking fins or scales (Figure 4-19). These taboos arose partially from concern for the environment by the Hebrews, who lived as pastoral nomads in lands bordering the eastern Mediterranean. The pig, for example, is prohibited in part because it is more suited to sedentary farming than pastoral nomadism and in part because its meat spoils relatively quickly in hot climates, such as the Mediterranean. These biblical taboos were developed through oral tradition and by rabbis into the kosher laws observed today by some Jews.

Similarly, Muslims embrace the taboo against pork because pigs are unsuited for the dry lands of the Arabian Peninsula. Pigs would compete with humans for food and water, without offering compensating benefits, such as being able to pull a plow, carry loads, or provide milk and wool. Widespread raising of pigs would be an ecological disaster in Islam's hearth.

Hindu taboos against consuming cattle can also be partly explained by environmental reasons. Cows are the source of oxen (castrated male bovine), the traditional choice for pulling plows as well as carts. A large supply of oxen must be maintained in India because every field has to be plowed at approximately the same time—when the monsoon rains arrive. Religious sanctions have kept India's cattle population large as a form of insurance against the loss of oxen and increasing population.

But the taboo against consumption of meat among many people, including Muslims, Hindus, and Jews, cannot be explained primarily by environmental factors. Social values must influence the choice of diet because people in similar climates and with similar levels of income consume different foods. The biblical food taboos were established in part to set the Hebrew people apart from others. That Christians ignore the biblical food injunctions reflects their desire to distinguish themselves from Jews. Furthermore, as a universalizing religion, Christianity was less tied to taboos that originated in the Middle East (see Chapter 6).

▼ **FIGURE 4-19 KOSHER RESTAURANT, ROME**

Pause and Reflect 4.2.2

What foods do you avoid? Do you avoid foods because of taboos or for other reasons?

POPULAR FOOD CULTURE

Learning Outcome 4.2.3
Describe regional variations in popular food preferences.

In the popular culture of twenty-first century America, food preferences seem far removed from folk traditions. Popular food preferences are influenced more by cultural values than by environmental features. Still, some regional variations can be observed, and environmental influences remain important in selected items.

DIFFERENCES AMONG COUNTRIES. Why do Coca-Cola and Pepsi have different sales patterns (Figure 4-20)? The two beverages are similar, and many people are unable to taste the difference. Yet consumers prefer Coke in some countries and Pepsi in others.

Coca-Cola accounts for more than one-half of the world's cola shares, and Pepsi for another one-fourth. Coca-Cola is the sales leader in most of the Western Hemisphere. The principal exception is Canada's French-speaking province of Québec, where Pepsi is preferred. Pepsi won over the Québécois with advertising that tied Pepsi to elements of uniquely French Canadian culture. The major indoor arena in Québec City is named the Colisée Pepsi (Pepsi Coliseum).

Cola preferences are influenced by politics in Russia. Under communism, government officials made a deal with Pepsi to allow that cola to be sold in the Soviet Union. With the breakup of the Soviet Union and the end of communism, Coke entered the Russian market. Russians quickly switched their preference to Coke because Pepsi was associated with the discredited Communist government.

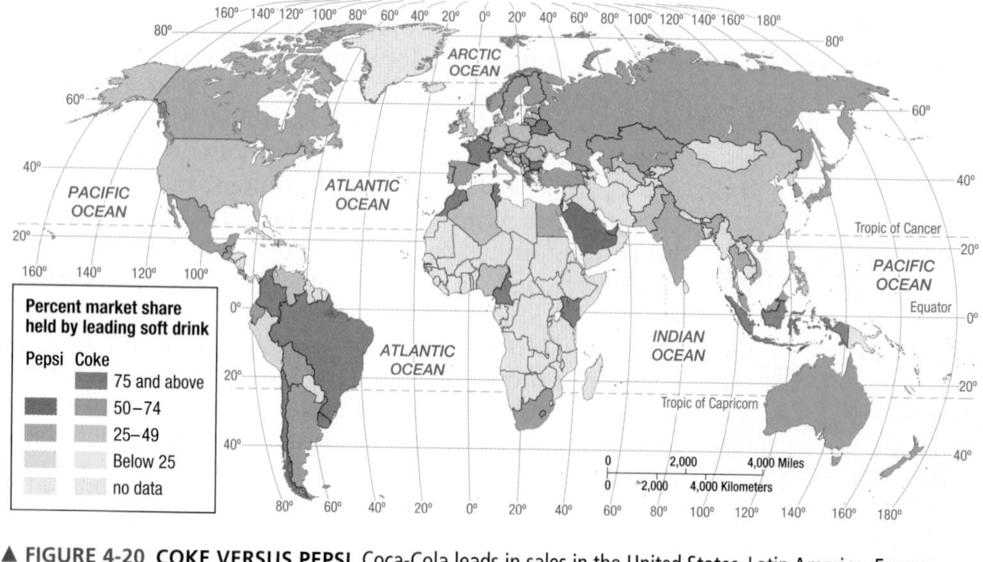

▲ **FIGURE 4-20 COKE VERSUS PEPSI** Coca-Cola leads in sales in the United States, Latin America, Europe, and Russia. Pepsi leads in Canada and South and Southwest Asia.

In Southwest Asia, religion influences cola preferences. At one time, the region's predominantly Muslim countries boycotted products that were sold in predominantly Jewish Israel. Because Coke but not Pepsi was sold in Israel, in most of Israel's neighbors Pepsi was preferred.

REGIONAL DIFFERENCES WITHIN THE UNITED STATES. Some of the leading franchised fast-food restaurants display regional variations in popularity. Lexicalist (www.lexicalist.com) reads through millions of words on Twitter and other social media to see what kind of words are being used throughout the United States. Based on the frequency of referring to particular fast-food restaurants, Lexicalist concludes that Dunkin' Donuts is especially popular in the Northeast, Krispy Kreme in the Southeast, White Castle in the Midwest, and In-N-Out Burger in the Southwest (Figure 4-21).

Americans may choose particular beverages or snacks in part on the basis of preference for what is produced, grown, or imported locally:

- Wine consumption is relatively high in California, where most of the U.S. production is concentrated, and beer consumption is relatively low there. Beer and spirits consumption are relatively high in the upper Midwest, where much of the grain is grown. Consumption of wine is low in that part of the country, where few grapes are grown.
- Southerners may prefer pork rinds because more hogs are raised there, and northerners may prefer popcorn and potato chips because more corn and potatoes are grown there.

Cultural backgrounds affect the amount and types of alcohol and snack foods consumed:

- Utah has a low rate of consumption of all types of alcohol because of a concentration there of members of the Church of Jesus Christ of Latter-day Saints, who abstain from all alcohol consumption. The adjacent state of Nevada has a high rate of consumption of all types of alcohol because of the heavy concentration of gambling and other resort activities there. Alcohol consumption relates partially to religious backgrounds and partially to income and advertising.
- Texans may prefer tortilla chips because of the large number of Hispanic Americans there. Westerners may prefer multigrain chips because of greater concern for the nutritional content of snack foods.

Geographers cannot explain most of the regional variations in food preferences. Variations

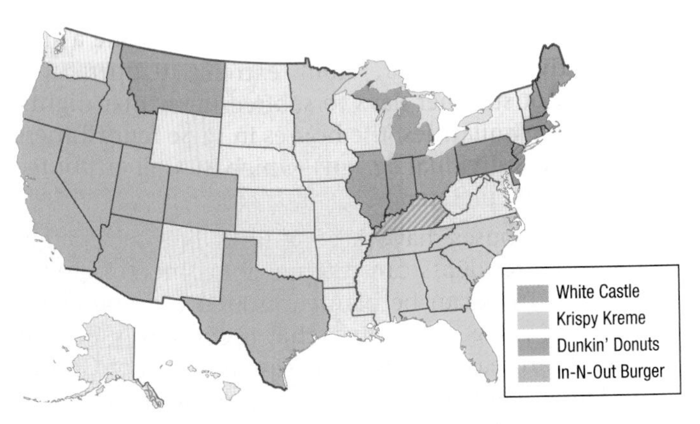

▲ **FIGURE 4-21 FAST-FOOD RESTAURANT PREFERENCES** Dunkin' Donuts is especially popular in the Northeast, Krispy Kreme in the Southeast, White Castle in the Midwest, and In-N-Out Burger in the Southwest.

within the United States are much less significant than differences between the United States and developing countries in Africa and Asia.

Pause and Reflect 4.2.3
Do your food preferences match the predominant ones in your region?

WINE PRODUCTION: ENVIRONMENTAL FACTORS. The spatial distribution of wine production demonstrates that environmental factors can be of some influence in the distribution of popular food customs. The distinctive character of a wine derives from a vineyard's terroir—the unique combination of soil, climate, and other physical characteristics at the place where the grapes are grown:

- **Climate.** Vineyards are best cultivated in temperate climates of moderately cold, rainy winters and fairly long, hot summers. Hot, sunny weather is necessary in the summer for the fruit to mature properly, whereas winter is the preferred season for rain because plant diseases that cause the fruit to rot are more active in hot, humid weather.
- **Topography.** Vineyards are planted on hillsides, if possible, to maximize exposure to sunlight and to facilitate drainage. A site near a lake or river is also desirable because water can temper extremes of temperature.
- **Soil.** Grapes can be grown in a variety of soils, but the best wine tends to be produced from grapes grown in soil that is coarse and well drained—a soil that is not necessarily fertile for other crops.

WINE PRODUCTION: CULTURAL FACTORS. Although grapes can be grown in a wide variety of locations, the production of wine is based principally on cultural values, both historical and contemporary. The distribution of wine production shows that the diffusion of popular customs depends less on the distinctive environment of a location than on the presence of beliefs, institutions, and material traits conducive to accepting those customs (Figure 4-22). Wine is made today primarily in locations that have a tradition of excellence in making it and people who like to drink it and can afford to purchase it.

The social custom of wine production in much of France and Italy extends back at least to the Roman Empire. Wine consumption declined after the fall of Rome, and many vineyards were destroyed. Monasteries preserved the wine-making tradition in medieval Europe for both sustenance and ritual. Wine consumption has become extremely popular again in Europe in recent centuries, as well as in the Western Hemisphere, which was colonized by Europeans. Vineyards are now typically owned by private individuals and corporations rather than religious organizations.

Wine production is discouraged in regions of the world dominated by religions other than Christianity. Hindus and Muslims in particular avoid alcoholic beverages. Thus wine production is limited in the Middle East (other than Israel) and southern Asia primarily because of cultural values, especially religion.

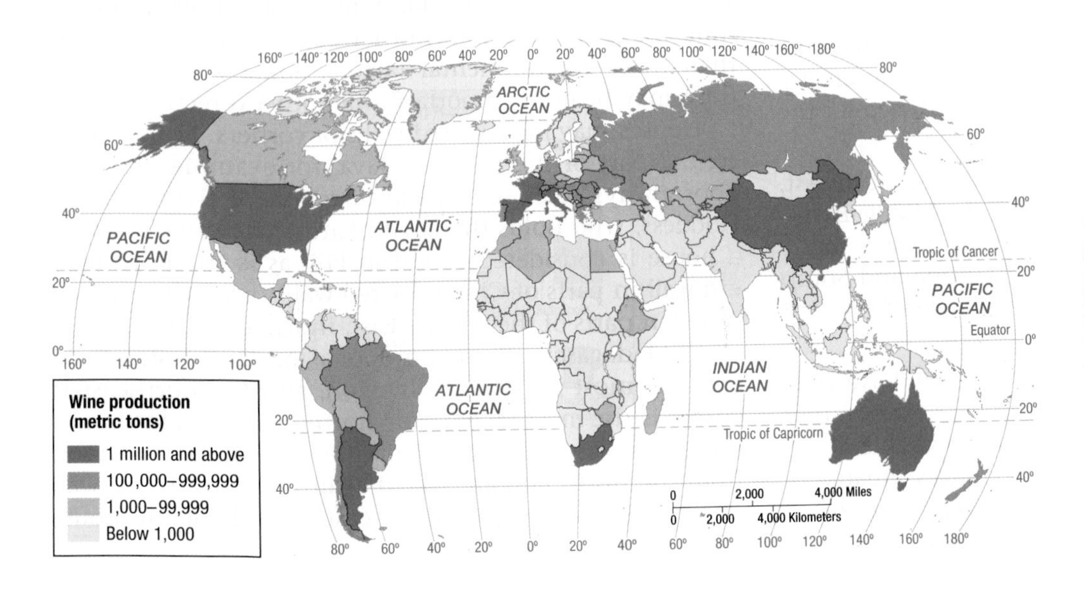

◀ **FIGURE 4-22 WINE PRODUCTION** The distribution of wine production is influenced in part by the physical environment and in part by social customs. Most grapes used for wine are grown near the Mediterranean Sea or in areas of similar climate. Income, preferences, and other social customs also influence the distribution of wine consumption, as seen in the lower production levels of predominantly Muslim countries south of the Mediterranean.

KEY ISSUE 3

Why Is Access to Folk and Popular Culture Unequal?

- **Electronic Diffusion of Popular Culture**
- **Challenges in Accessing Electronic Media**

Learning Outcome 4.3.1
Describe the origin, diffusion, and distribution of TV around the world.

Popular culture diffuses rapidly around the world, primarily through electronic media. The latest fashions in material culture and leisure activities can be viewed by anyone in the world who has access to one or more forms of electronic media. Electronic media increase access to popular culture for people who embrace folk culture and at the same time increase access to folk culture for people who are part of the world's popular culture scene.

The principal obstacle to popular culture is lack of access to electronic media. Access is limited primarily by lack of income. In some developing countries access is also limited by lack of electricity, cell phone service, and other electronic media.

Electronic Diffusion of Popular Culture

The world's most important electronic media format by far is TV. TV supplanted other formats, notably radio and telegraph, during the twentieth century. Into the twenty-first century, other formats have become popular, but they have not yet supplanted TV worldwide.

Watching TV remains especially important for popular culture for two reasons:

- Watching TV is the most popular leisure activity in the world. The average human watched 3.1 hours of TV per day in 2009, and the average American watched 4.6 hours.
- TV has been the most important mechanism by which popular culture, such as professional sports, rapidly diffuses across Earth.

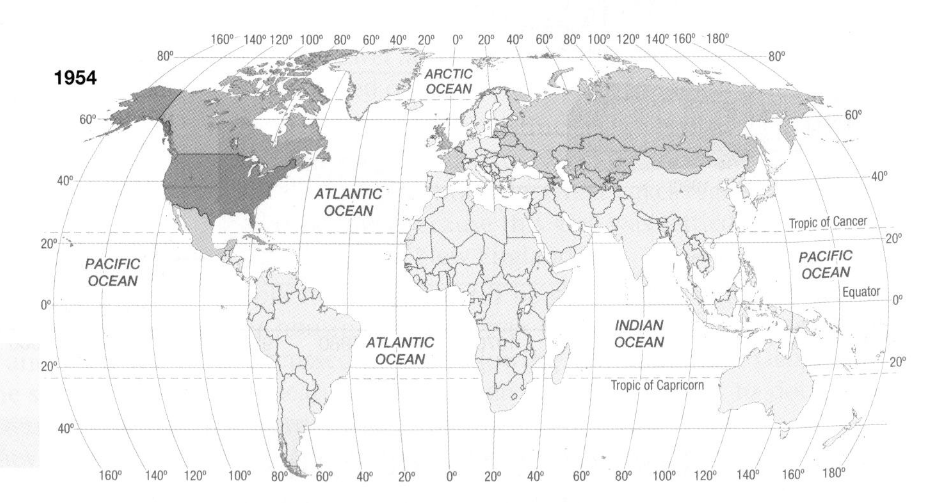

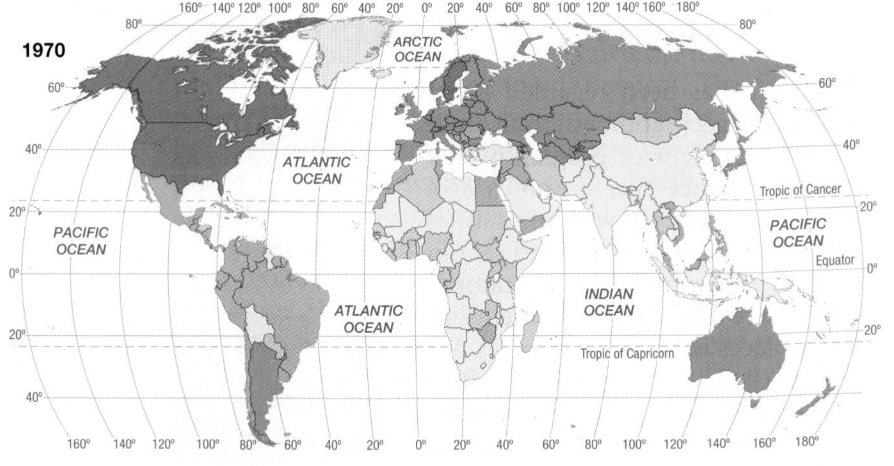

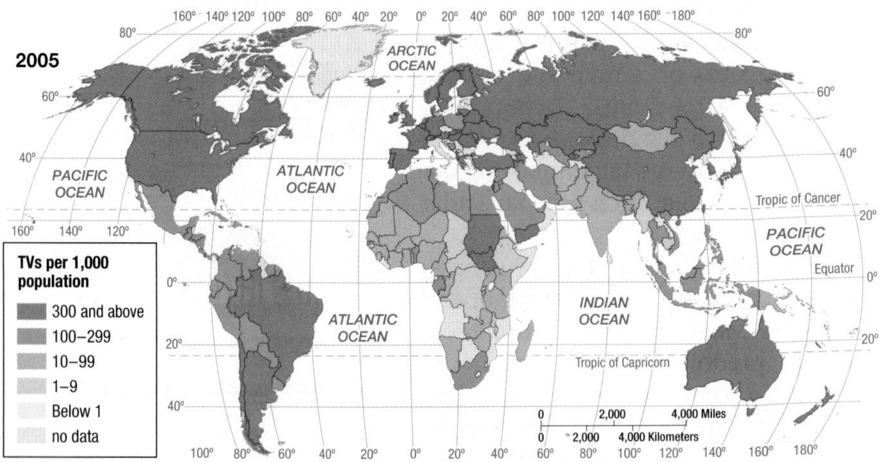

▲ **FIGURE 4-28 DIFFUSION OF TV** Televisions per 1,000 inhabitants in (top) 1954, (middle) 1970, and (bottom) 2005. Television has diffused from North America and Europe to other regions of the world. The United States and Canada had far more TV sets per capita than any other country as recently as the 1970s, but several European countries now have higher rates of ownership.

▲ **FIGURE 4-29 TV HEARTH** One of the first experimental TV broadcasts was by German engineers at the 1936 Olympics in Berlin.

DIFFUSION OF TV: MID-TWENTIETH CENTURY

Through the second half of the twentieth century, television diffused from the United States to Europe and other developed countries and then to developing countries (Figure 4-28):

- **Early twentieth century: Multiple hearths.** Television technology was developed simultaneously in the United Kingdom, France, Germany, Japan, and the Soviet Union, as well as in the United States, though in the early years of broadcasting the United States held a near monopoly (Figure 4-29).
- **Mid-twentieth century: United States dominates.** In 1954, the first year that the United Nations published data on the subject, the United States had 86 percent of the world's 37 million TV sets.
- **Late twentieth century: Diffusion to Europe.** Rapid growth of ownership in Europe meant that the share of the world's sets in the United States declined to one-fourth. Still, in 1970, half of the countries in the world, including most of those in Africa and Asia, had little if any TV broadcasting.
- **Early twenty-first century: Near-universal access.** By 2005, ownership rates climbed sharply in many developing countries, diminishing international differences (Figure 4-30).

Despite diffusion of TV sets around the world, the United States remains the country where people are most likely to watch it. According to the U.S. Time Use Survey, the average American male spent around 7 hours on leisure and recreation in a typical weekend in 2010, and TV watching took up 51 percent of the time. Women spent around 6 hours on leisure in a typical weekend and watched TV for 49 percent of the time (Figure 4-31).

Although people around the world spend a lot of time watching TV, they don't all watch the same programs. Sports are the most popular programs in North America, entertainment programs such as reality shows in most of Europe and China, fictional programs in South Asia, and news programs in Russia.

▲ **FIGURE 4-30 TV DIFFUSES WORLDWIDE** Uruguayan fans watch Uruguay play South Korea in 2010 World Cup.

The technology by which TV is delivered to viewers has changed. Between 2006 and 2013, the share of viewers around the world receiving programs over the air declined from 44 percent to 33 percent, and the share using cable increased modestly, from 35 percent to 37 percent. On the other hand, the share receiving programs through a satellite dish increased from 20 percent to 26 percent, and the share receiving TV programs through the Internet increased from less than 1 percent to 5 percent.

Pause and Reflect 4.3.1

How much TV do you watch? Which types of programs do you watch? Do you watch on a traditional TV set, or do you watch on a computer, tablet, or smartphone?

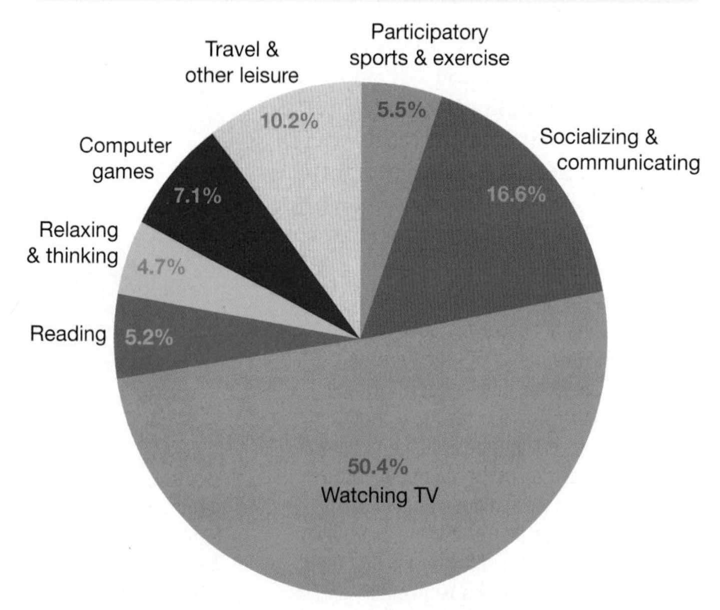

▲ **FIGURE 4-31 HOW AMERICANS SPEND THEIR WEEKENDS** Watching TV is by far the most common leisure activity for Americans.

DIFFUSION OF THE INTERNET: LATE TWENTIETH CENTURY

Learning Outcome 4.3.2
Compare the diffusion of the Internet and social media with the diffusion of TV.

The diffusion of Internet service follows the pattern established by television a generation earlier, but at a more rapid pace (Figure 4-32):

- In 1995, there were 40 million Internet users worldwide, including 25 million in the United States, and Internet service had not yet reached most countries.
- Between 1995 and 2000, Internet usage increased rapidly in the United States, from 9 percent to 44 percent of the population. But the worldwide increase was much greater, from 40 million Internet users in 1995 to 361 million in 2000. As Internet usage diffused rapidly, the U.S. percentage share declined rapidly in five years, from 62 to 31 percent.
- Between 2000 and 2011, Internet usage continued to increase rapidly in the United States, to 77 percent of the population. Again, the increase was more modest than in the rest of the world, and the share of the world's Internet users found in the United States continued to decline, to 10 percent in 2011.

Note that all six maps in Figures 4-28 and 4-32 use the same intervals. For example, the highest class in all maps is 300 or more per 1,000. What is different is the time interval period. The diffusion of television from the United States to the rest of the world took a half-century, whereas the diffusion of the Internet took only a decade. Given the history of television, the Internet is likely to diffuse further in the years ahead at a rapid rate (Figure 4-33).

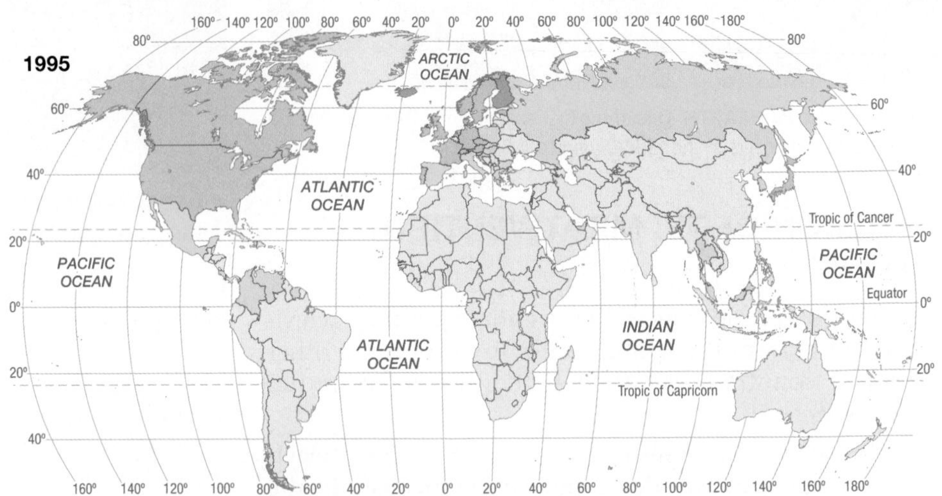

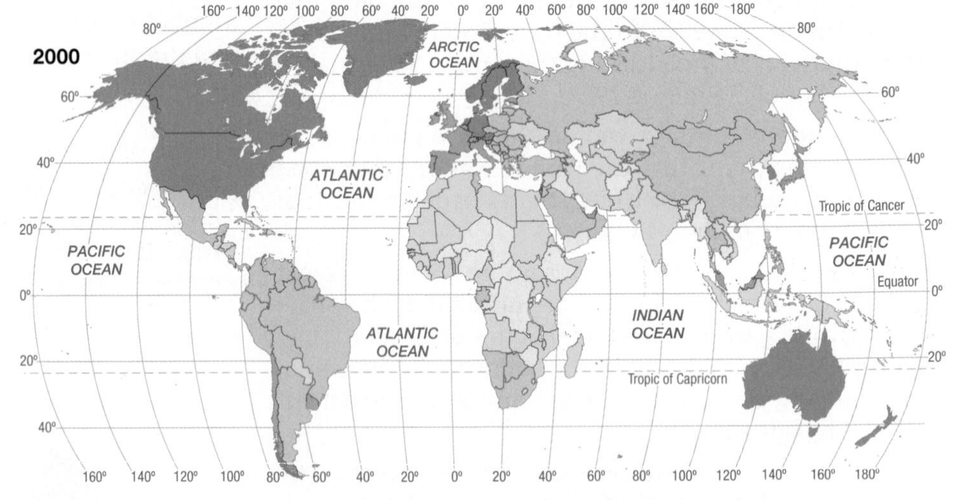

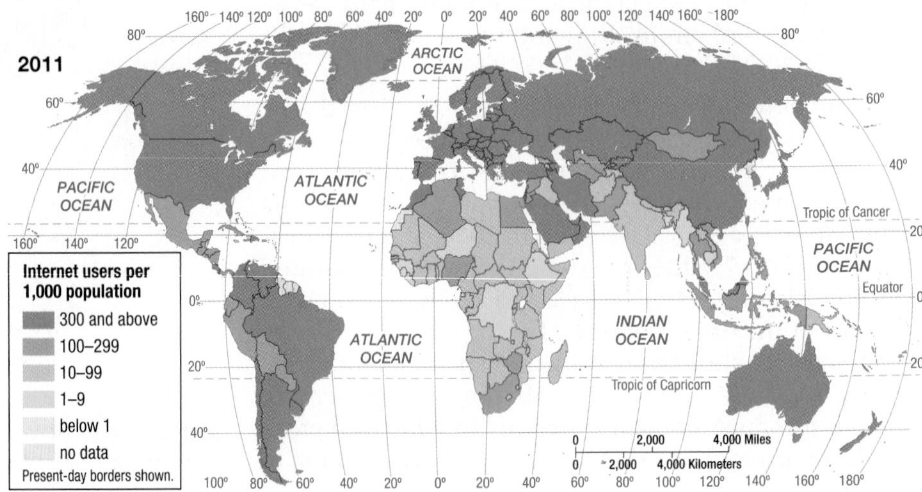

▶ **FIGURE 4-32 DIFFUSION OF THE INTERNET** Internet users per 1,000 inhabitants in (top) 1995, (middle) 2000, and (bottom) 2011. Compare to the diffusion of TV (Figure 4-28). Internet service is following a pattern in the twenty-first century similar to the pattern of diffusion of television in the twentieth century. The United States started out with a much higher rate of usage than elsewhere, until other countries caught up. The difference is that the diffusion of television took a half-century and the diffusion of the Internet only a decade.

DIFFUSION OF SOCIAL MEDIA: TWENTY-FIRST CENTURY

The familiar pattern has repeated in the twenty-first century. People based in the United States have dominated the

▲ **FIGURE 4-33 DIFFUSION OF THE INTERNET TO INDIA** Access to the Internet is available even in many rural areas of many LDCs.

use of social media during the early years. In the future, will U.S. dominance be reduced and perhaps disappear altogether, as occurred in the twentieth century with TV?

DIFFUSION OF FACEBOOK. Facebook, founded in 2004 by Harvard University students, has begun to diffuse rapidly. As with the first few years of TV and the Internet, once again the United States started out with far more Facebook users than any other country. In 2008, four years after Facebook's founding, the United States had one-third of all users worldwide. As Facebook has diffused to other countries, the share of users in the United States has declined, to one-fifth of the worldwide total in 2011 (Figure 4-34). In the years ahead, Facebook is likely to either diffuse to other parts of the world or be overtaken by other forms of electronic social networking and be relegated to a footnote in the continuous repeating pattern of diffusing electronic communications.

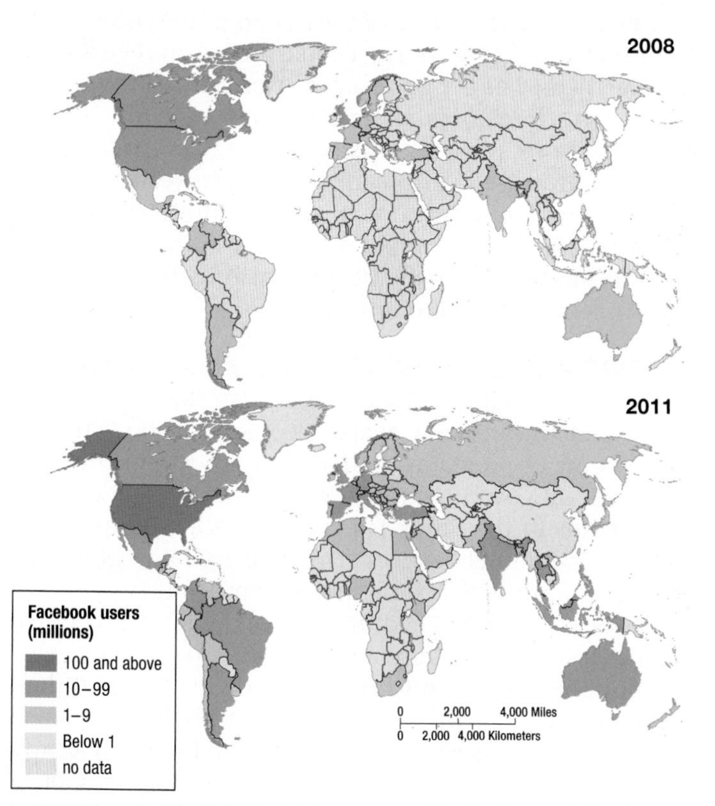

▲ **FIGURE 4-34 DIFFUSION OF FACEBOOK** Facebook users in (top) 2008 and (bottom) 2011.

DIFFUSION OF TWITTER. The United States was the source of one-third of all Twitter messages in 2011. Another one-third originated in six other countries—India, Japan, Germany, the United Kingdom, Brazil, and Canada (Figure 4-35). In the case of Twitter, the second leading Twitter country is one of the world's poorest, India. This may be a preview of future trends, in which electronic communications advances diffuse rapidly to developing countries, not just to other developed countries.

Americans or U.S.-based sources dominate the most popular Twitter postings. Nineteen of the 20 Twitter posters with the largest followings in 2010 were American, led by Ashton Kutcher, Britney Spears, Ellen DeGeneres, Barack Obama, and Lady Gaga. The only exception in the top 20 in 2010 was the UK band Coldplay.

DIFFUSION OF YOUTUBE. Again, the United States accounted for 30 percent of worldwide users in the early years of YouTube. Seventeen other countries, mostly in Europe, accounted for the remainder in 2011. Most countries of the world did not have YouTube users as of 2011 (Figure 4-36).

Pause and Reflect 4.3.2
Which social media do you prefer to use? Why?

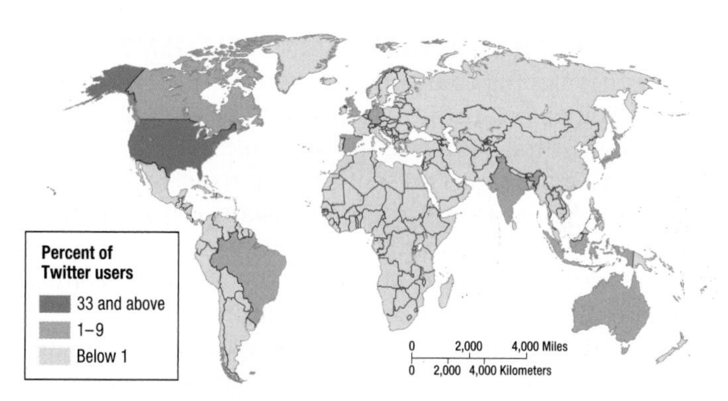

▲ **FIGURE 4-35 DISTRIBUTION OF TWITTER USERS** More than one-third of the world's Twitter users were in the United States in 2011.

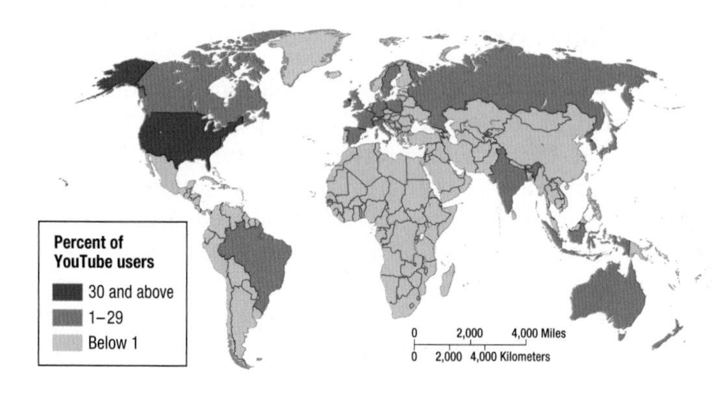

▲ **FIGURE 4-36 DISTRIBUTION OF YOUTUBE USERS** Nearly one-third of the world's YouTube users were in the United States in 2011.

Challenges in Accessing Electronic Media

Learning Outcome 4.3.3

Understand external and internal threats to folk culture posed by electronic media.

People in developing countries who embrace folk culture are challenged by the diffusion of popular culture through electronic media. On the one hand, they welcome the opportunity to view the Olympics or the latest fashions. On the other hand, increased availability of electronic media poses threats to the future of folk culture.

The threat to folk culture can be either external or internal. The external threat is that most of the content diffused through electronic media originates in a handful of developed countries. The internal threat is that the latest forms of social media enable people in developing countries to originate the content themselves—as long as they can afford the cost of access.

EXTERNAL THREAT: DEVELOPED COUNTRIES CONTROL THE MEDIA

Three developed countries dominate the television industry in developing countries—Japan, the United Kingdom, and the United States. These three countries are also the major exporters of programs.

ENTERTAINMENT, OR CULTURAL IMPERIALISM? Leaders of many developing countries view control of TV by a handful of developed countries— especially the United States— as a new method of economic and cultural imperialism. American TV programs present characteristically American beliefs and social forms, such as upward social mobility, relative freedom for women, glorification of youth, and stylized violence. These attractive themes may conflict with and drive out traditional folk culture.

To avoid offending traditional folk culture, many satellite and cable providers in developing countries block offending networks such as MTV and censor unacceptable programs. The entertainment programs that are substituted emphasize family values and avoid controversial or edgy cultural, economic, and political content.

NEWS—FAIR OR BIASED? Developing countries fear the threat of the news-gathering capability of the media even more than their entertainment function. The news media in most developing countries are dominated by the government, which typically runs the radio and TV service as well as the domestic news-gathering agency.

Sufficient funds are not available to establish an independent news service in developing countries. The process of gathering news worldwide is expensive, and most broadcasters and newspapers are unable to afford their own correspondents. Instead, they buy the right to use the dispatches of one or more of the main news organizations. The diffusion of information to newspapers around the world is dominated by the Associated Press (AP) and Reuters, which are owned by American and British companies, respectively. The AP and Reuters also supply most of the world's television news video. The world's 25 largest media companies are all based in developed countries: including 15 in the United States, 4 in the United Kingdom (including the parent company of the publisher of this book), and 2 each in France, Germany, and Japan.

NEWS COVERAGE AND PRESS FREEDOM. Many African and Asian government officials criticize the Western concept of freedom of the press. They argue that the American news organizations reflect American values and do not provide a balanced, accurate view of other countries. U.S. news-gathering organizations are more interested in covering earthquakes, hurricanes, and other sensational disasters than more meaningful but less visual and dramatic domestic stories, such as birth-control programs, health-care innovations, and construction of new roads.

Pause and Reflect 4.3.3

What would be a specific example of a distinctively American perspective on a U.S. TV show?

INTERNAL THREAT: SOCIAL MEDIA

George Orwell's novel *1984*, published in 1949, anticipated that TV—then in its infancy—would play a major role in the ability of undemocratic governments to control people's daily lives. In fact, many governments viewed TV as an important tool for fostering cultural integration. TV could extol the exploits of the leaders or the accomplishments of the political system. People turned on their TV sets and watched what the government wanted them to see.

Blocking foreign programming was easy for governments when TV service consisted of only a few over-the-air channels. Because over-the-air TV signals weakened with distance and were strong only up to roughly 100 kilometers (60 miles), few people could receive TV from other countries, so most were totally dependent on what their own government preferred to broadcast.

LIMITING ACCESS TO TV. Changing technology has made TV a force for political change rather than stability. Satellite dishes and the Internet enable people to choose from a wide variety of programs produced in other countries, not just the local government-controlled station. The delivery of programs in the future is likely to be closely integrated with other Internet services. This will facilitate people in different countries watching the same program.

Governments have had little success in shutting down satellite technology. Despite the threat of heavy fines, several hundred thousand Chinese still own satellite dishes. Consumers can outwit the government because the small size of satellite dishes makes them easy to smuggle into the country and erect out of sight, perhaps behind a brick wall or under a canvas tarpaulin. A dish may be expensive by local standards—twice the annual salary of a typical

Chinese worker, for example—but several neighbors can share the cost and hook up all of their TV sets to it.

LIMITING ACCESS TO THE INTERNET. As with television, governments try to limit Internet content. According to OpenNet Initiative, countries limit access to four types of Internet content (Figure 4-37):

1. Political content that expresses views in opposition to those of the current government or that is related to human rights, freedom of expression, minority rights, and religious movements.
2. Social content related to sexuality, gambling, and illegal drugs and alcohol, as well as other topics that may be socially sensitive or perceived as offensive.
3. Security content related to armed conflicts, border disputes, separatist movements, and militant groups.
4. Internet tools, such as e-mail, Internet hosting, and searching.

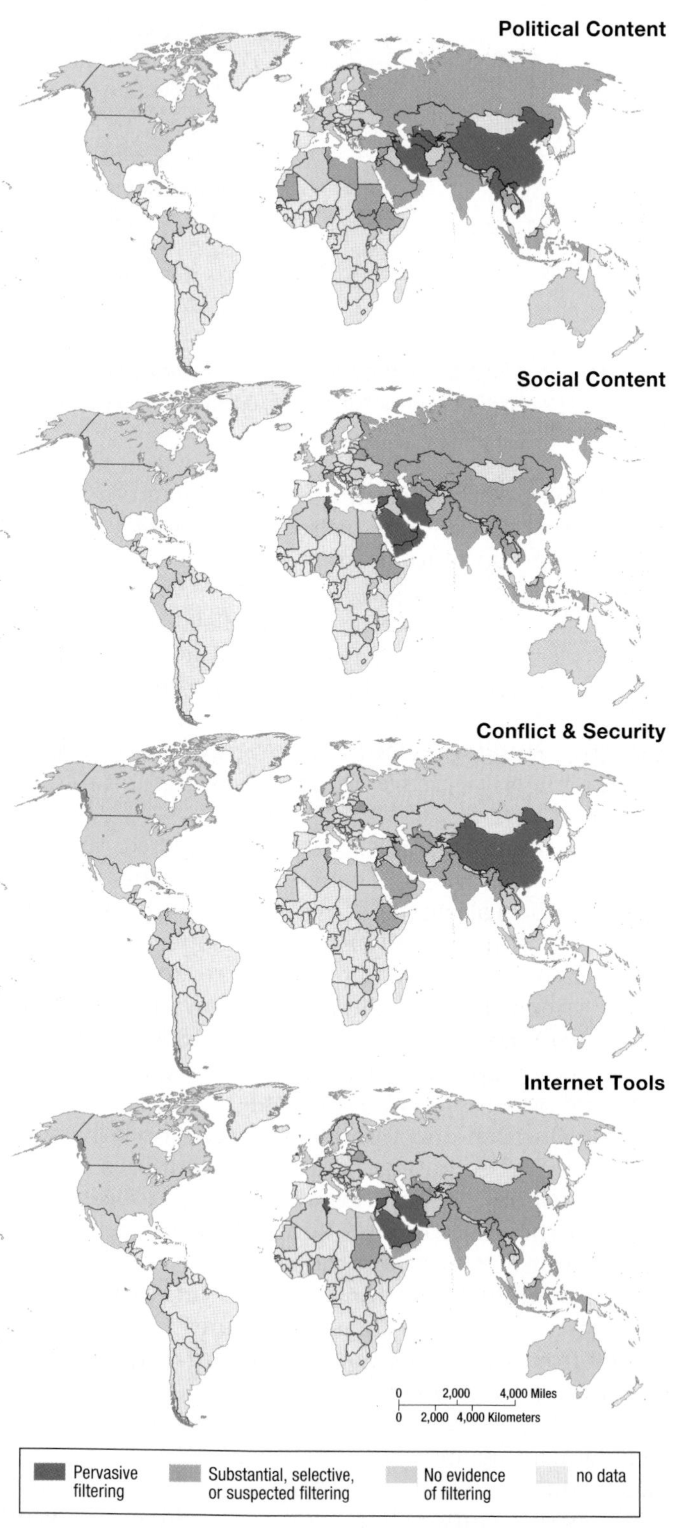

▲ **FIGURE 4-37 LIMITING FREEDOM ON THE INTERNET** Countries limit access to four types of Internet content: (top) political content, (second) social content, (third) security content, (bottom) Internet tools.

ELUDING CONTROL: NEW TECHNOLOGIES AND SOCIAL MEDIA. Social media have started to play a significant role in breaking the monopoly of government control over diffusion of information. As difficult as it is for governments to block satellite and Internet communications, it is even harder to block individual social media. Popular uprisings against undemocratic governments in Egypt, Libya, and other countries in Southwest Asia and North Africa in 2011 relied on individuals sending information through cell phones, Twitter, blogs, and other social media (Figure 4-38).

CHECK-IN: KEY ISSUE 3

Why Is Access to Folk and Popular Culture Unequal?

✓ **Popular culture diffuses primarily through electronic media, especially TV, as well as increasingly through other formats.**

✓ **Electronic media can pose a combination of external and internal threats to developing countries.**

▲ **FIGURE 4-38 PROTESTORS SHARING INFORMATION DURING ARAB SPRING** Two Egyptian protesters took photographs with their mobile phones when Egyptian riot police fired tear gas during an Arab Spring protest in 2011.

KEY ISSUE 4

Why Do Folk and Popular Culture Face Sustainability Challenges?

- **Sustainability Challenges for Folk Culture**
- **Sustainability Challenges for Popular Culture**

Learning Outcome 4.4.1
Summarize challenges for folk culture from diffusion of popular culture.

Elements of folk and popular culture face challenges in maintaining identities that are sustainable into the future. For folk culture, the challenges are to maintain unique local landscapes in an age of globalization. For popular culture, the challenges derive from the sustainability of practices designed to promote uniform landscapes.

Many fear the loss of folk culture, especially because rising incomes can fuel demand for the possessions typical of popular culture. When people turn from folk to popular culture, they may also turn away from the society's traditional values. And the diffusion of popular culture from developed countries can lead to dominance of Western perspectives.

Sustainability Challenges for Folk Culture

For folk culture, increased connection with popular culture can make it difficult to maintain centuries-old practices. The Amish in the United States and marriage customs in India are two examples.

THE AMISH: PRESERVING CULTURAL IDENTITY

Although the Amish number only about one-quarter million, their folk culture remains visible on the landscape in at least 19 U.S. states (Figure 4-39). Shunning mechanical and electrical power, the Amish still travel by horse and buggy and continue to use hand tools for farming. The Amish have distinctive clothing, farming, religious practices, and other customs.

The distribution of Amish folk culture across a major portion of the U.S. landscape is explained by examining

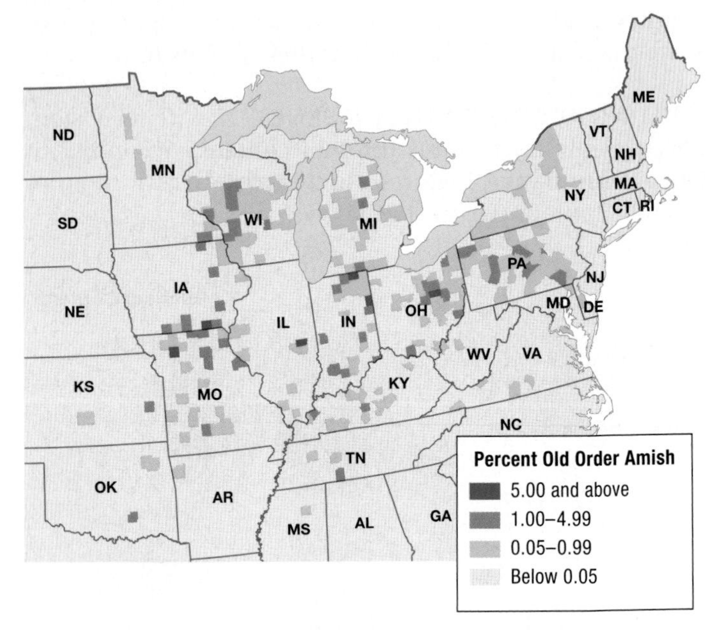

▲ FIGURE 4-39 **DISTRIBUTION OF AMISH** Amish settlements are distributed throughout the northeastern United States. Amish farmers minimize the use of mechanical devices.

the diffusion of their culture through migration. In the 1600s, a Swiss Mennonite bishop named Jakob Ammann gathered a group of followers who became known as the Amish. The Amish originated in Bern, Switzerland; Alsace in northeastern France; and the Palatinate region of southwestern Germany. They migrated to other portions of Northwestern Europe in the 1700s, primarily for religious freedom. In Europe, the Amish did not develop distinctive language, clothing, or farming practices, and they gradually merged with various Mennonite church groups.

Several hundred Amish families migrated to North America in two waves. The first group, primarily from Bern and the Palatinate, settled in Pennsylvania in the early 1700s, enticed by William Penn's offer of low-priced land. Because of lower land prices, the second group, from Alsace, settled in Ohio, Illinois, and Iowa in the United States and Ontario, Canada, in the early 1800s. From these core areas, groups of Amish migrated to other locations where inexpensive land was available.

Living in rural and frontier settlements relatively isolated from other groups, Amish communities retained their traditional customs, even as other European immigrants to the United States adopted new ones. We can observe Amish customs on the landscape in such diverse areas as southeastern Pennsylvania, northeastern Ohio, and east-central Iowa. These communities are relatively isolated from each other but share cultural traditions distinct from those of other Americans.

Amish folk culture continues to diffuse slowly through interregional migration within the United States. In recent years, a number of Amish families have sold their farms in Lancaster County, Pennsylvania—the oldest and at one time largest Amish community in the United States—and migrated to Christian and Todd counties in southwestern

▲ **FIGURE 4-40 AMISH AND TOURISTS** An Amish man demonstrates a cow milking machine to tourists in Shipshewana, Indiana.

Kentucky. According to Amish tradition, every son is given a farm when he is an adult, but land suitable for farming is expensive and hard to find in Lancaster County because of its proximity to growing metropolitan areas. With the average price of farmland in southwestern Kentucky less than one-fifth that in Lancaster County, an Amish family can sell its farm in Pennsylvania and acquire enough land in Kentucky to provide adequate farmland for all the sons. Amish families are also migrating from Lancaster County to escape the influx of tourists who come from the nearby metropolitan areas to gawk at the distinctive folk culture (Figure 4-40).

Pause and Reflect 4.4.1
In what ways might Amish people need to interact with popular culture?

MARRIAGE IN INDIA: CHALLENGING CULTURAL VALUES

Rapid changes in long-established cultural values can lead to instability, and even violence, in a society. This threatens not just the institutions of folk culture but the sustainability of the society as a whole.

The global diffusion of popular culture has challenged the subservience of women to men that is embedded in some folk customs. Women may have been traditionally relegated to performing household chores, such as cooking and cleaning, and to bearing and raising large numbers of children. Those women who worked outside the home were likely to be obtaining food for the family, either through agricultural work or by trading handicrafts.

At the same time, contact with popular culture has also had negative impacts for women in developing countries. Prostitution has increased in some developing countries to serve men from developed countries traveling on "sex tours." These tours, primarily from Japan and Northern Europe (especially Norway, Germany, and the Netherlands), include airfare, hotels, and the use of a predetermined number of women. Leading destinations include the Philippines, Thailand, and South Korea. International prostitution is encouraged in these countries as a major source of foreign currency. Through this form of global interaction, popular culture may regard women as essentially equal at home but as objects that money can buy in foreign folk societies.

Global diffusion of popular social customs has had an unintended negative impact for women in India: an increase in demand for dowries. Traditionally, a dowry was a "gift" from one family to another, as a sign of respect. In the past, the local custom in much of India was for the groom to provide a small dowry to the bride's family. In the twentieth century, the custom reversed, and the family of a bride was expected to provide a substantial dowry to the husband's family (Figure 4-41).

The government of India enacted anti-dowry laws in 1961, but the ban is widely ignored. In fact, dowries have become much larger in modern India and an important source of income for the groom's family. A dowry can take the form of either cash or expensive consumer goods, such as cars, electronics, and household appliances.

The government has tried to ban dowries because of the adverse impact on women. If the bride's family is unable to pay a promised dowry or installments, the groom's family may cast the bride out on the street, and her family may refuse to take her back. Husbands and in-laws angry over the small size of dowry payments killed 8,391 women in India in 2010, and disputes over dowries led to 90,000 cases of torture and cruelty toward women by men.

To raise awareness of dowry abuses, Shaadi.com, an Indian matrimonial web site with 2 million members, created an online game called Angry Brides. Each groom has a price tag, starting at 1.5 million rupees ($29,165). Every time the player hits a groom, his value decreases, and money is added to the player's Anti-Dowry Fund, which is shown on her Facebook page.

▼ **FIGURE 4-41 INDIA DOWRY** The photograph is held by the sister of a woman murdered by her husband for not meeting his dowry demands.

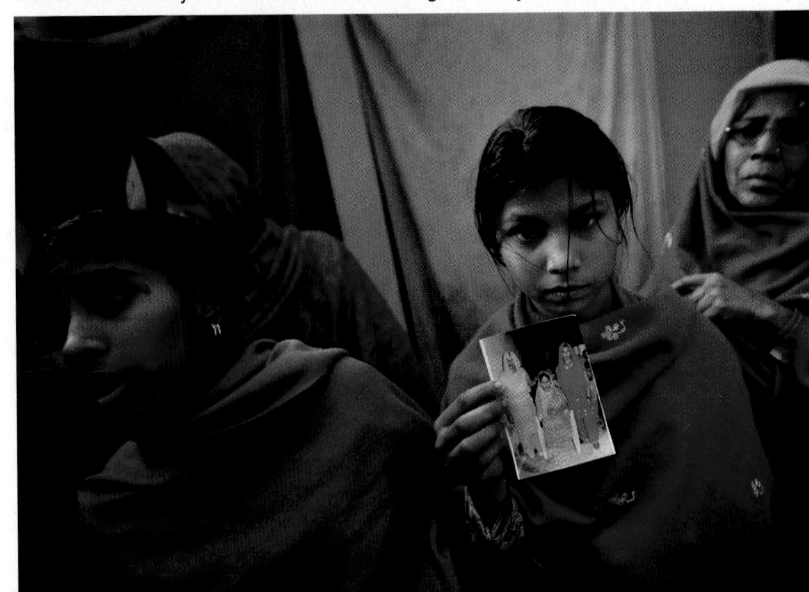

Sustainability Challenges for Popular Culture

Learning Outcome 4.4.2
Summarize the two principal ways that popular culture can adversely affect the environment.

Popular culture can significantly modify or control the environment. It may be imposed on the environment rather than spring forth from it, as with many folk customs. For many popular customs the environment is something to be modified to enhance participation in a leisure activity or to promote the sale of a product. Even if the resulting built environment looks "natural," it is actually the deliberate creation of people in pursuit of popular social customs.

The diffusion of some popular customs can adversely impact environmental quality in two ways:

- Pollution of the landscape
- Depletion of scarce natural resources

LANDSCAPE POLLUTION

Popular culture can pollute the landscape by modifying it with little regard for local environmental conditions, such as climate and soil. To create a uniform landscape, hills may be flattened and valleys filled in. The same building and landscaping materials may be employed regardless of location. Features such as golf courses consume large quantities of land and water; nonnative grass species are planted, and fertilizers and pesticides are laid on the grass to ensure an appearance considered suitable for the game.

UNIFORM LANDSCAPES. The distribution of popular culture around the world tends to produce more uniform landscapes. The spatial expression of a popular custom in one location will be similar to another. In fact, promoters of popular culture want a uniform appearance to generate "product recognition" and greater consumption (Figure 4-42).

▼ **FIGURE 4-42 UNIFORM LANDSCAPE** Route 66 in Springfield, Illinois.

The diffusion of fast-food restaurants is a good example of such uniformity. Such restaurants are usually organized as franchises. A franchise is a company's agreement with businesspeople in a local area to market that company's product. The franchise agreement lets the local outlet use the company's name, symbols, trademarks, methods, and architectural styles. To both local residents and travelers, the buildings are immediately recognizable as part of a national or multinational company. A uniform sign is prominently displayed.

Much of the attraction of fast-food restaurants comes from the convenience of the product and the use of the building as a low-cost socializing location for teenagers or families with young children. At the same time, the success of fast-food restaurants depends on large-scale mobility: People who travel or move to another city immediately recognize a familiar place. Newcomers to a particular place know what to expect in the restaurant because the establishment does not reflect strange and unfamiliar local customs that could be uncomfortable.

Fast-food restaurants were originally developed to attract people who arrived by car. The buildings generally were brightly colored, even gaudy, to attract motorists. Recently built fast-food restaurants are more subdued, with brick facades, pseudo-antique fixtures, and other stylistic details. To facilitate reuse of the structure in case the restaurant fails, company signs are often free-standing rather than integrated into the building design.

Uniformity in the appearance of the landscape is promoted by a wide variety of other popular structures in North America, such as gas stations, supermarkets, and motels. These structures are designed so that both local residents and visitors immediately recognize the purpose of the building, even if not the name of the company.

Physical expression of uniformity in popular culture has diffused from North America to other parts of the world. American motels and fast-food chains have opened in other countries. These establishments appeal to North American travelers, yet most customers are local residents who wish to sample American customs they have seen on television.

Pause and Reflect 4.4.2
How might fast-food restaurants reduce adverse impacts on the environment?

GOLF COURSES. Golf courses, because of their large size (80 hectares, or 200 acres), provide a prominent example of imposing popular culture on the environment. A surge in U.S. golf popularity spawned construction of several hundred courses during the late twentieth century. Geographer John Rooney attributed this to increased income and leisure time, especially among recently retired older people and younger people with flexible working hours. This trend slowed into the twenty-first century because of the severe recession.

The distribution of golf courses is not uniform across the United States. Although golf is perceived as a warm-weather sport, the number of golf courses per person is actually greatest in north-central states (Figure 4-43). People

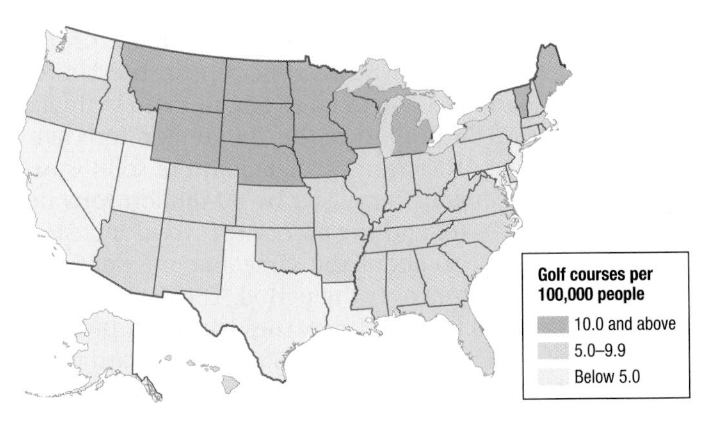

▲ **FIGURE 4-43 GOLF COURSES** The highest concentration of golf courses is in the upper Midwest.

in these regions have a long tradition of playing golf, and social clubs with golf courses are important institutions in the fabric of the regions' popular customs.

In contrast, access to golf courses is more limited in the South, in California, and in the heavily urbanized Middle Atlantic region between New York City and Washington, D.C. Rapid population growth in the South and West and lack of land on which to build in the Middle Atlantic region have reduced the number of courses per capita in those regions. Selected southern and western areas, such as coastal South Carolina, southern Florida, and central Arizona, have high concentrations of golf courses as a result of the arrival of large numbers of golf-playing northerners, either as vacationers or as permanent residents.

Golf courses are designed partially in response to local physical conditions. Grass species are selected to thrive in the local climate and still be suitable for the needs of greens, fairways, and roughs. Existing trees and native vegetation are retained if possible. (Few fairways in Michigan are lined by palms.) Yet, as with other popular customs, golf courses remake the environment—creating or flattening hills, cutting grass or letting it grow tall, carting in or digging up sand for traps, and draining or expanding bodies of water to create hazards. Ironically, golf originated as part of folk culture, as you can read in the following Sustainability and Inequality in Our Global Village feature.

ENVIRONMENTAL CAPACITY. The environment can accept and assimilate some level of waste from human activities. But popular culture generates a high volume of waste—solids, liquids, and gases—that must be absorbed into the environment. Although waste is discharged in all three forms, the most visible is solid waste—cans, bottles, old cars, paper, and plastics. These products are often discarded rather than recycled. With more people adopting popular customs worldwide, this problem grows.

Folk culture, like popular culture, can also cause environmental damage, especially when natural processes are ignored. A widespread belief exists that indigenous peoples of the Western Hemisphere practiced more "natural," ecologically sensitive agriculture before the arrival of Columbus and other Europeans. Geographers increasingly question this idea. In reality, pre-Columbian folk customs included burning grasslands for planting and hunting, cutting extensive forests, and overhunting some species. Very high rates of soil erosion have been documented in Central America from the practices of folk cultures.

SUSTAINABILITY AND INEQUALITY IN OUR GLOBAL VILLAGE

Golf: Folk or Popular Culture?

The modern game of golf originated as a folk custom in Scotland in the fifteenth century or earlier and diffused to other countries during the nineteenth century. In this respect, the history of golf is not unlike that of soccer, described earlier in this chapter. Early Scottish golf courses were primarily laid out on sand dunes adjacent to bodies of water (Figure 4-44). Largely because of golf's origin as a local folk custom, golf courses in Scotland do not modify the environment to the same extent as those constructed in more recent years in the United States and other countries, where hills, sand, and grass are imported, often with little regard for local environmental conditions. Modern golf also departs from its folk culture roots by being a relatively expensive sport to play in most places.

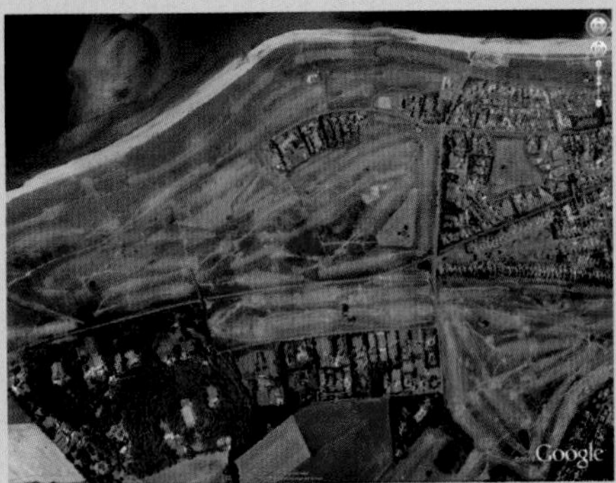

▲ **FIGURE 4-44 SCOTLAND AND U.S. GOLF COURSES** The Congressional Country Club golf course in Bethesda, Maryland (left), made substantial alterations to the landscape. Scotland's Royal Troon Golf Club was built into a seaside dune with little alteration of the landscape.

RESOURCE DEPLETION

Learning Outcome 4.4.3
Summarize major sources of waste and the extent to which each is recycled.

Increased demand for the products of popular culture can strain the capacity of the environment. Diffusion of some popular customs increases demand for animal products, ranging from rare wildlife to common domesticated animals, and for raw materials, such as minerals and other substances found beneath Earth's surface. The depletion of resources used to produce energy, especially petroleum, is discussed in Chapter 9.

DEMAND FOR ANIMAL PRODUCTS. Popular culture may demand a large supply of certain animals, resulting in depletion or even extinction of some species. For example, some animals are killed for their skins, which can be shaped into fashionable clothing and sold to people living thousands of kilometers from the animals' habitat. The skins of the mink, lynx, jaguar, kangaroo, and whale have been heavily consumed for various articles of clothing, to the point that the survival of these species is endangered. This unbalances ecological systems of which the animals are members. Folk culture may also encourage the use of animal skins, but the demand is usually smaller than for popular culture.

Increased meat consumption in popular culture has not caused extinction of cattle and poultry—we simply raise more. But animal consumption is an inefficient way for people to acquire calories—90 percent less efficient than if people simply ate grain directly. To produce 1 kilogram (2.2 pounds) of beef sold in the supermarket, nearly 10 kilograms (22 pounds) of grain are consumed by the animal. For every kilogram of chicken, nearly 3 kilograms (6.6 pounds) of grain are consumed by the fowl. This grain could be fed to people directly, bypassing the inefficient meat step. With a large percentage of the world's population undernourished, some question this inefficient use of grain to feed animals for eventual human consumption.

RECYCLING OF RESOURCES. The developed countries that produce endless supplies of consumer products for popular culture have created the technological capacity both to create large-scale environmental damage and to control it. However, a commitment of time and money must be made to control the damage.

Unwanted by-products are usually "thrown away," perhaps in a "trash can." Recycling is the separation, collection, processing, marketing, and reuse of the unwanted material. Recycling increased in the United States from 7 percent of all solid waste in 1970 to 10 percent in 1980, 17 percent in 1990, and 34 percent in 2010 (Figure 4-45).

As a result of recycling, about 85 million of the 250 million tons of solid waste generated in the United States in 2010 did not have to go to landfills and incinerators, compared to 34 million of the 200 million tons generated in 1990. In other words, the amount of solid waste generated by Americans increased by 50 million tons between 1990 and 2010, and the amount recycled increased by 51 million tons, so about the same amount went into landfills or incinerators over the period. The percentage of materials recovered by recycling varies widely by product: 63 percent of paper products and 58 percent of yard waste are recycled, compared to only 8 percent of plastic and 3 percent of food scraps (Figure 4-46).

RECYCLING COLLECTION. Recycling involves two main series of activities:

1. **Pick-up and processing.** Materials that would otherwise be "thrown away" are collected and sorted, in four principal ways:
   - **Curbside programs.** Recyclables can often be placed at the curb in a container separate from the nonrecyclable trash at a specified time each week, either at the same or different time as the other trash. The trash collector usually supplies homes with specially marked containers for the recyclable items.
   - **Drop-off centers.** Drop-off centers are sites, typically with several large containers placed at a central location, for individuals to leave recyclable materials. A separate container is designated for each type of recyclable material, and the containers are periodically emptied by a processor or recycler but are otherwise left unattended.
   - **Buy-back centers.** Commercial operations sometimes pay consumers for recyclable materials, especially aluminum cans, but also sometimes plastic containers and glass bottles. These materials are usually not processed at the buy-back center.
   - **Deposit programs.** Glass and aluminum containers can sometimes be returned to retailers. The price a

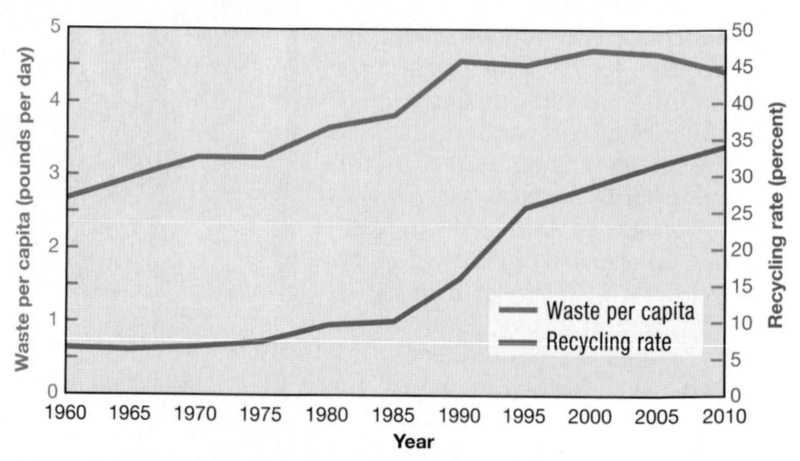

▲ **FIGURE 4-45 RECYCLING IN THE UNITED STATES**
Recycling has increased substantially in the United States. As a result, the amount of waste generated per person has not changed much.

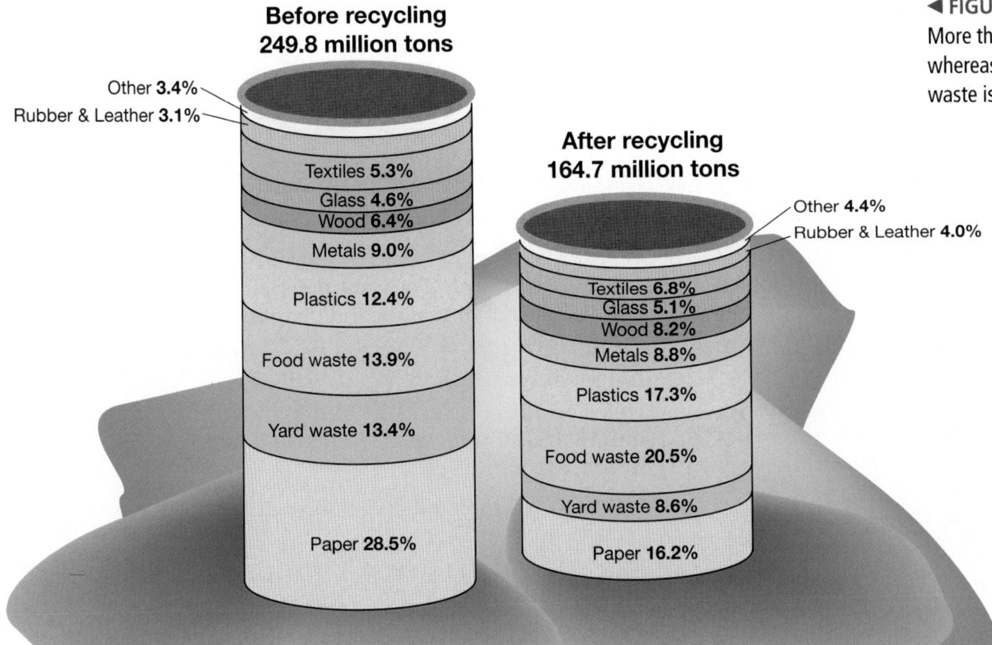

◀ **FIGURE 4-46 SOURCES OF SOLID WASTE**
More than one-half of paper and yard waste is recycled, whereas only a small percentage of plastic and food waste is recycled.

consumer pays for a beverage may include a deposit fee of 5¢ or 10¢ that the retailer refunds when the container is returned.

Regardless of the collection method, recyclables are sent to a materials recovery facility to be sorted and prepared as marketable commodities for manufacturing. Recyclables are bought and sold just like any other commodity; typical prices in recent years have been 30¢ per pound for plastic, $30 per ton for clear glass, and $90 per ton for corrugated paper. Prices for the materials change and fluctuate with the market.

Pause and Reflect 4.4.3
Which, if any, recycling systems operate in your community?

2. **Manufacturing.** Materials are manufactured into new products for which a market exists. Important inputs into manufacturing include recycled paper, plastic, glass, and aluminum:
   - **Paper.** Most types of paper can be recycled. Newspapers have been recycled profitably for decades, and recycling of other paper, especially computer paper, is growing. Rapid increases in virgin paper pulp prices have stimulated construction of more plants capable of using waste paper. The key to recycling is collecting large quantities of clean, well-sorted, uncontaminated, and dry paper.
   - **Plastic.** Different plastic types must not be mixed, as even a small amount of the wrong type of plastic can ruin the melt. Because it is impossible to tell one type from another by sight or touch, the plastic industry has developed a system of numbers marked inside triangles on the bottom of containers. Types 1 and 2 are commonly recycled, and the others generally are not.
   - **Glass.** Glass can be used repeatedly with no loss in quality and is 100 percent recyclable. The process of creating new glass from old is extremely efficient, producing virtually no waste or unwanted by-products. Though unbroken clear glass is valuable, mixed-color glass is nearly worthless, and broken glass is hard to sort.
   - **Aluminum.** The principal source of recycled aluminum is beverage containers. Aluminum cans began to replace glass bottles for beer during the 1950s and for soft drinks during the 1960s. Aluminum scrap is readily accepted for recycling, although other metals are rarely accepted.

Four major manufacturing sectors accounted for more than half of the recycling activity—paper mills, steel mills, plastic converters, and iron and steel foundries. Common household items that contain recycled materials include newspapers and paper towels; aluminum, plastic, and glass soft-drink containers; steel cans; and plastic laundry detergent bottles. Recycled materials are also used in such industrial applications as recovered glass in roadway asphalt ("glassphalt") and recovered plastic in carpet, park benches, and pedestrian bridges.

CHECK-IN: KEY ISSUE 4

Why Do Folk and Popular Culture Face Sustainability Challenges?

✓ **Folk culture faces loss of traditional values in the face of rapid diffusion of popular culture.**

✓ **Popular culture can cause two environmental concerns—pollution of the landscape and depletion of scarce resources.**

Summary and Review

KEY ISSUE 1

Where Are Folk and Popular Leisure Activities Distributed?

Culture can be divided into folk and popular culture. Leisure activities, such as music and sports, can be classified as folk or popular, depending on their characteristics.

LEARNING OUTCOME 4.1.1: Compare the origin, diffusion, and distribution of folk and popular culture.

- Folk culture is more likely to have an anonymous origin and to diffuse slowly through migration, whereas popular culture is more likely to be invented and diffuse rapidly with the use of modern communications.

LEARNING OUTCOME 4.1.2: Compare the characteristics of folk and popular music.

- Popular music has wide global distribution because of connections among artists and styles.

LEARNING OUTCOME 4.1.3: Describe how sports have been transformed from folk to popular culture.

- Sports that originated as isolated folk customs have been organized into popular culture with global distribution.

THINKING GEOGRAPHICALLY 4.1: In what ways might gender affect the distribution of leisure activities in folk or popular culture?

GOOGLE EARTH 4.1: Connections among Nepal's diverse folk culture groups are hindered by what feature of the physical environment? What does the white represent in the image of Nepal?

KEY ISSUE 2

Where Are Folk and Popular Material Culture Distributed?

Important elements of material culture include clothing, food, and shelter. Folk and popular material culture have different origins, patterns of diffusion, and distribution.

LEARNING OUTCOME 4.2.1: Compare reasons for distribution of clothing styles in folk and popular culture.

- Folk clothing is more likely to respond to environmental conditions and cultural values, whereas clothing styles vary more in time than in place.

LEARNING OUTCOME 4.2.2: Understand reasons for folk food preferences and taboos.

- Folk food culture is especially strongly embedded in environmental conditions.

LEARNING OUTCOME 4.2.3: Describe regional variations in popular food preferences.

- Popular food culture can display some regional variations.

LEARNING OUTCOME 4.2.4: Understand factors that influence patterns of folk housing.

- Folk housing styles, like other folk material culture, respond to environmental and cultural factors.

LEARNING OUTCOME 4.2.5: Understand variations in time and space of housing in the United States.

- U.S. housing has roots in folk culture, but newer housing displays features of popular culture.

THINKING GEOGRAPHICALLY 4.2: Which elements of material culture do countries depict in campaigns to promote tourism?

GOOGLE EARTH 4.2: Rüdesheim, Germany, a wine-producing community, is surrounded by hillside vineyards. Towards which direction (east, west, north, or south) do most of these vineyards slope, and how does this orientation maximize exposure to sunlight?

Key Terms

Custom (p. 109) The frequent repetition of an act, to the extent that it becomes characteristic of the group of people performing the act.

Folk culture (p. 108) Culture traditionally practiced by a small, homogeneous, rural group living in relative isolation from other groups.

Habit (p. 109) A repetitive act performed by a particular individual.

Popular culture (p. 108) Culture found in a large, heterogeneous society that shares certain habits despite differences in other personal characteristics.

Taboo (p. 118) A restriction on behavior imposed by social custom.

Terroir (p. 118) The contribution of a location's distinctive physical features to the way food tastes.

KEY ISSUE 3

Why Is Access to Folk and Popular Culture Unequal?

Popular culture is diffused around the world through electronic media. TV was the dominant format in the twentieth century. Social media formats are expanding in the twenty-first century.

LEARNING OUTCOME 4.3.1: Describe the origin, diffusion, and distribution of TV around the world.

- TV diffused during the twentieth century from the United States to Europe and then to developing countries.

LEARNING OUTCOME 4.3.2: Compare the diffusion of the Internet and social media with the diffusion of TV.

- Diffusion of the Internet and of social media has followed the pattern of TV, but at a much faster rate.

LEARNING OUTCOME 4.3.3: Understand external and internal threats to folk culture posed by electronic media.

- Folk culture may be threatened by the dominance of popular culture in the media and by decreasing ability to control people's access to the media.

THINKING GEOGRAPHICALLY 4.3: Which elements of the physical environment are emphasized in the portrayal of places on TV?

GOOGLE EARTH 4.3: Kathmandu, Nepal, situated at the foot of rugged mountains, is one of the world's most physically isolated capitals. TripAdvisor considers BoudhaNath Stupa to be the top attraction in Kathmandu. Using the Find Business and ruler features of Google Earth, how far is it from the stupa to the nearest Internet café?

KEY ISSUE 4

Why Do Folk and Popular Culture Face Sustainability Challenges?

LEARNING OUTCOME 4.4.1: Summarize challenges for folk culture from diffusion of popular culture.

- Popular culture threatens traditional elements of cultural identity in folk culture.

LEARNING OUTCOME 4.4.2: Summarize the two principal ways that popular culture can adversely affect the environment.

- Popular culture can deplete scarce resources and pollute the landscape.

LEARNING OUTCOME 4.4.3: Summarize major sources of waste and the extent to which each is recycled.

- Paper is the principal source of solid waste before recycling, but plastics and food waste are the leading sources after recycling.

THINKING GEOGRAPHICALLY 4.4: Are there examples of groups in North America besides the Amish that have successfully resisted the diffusion of popular culture?

GOOGLE EARTH 4.4: Paradise, Pennsylvania, is in the heart of Amish country. If you fly to 269 Old Leacock Road in Paradise and drag to street view, what distinctive feature of Amish culture is visible?

Chapter

5 Languages

Why did someone spread graffiti on this sign? Page 163.

Why is this sign in four languages – but not English? Page 165.

KEY ISSUE 1

Where Are Languages Distributed?

PLEASE DO NOT FEED THE PIGEONS
請勿餵白鴿
POR FAVOR NO DE COMIDA A LAS PALOMAS

A World of Languages p. 143

Languages are like leaves growing from language families and branches.

KEY ISSUE 2

Why Is English Related to Other Languages?

Languages Are Families p. 150

Most languages can be classified into a handful of families.

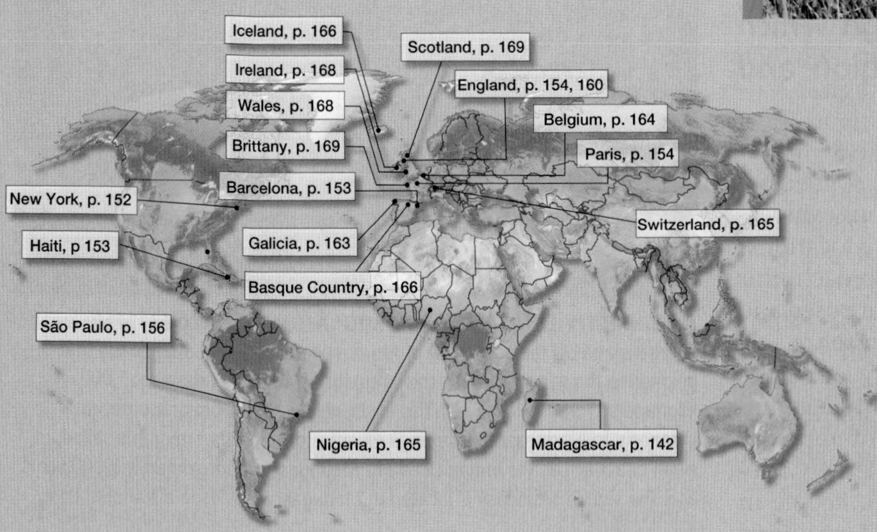

▲ No fishing in Israel's Alexander River. Can you identify the other three languages on the sign? Can you read any of them? Two-thirds of Americans believe that it is important to learn a second language, but only about one-fourth of Americans can hold a conversation in a second language.

KEY ISSUE 3

Why Do Individual Languages Vary among Places?

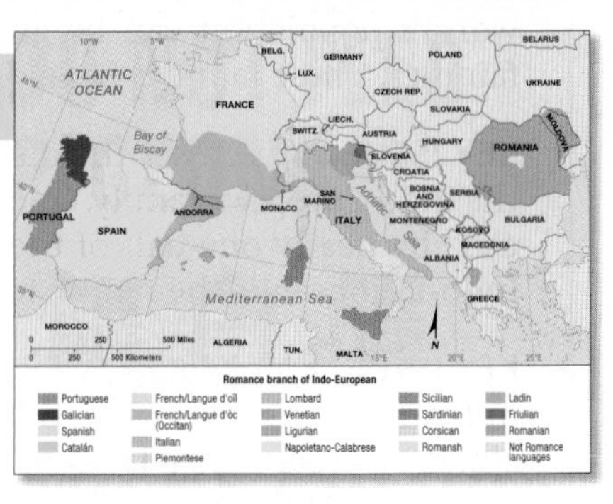

It Doesn't Sound Like English p. 158

The English language can sound very different in the United Kingdom than in the United States. Patterns of migration explain differences.

KEY ISSUE 4

Why Do People Preserve Local Languages?

What Does This Say? p. 164

Many languages have disappeared, meaning that no one who speaks them is alive today. But some, like Welsh, are being preserved.

Classification of Languages

Learning Outcome 5.1.1
Name the largest language families.

The several thousand spoken languages can be organized logically into a small number of language families. Larger language families can be further divided into language branches and language groups.

Figure 5-3 depicts differences among language families, branches, groups, and individual languages:

- Language families form the trunks of the trees.
- Individual languages are displayed as leaves.
- Some trunks divide into several branches, which logically represent language branches.
- The branches representing Germanic, Balto-Slavic, and Indo-Iranian in Figure 5-3 divide a second time into language groups.

The larger the trunks and leaves are, the greater the number of speakers of those families and languages.

Two-thirds of the people in the world speak a language that belongs to the Indo-European or Sino-Tibetan language family. Seven other language families are used by between 2 and 6 percent of the world (Figure 5-4). The remaining 5 percent of the world's people speak a language belonging to one of 100 smaller families.

Figure 5-3 displays each language family as a separate tree at ground level because differences among families predate recorded history. Some linguists speculate that language families were joined together as a handful of superfamilies tens of thousands of years ago. Superfamilies are shown as roots below the surface because their existence is highly controversial and speculative. A researcher in New Zealand, Quentin Atkinson, carries the speculation further, arguing that all languages can be ultimately traced to Africa. According to Atkinson, languages are most complex and diverse in Africa. Atkinson thinks humans outside Africa display less linguistic diversity because their languages have had a shorter time in which to evolve into new languages than have African languages.

Pause and Reflect 5.1.1
Based on Figure 5-3, what are the language family, branch, and group to which English belongs?

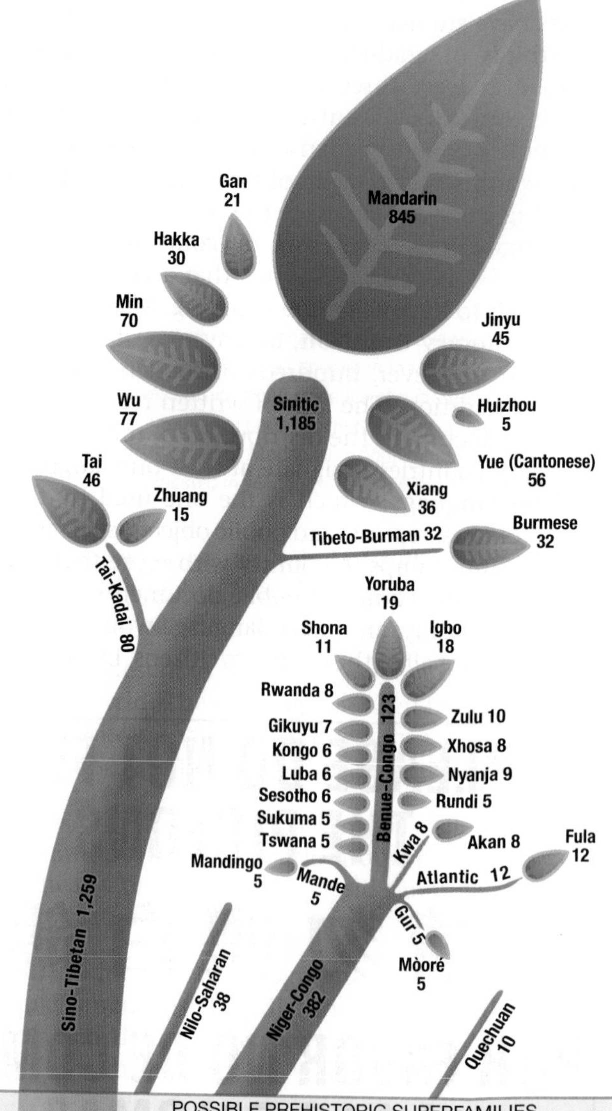

▶ **FIGURE 5-3 LANGUAGE FAMILY TREE** Language families with at least 10 million speakers according to *Ethnologue* are shown as trunks of trees. Some language families are divided into branches and groups. Individual languages that have more than 5 million speakers are shown as leaves. Below ground level, the language tree's "roots" are shown, but these are speculative because they predated recorded history.

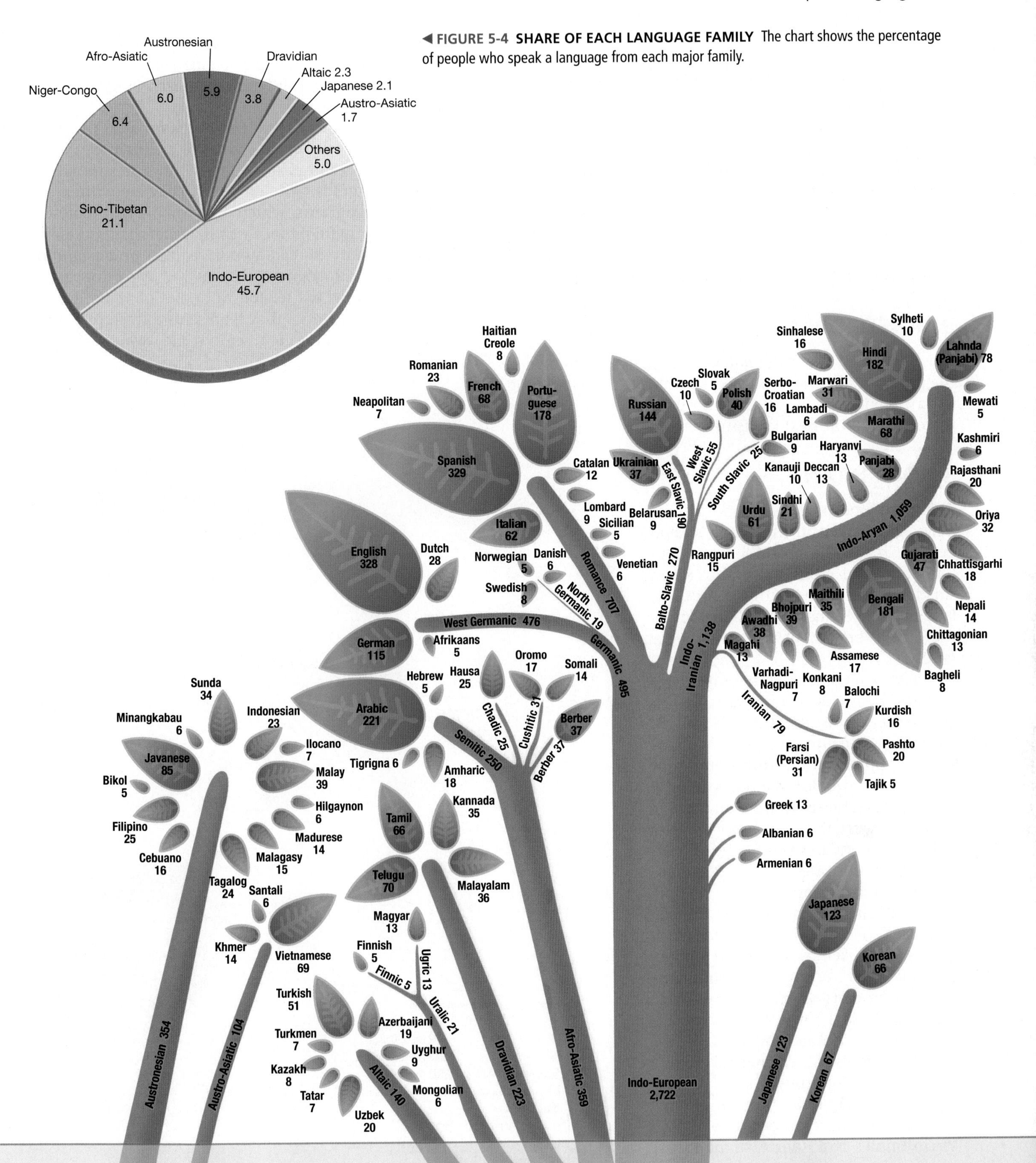

◀ **FIGURE 5-4 SHARE OF EACH LANGUAGE FAMILY** The chart shows the percentage of people who speak a language from each major family.

Distribution of Language Families

Learning Outcome 5.1.2

Identify the names and distribution of the two largest language families.

Language families with at least 10 million native speakers are shown in Figure 5-5. Individual languages with at least 50 million speakers are named on the map.

INDO-EUROPEAN

Indo-European, the most widely used language family, is the predominant one in Europe, South Asia, and North and Latin America. Its origin and distribution are discussed in more detail in the next key issue.

SINO-TIBETAN

The Sino-Tibetan family encompasses languages spoken in the People's Republic of China—the world's most populous state, at 1.3 billion—and in several smaller countries in Southeast Asia. The languages of China generally belong to the Sinitic branch of the Sino-Tibetan family.

There is no single Chinese language. Rather, the most commonly used is Mandarin (or, as the Chinese call it, Putonghua—"common speech"). Spoken by approximately three-fourths of the Chinese people, Mandarin is by a wide margin the most-used language in the world. Once the language of emperors in Beijing, Mandarin is now the official language of both the People's Republic of China and Taiwan, and it is one of the six official languages of the United Nations. Seven other Sinitic branch languages are spoken by at least 20 million each in China, mostly in the southern and eastern parts of the country—Wu, Min, Yue (also known as Cantonese), Jinyu, Xiang, Hakka, and Gan. However, the Chinese government is imposing Mandarin countrywide.

The relatively small number of languages in China (compared to India, for example) is a source of national strength and unity. Unity is also fostered by a consistent written form for all Chinese languages. Although the words are pronounced differently in each language, they are written the same way.

You already know the general structure of Indo-European quite well because you are a fluent speaker of at least one Indo-European language. But the structure of Chinese languages is quite different. They are written primarily with **logograms**, which are symbols that represent words, or meaningful parts of words, rather than sounds (as in English). Ability to read a book requires understanding several thousand logograms (Figure 5-6). Most logograms are compounds; words related to bodies of water, for example, include a symbol that represents a river, plus additional strokes that alter the river in some way.

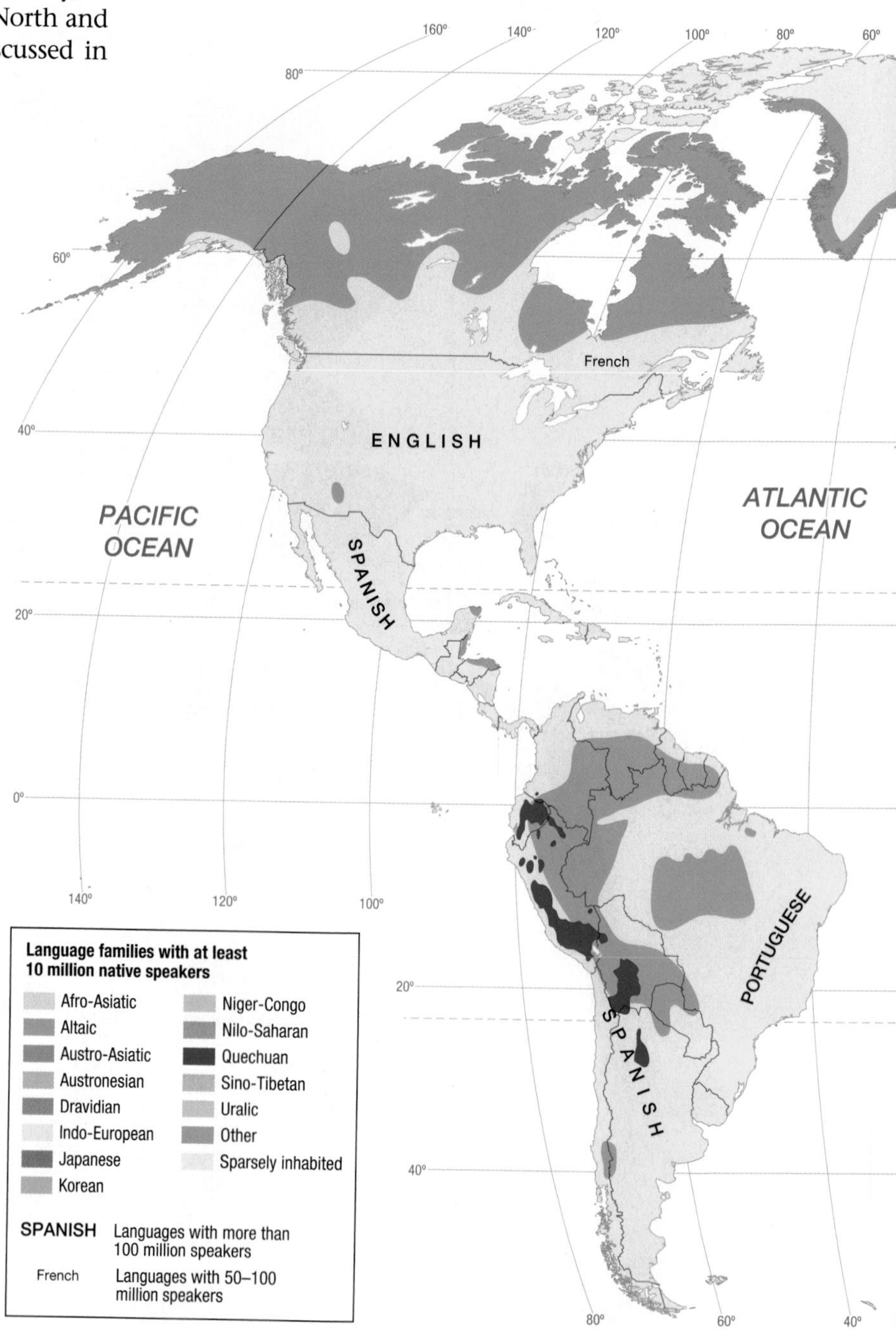

▶ **FIGURE 5-5 DISTRIBUTION OF LANGUAGE FAMILIES** Most language can be classified into one of a handful of language families.

River 河

River bed 河床

Lake 湖

Stream 流

Riptide 沖

◀ **FIGURE 5-6 CHINESE** Similar logograms represent various water-related words.

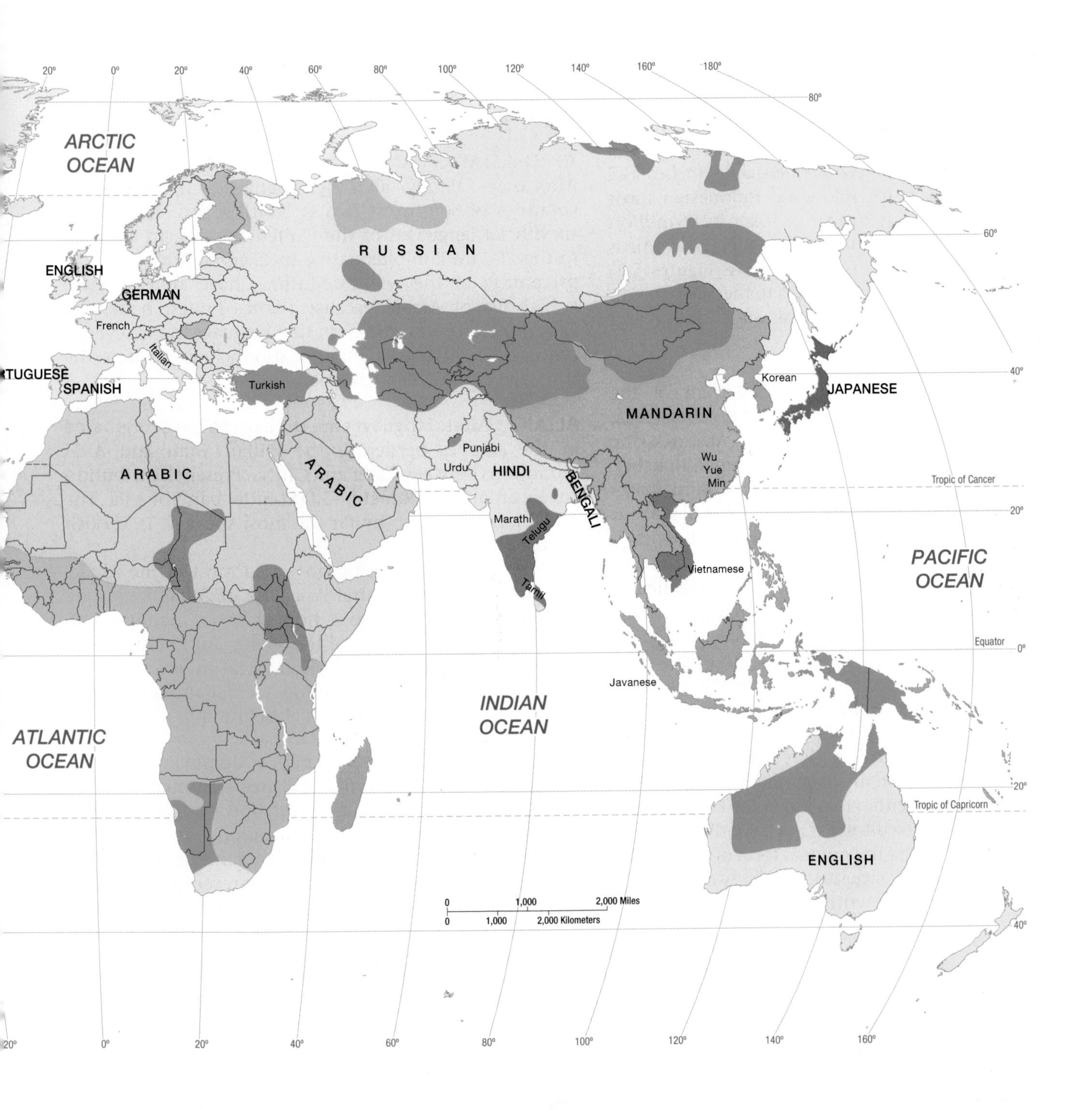

OTHER ASIAN LANGUAGE FAMILIES

Learning Outcome 5.1.3

Identify the names and distribution of the largest language families in addition to Indo-European and Sino-Tibetan.

In addition to Sino-Tibetan, several other language families spoken by large numbers of people can be found in East and Southeast Asia. If you look at their distribution in Figure 5-5, you can see a physical reason for their independent development: These language families are clustered on either islands or peninsulas.

AUSTRONESIAN. Austronesian languages are spoken by about 6 percent of the world's people, who are mostly in Indonesia, the world's fourth-most-populous country. With its inhabitants dispersed among thousands of islands, Indonesia has an extremely large number of distinct languages and dialects; *Ethnologue* identifies 722 actively used languages in Indonesia. Indonesia's most widely used first language is Javanese, spoken by 85 million people, mostly on the island of Java, where two-thirds of the country's population is clustered. As Figures 5-1, 5-3, and 5-5 show, Malagasy also belongs to the Austronesian family because of migration from Indonesia to Madagascar 2,000 years ago.

AUSTRO-ASIATIC. Spoken by about 2 percent of the world's population, Austro-Asiatic is based in Southeast Asia. Vietnamese, the most-spoken tongue of the Austro-Asiatic language family, is written with our familiar Roman alphabet, with the addition of a large number of diacritical marks above the vowels. The Vietnamese alphabet was devised in the seventeenth century by Roman Catholic missionaries.

TAI KADAI. The Tai Kadai family was once classified as a branch of Sino-Tibetan. The principal languages of this family are spoken in Thailand and neighboring portions of China. Similarities with the Austronesian family have led some linguistic scholars to speculate that people speaking these languages may have migrated from the Philippines.

JAPANESE. Written in part with Chinese logograms, Japanese also uses two systems of phonetic symbols, like Western languages, used either in place of the logograms or alongside them. Chinese cultural traits have diffused into Japanese society, including the original form of writing the Japanese language. But the structures of the two languages differ. Foreign terms may be written with one of these sets of phonetic symbols.

KOREAN. Unlike Sino-Tibetan languages and Japanese, Korean is written not with logograms but in a system known as hankul (also called hangul or onmun). In this system, each letter represents a sound, as in Western languages. More than half of the Korean vocabulary derives from Chinese words. In fact, Chinese and Japanese words are the principal sources for creating new words to describe new technology and concepts.

Pause and Reflect 5.1.3

If you are trying to recall where in the world language families are distributed, keep in mind that most of them are named for regions or countries. Based on their names, how would you expect the language families Austronesian and Austro-Asiatic to differ in their distribution?

LANGUAGES OF SOUTHWEST ASIA & NORTH AFRICA AND CENTRAL ASIA

The two largest language families in Southwest Asia & North Africa and Central Asia are Afro-Asiatic and Altaic. Uralic languages were once classified with Altaic.

AFRO-ASIATIC. Arabic is the major language of the Afro-Asiatic family, an official language in two dozen countries of Southwest Asia & North Africa, and one of six official languages of the United Nations. In addition to the 200-million-plus native speakers of Arabic, a large percentage of the world's Muslims have at least some knowledge of Arabic because Islam's holiest book, the Quran (Koran), was written in that language in the seventh century. The Afro-Asiatic family also includes Hebrew, the language of the Judeo-Christian Bible.

ALTAIC. Altaic languages are thought to have originated in the steppes bordering the Qilian Shan and Altai mountains between Tibet and China. Present distribution covers an 8,000-kilometer (5,000-mile) band of Asia. The Altaic language with by far the most speakers is Turkish (Figure 5-7).

When the Soviet Union governed most of the Altaic-speaking region of Central Asia, use of Altaic languages was suppressed to create a homogeneous national culture. With the dissolution of the Soviet Union in the early 1990s, Altaic languages became official in several newly independent countries, including Azerbaijan, Kazakhstan, Kyrgyzstan, Turkmenistan, and Uzbekistan.

URALIC. Every European country is dominated by Indo-European speakers, except for three—Estonia, Finland, and Hungary. The Estonians, Finns, and Hungarians speak languages that belong to the Uralic family.

The Altaic and Uralic language families were once thought to be linked as one family, but recent studies point to geographically distinct origins. Uralic languages are traceable back to a common language, Proto-Uralic, first used 7,000 years ago by people living in the Ural Mountains of present-day Russia, north of the Kurgan homeland. Migrants carried the Uralic languages to Europe, carving out homelands for themselves in the midst of Germanic- and Slavic-speaking peoples and retaining their language as a major element of cultural identity.

AFRICAN LANGUAGE FAMILIES

No one knows the precise number of languages spoken in Africa, and scholars disagree on classifying those known into families. In the 1800s, European missionaries and colonial officers began to record African languages using the Roman or Arabic alphabet. More than 1,000 distinct languages and several thousand named dialects have been documented. Most lack a written tradition.

NIGER-CONGO. More than 95 percent of the people in sub-Saharan Africa speak languages of the Niger-Congo family (Figure 5-8). One of these languages—Swahili—is the first language of only 800,000 people and an official language in only one country (Tanzania), but it is spoken as a second language by approximately 30 million Africans.

Especially in rural areas, the local language is used to communicate with others from the same village, and Swahili is used to communicate with outsiders. Swahili originally developed through interaction among African groups and Arab traders, so its vocabulary has strong Arabic influences. Also, Swahili is one of the few African languages with an extensive literature.

NILO-SAHARAN. Languages of the Nilo-Saharan family are spoken by a few million people in north-central Africa, immediately north of the Niger-Congo language region.

▼ FIGURE 5-7 TURKISH In 1928, Turkey's leader Kemal Ataturk ordered Turks to write Turkish with Roman letters instead of Arabic. Ataturk believed that using Roman letters would help modernize Turkey's economy and culture through increased communications with European countries. This painting depicts Ataturk writing with Roman letters.

Divisions within the Nilo-Saharan family exemplify the problem of classifying African languages. Despite have relatively few speakers, the Nilo-Saharan family is divided into six branches, plus numerous groups and subgroups. The total number of speakers of each individual Nilo-Saharan language is extremely small.

KHOISAN. A distinctive characteristic of the Khoisan languages is the use of clicking sounds. Upon hearing this, whites in southern Africa derisively and onomatopoeically named the most important Khoisan language Hottentot.

▼ FIGURE 5-8 AFRICA'S LANGUAGE FAMILIES More than 1,000 languages have been identified in Africa, and experts do not agree on how to classify them into families, especially languages in central Africa. Languages with more than 5 million speakers are named on the map. The great number of languages results from at least 5,000 years of minimal interaction among the thousands of cultural groups inhabiting the African continent. Each group developed its own language, religion, and other cultural traditions in isolation from other groups.

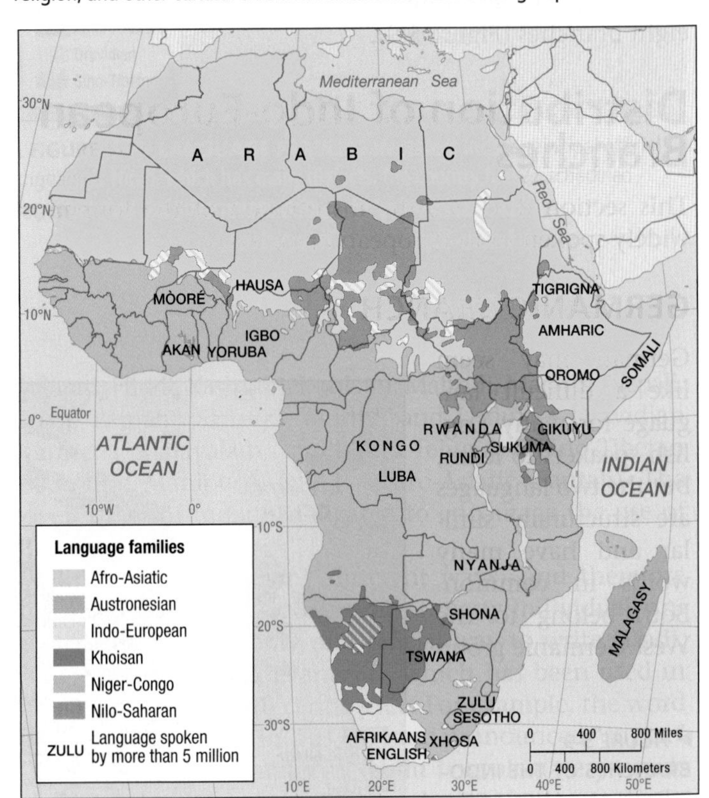

CHECK-IN: KEY ISSUE 1

Where Are Languages Distributed?

- ✓ **Languages can be classified into families and branches.**
- ✓ **The two largest families are Indo-European, which is found primarily in Europe, North America, Latin America, and South Asia, and Sino-Tibetan, which is clustered primarily in East Asia.**

BALTO-SLAVIC BRANCH

Learning Outcome 5.2.2
Learn the distribution of the Balto-Slavic and Romance branches of Indo-European.

Slavic was once a single language, but differences developed in the seventh century A.D. when several groups of Slavs migrated from Asia to different areas of Eastern Europe and thereafter lived in isolation from one other. As a result, this branch can be divided into East, West, and South Slavic groups as well as a Baltic group. Figure 5-9 shows the widespread area populated with Balto-Slavic speakers.

EAST SLAVIC AND BALTIC GROUPS. The most widely used Slavic languages are the eastern ones, primarily Russian, which is spoken by more than 80 percent of Russian people (Figure 5-12). Russian is one of the six official languages of the United Nations.

The importance of Russian increased with the Soviet Union's rise to power after the end of World War II in 1945. Soviet officials forced native speakers of other languages to learn Russian as a way of fostering cultural unity among the country's diverse peoples. In Eastern European countries that were dominated politically and economically by the Soviet Union, Russian was taught as the second language. The presence of so many non-Russian speakers was a measure of cultural diversity in the Soviet Union, and the desire to use languages other than Russian was a major drive in its breakup. With the demise of the Soviet Union, the newly independent republics adopted official languages other than Russian, although Russian remains the language for communication among officials in the countries that were formerly part of the Soviet Union.

▼ **FIGURE 5-12 RUSSIAN** New York City's Brighton Beach neighborhood is home to many Russian immigrants, including this bookseller. The red words say "Christian Library," with the text below announcing Saturday services at the Russian Evangelical Baptist Church.

After Russian, Ukrainian and Belarusan are the two most commonly used East Slavic languages and are the official languages in Ukraine and Belarus. Ukraine is a Slavic word meaning "border," and *bela-* means "white."

WEST AND SOUTH SLAVIC GROUPS. The most spoken West Slavic language is Polish, followed by Czech and Slovak. The latter two are quite similar, and speakers of one can understand the other.

The government of the former state of Czechoslovakia tried to balance the use of the two languages, even though the country contained twice as many Czechs as Slovaks. For example, the announcers on televised sports events used one of the languages during the first half and switched to the other for the second half. These balancing measures were effective in promoting national unity during the Communist era, but in 1993, four years after the fall of communism, Slovakia split from the Czech Republic. Slovaks rekindled their long-suppressed resentment of perceived dominance of the national culture by the Czech ethnic group.

The most widely used South Slavic language is the one spoken in Bosnia & Herzegovina, Croatia, Montenegro, and Serbia. When Bosnia & Herzegovina, Croatia, Montenegro, and Serbia were all part of Yugoslavia, the language was called Serbo-Croatian. This name now offends Bosnians and Croatians because it recalls when they were once in a country that was dominated by Serbs. Instead, the names Bosnian, Croatian, and Serbian are preferred by people in these countries, to demonstrate that each language is unique, even though linguists consider them one. Bosnians and Croats write the language in the Roman alphabet (what you are reading now), whereas Montenegrans and Serbs use the Cyrillic alphabet (for example, Serbia is written **Србија**).

Differences have crept into the South Slavic languages. Bosnian Muslims have introduced Arabic words used in their religion, and Croats have replaced words regarded as having a Serbian origin with words considered to be purely Croatian. For example, the Serbo-Croatian word for martyr or hero—*junak*—has been changed to *heroj* by Croats and *shahid* by Bosnian Muslims. In the future, after a generation of isolation and hostility among Bosnians, Croats, and Serbs, the languages spoken by the three may be sufficiently different to justify their classification as distinct languages.

In general, differences among all of the Slavic languages are relatively small. A Czech, for example, can understand most of what is said or written in Slovak and could become fluent without much difficulty. However, because language is a major element in a people's cultural identity, relatively small differences among Slavic as well as other languages are being preserved and even accentuated in recent independence movements.

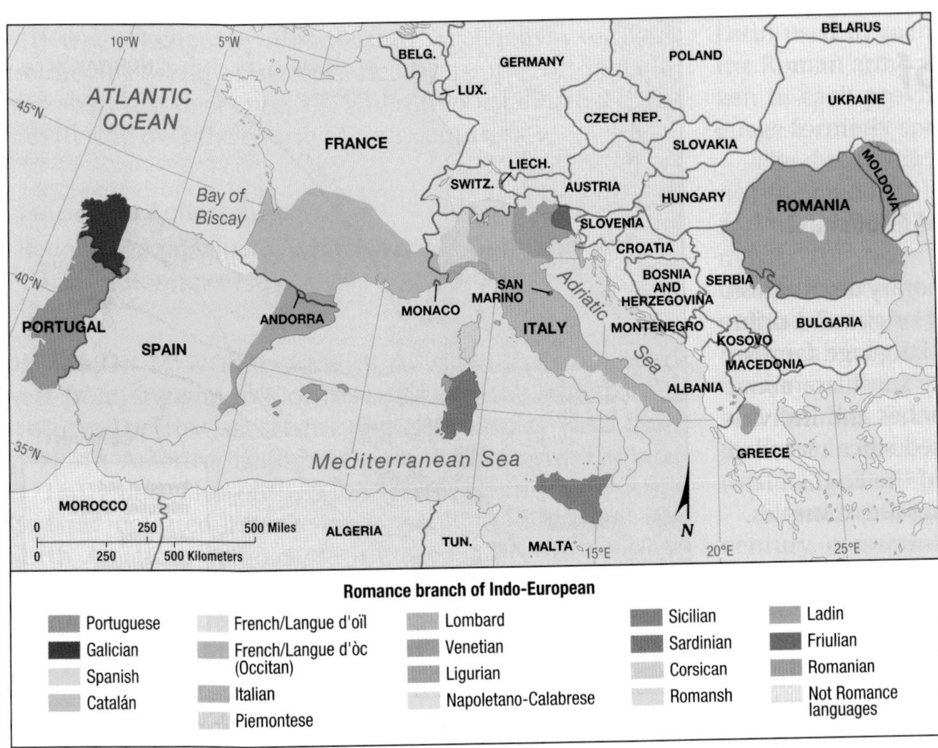

▲ **FIGURE 5-13 ROMANCE BRANCH OF THE INDO-EUROPEAN LANGUAGE FAMILY** Romance branch languages predominate in southwestern Europe.

Pause and Reflect 5.2.2
On the map of Europe, which branch predominates to the north, which to the south, and which to the east?

ROMANCE BRANCH

The Romance language branch evolved from the Latin language spoken by the Romans 2,000 years ago. The four most widely used contemporary Romance languages are Spanish, Portuguese, French, and Italian (Figure 5-13). Spanish and French are two of the six official languages of the United Nations.

The European regions in which these four languages are spoken correspond somewhat to the boundaries of the modern states of Spain, Portugal, France, and Italy. Rugged mountains serve as boundaries among these four countries. France is separated from Italy by the Alps and from Spain by the Pyrenees, and several mountain ranges mark the border between Spain and Portugal. Physical boundaries such as mountains are strong intervening obstacles, creating barriers to communication between people living on opposite sides.

The fifth most widely used Romance language, Romanian, is the principal language of Romania and Moldova. It is separated from the other Romance-speaking European countries by Slavic-speaking peoples.

The distribution of Romance languages shows the difficulty in trying to establish the number of distinct languages in the world. In addition to the five languages already mentioned, two other official Romance languages are Romansh and Catalán. Romansh is one of four official languages of Switzerland, although it is spoken by only 40,000 people. Catalán is the official language of Andorra, a tiny country of 70,000 inhabitants situated in the Pyrenees Mountains between Spain and France. Catalán is also spoken by 6 million people in eastern Spain and is the official language of Spain's highly autonomous Catalonia province, centered on the city of Barcelona (Figure 5-14). A third Romance language, Sardinian—a mixture of Italian, Spanish, and Arabic—was once the official language of the Mediterranean island of Sardinia.

In addition to these official languages, several other Romance languages have individual literary traditions. In Italy, Ladin (not Latin) is spoken by 30,000 people living in the South Tyrol, and Friulian is spoken by 800,000 people in the northeast. Ladin and Friulian (along with the official Romansh) are dialects of Rhaeto-Romanic.

A Romance tongue called Ladino—a mixture of Spanish, Greek, Turkish, and Hebrew—is spoken by 100,000 Sephardic Jews, most of whom now live in Israel. None of these languages have an official status in any country, although they are used in literature.

▲ **FIGURE 5-14 CATALÁN** The sign says "Passage is restricted to workers." The sign is in front of Sagrada Familia church, designed by Antoni Gaudi, in Barcelona, Spain.

Origin and Diffusion of Indo-European

Learning Outcome 5.2.4
Understand the two theories of the origin and diffusion of Indo-European.

If Germanic, Romance (Figure 5-17), Balto-Slavic, and Indo-Iranian languages are all part of the same Indo-European language family, then they must be descended from a single common ancestral language. Unfortunately, the existence of a single ancestor—which can be called Proto-Indo-European—cannot be proved with certainty because it would have existed thousands of years before the invention of writing or recorded history.

The evidence that Proto-Indo-European once existed is "internal," derived from the physical attributes of words themselves in various Indo-European languages. For example, the words for some animals and trees in modern Indo-European languages have common roots, including beech, oak, bear, deer, pheasant, and bee. Because all Indo-European languages share these similar words, linguists believe the words must represent things experienced in the daily lives of the original Proto-Indo-European speakers. In contrast, words for other features, such as elephant, camel, rice, and bamboo, have different roots in the various Indo-European languages. Such words therefore cannot be traced back to a common Proto-Indo-European ancestor and must have been added later, after the root language split into many branches. Individual Indo-European languages share common root words for winter and snow but not for ocean. Therefore, linguists conclude that original Proto-Indo-European speakers probably lived in a cold climate, or one that had a winter season, but did not come in contact with oceans.

▼ **FIGURE 5-17 A ROMANCE LANGUAGE: PORTUGUESE** The Museum of Portuguese Language in São Paulo, Brazil, has exhibits related to the Portuguese language, such as authors who have written in Portuguese.

Linguists and anthropologists generally accept that Proto-Indo-European must have existed, but they disagree on when and where the language originated and the process and routes by which it diffused. The debate over place of origin and paths of diffusion is significant; one theory argues that language diffused primarily through warfare and conquest, and another theory argues that the diffusion resulted from peaceful sharing of food. So where did Indo-European originate? Not surprisingly, scholars disagree on where and when the first speakers of Proto-Indo-European lived.

NOMADIC WARRIOR HYPOTHESIS. One influential hypothesis, espoused by Marija Gimbutas, is that the first Proto-Indo-European speakers were the Kurgan people, whose homeland was in the steppes near the border between present-day Russia and Kazakhstan. The earliest archaeological evidence of the Kurgans dates to around 4300 B.C.

The Kurgans were nomadic herders. Among the first people to domesticate horses and cattle, they migrated in search of grasslands for their animals. This took them westward through Europe, eastward to Siberia, and southeastward to Iran and South Asia. Between 3500 and 2500 B.C., Kurgan warriors, using their domesticated horses as weapons, conquered much of Europe and South Asia (Figure 5-18).

SEDENTARY FARMER HYPOTHESIS. Archaeologist Colin Renfrew argues that the first speakers of Proto-Indo-European lived 2,000 years before the Kurgans, in eastern Anatolia, part of present-day Turkey (Figure 5-19). Biologist Russell D. Gray supports the Renfrew position but dates the first speakers even earlier, at around 6700 B.C.

Renfrew believes they diffused from Anatolia westward to Greece (the origin of the Greek language branch) and from Greece westward toward Italy, Sicily, Corsica, the Mediterranean coast of France, Spain, and Portugal (the origin of the Romance language branch). From the Mediterranean coast, the speakers migrated northward toward central and northern France and on to the British Isles (perhaps the origin of the Celtic language branch).

Indo-European is also said to have diffused northward from Greece toward the Danube River (Romania) and westward to central Europe, according to Renfrew. From there the language diffused northward toward the Baltic Sea (the origin of the Germanic language branch) and eastward toward the Dnestr River near Ukraine (the origin of the Slavic language branch). From the Dnestr River, speakers migrated eastward to the Dnepr River (the homeland of the Kurgans).

The Indo-Iranian branch of the Indo-European language family originated either directly through migration from Anatolia along the

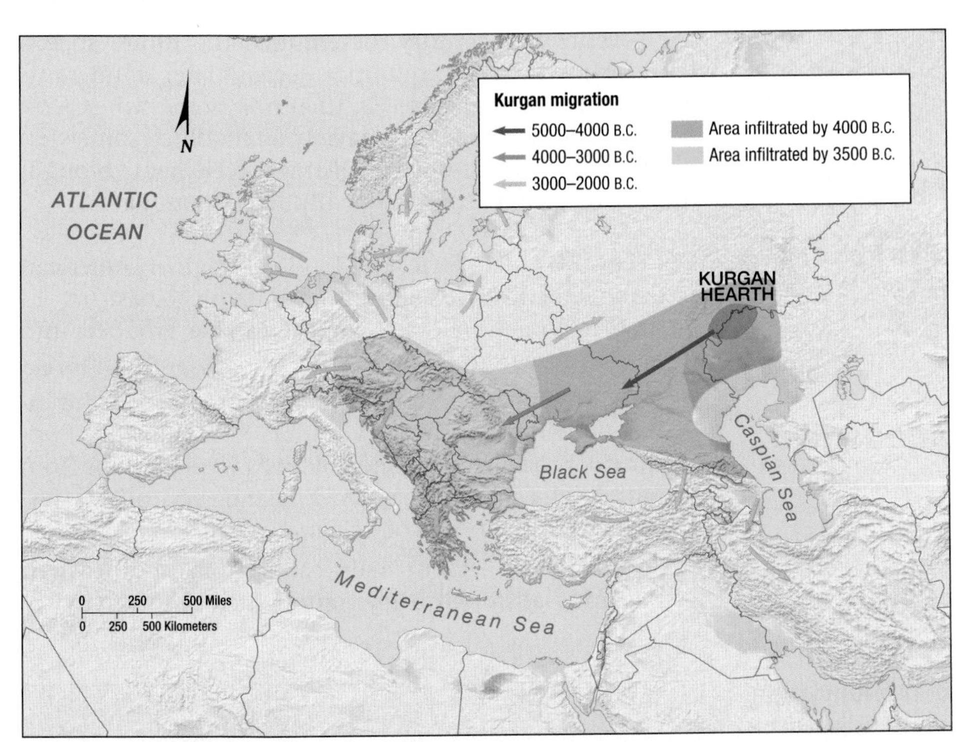

◀ FIGURE 5-18 ORIGIN AND DIFFUSION OF INDO-EUROPEAN (NOMADIC WARRIOR THEORY) The Kurgan homeland was north of the Caspian Sea, near the present-day border between Russia and Kazakhstan. According to this theory, the Kurgans may have infiltrated into Eastern Europe beginning around 4000 B.C. and into central Europe and Southwest Asia beginning around 2500 B.C.

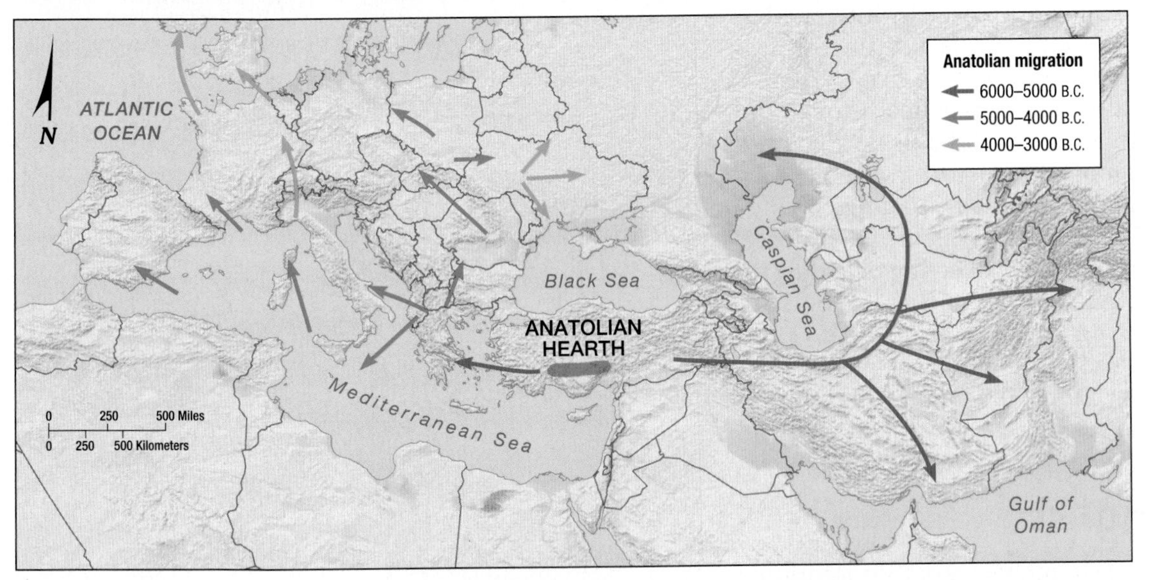

◀ FIGURE 5-19 ORIGIN AND DIFFUSION OF INDO-EUROPEAN (SEDENTARY FARMER THEORY) Indo-European may have originated in present-day Turkey 2,000 years before the Kurgans. According to this theory, the language diffused along with agricultural innovations west into Europe and east into Asia.

south shores of the Black and Caspian seas by way of Iran and Pakistan, or indirectly by way of Russia north of the Black and Caspian seas.

Renfrew argues that Indo-European diffused into Europe and South Asia along with agricultural practices rather than by military conquest. The language triumphed because its speakers became more numerous and prosperous by growing their own food instead of relying on hunting.

Regardless of how Indo-European diffused, communication was poor among different peoples, whether warriors or farmers. After many generations of complete isolation, individual groups evolved increasingly distinct languages.

Pause and Reflect 5.2.4

Which hypothesis appeals more to you: the "war" or the "peace" hypothesis? Why?

CHECK-IN: KEY ISSUE 2

Why Is English Related to Other Languages?

- ✓ **The Indo-European family has four widely spoken branches.**
- ✓ **Individual languages, such as English and languages of the Romance branch, have documented places of origin and patterns of diffusion.**
- ✓ **The origin and early diffusion of language families such as Indo-European is speculative because these language families existed before recorded history.**

KEY ISSUE 3

Why Do Individual Languages Vary among Places?

- Dialects of English
- Distinguishing between Languages and Dialects

Learning Outcome 5.3.1
Describe the main dialects in the United States.

A **dialect** is a regional variation of a language distinguished by distinctive vocabulary, spelling, and pronunciation. Generally, speakers of one dialect can understand speakers of another dialect. Geographers are especially interested in differences in dialects because they reflect distinctive features of the environments in which groups live.

The distribution of dialects is documented through the study of particular words. Every word that is not used nationally has some geographic extent within the country and therefore has boundaries. Such a word-usage boundary, known as an **isogloss**, can be constructed for each word. Isoglosses are determined by collecting data directly from people, particularly natives of rural areas. People are shown pictures to identify or are given sentences to complete with a particular word. Although every word has a unique isogloss, boundary lines of different words coalesce in some locations to form regions.

Dialects of English

When speakers of a language migrate to other locations, various dialects of that language may develop. This was the case with the migration of English speakers to North America several hundred years ago. Because of its large number of speakers and widespread distribution, English has an especially large number of dialects. North Americans are well aware that they speak English differently from the British, not to mention people living in India, Pakistan, Australia, and other English-speaking countries. Further, English varies by regions within individual countries. In both the United States and England, northerners sound different from southerners.

DIALECTS IN THE UNITED STATES

Major differences in U.S. dialects originated because of differences in dialects among the original settlers. The English dialect spoken by the first colonists, who arrived in the seventeenth century, determined the future speech patterns for their communities because later immigrants adopted the language used in their new homes when they arrived. The language may have been modified somewhat by the new arrivals, but the distinctive elements brought over by the original settlers continued to dominate.

SETTLEMENT IN THE EAST. The original American settlements stretched along the Atlantic Coast in 13 separate colonies. The settlements can be grouped into three dialect regions (Figure 5-20):

- **New England.** These colonies were established and inhabited almost entirely by settlers from England. Two-thirds of the New England colonists were Puritans from East Anglia in southeastern England, and only a few came from the north of England.
- **Southeastern.** About half came from southeastern England, although they represented a diversity of

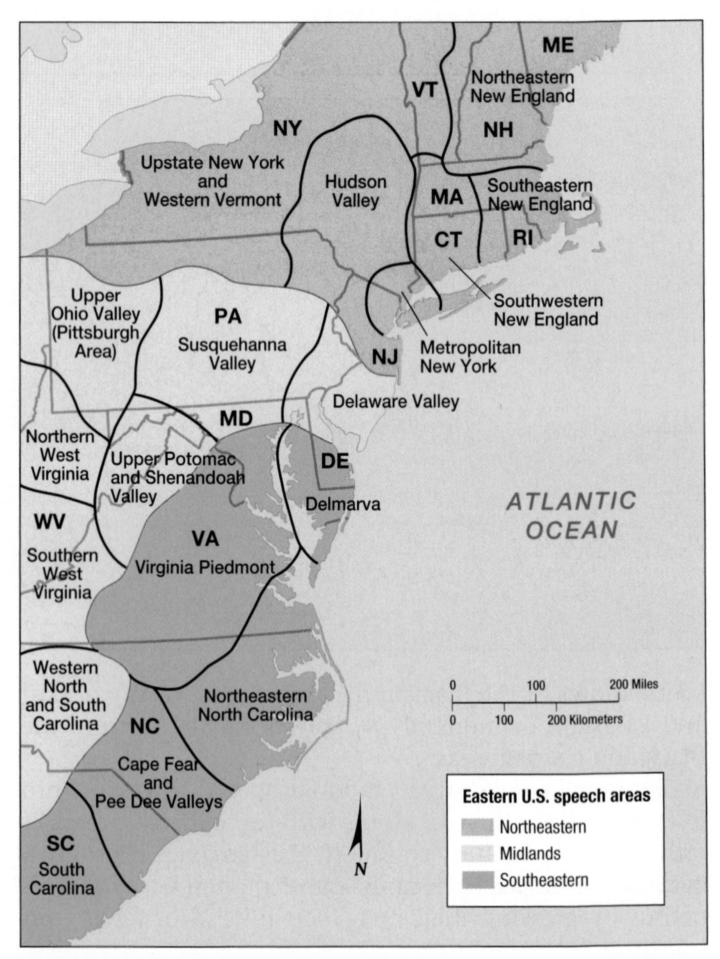

▲ **FIGURE 5-20 DIALECTS IN THE EASTERN UNITED STATES** The most comprehensive classification of dialects in the United States was made by Hans Kurath in 1949. He found the greatest diversity of dialects in the eastern part of the country, especially in vocabulary used on farms. Kurath divided the eastern United States into three major dialect regions—Northern, Midlands, and Southern—each of which contained a number of important subareas. Compare this to the map of source areas of U.S. house types (Figure 4-25). As Americans migrated west, they took with them distinctive house types as well as distinctive dialects.

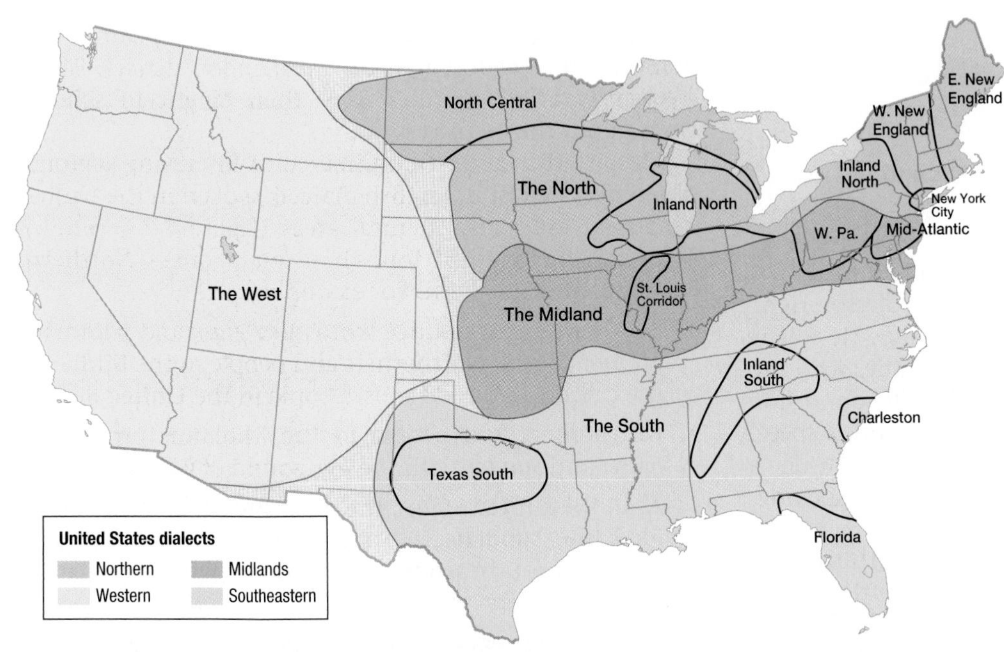

▲ **FIGURE 5-21 U.S. DIALECTS AND SUBDIALECTS** The four major U.S. dialect regions are Northern, Southern, Midlands, and West.

social-class backgrounds, including deported prisoners, indentured servants, and political and religious refugees.

- Midlands. These immigrants were more diverse. The early settlers of Pennsylvania were predominantly Quakers from the north of England. Scots and Irish also went to Pennsylvania, as well as to New Jersey and Delaware. The Middle Atlantic colonies also attracted many German, Dutch, and Swedish immigrants who learned their English from the English-speaking settlers in the area.

The English dialects now spoken in the U.S. Southeast and New England are easily recognizable. The dialects spoken in the former Midland colonies differ significantly from those spoken farther north and south because most of the settlers came from the north rather than the south of England or from other countries.

CURRENT DIALECT DIFFERENCES IN THE EAST. Major dialect differences continue to exist within the United States. The three major East Coast dialect regions have been joined by a fourth that developed in the West (Figure 5-21).

Many words that were once regionally distinctive are now national in distribution. Mass media, especially television and radio, influence the adoption of the same words throughout the country. Nonetheless, regional dialect differences persist in the United States. For example, the word for *soft drink* varies. Most people in the Northeast and Southwest, as well as the St. Louis area, use *soda* to describe a soft drink. Most people in the Midwest, Great Plains, and Northwest prefer *pop*. Southerners refer to all soft drinks as *coke* (Figure 5-22).

PRONUNCIATION DIFFERENCES. Regional pronunciation differences are more familiar to us than word differences, although it is harder to draw precise isoglosses for them:

- The southeastern dialect includes making such words as *half* and *mine* into two syllables ("ha-af" and "mi-yen").
- The northeastern dialect is well known for dropping the /r/ sound, so that *heart* and *lark* are pronounced "hot" and "lock." This characteristic dropping of the /r/ sound is shared with speakers from the south of England and reflects the place of origin of most New England colonists.

  It also reflects the relatively high degree of contact between the two groups. Residents of Boston, the Northeast's main port city, maintained especially close ties to the important ports of southern England.

The diffusion of particular English dialects is a result of the westward movement of colonists from the three East Coast dialect regions. The northeastern and southeastern accents sound unusual to the majority of Americans because the standard pronunciation throughout the American West comes from the Midlands rather than the northeastern and southeastern regions. This pattern occurred because most western settlers came from the Midlands. The three eastern dialect regions can also be divided into several subdialects, several of which are shown in Figure 5-21.

Pause and Reflect 5.3.1
Does your English fall into one of these dialects? Why or why not?

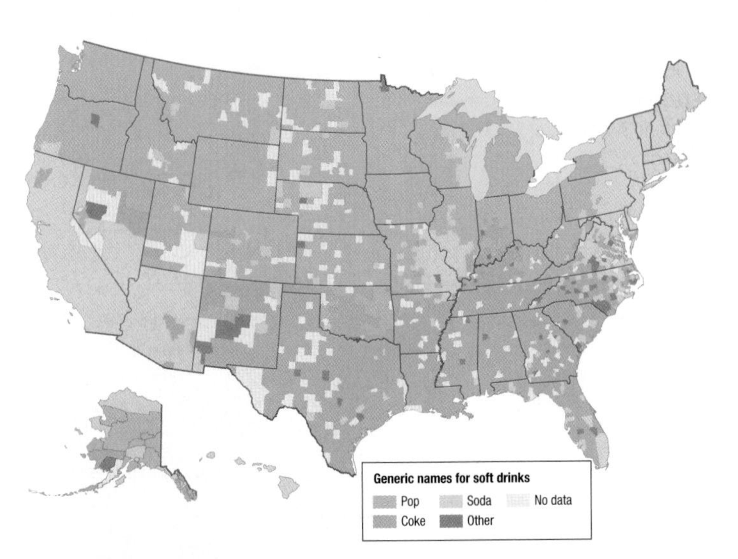

▲ **FIGURE 5-22 SOFT-DRINK DIALECTS** Soft drinks are called *soda* in the Northeast and Southwest, *pop* in the Midwest and Northwest, and *Coke* in the South. The map reflects voting at www.popvssoda.com.

DIALECTS IN THE UNITED KINGDOM

Learning Outcome 5.3.2

Understand the main ways that British and U.S. English dialects vary.

"If you use proper English, you're regarded as a freak; why can't the English learn to speak?" asked Professor Henry Higgins in the Broadway musical *My Fair Lady*. He was referring to the Cockney-speaking Eliza Doolittle, who pronounced *rain* like "rine" and dropped the /*h*/ sound from the beginning of words like *happy*. Eliza Doolittle's speech illustrates that English, like other languages, has a wide variety of dialects that use different pronunciations, spellings, and meanings for particular words.

As already discussed, English originated with three invading groups from Northern Europe who settled in different parts of Britain—the Angles in the north, the Jutes in the southeast, and the Saxons in the southwest. The language each spoke was the basis of distinct regional dialects of Old English—Kentish in the southeast, West Saxon in the southwest, Mercian in the center of the island, and Northumbrian in the north (Figure 5-23).

In a language with multiple dialects, one dialect may be recognized as the **standard language**, which is a dialect that is well established and widely recognized as the most acceptable for government, business, education, and mass communication. In the case of England, the standard language is known as **Received Pronunciation (RP).** It is well known around the world as the dialect commonly used by politicians, broadcasters, and actors.

RP was the dialect used by upper-class residents in the capital city of London and the two important university cities of Cambridge and Oxford. The diffusion of the upper-class London and university dialects was encouraged by the introduction of the printing press to England in 1476. Grammar books and dictionaries printed in the eighteenth century established rules for spelling and grammar that were based on the London dialect. These frequently arbitrary rules were then taught in schools throughout the country.

Despite the current dominance of RP, strong regional differences persist in English dialects spoken in the United Kingdom, especially in rural areas (Figure 5-23, center). They can be grouped into three main ones—Northern, Midland, and Southern. For example:

- Southerners pronounce words like *grass* and *path* with an /*ah*/ sound; Northerners and people in the Midlands use a short /*a*/, as do most people in the United States.
- Northerners and people in the Midlands pronounce *butter* and *Sunday* with the /*oo*/ sound of words like *boot*.

As in the United States, the main British dialects can be divided into subdialects. For example, distinctive southwestern and southeastern accents occur within England's Southern dialect:

- Southwesterners pronounce *thatch* and *thing* with the /*th*/ sound of *then* rather than *thin*. *Fresh* and *eggs* have an /*ai*/ sound.
- Southeasterners pronounce the /*a*/ in *apple* and *cat* like the short /*e*/ in *bet*.

The isoglosses between English dialects have been moving (Figure 5-23, right). The changes reflect patterns of migration. The emergence of a subdialect in London reflects migration of people from other countries into the capital city, and the northern expansion of the southeastern subdialect reflects the outmigration of Londoners.

BRITISH AND AMERICAN ENGLISH DIALECTS

Why don't Americans speak RP? The English language was brought to the North American continent by colonists from England who settled along the Atlantic Coast beginning in the seventeenth century. The early colonists naturally spoke

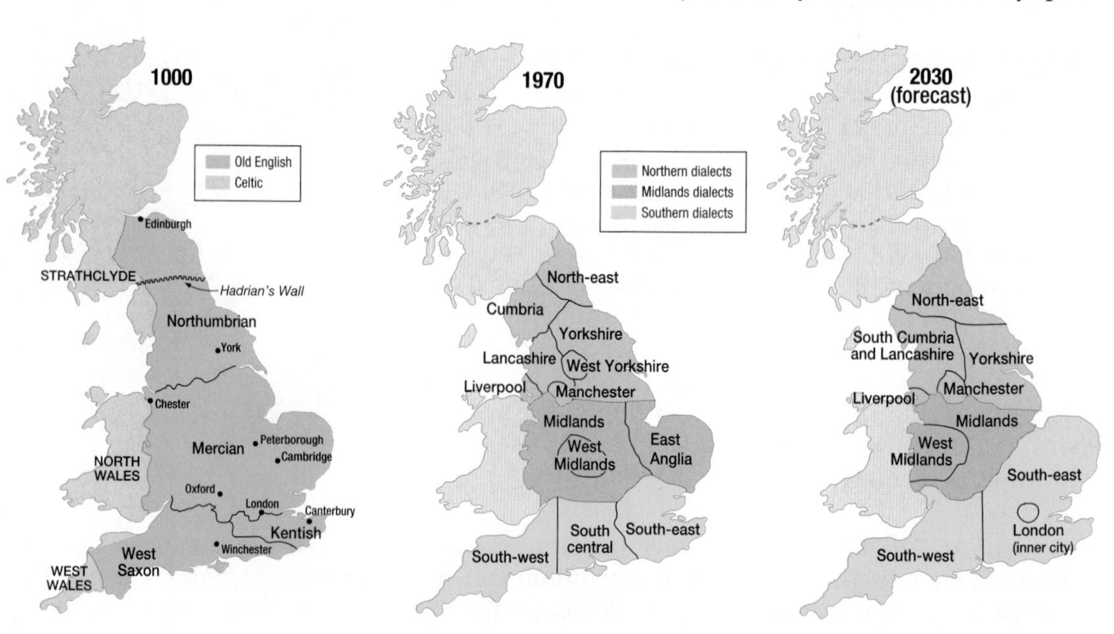

▶ **FIGURE 5-23 DIALECTS IN ENGLAND** Isoglosses between England's dialects of English are changing.

the language they had been using in England at the time.

Later immigrants from other countries found English already implanted here. Although they made significant contributions to American English, they became acculturated into a society that already spoke English. Therefore, the earliest colonists were most responsible for the dominant language patterns that exist today in the English-speaking part of the Western Hemisphere.

Why is the English language in the United States so different from that in England? As is so often the case with languages, the answer is isolation. Separated by the Atlantic Ocean, English in the United States and in England evolved independently during the eighteenth and nineteenth centuries, with little influence on one another. Few residents of one country could visit the other, and the means to transmit the human voice over long distances would not become available until the twentieth century.

U.S. English differs from the English of England in three significant ways—vocabulary, spelling, and pronunciation.

BRITISH American

PETROL Gas
LORRY Truck
SLEEPING POLICEMAN Speed Bump
CAR PARK Parking Lot
CAR JOURNEY Road Trip
ZEBRA CROSSING Crosswalk
MOTORWAY Freeway
SALOON Sedan
PETROL STATION Gas Station
BONNET Hood
WINDSCREEN Windshield
BOOT Trunk
REVERSING LIGHTS Back-up Lights
EXHAUST PIPE Tail Pipe
DUAL CARRIAGEWAY Divided Highway
NUMBER PLATE License Plate
FLYOVER Overpass
MULTI-STOREY CAR PARK Parking Garage
CAT'S EYE RAISED Pavement Marker
CARAVAN/CAMPERVAN RV
PAVEMENT Sidewalk
ESTATE CAR Station Wagon
MANUAL CAR Stickshift Car
GEAR STICK Stick
INDICATORS Turn Signal
TRAFFIC LIGHTS Stoplight
AMBER LIGHT (TRAFFIC LIGHTS) Yellow Light

▲ FIGURE 5-24 **DIFFERENCES BETWEEN BRITISH AND AMERICAN ENGLISH** Numerous features related to a car are identified by different words in American and British English dialects.

VOCABULARY. The vocabulary of U.S. English differs from the English of England largely because settlers in America encountered many new objects and experiences. The new continent contained physical features, such as large forests and mountains, that had to be given new names.

New animals were encountered, including the moose, raccoon, and chipmunk, all of which were given names borrowed from Native Americans. Indigenous American "Indians" also enriched American English with names for objects such as canoe, moccasin, and squash.

As new inventions appeared, they acquired different names on either side of the Atlantic. For example, the elevator is called a *lift* in England, and the flashlight is known as a *torch*. The British call the hood of a car the *bonnet* and the trunk the *boot* (Figure 5-24).

SPELLING. American spelling diverged from the British standard because of a strong national feeling in the United States for an independent identity. Noah Webster, the creator of the first comprehensive American dictionary and grammar books, was not just a documenter of usage; he had an agenda.

Webster was determined to develop a uniquely American dialect of English. He either ignored or was unaware of recently created rules of grammar and spelling developed in England. Webster argued that spelling and grammar reforms would help establish a national language, reduce cultural dependence on England, and inspire national pride. The spelling differences between British and American English, such as the elimination of the *u* from the British spelling of words such as *honour* and *colour* and the substitution of *s* for *c* in *defence*, are due primarily to the diffusion of Webster's ideas inside the United States.

PRONUNCIATION. From the time of their arrival in North America, colonists began to pronounce words differently from the British. Such divergence is normal, for interaction between the two groups was largely confined to exchange of letters and other printed matter rather than direct speech.

Such words as *fast*, *path*, and *half* are pronounced in England like the /*ah*/ in *father* rather than the /*a*/ in *man*. The British also eliminate the *r* sound from pronunciation except before vowels. Thus *lord* in British pronunciation sounds like *laud*.

Americans pronounce unaccented syllables with more clarity than do British English speakers. The words *secretary* and *necessary* have four syllables in American English but only three in British (*secret'ry* and *necess'ry*).

Surprisingly, pronunciation has changed more in England than in the United States. The letters *a* and *r* are pronounced in the United States closer to the way they were pronounced in Britain in the seventeenth century, when the first colonists arrived. A single dialect of Southern English did not emerge as the British national standard until the late eighteenth century, after the American colonies had declared independence and were politically as well as physically isolated from England. Thus people in the United States do not speak "proper" English because when the colonists left England, "proper" English was not what it is today. Furthermore, few colonists were drawn from the English upper classes.

Pause and Reflect 5.3.2

In British English dialect, circus has a second meaning in addition to a carnival with clowns. What is it?

Distinguishing between Languages and Dialects

Learning Outcome 5.3.3

Understand why it is sometimes difficult to distinguish between a language and a dialect.

Dialects are not confined to English; other languages, such as those in the Romance branch, have dialects. The Romance branch also demonstrates difficulties in distinguishing between dialects and distinct languages.

ROMANCE BRANCH DIALECTS

Distinct Romance languages did not suddenly appear in the former Roman Empire. As with other languages, they evolved over time. Numerous dialects existed within each province, and many of them are still spoken today. The creation of standard national languages, such as French and Spanish, occurred relatively recently.

SPANISH AND PORTUGUESE. Spain, like France, contained many dialects during the Middle Ages. One dialect, known as Castilian, arose during the ninth century in Old Castile, located in the north-central part of the country. The dialect spread southward over the next several hundred years, as independent kingdoms were unified into one large country.

Spain grew to its approximate present boundaries in the fifteenth century, when the Kingdom of Castile and Léon merged with the Kingdom of Aragón. At that time, Castilian became the official language for the entire country. Regional dialects, such as Aragón, Navarre, Léon, Asturias, and Santander, survived only in secluded rural areas. The official language of Spain is now called Spanish, although the term Castilian is still used in Latin America. Portuguese developed as a separate language because of Portugal's relative isolation on the west coast of the Iberian peninsula, especially after the fall of the Roman Empire.

Spanish and Portuguese have achieved worldwide importance because of the colonial activities of their European speakers. Approximately 90 percent of the speakers of these two languages live outside Europe, mainly in Central and South America. Spanish is the official language of 18 Latin American states, and Portuguese is spoken in Brazil, which has as many people as all the other South American countries combined and 18 times more people than Portugal itself.

These two Romance languages were diffused to the Americas by Spanish and Portuguese explorers. The division of Central and South America into Portuguese- and Spanish-speaking regions resulted from a 1493 decision by Pope Alexander VI to give the western portion of the New World to Spain and the eastern part to Portugal. The Treaty of Tordesillas, signed a year later, carried out the papal decision.

The Portuguese and Spanish languages spoken in the Western Hemisphere differ somewhat from their European versions, as is the case with English. The members of the Spanish Royal Academy meet every week in a mansion in Madrid to clarify rules for the vocabulary, spelling, and pronunciation of the Spanish language around the world. The academy's official dictionary, published in 1992, has added hundreds of "Spanish" words that originated either in the regional dialects of Spain or the Indian languages of Latin America.

Brazil, Portugal, and several Portuguese-speaking countries in Africa agreed in 1994 to standardize the way their common language is written. Many people in Portugal are upset that the new standard language more closely resembles the Brazilian version, which eliminates some of the accent marks—such as tildes (as in São Paulo), cedillas (as in Alcobaça), circumflexes (as in Estância), and hyphens—and the agreement recognizes as standard thousands of words that Brazilians have added to the language, such as flowers, animals, and other features of the natural environment found in Brazil but not in Portugal.

The standardization of Portuguese is a reflection of the level of interaction that is possible in the modern world between groups of people who live tens of thousands of kilometers apart. Books and television programs produced in one country diffuse rapidly to other countries where the same language is used. Refer to Figure 5-17, which shows an exhibit at the Museum of Portuguese Language in São Paulo, Brazil.

Pause and Reflect 5.3.3

Five hundred years from now, why might Spanish tourists in Peru not be easily understood by Peruvians if they speak their own version of Spanish?

DIALECT OR LANGUAGE?

Difficulties arise in determining whether two languages are distinct or whether they are two dialects of the same language. Here are several examples from Romance languages.

LANGUAGES OF ITALY. Several languages in Italy that have been traditionally classified as dialects of Italian are now viewed by *Ethnologue* as sufficiently different to merit consideration as languages distinct from Italian (number of speakers in parentheses):

- Emiliano-Romagnolo (2 million)
- Liguria (2 million)
- Lombard (9 million)
- Napoletano-Calebrese (7 million)
- Piemontese (3 million)
- Sicilian (5 million)
- Venetian (2 million)

Refer to Figure 5-13 for the distribution of these languages (or dialects) within Italy.

▲ **FIGURE 5-25 CATALÁN** This sign warning that this is private property was written in Spanish. The graffiti is in Catalán.

CATALÁN-VALENCIAN-BALEAR. Catalán was once regarded as a dialect of Spanish, but linguists now agree that it is a separate Romance language (refer to Figure 5-13). Like other Romance languages, Catalán can be traced to Vulgar Latin, and it developed as a separate language after the collapse of the Roman Empire (Figure 5-25).

As the status of Catalán as a separate language is settled, linguists are identifying its principal dialects. Linguists agree that Balear is a dialect of Catalán that is spoken in the Balearic Islands, which include Ibiza and Majorca. More controversial is the status of Valencian, which is spoken mostly in and around the city of Valencia. Most linguists consider Valencian a dialect of Catalán. However, many in Valencia, including the Valencian Language Institute, consider Valencian a separate language, because it contains words derived from people who lived in the region before the Roman conquest. *Ethnologue* now calls the language Catalán -Valencian-Balear.

GALICIAN. Whether Galician, which is spoken in northwestern Spain and northeastern Portugal, is a dialect of Portuguese or a distinct language is debated among speakers of Galician. The Academy of Galician Language considers it a separate language and a symbol of cultural independence. The Galician Association of the Language prefers to consider it a dialect because as a separate language, it would be relegated to a minor and obscure status, whereas as a dialect of Portuguese it can help to influence one of the world's most widely used languages.

MOLDOVAN. Generally classified as a dialect of Romanian, Moldovan is the official language of Moldova. Moldovan is written, like Russian, in Cyrillic letters, a legacy of Moldova being a part of the Soviet Union, whereas Romanian is written in Roman letters.

CREOLE LANGUAGES. Romance languages spoken in some former colonies can also be classified as separate languages because they differ substantially from the original introduced by European colonizers. Examples include French Creole in Haiti, Papiamento (creolized Spanish) in Netherlands Antilles (West Indies), and Portuguese Creole in the Cape Verde Islands off the African coast. A **creole**, or **creolized language**, is defined as a language that results from the mixing of the colonizer's language with the indigenous language of the people being dominated (Figure 5-26). A creolized language forms when the colonized group adopts the language of the dominant group but makes some changes, such as simplifying the grammar and adding words from the former language.

CHECK-IN: KEY ISSUE 3

Why Do Individual Languages Vary among Places?

- ✓ **A dialect is a regional variation of a language; the United States has several major dialects.**
- ✓ **Dialects vary based on vocabulary, spelling, and pronunciation.**
- ✓ **The distinction between a dialect and an entirely different language is not always clear-cut.**

▼ **FIGURE 5-26 CREOLE LANGUAGE** This note, written in French Creole in Haiti, shortly after a devastating earthquake killed 40,000 in January 2010, is the beginning of 2 Timothy 3:16, "All Scripture is given by inspiration of God."

KEY ISSUE 4

Why Do People Preserve Local Languages?

- Language Diversity
- Global Dominance of English

Learning Outcome 5.4.1
Understand how several countries peacefully embrace more than one language.

The distribution of a language is a measure of the fate of a cultural group. English has diffused around the world from a small island in northwestern Europe because of the dominance of England and the United States over other territory on Earth's surface. Icelandic remains a little-used language because of the isolation of the Icelandic people.

As in other cultural traits, language displays the two competing geographic trends of globalization and local diversity. English has become the principal language of communication and interaction for the entire world. At the same time, local languages endangered by the global dominance of English are being protected and preserved.

Language Diversity

In some countries, multiple languages coexist, with varying degrees of success. Other countries maintain the use of languages that have little if any relationship to other languages.

MULTILINGUAL STATES

Difficulties can arise at the boundary between two languages. Belgium, Switzerland, and Nigeria offer examples of varying degrees of difficulties.

BELGIUM. Note in Figures 5-9 (Indo-European languages) and 5-10 (Germanic languages) that the boundary between the Romance and Germanic branches runs through the middle of two small European countries, Belgium and Switzerland. Belgium has had more difficulty than Switzerland in reconciling the interests of the different language speakers.

Southern Belgians (known as Walloons) speak French, whereas northern Belgians (known as Flemings) speak Flemish, a dialect of the Germanic language Dutch (Figure 5-27). The language boundary sharply divides the country into two regions. Antagonism between the Flemings and Walloons is aggravated by economic and political

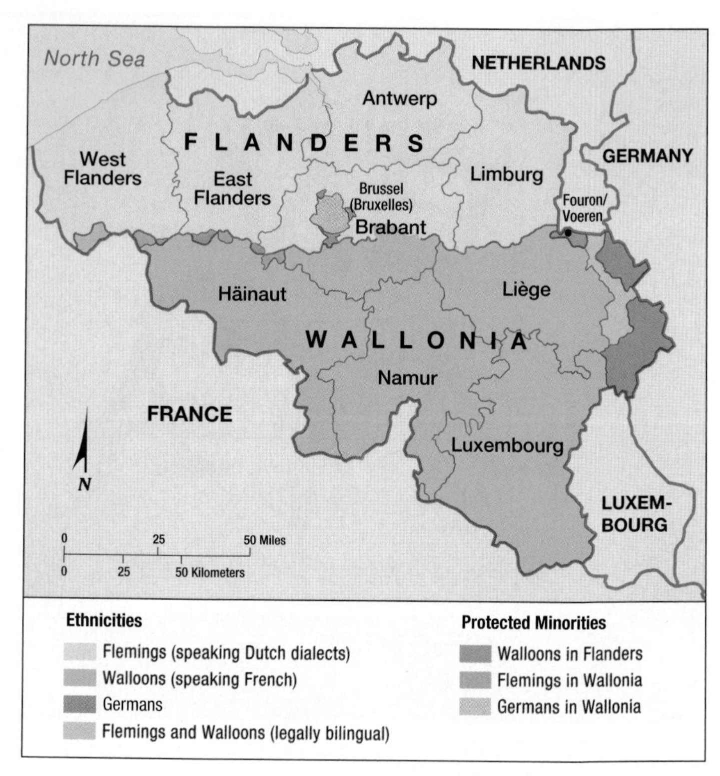

▲ **FIGURE 5-27 LANGUAGES IN BELGIUM** Flemings in the north speak Flemish, a Dutch dialect. Walloons in the south speak French. The two groups have had difficulty sharing national power.

differences. Historically, the Walloons dominated Belgium's economy and politics, and French was the official state language. Brussels, the capital city, is officially bilingual, and signs are in both French and Flemish (Figure 5-28).

In response to pressure from Flemish speakers, Belgium has been divided into two autonomous regions, Flanders and Wallonia. Each elects an assembly that controls cultural affairs, public health, road construction, and urban development in its region. But for many in Flanders, regional autonomy is not enough. They want to see Belgium divided into two independent countries. Were that to occur, Flanders would be one of Europe's richest countries and Wallonia one of the poorest.

▼ **FIGURE 5-28 LANGUAGE DIVERSITY IN BELGIUM** Delhaize, a supermarket chain in Belgium, advertises on adjacent posters "the best at the best prices" (left) in French and (right) in Flemish.

SWITZERLAND. In contrast with Belgium, Switzerland peacefully exists with multiple languages. The key is a long tradition of decentralized government, in which local authorities hold most of the power, and decisions are frequently made by voter referenda. Switzerland has four official languages—German (used by 65 percent of the population), French (18 percent), Italian (10 percent), and Romansh (1 percent). Swiss voters made Romansh an official language in a 1938 referendum, despite the small percentage of people who use the language.

Switzerland is divided into four main linguistic regions, as shown in Figure 5-29, but people living in individual communities, especially in the mountains, may use a language other than the prevailing local one. The Swiss, relatively tolerant of citizens who speak other languages, have institutionalized cultural diversity by creating a form of government that places considerable power in small communities.

NIGERIA. Africa's most populous country, Nigeria, displays problems that can arise from the presence of many speakers of many languages. Nigeria has 527 distinct languages, according to *Ethnologue*, only three of which have widespread use—Hausa, Yoruba, and Igbo, each spoken by one-eighth of the population (Figure 5-30).

Groups living in different regions of Nigeria have often battled. The southern Igbos attempted to secede from Nigeria during the 1960s, and northerners have repeatedly claimed that the Yorubas discriminate against them. To reduce these regional tensions, the government has moved the capital from Lagos in the Yoruba-dominated southwest to Abuja in the center of Nigeria.

Nigeria reflects the problems that can arise when great cultural diversity—and therefore language diversity—is packed into a relatively small region. Nigeria also illustrates the importance of language in identifying distinct cultural groups at a local scale. Speakers of one language are unlikely to understand any of the others in the same language family, let alone languages from other families.

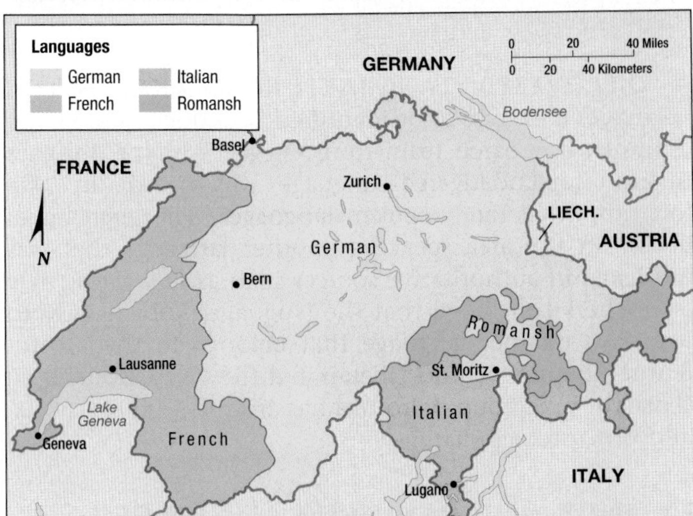

▲ **FIGURE 5-29 LANGUAGE DIVERSITY IN SWITZERLAND** The map shows Switzerland's four official languages. The photo shows a sign that prevents hikers, vehicles, and horses from entering the forest because of timber cutting. German is top left, French top right, Italian lower left, and Romansh lower right. Switzerland lives peacefully with four official languages, including Romansh, which is used by only 1 percent of the population.

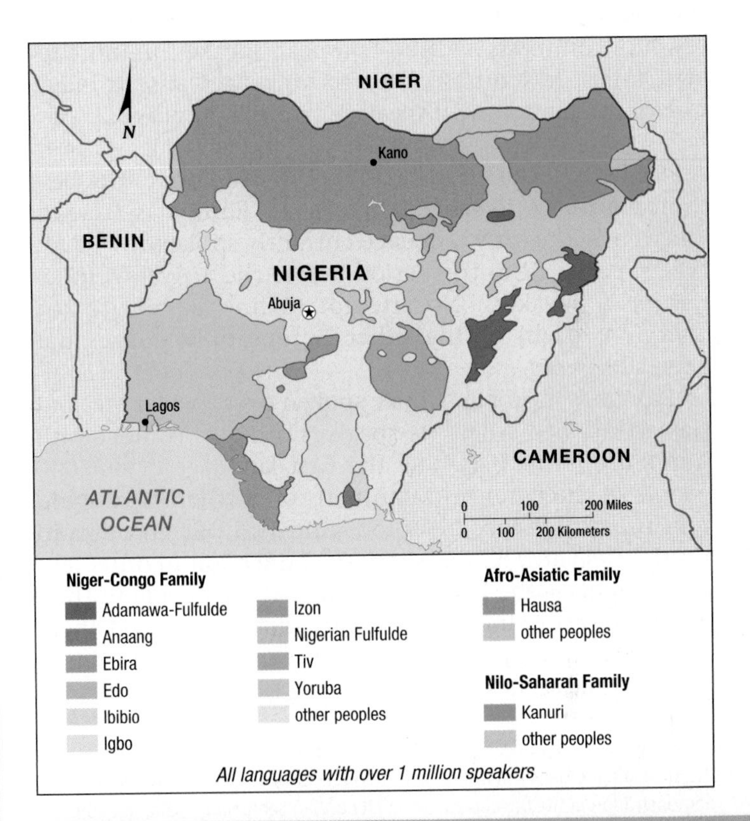

▲ **FIGURE 5-30 LANGUAGE DIVERSITY IN NIGERIA** The map shows Nigeria's principal languages. The photo shows Nigeria's capital city Abuja, which was built in the center of the country, where none of the three largest languages dominates. The city skyline includes a cathedral (left), national bank (center), and mosque (right).

ISOLATED LANGUAGES

Learning Outcome 5.4.2
Understand what is meant by an isolated language and an extinct language.

An **isolated language** is a language unrelated to any other and therefore not attached to any language family. Similarities and differences between languages—our main form of communication—are a measure of the degree of interaction among groups of people.

The diffusion of Indo-European languages demonstrates that a common ancestor dominated much of Europe before recorded history. Similarly, the diffusion of Indo-European languages to the Western Hemisphere is a result of conquests by Indo-European speakers in more recent times. In contrast, isolated languages arise through lack of interaction with speakers of other languages.

A PRE-INDO-EUROPEAN SURVIVOR: BASQUE. The best example of an isolated language in Europe is Basque, apparently the only language currently spoken in Europe that survives from the period before the arrival of Indo-European speakers. No attempt to link Basque to the common origin of the other European languages has been successful.

Basque was probably once spoken over a wider area but was abandoned where its speakers came in contact with Indo-Europeans. It is now the first language of 666,000 people in the Pyrenees Mountains of northern Spain and southwestern France (refer to Figure 5-13, the gray area in northern Spain). Basque's lack of connection to other languages reflects the isolation of the Basque people in their mountainous homeland. This isolation has helped them preserve their language in the face of the wide diffusion of Indo-European languages (Figure 5-31).

▼ **FIGURE 5-31 BASQUE** Protestors hold banners that say, in Basque, "Stop the state of emergency; self-determination for Basque Country," during a demonstration in the Basque-speaking city of San Sebastian, Spain, in 2009.

▲ **FIGURE 5-32 ICELANDIC** The warning sign in Icelandic and English is located in Hveragerdi, Iceland.

AN UNCHANGING LANGUAGE: ICELANDIC. Icelandic is related to other languages in the North Germanic group of the Germanic branch of the Indo-European family (Figure 5-32). Icelandic's significance is that over the past 1,000 years, it has changed less than any other language in the Germanic branch. As was the case with England, people in Iceland speak a Germanic language because their ancestors migrated to the island from the east, in this case from Norway. Norwegian settlers colonized Iceland in A.D. 874.

When an ethnic group migrates to a new location, it takes along the language spoken in the former home. The language spoken by most migrants—such as the Germanic invaders of England—changes in part through interaction with speakers of other languages. But in the case of Iceland, the Norwegian immigrants had little contact with speakers of other languages when they arrived in Iceland, and they did not have contact with speakers of their language back in Norway. After centuries of interaction with other Scandinavians, Norwegian and other North Germanic languages had adopted new words and pronunciation, whereas the isolated people of Iceland had less opportunity to learn new words and no reason to change their language.

A "DISCOVERED" LANGUAGE: KORO AKA. Isolated languages continue to be identified and documented. For example, a research team from Oregon's Living Tongues Institute for Endangered Languages was in India in 2008 to study other rarely spoken languages. The team heard people in the area speaking another language that was not listed in authoritative sources such as *Ethnologue*. The researchers concluded that the language, known as Koro Aka, is a distinct language that belongs to the Tibeto-Burman branch of Sino-Tibetan, but they were not able to classify it in a group. Koro Aka has around 1,000 speakers, in northeastern India.

EXTINCT AND REVIVED LANGUAGES

Thousands of languages are **extinct languages** that were once in use—even in the recent past—but are no longer spoken or read in daily activities by anyone in the world. *Ethnologue* considers 473 languages to be nearly extinct because only a few older speakers are still living, and they are not teaching the languages to their children. According to *Ethnologue*, 46 of these nearly extinct languages are in Africa, 182 in the Americas, 84 in Asia, 9 in Europe, and 152 in the Pacific.

MANY EXTINCT LANGUAGES: NATIVE AMERICANS. When Spanish missionaries reached the eastern Amazon region of Peru in the sixteenth century, they found more than 500 languages. Only 92 survive today, according to *Ethnologue*, and 14 of these face immediate extinction because fewer than 100 speakers remain. Of Peru's 92 surviving indigenous languages, only Cusco, a Quechuan language, is currently used by more than 1 million people.

Ethnologue lists 74 languages based in the United States that are now extinct. These are languages once spoken by groups of Native Americans, especially in the West (Figure 5-33).

AN EXTINCT LANGUAGE: GOTHIC. Gothic was widely spoken by people in Eastern and Northern Europe in the third century. Not only is Gothic extinct but so is the entire language group to which it belonged, the East Germanic group of the Germanic branch of Indo-European. The last speakers of Gothic lived in the Crimea in Russia in the sixteenth century.

The Gothic language died because the descendants of the Goths were converted to other languages through processes of integration, such as political dominance and cultural preference. For example, many Gothic people switched to speaking the Latin language after their conversion to Christianity. Similarly, indigenous languages are disappearing in Peru as speakers switch to Spanish.

REVIVING AN EXTINCT LANGUAGE: HEBREW. Hebrew is a rare case of an extinct language that has been revived (Figure 5-34). Most of the Jewish Bible (Christian Old Testament) was written in Hebrew. (A small part of it was written in another Afro-Asiatic language, Aramaic.) A language of daily activity in biblical times, Hebrew diminished in use in the fourth century B.C. and was thereafter retained only for Jewish religious services. At the time of Jesus, people in present-day Israel generally spoke Aramaic, which in turn was replaced by Arabic.

When Israel was established as an independent country in 1948, Hebrew became one of the new country's two official languages, along with Arabic. Hebrew was chosen because the Jewish population of Israel consisted of refugees and migrants from many countries who spoke many languages. Because Hebrew was still used in Jewish prayers, no other language could so symbolically unify the disparate cultural groups in the new country.

The task of reviving Hebrew as a living language was formidable. Words had to be created for thousands of objects and inventions unknown in biblical times, such as telephones, cars, and electricity. The revival effort was initiated by Eliezer Ben-Yehuda, who lived in Palestine before the creation of the state of Israel and who refused to speak any language other than Hebrew. Ben-Yehuda is credited with the invention of 4,000 new Hebrew words—related when possible to ancient ones—and the creation of the first modern Hebrew dictionary.

Pause and Reflect 5.4.2
Can you think of other words that would not have existed in ancient times?

▼ **FIGURE 5-33 ALGONQUIN** Student in Chisasibi, Québec, writes in Cree, an Algonquian language.

▼ **FIGURE 5-34 HEBREW** The road signs are in (top) Hebrew, (middle) Arabic, and English.

PRESERVING ENDANGERED LANGUAGES: CELTIC

Learning Outcome 5.4.3
Understand why the number of Celtic speakers has declined and how the languages are being preserved.

Some endangered languages are being preserved. Nonetheless, linguists expect that hundreds of languages will become extinct during the twenty-first century and that only about 300 languages are clearly safe from extinction because they have sufficient speakers and official government support.

The Celtic branch of Indo-European is of particular interest to English speakers because it was the major language in the British Isles before the Germanic Angles, Jutes, and Saxons invaded. Two thousand years ago, Celtic languages were spoken in much of present-day Germany, France, and northern Italy, as well as in the British Isles. Today, Celtic languages survive only in remote parts of Scotland, Wales, and Ireland and on the Brittany peninsula of France.

The Celtic language branch is divided into Goidelic (Gaelic) and Brythonic groups. Two Goidelic languages survive—Irish Gaelic and Scottish Gaelic. Speakers of Brythonic (also called Cymric or Britannic) fled westward during the Germanic invasions to Wales, southwestward to Cornwall, or southward across the English Channel to the Brittany peninsula of France. Recent efforts have prevented the disappearance of Celtic languages and others in Europe. The fate of five Celtic languages is described here, in order of number of speakers.

WELSH (BRYTHONIC). Wales—the name derived from the Germanic invaders' word for *foreign*—was conquered by the English in 1283. Welsh remained dominant in Wales until the nineteenth century, when many English speakers migrated there to work in coal mines and factories. A 2004 survey found 611,000 Welsh speakers in Wales, 22 percent of the population. In some isolated communities in the northwest, especially in the county of Gwynedd, two-thirds speak Welsh.

Cymdeithas yr Iaith Gymraeg (Welsh Language Society) has been instrumental in preserving the language. Britain's 1988 Education Act made Welsh language training a compulsory subject in all schools in Wales, and Welsh history and music have been added to the curriculum. All local governments and utility companies are now obliged to provide services in Welsh. Welsh-language road signs have been posted throughout Wales, and the British Broadcasting Corporation (BBC) produces Welsh-language television and radio programs (Figure 5-35). Knowledge of Welsh is now required for many jobs, especially in public service, media, culture, and sports.

▲ **FIGURE 5-35 WELSH** Members of the Welsh Language Society protest closure of small rural schools; the signs say "save our Welsh-speaking village schools."

IRISH. Irish Gaelic and English are the Republic of Ireland's two official languages. Irish is spoken by 350,000 people on a daily basis, and 1.5 million say that they can speak it (Figure 5-36). An Irish-language TV station began broadcasting in 1996. English road signs were banned from portions of western Ireland in 2005. The revival is being led by young Irish living in other countries who wish to distinguish themselves from the English (in much the same way that Canadians traveling abroad often make efforts to distinguish themselves from U.S. citizens). Irish singers, including many rock groups (although not U2), have begun to record and perform in Gaelic.

In the 1300s, the Irish were forbidden to speak their own language in the presence of their English masters. By the nineteenth century, Irish children were required

▼ **FIGURE 5-36 IRISH** The name of the pub means "the little bridge" in Irish.

to wear "tally sticks" around their necks at school. The teacher carved a notch in the stick every day the child used an Irish word, and at the end of the day meted out punishment based on the number of tallies. Parents encouraged their children to learn English so that they could compete for jobs.

Pause and Reflect 5.4.3
Use Google Translate to type something in English and see its translation in Irish and in Welsh. Do Irish and Welsh appear similar or very different?

BRETON. In Brittany—like Cornwall, an isolated peninsula that juts out into the Atlantic Ocean—around 250,000 people speak Breton regularly. Breton differs from the other Celtic languages in that it has more French words (Figure 5-37).

SCOTTISH. In Scotland 59,000, or 1 percent of the people, speak Scottish Gaelic (Figure 5-38). An extensive body of literature exists in Gaelic languages, including the Robert Burns poem *Auld Lang Syne* ("old long since"), the basis for the popular New Year's Eve song. Gaelic was carried from Ireland to Scotland about 1,500 years ago.

CORNISH. Cornish became extinct in 1777, with the death of the language's last known native speaker, Dolly Pentreath, who lived in Mousehole (pronounced "muzzle"). Before Pentreath died, an English historian wrote down as much of her speech as possible so that future generations could study the Cornish language. One of her last utterances was later translated as "I will not speak English . . . you ugly, black toad!"

▲ **FIGURE 5-38 SCOTTISH** The sign over the door says that this is a florist. Eilean Iarmain is the Scottish name for the village of Isleornsay, Scotland.

▼ **FIGURE 5-37 BRETON** Sign for the town is in French and Breton. In the background is the world's largest collections of ancient stones, which were erected more than 5,000 years ago by people who inhabited Brittany before the Celts.

A few hundred people have become fluent in the formerly extinct Cornish language, which was revived in the 1920s. Cornish is taught in grade schools and adult evening courses and is used in some church services. Some banks accept checks written in Cornish. See the Sustainability and Inequality in Our Global Village box for more on the revival of Cornish. After years of dispute over how to spell the revived language, various groups advocating for the revival of Cornish reached an agreement in 2008 on a standard written version of the language. Because the language became extinct, it is impossible to know precisely how to pronounce Cornish words.

The long-term decline of languages such as Celtic provides an excellent example of the precarious struggle for survival that many languages experience. Faced with the diffusion of alternatives used by people with greater political and economic strength, speakers of Celtic and other languages must work hard to preserve their linguistic cultural identity.

PRESERVING ABORIGINAL AND MAORI IN AUSTRALIA AND NEW ZEALAND

English is the most widely used language in Australia and New Zealand as a result of British colonization during the early nineteenth century. Settlers in Australia and New Zealand established and maintained outposts of British culture, including use of the English language.

Though English remains the dominant language of Australia and New Zealand, the languages that predate British settlement survive in both countries. However, the two countries have adopted different policies with regard to preserving indigenous languages. Australia regards English as a tool for promoting cultural diversity, whereas New Zealand regards linguistic diversity as an important element of cultural diversity.

AUSTRALIA. In Australia, 1 percent of the population is Aboriginal. Many elements of Aboriginal culture are now being preserved. But education is oriented toward teaching English rather than maintaining local languages. English is the language of instruction throughout Australia, and others are relegated to the status of second language.

An essential element in maintaining British culture was restriction of immigration from non-English-speaking places during the nineteenth and early twentieth centuries. Fear of immigration was especially strong in Australia because of its proximity to other Asian countries. Under a "White Australia" policy, every prospective immigrant was required to write 50 words of a European language dictated by an immigration officer. The dictation test was not eliminated until 1957. The Australian government now merely requires that immigrants learn English.

NEW ZEALAND. In New Zealand, more than 10 percent of the population is Maori, descendents of Polynesian people who migrated there around 1,000 years ago (Figure 5-39). In contrast with Australia, New Zealand has adopted policies to preserve the Maori language. Most notably, Maori has became one of New Zealand's three official languages, along with English and sign language. A Maori Language Commission was established to preserve the language. Despite official policies, only 1 percent of New Zealanders are fluent in Maori, most of whom are over age 50. Preserving the language requires skilled teachers and the willingness to endure inconvenience compared to using the world's lingua franca, English.

On the other hand, New Zealand's language requirement for immigrants is more stringent than Australia's: In most circumstances, immigrants must already be fluent in English, although free English lessons are available to immigrants for the exceptions. More remote from Asian landmasses, New Zealand has attracted fewer Asian immigrants.

Pause and Reflect 5.4.6
Which language policy do you favor, Australia's or New Zealand's? Why?

PRESERVING OCCITAN IN FRANCE

The most important linguistic difference within France is between the north and the south (refer to Figure 5-13). In the north, the most commonly spoken language is what is now known as French. The standard form of French derives from Francien, which was once a dialect of the Île-de-France region of the country.

Francien became the standard form of French because the region included Paris, which became the capital and largest city of France. Francien French became the country's official language in the sixteenth century, and local dialects tended to disappear as a result of the capital's longtime dominance over French political, economic, and social life.

Occitan is spoken by about 2 million people in southern France and adjacent countries. The name derives from the French region of Aquitaine, which in French has a similar

▲ **FIGURE 5-39 MAORI LANGUAGE, NEW ZEALAND** The sign, in Maori, is the name of this place, and is said to be the world's second-longest place name, at 85 letters. The Maori translates as "the summit where Tamatea, the man with the big knees, the climber of mountains, the land-swallower who travelled about, played his nose flute to his loved one."

SUSTAINABILITY AND INEQUALITY IN OUR GLOBAL VILLAGE

Preserving Lesser-Used Languages

The sustainability of any language depends on the political and military strength of its speakers. The Celtic languages declined because the Celts lost most of the territory they once controlled to speakers of other languages. Most remaining Celtic speakers also know the language of their English or French conquerors.

In 1982, the European Union established the European Bureau for Lesser Used Languages (EBLUL), based in Dublin, Ireland, to provide financial support for the preservation of several dozen indigenous, regional, and minority languages spoken by 46 million Europeans. The Celtic languages received a lot of attention from EBLUL; for example, in 2002, EBLUL granted Cornish official status within the European Union (Figure 5-40).

▲ **FIGURE 5-40 CORNISH** Sign for the town is in English and Cornish. The literal English translation of the Cornish version is "Cornwall welcomes you."

The European Union cut off funding for EBLUL in 2010, and the office was closed. Local individually based organizations such as *Cymdeithas yr Iaith Gymraeg* are expected to carry the responsibility of preserving lesser-used languages.

pronunciation to Occitan. Numerous dialects of Occitan are spoken, including Auvergnat, Gascon, and Provençal

French dialects of northern France are sometimes known by the French phrases langue d'oïl and the southern as langue d'òc. It is worth exploring these terms, for they provide insight into how languages evolve. These names derive from different ways in which the word for "yes" was said. One Roman term for "yes" was hoc illud est, meaning "that is so." In the south, the phrase was shortened to hoc, or òc, because the /h/ sound was generally dropped, just as we drop it on the word honor today. Northerners shortened the phrase to o-il after the first sound in the first two words of the phrase, again with the initial /h/ suppressed. If the two syllables of o-il are spoken very rapidly, they are combined into a sound like the English word wheel. Eventually, the final consonant was eliminated, as in many French words, giving a sound for "yes" like the English we, spelled in French oui.

The French government has established bilingual elementary and high schools called calandretas in the Occitan region. These schools teach both French and Occitan, according to a curriculum established by the national ministry of education. Still, many people living in southern France want to see more efforts by the government of France to encourage the use of Occitan (Figure 5-41).

▲ **FIGURE 5-41 PROVENCAL** People demonstrate in Beaucaire, France, for the preservation of the Provencal language *langue d'òc*.

Global Dominance of English

Learning Outcome 5.4.4
Understand the concept of a lingua franca.

One of the most fundamental needs in a global society is a common language for communication. Increasingly in the modern world, the language of international communication is English. A Polish airline pilot who flies over Spain speaks to the traffic controller on the ground in English. Swiss bankers speak a dialect of German among themselves, but with German bankers they prefer to speak English rather than German. English is the official language at an aircraft factory in France and an appliance company in Italy.

The dominance of English as an international language has facilitated the diffusion of popular culture and science and the growth of international trade. However, people who forsake their native language must weigh the benefits of using English against the cost of losing a fundamental element of local cultural identity.

English is the first language of 328 million people and is spoken fluently by another estimated ½ to 1 billion people (Figure 5-42). English is an official language in 57 countries, more than any other language, and is the predominant language in 2 more (Australia and the United States). Two billion people—one-third of the world—live in a country where English is an official language, even if they cannot speak it (Figure 5-43).

ENGLISH: AN EXAMPLE OF A LINGUA FRANCA

A language of international communication, such as English, is known as a **lingua franca.** To facilitate trade, speakers of two different languages create a lingua franca by mixing elements of the two languages into a simple common language. The term, which means *language of the Franks,* was originally applied by Arab traders during the Middle Ages to describe the language they used to communicate with Europeans, whom they called Franks.

People in smaller countries need to learn English to participate more fully in the global economy and culture. All children learn English in the schools of countries such as the Netherlands and Sweden to facilitate international communication. This may seem culturally unfair, but obviously it is more likely that several million Dutch people will learn English than that a half-billion English speakers around the world will learn Dutch.

The rapid growth in importance of English is reflected in the percentage of students learning English as a second language in school. More than 90 percent of students in the European Union learn English in middle or high school, not just in smaller countries such as Denmark and the Netherlands but also in populous countries such as France, Germany, and Spain. The Japanese government, having determined that fluency in English is mandatory in a global economy, has even considered adding English as a second official language.

Foreign students increasingly seek admission to universities in countries that teach in English rather than in

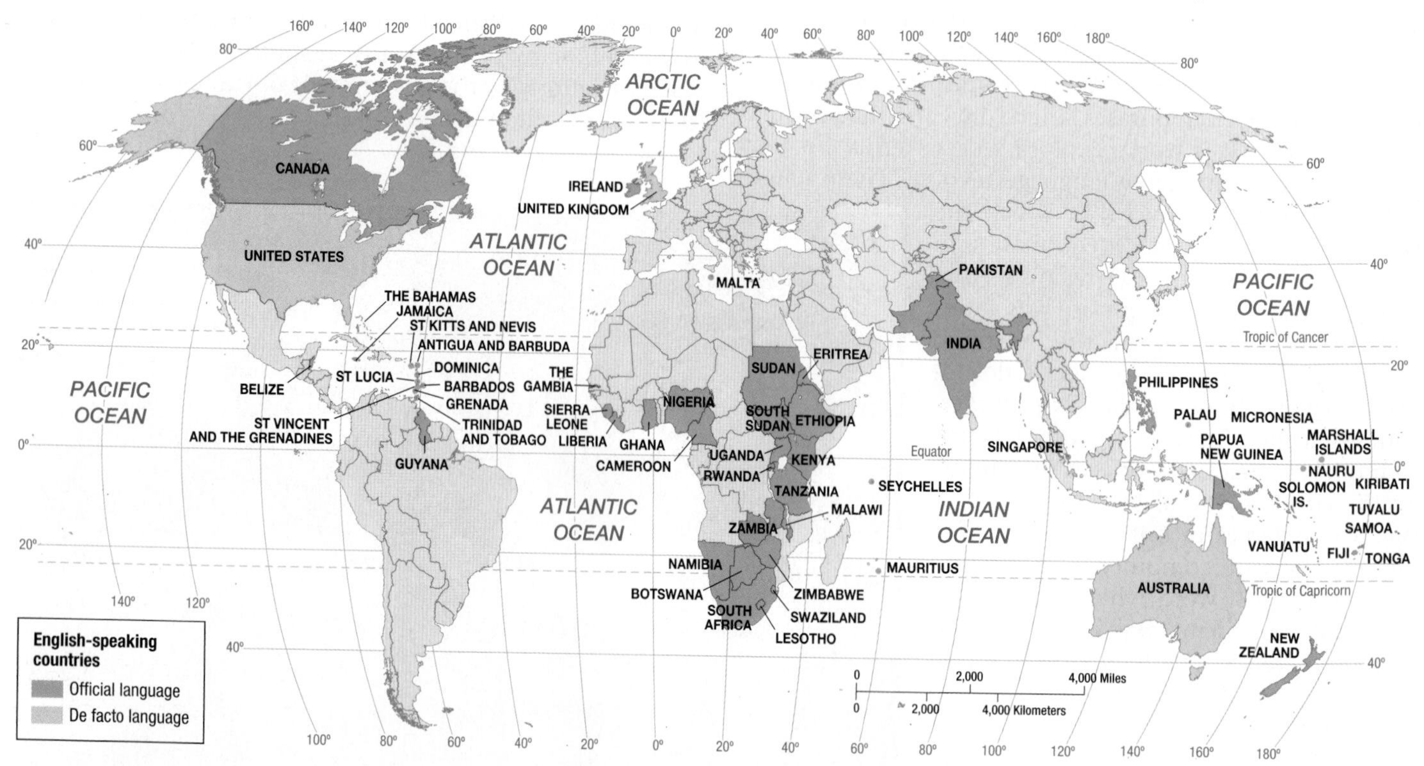

▲ **FIGURE 5-42 ENGLISH-SPEAKING COUNTRIES** English is an official language in 56 countries. English is also the predominant language in the United Kingdom, United States, and Australia, although these countries have declared it to be the official language.

▲ **FIGURE 5-43 TEACHING ENGLISH** English is widely taught around the world, including this school in China.

German, French, or Russian. Students around the world want to learn in English because they believe it is the most effective way to work in the global economy and participate in the global culture.

A group that learns English or another lingua franca may learn a simplified form, called a **pidgin language.** To communicate with speakers of another language, two groups construct a pidgin language by learning a few of the grammar rules and words of a lingua franca, mixing in some elements of their own languages. A pidgin language has no native speakers; it is always spoken in addition to one's native language.

Other than English, modern lingua franca languages include Swahili in East Africa, Hindi in South Asia, Indonesian in Southeast Asia, and Russian in the former Soviet Union. A number of African and Asian countries that became independent in the twentieth century adopted English or Swahili as an official language for government business, as well as for commerce, even if the majority of the people couldn't speak it.

In view of the global dominance of English, many U.S. citizens do not recognize the importance of learning other languages. One of the best ways to learn about the beliefs, traits, and values of people living in other regions is to learn their language. The lack of effort by Americans to learn other languages is a source of resentment among people elsewhere in the world, especially when Americans visit or work in other countries. The inability to speak other languages is also a handicap for Americans who try to conduct international business. Successful entry into new overseas markets requires knowledge of local culture, and officials who can speak the local language are better able to obtain important information. Japanese businesses that wish to expand in the United States send English-speaking officials, but American businesses that wish to sell products to the Japanese are rarely able to send a Japanese-speaking employee.

CONTEMPORARY GEOGRAPHIC TOOLS

The Death of English as a Lingua Franca?

English will disappear as a lingua franca, claims Nicholas Ostler, who heads the United Kingdom's Foundation for Endangered Languages, and no other language will replace it. Advances in technology enable people to continue speaking their native language while using the computer and speech recognition devices to translate between it and English.

Figure 5-44 is an excerpt from the Welsh language version of the 2011 UK census form. What are questions 18 and 19 asking? Use an online translation service, such as Google translator, at http://translate.google.com. Set the left box for Welsh and the right box for English, and type the Welsh from the census form into the left box.

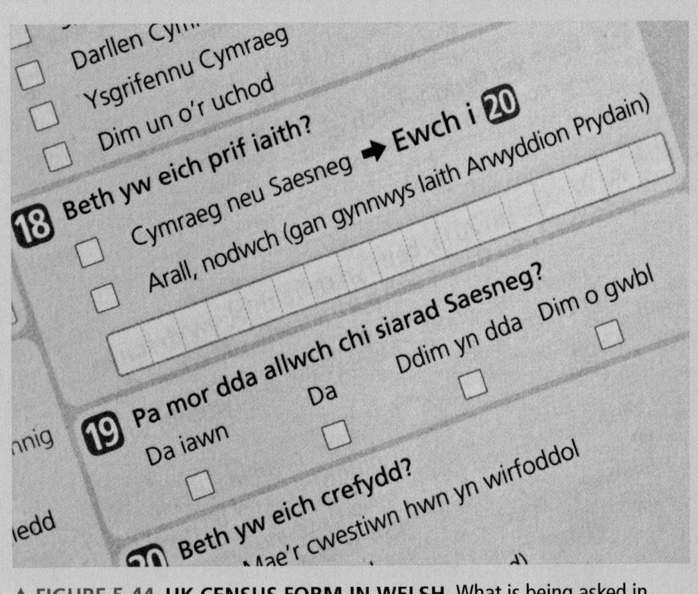
Ysgrifennu Cymraeg
Dim un o'r uchod
18 Beth yw eich prif iaith?
Cymraeg neu Saesneg ➜ Ewch i 20
Arall, nodwch (gan gynnwys Iaith Arwyddion Prydain)
19 Pa mor dda allwch chi siarad Saesneg?
Da iawn
Da
Ddim yn dda
Dim o gwbl
20 Beth yw eich crefydd?
Mae'r cwestiwn hwn yn wirfoddol

▲ **FIGURE 5-44 UK CENSUS FORM IN WELSH** What is being asked in questions 18 and 19?

EXPANSION DIFFUSION OF ENGLISH

Learning Outcome 5.4.5
Understand how English has diffused to other languages.

In the past, a lingua franca achieved widespread distribution through migration and conquest. Two thousand years ago, use of Latin spread through Europe along with the Roman Empire. In recent centuries, use of English spread around the world primarily through the British Empire.

In contrast, the recent growth in the use of English is an example of expansion diffusion, the spread of a trait through the snowballing effect of an idea rather than through the relocation of people. Expansion diffusion has occurred in two ways with English:

1. English is changing through diffusion of new vocabulary, spelling, and pronunciation.
2. English words are fusing with other languages.

For a language to remain vibrant, new words and usage must be coined to deal with new situations. Unlike most examples of expansion diffusion, recent changes in English have percolated up from common usage and ethnic dialects rather than being directed down to the masses by elite people. Examples include dialects spoken by African Americans and residents of Appalachia.

AFRICAN AMERICAN ENGLISH. Some African Americans speak a dialect of English heavily influenced by the group's distinctive heritage of forced migration from Africa during the eighteenth century to be slaves in the southern colonies. African American slaves preserved a distinctive dialect in part to communicate in a code not understood by their white masters. Black dialect words such as *gumbo* and *jazz* have long since diffused into the standard English language.

In the twentieth century, many African Americans migrated from the South to the large cities in the Northeast and Midwest (see Chapter 7). Living in racially segregated neighborhoods within northern cities and attending segregated schools, many African Americans preserved their distinctive dialect. That dialect has been termed African American Vernacular English (AAVE). Since 1996, the term **Ebonics**, a combination of *ebony* and *phonics*, has sometimes been used as a synonym for AAVE.

The American Speech, Language and Hearing Association classifies AAVE as a distinct dialect, with a recognized vocabulary, grammar, and word meaning. Among the distinctive elements of Ebonics are the use of double negatives, such as "I ain't going there no more," and such sentences as "She be at home" instead of "She is usually at home."

Use of AAVE is controversial within the African American community. On one hand, some regard it as substandard, a measure of poor education, and an obstacle to success in the United States. Others see AAVE as a means for preserving a distinctive element of African American culture and an effective way to teach African Americans who otherwise perform poorly in school.

Pause and Reflect 5.4.5
Should AAVE be taught in schools? Why or why not?

APPALACHIAN ENGLISH. Natives of Appalachian communities, such as in rural West Virginia, also have a distinctive dialect, pronouncing *hollow* as "holler" and *creek* as "crick." Distinctive grammatical practices include the use of the double negative as in Ebonics and adding "a" in front of verbs ending in "ing," such as *a-sitting*.

As with Ebonics, speaking an Appalachian dialect produces both pride and problems. An Appalachian dialect is a source of regional identity but has long been regarded by other Americans as a sign of poor education and an obstacle to obtaining employment in other regions of the United States. Some Appalachian residents are "bidialectic": They speak "standard" English outside Appalachia and slip back into their regional dialect at home.

DIFFUSION TO OTHER LANGUAGES

English words have become increasingly integrated into other languages. Many French speakers regard the invasion of English words with alarm, but Spanish speakers may find the mixing of the two languages stimulating.

FRANGLAIS. Traditionally, language has been an especially important source of national pride and identity in France. The French are particularly upset with the increasing worldwide domination of English, especially the invasion of their language by English words and the substitution of English for French as the most important language of international communications.

French is an official language in 29 countries and for hundreds of years served as the lingua franca for international diplomats. Many French are upset that English words such as *cowboy*, *hamburger*, *jeans*, and *T-shirt* were allowed to diffuse into the French language and destroy the language's purity. The widespread use of English in the French language is called **Franglais**, a combination of *français* and *anglais*, the French words for *French* and *English*. (Figure 5-45)

Since 1635, the French Academy has been the supreme arbiter of the French language. In modern times, it has promoted the use of French terms in France, such as *stationnement* rather than *parking*, *fin du semaine* rather than *le weekend*, *logiciel* rather than *software*, and *arrosage* rather than *spam*. France's highest court, however, ruled in 1994 that most of the country's laws banning Franglais were illegal.

SPANGLISH. English is diffusing into the Spanish language spoken by 34 million Hispanics in the United States to create **Spanglish**, a combination of Spanish and English (Figure 5-46). In Miami's large Cuban American community, Spanglish is sometimes called Cubonics, a combination of Cuban and phonetics.

▲ **FIGURE 5-45 FRANGLAIS** A restaurant awning mixes French (*dejeuner*, *salades*, and *pâtes*), English (*burgers* and *bagels*), and Franglais (*club sandwichs*).

As with Franglais, Spanglish involves converting English words to Spanish forms. Some of the changes modify the spelling of English words to conform to Spanish preferences and pronunciations, such as dropping final consonants and replacing *v* with *b*. For example, *shorts* (pants) becomes *chores*, and *vacuum cleaner* becomes *bacuncliner*. In other cases, awkward Spanish words or phrases are dropped in favor of English words. For example, *parquin* is used rather than *estacionamiento* for "parking," and *taipear* is used instead of *escribir a máquina* for "to type."

Spanglish is a richer integration of English with Spanish than the mere borrowing of English words. New words have been invented in Spanglish that do not exist in English but would be useful if they did. For example, *textear* is a verb derived from the English text, and is less awkward than the Spanish *mandar un mensajito*; *i-meiliar* is a verb that means "to e-mail someone." Spanglish also mixes English and Spanish words in the same phrase. For example, a magazine article is titled "When he says *me voy* . . . what does he really mean?" (*me voy* means "I'm leaving").

▼ **FIGURE 5-46 SPANGLISH** A restaurant in Santa Ana, California, mixes Spanish and English.

Spanglish has become especially widespread in popular culture, such as song lyrics, television, and magazines aimed at young Hispanic women, but it has also been adopted by writers of serious literature. Inevitably, critics charge that Spanglish is a substitute for rigorously learning the rules of standard English and Spanish. And Spanglish has not been promoted for use in schools, as has Ebonics. Rather than a threat to existing languages, Spanglish is generally regarded as enriching both English and Spanish by adopting the best elements of each—English's ability to invent new words and Spanish's ability to convey nuances of emotion. Many Hispanic Americans like being able to say *Hablo un mix de los dos languages.*

DENGLISH. The diffusion of English words into German is called **Denglish**, with the "D" for *Deutsch*, the German word for *German* (Figure 5-47). In Germany, airlines, car dealers, and telephone companies use English slogans in advertising. For many Germans, wishing someone "happy birthday" sounds more melodic than the German *Herzlichen Glückwunsch zum Geburtstag.*

The German telephone company Deutsche Telekom uses the German word *Deutschlandverbindungen* for "long distance" and the Denglish word *Cityverbindungen* for "local" (rather than the German word *Ortsverbindungen*). The telephone company originally wanted to use the English "German calls" and "city calls" to describe its long-distance and local services, but the Institute for the German Language, which defines rules for the use of German, protested, so Deutsche Telekom compromised with one German word and one Denglish word.

English has diffused into other languages as well. The Japanese, for example, refer to *beisboru* ("baseball"), *naifu* ("knife"), and *sutoroberi keki* ("strawberry cake").

▼ **FIGURE 5-47 DENGLISH** An ad for a radio station in Berlin, Germany, mixes German and English.

SPANISH AND FRENCH IN THE UNITED STATES AND CANADA

Learning Outcome 5.4.6
Understand the role of Spanish and French in North America.

North America is dominated by English speakers. However, other languages, especially French in Canada and Spanish in the United States, are becoming increasingly prominent.

SPANISH-SPEAKING UNITED STATES. Linguistic unity is an apparent feature of the United States, a nation of immigrants who learn English to become Americans. However, the diversity of languages in the United States is greater than it first appears. In 2008, a language other than English was spoken at home by 56 million Americans over age 5, 20 percent of the population. Spanish was spoken at home by 35 million people in the United States. More than 2 million spoke Chinese; at least 1 million each spoke French, German, Korean, Tagalog, and Vietnamese.

Spanish has become an increasingly important language in recent years because of large-scale immigration from Latin America. In some communities, public notices, government documents, and advertisements are printed in Spanish. Several hundred Spanish-language newspapers and radio and television stations operate in the United States, especially in southern Florida, the Southwest, and large northern cities, where most of the 35 million Spanish-speaking people live (Figure 5-48).

Promoting the use of English symbolizes that language is the chief cultural bond in the United States in an otherwise heterogeneous society. With the growing dominance of the English language in the global economy and culture, knowledge of English is important for people around the world, not just inside the United States. At the same time, the increasing use of other languages in the United States is a reminder of the importance that groups place on preserving cultural identity and the central role that language plays in maintaining that identity.

In reaction against the increasing use of Spanish in the United States, 30 states and a number of localities have laws making English the official language. (Hawaii has two official languages, English and Hawaiian, which is in the Austronesian language family.) Some courts have judged these laws to be unconstitutional restrictions on free speech. The U.S. Congress has debated enacting similar legislation. For a state such as Montana, the law is symbolic, because it has few non-English speakers. But for states such as California and Florida, with large Hispanic populations, the debate affects access to jobs, education, and social services.

FRENCH-SPEAKING CANADA. French is one of Canada's two official languages, along with English. French speakers comprise one-fourth of the country's population. Most French-speaking Canadians are clustered in Québec, where they account for more than three-fourths of the province's speakers (Figure 5-49). Colonized by the French in the seventeenth century, Québec was captured by the British in 1763, and in 1867 it became one of the provinces in the Confederation of Canada.

Until recently, Québec was one of Canada's poorest and least-developed provinces. Its economic and political activities were dominated by an English-speaking minority, and the province suffered from cultural isolation and lack of French-speaking leaders.

When French President Charles de Gaulle visited Québec in 1967, he encouraged the development of an independent Québec by shouting in his speech, *"Vive le Québec libre!"* ("Long live free Québec!") Voters in Québec have thus far rejected separation from Canada, but by a slim majority.

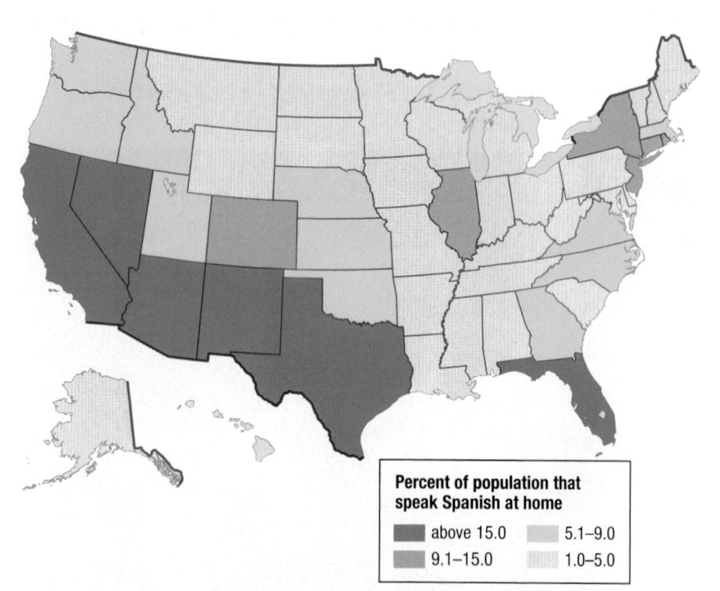

▲ **FIGURE 5-48 SPANISH SPEAKERS IN THE UNITED STATES** The largest percentages of Spanish speakers are in the Southwest and in Florida.

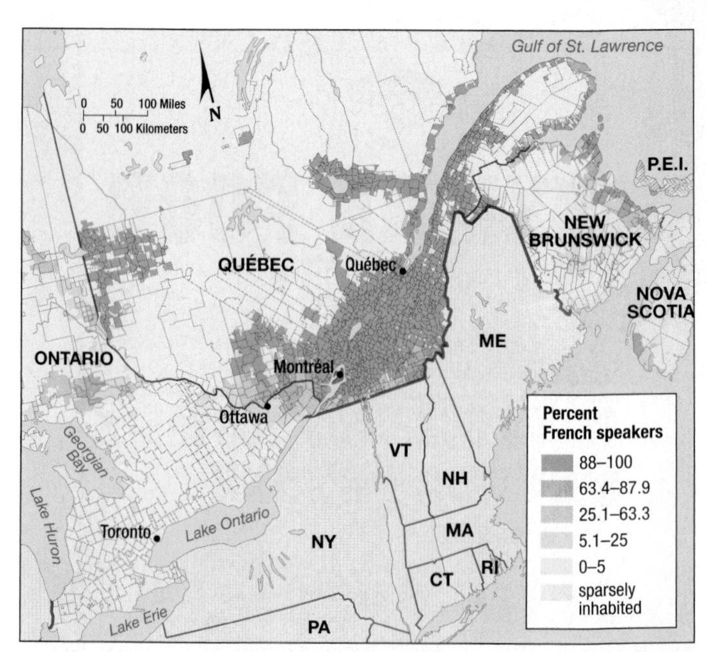

▲ **FIGURE 5-49 CANADA'S FRENCH–ENGLISH LANGUAGE BOUNDARY** French is the first language of 81 percent living in the province of Québec and 8 percent of Canadians living elsewhere in the country.

The Québec government has made the use of French mandatory in many daily activities. Québec's Commission de Toponymie has renamed towns, rivers, and mountains that have names with English-language origins. French must be the predominant language on all commercial signs, and the legislature passed a law banning non-French outdoor signs altogether. (However, the Canadian Supreme Court ruled this legislation unconstitutional.)

Confrontation during the 1970s and 1980s has been replaced in Québec by increased cooperation between French and English speakers. The neighborhoods of Montréal, Québec's largest city, were once highly segregated between French-speaking residents on the east and English-speaking residents on the west, but in recent years they have become more linguistically mixed. One-third of Québec's native English speakers have married French speakers in recent years. Children of English speakers are increasingly likely to be bilingual.

Although French dominates over English, Québec faces a fresh challenge of integrating a large number of immigrants from Europe, Asia, and Latin America who don't speak French. Many immigrants would prefer to use English rather than French as their lingua franca but are prohibited from doing so by the Québec government. Even immigrants who learn to speak French charge that they face discrimination because of their accents.

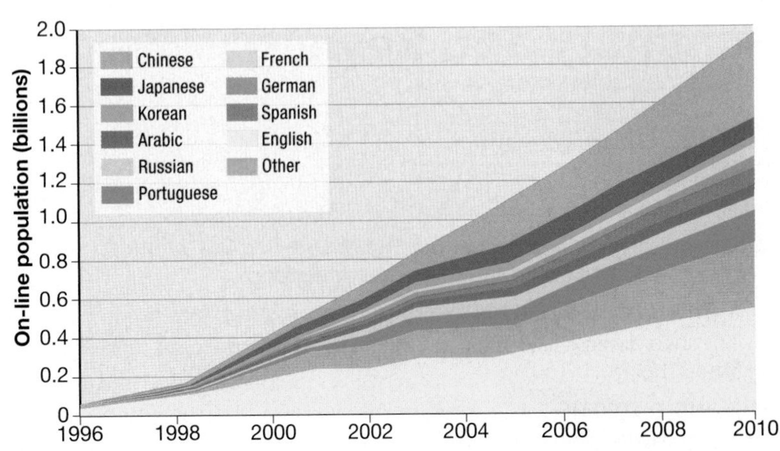

▲ **FIGURE 5-50 LANGUAGES OF ONLINE SPEAKERS** English remains the most widely used language on the Internet, but Chinese is growing more rapidly.

ENGLISH ON THE INTERNET

The emergence of the Internet as an important means of communication has further strengthened the dominance of English. Because a majority of the material on the Internet is in English, knowledge of English is essential for Internet users around the world. English was the dominant language of the Internet during the 1990s. In 1998, 71 percent of people online were using English (Figure 5-50). The early dominance of English on the Internet was partly a reflection of the fact that the most populous English-speaking country, the United States, had a head start on the rest of the world in making the Internet available to most of its citizens (refer to Figure 4-32).

English continued as the leading Internet language in the first years of the twenty-first century, but it was far less dominant. The percentage of English-language online users declined from 46 percent in 2000 to 27 percent in 2010. Chinese (Mandarin) language online users increased from 2 percent of the world total in 1998 to 22 percent in 2010, and Mandarin will probably replace English as the most-frequently used online language before 2020.

English may be less dominant as the language of the Internet in the twenty-first century. But the United States—and with it the English language—remains the Internet leader in key respects. The United States created the English-language nomenclature for the Internet that the rest of the world has followed. The designation "www," which English speakers recognize as an abbreviation of "World Wide Web," is awkward in other languages, most of which do not have an equivalent sound to the English *w*. In French, for example, *w* is pronounced "doo-blah-vay."

The U.S.-based Internet Corporation for Assigned Names and Numbers (ICANN) has been responsible for assigning domain names and for the suffixes following the dot, such as "com" and "edu." Domain names in the rest of the world include a two-letter suffix for the country, such as "fr" for France and "jp" for Japan, whereas U.S.-based domain names don't need the suffix. Reflecting the globalization of the languages of the Internet, ICANN agreed in 2009 to permit domain names in characters other than Latin. Arabic, Chinese, and other characters may now be used.

U.S.-based companies provide the principal search engines for Internet users everywhere. U.S.-based Google was used for 83 percent of all searches worldwide in 2011. Google, which offers search engines in languages other than English, was heavily criticized when its Mandarin-language Google.cn was designed to block web sites that China's government deemed unsuitable. A distant second was another U.S.-based company, Yahoo!, with 6 percent. A Chinese-language service Baidu was in third place in 2011, at 5 percent worldwide.

Pause and Reflect 5.4.4

Go to the home page of Google in a language other than English. How similar or different does it appear from the familiar English version?

CHECK-IN: KEY ISSUE 4

Why Do People Preserve Local Languages?

- ✓ **Some countries peacefully embrace multiple languages.**
- ✓ **Some languages survive in isolation from others, while some languages become extinct.**
- ✓ **Some endangered languages are being preserved.**
- ✓ **English has increasingly become the world's most important lingua franca, but Mandarin is catching up.**

Summary and Review

KEY ISSUE 1

Where Are Languages Distributed?

Languages can be classified as belonging to particular families. Some families are divided into branches and groups.

LEARNING OUTCOME 5.1.1 Name the largest language families.

- The two largest language families are Indo-European and Sino-Tibetan.

LEARNING OUTCOME 5.1.2: Identify the names and distribution of the two largest language families.

- Indo-European is the predominant language family of Europe, Latin America, North America, South Asia, and South Pacific. Sino-Tibetan is the predominant language family of East Asia.

LEARNING OUTCOME 5.1.3: Identify the names and distribution of the largest language families in addition to Indo-European and Sino-Tibetan.

- In addition to Indo-European and Sino-Tibetan, most of the world's remaining major language families are centered in Asia.

THINKING GEOGRAPHICALLY 5.1: What features of the Sino-Tibetan family make it especially difficult to learn to speak and to write the languages?

GOOGLE EARTH 5.1: Asakusa Shin-Nakamise is a shopping area in Tokyo. What are examples of English-language signs in the area?

KEY ISSUE 2

Why Is English Related To Other Languages?

English is in the Germanic branch of the Indo-European language family. Nearly one-half of humans currently speak a language in the Indo-European family. All Indo-European languages can be traced to a common ancestor.

LEARNING OUTCOME 5.2.1: Learn the distribution of the Germanic and Indo-Iranian branches of Indo-European.

- The four largest branches of Indo-European are Indo-Iranian, Romance, Germanic, and Balto-Slavic.

LEARNING OUTCOME 5.2.2: Learn the distribution of the Balto-Slavic and Romance branches of Indo-European.

- Balto-Slavic predominates in Eastern Europe, Romance in Southern Europe and Latin America, Germanic in Northern Europe and North America, and Indo-Iranian in South Asia and Central Asia.

LEARNING OUTCOME 5.2.3: Understand the origin and diffusion of English.

- English is a Germanic branch language because German-speaking tribes invaded England more than 1,500 years ago. Romance branch words entered English after French-speaking Normans invaded England nearly 1,000 years ago.

LEARNING OUTCOME 5.2.4: Understand the two theories of the origin and diffusion of Indo-European.

- Indo-European originated before recorded history; two competing theories disagree on whether origin and diffusion occurred primarily because of conquest or agriculture.

THINKING GEOGRAPHICALLY 5.2: Should the United States make English the official language? Why or why not? Should more than one language be made official? If so, which ones?

GOOGLE EARTH 5.2: Fly to Kutuluk, Russia, near the northern shore of the Caspian Sea and switch to ground-level view. Does the ancient homeland of the Kurgan warriors appear flat or mountainous? Grasslands or forests?

Key Terms

Creole, or creolized language (p. 163) A language that results from the mixing of a colonizer's language with the indigenous language of the people being dominated.

Denglish (p. 175) A combination of German and English.

Dialect (p. 158) A regional variety of a language distinguished by vocabulary, spelling, and pronunciation.

Ebonics (p. 174) A dialect spoken by some African Americans.

Extinct language (p. 166) A language that was once used by people in daily activities but is no longer used.

Franglais (p. 174) A term used by the French for English words that have entered the French language; a combination of *français* and *anglais*, the French words for *French* and *English*, respectively.

Isogloss (p. 158) A boundary that separates regions in which different language usages predominate.

Isolated language (p. 166) A language that is unrelated to any other languages and therefore not attached to any language family.

Language (p. 143) A system of communication through the use of speech, a collection of sounds understood by a group of people to have the same meaning.

Language branch (p. 143) A collection of languages related through a common ancestor that existed several thousand years ago. Differences are not as extensive or as old as with language families, and archaeological evidence can confirm that the branches derived from the same family.

Language family (p. 143) A collection of languages related to each other through a common ancestor long before recorded history.

Language group (p. 143) A collection of languages within a branch that share a common origin in the relatively recent past and display relatively few differences in grammar and vocabulary.

Lingua franca (p. 172) A language mutually understood and commonly used in trade by people who have different native languages.

Literary tradition (p. 143) A language that is written as well as spoken.

Logogram (p. 146) A symbol that represents a word rather than a sound.

KEY ISSUE 3

Why Do Individual Languages Vary Among Places?

A dialect is a regional variation of a language.

LEARNING OUTCOME 5.3.1: Describe the main dialects in the United States.

- U.S. English is divided into four main dialects. Differences can be traced to patterns of migration to the American colonies from various parts of England.

LEARNING OUTCOME 5.3.2: Understand the main ways that British and U.S. English dialects vary

- British and American dialects vary by vocabulary, spelling, and pronunciation.

LEARNING OUTCOME 5.3.3: Understand why it is sometimes difficult to distinguish between a language and a dialect.

- The distinction is often based on political decisions rather than the actual characteristics of the languages or dialects.

THINKING GEOGRAPHICALLY 5.3: Based on a comparison of Figure 5-23 center and right, which dialects are forecast to expand, and which are expected to contract by 2030? What geographic factor would account for this changing distribution?

GOOGLE EARTH 5.3: *Circus*, such as Piccadilly Circus in London, is an example of a British word that differs from American usage. The name *Piccadilly* can be traced to a house by that name built around 1612 by Robert Baker, a tailor who had a shop that sold stiff collars known as *piccadills*. Based on the feature visible here in the middle of Piccadilly Circus what would be the American equivalent of a *circus*?

Official language (p. 143) The language adopted for use by the government for the conduct of business and publication of documents.

Pidgin language (p. 173) A form of speech that adopts a simplified grammar and limited vocabulary of a lingua franca; used for communications among speakers of two different languages.

Received Pronunciation (RP) (p. 160) The dialect of English associated with upper-class Britons living in London and now considered standard in the United Kingdom.

Spanglish (p. 174) A combination of Spanish and English spoken by Hispanic Americans.

Standard language (p. 160) The form of a language used for official government business, education, and mass communications.

Vulgar Latin (p. 155) A form of Latin used in daily conversation by ancient Romans, as opposed to the standard dialect, which was used for official documents.

KEY ISSUE 4

Why Do People Preserve Local Languages?

English has become the most important language for international communication in culture and business.

LEARNING OUTCOME 5.4.1: Understand how several countries peacefully embrace more than one language.

- Switzerland, Belgium, and Nigeria have varying approaches to multilingual societies.

LEARNING OUTCOME 5.4.2: Understand what is meant by an isolated language and an extinct language.

- Thousands of languages once in use are now extinct. Some isolated languages survive that are unrelated to any other.

LEARNING OUTCOME 5.4.3: Understand why the number of Celtic speakers has declined and how the languages are being preserved.

- Celtic languages are being preserved through the efforts of advocacy groups and government agencies.

LEARNING OUTCOME 5.4.4: Understand the concept of a lingua franca.

- A lingua franca is a language of international communication.
- English is currently the world's most widely used lingua franca.

LEARNING OUTCOME 5.4.5: Understand how English has diffused to other languages.

- English is being combined with other languages, such as French and Spanish.

LEARNING OUTCOME 5.4.6: Understand the role of Spanish and French in North America.

- French is widely used in Canada, especially in Québec. Spanish is widely used in the United States.

THINKING GEOGRAPHICALLY 5.4: Because of Québec's French language and culture, some in the province have advocated Québec's separating from Canada and becoming an independent nation. Is a monolingual nation preferable to a bilingual one? State your argument for or against Québec's independence.

GOOGLE EARTH 5.4: Fly to 47 Mostyn St, Llandudno, Wales. In what language are most of the shop signs? In what language are most of the street signs and the sign in front of the church?

Chapter 6 Religions

Why are all of these clothed people standing in the Ganges River? Page 204

Why is this rock important to many religions? Page 221

KEY ISSUE 1

Where Are Religions Distributed?

A World of Religions p. 183

Only a few religions can claim the adherence of large numbers of people.

KEY ISSUE 2

Why Do Religions Have Different Distributions?

Origin and Diffusion p. 192

Some religions have known origins and diffusion, and some are shrouded in mystery.

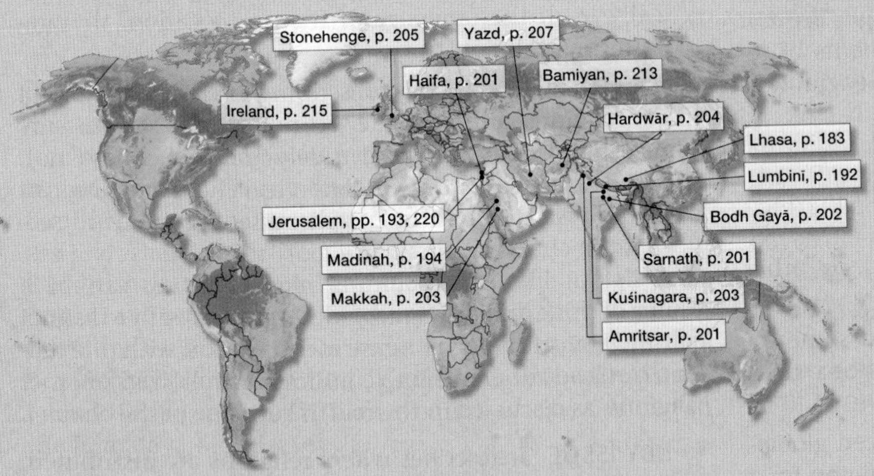

▲ The Dalai Lama, the spiritual leader of Tibetan Buddhists, is as important to that religion as the Pope is to Roman Catholics. The Dalai Lama has become an articulate spokesperson for religious freedom, and in 1989 he was awarded the world's most prestigious award for peace, the Nobel Prize. Despite the efforts of the Dalai Lama and other Buddhists, though, when the current generation of priests dies, many Buddhist traditions in Tibet may be lost forever.

KEY ISSUE 3

Why Do Religions Organize Space in Distinctive Patterns?

Religions in the Environment p. 200

Religions leave their mark on the landscape in many ways.

KEY ISSUE 4

Why Do Territorial Conflicts Arise Among Religious Groups?

Warring Religions p. 212

Followers of one religion often clash with followers of other religions, or with nonreligious institutions.

Distribution of Religions

Learning Outcome 6.1.1
Describe the distribution of the major religions.

Only a few religions can claim the adherence of large numbers of people. Each of these faiths has a distinctive distribution across Earth's surface.

Geographers distinguish two types of religions:

- **Universalizing religions** attempt to be global, to appeal to all people, wherever they may live in the world, not just to those of one culture or location.
- **Ethnic religions** appeal primarily to one group of people living in one place.

Statistics on the number of followers of religions can be controversial. No official count of religious membership is taken in the United States or in many other countries. Most statistics in this chapter come from Adherents.com, an organization not affiliated with any religion. According to Adherents.com, approximately 58 percent of the world's population practice a universalizing religion, 26 percent an ethnic religion, and 16 percent no religion.

The three universalizing religions with the largest numbers of adherents are Christianity, Islam, and Buddhism. According to Adherents.com, there are 2.1 billion Christians, 1.5 billion Muslims, and 376 million Buddhists in the world. Each of these religions has a distinctive distribution (Figure 6-3). The next three largest universalizing religions are Sikhism (23 million adherents), Bahá'í (7 million), and Zoroastrianism (3 million).

Hinduism is the ethnic religion with by far the largest number of adherents—900 million. Three other ethnic religions have at least 100 million adherents: Chinese traditional (394 million), Asian primal-indigenous (300 million), and African traditional religions (100 million). Three others—Juchte, Spiritism, and Judaism—have between 14 million and 19 million adherents each.

The nonreligious category consists primarily of people who express no religious interest or preference and don't participate in any organized religious activity. Some people in this group espouse **atheism**, which is belief that God does not exist, or **agnosticism**, which is belief that nothing can be known about whether God exists. According to Adherents.com, most people in this category affirm neither belief nor lack of belief in God or some other Higher Power.

Figure 6-4 shows the worldwide percentage of people adhering to the various religions. The small pie charts in Figure 6-3 show the overall proportion of the world's religions in each world region. Table 6-1 shows the distribution of religions in the United States.

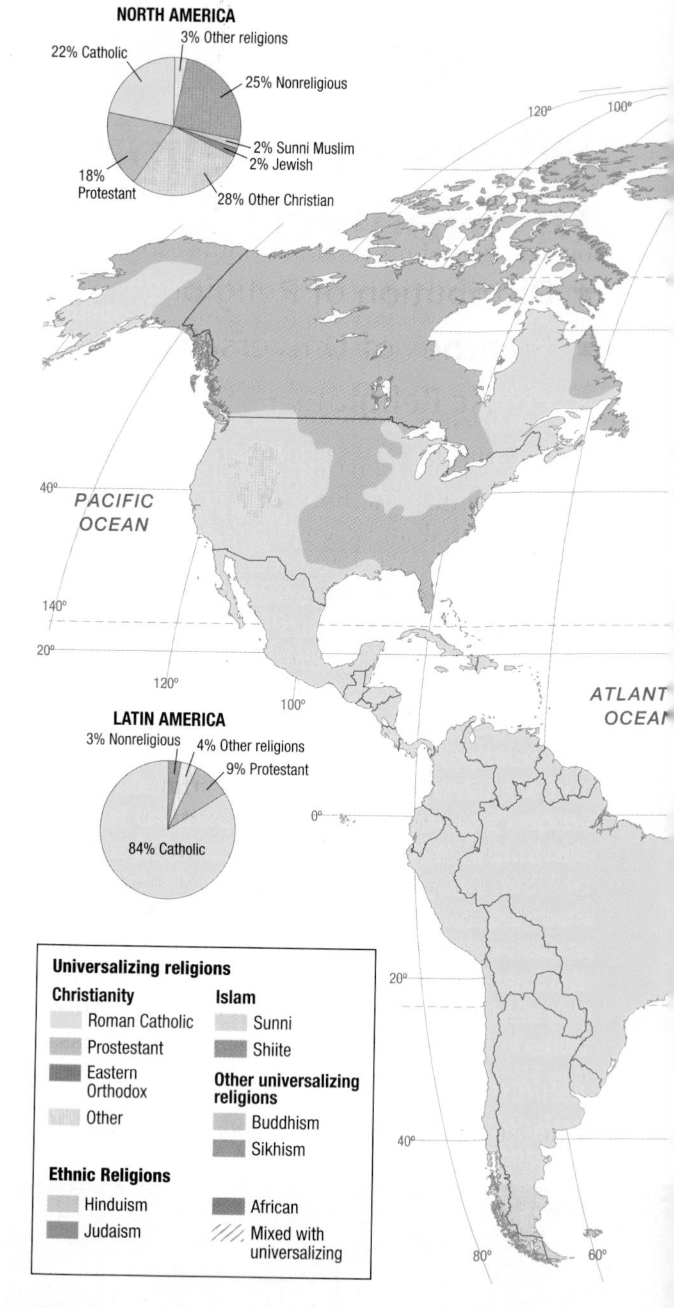

▲ **FIGURE 6-3 WORLD DISTRIBUTION OF RELIGIONS** The pie charts show the share of major religions in each world region.

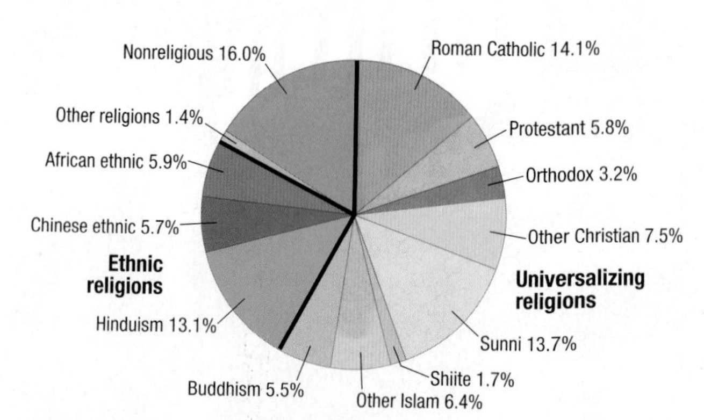

▲ **FIGURE 6-4 ADHERENTS OF WORLD RELIGIONS** Nonreligious includes atheists and agnostics.

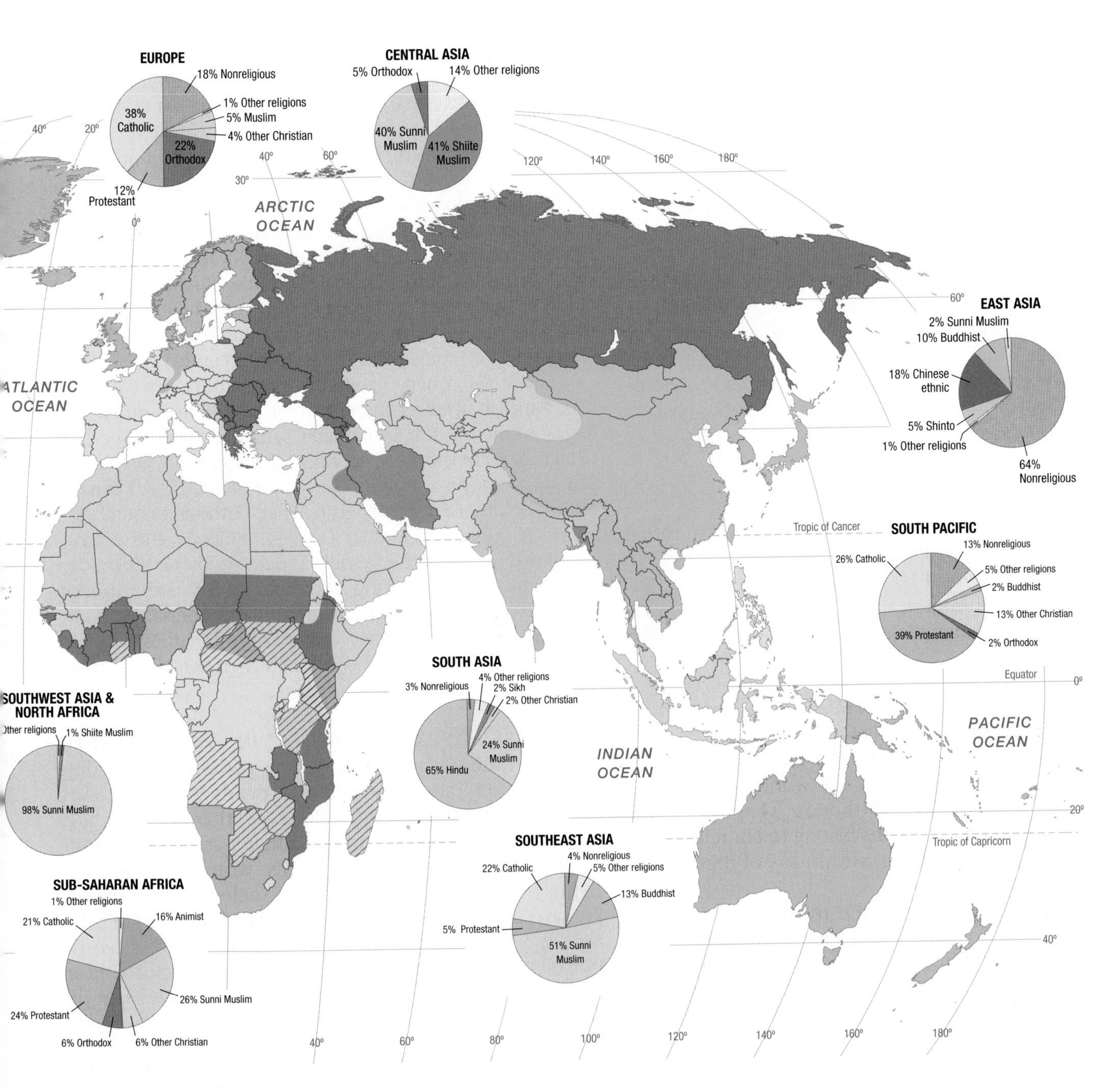

TABLE 6.1 RELIGIONS OF THE UNITED STATES

- **Nonreligious or atheist: 30 million**
- **Christians: 161 million**
  - Roman Catholics: 66 million
  - Protestants: 82 million
    - A Baptist church: 37 million
      - A Southern Baptist Convention church: 17 million
      - A National Baptist Convention, U.S.A., church: 8 million
      - A National Baptist Convention of America church: 4 million
      - A National Missionary Baptist Convention of America church: 3 million
      - A Progressive National Baptist Convention church: 3 million
      - An American Baptist Church, USA: 2 million
      - Another Baptist church: 3 million
    - A Methodist church: 13 million
      - A United Methodist church: 8 million
      - An African Methodist Episcopal or Episcopal Zion church: 4 million
    - A Pentecostal church: 11 million
      - A Church of God in Christ: 6 million
      - One of the Assemblies of God churches: 3 million
      - One of the Pentecostal Assemblies of the world churches: 2 million
    - A Lutheran church: 8 million
      - An Evangelical Lutheran Church in America: 5 million
      - One of the Lutheran Church Missouri Synod churches: 3 million
    - A Presbyterian Church U.S.A.: 4 million
    - A Reformed church: 2 million
      - A United Church of Christ: 1 million
      - Another Reformed Church: 1 million
    - An Episcopal church: 2 million
    - One of the Churches of Christ: 3 million
    - A Christian Church (Disciples of Christ): 1 million
    - A Seventh Day Adventist church: 1 million
  - Orthodox: 3 million
    - A church of the Greek Orthodox Archdiocese of America: 2 million
    - Another Orthodox church: 1 million
  - Other Christians: 10 million
    - A Church of Jesus Christ of Latter-Day Saints: 6 million
    - A Jehovah's Witness church: 1 million
    - Other Christians: 3 million
- **Buddhists: 1 million**
- **Hindus: 1 million**
- **Jews: 3 million**
- **Muslims: 1 million**
- **Other faiths: 6 million**

Branches of Universalizing Religions

Learning Outcome 6.1.2
Describe the distribution of the major branches of Christianity.

The three principal universalizing religions are divided into branches, denominations, and sects. A **branch** is a large and fundamental division within a religion. A **denomination** is a division of a branch that unites a number of local congregations in a single legal and administrative body. A **sect** is a relatively small group that has broken away from an established denomination.

CHRISTIANITY

Christianity has more than 2 billion adherents, more than any other world religion, and it also has the most widespread distribution. It is the predominant religion in North America, South America, Europe, and Australia, and countries with a Christian majority exist in Africa and Asia as well (Figure 6-5).

BRANCHES OF CHRISTIANITY IN EUROPE. Christianity has three major branches—Roman Catholic, Protestant, and Orthodox. According to *Encyclopaedia Britannica*, Roman Catholics comprise 51 percent of the world's Christians, Protestants 24 percent, and Orthodox 11 percent. In addition, 14 percent of Christians belong to churches that do not consider themselves to be within any of these three branches.

In Europe, Roman Catholicism is the dominant Christian branch in the southwest and east, Protestantism in the northwest, and Orthodoxy in the east and southeast. The regions of Roman Catholic and Protestant majorities frequently have sharp boundaries, even when they run through the middle of countries. For example, the Netherlands and Switzerland have approximately equal percentages of Roman Catholics and Protestants, but the Roman Catholic populations are concentrated in the south of these countries and the Protestant populations in the north.

The Orthodox branch of Christianity (often called Eastern Orthodox) is a collection of 14 self-governing churches in Eastern Europe and the Middle East. More than 40 percent of all Orthodox Christians belong to the Russian Orthodox Church, the largest of these 14 churches. Christianity came to Russia in the tenth century, and the Russian Orthodox Church was established in the sixteenth century.

Nine of the other 13 self-governing churches were established in the nineteenth century or twentieth century. The largest of these 9, the Romanian Church, includes 20 percent of all Eastern Orthodox Christians. The Bulgarian, Greek, and Serbian Orthodox churches have approximately 10 percent each. The other 5 recently established Orthodox churches—those of Albania, Cyprus, Georgia, Poland, and Sinai—combined account for about 2 percent of all Orthodox Christians. The remaining 4 of the 14 Eastern Orthodox churches—those of Constantinople, Alexandria, Antioch, and Jerusalem—trace their origins to the earliest days of Christianity. They have a combined membership of about 3 percent of all Orthodox Christians.

BRANCHES OF CHRISTIANITY IN THE WESTERN HEMISPHERE. The overwhelming percentage of people living in the Western Hemisphere—nearly 90 percent—are Christian. About 5 percent belong to other religions, and the remaining 6 percent profess adherence to no religion.

A fairly sharp boundary exists within the Western Hemisphere in the predominant branches of Christianity. Roman Catholics comprise 93 percent of Christians in Latin America, compared with 40 percent in North America. Within North America, Roman Catholics are clustered in the southwestern and northeastern United States and the Canadian province of Québec.

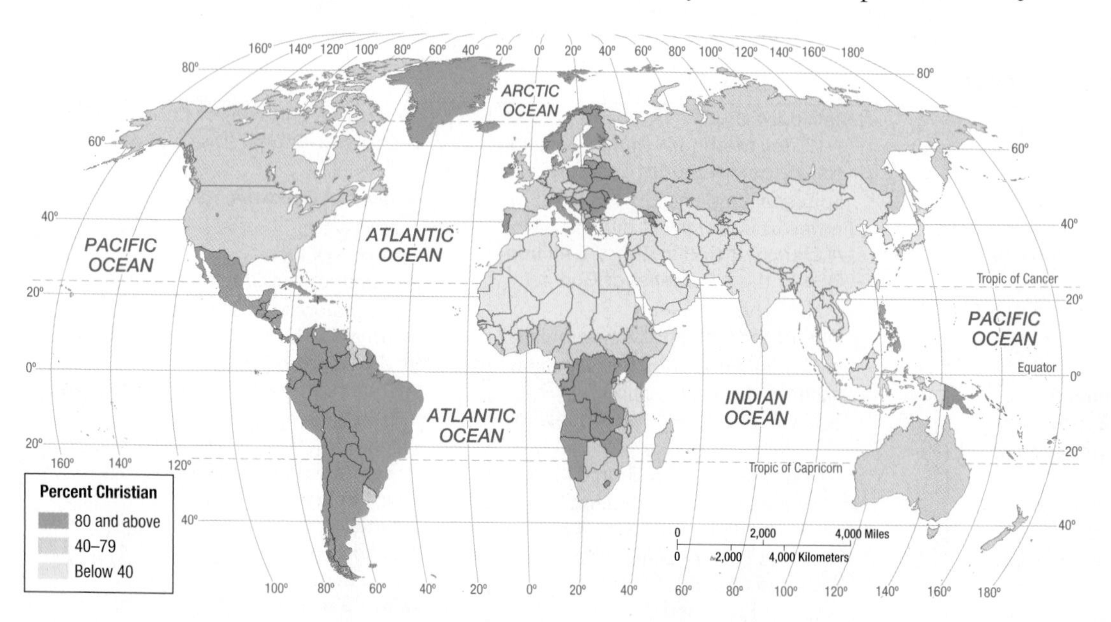

◀ **FIGURE 6-5 DISTRIBUTION OF CHRISTIANS** At least 80 percent of the population adheres to Christianity in Europe, the Western Hemisphere, the South Pacific, and selected countries in sub-Saharan Africa.

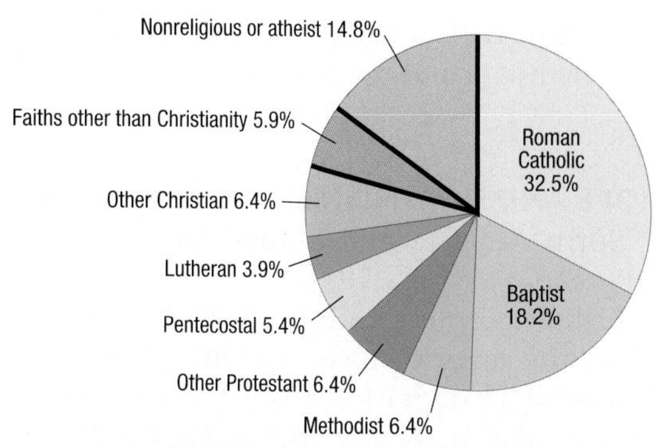

▲ **FIGURE 6-6 PERCENTAGE OF FAITHS IN THE UNITED STATES** Approximately 6 percent of the U.S. population adhere to a religion other than Christianity, and 15 percent adhere to no religion.

Protestant churches have approximately 82 million members, or about 28 percent of the U.S. population over age five (Figure 6-6). Baptist churches have the largest number of adherents in the United States, about 37 million combined over age five (refer to Table 6-1). Membership in some Protestant churches varies by region of the United States. Baptists, for example, are highly clustered in the southeast, and Lutherans in the upper Midwest. Other Christian denominations are more evenly distributed around the country (Figure 6-7).

OTHER CHRISTIANS. Several other Christian churches developed independently of the three main branches. Many of these Christian communities were isolated from others at an early point in the development of Christianity, partly because of differences in doctrine and partly as a result of Islamic control of intervening territory in Southwest Asia and North Africa.

Two small Christian churches survive in northeast Africa—the Coptic Church of Egypt and the Ethiopian Church. The Ethiopian Church, with perhaps 10 million adherents, split from the Egyptian Coptic Church in 1948, although it traces its roots to the fourth century, when two shipwrecked Christians, who were taken as slaves, ultimately converted the Ethiopian king to Christianity.

The Armenian Church originated in Antioch, Syria, and was important in diffusing Christianity to South Asia and East Asia between the seventh and thirteenth centuries. The church's few present-day adherents are concentrated in Lebanon and Armenia, as well as in northeastern Turkey and western Azerbaijan. Despite the small number of adherents, the Armenian Church, like other small sects, plays a significant role in regional conflicts. For example, Armenian Christians have fought for the independence of Nagorno-Karabakh, a portion of Azerbaijan, because Nagorno-Karabakh is predominantly Armenian, whereas the remainder of Azerbaijan is overwhelmingly Shiite Muslim (see Chapter 7).

The Maronites are another example of a small Christian sect that plays a disproportionately prominent role in political unrest. They are clustered in Lebanon, which has suffered through a long civil war fought among religious groups (see Chapter 7).

In the United States, members of The Church of Jesus Christ of Latter-day Saints (Mormons) regard their church as a branch of Christianity separate from other branches. About 3 percent of Americans are members of the Latter-day Saints, and a large percentage is clustered in Utah and surrounding states.

Pause and Reflect 6.1.2

Based on what you see in Figure 6-6 or Table 6-1, what are some of the largest Christian denominations in the United States that do not have highly clustered distributions in Figure 6-7?

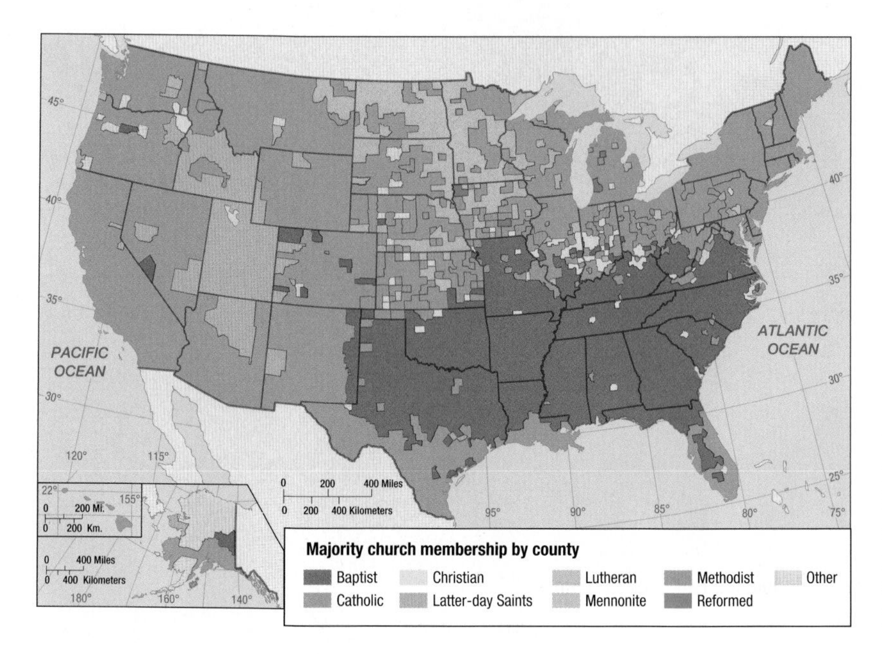

◀ **FIGURE 6-7 DISTRIBUTION OF CHRISTIANS IN THE UNITED STATES** The shaded areas are U.S. counties in which more than 50 percent of church membership is concentrated in either Roman Catholicism or one Protestant denomination. The distinctive distribution of religious groups within the United States results from patterns of migration, especially from Europe in the nineteenth century and from Latin America in recent years.

ISLAM

Learning Outcome 6.1.3
Identify the major branches of Islam and Buddhism.

Islam, the religion of 1.5 billion people, is the predominant religion of the Middle East from North Africa to Central Asia (Figure 6-8). Half of the world's Muslims live in four countries outside the Middle East—Indonesia, Pakistan, Bangladesh, and India.

The word *Islam* in Arabic means "submission to the will of God," and it has a similar root to the Arabic word for *peace*. An adherent of the religion of Islam is known as a *Muslim*, which in Arabic means "one who surrenders to God."

BRANCHES OF ISLAM. Islam is divided into two important branches:

- **Sunni.** The word *Sunni* comes from the Arabic for "people following the example of Muhammad." Sunnis comprise 83 percent of Muslims and are the largest branch in most Muslim countries in Southwest Asia and North Africa.
- **Shiite.** The word *Shiite* comes from the Arabic word for "sectarian." Shiites (sometimes called Shias in English) comprise 16 percent of Muslims, clustered in a handful of countries. Nearly 30 percent of all Shiites live in Iran, 15 percent in Pakistan, and 10 percent in Iraq. Shiites comprise nearly 90 percent of the population in Iran and more than half of the population in Azerbaijan, Iraq, and the less populous countries of Oman and Bahrain.

ISLAM IN EUROPE AND NORTH AMERICA. The Muslim population of North America and Europe has increased rapidly in recent years.

In Europe, Muslims account for 5 percent of the population. France has the largest Muslim population, about 4 million, a legacy of immigration from predominantly Muslim former colonies in North Africa. Germany has about 3 million Muslims, also a legacy of immigration, in Germany's case primarily from Turkey. In Southeast Europe, Albania, Bosnia, and Serbia each have about 2 million Muslims.

Estimates of the number of Muslims in North America vary widely, from 1 million to 5 million, but in any event, the number has increased dramatically from only a few hundred thousand in 1990. Muslims in the United States come from a variety of backgrounds. According to the U.S. State Department, approximately one-third of U.S. Muslims trace their ancestry to Pakistan and other South Asian countries and one-fourth from Arab countries of Southwest Asia and North Africa. Many of these Muslims immigrated to the United States during the 1990s. Another one-fourth are African Americans, who have converted to Islam.

Islam also has a presence in the United States through the Nation of Islam, also known as Black Muslims, founded in Detroit in 1930 and led for more than 40 years by Elijah Muhammad, who called himself "the messenger of Allah."

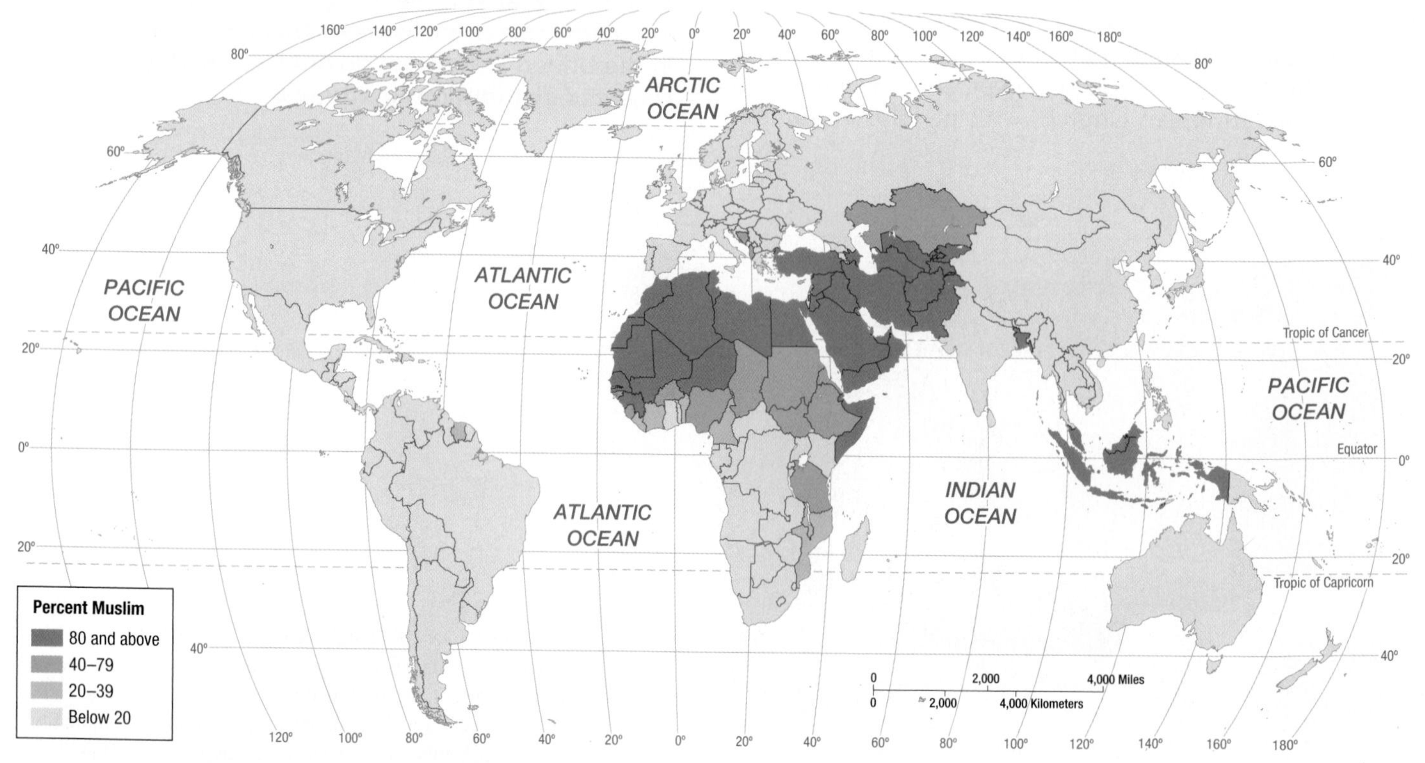

▲ **FIGURE 6-8 DISTRIBUTION OF MUSLIMS** At least 80 percent of the population adheres to Islam in Southwest Asia & North Africa and selected countries in Southeast Asia.

Black Muslims lived austerely and advocated a separate autonomous nation within the United States for their adherents. Tension between Muhammad and a Black Muslim minister, Malcolm X, divided the sect during the 1960s. After a pilgrimage to Makkah in 1963, Malcolm X converted to orthodox Islam and founded the Organization of Afro-American Unity. He was assassinated in 1965. After Muhammad's death, in 1975, his son Wallace D. Muhammad led the Black Muslims closer to the principles of orthodox Islam, and the organization's name was changed to the American Muslim Mission. A splinter group adopted the original name, Nation of Islam, and continues to follow the separatist teachings of Elijah Muhammad.

BUDDHISM

Buddhism, the third of the world's major universalizing religions, is clustered primarily in East Asia and Southeast Asia. Like the other two universalizing religions, Buddhism split into more than one branch, as followers disagreed on interpreting statements by the founder, Siddhartha Gautama. The three main branches are (Figure 6-9):

- **Mahayana.** Mahayanists account for about 56 percent of Buddhists, primarily in China, Japan, and Korea.
- **Theravada.** Theravadists comprise about 38 percent of Buddhists, especially in Cambodia, Laos, Myanmar, Sri Lanka, and Thailand.
- **Vajrayana.** Vajrayanists, also known as Lamaists and Tantrayanists, comprise about 6 percent and are found primarily in Tibet and Mongolia.

An accurate count of Buddhists is especially difficult because only a few people participate in Buddhist institutions. Religious functions are performed primarily by monks rather than by the general public. The number of Buddhists is also difficult to count because Buddhism, although a universalizing religion, differs in significant respects from the Western concept of a formal religious system. Someone can be both a Buddhist and a believer in other Eastern religions, whereas Christianity and Islam both require exclusive adherence. Most Buddhists in China and Japan, in particular, believe at the same time in an ethnic religion.

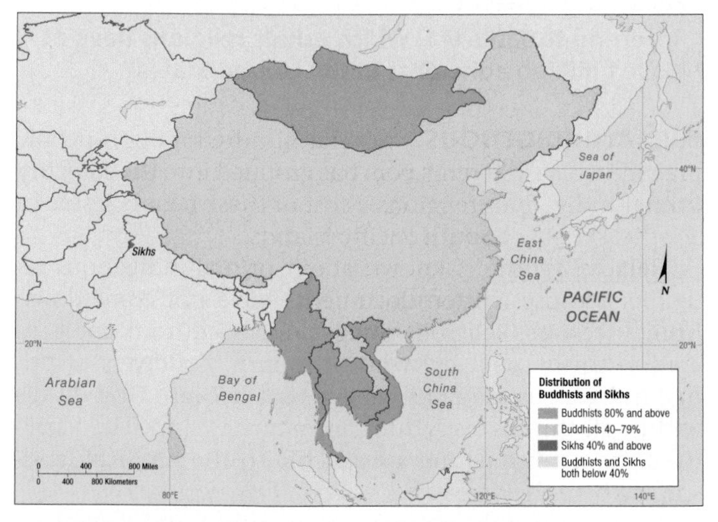

▲ **FIGURE 6-9 DISTRIBUTION OF BUDDHISTS AND SIKHS**
At least 40 percent of the population adheres to Buddhism in East Asia and Southeast Asia. At least 40 percent of the population adheres to Sikhism in northwestern India.

OTHER UNIVERSALIZING RELIGIONS

Sikhism and Bahá'í are the two universalizing religions other than Christianity, Islam, and Buddhism that have the largest numbers of adherents. There are an estimated 23 million Sikhs and 7 million Bahá'ís. All but 3 million Sikhs are clustered in the Punjab region of India Bahá'ís are dispersed among many countries, primarily in Africa and Asia.

Sikhism's first guru (religious teacher or enlightener) was Nanak (1469–1538), who lived in a village near the city of Lahore, in present-day Pakistan. God was revealed to Guru Nanak as The One Supreme Being, or Creator, who rules the universe by divine will. Only God is perfect, but people have the capacity for continual improvement and movement toward perfection by taking individual responsibility for their deeds and actions on Earth, such as heartfelt adoration, devotion, and surrender to the one God. Sikhism's most important ceremony, introduced by the tenth guru, Gobind Singh (1666–1708), is the Amrit (or Baptism), in which Sikhs declare they will uphold the principles of the faith. Gobind Singh also introduced the practice of men wearing turbans on their heads and never cutting their beards or hair. Wearing a uniform gave Sikhs a disciplined outlook and a sense of unity of purpose.

The Bahá'í religion is even more recent than Sikhism. It grew out of the Bábi faith, which was founded in Shíráz, Iran, in 1844 by Siyyid 'Ali Muhammad, known as the Báb (Persian for "gateway"). Bahá'ís believe that one of the Báb's disciples, Husayn 'Ali Nuri, known as Bahá'u'lláh (Arabic for "Glory of God"), was the prophet and messenger of God. Bahá'u'lláh's function was to overcome the disunity of religions and establish a universal faith through abolition of racial, class, and religious prejudices.

Pause and Reflect 6.1.3

Refer to the small pie charts in Figure 6-3. Which regions have enough adherents of each of the three universalizing religions so that all three appear on the pie charts?

Ethnic Religions

Learning Outcome 6.1.4
Describe the distribution of the largest ethnic religions.

In contrast to universalizing religions, which often spread from one culture to another, ethnic religions tend to remain within the culture where they originated. Ethnic religions typically have much more clustered distributions than do universalizing religions. The ethnic religion with by far the largest number of followers is Hinduism. Ethnic religions in Asia and Africa comprise most of the remainder.

HINDUISM

Hinduism is the world's third-largest religion, with 900 million adherents, but 90 percent of Hindus are concentrated in one country, India, and most of the remainder can be found in India's neighbors Bangladesh and Nepal. Hindus comprise more than 80 percent of the population of India and Nepal, about 9 percent in Bangladesh, and a small minority in every other country (Figure 6-10).

The average Hindu has allegiance to a particular god or concept within a broad range of possibilities. The manifestation of God with the largest number of adherents—an estimated 80 percent—is Vaishnavism, which worships the god Vishnu, a loving god incarnated as Krishna. The second-largest is Sivaism, dedicated to Siva, a protective and destructive god.

OTHER ASIAN ETHNIC RELIGIONS

Three religions based in East Asia and Southeast Asia show the difficulty of classifying ethnic religions and counting adherents. Chinese traditional religions are **syncretic**, which means they combine several traditions. Primal-indigenous religions are especially difficult to document because they are based on oral traditions rather than writing. Juchte is classified as a religion by Adherents.com but not by other sources.

CHINESE TRADITIONAL. Adherents.com considers Chinese traditional religion to be a combination of Buddhism (a universalizing religion) with Confucianism, Taoism, and other traditional Chinese practices. Most Chinese who consider themselves religious blend together the religious cultures of these multiple traditions:

- **Confucianism.** Confucius (551–479 B.C.) was a philosopher and teacher in the Chinese province of Lu. His sayings, which were recorded by his students, emphasized the importance of the ancient Chinese tradition of *li*, which can be translated roughly as "propriety" or "correct behavior." Confucianism prescribed a series of ethical principles for the orderly conduct of daily life in China, such as following traditions, fulfilling obligations, and treating others with sympathy and respect. These rules applied to China's rulers as well as to their subjects.
- **Taoism.** Lao-Zi (604–531? B.C., also spelled Lao Tse) organized Taoism. Although a government administrator by profession, Lao-Zi's writings emphasized the mystical and magical aspects of life rather than the importance of public service, which Confucius had emphasized. Tao, which means "the way" or "the path," cannot be comprehended by reason and knowledge because not everything is knowable. (Figure 6-11).

Commingling of diverse philosophies is not totally foreign to Americans. The tenets of a religion such as Christianity, the wisdom of the ancient Greek philosophers, and the ideals of the Declaration of Independence can all be held dear without doing grave injustice to the others.

Pause and Reflect 6.1.4
Referring to Table 6-1, which ethnic religions have at least 1 million adherents in the United States?

PRIMAL-INDIGENOUS. Several hundred million people practice what Adherents.com has grouped into the category primal-indigenous religions. Most of these people reside in Southeast Asia or South Pacific islands.

Relatively little is known about primal-indigenous religions because written documents have not come down from ancestors. Religious rituals are passed from one generation to the next by word of mouth. Followers of primal-indigenous religions believe that because God dwells within all things, everything in nature is spiritual. Narratives concerning nature are specific to the physical landscape where they are told.

Included in this group are Shamanism and Paganism. According to Shamans, invisible forces or spirits affect the lives of the living. **Pagan** used to refer to the practices of ancient peoples, such as the Greeks and Romans, who had multiple gods with human forms. The term is currently used to refer to beliefs that originated with religions that predated Christianity and Islam.

▼ **FIGURE 6-10**
DISTRIBUTION OF HINDUS All but 10 percent of the world's Hindus live in India.

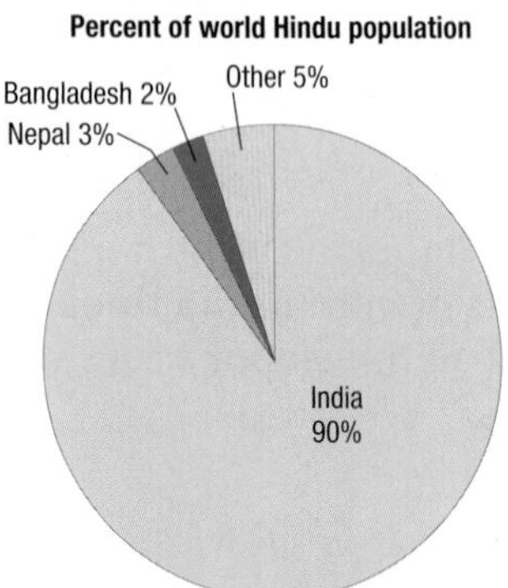

▶ **FIGURE 6-11**
DISTRIBUTION OF TAOISTS All but 4 percent of the world's Taoists live in China.

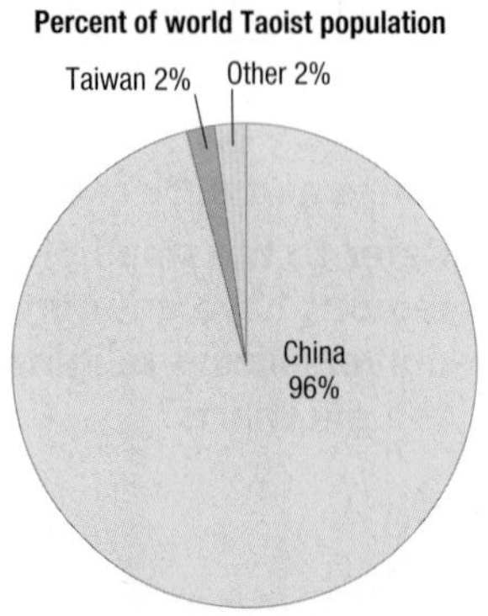

Percent of world Taoist population
Taiwan 2%
Other 2%
China 96%

JUCHTE. Most North Koreans are classified by Adherents.com as following Juchte, which is a Korean word meaning "self-reliance." Juchte was organized by Kim Il-sung, the leader of North Korea between 1948 and his death in 1994. Rather than a religion, Juchte is widely regarded as a government ideology or philosophy.

OTHER ETHNIC RELIGIONS

Outside Asia, the principal ethnic religions are African. Spiritism and Judaism are other ethnic religions that have at least 10 million adherents worldwide.

AFRICAN TRADITIONAL. Approximately 100 million Africans, 12 percent of the continent's people, follow traditional ethnic religions sometimes called **animism**. Animists believe that inanimate objects such as plants and stones, or natural events such as thunderstorms and earthquakes, are "animated," or have discrete spirits and conscious life.

African animist religions are apparently based on monotheistic concepts, although below the supreme god there is a hierarchy of divinities. These divinities may be assistants to the supreme god or personifications of natural phenomena, such as trees or rivers.

Africa is 46 percent Christian—split about evenly among Roman Catholic, Protestant, and other—and another 40 percent are Muslims (Figure 6-12). The growth in the two universalizing refer to at the expense of ethnic religions reflects fundamental geographic differences between the two types of religions, discussed in the next key issue.

SPIRITISM. Spiritism is the belief that the human personality continues to exist after death and can communicate with the living through the agency of a medium or psychic. Most Spiritists reside in Brazil.

JUDAISM. Roughly two-fifths of the world's 14 million Jews live in the United States and another two-fifths in Israel. The name *Judaism* derives from Judah, one of the patriarch Jacob's 12 sons; Israel is another biblical name for Jacob. The Bible recounts the ancient history of the Jewish people. Judaism plays a more substantial role in Western civilization than its number of adherents would suggest:

- Judaism is the first recorded religion to espouse **monotheism**, belief that there is only one God. Fundamental to Judaism is belief in one all-powerful God. Judaism offered a sharp contrast to the **polytheism** practiced by neighboring people, who worshipped a collection of gods.
- Two of the three main universalizing religions—Christianity and Islam—find some of their roots in Judaism. About 4,000 years ago Abraham, considered the patriarch or father of Judaism, migrated from present-day Iraq to present-day Israel, along a route known as the Fertile Crescent (see discussion of the Fertile Crescent in Chapter 8 and Figure 8-10). About 2,000 years after Abraham, Jesus was born a Jew, and about 500 years after Jesus Muhammad traced his ancestry to Abraham.

CHECK-IN: KEY ISSUE 1

Where Are Religions Distributed?

✓ **Religions can be classified into universalizing and ethnic.**

✓ **The three largest universalizing religions are Christianity, which is found primarily in Europe, North America, and Latin America; Islam, which is found primarily in Southeast, Central, and Southwest Asia, as well as North Africa; and Buddhism, which is found primarily in East Asia.**

✓ **The largest ethnic religion is Hinduism, which is found primarily in South Asia.**

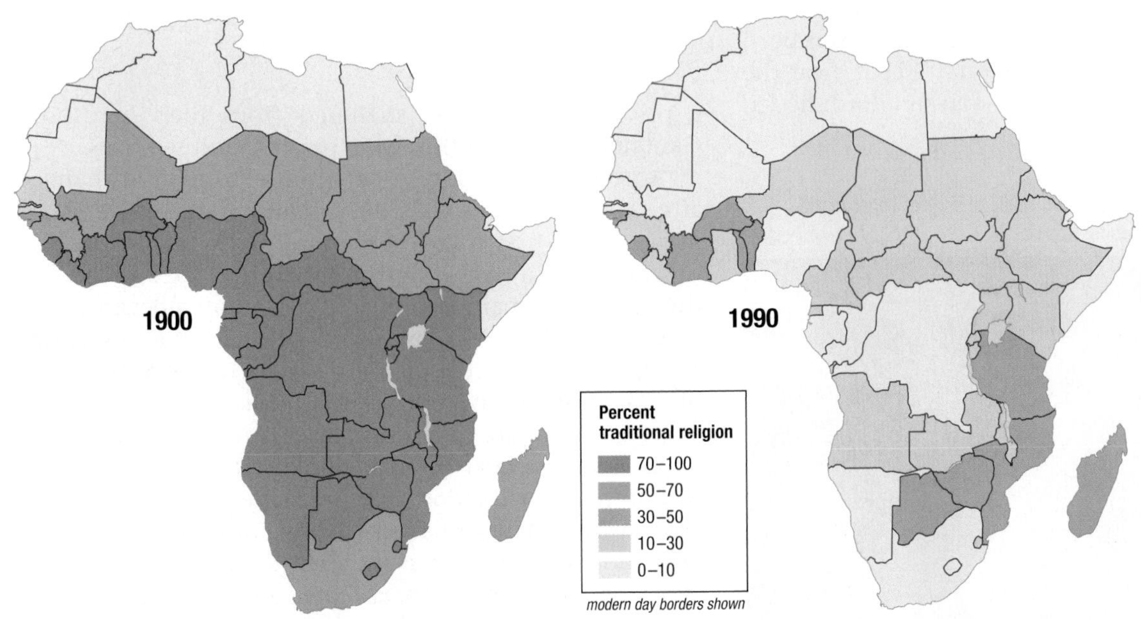

◀ **FIGURE 6-12 DISTRIBUTION OF AFRICAN TRADITIONAL RELIGIONS** The percentage of animists in sub-Saharan Africa has declined from more than 70 percent in 1900 to around 12 percent in 2010. As recently as 1980, some 200 million Africans—half the population of the region at the time—were classified as animists. Followers of traditional African religions now constitute a clear majority of the population only in Botswana. The rapid decline in animists in Africa has been caused by increases in the numbers of Christians and Muslims.

KEY ISSUE 2

Why Do Religions Have Different Distributions?

- Origin of Religions
- Diffusion of Religions

Learning Outcome 6.2.1
Describe the origin of universalizing religions.

We can identify several major geographic differences between universalizing and ethnic religions. These differences include the locations where the religions originated, the processes by which they diffused from their place of origin to other regions, the types of places that are considered holy, the calendar dates identified as important holidays, and attitudes toward modifying the physical environment.

Origin of Religions

Universalizing religions have precise places of origin based on events in the life of a man. Ethnic religions have unknown or unclear origins, not tied to single historical individuals.

Each of the three universalizing religions can be traced to the actions and teachings of a man who lived since the start of recorded history. The beginnings of Buddhism go back about 2,500 years, Christianity 2,000 years, and Islam 1,500 years. Specific events also led to the division of the universalizing religions into branches.

ORIGIN OF BUDDHISM

The founder of Buddhism, Siddhartha Gautama, was born about 563 B.C. in Lumbinī in present-day Nepal, near the border with India (Figure 6-13). The son of a lord, he led a privileged existence, sheltered from life's hardships. Gautama had a beautiful wife, palaces, and servants.

According to Buddhist legend, Gautama's life changed after a series of four trips. He encountered a decrepit old man on the first trip, a disease-ridden man on the second trip, and a corpse on the third trip. After witnessing these scenes of pain and suffering, Gautama began to feel he could no longer enjoy his life of comfort and security. Then, on a fourth trip, Gautama saw a monk, who taught him about withdrawal from the world.

At age 29 Gautama left his palace one night and lived in a forest for the next 6 years, thinking and experimenting with forms of meditation. Gautama emerged as the Buddha, the "awakened or enlightened one," and spent 45 years preaching his views across India. In the process, he trained monks, established orders, and preached to the public.

▼ **FIGURE 6-13 ORIGIN OF BUDDHISM** Ruins of shrines constructed around Buddha's birthplace in Lumbinī, Nepal.

CONTEMPORARY GEOGRAPHIC TOOLS

Counting Religious Adherents

An accurate count of the number of adherents to various religions, branches, denominations, and sects is impossible in the United States because the census does not ask questions about religion. Other countries do ask about religion.

In Canada, for example, the census asks:

> "What is this person's religion? Indicate a specific denomination or religion even if this person is not currently a practising member of that group. For example, Roman Catholic, Ukrainian Catholic, United Church, Anglican, Lutheran, Baptist, Coptic Orthodox, Greek Orthodox, Jewish, Islam, Buddhist, Hindu, Sikh, etc."
>
> ☐ Specify one denomination or religion only ________
>
> ☐ No religion

Critics charged that including the phrase "even if this person is not currently a practising member" inflated the number of people who were thought to be religious in Canada.

Nonetheless, an analysis of recent census data by mathematicians showed that the fastest-growing response to census questions about religion in a number of countries is "unaffiliated." When they extrapolated the growth of "unaffiliated" into the future, the mathematicians concluded that religion would become extinct during the twenty-first century in nine countries: Australia, Austria, Canada, the Czech Republic, Finland, Ireland, the Netherlands, New Zealand, and Switzerland.

The foundation of Buddhism is represented by these concepts, known as the Four Noble Truths:

1. All living beings must endure suffering.
2. Suffering, which is caused by a desire to live, leads to reincarnation (repeated rebirth in new bodies or forms of life).
3. The goal of all existence is to escape suffering and the endless cycle of reincarnation into Nirvana (a state of complete redemption), which is achieved through mental and moral self-purification.
4. Nirvana is attained through an Eightfold Path, which includes rightness of belief, resolve, speech, action, livelihood, effort, thought, and meditation.

THERAVADA BUDDHISM. Theravada is the older of the two largest branches of Buddhism. The word means "the way of the elders," indicating the Theravada Buddhists' belief that they are closer to Buddha's original approach. Theravadists believe that Buddhism is a full-time occupation, so to become a good Buddhist, one must renounce worldly goods and become a monk.

MAHAYANA BUDDHISM. Mahayana split from Theravada Buddhism about 2,000 years ago. *Mahayana* is translated as the great vehicle, and Mahayanists call Theravada Buddhism by the name *Hinayana*, or the inferior vehicle. Mahayanists claim that their approach to Buddhism can help more people because it is less demanding and all-encompassing. Theravadists emphasize Buddha's life of self-help and years of solitary introspection, and Mahayanists emphasize Buddha's later years of teaching and helping others. Theravadists cite Buddha's wisdom and Mahayanists his compassion.

VAJRAYANA BUDDHISM. Vajrayanas emphasize the practice of rituals, known as Tantras, which have been recorded in texts. Vajrayanas believe that Buddha began to practice Tantras during his lifetime, although other Buddhists regard Vajrayana as an approach to Buddhism that evolved from Mahayana Buddhism several centuries later.

ORIGIN OF CHRISTIANITY

Christianity was founded upon the teachings of Jesus, who was born in Bethlehem between 8 and 4 B.C. and died on a cross in Jerusalem about A.D. 30. Raised as a Jew, Jesus gathered a small band of disciples and preached the coming of the Kingdom of God. The four Gospels of the Christian Bible—Matthew, Mark, Luke, and John—document miracles and extraordinary deeds that Jesus performed. He was referred to as Christ, from the Greek word for the Hebrew word *messiah*, which means "anointed."

In the third year of his mission, Jesus was betrayed to the authorities by one of his companions, Judas Iscariot. After sharing the Last Supper (the Jewish Passover Seder) with his disciples in Jerusalem, Jesus was arrested and put to death as an agitator. On the third day after his death, his tomb was found empty (Figure 6-14). Christians believe that Jesus died to atone for human sins, that he was raised from the dead by God, and that his Resurrection from the dead provides people with hope for salvation.

▲ **FIGURE 6-14 ORIGIN OF CHRISTIANITY** This tomb in the center of the Church of the Holy Sepulchre in Jerusalem was erected on the site where Jesus is thought to have been buried and resurrected.

ROMAN CATHOLIC BRANCH. Roman Catholics accept the teachings of the Bible, as well as the interpretation of those teachings by the Church hierarchy, headed by the Pope. Roman Catholics recognize the Pope as possessing a universal primacy or authority, and they believe that the Church is infallible in resolving theological disputes. According to Roman Catholic belief, God conveys His grace directly to humanity through seven sacraments: Baptism, the Eucharist (the partaking of bread and wine that repeats the actions of Jesus at the Last Supper), Penance, Confirmation, Matrimony, Holy Orders, and Anointing the Sick.

ORTHODOX BRANCH. Orthodoxy comprises the faith and practices of a collection of churches that arose in the eastern part of the Roman Empire. The split between the Roman and Eastern churches dates to the fifth century, as a result of rivalry between the Pope of Rome and the Patriarchy of Constantinople, which was especially intense after the collapse of the Roman Empire. The split between the two churches became final in 1054, when Pope Leo IX condemned the Patriarch of Constantinople. Orthodox Christians accept the seven sacraments but reject doctrines that the Roman Catholic Church added since the eighth century.

PROTESTANT BRANCH. Protestantism originated with the principles of the Reformation in the sixteenth century. The Reformation movement is regarded as beginning when Martin Luther (1483–1546) posted 95 theses on the door of the church at Wittenberg on October 31, 1517. According to Luther, individuals have primary responsibility for achieving personal salvation through direct communication with God. Grace is achieved through faith rather than through sacraments performed by the Church.

ORIGIN OF ISLAM

Learning Outcome 6.2.2
Understand differences in the origin of universalizing and ethnic religions.

Like other universalizing religions, Islam arose from the teachings of a historical founder. The core of Islamic belief involves performing five acts, known as five pillars of faith:

1. *Shahadah*, which means frequent recitation that there is no god worthy of worship except the one God, the source of all creation, and Muhammad is the messenger of God.
2. *Salat*, which means that five times daily, a Muslim prays, facing the city of Makkah (Mecca), as a direct link to God.
3. *Zakat*, which means that a Muslim gives generously to charity as an act of purification and growth.
4. *Sawm of Ramadan*, which means that a Muslim fasts during the month of Ramadan as an act of self-purification.
5. *Hajj*, which means that if physically and financially able, a Muslim makes a pilgrimage to Makkah.

Islam traces its origin to the same narrative as Judaism and Christianity. All three religions consider Adam to have been the first man and Abraham to have been one of his descendants. According to the biblical narrative:

- Abraham married Sarah, who did not bear children; as polygamy was a custom of the culture, Abraham then married Hagar, who bore a son, Ishmael.
- Sarah's fortunes changed, and she bore a son, Isaac.
- Jews and Christians trace their story through Abraham's original wife Sarah and her son Isaac.
- Muslims trace their story through his second wife, Hagar, and her son Ishmael; the Islamic tradition tells that Abraham brought Hagar and Ishmael to Makkah (spelled Mecca on many English-language maps), in present-day Saudi Arabia.
- Centuries later, one of Ishmael's descendants, Muhammad, became the Prophet of Islam.

PROPHET MUHAMMAD. Muhammad was born in Makkah about 570. At age 40, while engaged in a meditative retreat, Muhammad received his first revelation from God through the Angel Gabriel. The Quran, the holiest book in Islam, is a record of God's words, as revealed to the Prophet Muhammad through Gabriel. Arabic is the lingua franca, or language of communication, within the Muslim world, because it is the language in which the Quran is written.

As he began to preach the truth that God had revealed to him, Muhammad suffered persecution, and in 622 he was commanded by God to emigrate. His migration from Makkah to the city of Yathrib—an event known as the *Hijra* (from the Arabic word for "migration," sometimes spelled *hegira*)—marks the beginning of the Muslim calendar. Yathrib was subsequently renamed Madinah, Arabic for "the City of the Prophet" (Figure 6-15). After several years, Muhammad and his followers returned to Makkah and established Islam as the city's religion. By Muhammad's death, in 632 at about age 63, Islam had spread through most of present-day Saudi Arabia.

▲ **FIGURE 6-15 ORIGIN OF ISLAM** Muhammad is buried under the green dome in the Mosque of the Prophet in Madinah, Saudi Arabia. The mosque, built on the site of Muhammad's house, is the second holiest in Islam and the second largest mosque in the world.

SHIITES VERSUS SUNNIS. Differences between the two main branches of Islam—Shiites and Sunnis—go back to the earliest days of the religion and basically reflect disagreement over the line of succession in Islamic leadership. Muhammad had no surviving son and no follower of comparable leadership ability. His successor was his father-in-law, Abu Bakr (573–634), an early supporter from Makkah, who became known as *caliph* ("successor of the prophet"). The next two caliphs, Umar (634–644) and Uthman (644–656), expanded the territory under Muslim influence to Egypt and Persia.

Uthman was a member of a powerful Makkah clan that had initially opposed Muhammad before the clan's conversion to Islam. The more ardent converts criticized Uthman for seeking compromises with other formerly pagan families in Makkah. Uthman's opponents found a leader in Ali (600?–661), a cousin and son-in-law of Muhammad, and thus Muhammad's nearest male heir. When Uthman was murdered, in 656, Ali became caliph, although five years later he, too, was assassinated (Figure 6-16).

Ali's descendants claim leadership of Islam, and Shiites support this claim. But Shiites disagree among themselves about the precise line of succession from Ali to modern times. They acknowledge that the chain of leadership was broken, but they dispute the date and events surrounding the disruption.

During the 1970s both the shah (king) of Iran and an ayatollah (religious scholar) named Khomeini claimed to be the divinely appointed interpreter of Islam for the

Shiites. The allegiance of the Iranian Shiites switched from the shah to the ayatollah largely because the ayatollah made a more convincing case that he was more faithfully adhering to the rigid laws laid down by Muhammad in the Quran.

ORIGIN OF OTHER UNIVERSALIZING RELIGIONS

Sikhism and Bahá'í were founded more recently than the three large universalizing religions. The founder of Sikhism, Guru Nanak, traveled widely through South Asia around 500 years ago, preaching his new faith, and many people became his *Sikhs*, which is the Hindi word for "disciples." Nine other gurus succeeded Guru Nanak. Arjan, the fifth guru, compiled and edited in 1604 the *Guru Granth Sahib* (the Holy Granth of Enlightenment), which became the book of Sikh holy scriptures.

When it was established in Iran during the nineteenth century, Bahá'í provoked strong opposition from Shiite Muslims. The Báb was executed in 1850, as were 20,000 of his followers. Bahá'u'lláh, the prophet of Bahá'í, was also arrested but was released in 1853 and exiled to Baghdad. In 1863, his claim that he was the messenger of God anticipated by the Báb was accepted by other followers. Before he died in 1892, Bahá'u'lláh appointed his eldest son 'Abdu'l-Bahá (1844–1921) to be the leader of the Bahá'í community and the authorized interpreter of his teachings.

UNKNOWN ORIGIN OF HINDUISM

Unlike the universalizing religions, Hinduism did not originate with a specific founder. The origins of Christianity, Islam, and Buddhism are recorded in the relatively recent past, but Hinduism existed prior to recorded history.

The word *Hinduism* originated in the sixth century B.C. to refer to people living in what is now India. The earliest surviving Hindu documents were written around 1500 B.C., although archaeological explorations have unearthed objects relating to the religion from 2500 B.C. Aryan tribes from Central Asia invaded India about 1400 B.C. and brought with them Indo-European languages, as discussed in Chapter 5. In addition to their language, the Aryans brought their religion. The Aryans first settled in the area now called the Punjab in northwestern India and later migrated east to the Ganges River valley, as far as Bengal. Centuries of intermingling with the Dravidians already living in the area modified their religious beliefs (Figure 6-17).

▲ **FIGURE 6-16 ORIGIN OF SHIITE ISLAM** The shrine of Imam Ali, in Najaf, Iran, contains the tomb of Ali, from whom traces the Shiite branch of Islam.

Pause and Reflect 6.2.2
What is the significance of Abraham in Judaism, Christianity, and Islam?

◀ **FIGURE 6-17 ORIGIN OF HINDUISM** Mount Kailās (also spelled Kailash) in Tibet is a place of eternal bliss in Hinduism, as well as several other religions. Because of its importance, no human in recorded history has ever climbed to its summit. Hindus believe that this mountain is home of Lord Siva (also spelled Shiva), who is the destroyer of evil and sorrow.

KEY ISSUE 3

Why Do Religions Organize Space in Distinctive Patterns?

- Sacred Space
- The Calendar
- Administration of Space

Learning Outcome 6.3.1
Compare the role of places of worship in various religions.

Geographers study the major impact on the landscape made by all religions, regardless of whether they are universalizing or ethnic. In large cities and small villages around the world, regardless of the region's prevailing religion, the tallest, most elaborate buildings are often religious structures.

Sacred Space

The distribution of religious elements on the landscape reflects the importance of religion in people's values. The impact of religion on the landscape is particularly profound, for many religious people believe that their life on Earth ought to be spent in service to God.

The impact of religion is clearly seen in the arrangement of human activities on the landscape at several scales, from relatively small parcels of land to entire communities. How each religion distributes its elements on the landscape depends on its beliefs. Important religious land uses include burial of the dead and religious settlements.

PLACES OF WORSHIP IN UNIVERSALIZING RELIGIONS

Church, basilica, mosque, temple, pagoda, and synagogue are familiar names that identify places of worship in various religions. Sacred structures are physical "anchors" of religion. All major religions have structures, but the functions of the buildings influence the arrangement of the structures across the landscape. They may house shrines or be places where people assemble for worship. Some religions require a relatively large number of elaborate structures, whereas others have more modest needs.

CHRISTIAN CHURCHES. The Christian landscape is dominated by a high density of churches. The word *church* derives from a Greek term meaning "lord," "master," and "power." *Church* also refers to a gathering of believers, as well as the building at which the gathering occurs.

The church plays a more critical role in Christianity than do buildings in other religions, in part because the structure is an expression of religious principles, an environment in the image of God. The church is also more prominent in Christianity because attendance at a collective service of worship is considered extremely important.

The prominence of churches on the landscape also stems from their style of construction and location. In some communities, the church was traditionally the largest and tallest building and was placed at an important square or other prominent location. Although such characteristics may no longer apply in large cities, they are frequently still true for small towns and neighborhoods within cities.

Since Christianity split into many denominations, no single style of church construction has dominated. Churches reflect both the cultural values of the denomination and the region's architectural heritage. Orthodox churches follow an architectural style that developed in the Byzantine Empire during the fifth century. Byzantine-style Orthodox churches tend to be highly ornate, topped by prominent domes. Many Protestant churches in North America, on the other hand, are simple, with little ornamentation. This austerity is a reflection of the Protestant conception of a church as an assembly hall for the congregation.

Availability of building materials also influences church appearance. In the United States, early churches were most frequently built of wood in the Northeast, brick in the Southeast, and adobe in the Southwest. Stucco and stone predominated in Latin America. This diversity reflected differences in the most common building materials found by early settlers.

MUSLIM MOSQUES. Religious buildings are highly visible and important features of the landscapes in regions dominated by religions other than Christianity. But unlike Christianity, other major religions do not consider their important buildings sanctified places of worship.

Muslims consider a mosque to be a space for community assembly. Unlike a church, a mosque is not viewed as a sanctified place but rather as a location for the community to gather together for worship. Mosques are found primarily in larger cities of the Muslim world; simple structures may serve as places of prayer in rural villages.

A mosque is organized around a central courtyard—traditionally open-air, although it may be enclosed in harsher climates. The pulpit is placed at the end of the courtyard facing Makkah, the direction toward which all Muslims pray. Surrounding the courtyard is a cloister used for schools and nonreligious activities. A distinctive feature of the mosque is the *minaret*, a tower where a man known as a *muezzin* summons people to worship.

BUDDHIST PAGODAS. The pagoda is a prominent and visually attractive element of the Buddhist landscape. Frequently elaborate and delicate in appearance, pagodas typically include tall, many-sided towers arranged in a series of tiers, balconies, and slanting roofs. Pagodas contain relics that Buddhists believe to be a portion of Buddha's body or clothing (Figure 6-23). After Buddha's death, his followers scrambled to obtain these relics. As part of the process of diffusing the religion, Buddhists carried these relics to other countries and built pagodas for them. Pagodas are not designed for congregational worship. Individual prayer or meditation is more likely to be undertaken at an adjacent temple, at a remote monastery, or in a home.

BAHÁ'Í HOUSES OF WORSHIP. Bahá'ís have built Houses of Worship in Wilmette, Illinois, in 1953; Sydney, Australia, and Kampala, Uganda, both in 1961; Lagenhain, near Frankfurt, Germany, in 1964; Panama City, Panama, in 1972; Tiapapata, near Apia, Samoa, in 1984; and New Delhi, India, in 1986 (Figure 6-24). The first Bahá'í House of Worship, built in 1908 in Ashgabat, Russia, now the capital of Turkmenistan, was turned into a museum by the Soviet Union and demolished in 1962 after a severe earthquake. Additional Houses of Worship are planned in Tehran, Iran; Santiago, Chile; and Haifa, Israel.

The locations have not been selected because of proximity to clusters of Bahá'ís. Instead, the Houses of Worship have been dispersed to different continents to dramatize Bahá'í as a universalizing religion with adherents all over the world. The Houses of Worship are open to adherents of all religions, and services include reciting the scriptures of various religions.

▲ FIGURE 6-23 **BUDDHIST PAGODA, SARNATH, INDIA** The Dhamek pagoda, in Deer Park, Sarnath, was built in the third century B.C., and is probably the oldest surviving Buddhist structure in the world.

▲ FIGURE 6-24 **BAHÁ'Í HOUSES OF WORSHIP** Shrine of the Báb, Haifa, Israel.

SIKHISM'S GOLDEN TEMPLE OF AMRITSAR. Sikhism's most holy structure, the Darbar Sahib, or Golden Temple, was built at Amritsar, in the Punjab, by Arjan, the fifth guru, during the sixteenth century (Figure 6-25). The holiest book in Sikhism, the Guru Granth Sahib, is kept there.

Militant Sikhs used the Golden Temple at Amritsar as a base for launching attacks in support of greater autonmy for the Punjab during the 1980s. In 1984, the Indian army attacked the Golden Temple at Amritsar and killed between 500 and 1,500 Sikhs defending the temple. In retaliation later that year, India's Prime Minister Indira Gandhi was assassinated by two of her guards, who were Sikhs.

Pause and Reflect 6.3.1

What is the purpose of the main religious structure in Christianity, Islam, Buddhism, Bahá'í, and Sikhism?

▼ FIGURE 6-25 **SIKH GOLDEN TEMPLE OF AMRITSAR, INDIA** The Darbar Sahib, or Gold Temple, at Amritsar, is the most holy structure for Sikhs, most of whom live in northwestern India.

SACRED PLACES IN UNIVERSALIZING RELIGIONS

Learning Outcome 6.3.2
Explain why places are sacred in universalizing religions.

Religions may elevate particular places to holy positions. Universalizing and ethnic religions differ on the types of places that are considered holy:

- An ethnic religion typically has a less widespread distribution than a universalizing one in part because its holy places derive from the distinctive physical environment of its hearth, such as mountains, rivers, or rock formations.
- A universalizing religion endows with holiness cities and other places associated with the founder's life. Its holy places do not necessarily have to be near each other, and they do not need to be related to any particular physical environment.

Buddhism and Islam are the universalizing religions that place the most emphasis on identifying shrines. Places are holy because they are the locations of important events in the life of Buddha or Muhammad. Making a **pilgrimage** to these holy places—a journey for religious purposes to a place considered sacred—is incorporated into the rituals of some religions. Hindus and Muslims are especially encouraged to make pilgrimages to visit holy places in accordance with recommended itineraries.

BUDDHIST SHRINES. Eight places are holy to Buddhists because they were the locations of important events in Buddha's life (Figure 6-26). The four most important of the eight places are concentrated in a small area of northeastern India and southern Nepal:

- Lumbinī in southern Nepal, where Buddha was born around 563 B.C., is most important. Many sanctuaries and monuments were built there, but all are in ruins today.
- Bodh Gayā, 250 kilometers (150 miles) southeast of Buddha's birthplace, is the site of the second great event in his life, where he reached perfect wisdom. A temple has stood near the site since the third century B.C., and part of the surrounding railing built in the first century A.D. still stands. Because Buddha reached perfect Enlightenment while sitting under a bo tree, that tree has become a holy object as well (Figure 6-27). To honor Buddha, the bo tree has been diffused to other Buddhist countries, such as China and Japan.
- Deer Park in Sarnath, where Buddha gave his first sermon, is the third important location. The Dhamek pagoda at Sarnath, built in the third century B.C., is probably the oldest surviving structure in India (refer to Figure 6-23). Nearby is an important library of Buddhist literature, including many works removed from Tibet when Tibet's Buddhist leader, the Dalai Lama, went into exile.
- Kuśinagara, the fourth holy place, is where Buddha died at age 80 and passed into Nirvana, a state of peaceful extinction (Figure 6-28). Temples built at the site are currently in ruins.

Four other sites in northeastern India are particularly sacred because they were the locations of Buddha's principal miracles:

- Srāvastī is where Buddha performed his greatest miracle. Before an assembled audience of competing religious leaders, Buddha created multiple images of himself and

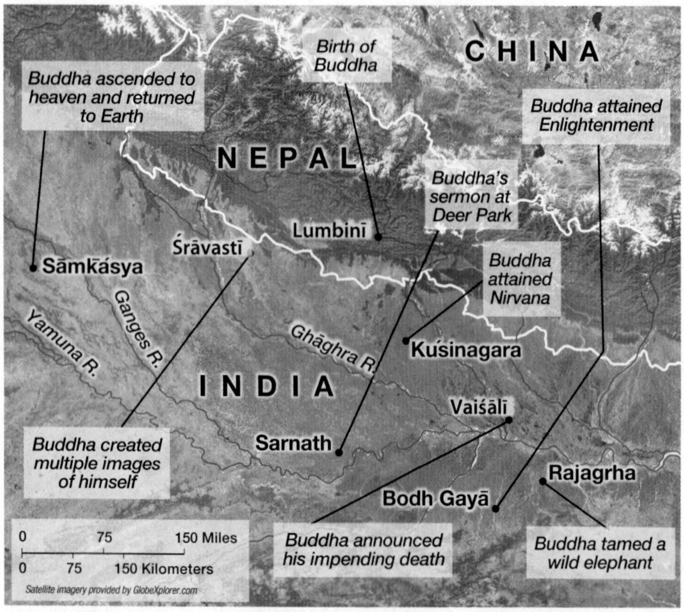

▲ **FIGURE 6-26 HOLY PLACES IN BUDDHISM** Most are clustered in northeastern India and southern Nepal because they were the locations of important events in Buddha's life.

▲ **FIGURE 6-27 BO TREE, BODH GAYĀ, INDIA** A Buddhist monk sits under a Bo tree at Bodh Gayā, the place where Buddha reached perfect wisdom.

▲ **FIGURE 6-28 SLEEPING BUDDHA, KUŚINAGARA** The statue of Buddha asleep marks the location where Buddha is thought to have attained nirvana.

visited heaven. Srāvastī became an active center of Buddhism, and one of the most important monasteries was established there.

- Sāmkāśya, the second miracle site, is where Buddha is said to have ascended to heaven, preached to his mother, and returned to Earth.
- Rajagrha, the third site, is holy because Buddha tamed a wild elephant there, and shortly after Buddha's death, it became the site of the first Buddhist Council.
- Vaisālā, the fourth location, is the site of Buddha's announcement of his impending death and the second Buddhist Council.

All four miracle sites are in ruins today, although excavation activity is under way.

HOLY PLACES IN ISLAM. The holiest locations in Islam are in cities associated with the life of the Prophet Muhammad. The holiest city for Muslims is Makkah (Mecca), the birthplace of Muhammad. The word *mecca* now has a general meaning in the English language as a goal sought or a center of activity.

Now a city of 1.3 million inhabitants, Makkah contains the holiest object in the Islamic landscape, namely al–Ka'ba, a cubelike structure encased in silk, which stands at the center of the Great Mosque, Masjid al-Haram, Islam's largest mosque (Figure 6-29). The Ka'ba, thought to have been built by Abraham and Ishmael, contains a black stone given to Abraham by Gabriel as a sign of the covenant with Ishmael and the Muslim people.

The Ka'ba had been a religious shrine in Makkah for centuries before the origin of Islam. After Muhammad defeated the local people, he captured the Ka'ba, cleared it of idols, and rededicated it to the all-powerful Allah (God). The Masjid al-Haram mosque also contains the well of Zamzam, considered to have the same water source as that given to Hagar by the Angel Gabriel to quench the thirst of her infant, Ishmael.

▲ **FIGURE 6-29 MASJID AL-HARAM, MAKKAH, SAUDI ARABIA** The black cube-like Ka'ba at the center of Masjid al-Haram (Great Mosque) in Makkah is Islam's holiest object.

The second-most-holy geographic location in Islam is Madinah (Medina), a city of 1.3 million inhabitants, 350 kilometers (220 miles) north of Makkah. Muhammad received his first support from the people of Madinah and became the city's chief administrator. Muhammad's tomb is at Madinah, inside Islam's second-largest mosque (refer to Figure 6-15).

Every healthy Muslim who has adequate financial resources is expected to undertake a pilgrimage, called a *hajj*, to Makkah (Mecca). Regardless of nationality and economic background, all pilgrims dress alike, in plain white robes, to emphasize common loyalty to Islam and the equality of people in the eyes of Allah. A precise set of rituals is practiced, culminating in a visit to the Ka'ba. The *hajj* attracts millions of Muslims annually to Makkah. *Hajj* visas are issued by the government of Saudi Arabia according to a formula of 1 per 1,000 Muslims in a country. Roughly 80 percent come from Southwest Asia & North Africa and 20 percent from elsewhere in Asia. Although Indonesia is the country with the most Muslims, it has not sent the largest number of pilgrims to Makkah because of the relatively long travel distance.

Pause and Reflect 6.3.2

Based on the lives of the Buddha and the prophet Muhammad, what types of sites are likely to be goals of pilgrimage for the followers of a universalizing religion?

THE LANDSCAPE IN ETHNIC RELIGIONS

Learning Outcome 6.3.3

Analyze the importance of the physical geography in ethnic religions.

One of the principal reasons that ethnic religions are highly clustered is that they are closely tied to the physical geography of a particular place. Pilgrimages are undertaken to view these physical features.

HINDU LANDSCAPE. As an ethnic religion of India, Hinduism is closely tied to the physical geography of India. According to a survey conducted by the geographer Surinder Bhardwaj, the natural features most likely to rank among the holiest shrines in India are riverbanks and coastlines. Hindus consider a pilgrimage, known as a *tirtha*, to be an act of purification. Although not a substitute for meditation, the pilgrimage is an important act in achieving redemption.

Hindu holy places are organized into a hierarchy. Particularly sacred places attract Hindus from all over India, despite the relatively remote locations of some; less important shrines attract primarily local pilgrims. Because Hinduism has no central authority, the relative importance of shrines is established by tradition, not by doctrine. For example, many Hindus make long-distance pilgrimages to Mt. Kailās, located at the source of the Ganges in the Himalayas, which is holy because Siva lives there (refer to Figure 6-17). Other mountains may attract only local pilgrims: Local residents may consider a nearby mountain to be holy if Siva is thought to have visited it at one time.

Hindus believe that they achieve purification by bathing in holy rivers. The Ganges is the holiest river in India because it is supposed to spring forth from the hair of Siva, one of the main deities. Indians come from all over the country to Hardwār, the most popular location for bathing in the Ganges (Figure 6-30).

The remoteness of holy places from population clusters once meant that making a pilgrimage required major commitments of time and money as well as undergoing considerable physical hardship. Recent improvements in transportation have increased the accessibility of shrines. Hindus can now reach holy places in the Himalaya Mountains by bus or car, and Muslims from all over the world can reach Makkah by airplane.

▼ **FIGURE 6-30 BATHING IN THE GANGES, HARDWĀR, INDIA** Hindus bathe in the Ganges River to wash away their sins.

HINDU TEMPLES. Sacred structures for collective worship are relatively unimportant in Asian ethnic and universalizing religions. Instead, important religious functions are more likely to take place at home within the family. Temples are built to house shrines for particular gods rather than for congregational worship. The Hindu temple serves as a home to one or more gods, although a particular god may have more than one temple.

A typical Hindu temple contains a small, dimly lit interior room where a symbolic artifact or some other image of the god rests. Because congregational worship is not part of Hinduism, the temple does not need a large closed interior space filled with seats. The site of the temple, usually demarcated by a wall, may also contain a structure for a caretaker and a pool for ritual baths. Space may be devoted to ritual processions.

Wealthy individuals or groups usually maintain local temples. Size and number of temples are determined by local preferences and commitment of resources rather than standards imposed by religious doctrine.

COSMOGONY

Ethnic religions differ from universalizing religions in their understanding of relationships between human beings and nature. These differences derive from distinctive concepts of **cosmogony**, which is a set of religious beliefs concerning the origin of the universe. A variety of events in the physical environment are more likely to be incorporated into the principles of an ethnic religion. These events range from the familiar and predictable to unexpected disasters.

COSMOGONY IN ETHNIC RELIGIONS. Chinese ethnic religions, such as Confucianism and Taoism, believe that the universe is made up of two forces, yin and yang, which exist in everything. The yin force is associated with earth, darkness, female, cold, depth, passivity, and death. The yang force is associated with heaven, light, male, heat, height, activity, and life. Yin and yang forces interact with each other to achieve balance and harmony, but they are in a constant state of change. An imbalance results in disorder and chaos. The principle of yin and yang applies to the creation and transformation of all natural features.

COSMOGONY IN UNIVERSALIZING RELIGIONS. The universalizing religions that originated in Southwest Asia, notably Christianity and Islam, consider that God created the universe, including Earth's physical environment and human beings. A religious person can serve God by cultivating the land, draining wetlands, clearing forests, building new settlements, and otherwise making

▲ **FIGURE 6-31 STONEHENGE** Stonehenge, in southwestern England, was constructed between 4,000 and 5,000 years ago.

productive use of natural features that God created. As the very creator of Earth itself, God is more powerful than any force of nature, and if in conflict, the laws of God take precedence over laws of nature.

Christian and Islamic cosmogony differ in some respects. For example, Christians believe that Earth was given by God to humanity to finish the task of creation. Obeying the all-supreme power of God means independence from the tyranny of natural forces. Muslims regard humans as representatives of God on Earth, capable of reflecting the attributes of God in their deeds, such as growing food or doing other hard work to improve the land. But they believe that humans are not partners with God, who alone was responsible for Earth's creation.

In the name of God, some people have sought mastery over nature, not merely independence from it. Large-scale development of remaining wilderness is advocated by some religious people as a way to serve God. To those who follow this approach, failure to make full and complete use of Earth's natural resources is considered a violation of biblical teachings. Christians are more likely to consider floods, droughts, and other natural disasters to be preventable and may take steps to overcome the problem by modifying the environment. Some Christians regard natural disasters as punishment for human sins.

Adherents of ethnic religions do not attempt to transform the environment to the same extent. To animists, for example, God's powers are mystical, and only a few people on Earth can harness these powers for medical or other purposes. God can be placated, however, through prayer and sacrifice. Environmental hazards may be accepted as normal and unavoidable.

THE SOLSTICE. The **solstice** has special significance in some ethnic religions. A major holiday in some pagan religions is the winter solstice, December 21 or 22 in the Northern Hemisphere. The winter solstice is the shortest day and longest night of the year, when the Sun appears lowest in the sky, and the apparent movement of the Sun's path north or south comes to a stop before reversing direction (*solstice* comes from the Latin to "stand still"). Stonehenge, a collection of stones erected in southwestern England some 3,500 years ago (Figure 6-31), is a prominent remnant of a pagan structure apparently aligned so the Sun rises between two stones on the summer and winter solstices (Figure 6-32).

If you stand at the western facade of the U.S. Capitol in Washington at sunset on the summer solstice (June 21 or 22 in the Northern Hemisphere) and look down Pennsylvania Avenue, the Sun is directly over the center of the avenue. Similarly, at the winter solstice, sunset is directly aligned with the view from the Capitol down Maryland Avenue. Will archaeologists of the distant future think we erected the Capitol Building and aligned the streets as a religious ritual? Did the planner of Washington, Pierre L'Enfant, create the pattern accidentally or deliberately, and if so, why?

Pause and Reflect 6.3.3

How do adherents of universalizing religions such as Christianity and Islam and adherents of ethnic religions tend to differ in their attitudes toward Earth's physical environment?

▼ **FIGURE 6-32 SUNRISE ON THE SOLSTICE AT STONEHENGE** Stones were apparently aligned with regard for the solstice.

DISPOSING OF THE DEAD

Learning Outcome 6.3.4

Describe ways in which the landscape is used in religiously significant ways.

A prominent example of religiously inspired arrangement of land at a smaller scale is burial practices. Climate, topography, and religious doctrine combine to create differences in practices to shelter the dead.

BURIAL. Christians, Muslims, and Jews usually bury their dead in a specially designated area called a cemetery (Figure 6-33). The Christian burial practice can be traced to the early years of the religion. In ancient Rome, underground passages known as *catacombs* were used to bury early Christians (and to protect the faithful when the religion was still illegal).

After Christianity became legal, Christians buried their dead in the yard around the church. As these burial places became overcrowded, separate burial grounds had to be established outside the city walls. Public health and sanitation considerations in the nineteenth century led to public management of many cemeteries. Some cemeteries are still operated by religious organizations. The remains of the dead are customarily aligned in some traditional direction. Some Christians bury the dead with the feet toward Jerusalem so that they may meet Christ there on the Day of Judgment.

Cemeteries may consume significant space in a community, increasing the competition for scarce space. In congested urban areas, Christians and Muslims have traditionally used cemeteries as public open space. Before the widespread development of public parks in the nineteenth century, cemeteries were frequently the only green space in rapidly growing cities. Cemeteries are still used as parks in Muslim countries, where the idea faces less opposition than in Christian societies.

▼ **FIGURE 6-33 MUSLIM CEMETERY** Fatimid cemetery, in Aswan, Egypt, is approximately 1,000 years old.

Traditional burial practices in China have put pressure on agricultural land. By burying dead relatives, rural residents have removed as much as 10 percent of the land from productive agriculture. The government in China has ordered the practice discontinued, even urging farmers to plow over old burial mounds. Cremation is encouraged instead.

OTHER METHODS OF DISPOSING OF BODIES. Not all faiths bury their dead. Hindus generally practice cremation rather than burial (Figure 6-34). The body is washed with water from the Ganges River and then burned with a slow fire on a funeral pyre. Burial is reserved for children, ascetics, and people with certain diseases. Cremation is considered an act of purification, although it tends to strain India's wood supply.

Motivation for cremation may have originated from unwillingness on the part of nomads to leave their dead behind, possibly because of fear that the body could be attacked by wild beasts or evil spirits, or even return to life. Cremation could also free the soul from the body for departure to the afterworld and provide warmth and comfort for the soul as it embarked on the journey to the afterworld. Cremation was the principal form of disposing of bodies in Europe before Christianity. It is still practiced in parts of Southeast Asia, possibly because of Hindu influence.

To strip away unclean portions of the body, Zoroastrians (Parsis) traditionally exposed the dead to scavenging birds and animals. The ancient Zoroastrians did not want the body to contaminate the sacred elements of fire, earth, or water. The dead were exposed in a circular structure called a *dakhma*, or tower of silence (Figure 6-35). Tibetan Buddhists also practiced exposure for some dead, with cremation reserved for the most exalted priests.

Disposal of bodies at sea is used in some parts of Micronesia, but the practice is much less common than in the past. The bodies of lower-class people would be flung into the sea; elites could be set adrift on a raft or boat. Water burial was regarded as a safeguard against being contaminated by the dead.

▼ **FIGURE 6-34 HINDU CREMATION** Family members cover a body with wood for cremation, Agra, India.

▲ **FIGURE 6-35 ZOROASTRIAN TOWER OF SILENCE, YAZD, IRAN** Zoroastrians placed bodies in the pit in the center of the tower. The practice has been discontinued.

Pause and Reflect 6.3.4
What are some of the cultural or religious factors that influence methods of disposing of bodies other than burial?

RELIGIOUS SETTLEMENTS AND PLACE NAMES

Buildings for worship and burial places are smaller-scale manifestations of religion on the landscape, but there are larger-scale examples—entire settlements. Most human settlements serve an economic purpose (see Chapter 12), but some are established primarily for religious reasons.

UTOPIAN SETTLEMENTS. A *utopian settlement* is an ideal community built around a religious way of life. Buildings are sited and economic activities organized to integrate religious principles into all aspects of daily life. An early utopian settlement in the United States was Bethlehem, Pennsylvania, founded in 1741 by Moravians, Christians who had emigrated from the present-day Czech Republic. By 1858, some 130 different utopian settlements had begun in the United States, in conformance with a group's distinctive religious beliefs. Examples include Oneida, New York; Ephrata, Pennsylvania; Nauvoo, Illinois; and New Harmony, Indiana.

The culmination of the utopian movement in the United States was the construction of Salt Lake City by the Mormons, beginning in 1848. The layout of Salt Lake City is based on a plan of the city of Zion given to the church elders in 1833 by the Mormon prophet Joseph Smith. The city has a regular grid pattern, unusually broad boulevards, and church-related buildings situated at strategic points.

Most utopian communities declined in importance or disappeared altogether. Some disappeared because the inhabitants were celibate and could not attract immigrants; in other cases, residents moved away in search of better economic conditions. The utopian communities that have not been demolished are now inhabited by people who are not members of the original religious sect, although a few have been preserved as museums.

Although most colonial settlements were not planned primarily for religious purposes, religious principles affected many of the designs. Most early New England settlers were members of a Puritan Protestant denomination. The Puritans generally migrated together from England and preferred to live near each other in clustered settlements rather than on dispersed, isolated farms. Reflecting the importance of religion in their lives, New England settlers placed the church at the most prominent location in the center of the settlement, usually adjacent to a public open space known as a *common*, because it was for common use by everyone.

RELIGIOUS PLACE NAMES. Roman Catholic immigrants have frequently given religious place names, or *toponyms*, to their settlements in the New World, particularly in Québec and the U.S. Southwest. Québec's boundaries with Ontario and the United States clearly illustrate the difference between toponyms selected by Roman Catholic and Protestant settlers. Religious place names are common in Québec but rare in the two neighbors (Figure 6-36).

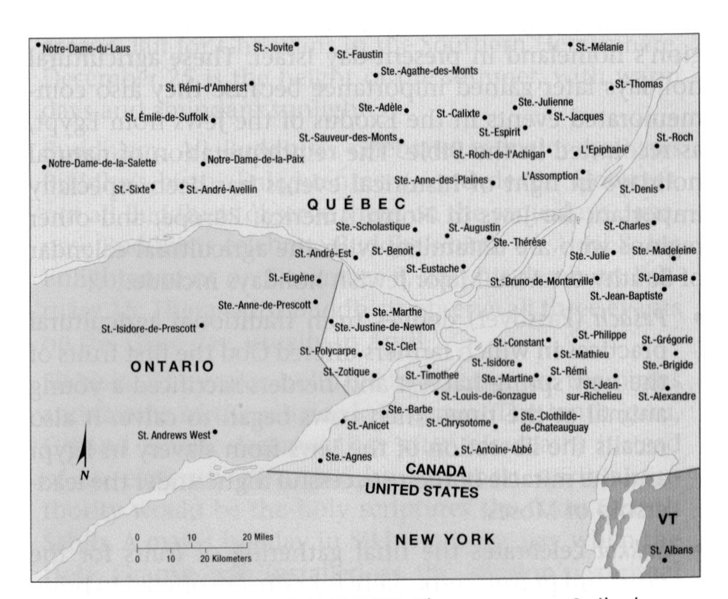

▲ **FIGURE 6-36 RELIGIOUS TOPONYMS** Place names near Québec's boundaries with Ontario and the United States show the impact of religion on the landscape. In Québec, a province with a predominantly Roman Catholic population, a large number of settlements are named for saints, where relatively few religious toponyms are found in predominantly Protestant Ontario, New York, and Vermont.

Administration of Space

Learning Outcome 6.3.6
Compare the administrative organization of hierarchical and locally autonomous religions.

Followers of a universalizing religion must be connected in order to ensure communication and consistency of doctrine. The method of interaction varies among universalizing religions, branches, and denominations. Ethnic religions tend not to have organized, central authorities.

HIERARCHICAL RELIGIONS

A **hierarchical religion** has a well-defined geographic structure and organizes territory into local administrative units. Roman Catholicism provides a good example of a hierarchical religion.

LATTER-DAY SAINTS. Latter-day Saints (Mormons) exercise strong organization of the landscape. The territory occupied by Mormons, primarily Utah and portions of surrounding states, is organized into wards, with populations of approximately 750 each. Several wards are combined into a stake of approximately 5,000 people. The highest authority in the Church—the board and president—frequently redraws ward and stake boundaries in rapidly growing areas to reflect the ideal population standards.

ROMAN CATHOLIC HIERARCHY. The Roman Catholic Church has organized much of Earth's inhabited land into an administrative structure ultimately accountable to the Pope in Rome (Figure 6-40). Here is the top hierarchy of Roman Catholicism:

- The *Pope* is also the bishop of the Diocese of Rome.
- *Archbishops* report to the Pope. Each heads a province, which is a group of several dioceses. The archbishop also is bishop of one diocese within the province, and some distinguished archbishops are elevated to the rank of cardinal.
- *Bishops* report to an archbishop. Each administers a diocese, which is the basic unit of geographic organization in the Roman Catholic Church. The bishop's headquarters, called a "see," is typically the largest city in the diocese.
- *Priests* report to bishops. A diocese is spatially divided into parishes, each headed by a priest.

Pause and Reflect 6.3.6
What are the different spatial units of administration in the Roman Catholic Church?

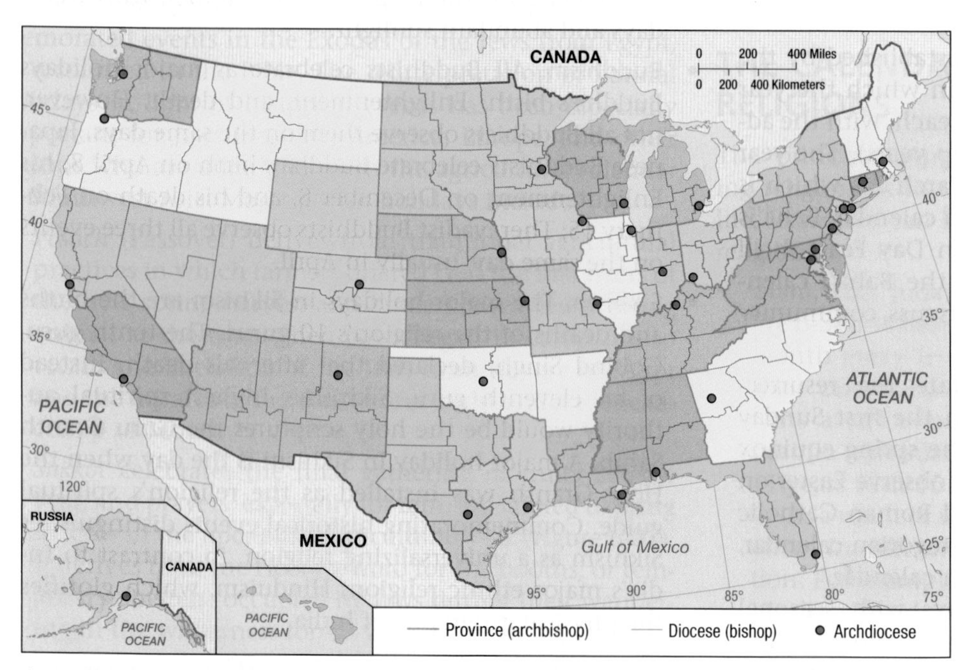

▲ **FIGURE 6-40 ROMAN CATHOLIC HIERARCHY IN THE UNITED STATES** The Roman Catholic Church divides the United States into provinces, each headed by an archbishop. Provinces are subdivided into dioceses, each headed by a bishop. The archbishop of a province also serves as the bishop of a diocese. Dioceses that are headed by archbishops are called archdioceses.

The area and population of parishes and dioceses vary according to historical factors and the distribution of Roman Catholics across Earth's surface. In parts of Europe, the overwhelming majority of the dense population is Roman Catholic. Consequently, the density of parishes is high. A typical parish may encompass only a few square kilometers and fewer than 1,000 people. At the other extreme, Latin American parishes may encompass several hundred square kilometers and 5,000 people. The more dispersed Latin American distribution is attributable partly to a lower population density than in Europe.

Because Roman Catholicism is a hierarchical religion, individual parishes must work closely with centrally located officials concerning rituals and procedures. If Latin America followed the European model of small parishes, many would be too remote for the priest to communicate with others in the hierarchy. The less intensive network of Roman Catholic institutions also results in part from colonial traditions, for both Portuguese and Spanish rulers discouraged parish development in Latin America.

The Roman Catholic population is growing rapidly in the U.S. Southwest and in suburbs of some large North American and European cities. Some of these areas have a low density of parishes and dioceses compared to the population, so the Church must adjust its territorial organization. New local administrative units can be created, although funds to provide the desired number of churches, schools, and other religious structures might be scarce. Conversely, the Roman Catholic population is declining in inner cities and rural areas. Maintaining services in these areas is expensive, but the process of combining parishes and closing schools is very difficult.

LOCALLY AUTONOMOUS RELIGIONS

Some universalizing religions are highly **autonomous religions**, or self-sufficient, and interaction among communities is confined to little more than loose cooperation and shared ideas. Islam and some Protestant denominations are good examples.

LOCAL AUTONOMY IN ISLAM. Among the three large universalizing religions, Islam provides the most local autonomy. Like other locally autonomous religions, Islam has neither a religious hierarchy nor a formal territorial organization. A mosque is a place for public ceremony, and a leader known as a *muezzin* calls the faithful to prayer (Figure 6-41), but everyone is expected to participate equally in the rituals and is encouraged to pray privately.

In the absence of a hierarchy, the only formal organization of territory in Islam is through the coincidence of religious territory with secular states. Governments in some predominantly Islamic countries include in their bureaucracy people who administer Islamic institutions. These administrators interpret Islamic law and run welfare programs.

▼ **FIGURE 6-41 CALLING MUSLIMS TO PRAYER, CAIRO, EGYPT**
Muslims are called to prayer by a muezzin, who recites the *shahadah*.

Strong unity in the Islamic world is maintained by a relatively high degree of communication and migration, such as the pilgrimage to Makkah. In addition, uniformity is fostered by Islamic doctrine, which offers more explicit commands than other religions.

PROTESTANT DENOMINATIONS. Protestant Christian denominations vary in geographic structure from extremely autonomous to somewhat hierarchical. The Episcopalian, Lutheran, and most Methodist churches have hierarchical structures, somewhat comparable to the Roman Catholic Church. Extremely autonomous denominations such as Baptists and United Church of Christ are organized into self-governing congregations. Each congregation establishes the precise form of worship and selects the leadership.

Presbyterian churches represent an intermediate degree of autonomy. Individual churches are united in a presbytery, several of which in turn are governed by a synod, with a general assembly as ultimate authority over all churches. Each Presbyterian church is governed by an elected board of directors with lay members.

ETHNIC RELIGIONS. Judaism and Hinduism also have no centralized structure of religious control. To conduct a full service, Judaism merely requires the presence of 10 adult males. (Females count in some Jewish communities.)

Hinduism is even more autonomous because worship is usually done alone or with others in the household. Hindus share ideas primarily through undertaking pilgrimages and reading traditional writings.

CHECK-IN: KEY ISSUE 3

Why Do Religions Organize Space in Distinctive Patterns?

- ✓ **Religious structures, such as churches and mosques, are prominent features of the landscape.**
- ✓ **Some religions encourage pilgrimages to holy places.**
- ✓ **Ethnic religions are more closely tied to their local physical environment than are universalizing religions.**
- ✓ **The calendar typically revolves around the physical environment in ethnic religions and the founder's life in universalizing religions.**
- ✓ **Some religions have hierarchical administrative structures, whereas others emphasize local autonomy.**

RELIGION VERSUS COMMUNISM

Learning Outcome 6.4.2
Summarize reasons for conflicts between religions.

Organized religion was challenged in the twentieth century by the rise of Communism in Eastern Europe and Asia. The three religions most affected were Orthodox Christianity, Islam, and Buddhism. Communist regimes generally discouraged religious belief and practice.

CHRISTIANITY AND ISLAM VERSUS THE FORMER SOVIET UNION. In 1721, Czar Peter the Great made the Russian Orthodox Church a part of the Russian government (Figure 6-45). The patriarch of the Russian Orthodox Church was replaced by a 12-member committee, known as the Holy Synod, nominated by the czar.

Following the 1917 Bolshevik Revolution, which overthrew the czar, the Communist government of the Soviet Union pursued antireligious programs. Karl Marx had called religion "the opium of the people," a view shared by V. I. Lenin and other early Communist leaders. Marxism became the official doctrine of the Soviet Union, so religious doctrine was a potential threat to the success of the revolution.

The Soviet government in 1918 eliminated the official church–state connection that Peter the Great had forged. All church buildings and property were nationalized and could be used only with local government permission. People's religious beliefs could not be destroyed overnight, but the role of organized religion in Soviet life could be reduced—and it was. The Orthodox religion retained adherents in the Soviet Union, especially among the elderly, but younger people generally had little contact with the church beyond attending a service perhaps once a year. With religious organizations prevented from conducting social and cultural work, religion dwindled in daily life.

The end of Communist rule in the late twentieth century brought a religious revival in Eastern Europe, especially where Roman Catholicism is the most prevalent branch of Christianity, including Croatia, the Czech Republic, Hungary, Lithuania, Poland, Slovakia, and Slovenia. Property confiscated by the Communist governments reverted to Church ownership, and attendance at church services increased.

In Central Asia, countries that were former parts of the Soviet Union—Kazakhstan, Kyrgyzstan, Tajikistan, Turkmenistan, and Uzbekistan—most people are Muslims. These newly independent countries are struggling to determine the extent to which laws should be rewritten to conform to Islamic custom rather than to the secular tradition inherited from the Soviet Union.

Pause and Reflect 6.4.2
How did the end of communism in the former Soviet Union and Eastern Europe affect religion?

BUDDHISM VERSUS SOUTHEAST ASIAN COUNTRIES. In Southeast Asia, Buddhists were hurt by the long Vietnam War—waged between the French and later by the Americans, on one side, and Communist groups on the other. Neither antagonist was particularly sympathetic to Buddhists. U.S. air raids in Laos and Cambodia destroyed many Buddhist shrines, and other shrines were vandalized by Vietnamese and by the Khmer Rouge Cambodian Communists. On a number of occasions, Buddhists immolated (burned) themselves to protest policies of the South Vietnamese government.

The current Communist governments in Southeast Asia have discouraged religious activities and permitted monuments to decay, most notably the Angkor Wat complex in Cambodia, considered one of the world's most beautiful Buddhist and Hindu structures (Figure 6-46). In any event, these countries do not have the funds necessary to restore the structures, although international organizations have helped.

▼ **FIGURE 6-45 ST BASIL'S, MOSCOW** A Russian Orthodox cathedral has stood at the center of Moscow since the sixteenth century. The communists turned it into a museum.

▲ **FIGURE 6-46 VANDALIZING RELIGIOUS SHRINES** Angkor Wat, Cambodia, considered one of the world's most important Hindu and Buddhist shrines, was vandalized by the Khmer Rouge.

Religion versus Religion

Refer to the map of world religions near the beginning of this chapter (Figure 6-3). Conflicts are most likely to occur where colors change, indicating a boundary between two religious groups.

Two long-standing conflicts involving religious groups are in Northern Ireland and Southwest Asia.

RELIGIOUS WARS IN IRELAND

The most troublesome religious boundary in Western Europe lies on the island of Eire (Ireland). The Republic of Ireland, which occupies five-sixths of the island, is 87 percent Roman Catholic, but the island's northern one-sixth, which is part of the United Kingdom rather than Ireland, is 46 percent Protestant and 40 percent Roman Catholic, according to the 2001 census. (The remaining 14 percent stated no religion or did not respond.)

The entire island was an English colony for many centuries and was made part of the United Kingdom in 1801. Agitation for independence from Britain increased in Ireland during the nineteenth century, especially after poor economic conditions and famine in the 1840s led to mass emigration. Following a succession of bloody confrontations, Ireland became a self-governing dominion within the British Empire in 1921. Complete independence was declared in 1937, and a republic was created in 1949. When most of Ireland became independent, a majority in six northern counties voted to remain in the United Kingdom. Protestants, who comprised the majority in Northern Ireland, preferred to be part of the predominantly Protestant United Kingdom rather than join the predominantly Roman Catholic Republic of Ireland (Figure 6-47).

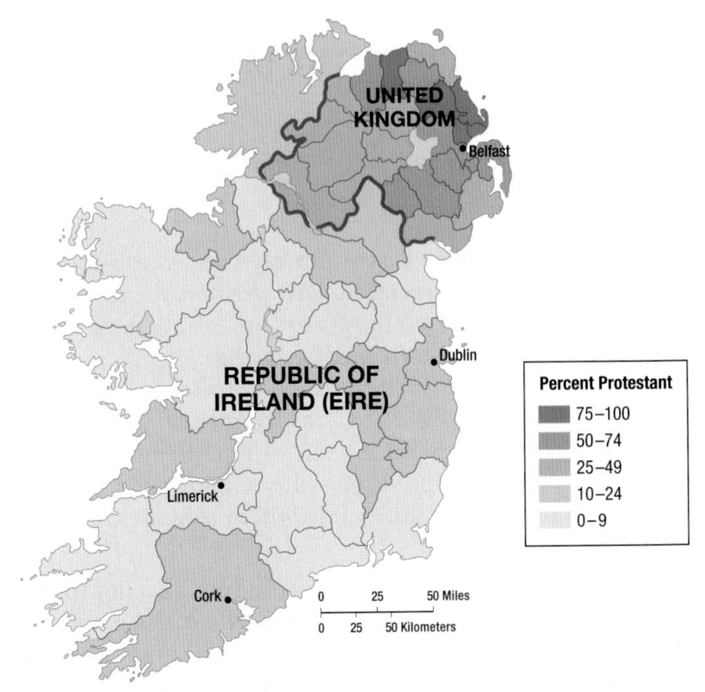

▲ **FIGURE 6-47 DISTRIBUTION OF CATHOLICS AND PROTESTANTS IN IRELAND, 1911** Long a colony of England, Ireland became a self-governing dominion within the British Empire in 1921. In 1937, it became a completely independent country, but 26 districts in the north of Ireland chose to remain part of the United Kingdom. The Republic of Ireland today is 87 percent Roman Catholic, whereas Northern Ireland has a Protestant majority. The boundary between Roman Catholics and Protestants does not coincide precisely with the international border, so Northern Ireland includes some communities that are predominantly Roman Catholic. This is the root of a religious conflict that continues today.

Roman Catholics in Northern Ireland have been victimized by discriminatory practices, such as exclusion from higher-paying jobs and better schools. The capital Belfast is highly segregated, with predominantly Catholic neighborhoods to the west and Protestant neighborhoods to the east (Figure 6-48). Demonstrations by Roman Catholics protesting discrimination began in 1968. Since then, more than 3,000 have been killed in Northern Ireland—both Protestants and Roman Catholics—in a continuing cycle of demonstrations and protests.

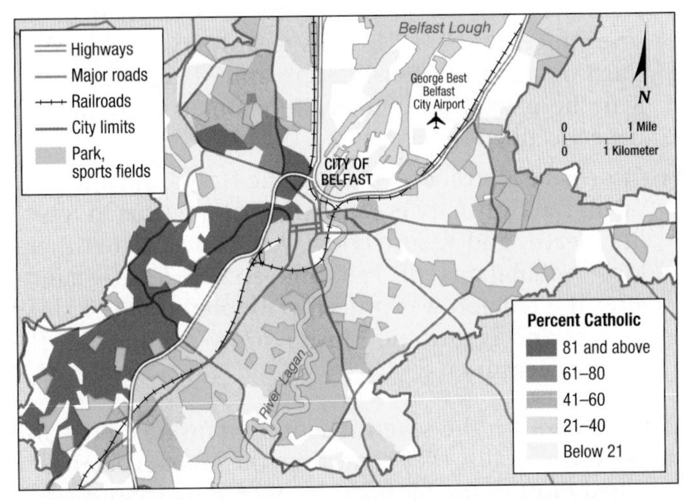

▲ **FIGURE 6-48 DISTRIBUTION OF CATHOLICS AND PROTESTANTS IN BELFAST** Belfast, Northern Ireland, is highly segregated. Most Roman Catholics live to the west, and Protestants to the east.

A small number of Roman Catholics in both Northern Ireland and the Republic of Ireland joined the Irish Republican Army (IRA), a militant organization dedicated to achieving Irish national unity by whatever means available, including violence. Similarly, a scattering of Protestants created extremist organizations to fight the IRA, including the Ulster Defense Force (UDF).

Although the overwhelming majority of Northern Ireland's Roman Catholics and Protestants are willing to live peacefully with the other religious group, extremists disrupt daily life for everyone and do well in elections. As long as most Protestants are firmly committed to remaining in the United Kingdom and most Roman Catholics are equally committed to union with the Republic of Ireland, peaceful settlement appears difficult. Peace agreements implemented in 1999 provided for the sharing of power, but the British government has suspended the arrangement several times because of violations.

RELIGIOUS WARS IN THE MIDDLE EAST

Learning Outcome 6.4.3

Analyze reasons for religious conflict in the Middle East.

Conflict in the Middle East is among the world's longest standing and most intractable. Jews, Christians, and Muslims have fought for 2,000 years to control the same small strip of land in the Eastern Mediterranean.

To some extent, the hostility among Christians, Muslims, and Jews in the Middle East stems from their similar heritage. All three groups trace their origins to Abraham in the Hebrew Bible narrative, but the religions diverged in ways that have made it difficult for them to share the same territory:

- *Judaism*, an ethnic religion, makes a special claim to the territory it calls the Promised Land. The major events in the development of Judaism took place there, and the religion's customs and rituals acquired meaning from the agricultural life of the ancient Hebrew tribe. Descendants of 10 of Jacob's sons, plus 2 of his grandsons, constituted the 12 tribes of Hebrews who emigrated from Egypt in the Exodus narrative. Each received a portion of Canaan. After the Romans gained control of the area, which they called the province of Palestine, they dispersed the Jews from Palestine, and only a handful were permitted to live in the region until the twentieth century.
- *Islam* became the most widely practiced religion in Palestine after the Muslim army conquered it in the seventh century A.D. Muslims regard Jerusalem as their third holiest city, after Makkah and Madinah, because it is the place from which Muhammad is thought to have ascended to heaven.
- *Christianity* considers Palestine the Holy Land and Jerusalem the Holy City because the major events in Jesus's life, death, and Resurrection were concentrated there. Most inhabitants of Palestine accepted Christianity after the religion was officially adopted by the Roman Empire and before the Muslim army conquest in the seventh century.

CRUSADES. In the seventh century, Muslims, now also called Arabs because they came from the Arabian peninsula, captured most of the Middle East, including Palestine and Jerusalem. The Arab Muslim presence the Arabic language across the Middle East and diffused subsequently converted most of the people from Christianity to Islam.

The Arab Muslims moved west across North Africa and invaded Europe at Gibraltar in A.D. 711 (see Figure 6-20). The army conquered most of the Iberian Peninsula, crossed the Pyrenees Mountains a few years later, and for a time occupied much of present-day France. Its initial advance in Europe was halted by the Franks (a West Germanic people), led by Charles Martel, at Poitiers, France, in 732. The Muslims made further gains in Europe in subsequent years and continued to control portions of present-day Spain until 1492, but Martel's victory ensured that Christianity rather than Islam would be Europe's dominant religion.

To the east, Ottoman Turks captured Eastern Orthodox Christianity's most important city, Constantinople (present-day Istanbul in Turkey), in 1453 and advanced a few years later into southeast Europe, as far north as present-day Bosnia & Herzegovina. The recent civil war in that country is a legacy of the fifteenth-century Muslim invasion (see Chapter 7).

To recapture the Holy Land from its Muslim conquerors, European Christians launched a series of military campaigns, known as Crusades, over a 150-year period. Crusaders captured Jerusalem from the Muslims in 1099 during the First Crusade, lost it in 1187 (which led to the Third Crusade), regained it in 1229 as part of a treaty ending the Sixth Crusade, and lost it again in 1244.

Pause and Reflect 6.4.3

Why is a narrow strip of land at the eastern end of the Mediterranean Sea so important in Judaism, Christianity, and Islam?

PARTITION OF PALESTINE. The Muslim Ottoman Empire controlled Palestine for most of the four centuries between 1516 and 1917. Upon the empire's defeat in World War I, the United Kingdom took over Palestine, under a mandate from the League of Nations, and later from the United Nations.

For a few years, the British allowed some Jews to return to Palestine, but immigration was restricted again during the 1930s, in response to intense pressure by Arabs in the region. As violence initiated by both Jewish and Muslim settlers escalated after World War II, the British announced their intention to withdraw from Palestine. The United Nations voted in 1947 to partition Palestine into two independent states, one Jewish and one Muslim (Figure 6-49, left). Jerusalem was to be an international city, open to all religions, and run by the United Nations.

WARS BETWEEN ISRAEL AND NEIGHBORS. When the British withdrew in 1948, Jews declared an independent state of Israel within the boundaries prescribed by the UN resolution. Over the next quarter-century, Israel fought four wars with its neighbors:

- **1948–1949 Independence War.** The day after Israel declared independence, the neighboring Arab Muslim states declared war. Israel survived the attack, and the combatants signed an armistice in 1949. Israel's boundaries were extended beyond the UN partition, including the western suburbs of Jerusalem. Jordan gained control of the West Bank and East Jerusalem, including the Old City, where holy places are clustered. Egypt gained the Gaza Strip.
- **1956 Suez War.** Egypt seized the Suez Canal, a key shipping route between Europe and Asia that had been

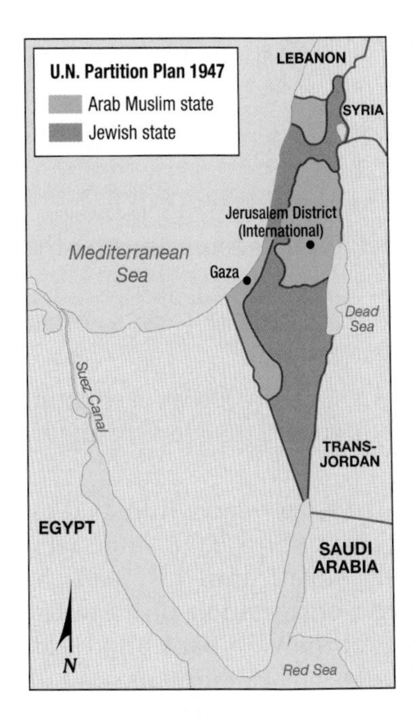

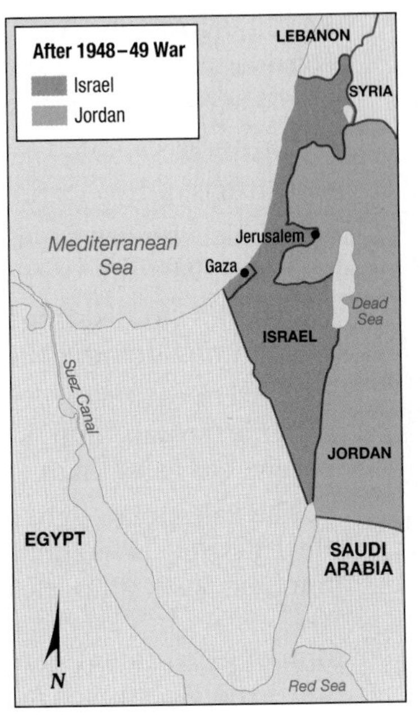

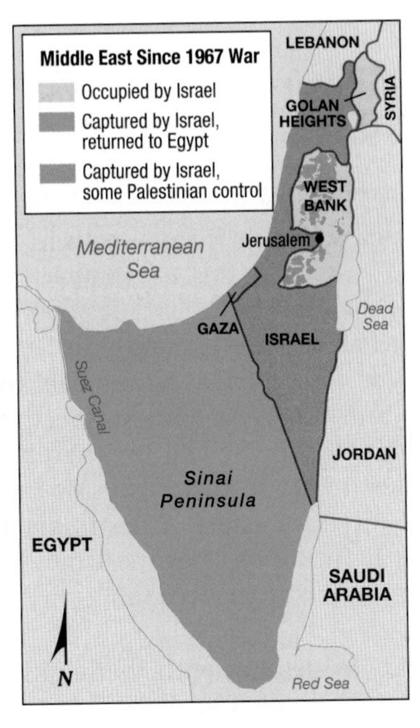

◀ FIGURE 6-49 **BOUNDARY CHANGES IN ISRAEL/PALESTINE** (left) The 1947 UN partition plan, (center) Israel after the 1948–1949 war, (right) Israel and its neighbors since the 1967 Six-Day War.

built and controlled up until then by France and the United Kingdom. Egypt also blockaded international waterways near its shores that Israeli ships were using. Israel, France, and the United Kingdom attacked Egypt and got the waterways reopened, although Egypt retained control of the Suez Canal.

- **1967 Six-Day War.** Israel's neighbors massed a quarter-million troops along the borders and again blocked Israeli ships from using international waterways. In retaliation, Israel launched a surprise attack, destroying the coalition's air forces. Israel captured territory:
  - From Jordan, the Old City of Jerusalem and the West Bank (the territory west of the Jordan River taken by Jordan in the 1948–1949 war) (Figure 6-50)
  - From Syria, the Golan Heights
  - From Egypt, the Gaza Strip and Sinai Peninsula
- **1973 Yom Kippur War.** A surprise attack on Israel by its neighbors took place on the holiest day of the year for Jews. The war ended without a change in boundaries.
- **1979 Peace Treaty.** Egypt's President Anwar Sadat and Israel's Prime Minister Menachem Begin signed a peace treaty in 1979, following a series of meetings with U.S. President Jimmy Carter at Camp David, Maryland. Israel returned the Sinai Peninsula to Egypt, and in return Egypt recognized Israel's right to exist. Sadat was assassinated by Egyptian soldiers, who were extremist Muslims opposed to compromising with Israel, but his successor Hosni Mubarak carried out the terms of the treaty. A half-century after the Six-Day War, the status of the other territories occupied by Israel has still not been settled.

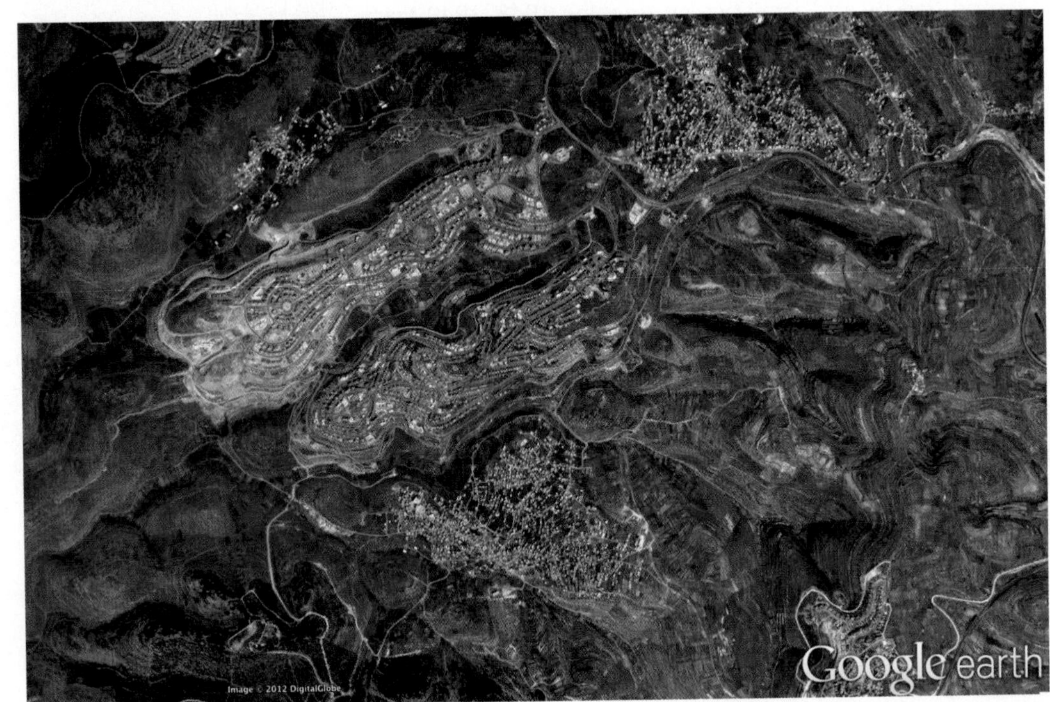

▶ FIGURE 6-50 **WEST BANK SETTLEMENT** In this Google Earth image from 2010, the Israeli settlement Betar Illit is under construction (top of the photo) in the West Bank, on a hillside overlooking the Palestinian villages Nahalin (bottom) and Husan (top right).

CONFLICTING PERSPECTIVES OF THE HOLY LAND

Learning Outcome 6.4.4
Describe differences in geographic frameworks in the Middle East.

After the 1973 war, the Palestinians emerged as Israel's principal opponent. Egypt and Jordan renounced their claims to the Gaza Strip and the West Bank, respectively, and recognized the Palestinians as the legitimate rulers of these territories. The Palestinians in turn also saw themselves as the legitimate rulers of Israel. Palestinian and Israeli perspectives over the future of Palestine/Israel have not been reconciled over the past four decades.

ISRAELI PERSPECTIVES. In dealing with its neighbors, Israel considers two elements of the local landscape especially meaningful:

- Israel is a very small country (smaller than New Hampshire), with a Jewish majority, surrounded by a region of hostile Muslim Arabs encompassing more than 25 million square kilometers (10 million square miles). Israel's people live extremely close to international borders, making them vulnerable to attack.
- Palestine is divided into three narrow, roughly parallel physical regions (Figure 6-51):
    - A coastal plain along the Mediterranean Sea
    - A series of hills reaching elevations above 1,000 meters (3,300 feet)
    - The Jordan River valley, much of which is below sea level

The UN plan for the partition of Palestine in 1947 (as modified by the armistice ending the 1948–1949 war) allocated most of the coastal plain to Israel, whereas Jordan took most of the hills between the coastal plain and the Jordan River valley, a region generally called the West Bank (of the Jordan River). Farther north, Israel's territory extended eastward to the Jordan River valley, but Syria controlled the highlands east of the valley, known as the Golan Heights.

Jordan and Syria used the hills between 1948 and 1967 as staging areas to attack Israeli settlements on the adjacent coastal plain and in the Jordan River valley. Israel captured these highlands during the 1967 war to stop attacks on the lowland population concentrations. Israel still has military control over the Golan Heights and West Bank a generation later, yet attacks by Palestinians against Israeli citizens have continued.

Israeli Jews were divided for many years between those who wished to retain the occupied territories and those who wished to make compromises with the Palestinians. In recent years, a large majority of Israelis have supported construction of a barrier to deter Palestinian attacks (refer to the Sustainability and Inequality in Our Global Village box).

PALESTINIAN PERSPECTIVES. Five groups of people consider themselves Palestinians:

- People living in the West Bank, Gaza, and East Jerusalem territories captured by Israel in 1967
- Citizens of Israel who are Muslims rather than Jews
- People who fled from Israel to other countries after the 1948–1949 war
- People who fled from the West Bank or Gaza to other countries after the 1967 Six-Day War
- Citizens of other countries, especially Jordan, Lebanon, Syria, Kuwait, and Saudi Arabia, who identify themselves as Palestinians

The Palestinian fight against Israel was coordinated by the Palestine Liberation Organization (PLO), under the longtime leadership of Yassir Arafat, until his death in 2004. Israel has permitted the organization of a limited form of government in much of the West Bank and Gaza, called the Palestinian Authority, but Palestinians are not satisfied with either the territory or the power they have received thus far.

The Palestinians have been divided by sharp differences, reflected in a struggle for power between the Fatah and Hamas parties. Some Palestinians, especially those aligned with the Fatah Party, are willing to recognize the state of Israel with its Jewish majority in exchange for return of all territory taken by Israel in the 1967 Six-Day War. Other Palestinians, especially those aligned with the Hamas Party, do not recognize the right of Israel to exist

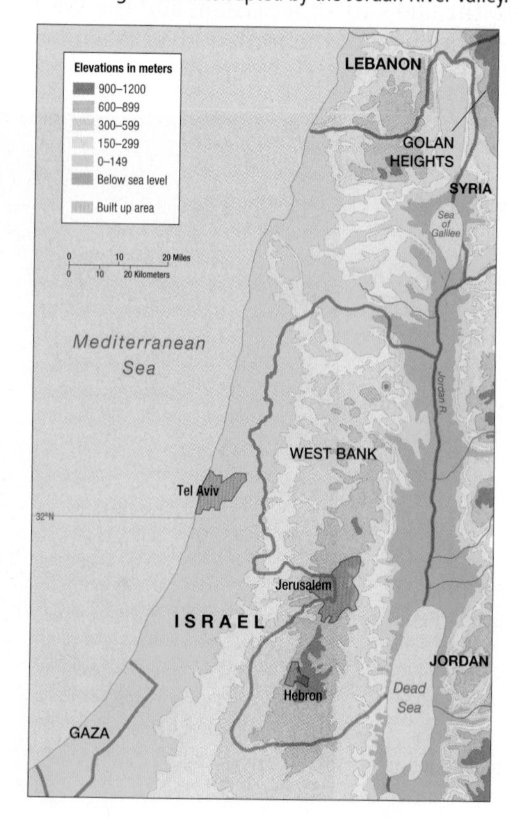

▼ **FIGURE 6-51 ISRAEL/PALESTINE PHYSICAL GEOGRAPHY** The physical geography of Israel/Palestine consists of narrow coastal lowlands and interior highlands interrupted by the Jordan River valley.

and want to continue fighting for control of the entire territory between the Jordan River and the Mediterranean Sea. The United States, European countries, and Israel consider Hamas to be a terrorist organization.

After capturing the West Bank from Jordan in 1967, Israel permitted Jewish settlers to construct more than 100 settlements in the territory (refer to Figure 6-51 in the Sustainability and Inequality in our Global Village feature). Some Israelis built settlements in the West Bank because they regarded the territory as an integral part of the biblical Jewish homeland, known as Judea and Samaria. Others migrated to the settlements because of a shortage of affordable housing inside Israel's pre-1967 borders. Jewish settlers comprise about 10 percent of the West Bank population, and Palestinians see their immigration as a hostile act. To protect the settlers, Israel has military control over most of the West Bank.

Pause and Reflect 6.4.4

What is the difference in elevation between Hebron (the largest city in the West Bank) and Tel Aviv (the largest city in Israel)?

SUSTAINABILITY AND INEQUALITY IN OUR GLOBAL VILLAGE

West Bank Barrier: Security Fence or Segregation Wall

Constructing a barrier to keep out the unwanted is one of the oldest of geographic tools. The United States is using this tool today, building a fence along the border with Mexico (refer to Figure 3-39 in Chapter 3).

To deter Palestinian suicide bombers from crossing into Israel, the Israeli government has constructed barriers along the West Bank and the Gaza Strip. The West Bank barrier is especially controversial because it places on Israel's side around 10 percent of the land, home to between 10,000 and 50,000 Palestinians, according to various sources (Figure 6-52).

According to Israel's government, the routes of the barrier were selected for two technical reasons:

- The area had to be wide enough to make construction of a barrier 60-meters (200 feet) wide feasible.
- High ground was placed on the Israeli side.

Critics charge that the circuitous route was chosen to encompass most of the 327,000 Israelis living in West Bank settlements that most other countries consider illegal.

Naming the structure is controversial. Israel calls the barrier a "security fence," and Palestinians call it a "racial segregation wall." Neutral sources call it a "separation barrier."

The Israel Supreme Court has twice declared portions of the route illegal because Palestinian rights were violated. The barrier made daily life unsustainable for some Palestinians: They could no longer reach their fields, water sources, and places of work.

▲ **FIGURE 6-52 WEST BANK SEPARATION BARRIER** (left) Route of the barrier. (right) The barrier separating Palestinian land (foreground) from Jewish settlement near Jerusalem (rear).

JERUSALEM: CONTESTED GEOGRAPHY

Learning Outcome 6.4.5
Explain the importance of Jerusalem to Jews and Muslims.

One of the most intractable obstacles to comprehensive peace in the Middle East is the status of Jerusalem (Figure 6-53). As long as any one religion—Jewish, Muslim, or Christian—maintains exclusive political control over Jerusalem, the other religious groups will not be satisfied. But Israelis have no intention of giving up control of the Old City of Jerusalem, and Palestinians have no intention of giving up their claim to it.

The geography of Jerusalem makes it difficult if not impossible to settle the long-standing religious conflicts. The difficulty is that the most sacred space in Jerusalem for Muslims was literally built on top of the most sacred space for Jews.

JUDAISM'S JERUSALEM. Jerusalem is especially holy to Jews as the location of the Temple, their center of worship in ancient times. The First Temple, built by King Solomon in approximately 960 B.C. was destroyed by the Babylonians in 586 B.C. After the Persian Empire, led by Cyrus the Great, gained control of Jerusalem in 614 B.C., Jews were allowed to build a Second Temple in 516 B.C. The Romans destroyed the Jewish Second Temple in A.D. 70. The Western Wall of the Temple survives.

Christians and Muslims call the Western Wall the Wailing Wall because for many centuries Jews were allowed to visit the surviving Western Wall only once a year to lament the Temple's destruction. After Israel captured the entire city of Jerusalem during the 1967 Six-Day War, it removed the barriers that had prevented Jews from visiting and living in the Old City of Jerusalem, including the Western Wall. The Western Wall soon became a site for daily prayers by observant Jews.

ISLAM'S JERUSALEM. The most important Muslim structure in Jerusalem is the Dome of the Rock, built in 691 (Figure 6-54). Muslims believe that the large rock beneath the building's dome is the place from which Muhammad ascended to heaven, as well as the altar on which Abraham prepared to sacrifice his son Isaac (according to Jews and Christians) or his son Ishmael (according to Muslims). Immediately south of the Dome of the Rock is the al-Aqsa Mosque. The challenge facing Jews and Muslims is that al-Aqsa Mosque was built on the site of the ruins of the Jewish Second Temple. Thus, the surviving Western Wall of the Jewish Temple is situated immediately beneath holy Muslim structures.

Israel allows Muslims unlimited access to that religion's holy structures in Jerusalem and some control over them. Ramps and passages patrolled by Palestinian guards provide Muslims access to the Dome of the Rock and the al-Aqsa Mosque without having to walk in front of the Western Wall, where Jews are praying. However, because the holy Muslim structures sit literally on top of the holy Jewish structure, the two sets of holy structures cannot be logically divided by a line on a map (Figure 6-55).

The ultimate obstacle to comprehensive peace in the Middle East is the status of Jerusalem. As long as any one religion—Jewish, Muslim, or Christian—maintains exclusive political control over Jerusalem, the other religious groups will not be satisfied. But Israelis have no intention of giving up control of the Old City of Jerusalem, and Palestinians have no intention of giving up their claim to it.

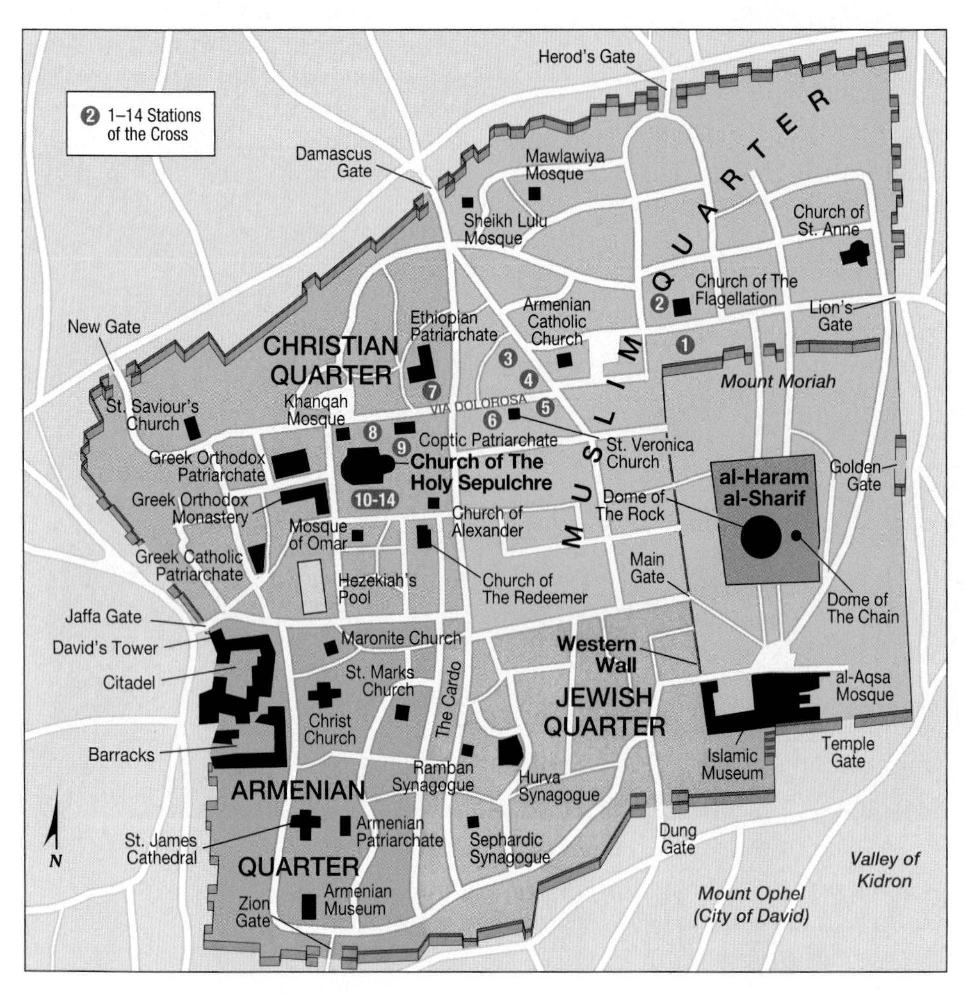

◀ **FIGURE 6-53 OLD CITY OF JERUSALEM** The Old City of Jerusalem is less than 1 square kilometer (0.4 square miles). It is divided into four quarters.

▲ **FIGURE 6-54 DOME OF THE ROCK** The large rock, which is under the golden dome of the Dome of the Rock is believed by Jews, Christians, and Muslims to be the place where Abraham was prepared to sacrifice his son. The son to be sacrificed was Isaac according to Jews and Christians, and Ishmael according to Muslims.

Pause and Reflect 6.4.5

Why are the Western Wall important in Judaism and the Dome of the Rock important in Islam?

CHECK-IN: KEY ISSUE 4

Why Do Territorial Conflicts Arise among Religious Groups?

- ✓ **Religious groups have opposed government policies, especially those of Communist governments.**
- ✓ **Religious principles seen as representing Western social values have been opposed by groups in Asia.**
- ✓ **An especially long-standing and intractable conflict among religious has been centered in Israel/Palestine, an area considered holy by Jews, Christians, and Muslims.**

▲ **FIGURE 6-55 WESTERN WALL AND DOME OF THE ROCK** A crowd of Jews are praying at the Western Wall (right), situated immediately below the mount containing Islam's Dome of the Rock (top left) and al-Aqsa Mosque (top right).

Summary and Review

KEY ISSUE 1

Where Are Religions Distributed?

Religions are classified as universalizing or ethnic. The world has three large universalizing religions—Christianity, Islam, and Buddhism, each of which is divided into branches and denominations. Hinduism is the largest ethnic religion.

LEARNING OUTCOME 6.1.1: Describe the distribution of the major religions.

- Christianity predominates in Europe and the Western Hemisphere, Buddhism in East Asia, Hinduism in South Asia, and Islam in other regions of Asia, as well as North Africa.

LEARNING OUTCOME 6.1.2: Describe the distribution of the major branches of Christianity.

- Christianity is divided into three main branches: Roman Catholic, which predominates in southwest Europe and Latin America; Protestant, which predominates in northwest Europe and North America; and Orthodox, which predominates in Eastern Europe.

LEARNING OUTCOME 6.1.3: Identify the major branches of Islam and Buddhism.

- Islam's two major branches are Sunni and Shiite. The two largest branches of Buddhism are Mahayana and Theravada.

LEARNING OUTCOME 6.1.4: Describe the distribution of the largest ethnic religions.

- Hinduism is clustered primarily in India. Other ethnic religions with the largest numbers of followers are clustered elsewhere in Asia.

THINKING GEOGRAPHICALLY 6.1: Islam seems strange and threatening to some people in predominantly Christian countries. To what extent is this attitude shaped by knowledge of the teachings of Muhammad and the Quran, and to what extent is it based on lack of knowledge of the religion?

GOOGLE EARTH 6.1: The large square in front of Saint Peter's Basilica, in the Vatican, is the length of approximately how many football fields?

KEY ISSUE 2

Why Do Religions Have Different Distributions?

A universalizing religion has a known origin and clear patterns of diffusion, whereas ethnic religions typically have unknown origins and little diffusion.

LEARNING OUTCOME 6.2.1: Describe the process of origin of universalizing religions.

- A universalizing religion originated with a single historical individual.

LEARNING OUTCOME 6.2.2: Understand differences in the origin of universalizing and ethnic religions.

- Ethnic religions typically have unknown origins.

LEARNING OUTCOME 6.2.3: Describe the process of diffusion of universalizing religions.

- Universalizing religions have diffused from their place of origin to other regions of the world.

LEARNING OUTCOME 6.2.4: Compare the diffusion of universalizing and ethnic religions.

- Ethnic religions typically do not diffuse far from their place of origin.

THINKING GEOGRAPHICALLY 6.2: People carry their religious beliefs with them when they migrate. Over time, change occurs in the regions from which most U.S. immigrants originate and in the U.S. regions where they settle. How has the distribution of U.S. religious groups been affected by these changes?

GOOGLE EARTH 6.2: Fly to 80 Ft Tall Lord Buddha, Bodhgaya, Bihar, India, click on 3D Buildings, and switch to ground-level view. Pan around the statue; what other Buddhist structure is visible in 3D?

Key Terms

Agnosticism (p. 184) Belief that nothing can be known about whether God exists.

Animism (p. 191) Belief that objects, such as plants and stones, or natural events, like thunderstorms and earthquakes, have a discrete spirit and conscious life.

Atheism (p. 184) Belief that God does not exist.

Autonomous religion (p. 211) A religion that does not have a central authority but shares ideas and cooperates informally.

Branch (p. 186) A large and fundamental division within a religion.

Caste (p. 213) The class or distinct hereditary order into which a Hindu is assigned, according to religious law.

Cosmogony (p. 204) A set of religious beliefs concerning the origin of the universe.

Denomination (p. 186) A division of a branch that unites a number of local congregations into a single legal and administrative body.

Ethnic religion (p. 184) A religion with a relatively concentrated spatial distribution whose principles are likely to be based on the physical characteristics of the particular location in which its adherents are concentrated.

Fundamentalism (p. 212) Literal interpretation and strict adherence to basic principles of a religion (or a religious branch, denomination, or sect).

Ghetto (p. 199) During the Middle Ages, a neighborhood in a city set up by law to be inhabited only by Jews; now used to denote a section of a city in which members of any minority group live because of social, legal, or economic pressure.

Hierarchical religion (p. 210) A religion in which a central authority exercises a high degree of control.

Missionary (p. 196) An individual who helps to diffuse a universalizing religion.

Monotheism (p. 191) The doctrine of or belief in the existence of only one god.

KEY ISSUE 3

Why Do Religions Organize Space in Distinctive Patterns?

Holy places and holidays in a universalizing religion are related to events in the life of its founder or prophet and are related to the local physical geography in an ethnic religion. Religions affect the landscape in other ways: Religious communities are built, religious toponyms mark the landscape, and extensive tracts are reserved for burying the dead.

LEARNING OUTCOME 6.3.1: Compare the role of places of worship in various religions.

- Religions have places of worship, but these places play differing roles for the various religions.

LEARNING OUTCOME 6.3.2: Explain why places are sacred in universalizing religions.

- In universalizing religions, holy places derive from events in the founder's life.

LEARNING OUTCOME 6.3.3: Analyze the importance of the physical geography in ethnic religions.

- In ethnic religions, holy places derive from the physical geography where the religion's adherents are clustered.

LEARNING OUTCOME 6.3.4: Describe ways in which the landscape is used in religiously significant ways.

- Religions have varying practices for handling the dead.

LEARNING OUTCOME 6.3.5: Compare the calendars and holidays of ethnic and universalizing religions.

- In ethnic religions, holidays derive from the physical geography where the religion is clustered.

LEARNING OUTCOME 6.3.6: Compare the administrative organization of hierarchical and locally autonomous religions.

- Religions can be divided into those that are administered through a hierarchy and those that are locally autonomous.

THINKING GEOGRAPHICALLY 6.3: Some Christians believe that they should be prepared to carry the word of God and the teachings of Jesus Christ to people who have not been exposed to them, at any time and at any place. Are missionary activities equally likely to occur at any time and at any place, or are some places more suited than others? Why?

GOOGLE EARTH 6.3: What is the physical environment around Badrinath Temple, one of Hindu's holiest temples, to Vishnu, in Badrinath, India?

Pagan (p. 190) A follower of a polytheistic religion.

Pilgrimage (p. 202) A journey to a place considered sacred for religious purposes.

Polytheism (p. 191) Belief in or worship of more than one god.

Sect (p. 186) A relatively small group that has broken away from an established denomination.

Solstice (p. 205) An astronomical event that happens twice each year, when the tilt of Earth's axis is most inclined toward or away from the Sun, causing the Sun's apparent position in the sky to reach it most northernmost or southernmost extreme, and resulting in the shortest and longest days of the year.

Syncretic (p. 190) A religion that combines several traditions.

Universalizing religion (p. 184) A religion that attempts to appeal to all people, not just those living in a particular location.

KEY ISSUE 4

Why Do Territorial Conflicts Arise among Religious Groups?

With Earth's surface dominated by four large religions, expansion of the territory occupied by one religion may reduce the territory of another. In addition, religions must compete for control of territory with nonreligious ideas, notably communism and economic modernization.

LEARNING OUTCOME 6.4.1: Understand reasons for religious conflicts arising from government policies.

- Religions can come into conflict with government policies, social changes, or other religions.

LEARNING OUTCOME 6.4.2: Summarize reasons for conflicts between religions.

- Conflicts among religions have been especially strong in Ireland and in the Middle East.

LEARNING OUTCOME 6.4.3: Analyze reasons for religious conflict in the Middle East.

- Religious conflict in the Middle East goes back thousands of years. Jews, Muslims, and Christians have fought for control of the Middle East land that is now part of Israel/Palestine.

LEARNING OUTCOME 6.4.4: Describe differences in geographic frameworks in the Middle East.

- Combatants in the Middle East have different perspectives on the division of land in the area.

LEARNING OUTCOME 6.4.5: Explain the importance of Jerusalem to Jews and Muslims.

- The most sacred space in Jerusalem for Muslims was built on top of the most sacred space for Jews.

THINKING GEOGRAPHICALLY 6.4: Sharp differences in demographic characteristics, such as natural increase, crude birth, and migration rates, can be seen among Jews, Christians, and Muslims in the Middle East and between Roman Catholics and Protestants in Northern Ireland. How might demographic differences affect future relationships among the groups in these two regions?

GOOGLE EARTH 6.4: The Abraj Al Bait (Royal Hotel Clock Tower) in Mecca, Saudi Arabia, the tallest hotel in the world, towers over what holy Muslim structure described in this chapter?

Chapter

7 Ethnicities

Why was this bridge blown up? Page 250

Why are these people burning torches on a mountain? Page 243

KEY ISSUE 1

Where Are Ethnicities Distributed?

A World of Ethnicities p. 227

Hispanics, African Americans, and Asian Americans are the most numerous U.S. ethnicities.

KEY ISSUE 2

Why Do Ethnicities Have Distinctive Distributions?

Ethnic Segregation p. 232

Migration of ethnicities can result in patterns of segregation, sometimes caused by discrimination.

▲ South Africa is a country of ethnic diversity. Between 1948 and 1994, the whites who controlled the government enacted laws known as apartheid that segregated the country's ethnicities. Most of the rights of people other than whites were taken away. The laws have been repealed, but many symbols of apartheid remain in South Africa, including these benches outside the law court in Cape Town.

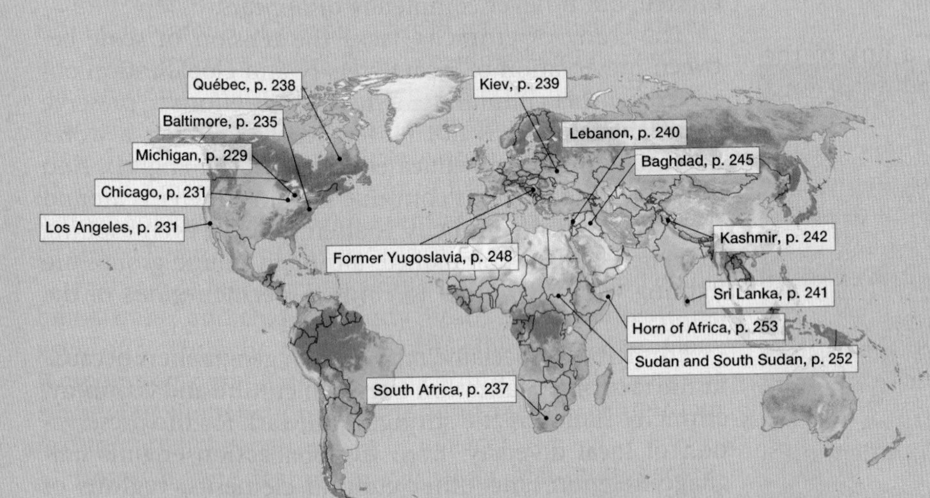

KEY ISSUE 3

Why Do Conflicts Arise among Ethnicities?

Ethnic Diversity p. 238

Ethnicities compete to control portions of Earth's surface.

KEY ISSUE 4

Why Do Ethnicities Engage in Ethnic Cleansing and Genocide?

Ethnic Cleansing p. 246

At its most extreme, competition among ethnic groups has led to atrocities.

Ethnicities in the United States

Learning Outcome 7.1.1
Identify and describe the major ethnicities in the United States.

The United States has always been defined, in part, by its ethnic diversity. Today, Americans are more diverse than ever before. Every 10 years, the U.S. Bureau of the Census asks people to classify themselves according to the ethnicity with which they most closely identify. Americans are asked to identify themselves by answering two questions:

- Check the box next to one or more of the following fifteen categories:
  - White
  - Black, African American, or Negro
  - American Indian or Alaska Native
  - Asian Indian
  - Chinese
  - Filipino
  - Other Asian
  - Japanese
  - Korean
  - Vietnamese
  - Native Hawaiian
  - Guamanian or Chamorro
  - Samoan
  - Other Pacific Islander
  - Other race
- Respond yes or no to being of Hispanic, Latino, or Spanish origin. If the response is yes, individuals are asked to pick one of these categories:
  - Mexican, Mexican Am., Chicano [the census uses the abbreviation "Am."]
  - Puerto Rican
  - Cuban
  - Another Hispanic, Latino, or Spanish origin

Respondents who select American Indian, Other Asian, Other Pacific Islander, Other race, or Another Hispanic are asked to write in the specific names on the census form.

Pause and Reflect 7.1.1
How would you answer the census questions about yourself?

Hispanic and *Hispanic American* are terms that the U.S. government chose in 1973 to describe the group because they are inoffensive labels that can be applied to all people from Spanish-speaking countries. Some Americans of Latin American descent have instead adopted the terms Latino (males) and Latina (females). A 1995 U.S. Census Bureau survey found that 58 percent of Americans of Latin American descent preferred the term *Hispanic* and 12 percent *Latino/Latina.*

Most Hispanics identify with a more specific ethnic or national origin. Around two-thirds come from Mexico and are sometimes called Chicanos (males) or Chicanas (females). Originally these terms were considered insulting, but in the 1960s Mexican American youths in Los Angeles began to call themselves Chicanos and Chicanas with pride.

In 2010 about 72 percent of Americans said on the census that they were white, 13 percent black or African American, 5 percent one of the seven Asian categories, 1 percent American Indian or Alaska Native, and 6 percent other. The census permits people to check more than one box, and 3 percent did that in 2010. Approximately 16 percent said they were Hispanic, and 84 percent said they were not.

The U.S. census shows the difficulty in distinguishing between ethnicity and race. Most of the census categories relate to ethnicity because they derive from places, such as African American or Asian Indian. However, the census also offers three race-related categories—black, white, and other race. The three most numerous U.S. ethnicities—Asian American, African American, and Hispanic American—further illustrate the difficulty. These three display distinct cultural traditions that originate at particular hearths but are regarded in different ways:

- *Asian American* as an ethnicity and *Asian* a race refer to the same group of people, which encompasses Americans from many countries in Asia (Figure 7-3).
- *African American* as an ethnicity and *black* as a race encompass different groups, although the 2010 census combines the two. Most black Americans are descended from African immigrants and therefore also belong to an African American ethnicity (Figure 7-4). Some American blacks, however, trace their cultural heritage to regions other than Africa, including Latin America, Asia, and Pacific islands. The term *African American* identifies a group with an extensive cultural tradition, whereas the term *black* in principle denotes nothing more than dark skin. Because many Americans make judgments about the values and behavior of others simply by observing

▼ **FIGURE 7-3 ASIAN AMERICANS** San Francisco's Chinatown.

▲ **FIGURE 7-4** **AFRICAN AMERICANS** New York's Harlem.

skin color, *black* is substituted for *African American* in daily language.

- *Hispanic* is an ethnicity but not a race, so Hispanics can identify with any race they wish. Hispanics have an especially difficult time doing so on the census. In 2010, 53 percent of Hispanics picked white, 37 percent other race, 6 percent more than one box, and 4 percent one of the 13 other categories (Figure 7-5).

Today, many Americans are of mixed ancestry and may or may not choose to identify with a single race or ethnicity. Other Americans trace their heritage to places in Europe, such as Ireland and Italy, that are not included in the two race and ethnicity census questions.

▲ **FIGURE 7-5** **HISPANICS** Chicago's Pilsen neighborhood.

ETHNIC CLUSTERING: STATE SCALE

The distinctive distribution of African Americans and Hispanics is especially noticeable at the state level. At the state level, African Americans comprise 85 percent of the population in the city of Detroit and only 7 percent in the rest of Michigan. Otherwise stated, Detroit contains less than one-tenth of Michigan's total population but more than one-half of the state's African American population (Figure 7-6). Similarly, Chicago is more than one-third African American, compared to one-twelfth in the rest of Illinois. Chicago has less than one-fourth of Illinois' total population and more than one-half of the state's African Americans.

The distribution of Hispanics is similar to that of African Americans in large northern cities. For example, New York City is more than one-fourth Hispanic, compared to one-sixteenth in the rest of New York State, and New York City contains two-fifths of the state's total population and three-fourths of its Hispanics.

In the states with the largest Hispanic populations—California and Texas—the distribution is mixed. In California, Hispanics comprise nearly half of Los Angeles's population, but the percentage of Hispanics in California's other large cities is less than or about equal to the overall state average. In Texas, El Paso and San Antonio—the two large cities closest to the Mexican border—are more than one-half Hispanic, but the state's other large cities have percentages below or about equal to the state's average of around one-third.

◀ **FIGURE 7-6** **DISTRIBUTION OF ETHNICITIES IN MICHIGAN** Most of Michigan's African Americans live in Detroit.

Distribution of Ethnicities in the United States

Learning Outcome 7.1.2
Describe the distribution of major U.S. ethnicities among regions and within urban areas.

Within a country, clustering of ethnicities can occur on two scales. Ethnic groups may live in particular regions of the country, and they may live in particular communities within cities and states. Within the United States, ethnicities are clustered at both scales.

ETHNIC CLUSTERING: REGIONAL SCALE

On a regional scale, ethnicities have distinctive distributions within the United States:

- **Hispanics.** Clustered in the Southwest, Hispanics exceed one-third of the population of Arizona, New Mexico, and Texas and one-quarter of California (Figure 7-7). California is home to one-third of all Hispanics, Texas one-fifth, and Florida and New York one-sixth each.
- **African Americans.** Clustered in the Southeast, African Americans comprise at least one-fourth of the population in Alabama, Georgia, Louisiana, Maryland, and South Carolina and more than one-third in Mississippi (Figure 7-8). Concentrations are even higher in selected counties. At the other extreme, nine states in upper New England and the West have less than 1 percent African Americans.
- **Asian Americans.** Clustered in the West, Asian Americans comprise more than 40 percent of the population of Hawaii (Figure 7-9). One-half of all Asian Americans live in California, where they comprise 12 percent of the population.

ETHNIC CLUSTERING: URBAN SCALE

African Americans and Hispanics are highly clustered in urban areas. Around 90 percent of these ethnicities live in metropolitan areas, compared to around 75 percent for all Americans. The clustering of ethnicities is especially pronounced on the scale of neighborhoods within cities. In the early twentieth century, Chicago, Cleveland, Detroit, and other Midwest cities attracted ethnic groups primarily from Southern and Eastern Europe to work in the rapidly growing steel, automotive, and related industries. For example, in 1910, when Detroit's auto production was expanding, three-fourths of the city's residents were immigrants and children of immigrants. Southern and Eastern European ethnic groups clustered in newly constructed neighborhoods that were often named for their predominant ethnicities, such as Detroit's Greektown and Poletown.

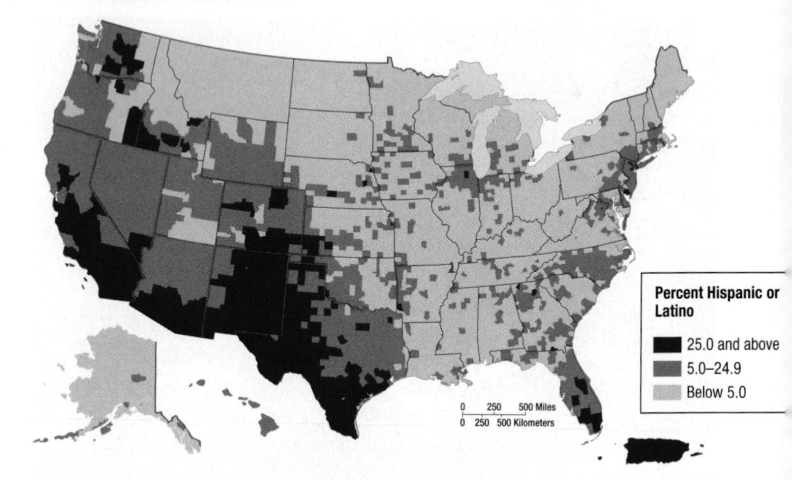

▲ **FIGURE 7-7 DISTRIBUTION OF HISPANICS IN THE UNITED STATES** The counties with the highest percentages in 2010 are in the Southwest, near the Mexican border, and in northern cities.

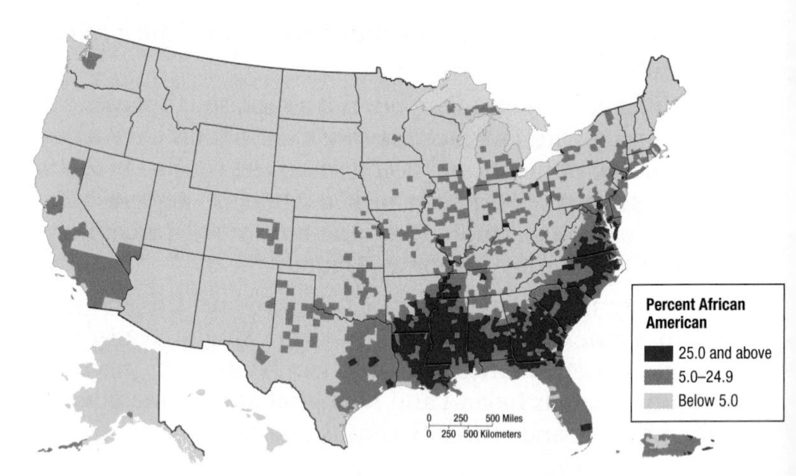

▲ **FIGURE 7-8 DISTRIBUTION OF AFRICAN AMERICANS IN THE UNITED STATES** The counties with the highest percentages of African Americans are in the rural South and in northern cities.

▲ **FIGURE 7-9 DISTRIBUTION OF ASIAN AMERICANS IN THE UNITED STATES** The counties with the highest percentages of Asian Americans are in Hawaii and California.

The children and grandchildren of European immigrants moved out of most of the original inner-city neighborhoods during the twentieth century. For descendants of European immigrants, ethnic identity is more likely to be retained through religion, food, and other cultural traditions than through location of residence. A visible remnant of early twentieth-century European ethnic neighborhoods is the clustering of restaurants in such areas as Little Italy and Greektown.

Ethnic concentrations in U.S. cities increasingly consist of African Americans who migrate from the South or immigrants from Latin America and Asia. In cities such as Detroit, African Americans now comprise the majority and live in neighborhoods originally inhabited by European ethnic groups. Chicago has extensive African American neighborhoods on the south and west sides of the city, but the city also contains a mix of neighborhoods inhabited by European, Latin American, and Asian ethnicities (Figure 7-10).

In Los Angeles, which contains large percentages of African Americans, Hispanics, and Asian Americans, the major ethnic groups are clustered in different areas (Figure 7-11). African Americans are located in south-central Los Angeles and Hispanics in the east. Asian Americans are located to the south and west, contiguous to the African American and Hispanic areas.

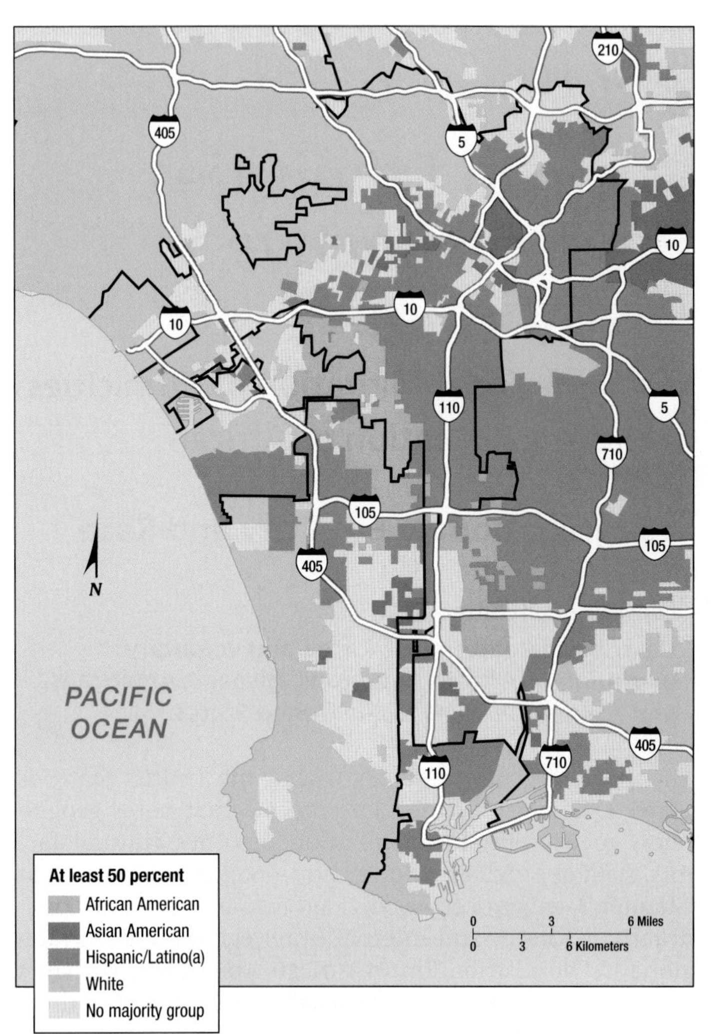

▲ FIGURE 7-11 **DISTRIBUTION OF ETHNICITIES IN LOS ANGELES** According to the 2010 Census, African Americans were clustered to the south of downtown Los Angeles and Hispanics to the east. Asian American neighborhoods were contiguous to the African American and Hispanic areas.

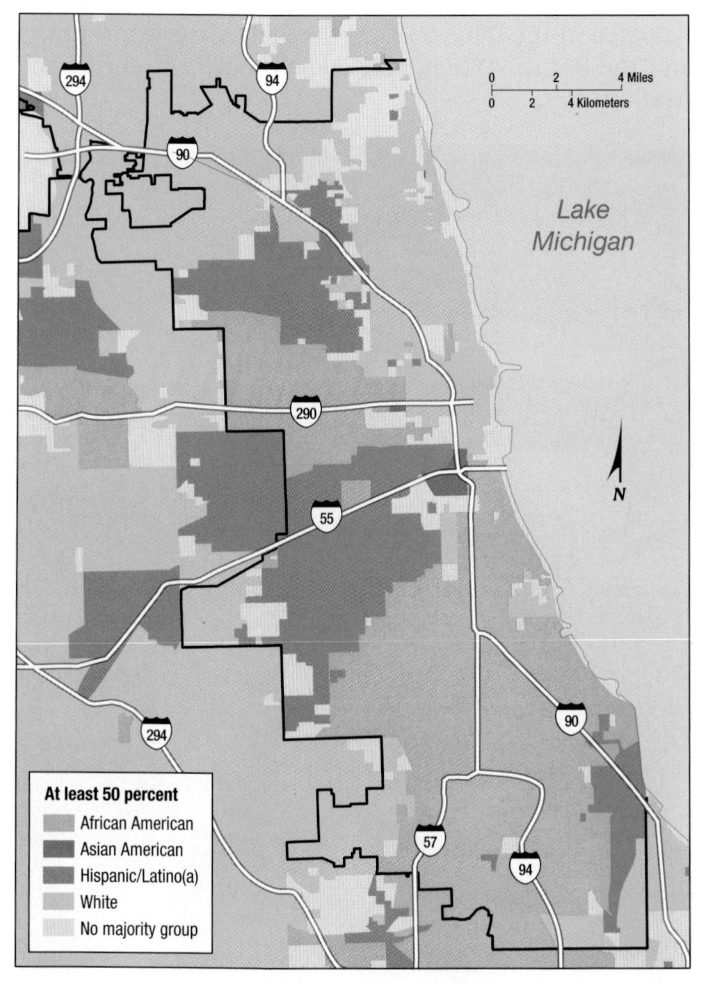

▲ FIGURE 7-10 **DISTRIBUTION OF ETHNICITIES IN CHICAGO** According to the 2010 Census, African Americans were clustered on the south and west sides, Hispanics on the northwest and southwest side, and whites on the north side.

CHECK-IN: KEY ISSUE 1

Where Are Ethnicities Distributed?

- ✓ **The most numerous ethnicities in the United States are Hispanic, African American, and Asian American.**
- ✓ **The three most numerous U.S. ethnicities have distinctive distributions at regional, state, and urban scales.**

Pause and Reflect 7.1.2

Where are the principal clusters of ethnic minorities found in your community?

KEY ISSUE 2

Why Do Ethnicities Have Distinctive Distributions?

- International Migration of Ethnicities
- Internal Migration of African Americans
- Segregation by Ethnicity and Race

Learning Outcome 7.2.1
Describe the patterns of forced and voluntary migration of African Americans, Hispanic Americans, and Asian Americans to the United States.

The clustering of ethnicities within the United States is partly a function of the same process that helps geographers to explain the distribution of other cultural factors, such as language and religion—namely migration. In Chapter 3, migration was divided into international (voluntary or forced) and internal (interregional and intraregional). The distribution of African Americans, Hispanic Americans, and Asian Americans demonstrates all of these migration patterns.

International Migration of Ethnicities

Most African Americans are descended from Africans forced to migrate to the Western Hemisphere as slaves during the eighteenth century. Most Asian Americans and Hispanics are descended from voluntary immigrants to the United States during the late twentieth and early twenty-first centuries, although some felt compelled for political reasons to come to the United States.

FORCED MIGRATION FROM AFRICA

Slavery is a system whereby one person owns another person as a piece of property and can force that slave to work for the owner's benefit. The first Africans brought to the American colonies as slaves arrived at Jamestown, Virginia, on a Dutch ship in 1619 (Figure 7-12). During the eighteenth century, the British shipped about 400,000 Africans to the 13 colonies that later formed the United States. In 1808 the United States banned bringing in additional Africans as slaves, but an estimated 250,000 were illegally imported during the next half-century.

Slavery was widespread during the time of the Roman Empire, about 2,000 years ago. During the Middle Ages, slavery was replaced in Europe by a feudal system, in which laborers working the land (known as serfs) were bound to the land and not free to migrate elsewhere. Serfs had to turn over a portion of their crops to the lord and provide other services, as demanded by the lord.

Although slavery was rare in Europe, Europeans were responsible for diffusing the practice to the Western Hemisphere. Europeans who owned large plantations in the Americas turned to African slaves as an abundant source of labor that cost less than paying wages to other Europeans.

At the height of the slave trade between 1710 and 1810, at least 10 million Africans were uprooted from their homes and sent on European ships to the Western Hemisphere for sale in the slave markets. During that period, the British and Portuguese each shipped about 2 million slaves to the Western Hemisphere, with most of the British slaves going to Caribbean islands and the Portuguese slaves to Brazil.

The forced migration began when people living along the east and west coasts of Africa, taking advantage of their superior weapons, captured members of other groups living farther inland and sold the captives to Europeans. Europeans in turn shipped the captured Africans to the Americas, selling them as slaves either on consignment or through auctions. The Spanish and Portuguese first participated in the slave trade in the early sixteenth century, and the British, Dutch, and French joined in during the next century.

▲ **FIGURE 7-12 SLAVE SHIP** This drawing made around 1845 for a French magazine shows the high density and poor conditions of Africans transported to the Western Hemisphere to become slaves.

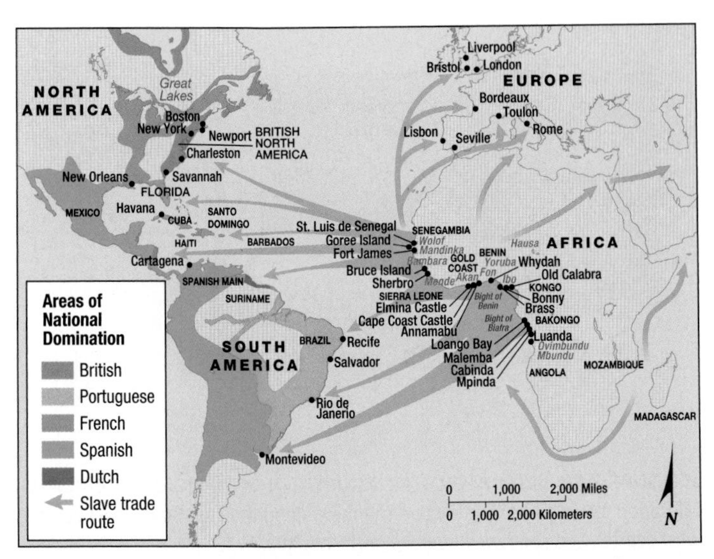

▲ **FIGURE 7-13 ORIGIN AND DESTINATION OF SLAVES** Most slaves were transported across the Atlantic from West Africa to the Americas.

Different European countries operated in various regions of Africa, each sending slaves to different destinations in the Americas (Figure 7-13). At the height of the eighteenth-century slave demand, a number of European countries adopted the **triangular slave trade**, an efficient triangular trading pattern (Figure 7-14).

The large-scale forced migration of Africans caused them unimaginable hardship, separating families and destroying villages. Traders generally seized the stronger and younger villagers, who could be sold as slaves for the highest price. The Africans were packed onto ships at extremely high density, kept in chains, and provided with minimal food and sanitary facilities. Approximately one-fourth died crossing the Atlantic.

In the 13 colonies that later formed the United States, most of the large plantations in need of labor were located in the South, primarily those growing cotton as well as tobacco. Consequently, nearly all Africans shipped to the 13 colonies ended up in the Southeast.

Attitudes toward slavery dominated U.S. politics during the nineteenth century. During the early 1800s, when new states were carved out of western territory, anti-slavery northeastern states and pro-slavery southeastern states bitterly debated whether to permit slavery in the new states. The Civil War (1861–1865) was fought to prevent 11 pro-slavery Southern states from seceding from the Union. In 1863, during the Civil War, Abraham Lincoln issued the Emancipation Proclamation, freeing the slaves in the 11 Confederate states. The Thirteenth Amendment to the Constitution, adopted 8 months after the South surrendered, outlawed slavery.

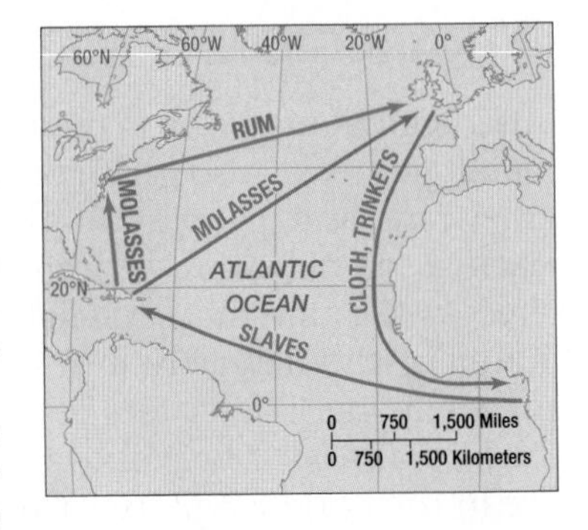

▶ **FIGURE 7-14 TRIANGULAR SLAVE TRADE**
- Ships left Europe for Africa with cloth and other trade goods, used to buy the slaves.
- They then transported slaves and gold from Africa to the Western Hemisphere, primarily to the Caribbean islands.
- To complete the triangle, the same ships then carried sugar and molasses from the Caribbean on their return trip to Europe.
- Some ships added another step, making a rectangular trading pattern, in which molasses was carried from the Caribbean to the North American colonies and rum from the colonies to Europe.

VOLUNTARY MIGRATION FROM LATIN AMERICA AND ASIA

Until the late twentieth century, quotas limited the number of people who could immigrate to the United States from Latin America and Asia, as discussed in Chapter 3. After the immigration laws were changed during the 1960s and 1970s, the population of Hispanics and Asian Americans in the United States increased rapidly. Initially, most Hispanics and Asian Americans were recent immigrants who came to the United States in search of work, but in the twenty-first century most Americans who identify themselves as Hispanics or Asian Americans are children or grandchildren of immigrants.

The rapid growth of Hispanics in the United States beginning in the 1970s was fueled primarily by immigration from Mexico and Puerto Rico (Figure 7-15).

Chinese comprise the largest share of Asian Americans, followed by Indians, Filipinos, Koreans, and Vietnamese (Figure 7-16). Most Asian Americans are either immigrants who arrived in the late twentieth and early twenty-first centuries or their offspring.

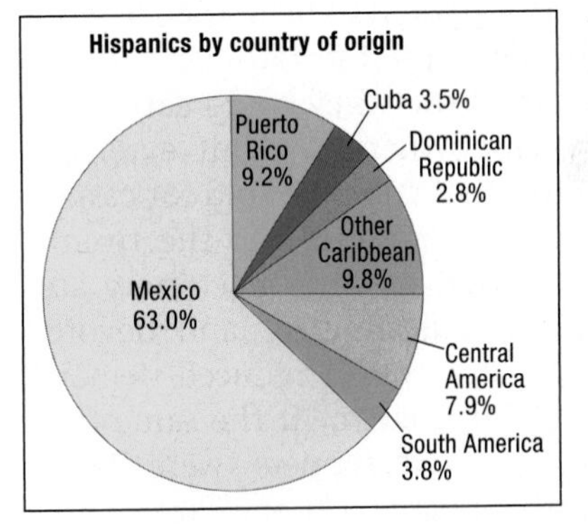

◀ **FIGURE 7-15 HISPANICS BY COUNTRY OF ORIGIN** Mexicans comprise nearly two-thirds of Hispanics in the United States.

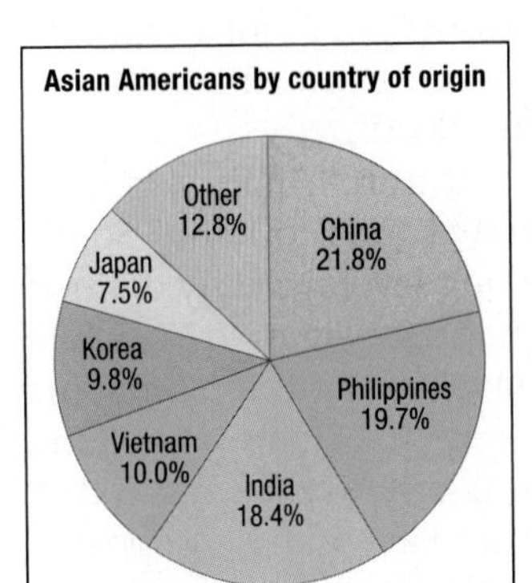

◀ **FIGURE 7-16 ASIAN AMERICANS BY COUNTRY OF ORIGIN** Chinese, Filipinos, and Indians comprise one-fifth each of Asian Americans in the United States.

Internal Migration of African Americans

Learning Outcome 7.2.2
Describe the patterns of migration of African Americans within the United States.

African Americans have displayed two distinctive internal migration patterns within the United States during the twentieth century:

- Interregional migration from the U.S. South to northern cities during the first half of the twentieth century.
- Intraregional migration from inner-city ghettos to outer city and inner suburban neighborhoods during the second half of the twentieth century.

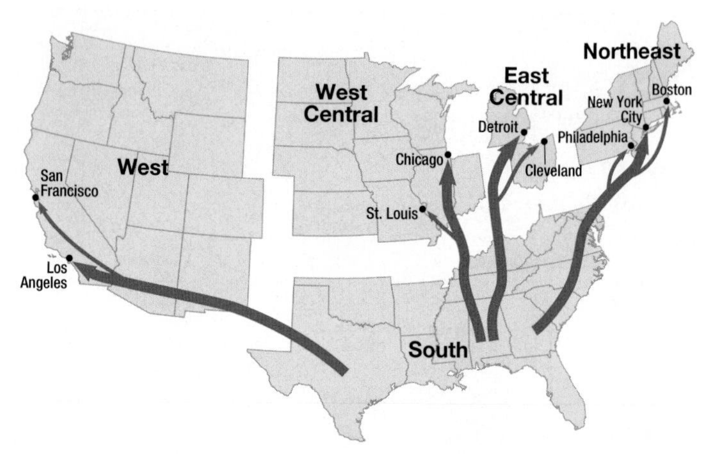

▲ **FIGURE 7-18 INTERREGIONAL MIGRATION OF AFRICAN AMERICANS** Migration followed four distinctive channels along the East Coast, east central, west central, and southwest regions of the country.

INTERREGIONAL MIGRATION

At the close of the Civil War, most African Americans were concentrated in the rural South. Today, as a result of interregional migration, many African Americans live in cities throughout the Northeast, Midwest, and West as well. Freed as slaves, most African Americans remained in the rural South during the late nineteenth century, working as sharecroppers (Figure 7-17). A **sharecropper** works fields rented from a landowner and pays the rent by turning over to the landowner a share of the crops. To obtain seed, tools, food, and living quarters, a sharecropper gets a line of credit from the landowner and repays the debt with yet more crops. The sharecropper system burdened poor African Americans with high interest rates and heavy debts. Instead of growing food that they could eat, sharecroppers were forced by landowners to plant extensive areas of crops such as cotton that could be sold for cash.

Sharecropping became less common into the twentieth century, as the introduction of farm machinery and a decline in land devoted to cotton reduced demand for labor. At the same time sharecroppers were being pushed off the farms, they were being pulled by the prospect of jobs in the booming industrial cities of the North.

▲ **FIGURE 7-17 SHARECROPPERS** Thirteen-year-old African American sharecropper plowing, 1937.

African Americans migrated out of the South along several clearly defined channels (Figure 7-18). Most traveled by bus and car along the major two-lane long-distance U.S. roads that were paved and signposted in the early decades of the twentieth century and have since been replaced by interstate highways:

- **East Coast.** From the Carolinas and other South Atlantic states north to Baltimore, Philadelphia, New York, and other northeastern cities, along U.S. Route 1 (parallel to present-day I-95).
- **East central.** From Alabama and eastern Tennessee north to either Detroit, along U.S. Route 25 (present-day I-75), or Cleveland, along U.S. Route 21 (present-day I-77).
- **West central.** From Mississippi and western Tennessee north to St. Louis and Chicago, along U.S. routes 61 and 66 (present-day I-55).
- **Southwest.** From Texas west to California, along U.S. routes 80 and 90 (present-day I-10 and I-20).

Southern African Americans migrated north and west in two main waves, the first in the 1910s and 1920s before and after World War I and the second in the 1940s and 1950s before and after World War II. The world wars stimulated expansion of factories in the 1910s and 1940s to produce war materiel, while the demands of the armed forces created shortages of factory workers. After the wars, during the 1920s and 1950s, factories produced steel, motor vehicles, and other goods demanded in civilian society.

INTRAREGIONAL MIGRATION

Intraregional migration—migration within cities and metropolitan areas—also changed the distribution of African Americans and people of other ethnicities. When they reached the big cities, African American immigrants clustered in the one or two neighborhoods where the small numbers who had arrived in the nineteenth century were already living. These areas became known as ghettos, after the term for neighborhoods in which Jews were forced to live in the Middle Ages (see Chapter 6).

EXPANSION OF THE GHETTO. African Americans moved from the tight ghettos into immediately adjacent neighborhoods during the 1950s and 1960s. Expansion of the ghetto typically followed major avenues that radiated out from the center of the city.

In Baltimore, for example, most of the city's quarter-million African Americans in 1950 were clustered in a 3-square-kilometer (1-square-mile) neighborhood northwest of downtown (Figure 7-19). The remainder were clustered east of downtown or in a large isolated housing project on the south side built for black wartime workers in port industries. Densities in the ghettos were high, with 40,000 inhabitants per square kilometer (100,000 per square mile) common. Contrast that density with the current level found in typical American suburbs of 2,000 inhabitants per square kilometer (5,000 per square mile). Because of the shortage of housing in the ghettos, families were forced to live in one room. Many dwellings lacked bathrooms, kitchens, hot water, and heat.

Baltimore's west side African American ghetto expanded from 3 square kilometers (1 square mile) in 1950 to 25 square kilometers (10 square miles) in 1970, and a 5-square-kilometer (2-square-mile) area on the east side became mainly populated by African Americans. Expansion of the ghetto continued to follow major avenues to the northwest and northeast in subsequent decades.

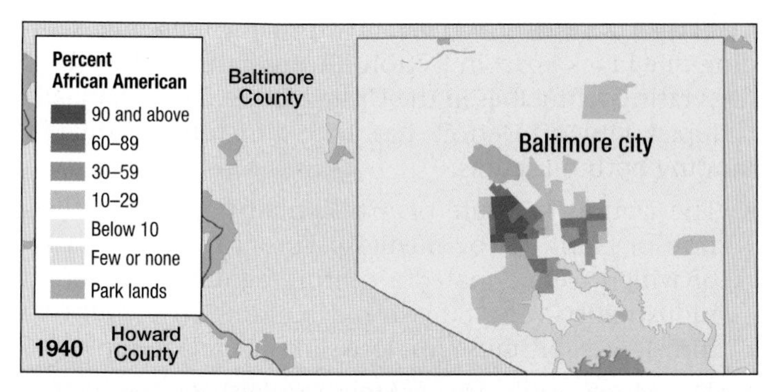

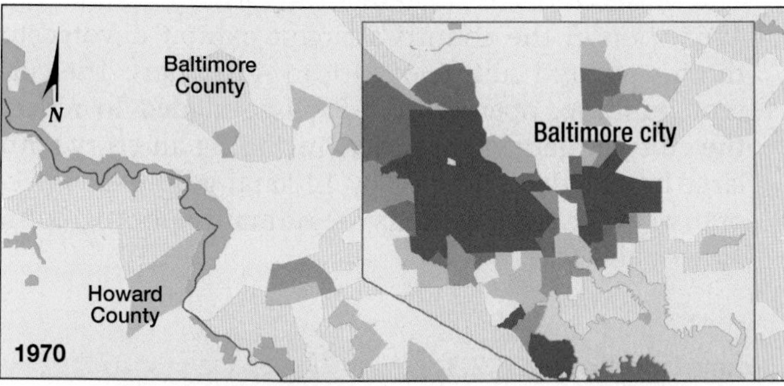

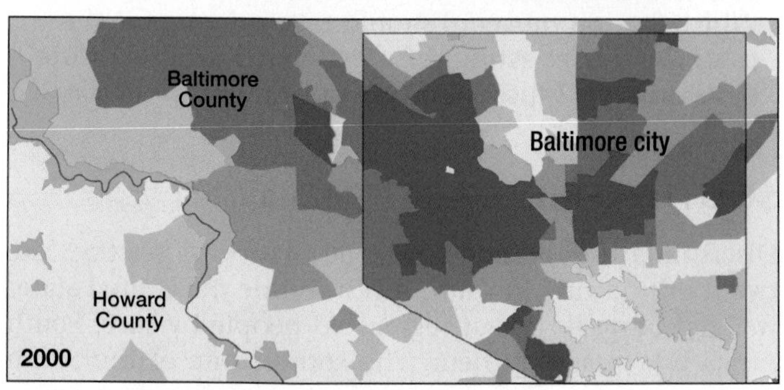

▲ **FIGURE 7-19 EXPANSION OF THE GHETTO IN BALTIMORE**
In 1950, most African Americans in Baltimore lived in a small area northwest of downtown. During the 1950s and 1960s, the African American area expanded to the northwest, along major radial roads, and a second node opened on the east side. The south-side African American area was an isolated public housing complex built for wartime workers in the nearby port industries.

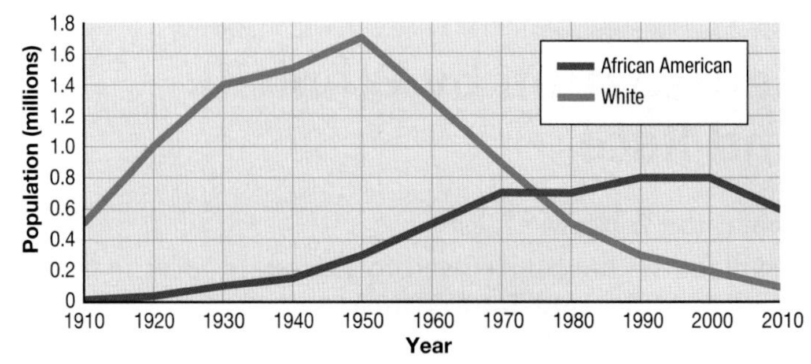

▲ **FIGURE 7-20 ETHNIC POPULATION CHANGE IN DETROIT**
Between 1950 and 2010, the white population of Detroit declined from 1.7 million to 100,000 today, whereas the African American population increased from 300,000 to 600,000.

"WHITE FLIGHT." The expansion of the black ghettos in American cities was made possible by "white flight," the emigration of whites from an area in anticipation of blacks immigrating into the area. Rather than integrate, whites fled.

Detroit provides a clear example. African Americans poured into Detroit in the early twentieth century. Many found jobs in the rapidly growing auto industry (Figure 7-20). Immigration into Detroit from the South subsided during the 1950s, but as legal barriers to integration crumbled, whites began to emigrate out of Detroit. Detroit's white population dropped by about 1 million between 1950 and 1975 and by another half million between 1975 and 2000. As a result, the overall population of Detroit declined from a historic peak of nearly 2 million in 1950 to around 700,000 in the early twenty-first century.

White flight was encouraged by unscrupulous real estate practices, especially blockbusting. Under **blockbusting**, real estate agents convinced white homeowners living near a black area to sell their houses at low prices, preying on their fears that black families would soon move into the neighborhood and cause property values to decline. The agents then sold the houses at much higher prices to black families desperate to escape the overcrowded ghettos. Through blockbusting, a neighborhood could change from all-white to all-black in a matter of months, and real estate agents could start the process all over again in the next white area.

The National Advisory Commission on Civil Disorders, known as the Kerner Commission, wrote in 1968 that U.S. cities were divided into two separate and unequal societies, one black and one white. A half-century later, despite serious efforts to integrate and equalize the two, segregation and inequality persist.

Pause and Reflect 7.2.2

Referring to Figure 7-20, which figure is higher in Detroit since 1950: the increasing number of African Americans or the decreasing number of whites?

Segregation by Ethnicity and Race

Learning Outcome 7.2.3
Explain the laws once used to segregate races in the United States and South Africa.

In explaining spatial regularities, geographers look for patterns of spatial interaction. A distinctive feature of ethnic relations in the United States and South Africa has been the strong discouragement of spatial interaction—in the past through legal means and today through cultural preferences or discrimination.

UNITED STATES: "SEPARATE BUT EQUAL"

The U.S. Supreme Court in 1896 upheld a Louisiana law that required black and white passengers to ride in separate railway cars. In *Plessy v. Ferguson*, the Supreme Court stated that Louisiana's law was constitutional because it provided separate, but equal, treatment of blacks and whites, and equality did not mean that whites had to mix socially with blacks.

SEGREGATION LAWS. Once the Supreme Court permitted "separate but equal" treatment of the races, southern states enacted a comprehensive set of laws to segregate blacks from whites as much as possible (Figure 7-21). These were called "Jim Crow" laws, named for a nineteenth-century song-and-dance act that depicted blacks offensively. Blacks had to sit in the backs of buses, and shops, restaurants, and hotels could choose to serve only whites. Separate schools were established for blacks and whites. This was equal, after all, white southerners argued, because the bus got blacks sitting in the rear to the destination at the same time as the whites in the front, some commercial establishments served only blacks, and all of the schools had teachers and classrooms.

Throughout the country, not just in the South, house deeds contained restrictive covenants that prevented the owners from selling to blacks, as well as to Roman Catholics or Jews in some places. Restrictive covenants kept blacks from moving into an all-white neighborhood. And because schools, especially at the elementary level, were located to serve individual neighborhoods, most were segregated in practice, even if not by legal mandate.

U.S. segregation laws were eliminated during the 1950s and 1960s. The landmark Supreme Court decision *Brown v. Board of Education of Topeka, Kansas*, in 1954, found that having separate schools for blacks and whites was unconstitutional because no matter how equivalent the facilities, racial separation branded minority children as inferior and therefore was inherently unequal. A year later, the Supreme Court further ruled that schools had to be desegregated "with all deliberate speed."

CULTURAL SEGREGATION. Two major museums standing one block apart in Detroit illustrate the challenges of integrating ethnicities in the United States. The financially strapped city of Detroit has had difficulty adequately funding both museums:

- The Detroit Institute of Arts contains a major collection of paintings by medieval European artists, many of which were donated a century ago by rich Detroit industrialists. The 80-year-old building, the country's fifth-largest art museum, looks like a Greek temple.
- The Museum of African American History, founded in 1965, houses the country's largest exhibit devoted to the history and culture of African Americans. The current building, opened in 1997, is designed to reflect the cultural heritage of Africa, including an entry with large bronze doors topped by 14-karat gold-plated decorative masks. The exhibits are primarily photographs, videos, and text.

▼ FIGURE 7-21 SEGREGATION IN THE UNITED STATES Until the 1960s in the U.S. South, whites and blacks had to use separate drinking fountains, as well as separate restrooms, bus seats, hotel rooms, and other public facilities.

Pause and Reflect 7.2.3
Which Detroit museum should take priority for the city's limited investment funds—the Detroit Institute of Arts or the Museum of African American History?

SOUTH AFRICA: APARTHEID

Discrimination by race reached its peak in the late twentieth century in South Africa. While the United States was repealing laws that segregated people by race, South Africa was enacting them. The cornerstone of the South African policy was the creation of a legal system called apartheid (Figure 7-22). **Apartheid** was the physical separation of different races into different geographic areas. Although South Africa's apartheid laws were repealed during the 1990s, it will take many years to erase the impact of those policies.

▲ **FIGURE 7-22 APARTHEID IN SOUTH AFRICA** South Africa's apartheid laws were designed to spatially segregate races as much as possible. This 1984 image of City Hall in Johannesburg shows that whites and nonwhites were required to use separate bathrooms.

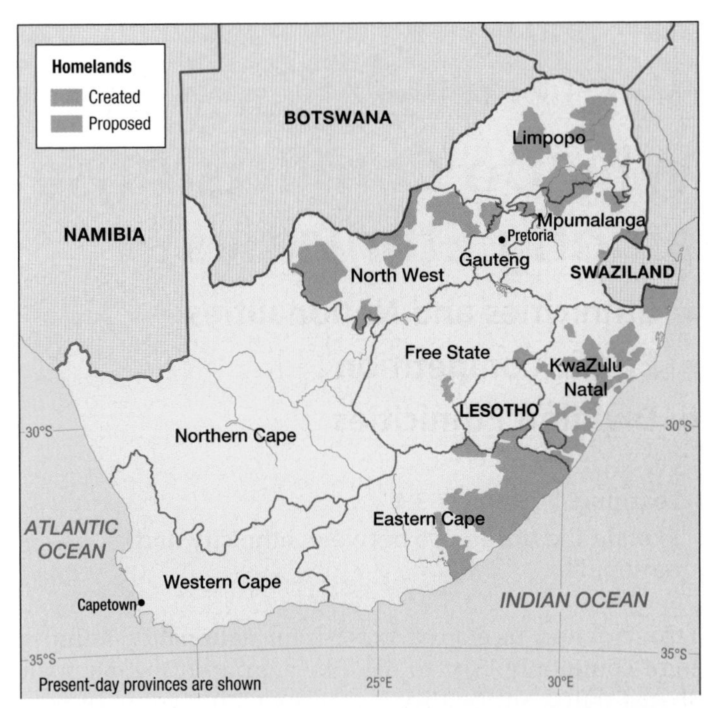

▲ **FIGURE 7-23 SOUTH AFRICA'S APARTHEID HOMELANDS** As part of its apartheid system, the government of South Africa designated 10 homelands, expecting that ultimately every black would become a citizen of one of them. South Africa declared 4 of these homelands to be independent states, but no other country recognized the action. With the end of apartheid and the election of a black majority government, the homelands were abolished, and South Africa was reorganized into 9 provinces.

In South Africa, under apartheid, a newborn baby was classified as being one of four races—black, white, colored (mixed white and black), or Asian. Under apartheid, each of the four races had a different legal status in South Africa. The apartheid laws determined where different races could live, attend school, work, shop, and own land. Blacks were restricted to certain occupations and were paid far lower wages than were whites for similar work. Blacks could not vote or run for political office in national elections. The apartheid system was created by descendants of whites who arrived in South Africa from the Netherlands in 1652 and settled in Cape Town, at the southern tip of the territory. They were known either as Boers, from the Dutch word for "farmer," or Afrikaners, from the word "Afrikaans," the name of their language, which is a dialect of Dutch.

The British seized the Dutch colony in 1795 and controlled South Africa's government until 1948, when the Afrikaner-dominated Nationalist Party won elections. The Afrikaners gained power at a time when colonial rule was being replaced in the rest of Africa by a collection of independent states run by the local black population. The Afrikaners vowed to resist pressures to turn over South Africa's government to blacks, and the Nationalist Party created the apartheid laws in the next few years to perpetuate white dominance of the country. To ensure geographic isolation of different races, the South African government designated 10 so-called homelands for blacks (Figure 7-23). The white minority government expected every black to become a citizen of one of the homelands and to move there. More than 99 percent of the population in the 10 homelands was black.

The white-dominated government of South Africa repealed the apartheid laws in 1991. The principal anti-apartheid organization, the African National Congress, was legalized, and its leader, Nelson Mandela, was released from jail after more than 27 years of imprisonment. When all South Africans were permitted to vote in national elections for the first time, in 1994, Mandela was overwhelmingly elected the country's first black president.

Now that South Africa's apartheid laws have been dismantled and the country is governed by its black majority, other countries have reestablished economic and cultural ties. However, the legacy of apartheid will linger for many years: South Africa's blacks have achieved political equality, but they are much poorer than white South Africans. Average income among white South Africans is about 10 times higher than that of blacks.

CHECK-IN: KEY ISSUE 2

Why Do Ethnicities Have Distinctive Distributions?

- ✓ **Ancestors of African Americans immigrated to the United States primarily as slaves.**
- ✓ **Large numbers of African Americans migrated from the South to the North and West during the early twentieth century.**
- ✓ **In the United States, as well as in South Africa, segregation of races was legal for much of the twentieth century.**

KEY ISSUE 3

Why Do Conflicts Arise among Ethnicities?

- Ethnicities and Nationalities
- Ethnic Competition
- Dividing Ethnicities

Learning Outcome 7.3.1
Explain the difference between ethnicity and nationality.

Ethnicity and race are distinct from nationality, another term commonly used to describe a group of people with shared traits. **Nationality** is identity with a group of people who share legal attachment and personal allegiance to a particular country. It comes from the Latin word *nasci*, which means "to have been born."

Ethnicities and Nationalities

Nationality and ethnicity are similar concepts in that membership in both is defined through shared cultural values. In principle, the cultural values shared with others of the same ethnicity derive from religion, language, and material culture, whereas those shared with others of the same nationality derive from voting, obtaining a passport, and performing civic duties.

NATIONALITIES IN NORTH AMERICA

In the United States, *nationality* is generally kept reasonably distinct from *ethnicity* and *race* in common usage:

- Nationality identifies citizens of the United States of America, including those born in the country and those who immigrated and became citizens.
- Ethnicity identifies groups with distinct ancestry and cultural traditions, such as African Americans, Hispanic Americans, Chinese Americans, or Polish Americans.
- Race distinguishes blacks and other persons of color from whites.

The United States forged a nationality in the late eighteenth century out of a collection of ethnic groups gathered primarily from Europe and Africa, not through traditional means of issuing passports (African Americans weren't considered citizens then) or voting (women and African Americans couldn't vote then), but through sharing the values expressed in the Declaration of Independence and the U.S. Constitution. To be an American meant believing in the "unalienable rights" of "life, liberty, and the pursuit of happiness."

▲ FIGURE 7-24 QUÉBEC INDEPENDENCE RALLY Supporters of independence for Québec march through the streets of Montréal prior to a 1995 referendum in which voters voted 50.6 percent to 49.4 percent to remain part of Canada.

In Canada, the Québécois are clearly distinct from other Canadians in language, religion, and other cultural traditions. But do the Québécois form a distinct ethnicity within the Canadian nationality or a second nationality separate altogether from Anglo-Canadian? The distinction is critical because if Québécois is recognized as a separate nationality from Anglo-Canadian, the Québec government would have a much stronger justification for breaking away from Canada to form an independent country (Figure 7-24).

ETHNICITIES AND NATIONALITIES IN THE UNITED KINGDOM

Outside North America, distinctions between ethnicity and nationality are even muddier. An example of the complexity is the British Isles, which comprise several thousand islands, including Ireland (called Eire in Irish) and Great Britain. The British Isles contain four principal ethnicities (Figure 7-25):

- **English.** The English are descendants of Germanic tribes who crossed the North Sea and invaded the country in the fifth century (see Chapter 5).

► FIGURE 7-25 ETHNICITIES AND NATIONALITIES IN THE UNITED KINGDOM AND IRELAND
The British Isles comprise two countries: the Republic of Ireland and the United Kingdom of Great Britain and Northern Ireland

- **Welsh.** The Welsh were Celtic people conquered by England in 1282 and formally united with England through the Act of Union of 1536. Welsh laws were abolished, and Wales became a local government unit.
- **Scots.** The Scots were Celtic people who had an independent country for more than 700 years, until 1603, when Scotland's King James VI also became King James I of England, thereby uniting the two countries. The Act of Union in 1707 formally merged the two governments, although Scotland was allowed to retain its own systems of education and local laws.
- **Irish.** The Irish were Celtic people who were ruled by England until the twentieth century, when most of the island became the independent country of Ireland.

Ireland and Great Britain are divided into two nationalities:

- **The United Kingdom** comprises Great Britain and Northern Ireland. The term *British* refers to the nationality.
- **The Republic of Ireland** comprises the southern 84 percent of the island of Ireland. The island of Ireland contains one predominant ethnicity—Irish—divided between two nationalities.

Within the United Kingdom, a strong element of ethnic identity comes from sports. Even though they are not separate countries, England, Scotland, Wales, and Northern Ireland field their own national soccer and compete separately in major international tournaments, such as the World Cup. The most important annual rugby tournament, known as the Six Nations' Championship, includes teams from England, Scotland, and Wales, as well as Ireland, Italy, and France. Given the history of English conquest, the other nationalities often root against England when it is playing teams from other countries.

Sorting out ethnicity and nationality can be challenging for many, including prominent sports stars. The golfer Rory McIlroy's ethnicity is Irish Catholic, and his nationality is United Kingdom, because Northern Ireland is part of the United Kingdom. But many Catholics in Northern Ireland feel closeness to the Republic of Ireland (see Chapter 6).

Tiger Woods has the reverse situation. His nationality is clearly the United States, but his ethnicity is less clear. His father was a mix of African American, Native American, and possibly Chinese, and his mother was a mix of Thai, Chinese, and Dutch. Woods describes his complex ethnicity as "Cablinasian."

Pause and Reflect 7.3.1

If Scotland becomes an independent country, how would the arrangement of nationalities in the British Isles change?

NATIONALISM

A nationality, once established, must hold the loyalty of its citizens to survive (Figure 7-26). Politicians and governments try to instill loyalty through **nationalism**, which is loyalty and devotion to a nationality. Nationalism typically promotes a sense of national consciousness that exalts one nation above all others and emphasizes its culture and interests as opposed to those of other nations. People display nationalism by supporting a country that preserves and enhances the culture and attitudes of their nationality.

▲ **FIGURE 7-26 NATIONALISM** Ukrainians celebrate independence day on August 24 by waving flags while walking along Khreshchatyk Street in the capital, Kiev. Ukraine declared its independence from the former Soviet Union on August 24, 1991.

States foster nationalism by promoting symbols of the country, such as flags and songs. The symbol of the hammer and sickle on a field of red was long synonymous with the beliefs of communism. After the fall of communism, one of the first acts in a number of Eastern European countries was to redesign flags without the hammer and sickle. Legal holidays were changed from dates associated with Communist victories to those associated with historical events that preceded Communist takeovers.

Nationalism can have a negative impact. The sense of unity within a nation-state is sometimes achieved through the creation of negative images of other nation-states. Travelers in southeastern Europe during the 1970s and 1980s found that jokes directed by one nationality against another recurred in the same form throughout the region, with only the name of the target changed. For example, "How many [fill in the name of a nationality] are needed to change a lightbulb?" Such jokes seemed harmless, but in hindsight reflected the intense dislike for other nationalities that led to conflict in the 1990s.

Nationalism is an important example of a **centripetal force**, which is an attitude that tends to unify people and enhance support for a state. (The word *centripetal* means "directed toward the center"; it is the opposite of *centrifugal*, which means "to spread out from the center.") Most countries find that the best way to achieve citizen support is to emphasize shared attitudes that unify the people.

Ethnic Competition

Learning Outcome 7.3.2

Identify and describe the principal ethnicities in Lebanon and Sri Lanka.

We have already seen in this chapter that identification with ethnicity and race can lead to discrimination and segregation. Confusion between ethnicity and nationality can lead to violent conflicts. Lebanon and Sri Lanka are examples of countries that have not successfully integrated diverse ethnicities.

ETHNIC COMPETITION IN LEBANON

Lebanon has 4 million people in an area of 10,000 square kilometers (4,000 square miles), a bit smaller and more populous than Connecticut. Once known as a financial and recreational center in the Middle East, Lebanon has been severely damaged by fighting among ethnicities since the 1970s.

Lebanon is divided between around 60 percent Muslims and 40 percent Christians (Figure 7-27). The precise distribution of religions in Lebanon is unknown because no census has been taken since 1932:

- **Christians.** Lebanon's most numerous Christian sect is Maronite, which split from the Roman Catholic Church in the seventh century. Maronites, ruled by the patriarch of Antioch, perform the liturgy in the ancient Syrian language. The second-largest Christian sect is Greek Orthodox, the Orthodox church that uses a Byzantine liturgy.
- **Muslims.** Most of Lebanon's Muslims belong to one of several Shiite sects. Sunnis, who are much more numerous than Shiites in the world, account for a minority of Lebanon's Muslims. Lebanon also has an important community of Druze, who were once considered to have a separate religion but now consider themselves Muslim. Many Druze rituals are kept secret from outsiders.

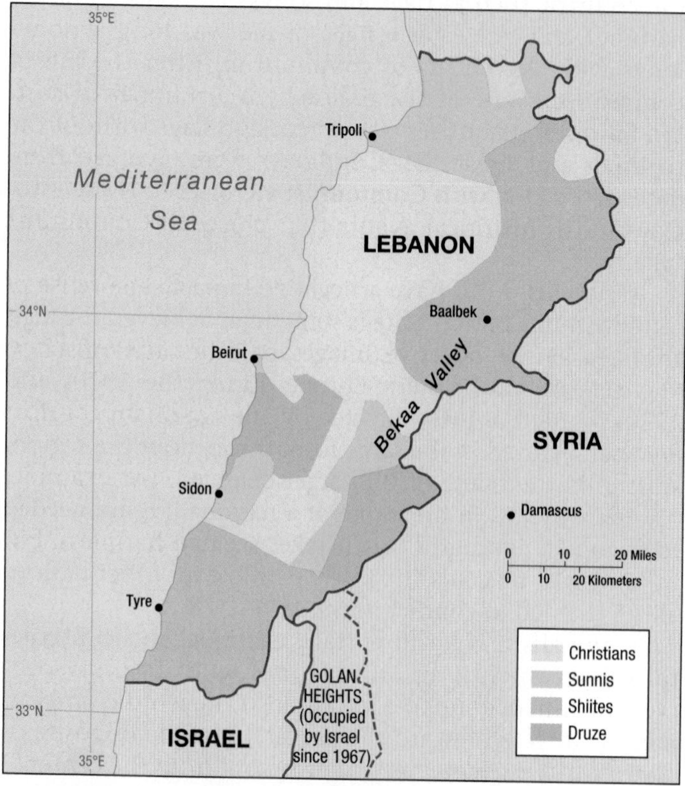

▲ **FIGURE 7-27 ETHNICITIES IN LEBANON** Christians dominate in the south and the northwest, Sunni Muslims in the far north, Shiite Muslims in the northeast and south, and Druze in the south-central and southeast.

Lebanon's diversity may appear to be religious rather than ethnic. But most of Lebanon's Christians consider themselves ethnically descended from the ancient Phoenicians who once occupied present-day Lebanon. In this way, Lebanon's Christians differentiate themselves from the country's Muslims, who are considered Arabs.

When Lebanon became independent in 1943, the constitution required that each religion be represented in the Chamber of Deputies according to its percentage in the 1932 census. By unwritten convention, the president of Lebanon was a Maronite Christian, the premier a Sunni Muslim, the speaker of the Chamber of Deputies a Shiite Muslim, and the foreign minister a Greek Orthodox Christian. Other cabinet members and civil servants were similarly apportioned among the various faiths.

Lebanon's religious groups have tended to live in different regions of the country. Maronites are concentrated in the west-central part, Sunnis in the northwest, and Shiites in the south and east. Beirut, the capital and largest city, has been divided between a Christian eastern zone and a Muslim western zone. During a civil war between 1975 and 1990, each religious group formed a private army or militia to guard its territory. The territory controlled by each militia changed according to results of battles with other religious groups.

When the governmental system was created, Christians constituted a majority and controlled the country's main businesses, but as Muslims became the majority, they demanded political and economic equality. The agreement ending the civil war in 1990 gave each religion one-half of the 128 seats in Parliament. Israel and the United States sent troops into Lebanon at various points in failed efforts to restore peace (Figure 7-28). The United States pulled out after 241 U.S. marines died in their barracks from a truck bomb in 1983. Lebanon was left under the control of neighboring Syria, which had a historical claim over the territory until it, too, was forced to withdraw its troops in 2005.

Pause and Reflect 7.3.2

What country borders Lebanon on the south? What conflict has been ongoing in that country, as described in Chapter 6?

ETHNIC DIVERSITY IN SRI LANKA

An island country of 19 million inhabitants off the Indian coast, Sri Lanka is inhabited by three principal ethnicities

▲ **FIGURE 7-28 ETHNIC CONFLICT IN LEBANON** U.S. Marines patrol the streets of Beirut, Lebanon, in 1983.

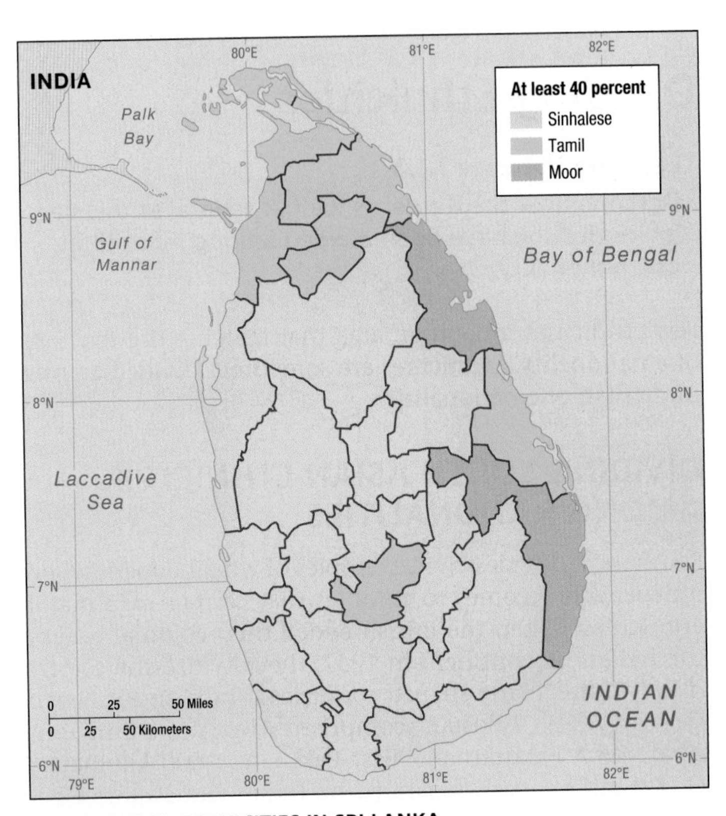

▲ **FIGURE 7-29 ETHNICITIES IN SRI LANKA**
The Sinhalese are Buddhists who speak an Indo-European language, whereas the Tamils are Hindus who speak a Dravidian language.

known as Sinhalese, Tamil, and Moors (Figure 7-29). War between the Sinhalese and Tamil erupted in 1983 and continued until 2009. During that period, 80,000 died in the conflict between the two ethnicities:

- **Sinhalese,** who comprise 74 percent of Sri Lanka's population, migrated from northern India in the fifth century B.C., occupying the southern two-thirds of the island. Three hundred years later, the Sinhalese were converted to Buddhism, and Sri Lanka became one of that religion's world centers. Sinhalese is an Indo-European language, in the Indo-Iranian branch.
- **Tamils,** who comprise 16 percent of Sri Lanka's population, migrated across the narrow 80-kilometer-wide (50-mile-wide) Palk Strait from India beginning in the third century B.C. and occupied the northern part of the island. Tamils are Hindus, and the Tamil language, in the Dravidian family, is also spoken by 60 million people in India.
- **Moors,** who comprise 10 percent of Sri Lanka's population, are ethnic Arabs, descended from traders from Southwest Asia who settled in Sri Lanka beginning in the eighth century A.D. Moors adhere to Islam but speak either Tamil or Sanhalese.

The dispute between Sri Lanka's two largest ethnicities extends back more than 2,000 years but was suppressed during 300 years of European control. Since the country gained independence in 1948, Sinhalese have dominated the government, military, and most of the commerce. Tamils feel that they suffer from discrimination at the hands of the Sinhalese-dominated government and have received support for a rebellion that began in 1983 from Tamils living in other countries.

The long war between the ethnicities ended in 2009, with the defeat of the Tamils (Figure 7-30). With their defeat, the Tamils fear that the future of Sri Lanka as a multinational state is jeopardized. Back in 1956, Sinhalese leaders made Buddhism the sole official religion and Sinhala the sole official language of Sri Lanka. The Tamils fear that their military defeat jeopardizes their ethnic identity again.

▼ **FIGURE 7-30 ETHNIC CONFLICT IN SRI LANKA** Tamils demonstrating in Switzerland for international support a few days before losing the war in 2009.

Dividing Ethnicities

Learning Outcome 7.3.3

Describe how the Kurds, as well as several ethnicities in South Asia, have been divided among more than one nationality.

Few ethnicities inhabit an area that matches the territory of a nationality. Ethnicities are sometimes divided among more than one nationality.

DIVIDING SOUTH ASIAN ETHNICITIES AMONG NATIONALITIES

South Asia provides vivid examples of what happens when independence comes to colonies that contain two major ethnicities. When the British ended their colonial rule of the Indian subcontinent in 1947, they divided the colony into two irregularly shaped countries—India and Pakistan (Figure 7-31). Pakistan comprised two noncontiguous areas, West Pakistan and East Pakistan, 1,600 kilometers (1,000 miles) apart, separated by India. East Pakistan became the independent state of Bangladesh in 1971. An eastern region of India was also practically cut off from the rest of the country, attached only by a narrow corridor north of Bangladesh that is less than 13 kilometers (8 miles) wide in some places.

The basis for separating West and East Pakistan from India was ethnicity. The people living in the two areas of Pakistan were predominantly Muslim; those in India were predominantly Hindu. Antagonism between the two religious groups was so great that the British decided to place the Hindus and Muslims in separate states. Hinduism has become a great source of national unity in India. In modern India, with its hundreds of languages and ethnic groups, Hinduism has become the cultural trait shared by the largest percentage of the population.

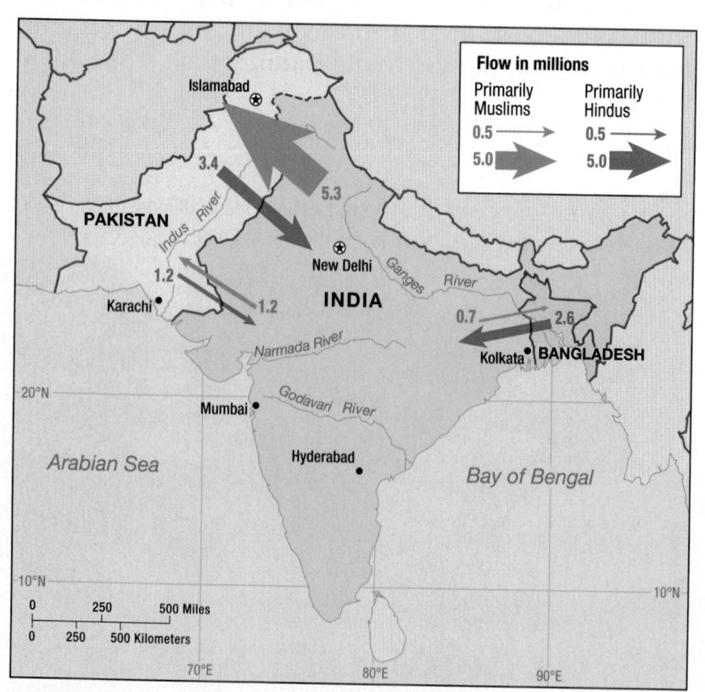

▲ **FIGURE 7-31 ETHNIC DIVISION OF SOUTH ASIA** In 1947, British India was partitioned into two independent states, India and Pakistan, which resulted in the migration of an estimated 17 million people. The creation of Pakistan as two territories nearly 1,600 kilometers (1,000 miles) apart proved unstable, and in 1971 East Pakistan became the independent country of Bangladesh.

Muslims have long fought with Hindus for control of territory, especially in South Asia. After the British took over India in the early 1800s, a three-way struggle began, with the Hindus and Muslims fighting each other as well as the British rulers. Mahatma Gandhi, the leading Hindu advocate of nonviolence and reconciliation with Muslims, was assassinated in 1948, ending the possibility of creating a single state in which Muslims and Hindus could live together peacefully.

The partition of South Asia into two states resulted in massive migration because the two boundaries did not correspond precisely to the territory inhabited by the two ethnicities. Approximately 17 million people caught on the wrong side of a boundary felt compelled to migrate during the late 1940s. Some 6 million Muslims moved from India to West Pakistan and about 1 million from India to East Pakistan. Hindus who migrated to India included approximately 6 million from West Pakistan and 3.5 million from East Pakistan. As they attempted to reach the other side of the new border, Hindus in Pakistan and Muslims in India were killed by people from the rival religion. Extremists attacked small groups of refugees traveling by road and halted trains to massacre the passengers.

Pakistan and India never agreed on the location of the boundary separating the two countries in the northern region of Kashmir (Figure 7-32). Since 1972, the two countries have maintained a "line of control" through the region, with Pakistan administering the northwestern portion and India the southeastern portion. Muslims, who comprise a majority in both portions, have fought a guerrilla war to secure reunification of Kashmir, either as part of Pakistan or as an independent country. India blames Pakistan for the unrest and vows to retain its portion of Kashmir. Pakistan argues that Kashmiris on both sides of the border should choose their own future in a vote, confident that the majority Muslim population would break away from India.

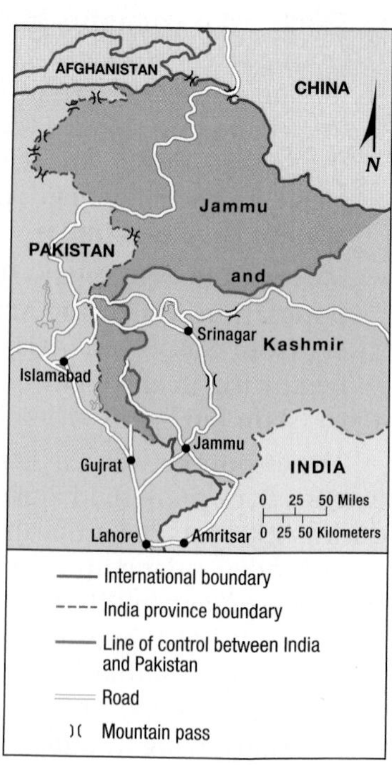

▲ **FIGURE 7-32 KASHMIR** India and Pakistan dispute the location of their border.

◀ **FIGURE 7-33 KURDS** Kurds in northern Iraq hold burning torches to celebrate their new year, which they call Newroz, on the first day of spring.

India's religious unrest is further complicated by the presence of 25 million Sikhs, who have long resented that they were not given their own independent country when India was partitioned (see Chapter 6). Although they constitute only 2 percent of India's total population, Sikhs comprise a majority in the Indian state of Punjab, situated south of Kashmir along the border with Pakistan. Sikh extremists have fought for more control over the Punjab or even complete independence from India.

DIVIDING THE KURDS AMONG NATIONALITIES

A prominent example of an ethnicity divided among several countries in western Asia is the Kurds, who live in the Caucasus Mountains (Figure 7-33). The Kurds are Sunni Muslims who speak a language in the Iranian group of the Indo-Iranian branch of Indo-European and have distinctive literature, dress, and other cultural traditions.

When the victorious European allies carved up the Ottoman Empire after World War I, they created an independent state of Kurdistan to the south and west of Van Gölü (Lake Van) under the 1920 Treaty of Sèvres. Before the treaty was ratified, however, the Turks, under the leadership of Mustafa Kemal (later known as Kemal Ataturk), fought successfully to expand the territory under their control beyond the small area the allies had allocated to them. The Treaty of Lausanne in 1923 established the modern state of Turkey, with boundaries nearly identical to the current ones. Kurdistan became part of Turkey and disappeared as an independent state.

Today the 30 million Kurds are split among several countries; 14 million live in eastern Turkey, 5 million in northern Iraq, 4 million in western Iran, 2 million in Syria, and the rest in other countries. Kurds comprise 19 percent of the population in Turkey, 16 percent in Iraq, 9 percent in Syria, and 6 percent in Iran (refer ahead to Figure 7-34 on the next page).

To foster the development of Turkish nationalism, the Turks have tried repeatedly to suppress Kurdish culture. Use of the Kurdish language was illegal in Turkey until 1991, and laws banning its use in broadcasts and classrooms remain in force. Kurdish nationalists, for their part, have waged a guerrilla war since 1984 against the Turkish army. Kurds in other countries have fared just as poorly as those in Turkey. Iran's Kurds secured an independent republic in 1946, but it lasted less than a year. Iraq's Kurds have made several unsuccessful attempts to gain independence, including in the 1930s, 1940s, and 1970s.

A few days after Iraq was defeated in the 1991 Gulf War, the country's Kurds launched another unsuccessful rebellion. The United States and its allies decided not to resume their recently concluded fight against Iraq on behalf of the Kurdish rebels, but after the revolt was crushed, they sent troops to protect the Kurds from further attacks by the Iraqi army. After the United States attacked Iraq and deposed Saddam Hussein in 2003, Iraqi Kurds achieved even more autonomy, but still not independence. Thus, despite their numbers, the Kurds are an ethnicity with no corresponding Kurdish state today. Instead, they are forced to live under the control of the region's more powerful nationalities.

Pause and Reflect 7.3.3

Refer ahead to Figure 7-34 on the next page. What is the largest ethnicity in Pakistan?

ETHNIC DIVERSITY IN WESTERN ASIA

Learning Outcome 7.3.4
Identify and describe the principal ethnicities in western Asia.

The lack of correspondence between the territory occupied by ethnicities and nationalities is especially severe in western Asia. Four nationalities in the region—Iraqi, Iranian, Afghan, and Pakistani—encompass dozens of ethnicities, most of whom inhabit more than one of the region's countries (Figure 7-34):

ETHNIC DIVERSITY IN IRAQ. Approximately three-fourths of Iraqis are Arabs, and one-sixth are Kurds. The Arab population is divided among Muslim branches, with two-thirds Shiite and one-third Sunni.

The United States led an attack against Iraq in 2003 that resulted in the removal and death of the country's longtime president, Saddam Hussein. U.S. officials justified removing Hussein because he ran a brutal dictatorship, created weapons of mass destruction, and allegedly had close links with terrorists (see Chapter 8).

Having invaded Iraq and removed Hussein from power, the United States expected an enthusiastic welcome from the Iraqi nation. Instead, the United States became embroiled in a complex and violent struggle among ethnic groups:

- Kurds welcomed the United States because they gained more security and autonomy than they had had under Hussein.
- Sunni Muslim Arabs opposed the U.S.-led attack because they feared loss of power and privilege given to them by Hussein, who was a Sunni.
- Shiite Muslim Arabs also opposed the U.S. presence. Although they had been treated poorly by Hussein and controlled Iraq's post-Hussein government, Shiites shared a long-standing hostility toward the United States with their neighbors in Shiite-controlled Iran.

Iraq's principal ethnic groups are split into regions, with Kurds in the north, Sunnis in the center, and Shiites in the south.

The capital, Baghdad, where one-fourth of the Iraqi people live, has some neighborhoods where virtually all residents are of one ethnicity, but most areas are mixed. In many of these historically mixed neighborhoods, the minority ethnicity has been forced to move away (Figure 7-35).

The major ethnicities are divided into numerous tribes and clans (Figure 7-36). Most Iraqis actually have stronger loyalty to a tribe or clan than to the nationality or a major ethnicity.

ETHNIC DIVERSITY IN IRAN. The most numerous ethnicity is Persian, but Azeri and Baluchi represent important minorities. Persians constitute the world's largest ethnic group that adheres to Shiite Islam. Persians

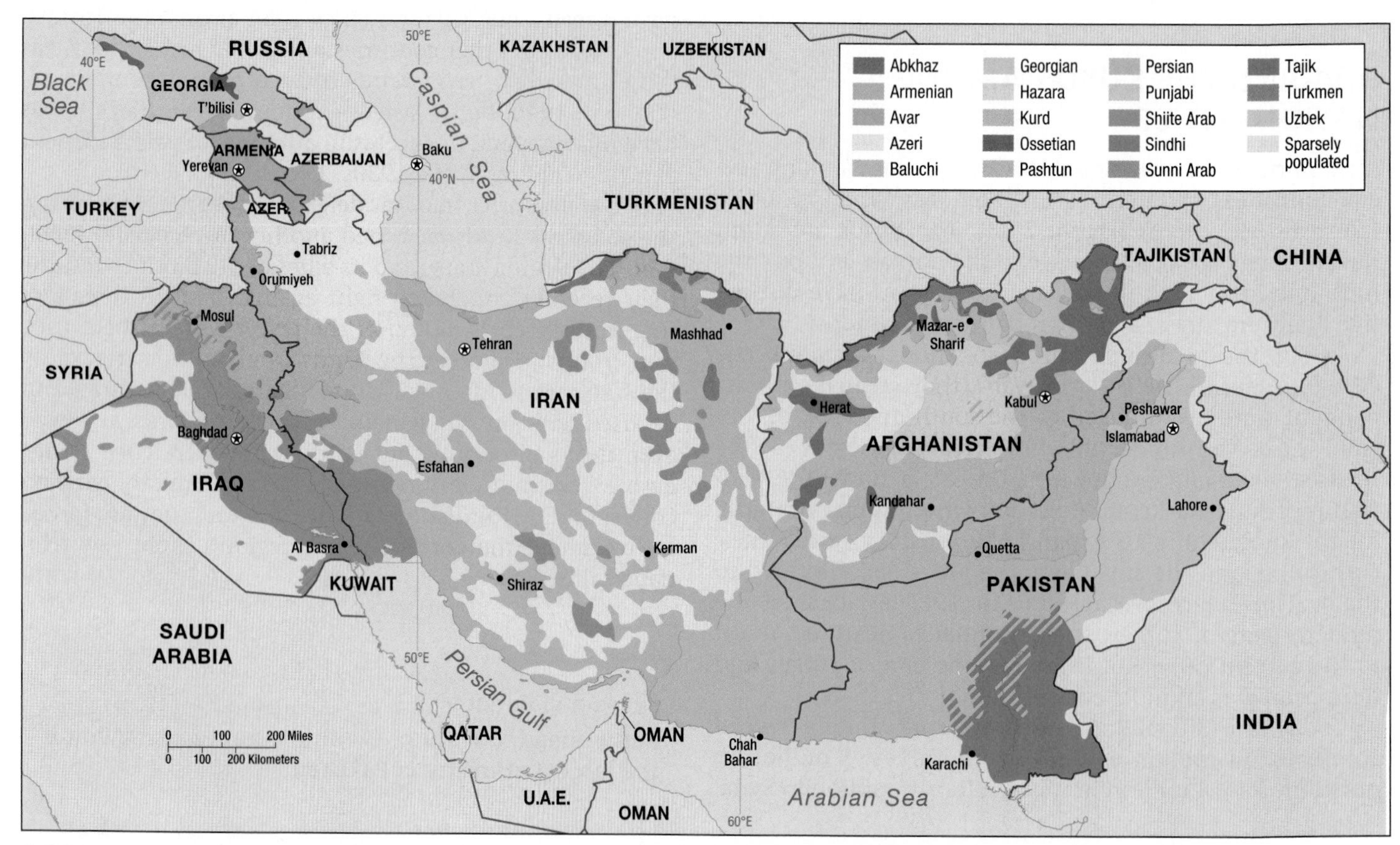

▲ **FIGURE 7-34 ETHNICITIES IN WESTERN ASIA** The complex distribution of ethnicities and nationalities across western Asia is a major source of conflict.

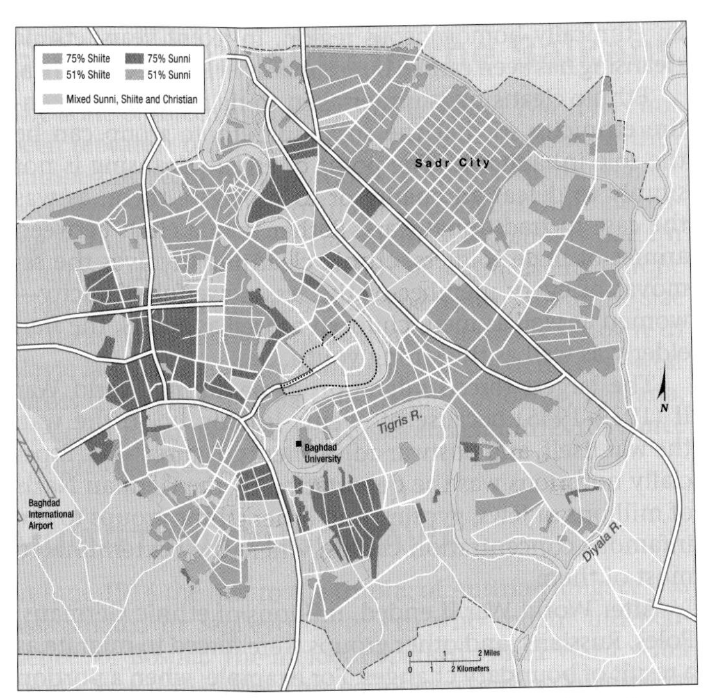

▲ **FIGURE 7-35 ETHNICITIES IN BAGHDAD** Baghdad contains a mix of Sunnis, Shiites, and other groups. Many neighborhoods were traditionally mixed, but in recent years the minority group has been forced to migrate.

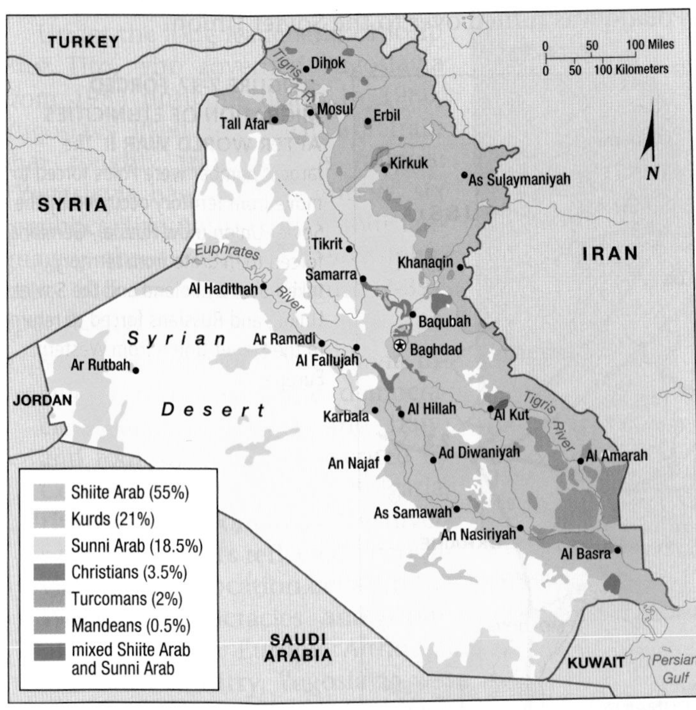

▲ **FIGURE 7-36 ETHNICITIES IN IRAQ** Iraq is home to around 150 distinct tribes. Some of the larger ones are shown on the map.

are believed to be descendants of the Indo-European tribes that began migrating from Central Asia into what is now Iran several thousand years ago (see Chapter 5). The Persian Empire extended from present-day Iran west as far as Egypt during the fifth and fourth centuries B.C. After the Muslim army conquered Persia in the seventh century, most Persians converted to Sunni Islam. The conversion to Shiite Islam came primarily in the fifteenth century.

ETHNIC DIVERSITY IN AFGHANISTAN. The most numerous ethnicities in Afghanistan are Pashtun, Tajik, and Hazara. The current unrest among Afghanistan's ethnicities dates from 1979, with the start of a rebellion by several ethnic groups against the government, which was being defended by more than 100,000 troops from the Soviet Union. Unable to subdue the rebellion, the Soviet Union withdrew its troops in 1989, and the Soviet-installed government in Afghanistan collapsed in 1992.

After several years of infighting among ethnicities, a faction of the Pashtun called the Taliban gained control over most of the country in 1995. The Taliban imposed very harsh, strict laws on Afghanistan, according to Islamic values as the Taliban interpreted them (see Chapter 6). The United States invaded Afghanistan in 2001 and overthrew the Taliban-led government because it was harboring terrorists (see Chapter 8). Removal of the Taliban unleashed a new struggle for control of Afghanistan among the country's many ethnic groups, including the Taliban.

ETHNIC DIVERSITY IN PAKISTAN. The most numerous ethnicity in Pakistan is Punjabi, but the border area with Afghanistan is principally Baluchi and Pashtun. The Punjabi have been the most numerous ethnicity since ancient times in what is now Pakistan. As with the neighboring Pashtun, the Punjabi converted to Islam after they were conquered by the Muslim army in the seventh century. The Punjabi remained Sunni Muslims rather than convert to Shiite Islam like their neighbors the Pashtun, who comprise Pakistan's second-largest ethnicity, especially along the border with Afghanistan. Fighting between Pakistan's army and supporters of the Taliban forced Pakistanis to leave their homes and move into camps, where they were fed by international relief organizations.

Pause and Reflect 7.3.4
How do the ethnic complexities of western Asia make it difficult to set up stable democratic governments?

CHECK-IN: KEY ISSUE 3

Why Do Conflicts Arise among Ethnicities?

- ✓ **Nationality is identity with a group of people who share legal attachment and personal allegiance to a particular country.**
- ✓ **Countries such as Lebanon and Sri Lanka have difficulty peacefully combining ethnicities into one nationality.**
- ✓ **Some ethnicities, such as the Kurds, are divided among more than one nationality.**
- ✓ **Lack of correspondence between ethnicities and nationalities is especially severe in western Asia.**

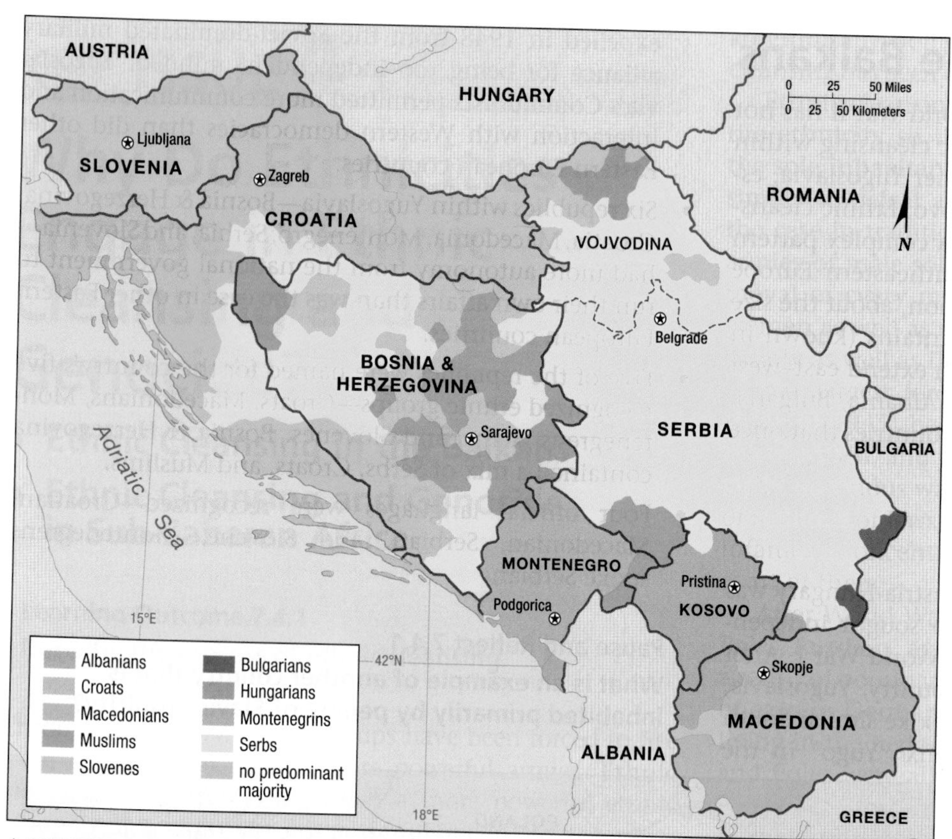

▲ **FIGURE 7-39 YUGOSLAVIA UNTIL ITS BREAKUP IN 1992** Yugoslavia comprised six republics (plus Kosovo and Vojvodina, autonomous regions within the Republic of Serbia).

Rivalries among ethnicities resurfaced in Yugoslavia during the 1980s after Tito's death, leading to the breakup of the country. Breaking away to form independent countries were Bosnia & Herzegovina, Croatia, Macedonia, and Slovenia during the 1990s and Montenegro in 2006. The breakup left Serbia standing on its own as well.

As long as Yugoslavia comprised one country, ethnic groups were not especially troubled by the division of the country into six republics. But when Yugoslavia's republics were transformed from local government units into five separate countries, ethnicities fought to redefine the boundaries. Not only did the boundaries of Yugoslavia's six republics fail to match the territory occupied by the five major nationalities, but the country contained other important ethnic groups that had not received official recognition as nationalities.

- Three major religions included Roman Catholic in the north, Orthodox in the east, and Islam in the south. Croats and Slovenes were predominantly Roman Catholic, Serbs and Macedonians predominantly Orthodox, and Bosnians and Montenegrens predominantly Muslim.
- Two of the four official languages—Croatian and Slovene—were written in the Roman alphabet; Macedonian and Serbian were written in Cyrillic. Most linguists outside Yugoslavia considered Serbian and Croatian to be the same language except with different alphabets.
- One, the refrain concluded, was the dinar, the national unit of currency. This meant that despite cultural diversity, common economic interests kept Yugoslavia's nationalities unified.

The Balkan Peninsula, a complex assemblage of ethnicities, has long been a hotbed of unrest (Figure 7-40). Northern portions were incorporated into the Austro-Hungarian Empire; southern portions were ruled by the Ottomans. Austria-Hungary extended its rule farther south in 1878 to include Bosnia & Herzegovina, where the majority of the people had been converted to Islam by the Ottomans.

The creation of Yugoslavia brought stability that lasted for most of the twentieth century. Old animosities among ethnic groups were submerged, and younger people began to identify themselves as Yugoslavs rather than as Serbs, Croats, or Montenegrens.

▲ **FIGURE 7-40 THE BALKANS IN 1914** At the outbreak of World War I, Austria-Hungary controlled the northern part of the region, including all or part of Croatia, Slovenia, and Romania. The Ottoman Empire controlled some of the south, although during the nineteenth century it had lost control of Albania, Bosnia & Herzegovina, Greece, Romania, and Serbia.

CONTEMPORARY GEOGRAPHIC TOOLS

Documenting Ethnic Cleansing

Early reports of ethnic cleansing by Serbs in the former Yugoslavia were so shocking that many people dismissed them as journalistic exaggeration or partisan propaganda. It took one of geography's most important analytic tools, aerial-photography interpretation, to provide irrefutable evidence of the process, as well as the magnitude, of ethnic cleansing.

A series of three photographs taken by NATO air reconnaissance over the village of Glodane, in western Kosovo, illustrated the four steps in ethnic cleansing. Figure 7-41 is the first of the three photos:

- Illustrating step 1, the red circles in Figure 7-41 show the location of Serb armored vehicles along the main street of the village. Figure 7-41 shows the village's houses and farm buildings clustered on the left side, with fields on the outskirts of the village, including the center and right portions of the photograph. As discussed in Chapter 12, rural settlements in most of the world have houses and farm buildings clustered together and surrounded by fields rather than in isolated, individual farms typical of North America.
- Illustrating step 2, the farm field immediately to the east of the main north–south road is filled with the villagers. At the scale that the photograph is reproduced in this book, the people appear as a dark mass. The white rectangles to the north of the people are civilian cars and trucks.
- Illustrating step 3, the second photograph of the sequence showed the same location a short time later, with one major change—the people and vehicles massed in the field in the first photograph are gone—no people and no vehicles.
- Illustrating step 4, the third photograph showed that the buildings in the village had been set on fire.

Aerial photographs such as these not only "proved" that ethnic cleansing was occurring but also provided critical evidence to prosecute Serb leaders for war crimes.

▲ FIGURE 7-41 **EVIDENCE OF ETHNIC CLEANSING IN KOSOVO** Ethnic cleansing by Serbs forced Albanians living in Kosovo to flee in 1999. The village of Glodane is on the west (left) side of the road. The villagers and their vehicles have been rounded up and placed in the field east of the road. The red circles show the locations of Serb armored vehicles.

ETHNIC CLEANSING IN BOSNIA

Learning Outcome 7.4.2
Explain the concept of ethnic cleansing in the Balkans.

The creation of a viable nationality has proved especially difficult in the case of Bosnia & Herzegovina. At the time of the breakup of Yugoslavia, the population of Bosnia & Herzegovina was 48 percent Bosnian Muslims, 37 percent Serbs, and 14 percent Croats. Bosnian Muslim was considered an ethnicity rather than a nationality. Rather than live in an independent multiethnic state with a Muslim plurality, Bosnia & Herzegovina's Serbs and Croats fought to unite the portions of the republic that they inhabited with Serbia and Croatia, respectively.

To strengthen their cases for breaking away from Bosnia & Herzegovina, Serbs and Croats engaged in ethnic cleansing of Bosnian Muslims (Figure 7-42). Ethnic cleansing ensured that areas did not merely have majorities of Bosnian Serbs and Bosnian Croats but were ethnically homogeneous and therefore better candidates for union with Serbia and Croatia. Ethnic cleansing by Bosnian Serbs against Bosnian Muslims was especially severe because much of the territory inhabited by Bosnian Serbs was separated from Serbia by areas with Bosnian Muslim majorities. By ethnically cleansing Bosnian Muslims from intervening areas, Bosnian Serbs created one continuous area of Bosnian Serb domination rather than several discontinuous ones.

Accords reached in Dayton, Ohio, in 1996 by leaders of the various ethnicities divided Bosnia & Herzegovina into three regions, one each dominated, respectively, by the Bosnian Croats, Muslims, and Serbs. The Bosnian Croat and Muslim regions were combined into a federation, with some cooperation between the two groups, but the Serb region has operated with almost complete independence in all but name from the others. In recognition of the success of their ethnic cleansing, Bosnian Serbs received nearly half of the country, although they comprised one-third of the population, and Bosnian Croats got one-fourth of the land, although they comprised one-sixth of the population. Bosnian Muslims, one-half of the population before the ethnic cleansing, got one-fourth of the land (Figure 7-43).

▲ **FIGURE 7-42 ETHNIC CLEANSING IN BOSNIA & HERZEGOVINA** (top) The Stari Most (old bridge), built by the Turks in 1566 across the Neretva River, was an important symbol and tourist attraction in the city of Mostar. (middle) The bridge was blown up by Croats in 1993, in an attempt to demoralize Bosnian Muslims as part of ethnic cleansing (bottom). With the end of the war in Bosnia & Herzegovina, the bridge was rebuilt in 2004.

Pause and Reflect 7.4.2
In which regions within Bosnia & Herzegovina did Serbs gain most of their territory?

ETHNIC CLEANSING IN KOSOVO

After the breakup of Yugoslavia, Serbia remained a multiethnic country. Particularly troubling was the province of Kosovo, where ethnic Albanians comprised 90 percent of the population. Under Tito, ethnic Albanians in Kosovo received administrative autonomy and national identity.

Serbia had a historical claim to Kosovo, having controlled it between the twelfth and fourteenth centuries. Serbs fought an important—though losing—battle in

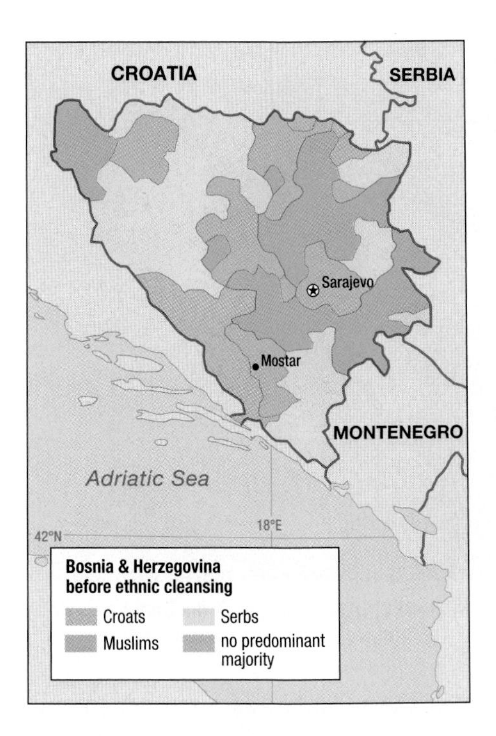

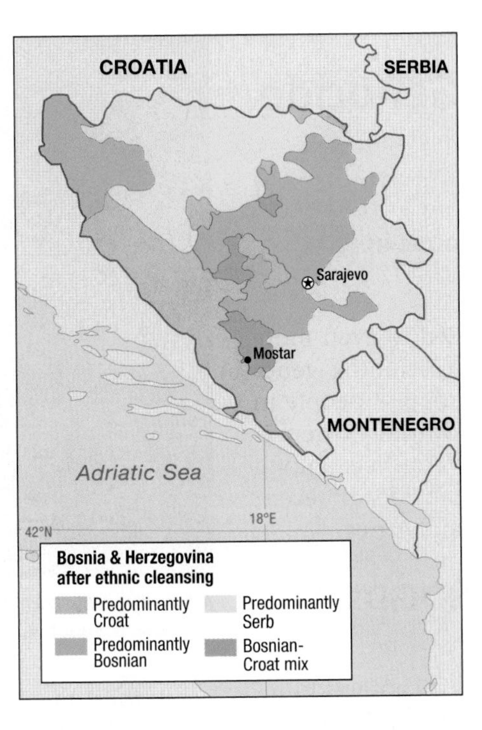

◀ **FIGURE 7-43 ETHNICITIES IN BOSNIA & HERZEGOVINA BEFORE AND AFTER ETHNIC CLEANSING** The territory occupied by Bosnian Muslims (left) was considerably reduced as a result of ethnic cleansing by Bosnian Serbs and Croats (right).

Kosovo against the Ottoman Empire in 1389. In recognition of its role in forming the Serb ethnicity, Serbia was given control of Kosovo when Yugoslavia was created in the early twentieth century.

With the breakup of Yugoslavia, Serbia took direct control of Kosovo and launched a campaign of ethnic cleansing of the Albanian majority. The process of ethnic cleansing involved four steps:

1. Move a large amount of military equipment and personnel into a village that has no strategic value (see the Contemporary Geographic Tools feature.
2. Round up all the people in the village. In Bosnia, Serbs often segregated men from women, children, and old people. The men were placed in detention camps or "disappeared"—undoubtedly killed—and the others were forced to leave the village. In Kosovo, men were herded together with the others rather than killed.
3. Force the people to leave the village. The villagers were typically forced into a convoy—some in the vehicles, others on foot—heading for the Albanian border.
4. Destroy the vacated village by setting it on fire.

At its peak in 1999, Serb ethnic cleansing had forced 750,000 of Kosovo's 2 million ethnic Albanian residents from their homes, mostly to camps in Albania. Outraged by the ethnic cleansing, the United States and Western European countries, operating through the North Atlantic Treaty Organization (NATO), launched an air attack against Serbia. The bombing campaign ended when Serbia agreed to withdraw all of its soldiers and police from Kosovo. Kosovo declared its independence from Serbia in 2008. Around 60 countries, including the United States, recognize Kosovo as an independent country, but Serbia and Russia oppose it.

BALKANIZATION

A century ago, the term **Balkanized** was widely used to describe a small geographic area that could not successfully be organized into one or more stable states because it was inhabited by many ethnicities with complex, long-standing antagonisms toward each other. World leaders at the time regarded **Balkanization**—the process by which a state breaks down through conflicts among its ethnicities—as a threat to peace throughout the world, not just in a small area. They were right: Balkanization led directly to World War I because the various nationalities in the Balkans dragged into the war the larger powers with which they had alliances.

After two world wars and the rise and fall of communism during the twentieth century, the Balkans have once again become Balkanized in the twenty-first century. Will the United States, Europe, and Russia once again be drawn reluctantly into conflict through entangled alliances in the Balkans? If peace comes to the Balkans, it will be because in a tragic way ethnic cleansing "worked." Millions of people were rounded up and killed or forced to migrate because they constituted ethnic minorities. Ethnic homogeneity may be the price of peace in areas that once were multiethnic.

Ethnic Cleansing and Genocide in Sub-Saharan Africa

Learning Outcome 7.4.3

Identify the principal episodes of genocide in northeastern Africa.

In some places, ethnic competition has led to even more extreme actions than ethnic cleansing, including genocide. **Genocide** is the mass killing of a group of people in an attempt to eliminate the entire group from existence. Sub-Saharan Africa has been plagued by conflicts among ethnic groups that have resulted in genocide in recent years, especially in northeastern and central Africa.

ETHNIC CLEANSING AND GENOCIDE IN NORTHEASTERN AFRICA

In northeastern Africa, three distinct ethnic conflicts in recent years have taken place in Sudan, Somalia, and Ethiopia.

SUDAN. In Sudan, several civil wars have raged since the 1980s between the Arab-Muslim dominated government in the north and other ethnicities in the south, west, and east (Figure 7-44):

- **South Sudan.** Black Christian and animist ethnicities resisted government attempts to convert the country from a multiethnic society to one nationality tied to Muslim traditions. A north–south war between 1983 and 2005 resulted in the death of an estimated 1.9 million Sudanese, mostly civilians. The war ended with the establishment of Southern Sudan as an independent state in 2011. However, fighting resumed as the governments of Sudan and South Sudan could not agree on boundaries between the two countries.

▲ **FIGURE 7-44 SUDAN AND SOUTH SUDAN** South Sudan became an independent country in 2011.

▲ **FIGURE 7-45 DARFUR REFUGEE CAMP** Refugees from Darfur are living in a camp in Adré, Chad.

- **Darfur.** As Sudan's religion-based civil war was winding down, an ethnic war erupted in Sudan's westernmost region, Darfur. Resenting discrimination and neglect by the national government, Darfur's black Africans launched a rebellion in 2003. Marauding Arab nomads, known as janjaweed, with the support of the Sudanese government, crushed Darfur's black population, made up mainly of settled farmers; 480,000 have been killed and another 2.8 million have been living in dire conditions in refugee camps in the harsh desert environment of Darfur (Figure 7-45). Actions of Sudan's government troops, including mass murders and rape of civilians, have been termed genocide by many other countries, and charges of war crimes have been filed against Sudan's leaders.
- **Eastern front.** Ethnicities in the east fought Sudanese government forces between 2004 and 2006, with the support of neighboring Eritrea. At issue was disbursement of profits from oil.

ETHIOPIA AND ERITREA. Eritrea, located along the Red Sea, became an Italian colony in 1890. Ethiopia, an independent country for more than 2,000 years, was captured by Italy during the 1930s. After World War II, Ethiopia regained its independence, and the United Nations awarded Eritrea to Ethiopia (Figure 7-46). The United Nations expected Ethiopia to permit Eritrea considerable authority to run its own affairs, but Ethiopia dissolved the Eritrean legislature and banned the use of Tigrinya, Eritrea's major local language. The Eritreans rebelled, beginning a 30-year fight for independence (1961–1991). During this civil war, an estimated 665,000 Eritrean refugees fled to neighboring Sudan.

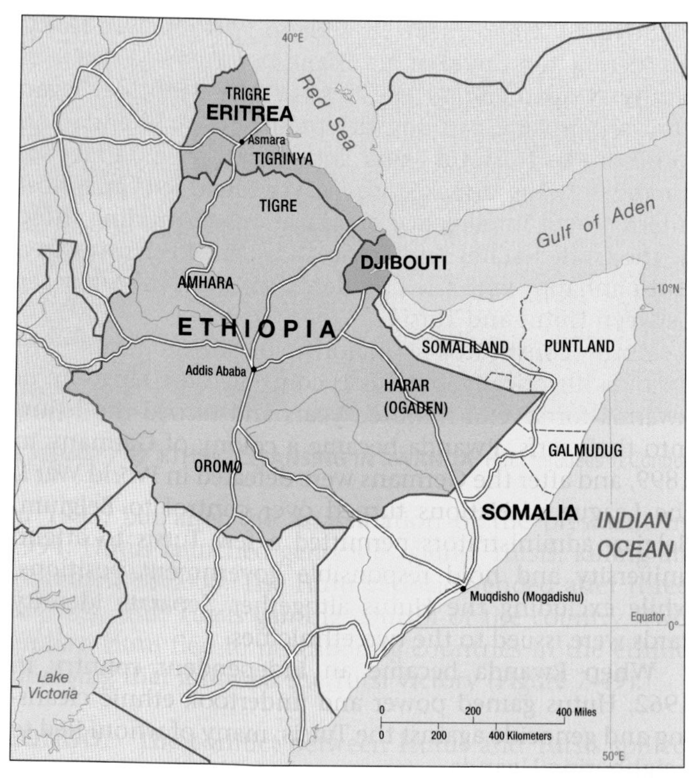

▲ **FIGURE 7-46 HORN OF AFRICA** Eritrea broke away from Ethiopia to become an independent country in the early 1990s. Somalia is divided into several territories controlled by various ethinic groups.

Eritrean rebels defeated the Ethiopian army in 1991, and 2 years later Eritrea became an independent state. But war between Ethiopia and Eritrea flared up again in 1998 because of disputes over the location of the border. Eritrea justified its claim through a 1900 treaty between Ethiopia and Italy, which then controlled Eritrea, but Ethiopia cited a 1902 treaty with Italy. Ethiopia defeated Eritrea in 2000 and took possession of the disputed areas. Battles along the border have continued (Figure 7-47).

▼ **FIGURE 7-47 ERITREA–ETHIOPIA BORDER** The border between Eritrea (background) and Ethiopia (foreground) is unmarked here.

A country of 5 million people split evenly between Christian and Muslim, Eritrea has two principal ethnic groups: Tigrinya and Tigre. At least in the first years of independence, a strong sense of national identity united Eritrea's ethnicities as a result of shared experiences during the 30-year war to break free of Ethiopia.

Even with the loss of Eritrea, Ethiopia remained a complex multiethnic state. From the late nineteenth century until the 1990s, Ethiopia was controlled by the Amharas, who are Christians. After the government defeat in the early 1990s, power passed to a combination of ethnic groups. The Oromo, who are Muslim fundamentalists from the south, are the largest ethnicity in Ethiopia, at 34 percent of the population. The Amhara, who comprise 27 percent of the population, had banned the use of languages other than Amharic, including Oromo.

SOMALIA. On the surface, Somalia should face fewer ethnic divisions than its neighbors in the Horn of Africa. Somalis are overwhelmingly Sunni Muslims and speak Somali. Most share a sense that Somalia is a nation-state, with a national history and culture.

Somalia's 9 million inhabitants are divided among several ethnic groups known as clans, each of which is divided into a large number of subclans. Traditionally, the major clans occupied different portions of Somalia. In 1991, a dictatorship that ran the country collapsed, and various clans and subclans claimed control over portions of the country. Clans have declared independent states of Somaliland in the north, Puntland in the northeast, Galmudug in the center, and Southwestern Somalia in the south.

The United States sent several thousand troops to Somalia in 1992, after an estimated 300,000 people, mostly women and children, died from famine and from warfare among clans. The purpose of the mission was to protect delivery of food by international relief organizations to starving Somali refugees and to reduce the number of weapons in the hands of the clan and subclan armies. After peace talks among the clans collapsed in 1994, U.S. troops withdrew.

Islamist militias took control of much of Somalia between 2004 and 2006. Neighboring countries were drawn into the conflict, Eritrea on the side of the Islamists and Ethiopia against them. Claiming that some of the leaders were terrorists, the United States also opposed the Islamists and launched air strikes in 2007. The fighting generated several hundred thousand refugees. Islamist militias withdrew from most of Somalia in 2006 but have since returned and again control much of the country. The ongoing conflict worsened the impact of a recent drought (see the Sustainability and Inequality in Our Global Village feature and Figure 7-48).

Pause and Reflect 7.4.3
Which countries with ethnic conflicts described in Key Issues 3 and 4 have had U.S. troops sent to try to restore the peace?

Summary and Review

KEY ISSUE 1

Where Are Ethnicities Distributed?

Ethnicity is identity with a group of people who share the cultural traditions of a particular homeland or hearth. Ethnicity is often confused with race, which is identity with a group of people who share a biological ancestor.

LEARNING OUTCOME 7.1.1: Identify and describe the major ethnicities in the United States.

- The three most numerous ethnicities are Hispanics, African Americans, and Asian Americans.

LEARNING OUTCOME 7.1.2: Describe the distribution of major U.S. ethnicities among states and within urban areas.

- Hispanics are clustered in the Southwest, African Americans in the Southeast, and Asian Americans in the West.
- African Americans and Hispanics are highly clustered in urban areas, especially in inner-city neighborhoods.

THINKING GEOGRAPHICALLY 7.1: A century ago European immigrants to the United States had much stronger ethnic ties than they do today, including clustering in specific neighborhoods. Discuss the rationale for retaining strong ethnic identity in the United States as opposed to full assimilation into the American nationality identity.

GOOGLE EARTH 7.1: Oldtown Mall in Baltimore is in a predominantly African American neighborhood. At Google Earth's ground-level view, does the mall look busy or quiet?

KEY ISSUE 2

Why Do Ethnicities Have Distinctive Distributions?

Ethnicities cluster within the United States as a result of distinctive patterns of migration.

LEARNING OUTCOME 7.2.1: Describe the patterns of forced and voluntary migration of African Americans, Hispanic Americans, and Asian Americans to the United States.

- Many African Americans trace their ancestry to forced migration from Africa for slavery.
- Many Hispanics and Asian Americans trace their heritage to people who migrated in the late twentieth century for economic prospects and political freedom.

LEARNING OUTCOME 7.2.2: Describe the patterns of migration of African Americans within the United States.

- African Americans migrated in large numbers from the South to the North and West in the early twentieth century.
- African Americans clustered in inner-city ghettos that have expanded in recent decades.

LEARNING OUTCOME 7.2.3: Explain the laws once used to segregate races in the United States and South Africa.

- Segregation of races was legal in the United States and South Africa until the late twentieth century.

THINKING GEOGRAPHICALLY 7.2: Despite the 1954 U.S. Supreme Court decision that racially segregated school systems are inherently unequal, most schools remain segregated, with virtually none or virtually all African American or Hispanic pupils. As long as most neighborhoods are segregated, how can racial integration in the schools be achieved?

GOOGLE EARTH 7.2: Mthatha (known until 2004 as Umtata), South Africa, is a city in one of the homelands established during apartheid. In Google Earth's ground-level view, what is the race of nearly all of the people?

Key Terms

Apartheid (p. 236) Laws (no longer in effect) in South Africa that physically separated different races into different geographic areas.

Balkanization (p. 251) A process by which a state breaks down through conflicts among its ethnicities.

Balkanized (p. 251) Descriptive of a small geographic area that could not successfully be organized into one or more stable states because it was inhabited by many ethnicities with complex, long-standing antagonisms toward each other.

Blockbusting (p. 235) A process by which real estate agents convince white property owners to sell their houses at low prices because of fear that persons of color will soon move into the neighborhood.

Centripetal force (p. 239) An attitude that tends to unify people and enhance support for a state.

Ethnic cleansing (p. 246) A process in which a more powerful ethnic group forcibly removes a less powerful one in order to create an ethnically homogeneous region.

Ethnicity (p. 227) Identity with a group of people that share distinct physical and mental traits as a product of common heredity and cultural traditions.

Genocide (p. 252) The mass killing of a group of people in an attempt to eliminate the entire group from existence.

Nationalism (p. 239) Loyalty and devotion to a particular nationality.

Nationality (p. 238) Identity with a group of people that share legal attachment and personal allegiance to a particular place as a result of being born there.

KEY ISSUE 3

Why Do Conflicts Arise among Ethnicities?

Conflicts can arise when a country contains several ethnicities competing with each other for control or dominance. Conflicts also arise when an ethnicity is divided among more than one country.

LEARNING OUTCOME 7.3.1: Explain the difference between ethnicity and nationality.

- Nationality is identity with a group of people who share legal attachment and personal allegiance to a particular country.
- Nationalism is loyalty and devotion to a nationality.

LEARNING OUTCOME 7.3.2: Identify and describe the principal ethnicities in Lebanon and Sri Lanka.

- Lebanon and Sri Lanka are examples of countries where ethnicities have not been able to live in peace.

LEARNING OUTCOME 7.3.3: Describe how the Kurds, as well as ethnicities in South Asia, have been divided among more than one nationality.

- Some ethnicities find themselves divided among more than one nationality.

LEARNING OUTCOME 7.3.4: Identify and describe the principal ethnicities in western Asia.

- The lack of correspondence between the territory occupied by ethnicities and nationalities is especially severe in western Asia.

THINKING GEOGRAPHICALLY 7.3: Ethnicities around the world seek the ability to be the majority in control of countries. What are some of the obstacles to multiple ethnicities sharing power in individual countries?

GOOGLE EARTH 7.3: Fly to Güven, Turkey to a village inhabited by Kurds. Turn on borders and labels; how far is Güven from Syria? From Iraq?

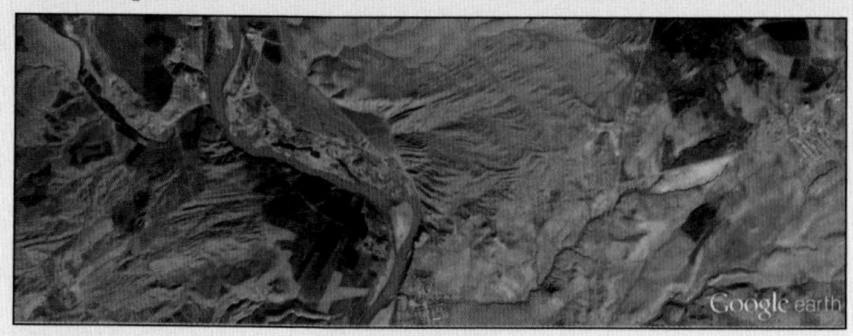

KEY ISSUE 4

Why Do Ethnicities Engage in Ethnic Cleansing and Genocide?

Ethnic cleansing is a process in which a more powerful ethnic group forcibly removes a less powerful one in order to create an ethnically homogeneous region.

LEARNING OUTCOME 7.4.1: Describe the process of ethnic cleansing.

- Ethnic cleansing has been undertaken in recent years in the Balkans.

LEARNING OUTCOME 7.4.2: Explain the concept of ethnic cleansing in the Balkans.

- Balkanization is a process by which a state breaks down through conflicts among its ethnicities.

LEARNING OUTCOME 7.4.3: Identify the principal episodes of genocide in northeastern Africa.

- Genocide is the mass killing of a group of people in an attempt to eliminate the entire group from existence.

LEARNING OUTCOME 7.4.4: Identify the principal episodes of genocide in central Africa.

- Genocide has been practiced in several places in Africa, including Sudan, Somalia, Rwanda, and the Democratic Republic of Congo.

THINKING GEOGRAPHICALLY 7.4: Sarajevo, the capital of Bosnia & Herzegovina, once was home to concentrations of many ethnic groups. In retaliation for ethnic cleansing by the Serbs and Croats, the Bosnian Muslims now in control of Sarajevo have been forcing other ethnic groups to leave the city, and Sarajevo is now inhabited overwhelmingly by Bosnian Muslims. Discuss the challenges in restoring Sarajevo as a multiethnic city.

GOOGLE EARTH 7.4: Gazi Husrev-beg Mosque in Sarajevo, Bosnia & Herzegovina, was heavily damaged during ethnic cleansing and since rebuilt. In ground-level view and 3D, pan around the mosque; what other religious structures are visible in 3D within 500 meters of the mosque?

Race (p. 227) Identity with a group of people descended from a biological ancestor.

Racism (p. 227) Belief that race is the primary determinant of human traits and capacities and that racial differences produce an inherent superiority of a particular race.

Racist (p. 227) A person who subscribes to the beliefs of racism.

Sharecropper (p. 234) A person who works fields rented from a landowner and pays the rent and repays loans by turning over to the landowner a share of the crops.

Triangular slave trade (p. 233) A practice, primarily during the eighteenth century, in which European ships transported slaves from Africa to Caribbean islands, molasses from the Caribbean to Europe, and trade goods from Europe to Africa.

Chapter

8 Political Geography

Why did Morocco build this wall across the Sahara Desert? Page 265

Who lived here? Page 293

KEY ISSUE 1

Where Are States Distributed?

A World of States p. 261

Earth is divided into approximately 200 states. This was not always the case.

KEY ISSUE 2

Why Are Nation-States Difficult to Create?

Nation-States and Multinational States p. 268

Dividing the world into states of single ethnicities has been difficult. States with multiple ethnicities are often in turmoil.

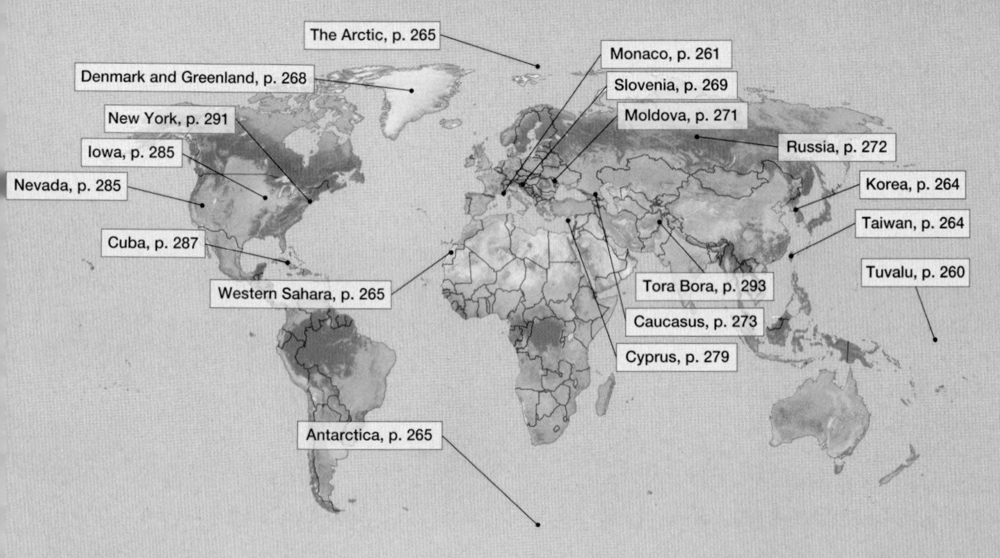

▲ This bicyclist is crossing the Rhine River on the Passerelle Mimram Pedestrian Bridge. He is heading from Strasbourg, France, to Kehl, Germany. France and Germany fought for centuries over control of Strasbourg and the Rhine. These former enemies are now allies, having joined with other European countries to eliminate passport checks and other border controls. Travel between France and Germany is now as easy as travel between two U.S. states.

KEY ISSUE 3

Why Do Boundaries Cause Problems?

Drawing a Line in the Sand (or Somewhere Else) p. 276

Boundaries between states and within states are hard to set and are often controversial.

KEY ISSUE 4

Why Do States Cooperate and Compete with Each Other?

States in War and Peace p. 286

States increasingly cooperate economically, but violence is increasingly led by terrorists.

Introducing
Political Geography

When looking at satellite images of Earth, we easily distinguish ***places***—landmasses and water bodies, mountains and rivers, deserts and fertile agricultural land, urban areas and forests. What we cannot see are where boundaries are located between countries.

To many, national boundaries are more meaningful than natural features. One of Earth's most fundamental cultural characteristics—one that we take for granted—is the division of our planet's surface into a collection of ***spaces*** occupied by individual countries.

During the Cold War (the late 1940s until the early 1990s), two superpowers—the United States and the Soviet Union—essentially "ruled" the world. As superpowers, they competed at a global ***scale***. Many countries belonged to one of two ***regions***, one allied with the former Soviet Union and the other allied with the United States.

With the end of the Cold War in the 1990s, the global political landscape changed fundamentally. In the post–Cold War era, the familiar division of the world into countries or states is crumbling. The United States is less dominant in the political landscape of the twenty-first century, and the Soviet Union no longer exists.

Wars have broken out in recent years—both between small neighboring states and among cultural groups within countries—over political control of territory. Old countries have been broken up into collections of smaller ones, some barely visible on world maps (Figure 8-1).

Geographic concepts help us to understand the altered political organization of Earth's surface. Geographers observe why this familiar division of the world is changing. We can also use geographic methods to examine the causes of political change and instability and to anticipate potential trouble spots around the world.

Today, globalization means more ***connections*** among states. Individual countries have transferred military, economic, and political authority to regional and worldwide collections of states. Power is exercised through connections among states created primarily for economic cooperation.

Despite (or perhaps because of) greater global political cooperation, local diversity has increased in political affairs, as individual cultural groups have demanded more control over the territory they inhabit. States have transferred power to local governments, but this has not placated cultural groups that seek complete independence.

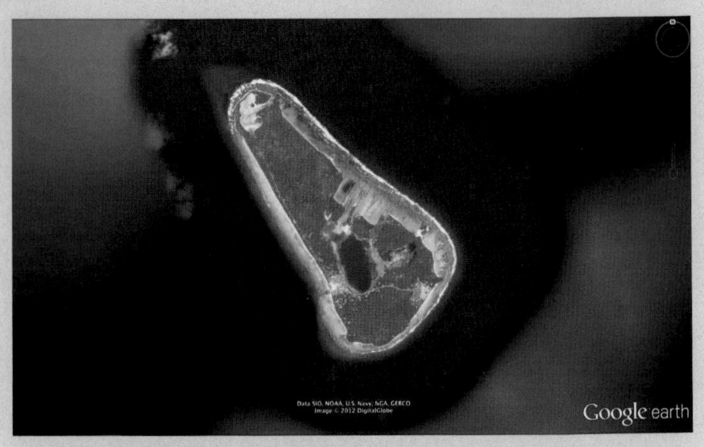

▲ **FIGURE 8-1 TUVALU** The island of Tuvalu, with 10,000 inhabitants, became an independent country in 1978. It is the world's fourth-smallest country.

No one can predict where the next war or terrorist attack will erupt, but political geography helps to explain the cultural and physical factors that underlie political unrest in the world. Political geographers study how people have organized Earth's land surface into countries and alliances, reasons underlying the observed arrangements, and the conflicts that result from the organization.

- **KEY ISSUE 1** describes ***where*** states are distributed. Nearly the entire land area of the world is divided into states, although what constitutes a state is not always clear-cut.
- **KEY ISSUE 2** explains ***why*** states can be difficult to create. ***Local diversity*** has increased in political affairs, as individual cultural groups have demanded more control over the territory they inhabit.
- **KEY ISSUE 3** looks at boundaries between states and within states. Boundary lines are not painted on Earth, but they might as well be, for these national divisions are very real.
- **KEY ISSUE 4** discusses competition and cooperation among states. Political conflicts during the twentieth century were dominated by the ***globalization*** of warfare, including two world wars involving most of the world's states and collections of allied states. Into the twenty-first century, the attacks against the United States on September 11, 2001, were initiated not by a hostile state but by a terrorist organization.

KEY ISSUE 1

Where Are States Distributed?

- A World of States
- Challenges in Defining States
- Development of State Concept

A **state** is an area organized into a political unit and ruled by an established government that has control over its internal and foreign affairs. It occupies a defined territory on Earth's surface and contains a permanent population. The term *country* is a synonym for *state*. A state has **sovereignty**, which means independence from control of its internal affairs by other states. Because the entire area of a state is managed by its national government, laws, army, and leaders, it is a good example of a formal or uniform region.

The term *state*, as used in political geography, does not refer to the 50 regional governments inside the United States. The 50 states of the United States are subdivisions within a single state—the United States of America.

How many of these states can you name? Old-style geography sometimes required memorization of countries and their capitals. Human geographers now emphasize a thematic approach. We are concerned with the location of activities in the world, the reasons for particular spatial distributions, and the significance of the arrangements. Despite this change in emphasis, you still need to know the locations of states. Without such knowledge, you lack a basic frame of reference—knowing where things are.

The land area occupied by the states of the world varies considerably. The largest state is Russia, which encompasses 17.1 million square kilometers (6.6 million square miles), or 11 percent of the world's entire land area. Other states with more than 5 million square kilometers (2 million square miles) include Canada, the United States, China, Brazil, and Australia.

At the other extreme are about two dozen **microstates**, which are states with very small land areas. If Russia were the size of this page, a microstate would be the size of a single letter on it. The smallest microstate in the United Nations—Monaco (Figure 8-2)—encompasses only 1.5 square kilometers (0.6 square miles).

Other UN member states that are smaller than 1,000 square kilometers (400 square miles) include Andorra, Antigua and Barbuda, Bahrain, Barbados, Dominica, Grenada, Kiribati, Liechtenstein, Maldives, Malta, Micronesia, Nauru, Palau, St. Kitts & Nevis, St. Lucia, St. Vincent & the Grenadines, San Marino, São Tomé e Príncipe, the Seychelles, Singapore, Tonga, and Tuvalu (refer to Figure 8-1). Many of the microstates are islands, which explains both their small size and sovereignty.

◀ FIGURE 8-2
MICROSTATE: MONACO The smallest microstate in the United Nations, Monaco is a principality, ruled by a prince.

A World of States

Learning Outcome 8.1.1

Explain the three eras of rapid growth in UN membership.

A map of the world shows that virtually all habitable land belongs to some country or other. But for most of history, until recently, this was not so. As recently as the 1940s, the world contained only about 50 countries, compared to approximately 200 today.

THE UNITED NATIONS

The most important global organization is the United Nations, created at the end of World War II by the victorious Allies. During this era of rapid changes in states and their relationships, the UN has provided a forum for the discussion of international problems. On occasion, the UN has intervened in conflicts between or within member states, authorizing military and peacekeeping actions. In addition, the UN seeks to promote international cooperation to address global economic problems, promote human rights, and provide humanitarian relief.

When it was organized in 1945, the UN had only 51 members, including 49 sovereign states plus Byelorussia (now Belarus) and Ukraine, then part of the Soviet Union (Figure 8-3). The number of UN members reached 193 in 2011.

The UN membership increased rapidly on three occasions (Figure 8-4):

- **1955.** Sixteen countries joined in 1955, mostly European countries that had been liberated from Nazi Germany during World War II.
- **1960.** Seventeen new members were added in 1960, all but one a former African colony of Britain or France. Only four African states were original members of the United Nations—Egypt, Ethiopia, Liberia, and South Africa—and only six more joined during the 1950s.
- **1990–1993.** Twenty-six countries were added between 1990 and 1993, primarily due to the breakup of the Soviet Union and Yugoslavia. UN membership also increased in the 1990s because of the admission of several microstates.

The United Nations was not the world's first attempt at international peacemaking. The UN replaced an earlier organization known as the League of Nations, which was established after World War I. The League of Nations was never an effective peacekeeping organization. The United States did not join it, despite the fact that President Woodrow Wilson initiated the idea, because the U.S. Senate refused to ratify the membership treaty. By the 1930s, Germany, Italy, Japan, and the Soviet Union had all withdrawn, and the League of Nations could not stop aggression by these states against neighboring countries.

UN members can vote to establish a peacekeeping force and request states to contribute military forces. The UN is playing an important role in trying to separate warring groups in a number of regions, especially in Eastern Europe, Central and Southwest Asia, and sub-Saharan Africa. However, any one of the five permanent members of the Security Council—China, France, Russia (formerly the Soviet Union), the United Kingdom, and the United States—can veto a peacekeeping operation. During the

▶ **FIGURE 8-3 UN MEMBERS** Nearly the entire land area of the world part of the UN.

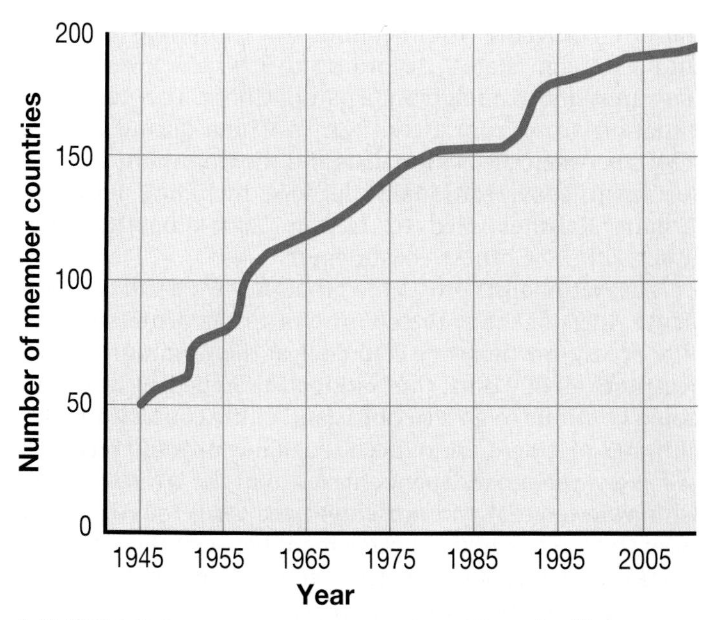

▲ **FIGURE 8-4 GROWTH IN UN MEMBERSHIP** UN membership has increased from 51 to 193.

Cold War era, the United States and the Soviet Union used the veto to prevent undesired UN intervention, and it was only after the Soviet Union's delegate walked out of a Security Council meeting in 1950 that the UN voted to send troops to support South Korea. More recently, the opposition of China and Russia has made it difficult for the international community to prevent Iran from developing nuclear weapons.

Because it must rely on individual countries to supply troops, the UN often lacks enough of them to keep peace effectively. The UN tries to maintain strict neutrality in separating warring factions, but this has proved difficult in places such as Bosnia & Herzegovina, where most of the world sees two ethnicities (Bosnia's Serbs and Croats) as aggressors undertaking ethnic cleansing against weaker victims (Bosnian Muslims). Despite its shortcomings, though, the UN represents a forum where, for the first time in history, virtually all states of the world can meet and vote on issues without resorting to war.

Pause and Reflect 8.1.1
How might UN membership substantially increase in the future beyond the current level?

Challenges in Defining States

Learning Outcome 8.1.2

Explain why it is difficult to determine whether some territories are states.

There is some disagreement about the actual number of sovereign states. This disagreement is closely tied to the history and geography of the places involved and most often involves neighboring states. In some disputes about sovereignty, multiple states lay claim to a territory. Among places that test the definition of a state are Korea, China, Kosovo, Western Sahara (Sahrawi Republic), and the polar regions of Antarctica and the Arctic Ocean.

KOREA: ONE STATE OR TWO?

A colony of Japan for many years, Korea was divided into two occupation zones by the United States and the former Soviet Union after they defeated Japan in World War II (Figure 8-5). The country was divided into northern and southern sections along 38° north latitude. The division of these zones became permanent in the late 1940s, when the two superpowers established separate governments and withdrew their armies. The new government of the Democratic People's Republic of Korea (North Korea) then invaded the Republic of Korea (South Korea) in 1950, touching off a three-year war that ended with a cease-fire line near the 38th parallel.

Both Korean governments are committed to reuniting the country into one sovereign state. Leaders of the two countries agreed in 2000 to allow exchange visits of families separated for a half century by the division and to increase economic cooperation. However, progress toward reconciliation was halted by North Korea's decision to build nuclear weapons, even though the country lacked the ability to provide its citizens with food, electricity, and other basic needs. Meanwhile, in 1992, North Korea and South Korea were admitted to the United Nations as separate countries.

▲ **FIGURE 8-5 NORTH AND SOUTH KOREA** A nighttime satellite image recorded by the U.S. Air Force Defense Meteorological Satellite Program shows the illumination of electric lights in South Korea, whereas North Korea has virtually no electric lights, a measure of its poverty and limited economic activity.

CHINA AND TAIWAN: ONE STATE OR TWO?

Are China and the island of Taiwan two sovereign states or one? Most other countries consider China (officially the People's Republic of China) and Taiwan (officially the Republic of China) as separate and sovereign states. According to China's government, Taiwan is not sovereign but a part of China. This confusing situation arose from a civil war in China during the late 1940s between the Nationalists and the Communists. After losing in 1949, Nationalist leaders, including President Chiang Kai-shek, fled to Taiwan, 200 kilometers (120 miles) off the Chinese coast (Figure 8-6).

The Nationalists proclaimed that they were still the legitimate rulers of the entire country of China. Until some future occasion when they could defeat the Communists and recapture all of China, the Nationalists argued, at least they could continue to govern one island of the country. In 1999 Taiwan's president announced that Taiwan would regard itself as a sovereign independent state, but the government of China viewed that announcement as a dangerous departure from the long-standing arrangement between the two.

The question of who constituted the legitimate government of China plagued U.S. officials during the 1950s and 1960s. The United States had supported the Nationalists during the civil war, so many Americans opposed acknowledging that China was firmly under the control of the Communists. Consequently, the United States continued to regard the Nationalists as the official government of China until 1971, when U.S. policy finally changed and the United Nations voted to transfer China's seat from the Nationalists to the Communists. Taiwan is now the most populous state not in the United Nations.

▲ **FIGURE 8-6 TAIWAN** Taiwanese wave flags at Chiang Kai-shek Memorial Hall in Taipei, Taiwan. The hall is named for the last Nationalist president of mainland China.

▲ **FIGURE 8-7 WESTERN SAHARA** Morocco built sand walls during the 1980s to isolate Polisario Front rebels fighting for independence.

WESTERN SAHARA (SAHRAWI REPUBLIC)

The Sahrawi Arab Democratic Republic, also known as Western Sahara, is considered by most African countries as a sovereign state. Morocco, however, claims the territory and to prove it has built a 2,700-kilometer (1,700-mile) wall around the territory to keep out rebels (Figure 8-7).

Spain controlled the territory on the continent's west coast between Morocco and Mauritania until withdrawing in 1976. An independent Sahrawi Republic was declared by the Polisario Front and recognized by most African countries, but Morocco and Mauritania annexed the northern and southern portions, respectively. Three years later Mauritania withdrew, and Morocco claimed the entire territory.

Morocco controls most of the populated area, but the Polisario Front operates in the vast, sparsely inhabited deserts, especially the one-fifth of the territory that lies east of Morocco's wall. The United Nations has tried but failed to reach a resolution among the parties.

POLAR REGIONS: MANY CLAIMS

The South Pole region contains the only large landmasses on Earth's surface that are not part of a state. Several states claim portions of the region, and some claims are overlapping and conflicting.

Several states, including Argentina, Australia, Chile, France, New Zealand, Norway, and the United Kingdom, claim portions of Antarctica (Figure 8-8). Argentina, Chile, and the United Kingdom have made conflicting, overlapping claims. The United States, Russia, and a number of other states do not recognize the claims of any country to Antarctica. The Antarctic Treaty, signed in 1959 by 47 states, provides a legal framework for managing Antarctica. States may establish research stations there for scientific investigations, but no military activities are permitted.

As for the Arctic, the 1982 United Nations Convention on the Law of the Sea permitted countries to submit claims inside the Arctic Circle by 2009 (Figure 8-9). The Arctic region is thought to be rich in energy resources.

Pause and Reflect 8.1.2

The polar ice caps are receding with the warming of Earth. How might this affect competing territorial claims?

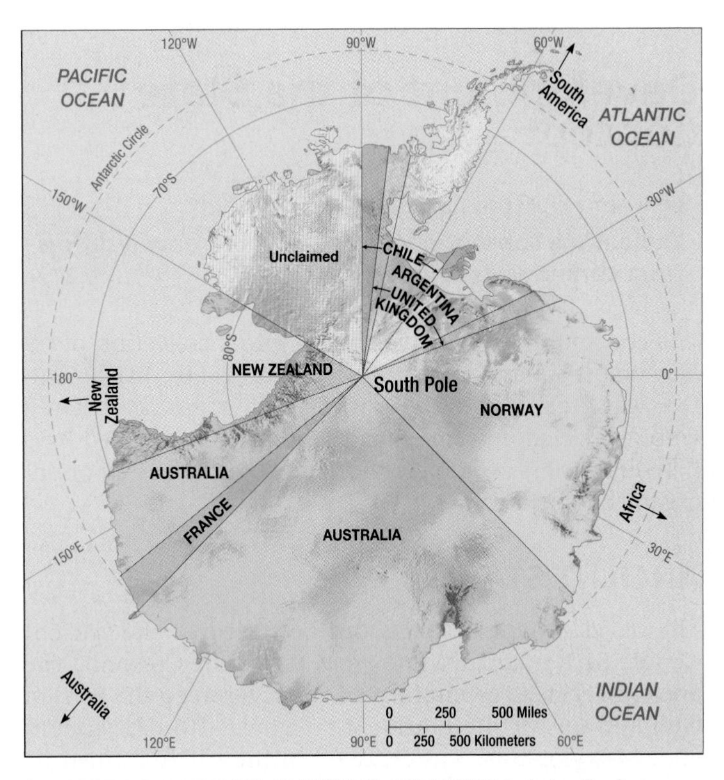

▲ **FIGURE 8-8 NATIONAL CLAIMS TO ANTARCTICA** Antarctica is the only large landmass in the world that is not part of a sovereign state. It comprises 14 million square kilometers (5.4 million square miles), which makes it 50 percent larger than Canada. Portions are claimed by Argentina, Australia, Chile, France, New Zealand, Norway, and the United Kingdom; claims by Argentina, Chile, and the United Kingdom are conflicting.

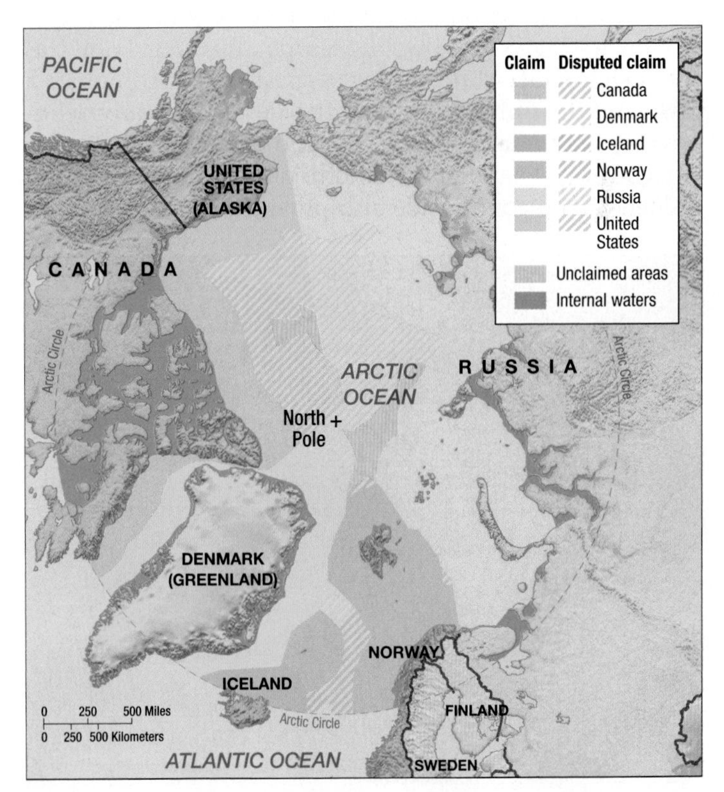

▲ **FIGURE 8-9 NATIONAL CLAIMS TO THE ARCTIC** Under the Law of the Sea Treaty of 1982, countries had until 2009 to submit claims to territory inside the Arctic Circle. Some of these claims overlap.

Development of the State Concept

Learning Outcome 8.1.3
Explain the concept of nation-state and how it differs from earlier ways to govern.

The concept of dividing the world into a collection of independent states is recent. Prior to the 1800s, Earth's surface was organized in other ways, such as into city-states, empires, kingdoms, and small land areas controlled by a hereditary class of nobles, and much of it consisted of unorganized territory.

ANCIENT STATES

The development of states can be traced to the ancient Middle East, in an area known as the Fertile Crescent. The ancient Fertile Crescent formed an arc between the Persian Gulf and the Mediterranean Sea (Figure 8-10). The eastern end, Mesopotamia, was centered in the valley formed by the Tigris and Euphrates rivers, in present-day Iraq. The Fertile Crescent then curved westward over the desert, turning southward to encompass the Mediterranean coast through present-day Syria, Lebanon, and Israel. The Nile River valley of Egypt is sometimes regarded as an extension of the Fertile Crescent. Situated at the crossroads of Europe, Asia, and Africa, the Fertile Crescent was a center for land and sea communications in ancient times.

The first states to evolve in Mesopotamia were known as city-states. A **city-state** is a sovereign state that comprises a town and the surrounding countryside. Walls clearly delineated the boundaries of the city, and outside the walls the city controlled agricultural land to produce food for urban residents. The countryside also provided the city with an outer line of defense against attack by other city-states. Periodically, one city or tribe in Mesopotamia would gain military dominance over the others and form an empire. Mesopotamia was organized into a succession of empires by the Sumerians, Assyrians, Babylonians, and Persians.

Pause and Reflect 8.1.3
What is the importance of the Fertile Crescent in the development of religions, as discussed in Chapter 6? How do you think the development of ancient states and religions in the region are related?

MEDIEVAL STATES

Political unity in the ancient world reached its height with the establishment of the Roman Empire, which controlled most of Europe, North Africa, and Southwest Asia, from modern-day Spain to Iran and from Egypt to England (Figure 8-11). At its maximum extent, the empire comprised 38 provinces, each using the same set of laws that had been created in Rome. Massive walls helped the Roman army defend many of the empire's frontiers.

The Roman Empire collapsed in the fifth century, after a series of attacks by people living on its frontiers and because of internal disputes. The European portion of the Roman Empire was fragmented into a large number of estates owned by competing kings, dukes, barons, and other nobles.

A handful of powerful kings emerged as rulers over large numbers of these European estates beginning about the year 1100. The consolidation of neighboring estates under the unified control of a king formed the basis for the development of such modern European states as England, France, and Spain (Figure 8-12). Much of Europe consolidated into a handful of empires, including Austrian, French, Ottoman, and Russian (Figure 8-13, top).

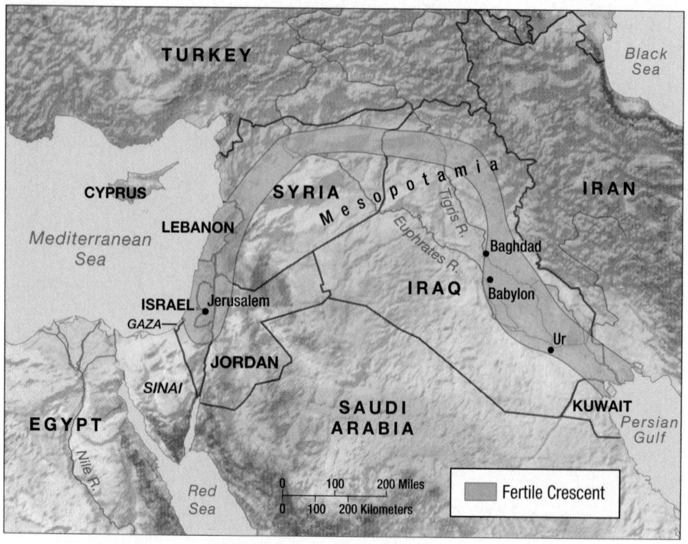

▲ **FIGURE 8-10 THE FERTILE CRESCENT** The crescent-shaped area of relatively fertile land was organized into a succession of empires starting several thousand years ago.

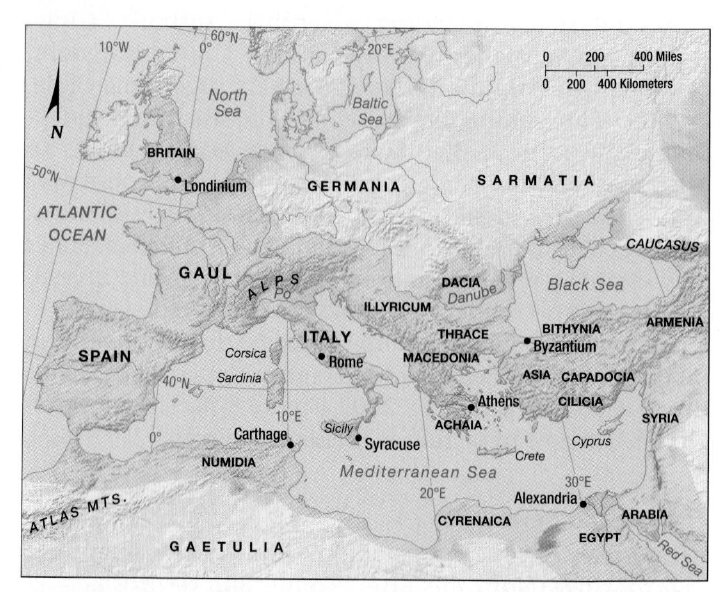

▲ **FIGURE 8-11 ROMAN EMPIRE, A.D. 100** At its height, the Roman Empire controlled much of Europe and Southwest Asia & North Africa.

▲ **FIGURE 8-12 EUROPE, 1300** Much of Europe was fragmented into small estates controlled by nobles.

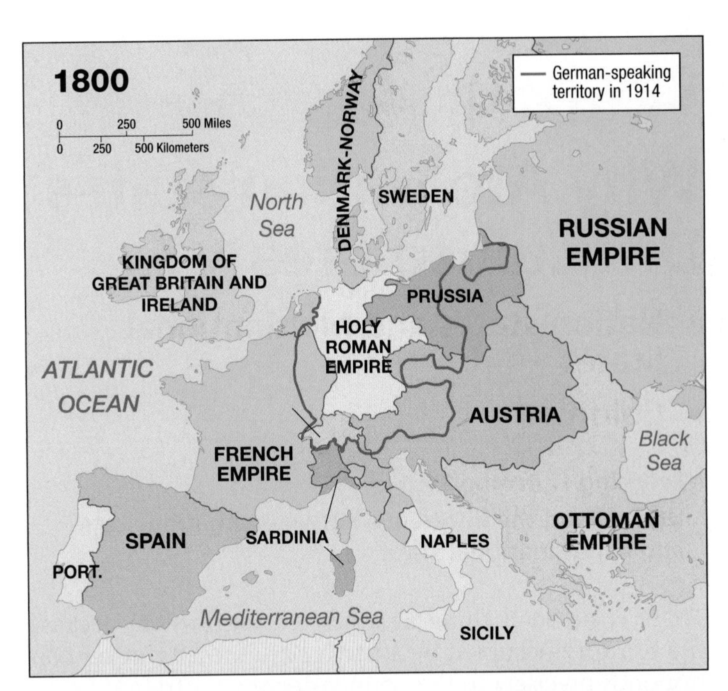

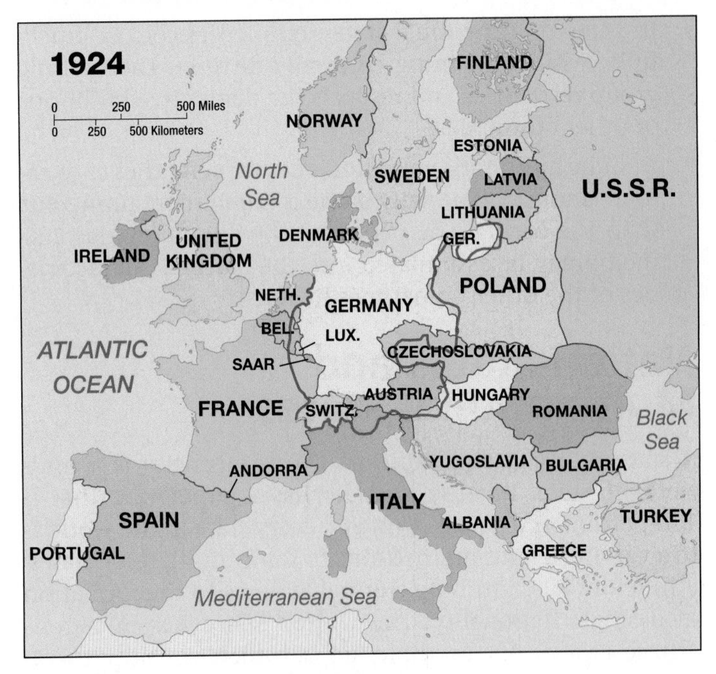

▲ **FIGURE 8-13 NATION-STATES IN EUROPE, 1800 AND 1924** (Top) In 1800, much of Europe was organized into empires. (bottom) After World War I, much of Europe was organized into nation-states.

NATION-STATES IN EUROPE

To preserve and enhance distinctive cultural characteristics, ethnicities seek to govern themselves without interference. A **nation-state** is a state whose territory corresponds to that occupied by a particular ethnicity. Ethnic groups have pushed to create nation-states because desire for self-rule is a very important shared attitude for many of them. The concept that ethnicities have the right to govern themselves is known as **self-determination**.

Some ethnicities were able to form nation-states in Europe during the nineteenth century, and by the early twentieth century most of Western Europe was made up of nation-states (Figure 8-13, bottom).

The movement to identify nationalities on the basis of language spread elsewhere in Europe during the twentieth century. After World War I, leaders of the victorious countries met at the Versailles Peace Conference to redraw the map of Europe. One of the chief advisers to President Woodrow Wilson, the geographer Isaiah Bowman, played a major role in the decisions. Language was the most important criterion the Allied leaders used to create new states in Europe and to adjust the boundaries of existing ones.

During the 1930s, German National Socialists (Nazis) claimed that all German-speaking parts of Europe constituted one nationality and should be unified into one state. After it was defeated in World War II, Germany was divided into two countries ((refer ahead to Figure 8-43). Two Germanys existed from 1949 until 1990.

With the end of communism, the German Democratic Republic ceased to exist, and its territory became part of the German Federal Republic. The present-day state of Germany, though, bears little resemblance to the territory occupied by German-speaking people prior to the upheavals of the twentieth century.

CHECK-IN: KEY ISSUE 1

Where Are States Distributed?

- ✓ The world is divided into approximately 200 states, all but a handful of which are members of the United Nations.
- ✓ It is not always clear-cut whether a territory can be defined as a single state.
- ✓ Organizing Earth into nation-states is a recent concept; other methods of controlling territory prevailed in the past.

KEY ISSUE 2

Why Are Nation-states Difficult to Create?

- Nation-states and Multinational States
- Colonies

Learning Outcome 8.2.1
Understand the difference between a nation-state and a multinational state.

There is no such thing as a perfect nation-state because the territory occupied by a particular ethnicity never corresponds precisely to the boundaries of countries:

- In some multinational states, ethnicities coexist peacefully, while remaining culturally distinct. Each ethnic group recognizes and respects the distinctive traditions of other ethnicities.
- In some multinational states, one ethnicity tries to dominate another, especially if one is much more numerous than the others. The people of the less numerous ethnicity may be assimilated into the cultural characteristics of the other, sometimes by force.

Nation-states and Multinational States

A state that contains more than one ethnicity is a **multiethnic** state. Because no state has a population that is 100 percent of a single ethnicity, every state in the world is to a varying degree multiethnic. In some multiethnic states, ethnicities all contribute cultural features to the formation of a single nationality. The United States has numerous ethnic groups, for example, all of which consider themselves as belonging to the American nationality.

A **multinational state** is a country that contains more than one ethnicity with traditions of self-determination. The Soviet Union was an especially prominent example of a multinational state until its collapse in the early 1990s. Russia, which comprised the largest portion of the Soviet Union, is now the world's largest multinational state. Relationships among ethnicities vary in multinational states.

NATION-STATES IN EUROPE

Two relatively clear examples of nation-states are Denmark and Slovenia, yet even these two are not perfect examples.

DENMARK. Ninety percent of the population of Denmark consists of ethnic Danes. The Danes have a strong sense of unity that derives from shared cultural characteristics and attitudes and a recorded history that extends back more than 1,000 years. Nearly all Danes speak the same language—Danish—and nearly all the world's speakers of Danish live in Denmark.

▲ **FIGURE 8-14 DENMARK AND GREENLAND** Greenland's official language is now Greenlandic, an Inuit language. Greenland is now officially known as Kalaallit Nunaat, and the capital city was changed from Godthaab to Nuuk.

However, 10 percent of Denmark's population consists of ethnic minorities. The two largest groups are guest workers from Turkey and refugees from ethnic cleansing in the former Yugoslavia. Further diluting the concept of a nation-state, Denmark controls two territories where few ethnic Danes live (Figure 8-14):

- Faeroe Islands, a group of 21 islands, has been ruled by Denmark for more than 600 years. The nearly 50,000 inhabitants of the Faeroe Islands speak Faeroese.
- Greenland, the world's largest island, is controlled by Denmark. Only 12 percent of Greenland's 58,000 residents are considered Danish; the remainder are native-born Greenlanders, primarily Inuit. Greenlanders control most of their own domestic affairs.

SLOVENIA. Slovenia was a republic within Yugoslavia that became an independent country in 1991 (Figure 8-15). Slovenes comprise 83 percent of the population of Slovenia, and nearly all the world's 2 million Slovenes live in Slovenia. The relatively close coincidence between the boundaries of the Slovene ethnic group and the country of Slovenia has promoted the country's relative peace and stability, compared to other former Yugoslavian republics, as discussed in Chapter 7.

A census in 1948 showed that Slovenes comprised 97 percent of Slovenia's population. The percentage has declined steadily since then. When it was part of Yugoslavia, Slovenia was the most prosperous republic, and it attracted migrants from other republics. Many of them remained in

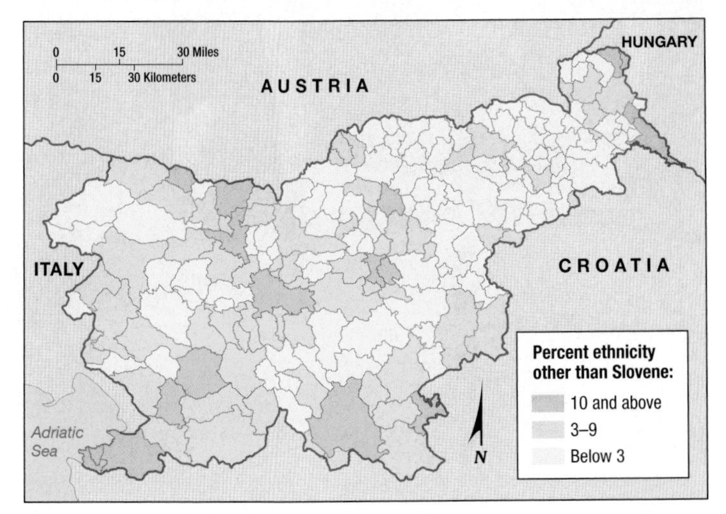

▲ **FIGURE 8-15 SLOVENIA** The percentage of ethnicities other than Slovene is higher in localities bordering neighboring countries, especially Hungary and Italy.

Slovenia after the country became independent. Slovenia's 90-member National Assembly reserves one seat each for Hungarian and Italian ethnic groups living in Slovenia. The province of Italy bordering Slovenia has a population that is approximately one-fifth Slovene. Boundary changes after World War II resulted in a number of Slovenes living in Italy and Italians living in Slovenia.

Pause and Reflect 8.2.1
Referring to Figure 7-40, where do the boundaries of Slovenia not match language boundaries?

NATION-STATES AND ETHNIC IDENTITY. Europeans thought that ethnicity had been left behind as an insignificant relic, such as wearing quaint costumes to amuse tourists. Karl Marx wrote that nationalism was a means for the dominant social classes to maintain power over workers, and he believed that workers would identify with other working-class people instead of with an ethnicity.

In the twenty-first century, ethnic identity has once again become important in the creation of nation-states in much of Europe. The breakup of the Soviet Union, Yugoslavia, and Czechoslovakia during the 1990s gave more-numerous ethnicities the opportunity to organize nation-states. But the less-numerous ethnicities found themselves existing as minorities in multinational states or divided among more than one of the new states. Especially severe problems have occurred in the Balkans, a rugged, mountainous region where nation-states could not be delineated peacefully.

Until they lost power around 1990, Communist leaders in Eastern Europe and the former Soviet Union used centripetal forces to discourage ethnicities from expressing their cultural uniqueness. Writers and artists were pressured to conform to a style known as "socialist realism," which emphasized Communist economic and political values. Use of the Russian language was promoted as a centripetal device throughout the former Soviet Union. It was taught as the second language in other Eastern European countries. The role of organized religion was minimized, suppressing a cultural force that competed with the government (Figure 8-16).

The Soviet Union, Yugoslavia, and Czechoslovakia were dismantled largely because minority ethnicities opposed the long-standing dominance of the most numerous ones in each country—Russians in the Soviet Union, Serbs in Yugoslavia, and Czechs in Czechoslovakia. The dominance was pervasive, including economic, political, and cultural institutions. No longer content to control a province or some other local government unit, ethnicities sought to be the majority in completely independent nation-states. Republics that once constituted local government units within the Soviet Union, Yugoslavia, and Czechoslovakia generally made peaceful transitions into independent countries—as long as their boundaries corresponded reasonably well with the territory occupied by a clearly defined ethnicity.

▲ **FIGURE 8-16 COMMUNIST ART** After the Communists took over Czechoslovakia in 1948, they altered this sixteenth-century clock in the city of Olomouc to conform to socialist realism art. Because they discouraged religion, the Communists removed the statues of the 12 apostles and replaced them with statues of workers.

INDEPENDENT NATION-STATES IN FORMER SOVIET REPUBLICS

Learning Outcome 8.2.2
Describe differences among states formerly in the Soviet Union.

For decades, the many ethnicities within the Soviet Union were unable to realize their nationalist aspirations and form independent nation-states. The Soviet Union consisted of 15 republics, based on its 15 largest ethnicities (Figure 8-17). The 15 republics that once constituted the Soviet Union are now independent states. These 15 states consist of five groups:

- Three Baltic states: Estonia, Latvia, and Lithuania
- Three European states: Belarus, Moldova, and Ukraine
- Five Central Asian states: Kazakhstan, Kyrgyzstan, Tajikistan, Turkmenistan, and Uzbekistan
- Three Caucasus states: Azerbaijan, Armenia, and Georgia
- Russia

Reasonably good examples of nation-states have been carved out of the Baltic, European, and some Central Asian states. On the other hand, peaceful nation-states have not been created in any of the small Caucasus states, and Russia is an especially prominent example of a state with major difficulties in keeping all its ethnicities contented. With the breakup of the Soviet Union into 15 independent countries, a number of these less-numerous ethnicities are now divided among these states.

▲ **FIGURE 8-17 STATES IN THE FORMER U.S.S.R.** The Union of Soviet Socialist Republics included 15 republics, named for the country's largest ethnicities. With the breakup of the Soviet Union, the 15 republics became independent states.

▲ **FIGURE 8-18 BORDER CROSSING BETWEEN ESTONIA AND LATVIA** This is the border between the towns of Valga, Estonia, and Valka, Latvia.

BALTIC STATES. Estonia, Latvia, and Lithuania are known as the Baltic states for their location on the Baltic Sea. They were independent countries between the end of World War I in 1918 and 1940, when the former Soviet Union annexed them under an agreement with Nazi Germany. These three small neighboring Baltic countries have clear cultural differences and distinct historical traditions:

- **Lithuania.** Of the three Baltic states, Lithuania most closely fits the definition of a nation-state because ethnic Lithuanians comprise 85 percent of its population. Most Lithuanians are Roman Catholic and speak a language of the Baltic group within the Balto-Slavic branch of the Indo-European language family (Figure 8-18).
- **Estonians.** In Estonia, ethnic Estonians comprise only 69 percent of the population. Most Estonians are Protestant (Lutheran) and speak a Uralic language related to Finnish.
- **Latvians.** In Latvia, only 59 percent are ethnic Latvians. Latvians are predominantly Lutheran, with a substantial Roman Catholic minority, and they speak a language of the Baltic group.

EUROPEAN STATES. To some extent, the former Soviet republics of Belarus, Moldova, and Ukraine now qualify as nation-states. Belarusians comprise 81 percent of the population of Belarus, Moldovans comprise 78 percent of the population of Moldova, and Ukrainians comprise 78 percent of the population of Ukraine. The ethnic distinctions among Belarusians, Ukrainians, and Russians are somewhat blurred. The three groups speak similar East Slavic languages, and all are predominantly Orthodox Christians (some western Ukrainians are Roman Catholics):

- **Belarus and Ukraine.** Belarusians and Ukrainians became distinct ethnicities because they were isolated from the main body of Eastern Slavs—the Russians—during the thirteenth and fourteenth centuries. This was the consequence of Mongolian invasions and conquests by Poles and Lithuanians. Russians conquered the Belarusian and Ukrainian homelands in the late 1700s, but after five centuries of exposure to non-Slavic influences, the three Eastern Slavic groups displayed

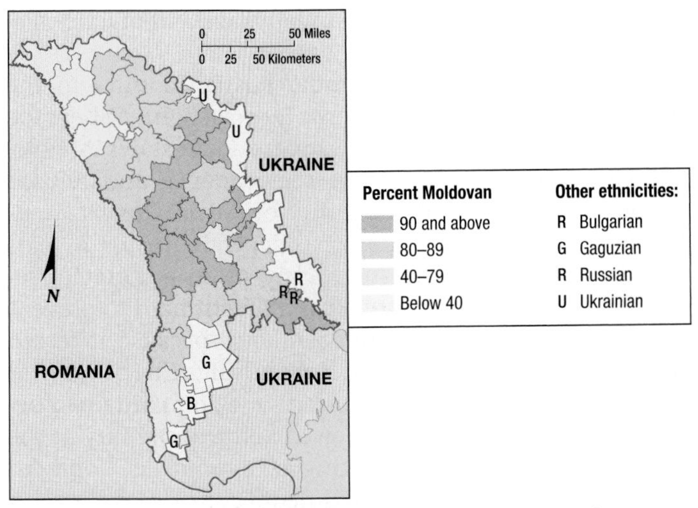

▲ **FIGURE 8-19 ETHNICITIES IN MOLDOVA** Ethnicities other than Moldovan predominate in the eastern portion of the country.

sufficient cultural diversity to consider themselves three distinct ethnicities.

- **Moldova.** Moldovans are ethnically indistinguishable from Romanians, and Moldova (then called Moldavia) was part of Romania until the Soviet Union seized it in 1940. When Moldova changed from a Soviet republic back to an independent country in 1992, many Moldovans pushed for reunification with Romania, both to reunify the ethnic group and to improve the region's prospects for economic development. But it was not to be that simple. When Moldova became a Soviet republic in 1940, its eastern boundary was the Dniester River. The Soviet government increased the size of Moldova by about 10 percent, transferring from Ukraine a 3,000-square-kilometer (1,200-square-mile) sliver of land on the east bank of the Dniester. The majority of the inhabitants of this area, known as Trans-Dniestria, are Ukrainian and Russian. They, of course, oppose Moldova's reunification with Romania (Figure 8-19).

Pause and Reflect 8.2.2
To what branches of Indo-European do the principal languages of Moldova belong? How might these linguistic differences affect politics in Moldova?

CENTRAL ASIAN STATES. The five states in Central Asia carved out of the former Soviet Union display varying degrees of conformance to the principles of a nation-state (Figure 8-20). Together the five provide an important reminder that multinational states can be more peaceful than nation-states:

- **Turkmenistan and Uzbekistan.** In Turkmenistan and Uzbekistan, the leading ethnic group has an overwhelming majority—85 percent Turkmen and 80 percent Uzbek, respectively. Both ethnic groups are Muslims who speak an Altaic language; they were conquered by Russia in the nineteenth century. Turkmen and Uzbeks are examples of ethnicities split into more than one country—Turkmen between Turkmenistan and Russia, and Uzbeks among Kyrgyzstan, Tajikistan, and Uzbekistan.
- **Kyrgyzstan.** Kyrgyzstan is 69 percent Kyrgyz, 15 percent Uzbek, and 9 percent Russian. The Kyrgyz—also Muslims who speak an Altaic language—resent the Russians for seizing the best farmland when they colonized this mountainous country early in the twentieth century.
- **Kazakhstan.** In principle, Kazakhstan, twice as large as the other four Central Asian countries combined, is a recipe for ethnic conflict. The country is divided between Kazakhs, who comprise 67 percent of the population, and Russians, at 18 percent. Kazakhs are Muslims who speak an Altaic language similar to Turkish, whereas the Russians are Orthodox Christians who speak an Indo-European language. Tensions exist between the two groups, but Kazakhstan has been peaceful, in part because it has a somewhat less depressed economy than its neighbors.
- **Tajikistan.** In contrast to Kazakhstan, Tajikistan—80 percent Tajik, 15 percent Uzbek, and 1 percent Russian—would appear to be a stable country, but it suffers from a civil war among the Tajik people, Muslims who speak a language in the Indic group of the Indo-Iranian branch of Indo-European language. The civil war has been between Tajiks, who are former Communists, and an unusual alliance of Muslim fundamentalists and Western-oriented intellectuals. Fifteen percent of the population has been made homeless by the fighting.

▲ **FIGURE 8-20 ETHNICITIES IN CENTRAL ASIA** The map shows the distribution of ethnicities in Central Asia.

THE LARGEST MULTINATIONAL STATE: RUSSIA

Learning Outcome 8.2.3

Describe patterns of distribution of ethnicities in Russia and the Caucasus.

Multinational states face complex challenges in maintaining unity and avoiding fragmentation as discontented ethnicities seek to break away and form new nation-states. Russia officially recognizes the existence of 39 ethnic groups as nationalities, many of which are eager for independence. Russia's ethnicities are clustered in two principal locations (Figure 8-21). Some are located along borders with neighboring states, including Buryats and Tuvinian near Mongolia, and Chechens, Dagestani, Kabardins, and Ossetians near the two former Soviet republics of Azerbaijan and Georgia. Overall, 20 percent of the country's population is non-Russian.

Other ethnicities are clustered in the center of Russia, especially between the Volga River basin and the Ural Mountains. Among the most numerous in this region are Bashkirs, Chuvash, and Tatars, who speak Altaic languages similar to Turkish, and Mordvins and Udmurts, who speak Uralic languages similar to Finnish. Most of these groups were conquered by the Russians in the sixteenth century, under the leadership of Ivan IV (Ivan the Terrible).

Independence movements are flourishing because Russia is less willing to suppress these movements forcibly than the Soviet Union once was. Particularly troublesome for the Russians are the Chechens, a group of Sunni Muslims who speak a Caucasian language and practice distinctive social customs.

Chechnya was brought under Russian control in the nineteenth century only after a 50-year fight. When the Soviet Union broke up into 15 independent states in 1991, the Chechens declared their independence and refused to join the newly created country of Russia. Russian leaders ignored the declaration of independence for 3 years but then sent in the Russian army in an attempt to regain control of the territory. Russia fought hard to prevent Chechnya from gaining independence because it feared that other ethnicities would follow suit. Chechnya was also important to Russia because the region contained deposits of petroleum. Russia viewed political stability in the area as essential for promoting economic development and investment by foreign petroleum companies.

TURMOIL IN THE CAUCASUS

The Caucasus region, an area about the size of Colorado, is situated between the Black and Caspian seas and gets its name from the mountains that separate Russia from Azerbaijan and Georgia. The region is home to several ethnicities, with Azeris, Armenians, and Georgians the most numerous (Figure 8-22). Other important ethnicities include Abkhazians, Chechens, Ingush, and Ossetians. Kurds and Russians—two ethnicities that are more numerous in other regions—are also represented in the Caucasus.

When the entire Caucasus region was part of the Soviet Union, the Soviet government promoted allegiance to communism and the Soviet state and quelled disputes among ethnicities, by force if necessary. With the breakup

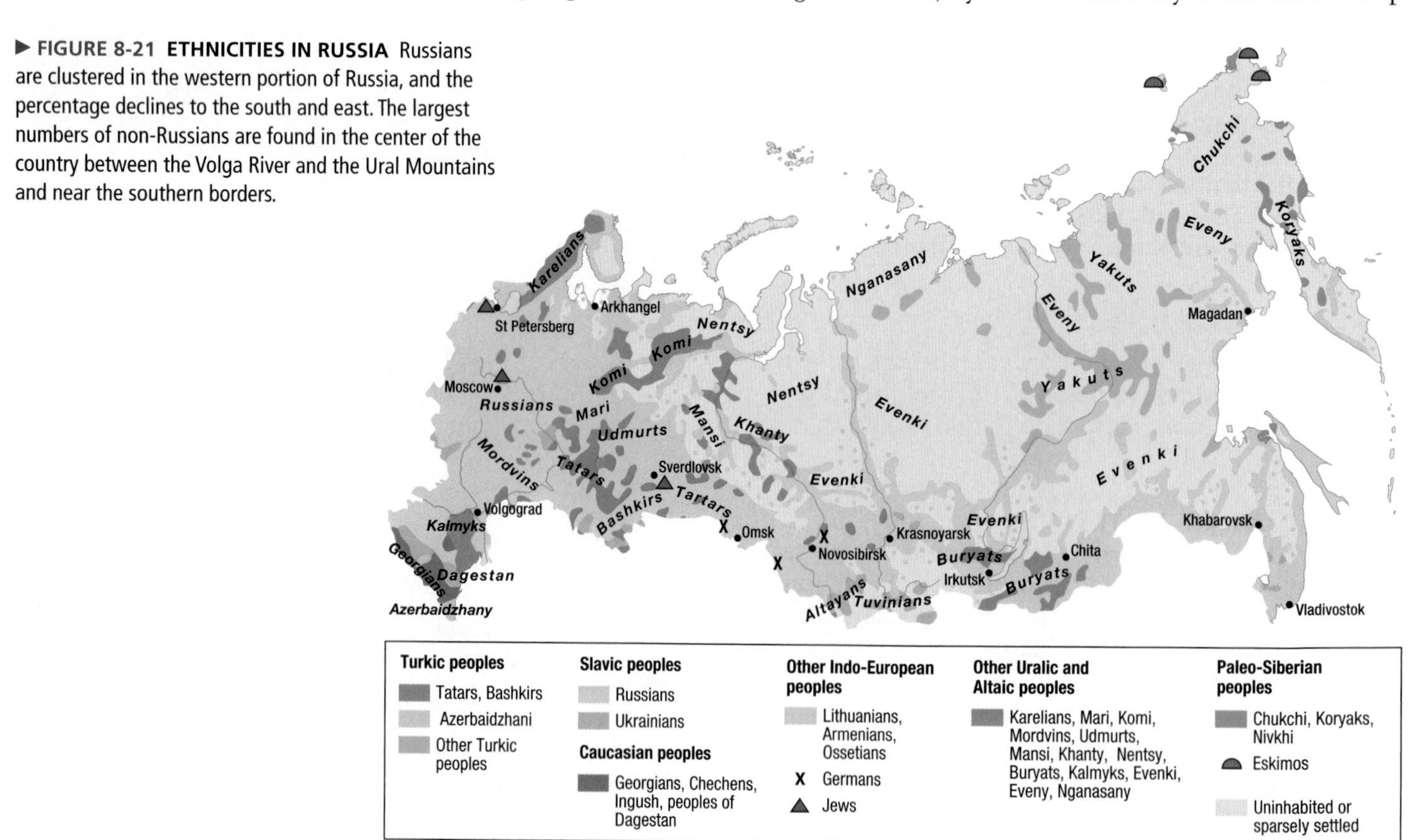

▶ **FIGURE 8-21 ETHNICITIES IN RUSSIA** Russians are clustered in the western portion of Russia, and the percentage declines to the south and east. The largest numbers of non-Russians are found in the center of the country between the Volga River and the Ural Mountains and near the southern borders.

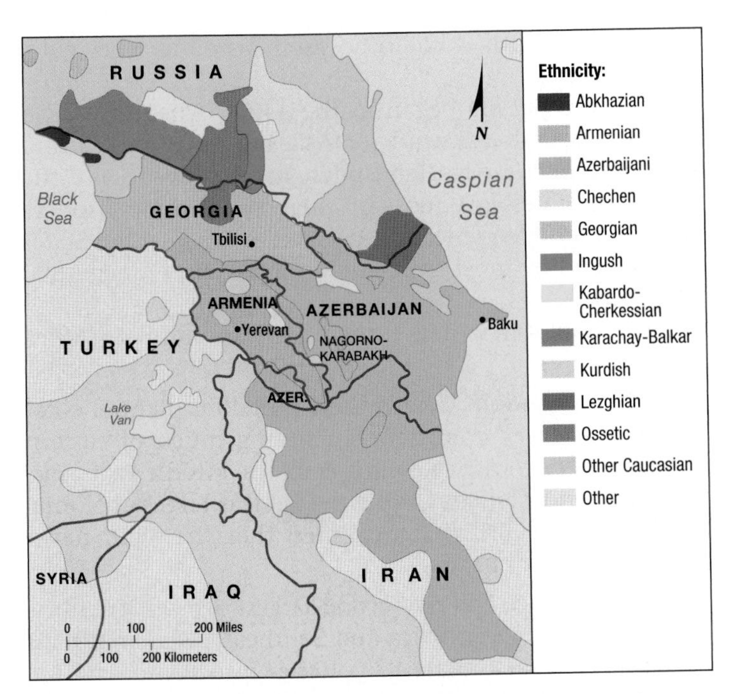

▲ **FIGURE 8-22 ETHNICITIES IN THE CAUCASUS** Armenians, Azeris, and Georgians are examples of ethnicities that were able to dominate new states during the 1990s, following the breakup of the Soviet Union. But the boundaries of the states of Armenia, Azerbaijan, and Georgia do not match the territories occupied by the Armenian, Azeri, and Georgian ethnicities. The Abkhazians, Chechens, Kurds, and Ossetians are examples of ethnicities in this region that have not been able to organize nation-states.

of the region into several independent countries, long-simmering conflicts among ethnicities have erupted into armed conflicts. Each ethnicity has a long-standing and complex set of grievances against others in the region. But from a political geography perspective, every ethnicity in the Caucasus has the same aspiration: to carve out a sovereign nation-state. The region's ethnicities have had varying degrees of success in achieving this objective, but none have fully achieved it.

AZERBAIJAN. Azeris (or Azerbaijanis) trace their roots to Turkish invaders who migrated from Central Asia in the eighth and ninth centuries and merged with the existing Persian population. An 1828 treaty allocated northern Azeri territory to Russia and southern Azeri territory to Persia (now Iran). In 1923, the Russian portion became the Azerbaijan Soviet Socialist Republic within the Soviet Union.

With the Soviet Union's breakup in 1991, Azerbaijan again became an independent country. The western part of the country, Nakhichevan (named for the area's largest city), is separated from the rest of Azerbaijan by a 40-kilometer (25-mile) corridor that belongs to Armenia.

More than 7 million Azeris now live in Azerbaijan, 91 percent of the country's total population. Another 16 million Azeris are clustered in northwestern Iran, where they constitute 24 percent of that country's population. Azeris hold positions of responsibility in Iran's government and economy, but Iran restricts teaching of the Azeri language.

ARMENIA. More than 3,000 years ago Armenians controlled an independent kingdom in the Caucasus. Converted to Christianity in 303, they lived for many centuries as an isolated Christian enclave under the rule of Turkish Muslims.

During the late nineteenth and early twentieth centuries, hundreds of thousands of Armenians were killed in a series of massacres organized by the Turks. Others were forced to migrate to Russia, which had gained possession of eastern Armenia in 1828.

After World War I the Allies created an independent state of Armenia, but it was soon swallowed by its neighbors. In 1921, Turkey and the Soviet Union agreed to divide Armenia between them. The Soviet portion became the Armenian Soviet Socialist Republic and then an independent country in 1991. Armenians comprise 98 percent of the population in Armenia, making it the most ethnically homogeneous country in the region.

Armenians and Azeris both have achieved long-held aspirations of forming nation-states, but after their independence from the Soviet Union the two went to war over the boundaries between them. The war concerned possession of Nagorno-Karabakh, a 5,000-square-kilometer (2,000-square-mile) enclave within Azerbaijan that is inhabited primarily by Armenians but placed under Azerbaijan's control by the Soviet Union during the 1920s. A 1994 cease-fire has left Nagorno-Karabakh technically part of Azerbaijan, but in reality it acts as an independent republic called Artsakh. Numerous clashes have occurred since then between Armenia and Azerbaijan.

GEORGIA. The population of Georgia is more diverse than that in Armenia and Azerbaijan. Ethnic Georgians comprise 71 percent of the population. The country also includes about 8 percent Armenian, 6 percent each Azeri and Russian, 3 percent Ossetian, and 2 percent each Abkhazian, Greek, and other ethnicities.

Georgia's cultural diversity has been a source of unrest, especially among the Ossetians and Abkhazians. During the 1990s, the Abkhazians fought for control of the northwestern portion of Georgia and have declared Abkhazia to be an independent state. In 2008, the Ossetians fought a war with the Georgians that resulted in the Ossetians declaring the South Ossetia portion of Georgia to be independent.

Russia has recognized Abkhazia and South Ossetia as independent countries and has sent troops there. Only a handful of other countries recognize the independence of Abkhazia and South Ossetia, although the two operate as if they were independent of Georgia.

Pause and Reflect 8.2.3

If Abkhazia, Artsakh, and South Ossetia were widely recognized independent states, how would they compare in size to microstates described earlier in this chapter?

Colonies

Learning Outcome 8.2.4

Explain the concept of colonies and describe their current distribution.

Although we live in an era when state creation has been a frequent phenomenon, some territories remain that have not achieved self-determination and statehood. A **colony** is a territory that is legally tied to a sovereign state rather than being completely independent. In some cases, a sovereign state runs only the colony's military and foreign policy. In others, it also controls the colony's internal affairs.

COLONIALISM

European states came to control much of the world through **colonialism**, which is an effort by one country to establish settlements in a territory and to impose its political, economic, and cultural principles on that territory (Figure 8-23). European states established colonies elsewhere in the world for three basic reasons:

- To promote Christianity.
- To extract useful resources and to serve as captive markets for their products.
- To establish relative power through the number of their colonies.

These three motives could be summarized as God, gold, and glory.

The colonial era began in the 1400s, when European explorers sailed westward for Asia but encountered and settled in the Western Hemisphere instead. Eventually, the European states lost most of their Western Hemisphere colonies: Independence was declared by the United States in 1776 and by most Latin American states between 1800 and 1824.

European states then turned their attention to Africa and Asia

- **United Kingdom.** The United Kingdom planted colonies on every continent, including much of eastern and southern Africa, South Asia, the Middle East, Australia, and Canada. With by far the largest colonial empire, the British proclaimed that the "Sun never set" on their empire.
- **France.** France had the second-largest overseas territory, primarily in West Africa and Southeast Asia. France attempted to assimilate its colonies into French culture and educate an elite group to provide local administrative leadership. After independence, most of these leaders retained close ties with France.

Most African and Asian colonies became independent after World War II. Only 15 African and Asian states were members of the United Nations when it was established in 1945, compared to 106 in 2012. The boundaries of the new states frequently coincide with former colonial provinces, although not always.

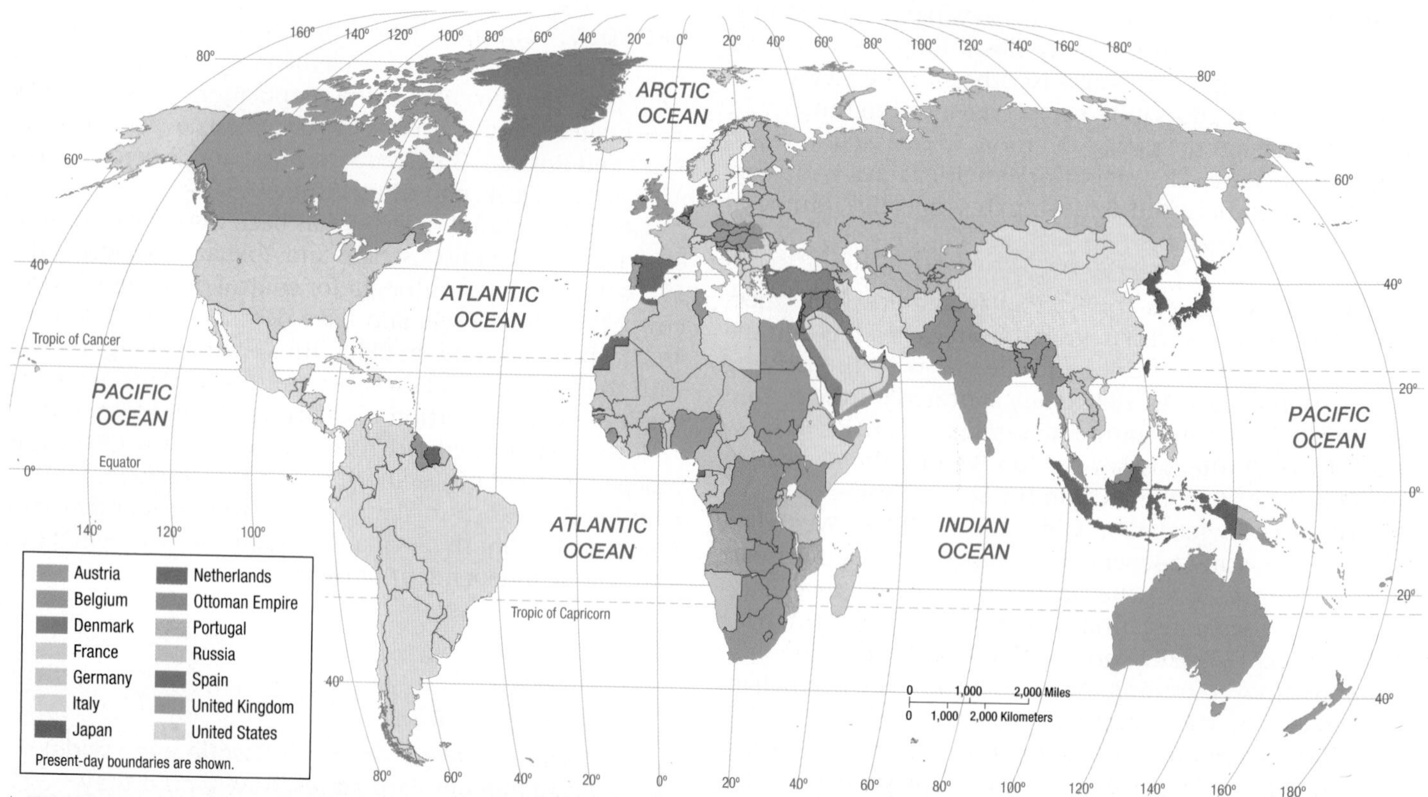

▲ **FIGURE 8-23 COLONIAL POSSESSIONS, 1914** At the outbreak of World War I in 1914, European states held colonies in much of the world, especially in Africa and Asia. Most of the countries in the Western Hemisphere were at one time colonized by Europeans but gained their independence in the eighteenth or nineteenth centuries.

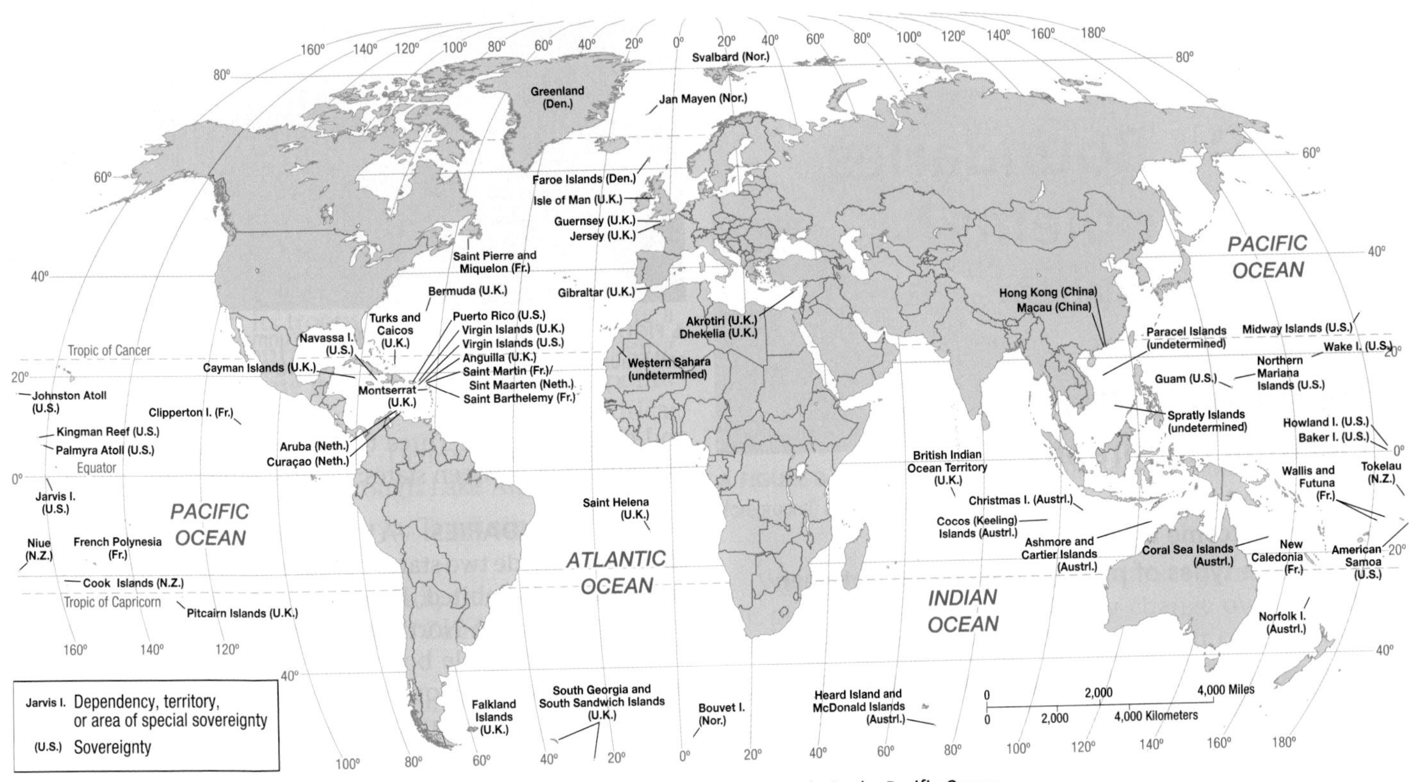

▲ **FIGURE 8-24 COLONIAL POSSESSIONS, 2012** Most remaining colonies are tiny specks in the Pacific Ocean and the Caribbean Sea, too small to appear on the map.

THE REMAINING COLONIES

At one time, colonies were widespread over Earth's surface, but only a handful remain today. The U.S. Department of State lists 68 places in the world that it calls dependencies and areas of special sovereignty (Figure 8-24). The list includes 43 with indigenous populations and 25 with no permanent population. Most current colonies are islands in the Pacific Ocean and Caribbean Sea.

The most populous is Puerto Rico, a commonwealth of the United States, with 4 million residents on an island of 8,870 square kilometers (3,500 square miles). Puerto Ricans are citizens of the United States, but they do not participate in U.S. elections or have a voting member of Congress.

One of the world's least-populated colonies is Pitcairn Island, a 47-square-kilometer (18-square-mile) possession of the United Kingdom. The island in the South Pacific was settled in 1790 by British mutineers from the ship *Bounty*, commanded by Captain William Bligh. Its 48 islanders survive by selling fish as well as postage stamps to collectors.

The U.S. State Department list does not include several inhabited islands considered by other sources to be colonies, including Australia's Lord Howe Island, Britain's Ascension Island, and Chile's Easter Island. On the other hand, the State Department list includes several entities that others do not classify as colonies:

- Greenland has a high degree of autonomy and self-rule and makes even foreign policy decisions independently of Denmark, as discussed earlier in the chapter. Greenland regards the Queen of Denmark as its head of state.
- Hong Kong and Macao, attached to the mainland of China, were colonies of the United Kingdom and Portugal, respectively. The British returned Hong Kong to China in 1997, and the Portuguese returned Macao to China in 1999. These two areas are classified as special administrative regions with autonomy from the rest of China in economic matters but not in foreign and military affairs.

Pause and Reflect 8.2.4
What would need to change for Puerto Rico to no longer be classified as a colony of the United States?

CHECK-IN: KEY ISSUE 2

Why Are Nation-States Difficult to Create?

- ✓ **Good examples of nation-states can be identified, though none are perfect.**
- ✓ **The Soviet Union was once the world's largest multinational state; with its breakup, Russia is now the largest.**
- ✓ **Much of Earth's land area once comprised colonies, but only a few colonies remain.**

CULTURAL BOUNDARIES

Learning Outcome 8.3.2
Describe types of cultural boundaries between states.

Two types of cultural boundaries are common: geometric and ethnic. Geometric boundaries are simply straight lines drawn on a map. Other boundaries between states coincide with differences in ethnicity, especially language and religion.

GEOMETRIC BOUNDARIES. Part of the northern U.S. boundary with Canada is a 2,100-kilometer (1,300-mile) straight line (more precisely, an arc) along 49° north latitude, running from Lake of the Woods between Minnesota and Manitoba to the Strait of Georgia between Washington State and British Columbia (Figure 8-29). This boundary was established in 1846 by a treaty between the United States and Great Britain, which still controlled Canada. The two countries share an additional 1,100-kilometer (700-mile) geometric boundary between Alaska and the Yukon Territory along the north–south arc of 141° west longitude.

Pause and Reflect 8.3.2
Where does the boundary between Canada and the United States follow physical features rather than geometry?

The 1,000-kilometer (600-mile) boundary between Chad and Libya is a straight line drawn across the desert in 1899 by the French and British to set the northern limit of French colonies in Africa (Figure 8-30). Libya claimed that the straight line should be 100 kilometers (60 miles) to the south. Citing an agreement between France and Italy in 1935, Libya seized the territory in 1973. In 1987, Chad

▼ **FIGURE 8-29 GEOMETRIC BOUNDARY BETWEEN CANADA AND THE UNITED STATES** Waterton-Glacier International Peace Park is located in both Canada and the United States. The international boundary between the United States (left) and Canada (right) is marked by the line of cut trees.

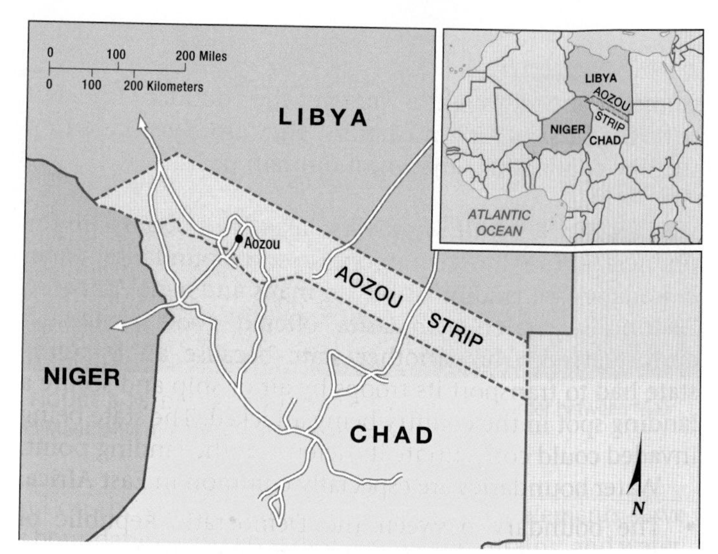

▲ **FIGURE 8-30 GEOMETRIC BOUNDARY BETWEEN CHAD AND LIBYA** The boundary between Chad and Libya is a straight line, drawn by European countries early in the twentieth century, when the area comprised a series of colonies. Libya, however, claims that the boundary should be located 100 kilometers (60 miles) to the south and that it should have sovereignty over the Aozou Strip.

expelled the Libyan army with the help of French forces and regained control of the strip.

ETHNIC BOUNDARIES. Boundaries between countries have been placed where possible to separate speakers of different languages or followers of different religions. Religious differences often coincide with boundaries between states, but in only a few cases has religion been used to select the actual boundary line.

The most notable example was in South Asia, when the British partitioned India into two states on the basis of religion. The predominantly Muslim portions were allocated to Pakistan, whereas the predominantly Hindu portions became the independent state of India (see Figure 7-31). Religion was also used to some extent to draw the boundary between two states on the island of Eire (Ireland). Most of the island became an independent country, but the northeast—now known as Northern Ireland—remained part of the United Kingdom. Roman Catholics comprise approximately 95 percent of the population in the 26 counties that joined the Republic of Ireland, whereas Protestants constitute the majority in the six counties of Northern Ireland (see Figure 6-47).

Language is an important cultural characteristic for drawing boundaries, especially in Europe. England, France, Portugal, and Spain are examples of European states that coalesced around distinctive languages before the nineteenth century. Germany and Italy emerged in the nineteenth century as states unified by language.

CYPRUS'S "GREEN LINE" BOUNDARY. Cyprus, the third-largest island in the Mediterranean Sea, contains two nationalities: Greek and Turkish (Figure 8-32). Although the island is physically closer to Turkey, Turks comprise only 18 percent of the country's population, whereas Greeks account for 78 percent. When Cyprus gained independence from Britain in 1960, its constitution guaranteed the Turkish minority a substantial share of elected offices and

CONTEMPORARY GEOGRAPHIC TOOLS

Demarcating Boundaries with GPS

GPS was defined in Chapter 1 as a system that determines the precise position of something on Earth. It is most commonly used for navigation, although GPS in a cell phone is used to identify the location of an individual. Surveyors are using the ability of GPS to pinpoint location to determine the precise boundary between North Carolina and South Carolina.

The original boundary between the two Carolina colonies, as decreed by the King of England in 1735, was drawn by eighteenth-century surveyors using the best technology then available—poles, chains, and compasses. The boundary was recorded with hatchet marks on trees, most of which have disappeared. The two states established a Joint Boundary Commission in 2010 to demarcate the boundary more precisely and mark it with stakes and stones. Surveyors found that nearly 100 properties thought to be in one state were actually in the other.

Shifting the boundary is not difficult on a map or on the ground, but the problems are considerable for the people and businesses suddenly shifted to the other state. In the U.S. system of federal government, taxes, services, and regulations vary considerably among states. The two state governments are trying to minimize the impact on the affected properties, essentially by ignoring the new precisely demarcated boundary (Figure 8-31).

▲ FIGURE 8-31 BOUNDARY BETWEEN NORTH CAROLINA AND SOUTH CAROLINA South of the Border is a large entertainment complex located on the South Carolina side of the border with North Carolina. After surveying, the complex remains on the South Carolina side.

▲ FIGURE 8-32 ETHNIC BOUNDARY BETWEEN GREEK AND TURKISH CYPRUS Since 1974, Cyprus has been divided into Greek and Turkish areas, separated by a United Nations buffer zone. The photo shows a crossing between the Greek side (foreground) and Turkish side (background), through the UN buffer zone (middle).

control over its own education, religion, and culture. But Cyprus has never peacefully integrated the Greek and Turkish nationalities.

Several Greek Cypriot military officers who favored unification of Cyprus with Greece seized control of the government in 1974. Shortly after the coup, Turkey invaded Cyprus to protect the Turkish Cypriot minority. The Greek coup leaders were removed within a few months, and an elected government was restored, but the Turkish army remained on Cyprus. The northern 36 percent of the island controlled by Turkey declared itself the independent Turkish Republic of Northern Cyprus in 1983, but only Turkey recognizes it as a separate state.

A wall was constructed between the two areas, and a buffer zone patrolled by the United Nations was delineated across the entire island. Traditionally, the Greek and Turkish Cypriots had mingled, but after the wall and buffer zone were established, the two nationalities became geographically isolated. The northern part of the island is now overwhelmingly Turkish, whereas the southern part is overwhelmingly Greek. Approximately one-third of the island's Greeks were forced to move from the region controlled by the Turkish army, whereas nearly one-fourth of the Turks moved from the region now regarded as the Greek side.

The two sides have been brought closer in recent years. A portion of the wall was demolished, and after three decades the two nationalities could again cross to the other side. The European Union accepted the entire island of Cyprus as a member in 2004. A UN Peace Plan for reunification was accepted by the Turkish side but rejected by the Greek side.

Shapes of States

Learning Outcome 8.3.3
Describe five shapes of states.

The shape of a state controls the length of its boundaries with other states. The shape therefore affects the potential for communication and conflict with neighbors. The shape also, as in the outline of the United States or Canada, is part of its unique identity. Beyond its value as a centripetal force, the shape of a state can influence the ease or difficulty of internal administration and can affect social unity.

Countries have one of five basic shapes—compact, prorupted, elongated, fragmented, or perforated—and examples of each can be seen in southern Africa (Figure 8-33). Each shape displays distinctive characteristics and challenges.

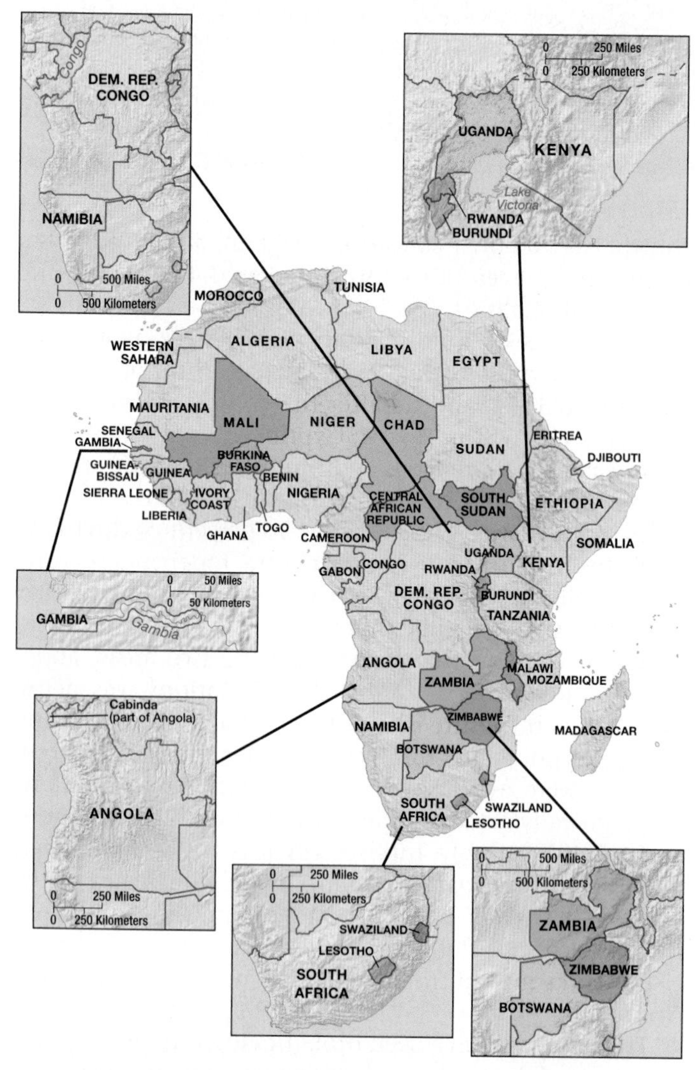

▲ **FIGURE 8-33 SHAPES OF STATES IN SOUTHERN AFRICA** Burundi, Kenya, Rwanda, and Uganda are examples of compact states. Malawi and Mozambique are elongated states. Namibia and the Democratic Republic of Congo are prorupted states. Angola and Tanzania are fragmented states. South Africa is a perforated state. The countries in color are landlocked African states, which must import and export goods by land-based transportation, primarily rail lines, to reach ocean ports in cooperating neighbor states.

COMPACT STATES: EFFICIENT

In a **compact state**, the distance from the center to any boundary does not vary significantly. The ideal theoretical compact state would be shaped like a circle, with the capital at the center and with the shortest possible boundaries to defend.

Compactness can be a beneficial characteristic for smaller states because good communications can be more easily established with all regions, especially if the capital is located near the center. However, compactness does not necessarily mean peacefulness, as compact states are just as likely as others to experience civil wars and ethnic rivalries.

ELONGATED STATES: POTENTIAL ISOLATION

A handful of **elongated states** have a long and narrow shape. Examples in sub-Saharan Africa include:

- Malawi, which measures about 850 kilometers (530 miles) north–south but only 100 kilometers (60 miles) east–west.
- Gambia, which extends along the banks of the Gambia River about 500 kilometers (300 miles) east–west but is only about 25 kilometers (15 miles) north–south.

Chile, a prominent example in South America, stretches north-south for more than 4,000 kilometers (2,500 miles) but rarely exceeds an east-west distance of 150 kilometers (90 miles). Chile is wedged between the Pacific Coast of South America and the rugged Andes Mountains, which rise more than 6,700 meters (20,000 feet).

Elongated states may suffer from poor internal communications. A region located at an extreme end of the elongation might be isolated from the capital, which is usually placed near the center.

PRORUPTED STATES: ACCESS OR DISRUPTION

An otherwise compact state with a large projecting extension is a **prorupted state**. Proruptions are created for two principal reasons:

- **To provide a state with access to a resource, such as water.** For example, in southern Africa, the Democratic Republic of Congo has a 500-kilometer (300-mile) proruption to the west along the Zaire (Congo) River. The Belgians created the proruption to give their colony access to the Atlantic.
- **To separate two states that otherwise would share a boundary.** For example, in southern Africa, Namibia has a 500-kilometer (300-mile) proruption to the east called the Caprivi Strip. When Namibia was a colony of Germany, the proruption disrupted communications among the British colonies of southern Africa. It also provided the Germans with access to the Zambezi, one of Africa's most important rivers.

Elsewhere in the world, the otherwise compact state of Afghanistan has a proruption approximately 300 kilometers (200 miles) long and as narrow as 20 kilometers (12 miles) wide. The British created the proruption to prevent Russia from sharing a border with Pakistan.

PERFORATED STATES: SOUTH AFRICA

A state that completely surrounds another one is a **perforated state**. In this situation, the state that is surrounded may face problems of dependence on, or interference from, the surrounding state. For example, South Africa completely surrounds the state of Lesotho. Lesotho must depend almost entirely on South Africa for the import and export of goods. Dependency on South Africa was especially difficult for Lesotho when South Africa had a government controlled by whites who discriminated against the black majority population. Elsewhere in the world, Italy surrounds the Holy See (the Vatican) and San Marino.

FRAGMENTED STATES: PROBLEMATIC

A **fragmented state** includes several discontinuous pieces of territory. Technically, all states that have offshore islands as part of their territory are fragmented. However, fragmentation is particularly significant for some states. There are two kinds of fragmented states, and both may face problems and costs associated with communications and maintaining national unity:

1. FRAGMENTED STATES SEPARATED BY WATER. An example in sub-Saharan Africa is Tanzania, which was created in 1964 as a union of the island of Zanzibar with the mainland territory of Tanganyika. Although home to different ethnic groups, the two entities agreed to join together because they shared common development goals and political priorities.

Elsewhere in the world, Indonesia comprises 13,677 islands that extend more than 5,000 kilometers (3,000 miles) between the Indian and Pacific oceans. Although more than 80 percent of the country's population live on two of the islands—Java and Sumatra—the fragmentation hinders communications and makes integration of people living on remote islands nearly impossible. To foster national integration, the Indonesian government has encouraged migration from the more densely populated islands to some of the sparsely inhabited ones.

Not all of the fragments joined Indonesia voluntarily. A few days after Timor-Leste (East Timor) gained its independence from Portugal in 1975, Indonesia invaded. A long struggle against Indonesia culminated in independence in 2002. West Papua, another fragment of Indonesia (the western portion of the island shared with Papua New Guinea), also claims that it should be an independent country. However, West Papua's attempt to break away from Indonesia gained less support from the international community.

2. FRAGMENTED STATES SEPARATED BY AN INTERVENING STATE. An example in sub-Saharan Africa is Angola, which is divided into two fragments by the Congo proruption described above. An independence movement is trying to detach Cabinda as a separate state from Angola, with the justification that its population belongs to distinct ethnic groups.

Elsewhere in the world, Russia has a fragment called Kaliningrad (Konigsberg), a 16,000-square-kilometer (6,000-square-mile) entity 400 kilometers (250 miles) west of the remainder of Russia, separated by the states of Lithuania and Belarus. The area was part of Germany until the end of World War II, when the Soviet Union seized it after the German defeat. The German population fled westward after the war, and virtually all of the area's 430,000 residents are Russians. Russia wants Kaliningrad because it has the country's largest naval base on the Baltic Sea.

Panama was a fragmented state for most of the twentieth century, divided in two parts by the canal built in 1914 by the United States. After the United States withdrew from the Canal Zone in 1999, Panama became an elongated state, 700 kilometers (450 miles) long and 80 kilometers (50 miles) wide.

LANDLOCKED STATES

A **landlocked state** lacks a direct outlet to a sea because it is completely surrounded by several other countries (or only one country, in the case of Lesotho). Landlocked states are most common in Africa, where 15 of the continent's 55 states have no direct ocean access (refer to the countries in colors on Figure 8-33). The prevalence of landlocked states in Africa is a remnant of the colonial era, when Britain and France controlled extensive regions. The European powers built railroads, mostly in the early twentieth century, to connect the interior of Africa with the sea. Railroads moved minerals from interior mines to seaports, and in the opposite direction, rail lines carried mining equipment and supplies from seaports to the interior.

Now that the British and French empires are gone, and former colonies have become independent states, some important colonial railroad lines pass through several independent countries. This has created new landlocked states, which must cooperate with neighboring states that have seaports. Direct access to an ocean is critical to states because it facilitates international trade. Bulky goods, such as petroleum, grain, ore, and vehicles, are normally transported long distances by ship. This means that a country needs a seaport where goods can be transferred between land and sea. To send and receive goods by sea, a landlocked state must arrange to use another country's seaport.

Pause and Reflect 8.3.3
Where outside of Africa is an example of a landlocked state?

Governing States

Learning Outcome 8.3.4
Describe differences among the three regime types.

A state has two types of government: a national government and local governments. At the national scale, a government can be more or less democratic. At the local scale, the national government can determine how much power to allocate to local governments.

NATIONAL SCALE: REGIME TYPES

National governments can be classified as democratic, autocratic, or anocratic (Figure 8-34). A **democracy** is a country in which citizens elect leaders and can run for office. An **autocracy** is a country that is run according to the interests of the ruler rather than the people. An **anocracy** is a country that is not fully democratic or fully autocratic, but rather displays a mix of the two types. According to the Center for Systemic Peace, democracies and autocracies differ in three essential elements:

Selection of Leaders:

- A democracy has institutions and procedures through which citizens can express effective preferences about alternative policies and leaders.
- An autocracy has leaders who are selected according to clearly defined (usually hereditary) rules of succession from within the established political elite.

Citizen Participation:

- A democracy has institutionalized constraints on the exercise of power by the executive.
- An autocracy has citizens' participation sharply restricted or suppressed.

Checks and Balances:

- A democracy has guarantees of civil liberties to all citizens in their daily lives and in acts of political participation.
- An autocracy has leaders who exercise power with no meaningful checks from legislative, judicial, or civil society institutions.

TREND TOWARD DEMOCRACY. In general, the world has become more democratic (Figure 8-35). The Center for Systemic Peace cites three reasons for this:

- The replacement of increasingly irrelevant and out-of-touch monarchies with elected governments that are able to regulate, tax, and mobilize citizens in exchange for broadening individual rights and liberties.
- The widening of participation in policy making to all citizens through universal rights to vote and to serve in government.
- The diffusion of democratic government structures created in Europe and North America to other regions of the world.

Pause and Reflect 8.3.4
What region of the world appears to have the greatest concentration of autocratic regimes?

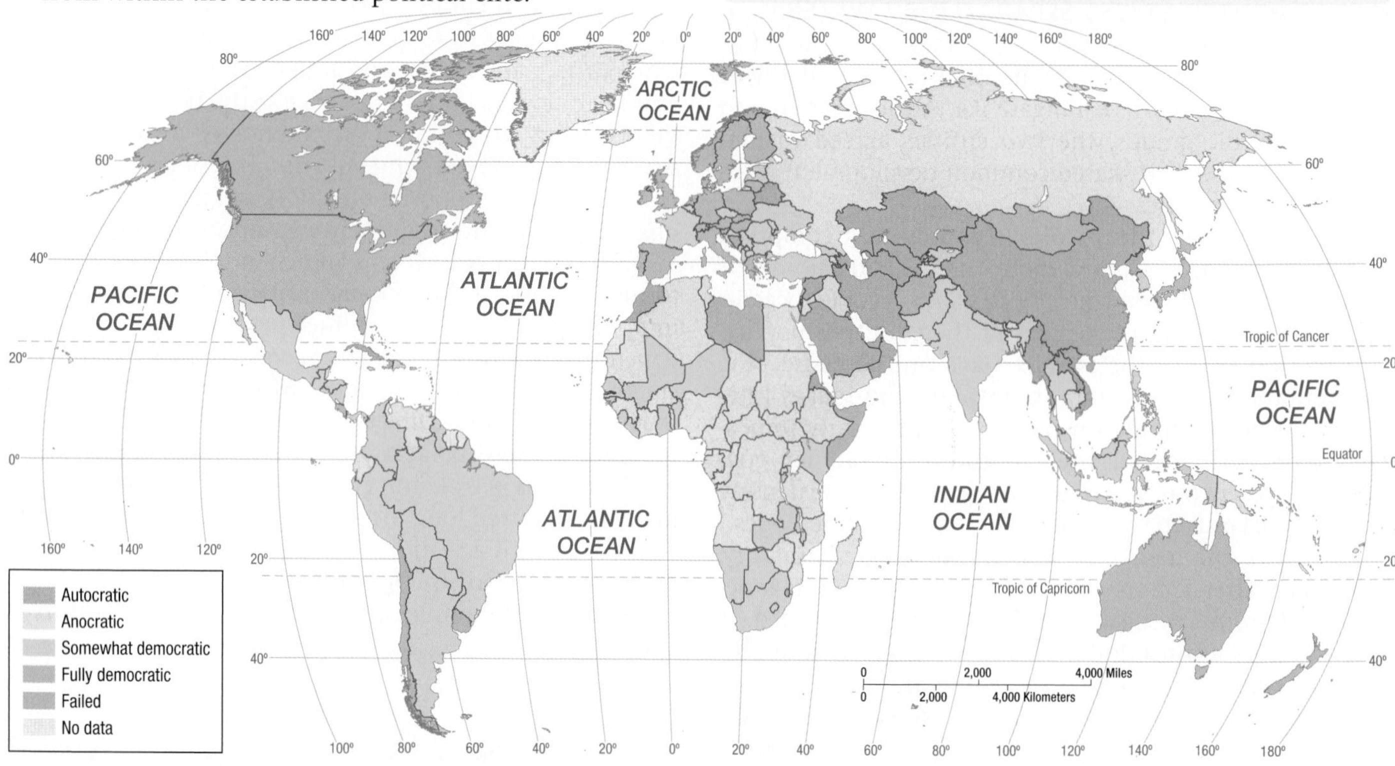

▲ **FIGURE 8-34 REGIME TYPE** Most states are either democratic, autocratic, or anocratic. In a few "failed" states, such as Somalia and Haiti, government institutions have broken down because of civil war, extreme poverty, or natural disasters—or some combination of the three.

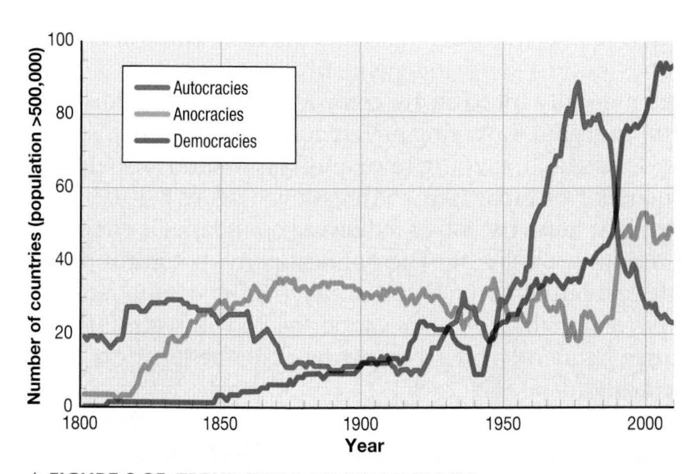

▲ **FIGURE 8-35 TREND TOWARD DEMOCRACY**
The number of autocracies has declined sharply since the late 1990s.

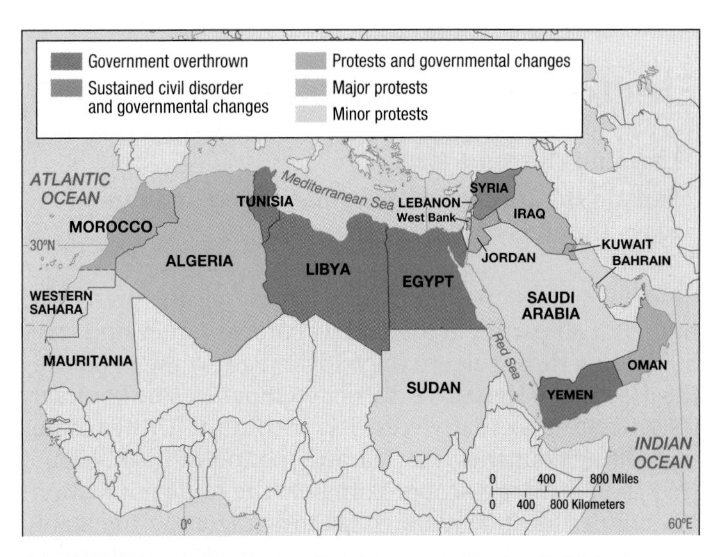

▲ **FIGURE 8-36 ARAB SPRING** Cell phones and other handheld devices were instrumental in rapidly diffusing information about uprisings despite government efforts to suppress the information.

ARAB SPRING. The most dramatic shift in governments in recent years has been Arab Spring, which began in late 2010 and reached its peak during spring 2011. Arab Spring consisted of major protests in a dozen countries in Southwest Asia and North Africa. The protests resulted in forcing from power autocratic rulers in Egypt, Libya, Tunisia, and Yemen (Figure 8-36).

The protests included demonstrations, rallies, strikes, and other forms of civil disobedience, many led by college-age people. Especially noteworthy was the use of social media and portable electronic devices to organize protests, communicate information, and distribute real-time images of events. Long-standing practices by autocratic regimes to suppress TV and newspaper coverage of opponents proved ineffective in the face of Facebook and Twitter, iPhones and iPads.

LOCAL SCALE: UNITARY AND FEDERAL STATES

The governments of states are organized according to one of two approaches:

- A **unitary state** places most power in the hands of central government officials.
- A **federal state** allocates strong power to units of local government within the country.

UNITARY STATES. In principle, the unitary government system works best in nation-states characterized by few internal cultural differences and a strong sense of national unity. Because the unitary system requires effective communications with all regions of the country, smaller states are more likely to adopt it. Unitary states are especially common in Europe.

Some multinational states have adopted unitary systems, so that the values of one nationality can be imposed on others. In Kenya and Rwanda, for instance, the mechanisms of a unitary state have enabled one ethnic group to extend dominance over weaker groups.

A good example of a nation-state, France has a long tradition of unitary government in which a very strong national government dominates local government decisions. Their basic local government unit is 96 *départements* (departments). A second tier of local government in France is the 36,686 *communes*. The French government has granted additional legal powers to the departments and communes in recent years. In addition, 22 regional councils that previously held minimal authority have been converted into full-fledged local government units, with elected councils and the power to levy taxes.

FEDERAL STATES. In a federal state, such as the United States, local governments possess considerable authority to adopt their own laws. Multinational states may adopt a federal system of government to empower different nationalities, especially if they live in separate regions of the country. Under a federal system, local government boundaries can be drawn to correspond with regions inhabited by different ethnicities.

The federal system is more suitable for very large states because the national capital may be too remote to provide effective control over isolated regions. Most of the world's largest states are federal, including Russia, Canada, the United States, Brazil, and India. However, the size of the state is not always an accurate predictor of the form of government: Tiny Belgium is a federal state (to accommodate the two main cultural groups, the Flemish and the Waloons, as discussed in Chapter 5), whereas China is a unitary state (to promote Communist values).

In recent years there has been a strong global trend toward federal government. Unitary systems have been sharply curtailed in a number of countries and scrapped altogether in others. In the face of increasing demands by ethnicities for more self-determination, states have restructured their governments to transfer some authority from the national government to local government units. An ethnicity that is not sufficiently numerous to gain control of the national government may be content with control of a regional or local unit of government.

Electoral Geography

Learning Outcome 8.3.5
Explain the concept of gerrymandering and three ways that it is done.

In democracies, politics must follow legally prescribed rules. But all parties to the political process often find ways of bending those rules to their advantage. A case in point is the drawing of legislative district boundaries. The boundaries separating legislative districts within the United States and other countries are redrawn periodically to ensure that each district has approximately the same population. Boundaries must be redrawn because migration inevitably results in some districts gaining population and others losing population. The 435 districts of the U.S. House of Representatives are redrawn every 10 years, following the Census Bureau's release of official population figures.

The process of redrawing legislative boundaries for the purpose of benefiting the party in power is called **gerrymandering**. The term *gerrymandering* was named for Elbridge Gerry (1744–1814), governor of Massachusetts (1810–1812) and vice president of the United States (1813–1814). As governor, Gerry signed a bill that redistricted the state to benefit his party. An opponent observed that an oddly shaped new district looked like a "salamander," whereupon another opponent responded that it was a "gerrymander." A newspaper subsequently printed a cartoon of a monster named "gerrymander" with a body shaped like the district. Gerrymandering takes three forms:

- *Wasted vote* spreads opposition supporters across many districts but in the minority (Figure 8-37).
- *Excess vote* concentrates opposition supporters into a few districts (Figure 8-38).
- *Stacked vote* links distant areas of like-minded voters through oddly shaped boundaries (Figure 8-39).

The job of redrawing boundaries in most European countries is entrusted to independent commissions. Commissions typically try to create compact homogeneous districts without regard for voting preferences or incumbents. A couple U.S. states, including Iowa and Washington, also use independent or bipartisan commissions (Figure 8-40), but in most U.S. states the job of redrawing boundaries is entrusted to the state legislature. The political party in control of the state legislature naturally attempts to redraw boundaries to improve the chances of its supporters to win seats. Political parties frequently offer competing plans designed to favor their candidates (Figure 8-41).

Stacked vote gerrymandering has been especially attractive for creating districts inclined to elect ethnic minorities. Because the two largest ethnic groups in the United States (African Americans and most Hispanics other than Cubans) tend to vote Democratic—in some elections more than 90 percent of African Americans vote Democratic—creating a majority African American district virtually guarantees election of a Democrat. Republicans support a "stacked" Democratic district because they are better able to draw boundaries that are favorable to their candidates in the rest of the state.

The U.S. Supreme Court ruled gerrymandering illegal in 1985 but did not require dismantling of existing oddly shaped districts, and a 2001 ruling allowed North Carolina to add another oddly shaped district that ensured the election of an African American Democrat. Through gerrymandering, only about one-tenth of congressional seats are competitive, making a shift of more than a few seats unlikely from one election to another in the United States, except in unusual circumstances.

Pause and Reflect 8.3.5
How was the city of Las Vegas treated in the two maps drawn by the political parties compared with the final map drawn by the court?

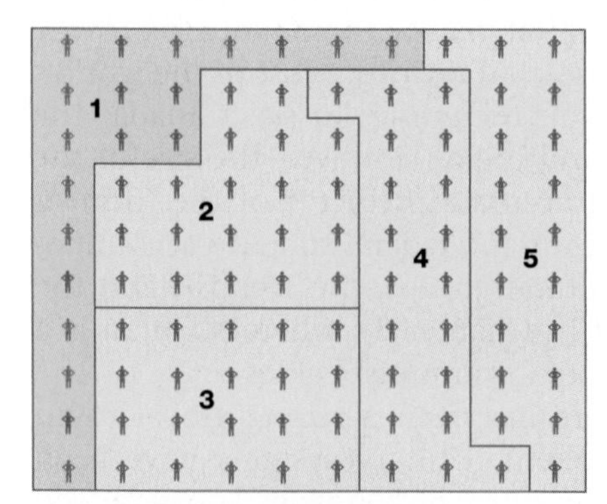

▲ **FIGURE 8-37 WASTED VOTE GERRYMANDERING** Wasted vote gerrymandering spreads opposition supporters across many districts as a minority. If the Blue Party controls the redistricting process, it could create a wasted vote gerrymander by creating four districts with a slender majority of Blue Party voters and one district (#1) with a strong majority of Red Party voters.

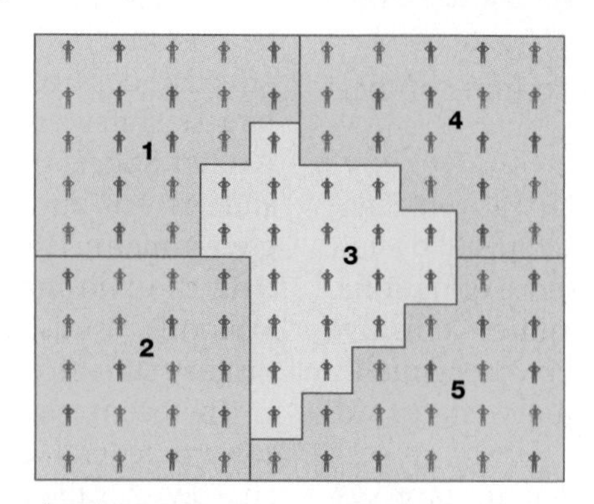

▲ **FIGURE 8-38 EXCESS VOTE GERRYMANDERING** Excess vote gerrymandering concentrates opposition supporters into a few districts. If the Red Party controls the redistricting process, it could create an excess vote gerrymander by creating four districts with a slender majority of Red Party voters and one district (#3) with an overwhelming majority of Blue Party voters.

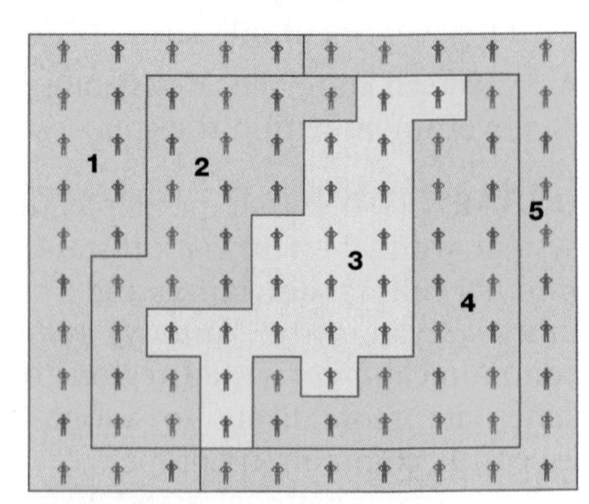

▲ **FIGURE 8-39 STACKED VOTE GERRYMANDERING** A stacked vote gerrymander links distant areas of like-minded voters through oddly shaped boundaries. In this example, the Red Party controls redistricting and creates five oddly shaped districts, four with a slender majority of Red Party voters and one (#3) with an overwhelming majority of Blue Party voters.

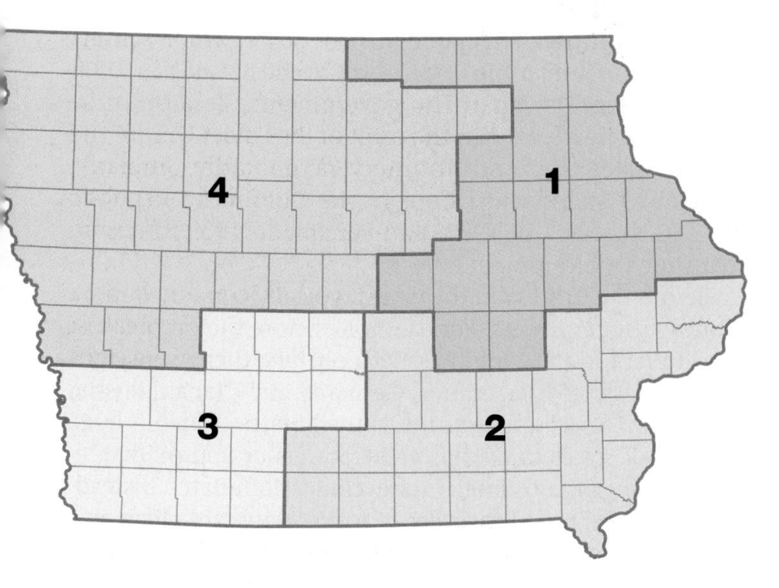

◀ **FIGURE 8-40 NO GERRYMANDERING: IOWA** Iowa does not have gerrymandered congressional districts. Each district is relatively compact, and boundaries coincide with county boundaries. A nonpartisan commission creates Iowa's districts each decade, without regard for past boundaries or impact on incumbents.

CHECK-IN: KEY ISSUE 3

Why Do Boundaries Cause Problems?

- ✓ **Two types of boundaries are physical and cultural.**
- ✓ **Deserts, mountains, and water can serve as physical boundaries between states.**
- ✓ **Geometry and ethnicity can create cultural boundaries between states.**
- ✓ **Five shapes of states are compact, elongated, prorupted, perforated, and fragmented.**
- ✓ **The governance of states can be classified as democratic, anocratic, or autocratic; democracies have been increasing.**
- ✓ **Boundaries dividing electoral districts within countries can be gerrymandered in several ways to favor one political party.**

Democratic proposal

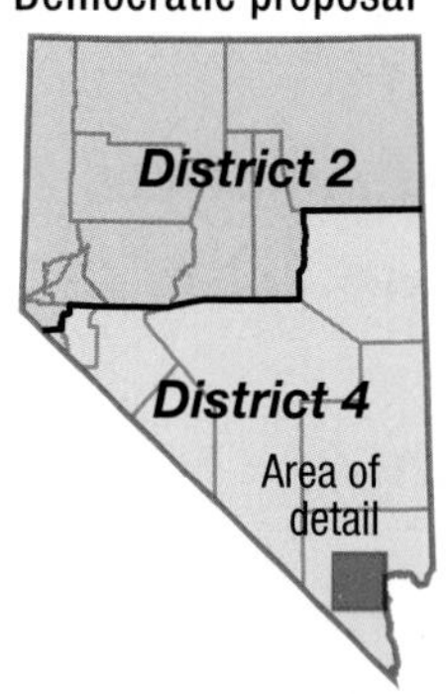

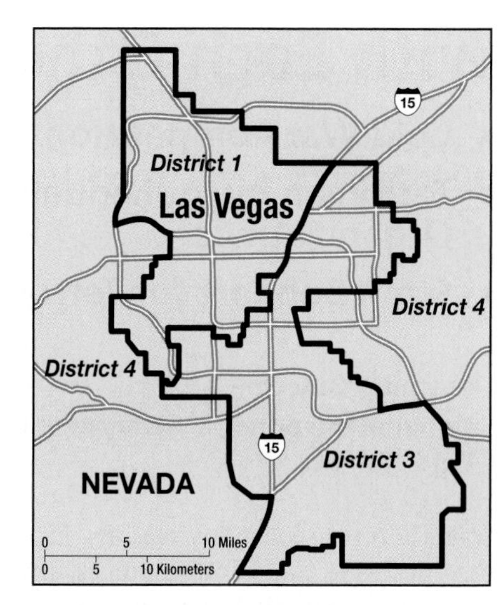

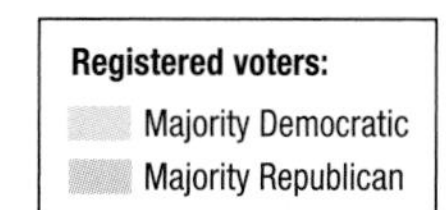

Republican proposal

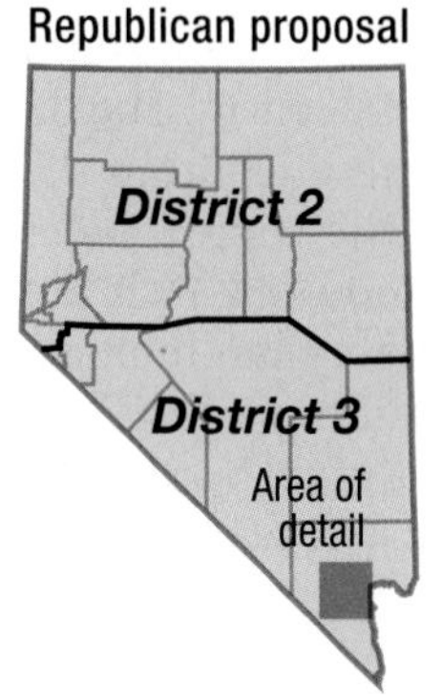

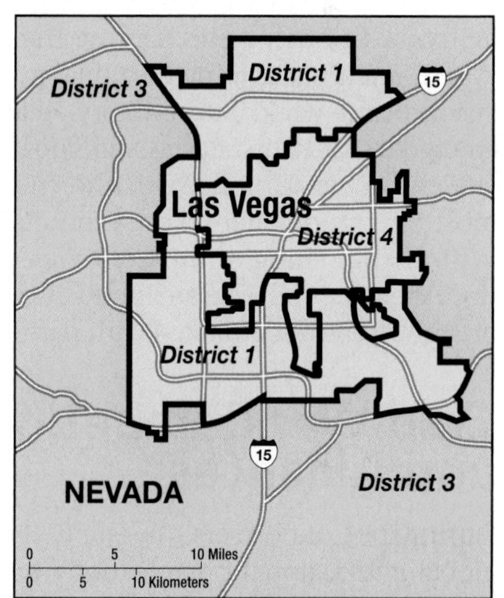

Court-imposed districts

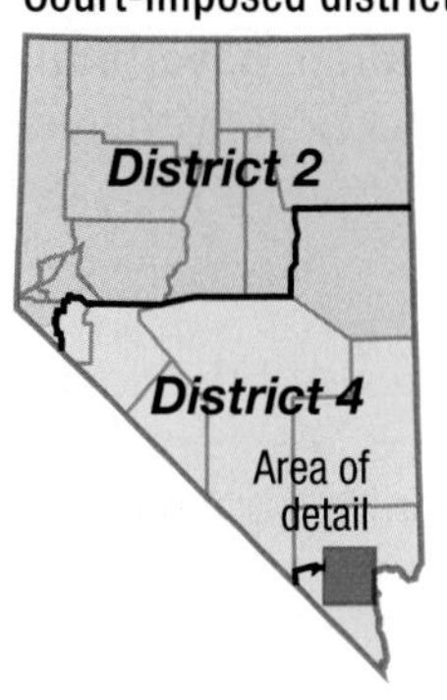

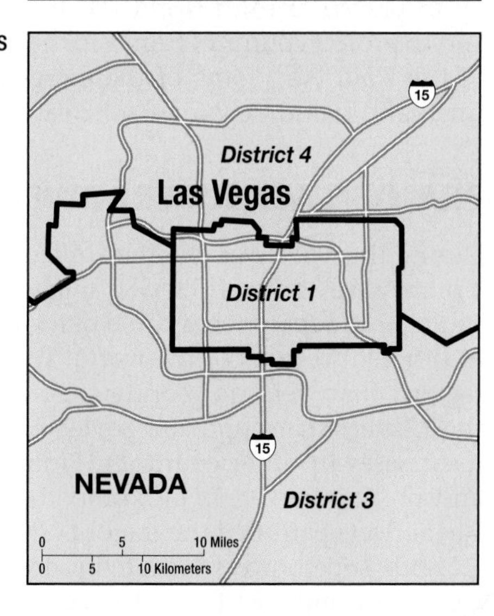

▶ **FIGURE 8-41 GERRYMANDERING: NEVADA**
Competing plans by Democrats and Republicans to draw boundaries for Nevada's four congressional districts illustrate all three forms of gerrymandering.

(top) Wasted vote gerrymander: The Democratic plan. Although Nevada as a whole has slightly more registered Democrats than Republicans (43 percent to 37 percent), the Democratic plan made Democrats more numerous than Republicans in three of the four districts.

(middle) Excess vote gerrymander: The Republican plan. By clustering a large share of the state's registered Democrats in District 4, the Republican plan gave Republicans the majority of registered voters in two of the four districts.

(both top and middle Stacked) vote gerrymander: In the Republican plan, District 4 has a majority Hispanic population and is surrounded by a C-shaped District 1. The Democratic plan created a long, narrow District 3.

(bottom) Nonpartisan plan without gerrymandering: The Nevada Court rejected both parties' maps and created regularly shaped districts that minimized gerrymandering. Three of the four districts happen to have more Democrats than Republicans, but District 3 is nearly even.

KEY ISSUE 4

Why Do States Cooperate and Compete with Each Other?

- Cold War Competition and Alliances
- Terrorism by Individuals and Organizations
- State Support for Terrorism

Learning Outcome 8.4.1
Describe the principal alliances in Europe during the Cold War era.

States compete for many reasons, including control of territory, access to trade and resources, and influence over other states. To further their competitive goals, states may form alliances with other states. During the Cold War, after World War II, many states joined regional military alliances. The division of the world into military alliances resulted from the emergence of two states as superpowers—the United States and the Soviet Union. With the end of the Cold War, the most important alliances are economic rather than military. With the lessening of the Cold War–era military confrontation, violence and wars are increasingly instigated by terrorist organizations not affiliated with particular states or alliances.

Cold War Competition and Alliances

During the Cold War era (the late 1940s until the early 1990s), global and regional organizations were established primarily to prevent a third world war in the twentieth century and to protect countries from a foreign attack. With the end of the Cold War, some of these organizations have flourished and found new roles, whereas others have withered.

ERA OF TWO SUPERPOWERS

During the Cold War era, the United States and the Soviet Union were the world's two superpowers. As very large states, both superpowers could quickly deploy armed forces in different regions of the world. To maintain strength in regions that were not contiguous to their own territory, the United States and the Soviet Union established military bases in other countries. From these bases, ground and air support were in proximity to local areas of conflict. Naval fleets patrolled the major bodies of water.

Both superpowers repeatedly demonstrated that they would use military force if necessary to prevent an ally from becoming too independent. The Soviet Union sent its armies into Hungary in 1956 and Czechoslovakia in 1968 to install more sympathetic governments. Because these states were clearly within the orbit of the Soviet Union, the United States chose not to intervene militarily. Similarly, the United States sent troops to the Dominican Republic in 1965, Grenada in 1983, and Panama in 1989 to ensure that they would remain allies.

Before the Cold War, the world typically contained more than two superpowers. For example, before the outbreak of World War I in the early twentieth century, there were eight great powers: Austria, France, Germany, Italy, Japan, Russia, the United Kingdom, and the United States. When a large number of states ranked as great powers of approximately equal strength, no single state could dominate. Instead, major powers joined together to form temporary alliances.

A condition of roughly equal strength between opposing alliances is known as a **balance of power**. In contrast, the post–World War II balance of power was bipolar between the United States and the Soviet Union. Because the power of these two states was so much greater than the power of all other states, the world comprised two camps, each under the influence of one of the superpowers. Other states lost the ability to tip the scales significantly in favor of one or the other superpower. They were relegated to a new role of either ally or satellite.

CUBAN MISSILE CRISIS. A major confrontation during the Cold War between the United States and Soviet Union came in 1962, when the Soviet Union secretly began to construct missile-launching sites in Cuba, less than 150 kilometers (90 miles) from U.S. territory. President John F. Kennedy went on national television to demand that the missiles be removed, and he ordered a naval blockade to prevent additional Soviet material from reaching Cuba.

At the United Nations, immediately after Soviet Ambassador Valerian Zorin denied that his country had placed missiles in Cuba, U.S. Ambassador Adlai Stevenson dramatically revealed aerial photographs taken by the U.S. Department of Defense, clearly showing preparations for them (see examples in Figure 8-42). Faced with irrefutable evidence that the missiles existed, the Soviet Union ended the crisis by dismantling them.

MILITARY COOPERATION IN EUROPE. After World War II, most European states joined one of two military alliances dominated by the superpowers—NATO (North Atlantic Treaty Organization) or the Warsaw Pact (Figure 8-43, left). NATO was a military alliance among 16 democratic states, including the United States and Canada plus 14 European states. The Warsaw Pact was a military agreement among Communist Eastern European countries to defend each other in case of attack. Eight members joined the Warsaw Pact when it was founded in 1955. Some of Hungary's leaders in 1956 asked for the help of Warsaw Pact troops to crush an uprising that threatened Communist control of the government. Warsaw Pact troops also invaded Czechoslovakia in 1968 to depose a government committed to reforms.

◀ **FIGURE 8-42 THE COLD WAR: 1962 CUBAN MISSILE CRISIS**
The U.S. Department of Defense took aerial photographs to show the Soviet buildup in Cuba. (top) Three Soviet ships with missile equipment are being unloaded at Mariel naval port in Cuba. (bottom) Within the outline box (enlarged below and rotated 90° clockwise) are Soviet missile transporters, fuel trailers, and oxidizer trailers (used to support the combustion of missile fuel).

NATO and the Warsaw Pact were designed to maintain a bipolar balance of power in Europe. For NATO allies, the principal objective was to prevent the Soviet Union from overrunning West Germany and other smaller countries. The Warsaw Pact provided the Soviet Union with a buffer of allied states between it and Germany to discourage a third German invasion of the Soviet Union in the twentieth century.

In a Europe no longer dominated by military confrontation between two blocs, the Warsaw Pact was disbanded, and the number of troops under NATO command was sharply reduced. NATO expanded its membership to include most of the former Warsaw Pact countries. Membership in NATO offered Eastern European countries an important sense of security against any future Russian threat, no matter how remote that might appear, as well as participation in a common united Europe (Figure 8-43, right).

Pause and Reflect 8.4.1
How does the map of military alliances in Europe during the Cold War compare to the map of regime types (Figure 8-34)?

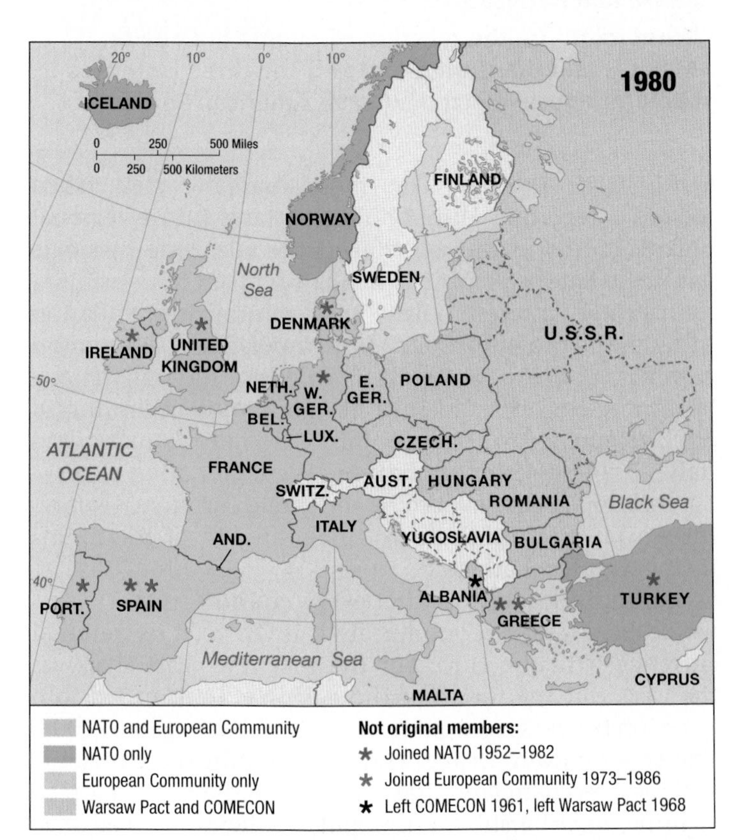

▲ **FIGURE 8-43 EUROPE MILITARY AND ECONOMIC ALLIANCES** (left) During the Cold War. Western European countries joined the European Union and the North Atlantic Treaty Organization (NATO), whereas Eastern European countries joined COMECON and the Warsaw Pact. (right) Post–Cold War. COMECON and the Warsaw Pact have been disbanded, whereas the European Union and NATO have accepted and plan to accept new members.

ECONOMIC ALLIANCES IN EUROPE

Learning Outcome 8.4.2
Describe the principal economic alliances in Europe in the period since World War II.

During the Cold War, two economic alliances formed in Europe:

- **European Union (EU).** The EU (formerly known as the European Economic Community, the Common Market, and the European Community), formed in 1958 with six members—Belgium, France, Italy, Luxembourg, the Netherlands, and the Federal Republic of Germany (West Germany). The EU was designed to heal Western Europe's scars from World War II (which had ended only 13 years earlier) when Nazi Germany, in alliance with Italy, conquered the other four countries.
- **Council for Mutual Economic Assistance (COMECON).** COMECON formed in 1949 with 10 members—the 8 Eastern European Communist states from the Warsaw Pact plus Cuba, Mongolia, and Vietnam. COMECON was designed to promote trade and sharing of natural resources.

With the end of the Cold War, economic cooperation throughout Europe has become increasingly important.

THE EU IN THE TWENTY-FIRST CENTURY. The EU expanded from its original 6 countries to 12 during the 1980s and 27 during the first decade of the twenty-first century. The most recent additions have been former members of COMECON, which disbanded in the 1990s, after the fall of communism. Future enlargements are likely: Croatia has begun negotiations to join, but the European Union has not yet set a timetable; Macedonia, Montenegro, and Serbia are candidates to join, but negotiations have not started; Iceland and Turkey are negotiating to become candidates; and Albania is considered a potential candidate.

The main task of the European Union is to promote development within the member states through economic and political cooperation:

- A European Parliament is elected by the people in each of the member states simultaneously.
- Subsidies are provided to farmers and to economically depressed regions.
- Most barriers to free trade have been removed; with a few exceptions, goods, services, capital, and people can move freely through Europe (Figure 8-44). For example, trucks can carry goods across borders without stopping, and a bank can open branches in any member country with supervision only by the bank's home country.

The effect of these actions has been to turn Europe into the world's wealthiest market.

▲ **FIGURE 8-44 TRAVEL IN EUROPE** Citizens of one EU state do not have to show passports to travel to other EU states.

Pause and Reflect 8.4.2
What might be the reaction of people in Canada, Mexico, and the United States if the three countries simultaneously elected a North American Parliament?

EUROZONE CRISIS. The most dramatic step taken toward integrating Europe's nation-states into a regional organization was the creation of the eurozone. A single bank, the European Central Bank, was given responsibility for setting interest rates and minimizing inflation throughout the eurozone. Most importantly, a common currency, the euro, was created for electronic transactions beginning in 1999 and in notes and coins beginning in 2002 (Figure 8-45). France's franc, Germany's mark, and Italy's lira—powerful symbols of sovereign nation-states—have disappeared, replaced by the single currency. Twenty-three countries use the euro, including 17 of the 27 EU members, plus 6 others.

European leaders bet that every country in the region would be stronger economically if it replaced its national currency with the euro. For the first few years that was the case, but the future of the euro has been called into question by the severe global recession that began in 2008. The economically weaker countries within the eurozone, such as Greece, Ireland, Italy, and Spain, have been forced to implement harsh and unpopular policies, such as drastically cutting services and raising taxes, whereas the economically strong countries, especially Germany, have been forced to subsidize the weaker states.

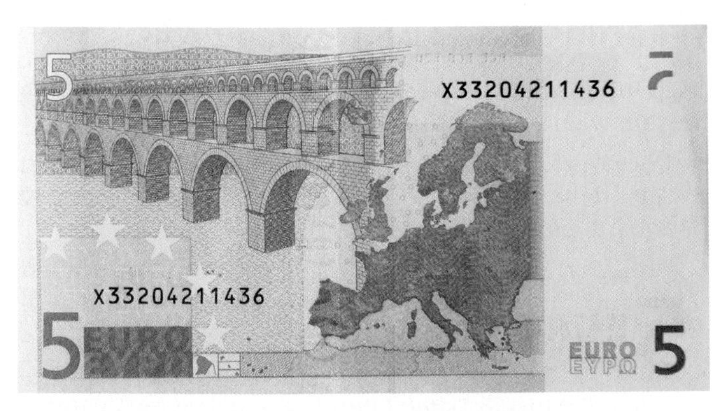

▲ FIGURE 8-45 **EURO** Euro paper money shows a map of Europe and a bridge on one side and architecture on the other side. Rather than actual structures, the bridges and architecture features are designed to represent a period in history; for example, they represent ancient times on the €5 note.

CULTURAL INTEGRATION IN EUROPE. Boundaries where hundreds of thousands of soldiers once stood guard now have little more economic significance in Europe than boundaries between states inside the United States. Crossing borders is a cultural rather than a political experience. For example, highways in the Netherlands are more likely than those in neighboring Belgium to be flanked by well-manicured vegetation and paths reserved for bicycles.

The most noticeable element of cultural diversity within Europe is language. Although English has rapidly become the principal language of business in the EU, much of the EU's budget is spent translating documents into other languages. Businesses must figure out how to effectively advertise their products in several languages. Rather than national boundaries, the most fundamental obstacle to European integration is the multiplicity of languages.

At the same time that residents of European countries are displaying increased tolerance for the cultural values of their immediate neighbors, opposition has increased to the immigration of people from the south and east, especially those who have darker skin and adhere to Islam. Immigrants from poorer regions of Europe, Africa, and Asia fill low-paying jobs (such as cleaning streets and operating buses) that Europeans are not willing to perform. Nonetheless, many Europeans fear that large-scale immigration will transform their nation-states into multiethnic societies.

Underlying this fear of immigration is recognition that natural increase rates are higher in most African and Asian countries than in Europe, as a result of higher crude birth rates. Many Europeans believe that Africans and Asians who immigrate to their countries will continue to maintain relatively high crude birth rates and consequently will constitute even higher percentages of the population in Europe in the future.

ALLIANCES IN OTHER REGIONS

Economic cooperation has been an important factor in the creation of international organizations that now can be found far beyond Western Europe. Other prominent regional organizations include:

- **Organization on Security and Cooperation in Europe (OSCE).** The OSCE's 56 members include the United States, Canada, and Russia, as well as most European countries. When founded in 1975, the Organization on Security and Cooperation was composed primarily of Western European countries and played only a limited role. With the end of the Cold War in the 1990s, the renamed OSCE expanded to include Warsaw Pact countries and became a more active forum for countries concerned with ending conflicts in Europe, especially in the Balkans and Caucasus. Although the OSCE does not directly command armed forces, it can call upon member states to supply troops, if necessary.
- **Organization of American States (OAS).** All 35 states in the Western Hemisphere are members of the OAS. Cuba is a member but was suspended from most OAS activities in 1962. The organization's headquarters, including the permanent council and general assembly, are located in Washington, D.C. The OAS promotes social, cultural, political, and economic links among member states.
- **African Union (AU).** Established in 2002, the AU encompasses 53 countries in Africa. The AU replaced an earlier organization called the Organization of African Unity, founded in 1963 primarily to seek an end to colonialism and apartheid in Africa. The new organization has placed more emphasis on promoting economic integration in Africa.
- **Commonwealth.** The Commonwealth includes the United Kingdom and 52 other states that were once British colonies, including Australia, Bangladesh, Canada, India, Nigeria, and Pakistan. Most other members are African states or island countries in the Caribbean or Pacific. Commonwealth members seek economic and cultural cooperation (Figure 8-46).

▲ FIGURE 8-46 **THE COMMONWEALTH** Commonwealth countries hold games every four years, including 2014, in Glasgow, Scotland. One of the venues in Glasgow is Clyde Auditorium.

Terrorism by Individuals and Organizations

Learning Outcome 8.4.3
Explain the concept of terrorism.

Terrorism is the systematic use of violence by a group in order to intimidate a population or coerce a government into granting its demands. Distinctive characteristics of terrorists include:

- Trying to achieve their objectives through organized acts that spread fear and anxiety among the population, such as bombing, kidnapping, hijacking, taking of hostages, and assassination.
- Viewing violence as a means of bringing widespread publicity to goals and grievances that are not being addressed through peaceful means.
- Believing in a cause so strongly that they do not hesitate to attack despite knowing they will probably die in the act.

The term *terror* (from the Latin "to frighten") was first applied to the period of the French Revolution between March 1793 and July 1794, known as the Reign of Terror. In the name of protecting the principles of the revolution, the Committee of Public Safety, headed by Maximilien Robespierre, guillotined several thousand of its political opponents. In modern times, the term *terrorism* has been applied to actions by groups operating outside government rather than to groups of official government agencies, although some governments provide military and financial support for terrorists.

Many political leaders have been assassinated, though this is not considered terrorism. For example:

- Four U.S. presidents—Lincoln (1865), Garfield (1881), McKinley (1901), and Kennedy (1963).
- Roman Emperor Julius Caesar (44 B.C.), vividly re-created for future generations through Shakespeare's play.
- Archduke Franz Ferdinand, heir to the throne of Austria-Hungary, by a Serb in Sarajevo (capital of present-day Bosnia and Herzegovina), June 28, 1914, which led directly to the outbreak of World War I.

Terrorism differs from assassination and other acts of political violence in that attacks are aimed at ordinary people rather than at military targets or political leaders. Other types of military action can result in civilian deaths—bombs can go astray, targets can be misidentified, or an enemy's military equipment can be hidden in civilian buildings—but average individuals are unintended victims rather than principal targets in most conflicts. A terrorist considers all citizens responsible for the actions he or she opposes, so they are therefore equally legitimate as victims.

TERRORISM AGAINST AMERICANS

The United States suffered several terrorist attacks during the late twentieth century:

- **December 21, 1988:** A terrorist bomb destroyed Pan Am Flight 103 over Lockerbie, Scotland, killing all 259 aboard, plus 11 on the ground.
- **February 26, 1993:** A car bomb parked in the underground garage damaged New York's World Trade Center, killing 6 and injuring about 1,000.
- **April 19, 1995:** A car bomb killed 168 people in the Alfred P. Murrah Federal Building in Oklahoma City.
- **June 25, 1996:** A truck bomb blew up an apartment complex in Dhahran, Saudi Arabia, killing 19 U.S. soldiers who lived there and injuring more than 100 people.
- **August 7, 1998:** U.S. embassies in Kenya and Tanzania were bombed, killing 190 and wounding nearly 5,000.
- **October 12, 2000:** The USS *Cole* was bombed while in the port of Aden, Yemen, killing 17 U.S. service personnel.

Some of the terrorists during the 1990s were American citizens operating alone or with a handful of others:

- Theodore J. Kaczynski, known as the Unabomber, was convicted of killing 3 people and injuring 23 others by sending bombs through the mail during a 17-year period. His targets were mainly academics in technological disciplines and executives in businesses whose actions he considered to be adversely affecting the environment.
- Timothy J. McVeigh was convicted and executed for the Oklahoma City bombing. For assisting McVeigh, Terry I. Nichols was convicted of conspiracy and involuntary manslaughter but not executed. McVeigh claimed that his terrorist act was provoked by rage against the U.S. government for such actions as the Federal Bureau of Investigation's 51-day siege of the Branch Davidian religious compound near Waco, Texas, culminating with an attack on April 19, 1993, that resulted in 80 deaths.

SEPTEMBER 11, 2001, ATTACKS

The most dramatic terrorist attacks against the United States came on September 11, 2001 (Figure 8-47). The tallest buildings in the United States, the 110-story twin towers of the World Trade Center in New York City, were destroyed (Figure 8-48), and the Pentagon, near Washington, D.C., was damaged. The attacks resulted in nearly 3,000 fatalities:

- 93 (5 terrorists, 77 other passengers, and 11 crew members) on American Airlines Flight 11, which crashed into World Trade Center Tower 1 (North Tower).
- 65 (5 terrorists, 51 other passengers, and 9 crew members) on United Airlines Flight 175, which crashed into World Trade Center Tower 2 (South Tower).
- 2,605 on the ground at the World Trade Center.

◀ **FIGURE 8-47 TERRORIST ATTACK ON THE WORLD TRADE CENTER**
On September 11, 2001, at 9:03 A.M., United Flight 175 approaches World Trade Center Tower 2 (left) and crashes into it (right). Tower 1 is already burning from the crash of American Flight 11 at 8:45 A.M.

- 64 (5 terrorists, 53 other passengers, and 6 crew members) on American Airlines Flight 77, which crashed into the Pentagon.
- 125 on the ground at the Pentagon.
- 44 (4 terrorists, 33 other passengers, and 7 crew members) on United Airlines Flight 93, which crashed near Shanksville, Pennsylvania, after passengers fought with terrorists on board, preventing an attack on another Washington, D.C., target.

Responsible or implicated in most of the anti-U.S. terrorism during the 1990s, as well as the September 11, 2001, attack, was the al-Qaeda network, founded by Osama bin Laden. His father, Mohammed bin Laden, a native of Yemen, established a construction company in Saudi Arabia and became a billionaire through close connections to the royal family. Osama bin Laden, one of about 50 children fathered by Mohammed with several wives, used his several-hundred-million-dollar inheritance to fund al-Qaeda (an Arabic word meaning "the foundation," or "the base") around 1990 to unite *opposition* fighters in Afghanistan, as well as supporters of bin Laden elsewhere in the Middle East.

▲ **FIGURE 8-48 AFTERMATH OF WORLD TRADE CENTER ATTACK**
Laser technology was used to create a topographic map of the World Trade Center site on September 19, 2001, eight days after the attack. Colors represent elevation above sea level (in green) or below sea level (in red) of the destroyed buildings. Rubble was piled more than 60 feet high where the twin towers once stood. The top of the image faces northeast. West Street runs across the foreground, and Liberty Street runs between the bottom center and the upper right. Tower 1 rubble is the square-shaped pile in the middle of the block facing West Street. The remains of Tower 2 face Liberty Street.

Bin Laden moved to Afghanistan during the mid-1980s to support the fight against the Soviet army and the country's Soviet-installed government. Calling the anti-Soviet fight a holy war, or *jihad*, bin Laden recruited militant Muslims from Arab countries to join the cause. After the Soviet Union withdrew from Afghanistan in 1989, bin Laden returned to Saudi Arabia, but he was expelled in 1991 for opposing the Saudi government's decision permitting the United States to station troops there during the 1991 war against Iraq. Bin Laden moved to Sudan but was expelled in 1994 for instigating attacks against U.S. troops in Yemen and Somalia, so he returned to Afghanistan, where he lived as a "guest" of the Taliban-controlled government.

Bin Laden issued a declaration of war against the United States in 1996 because of U.S. support for Saudi Arabia and Israel. In a 1998 *fatwa* ("religious decree"), bin Laden argued that Muslims have a duty to wage a holy war against U.S. citizens because the United States was responsible for maintaining the Saud royal family as rulers of Saudi Arabia and a state of Israel dominated by Jews. Destruction of the Saudi monarchy and the Jewish state of Israel would liberate from their control Islam's three holiest sites of Makkah (Mecca), Madinah, and Jerusalem.

Pause and Reflect 8.4.3
How has travel in the United States been affected by the 9/11 attacks?

AL-QAEDA

Learning Outcome 8.4.4
Describe ways that states have sponsored terrorism.

Al-Qaeda has been implicated in several attacks since 9/11:

- **May 8, 2002:** 13 died in a car bomb detonated outside the Sheraton Hotel in Karachi, Pakistan.
- **May 12, 2003:** 35 died (including 9 terrorists) in car bomb detonations at two apartment complexes in Riyadh, Saudi Arabia.
- **November 15, 2003:** Truck bombs killed 29 (including 2 terrorists) at two synagogues in Istanbul, Turkey.
- **November 20, 2003:** 32 (including 2 terrorists) were killed at the British consulate and British-owned HSBC Bank in Istanbul.
- **May 29, 2004:** 22 died in attacks on oil company offices in Khobar, Saudi Arabia.
- **July 7, 2005:** 56 died (including 4 terrorists) when several subway trains and buses were bombed in London, England.
- **July 23, 2005:** 88 died in bombings of resort hotels in Sharm-el-Sheikh, Egypt.
- **November 9, 2005:** 63 died in the bombing of three American-owned hotels in Amman, Jordan.
- **September 28, 2008:** 54 died in a truck bombing of a hotel in Islamabad, Pakistan.
- **December 25, 2009:** Al-Qaeda member Umar Farouk Abdulmatallab, a passenger on a flight from Amsterdam to Detroit, tried to detonate explosives sewn into his underwear. Passengers put out the flames from the failed detonation and restrained the operative until the plane landed. Abdulmatallab was sentenced to four consecutive life terms plus 50 years.

Al-Qaeda is not a single unified organization, and the number involved in al-Qaeda is unknown. Bin Laden was advised by a small leadership council, which has several committees that specialize in areas such as finance, military, media, and religious policy. In addition to the original organization founded by Osama bin Laden responsible for the World Trade Center attack, al-Qaeda also encompasses local franchises concerned with country-specific issues, as well as imitators and emulators ideologically aligned with al-Qaeda but not financially tied to it.

Jemaah Islamiyah is an example of an al-Qaeda franchise with local concerns, specifically with establishing fundamentalist Islamic governments in Southeast Asia. Jemaah Islamiyah terrorist activities have been concentrated in the world's most populous Muslim country, Indonesia:

- **October 12, 2002:** A nightclub in the resort town of Kuta on the island of Bali was bombed, killing 202.
- **August 5, 2003:** Car bombs killed 12 at a Marriott hotel in the capital Jakarta.
- **September 9, 2004:** Car bombs killed 3 at the Australian embassy, also in Jakarta.
- **October 1, 2005:** Attacks on a downtown square in Kuta as well as a food court in Jimbaran, also on Bali, killed 26.
- **July 17, 2009:** Bombs killed 9 at the Marriott and Ritz-Carlton hotels in Jakarta.

Other terrorist groups have been loosely associated with al-Qaeda. For example:

- November 28, 2002: A Somali terrorist group killed 10 Kenyan dancers and 3 Israeli tourists at a resort in Mombasa, Kenya, and fired two missiles at an Israeli airplane taking off from the Mombasa airport.
- March 11, 2004: A local terrorist group blew up several commuter trains in Madrid, Spain, killing 192.

Al-Qaeda's use of religion to justify attacks has posed challenges to Muslims and non-Muslims alike. For many Muslims, the challenge has been to express disagreement with the policies of governments in the United States and Europe yet disavow the use of terrorism. For many Americans and Europeans, the challenge has been to distinguish between the peaceful but unfamiliar principles and practices of the world's 1.3 billion Muslims and the misuse and abuse of Islam by a handful of terrorists.

State Support for Terrorism

Several states in the Middle East have provided support for terrorism in recent years, at three increasing levels of involvement:

- Providing sanctuary for terrorists wanted by other countries.
- Supplying weapons, money, and intelligence to terrorists.
- Planning attacks using terrorists.

SANCTUARY FOR TERRORISTS

Afghanistan and probably Pakistan have provided sanctuary for al-Qaeda terrorists.

AFGHANISTAN. The United States attacked Afghanistan in 2001, when its leaders, known as the Taliban, sheltered bin Laden and other al-Qaeda terrorists. During the battle of Tora Bora in December 2001, the United States overran positions held by al-Qaeda, but bin Laden escaped (Figure 8-49).

The Taliban had gained power in Afghanistan in 1995 and had imposed strict Islamic fundamentalist law on the population. Afghanistan's Taliban leadership treated women especially harshly. Women were prohibited from attending school, working outside the home, seeking health care, or driving a car. They were permitted to leave home only if fully covered by clothing and escorted by a male relative.

The six years of Taliban rule temporarily suppressed a civil war that has raged in Afghanistan on and off since the 1970s. The civil war began in 1973, when the king was

▼ **FIGURE 8-49 AFGHANISTAN** A mujahadeen fighter kneels by the entrance to a cave used by al-Qaeda fighters in the Tora Bora mountains of eastern Afghanistan, December 18, 2001.

overthrown in a bloodless coup led by Mohammed Daoud Khan. Daoud was murdered five years later and replaced by a government led by military officers sympathetic to the Soviet Union. The Soviet Union sent 115,000 troops to Afghanistan beginning in 1979, after fundamentalist Muslims, known as *mujahedeen*, or "holy warriors," started a rebellion against the pro-Soviet government.

Although heavily outnumbered by Soviet troops and possessing much less sophisticated equipment, the mujahedeen offset the Soviet advantage by waging a guerrilla war in the country's rugged mountains, where they were more comfortable than the Soviet troops and where Soviet air superiority was ineffective. Unable to subdue the mujahedeen, the Soviet Union withdrew its troops in 1989; the Soviet-installed government in Afghanistan collapsed in 1992. After several years of infighting among the factions that had defeated the Soviet Union, the Taliban gained control over most of the country.

Six years of Taliban rule came to an end in 2001, following the U.S. invasion. Destroying the Taliban was necessary in order for the United States to go after al-Qaeda leaders, including bin Laden, who were living in Afghanistan as guests of the Taliban. Removal of the Taliban unleashed a new struggle for control of Afghanistan among the country's many ethnic groups. When U.S. attention shifted to Iraq and Iran, the Taliban were able to regroup and resume an insurgency against the U.S.-backed Afghanistan government.

Pause and Reflect 8.4.4

Suspected terrorists captured primarily in Afghanistan have been detained at a detention camp run by the United States at Guantanamo Bay, Cuba. Do you think that suspected terrorists should be tried in a U.S. military court, brought to the United States for civilian trials, or sent back to the country from which they were captured?

PAKISTAN. The war on terrorism spilled over from Pakistan's western neighbor Afghanistan. Western Pakistan, along the border with Afghanistan, is a rugged, mountainous region inhabited by several ethnic minorities where the Taliban have been largely in control. U.S. intelligence and other experts thought that bin Laden was hiding out in the Taliban-controlled mountains of western Pakistan, but they were wrong. Navy SEALS killed bin Laden in a compound in the city of Abbottabad, only 120 kilometers (75 miles) from the capital.

The United States believed that Pakistan security had to be aware that bin Laden had been living in the compound for at least five years. The compound was heavily fortified, surrounded by high walls and barbed wire (Figure 8-50). Furthermore, the compound was located only 6 kilometers (4 miles) from the Pakistan Military Academy, the country's principal institution for training military officers, equivalent to the U.S. Military Academy in West Point. For their part, Pakistani officials were upset that the United States attacked the compound without their knowledge.

▲ **FIGURE 8-50 OSAMA BIN LADEN'S COMPOUND, PAKISTAN** While in this compound, Osama bin Laden was killed by U.S. Navy SEALS.

SUPPLYING TERRORISTS

Learning Outcome 8.4.5
Describe alleged sponsorship of terrorism in Iraq and Iran.

Iraq and Iran have both been accused of providing material and financial support for terrorists. The extent of their involvement in terrorism is controversial, especially in the case of Iraq.

IRAQ. U.S. claims of state-sponsored terrorism proved more controversial with regard to Iraq than to Afghanistan. The United States led an attack against Iraq in 2003 in order to depose Saddam Hussein, the country's longtime president. U.S. officials' justification for removing Hussein was that he had created biological and chemical weapons of mass destruction. These weapons could fall into the hands of terrorists, the U.S. government charged, because close links were said to exist between Iraq's government and al-Qaeda. The United Kingdom and a few other countries joined in the 2003 attack, but most countries did not offer support.

U.S. confrontation with Iraq predated the war on terror. From the time he became president of Iraq in 1979, Hussein's behavior had raised concern around the world. War with neighbor Iran, begun in 1980, ended 8 years later in stalemate. A nuclear reactor near Baghdad, where nuclear weapons to attack Israel were allegedly being developed, was destroyed in 1981 by Israeli planes. Hussein ordered the use of poison gas in 1988 against Iraqi Kurds, killing 5,000. Iraq's 1990 invasion of neighboring Kuwait, which Hussein claimed was part of Iraq, was opposed by the international community.

The 1991 U.S.-led Gulf War, known as Operation Desert Storm, drove Iraq out of Kuwait, but it failed to remove Hussein from power. Desert Storm was supported by nearly every country in the United Nations because the purpose was to end one country's unjustified invasion and attempted annexation of another. In contrast, few countries supported the U.S.-led attack in 2003; most did not agree with the U.S. assessment that Iraq still possessed weapons of mass destruction or intended to use them.

The U.S. assertion that Hussein had close links with al-Qaeda was also challenged by most other countries, as well as ultimately by U.S. intelligence agencies.

As the United States moved toward war with Iraq in 2003, Secretary of State Colin Powell scheduled a speech at the UN to present evidence to the world justifying military action against Iraq. Recalling the Cuban Missile Crisis (refer to Figure 8-42), Powell displayed a series of air photos designed to prove that Iraq possessed weapons of mass destruction. However, the photos did not provide clear evidence (Figure 8-51).

Lacking evidence of weapons of mass destruction and ties to al-Qaeda, the United States argued instead that Iraq needed a "regime change." Hussein's quarter-century record of brutality justified replacing him with a democratically elected government, according to U.S. officials.

Having invaded Iraq and removed Hussein from power, the United States expected an enthusiastic welcome from the Iraqi people. Instead, the United States became embroiled in a complex and violent struggle among these various religious sects and tribes.

IRAN. Hostility between the United States and Iran dates from 1979, when a revolution forced abdication of Iran's pro-U.S. Shah Mohammad Reza Pahlavi. Iran's majority Shiite population had demanded more democratic rule and opposed the Shah's economic modernization program that generated social unrest. Supporters of exiled fundamentalist Shiite Muslim leader Ayatollah Ruholiah Khomeini then proclaimed Iran an Islamic republic and rewrote the constitution to place final authority with the ayatollah. Militant supporters of the ayatollah seized the U.S. embassy on November 4, 1979, and held 62 Americans hostage until January 20, 1981.

Iran and Iraq fought a war between 1980 and 1988 over control of the Shatt al-Arab waterway, formed by the confluence of the Tigris and Euphrates rivers flowing into the Persian Gulf. Forced to cede control of the waterway to Iran in 1975, Iraq took advantage of Iran's revolution to seize the waterway in 1980, but Iran was not defeated outright, so an eight-year war began that neither side was able to win. An estimated 1.5 million died in the war, which ended when the two countries accepted a UN peace plan.

When the United States launched its war on terrorism after 9/11, Afghanistan was the immediate target, followed by Iraq. But after the election of Mahmoud Ahmadinejad as president in 2005, relations between the United States and Iran deteriorated. The United States accused Iran of harboring al-Qaeda members and of trying to gain influence in Iraq, where, as in Iran, the majority of the people were Shiites. More troubling to the international community was Iran's aggressive development of a nuclear program. Iran claimed that its nuclear program was for civilian purposes, but other countries believed that it was intended to develop weapons. Prolonged negotiations were undertaken to dismantle Iran's nuclear capabilities without resorting to yet another war in the Middle East.

Pause and Reflect 8.4.5
What events have occurred in Iran since this book was published?

STATE TERRORIST ATTACKS: LIBYA

The government of Libya was accused of sponsoring a 1986 bombing of a nightclub in Berlin, Germany, that was popular with U.S. military personnel then stationed there, killing three (including one U.S. soldier). U.S. relations with Libya had been poor since 1981, when U.S. aircraft shot down attacking Libyan warplanes while conducting

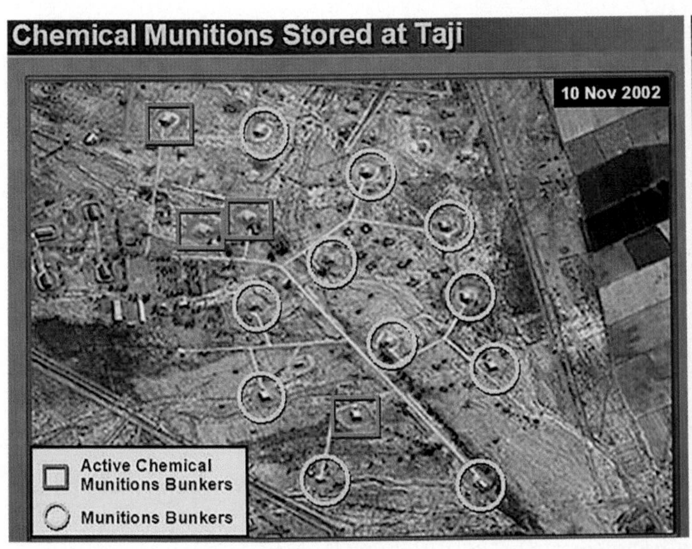

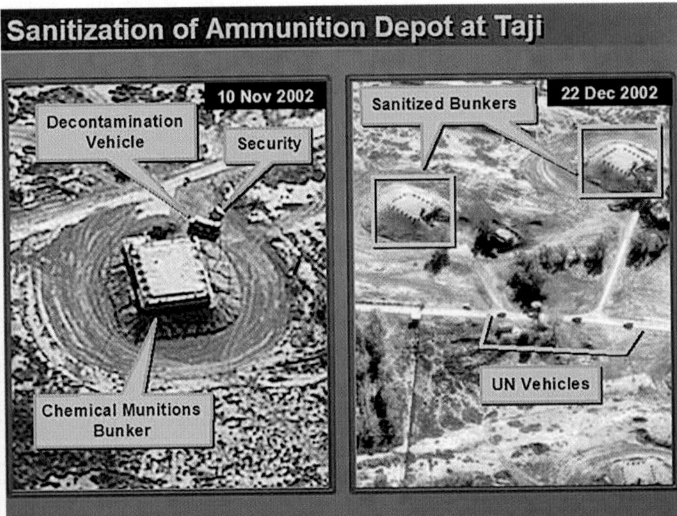

▲ **FIGURE 8-51 AIR PHOTOS ALLEGING IRAQ'S PREPARATIONS FOR CHEMICAL WARFARE** (left) U.S. satellite image purporting to show 15 munitions bunkers in Taji, Iraq. (center) Close-up of alleged munitions bunker outlined in red near the bottom of the left image. The truck labeled "decontamination vehicle" turned out to be a water truck. (right) Close-up of the two bunkers, outlined in red in the middle of the left image, allegedly sanitized.

exercises over waters in the Mediterranean Sea that the United States considered international but that Libya considered inside its territory. In response to the Berlin bombing, U.S. bombers attacked the Libyan cities of Tripoli and Benghazi in a failed attempt to kill Colonel Muammar el-Qaddafi.

Libyan agents were found to have planted bombs on Pan Am Flight 103 that killed 270 people in Lockerbie, Scotland, in 1988 (Figure 8-52), as well as 170 people on UTA Flight 772 over Niger in 1989. Following 8 years of UN economic sanctions, Qaddafi turned over suspects in the Lockerbie bombing for a trial that was held in the Netherlands under Scottish law. One of the two was acquitted; the other, Abdel Basset Ali al-Megrahi, was convicted and sentenced to life imprisonment, but he was released in 2009 after he was diagnosed with terminal cancer. Libya renounced terrorism in 2003 and has provided compensation for victims of Pan Am 103. UN sanctions have been lifted, and Libya is no longer considered a state sponsor of terrorism.

During Arab Spring, Qaddafi tried to crush protests with extreme violence, resulting in thousands of deaths and violations of human rights. To protect the protestors, the UN authorized member states to attack pro-Qaddafi forces. A coalition of 30 member states launched air and naval attacks that enabled the anti-Qaddafi forces to take the offensive and ultimately succeed. Qaddafi was captured and killed.

▼ **FIGURE 8-52 STATE-SPONSORED TERRORISM**
Libya authorized the bombing of Pan Am Flight 103, which blew up over Lockerbie, Scotland, in 1988, killing all 259 aboard, plus 11 on the ground.

CHECK-IN: KEY ISSUE 4

Why Do States Cooperate and Compete With Each Other?

- ✓ **During the Cold War, the world was divided into two alliances led by superpowers.**
- ✓ **With the end of the Cold War, economic alliances have become more important.**
- ✓ **Terrorism by individuals and organizations has included the 9/11 attacks on the United States.**
- ✓ **Some states have provided support for terrorism.**

Summary and Review

KEY ISSUE 1

Where Are States Distributed?

Earth's land area is divided into nearly 200 states. A state is a political unit, with an organized government and sovereignty.

LEARNING OUTCOME 8.1.1: Explain the three eras of rapid growth in UN membership.

- All but a handful of states are members of the UN.
- UN membership grew rapidly in 1955, 1960, and the 1990s.

LEARNING OUTCOME 8.1.2: Explain why it is difficult to determine whether some territories are states.

- Several places are not universally recognized as sovereign.
- Polar regions have not been organized into states, although neighboring states have competing claims on them.

LEARNING OUTCOME 8.1.3: Explain the concept of nation-state and how it differs from earlier ways to govern.

- Dividing the world into states is a modern concept.
- Historically, most of Earth's surface was organized in other ways, such as empires, or else unorganized.

THINKING GEOGRAPHICALLY 8.1: A century ago the British geographer Halford J. Mackinder identified a heartland in the interior of Eurasia (Europe and Asia) that was isolated by mountain ranges and the Arctic Ocean. Mackinder argued that whoever controlled the heartland would control Eurasia and hence the entire world. To what extent has Mackinder's theory been validated during the twentieth century by the creation and then the dismantling of the Soviet Union?

GOOGLE EARTH 8.1: The smallest state is the Holy See (Vatican). What is housed in the government building immediately to the west of St Peter's (identified in Google Earth 6.1)?

KEY ISSUE 2

Why Are Nation-States Difficult to Create?

A nation-state is a state whose territory matches that occupied by an ethnicity. It is impossible to find a perfect match between the boundaries of a state and the area inhabited by a single ethnicity.

LEARNING OUTCOME 8.2.1: Understand the difference between a nation-state and a multinational state.

- No perfect nation-state, exists, but some states come close.
- A multinational state contains multiple ethnicities rather than a single ethnicity.

LEARNING OUTCOME 8.2.2: Describe differences among states formerly in the Soviet Union.

- The U.S.S.R. was once the world's largest multinational state.
- The country's largest ethnicities were organized into 15 republics that are now independent states.

LEARNING OUTCOME 8.2.3: Describe patterns of distribution of ethnicities in Russia and the Caucasus.

- Russia is now the world's largest multinational state, with numerous ethnic groups.
- The Caucasus Mountain region contains a complex array of ethnicities divided among several small states.

LEARNING OUTCOME 8.2.4: Explain the concept of colonies and describe their current distribution.

- A colony is territory legally tied to a state. Into the twentieth century, much of the world consisted of colonies, but few remain.

THINKING GEOGRAPHICALLY 8.2: To what extent should a country's ability to provide its citizens with food, jobs, economic security, and material wealth, rather than the principle of self-determination, become the basis for dividing the world into independent countries?

GOOGLE EARTH 8.2: The boundary between what states run through the Caucasus Mountains?

Key Terms

Anocracy (p. 282) A country that is not fully democratic or fully autocratic, but rather displays a mix of the two types.

Autocracy (p. 282) A country that is run according to the interests of the ruler rather than the people.

Balance of power (p. 286) A condition of roughly equal strength between opposing countries or alliances of countries.

Boundary (p. 276) An invisible line that marks the extent of a state's territory.

City-state (p. 266) A sovereign state comprising a city and its immediately surrounding countryside.

Colonialism (p. 274) An attempt by one country to establish settlements and to impose its political, economic, and cultural principles in another territory.

Colony (p. 274) A territory that is legally tied to a sovereign state rather than completely independent.

Compact state (p. 280) A state in which the distance from the center to any boundary does not vary significantly.

Democracy (p. 282) A country in which citizens elect leaders and can run for office.

Elongated state (p. 280) A state with a long, narrow shape.

Federal state (p. 283) An internal organization of a state that allocates most powers to units of local government.

Fragmented state (p. 281) A state that includes several discontinuous pieces of territory.

Frontier (p. 276) A zone separating two states in which neither state exercises political control.

Gerrymandering (p. 284) The process of redrawing legislative boundaries for the purpose of benefiting the party in power.

Landlocked state (p. 281) A state that does not have a direct outlet to the sea.

Microstate (p. 261) A state that encompasses a very small land area.

Multiethnic state (p. 268) A state that contains more than one ethnicity.

Multinational state (p. 268) A state that contains two or more ethnic groups with traditions of self-determination that agree to

KEY ISSUE 3

Why Do Boundaries Cause Problems?

States are separated by boundaries, which are either physical or cultural. Boundaries affect the shape of a country and affect the ability of a country to live peacefully with its neighbors.

LEARNING OUTCOME 8.3.1: Describe the types of physical boundaries between states.

- Physical features used to delineate boundaries include deserts, mountains, and bodies of water.

LEARNING OUTCOME 8.3.2: Describe the types of cultural boundaries between states.

- Geometry and ethnicities can be used to delineate cultural boundaries between states.

LEARNING OUTCOME 8.3.3: Describe five shapes of states.

- States take five forms: compact, elongated, prorupted, perforated, and fragmented.

LEARNING OUTCOME 8.3.4: Describe differences among the three regime types.

- Regimes can be democratic, anocratic, or autocratic; the trend has been toward more democratic regimes.
- Local governments can be organized according to unitary or federal state principles; the trend has been toward more federal states.

LEARNING OUTCOME 8.3.5: Explain the concept of gerrymandering and three ways that it is done.

- Gerrymandering is the redrawing of electoral districts to benefit the party in power.
- Three forms of gerrymandering are wasted vote, excess vote, and stacked vote.

THINKING GEOGRAPHICALLY 8.3: Given the movement toward increased local government autonomy on the one hand and increased authority for international organizations on the other, what is the future of the nation-state? Have political and economic trends since the 1990s strengthened the concept of nation-state or weakened it?

GOOGLE EARTH 8.3: Portions of what four states can be seen near the Libyan Desert?

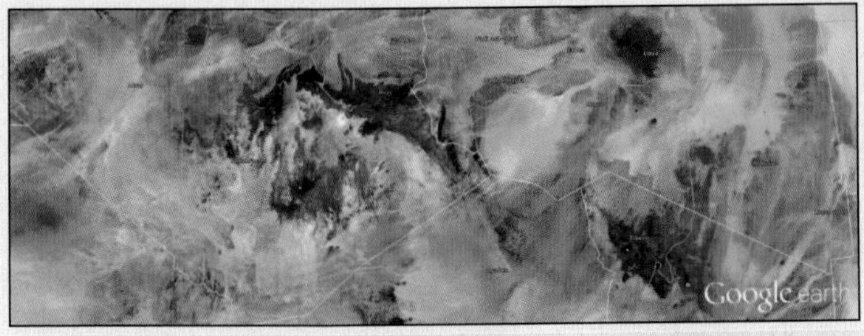

KEY ISSUE 4

Why Do States Cooperate and Compete with Each Other?

Competition among states has been replaced in some regions by economic alliances, especially in Europe. At the same time, violence has increased in the world because of terrorist attacks.

LEARNING OUTCOME 8.4.1: Describe the principal alliances in Europe during the Cold War era.

- States were allied with the two superpowers, the Soviet Union and the United States.

LEARNING OUTCOME 8.4.2: Describe the principal economic alliances in Europe in the period since World War II.

- With the end of the Cold War, economic alliances have replaced military alliances in importance, especially in Europe.

LEARNING OUTCOME 8.4.3: Explain the concept of terrorism.

- Terrorism is the systematic use of violence to intimidate a population or coerce a government.

LEARNING OUTCOME 8.4.4: Describe ways that states have sponsored terrorism.

- States have supported terrorism by providing sanctuary to terrorists, supplying them with weapons and intelligence, and planning state-sponsored attacks.

THINKING GEOGRAPHICALLY 8.4: In his book *1984*, George Orwell divided the world into three states, controlling people through technology. To what extent has Orwell's vision of a global political order been realized in an age of terrorism?

GOOGLE EARTH 8.4: If you zoom into the center of Abbotabad, Pakistan, where Osama bin Laden's hideout was located, turn on 3D, switch to ground-level view, and pan around, what is the only 3D building? Given the distribution of religions discussed in Chapter 6, why does this building seem out of place here?

coexist peacefully by recognizing each other as distinct nationalities.

Nation-state (p. 267) A state whose territory corresponds to that occupied by a particular ethnicity that has been transformed into a nationality.

Perforated state (p. 281) A state that completely surrounds another one.

Prorupted state (p. 280) An otherwise compact state with a large projecting extension.

Self-determination (p. 267) The concept that ethnicities have the right to govern themselves.

Sovereignty (p. 261) Ability of a state to govern its territory free from control of its internal affairs by other states.

State (p. 261) An area organized into a political unit and ruled by an established government that has control over its internal and foreign affairs.

Terrorism (p. 290) The systematic use of violence by a group in order to intimidate a population or coerce a government into granting its demands.

Unitary state (p. 283) An internal organization of a state that places most power in the hands of central government officials.

Chapter

9 Development

Why does India have so many bureaucrats? Page 328

Why is this coffee special? Page 336

KEY ISSUE 1

Why Does Development Vary among Countries?

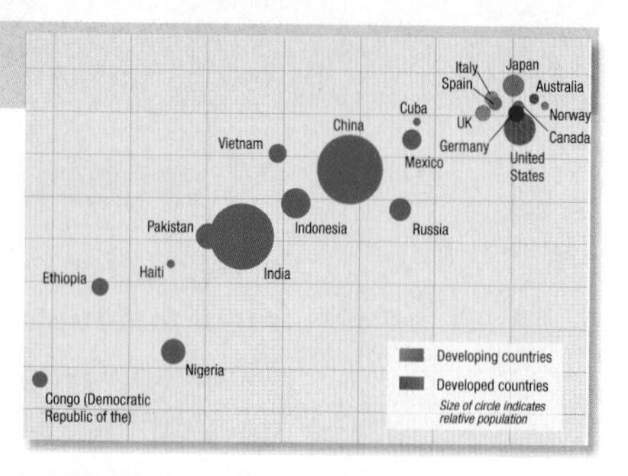

Developed and Developing Countries p. 301

Every country is at some level of development.

KEY ISSUE 2

Why Does Development Vary by Gender?

Gender Inequality p. 310

The UN says that men are treated better than women everywhere.

▲ This truck is traveling Trans-African Highway 8, which is the principal east–west road across Africa, 6,259 kilometers (3,890 miles) between Mombasa, Kenya, and Lagos, Nigeria. Aid from the United Nations and other international organizations has funded construction of the road. The eastern and western sections of the route are paved, but the central portion, through the Democratic Republic of Congo, is unpaved. The absence of good roads is one of many obstacles to development in sub-Saharan Africa.

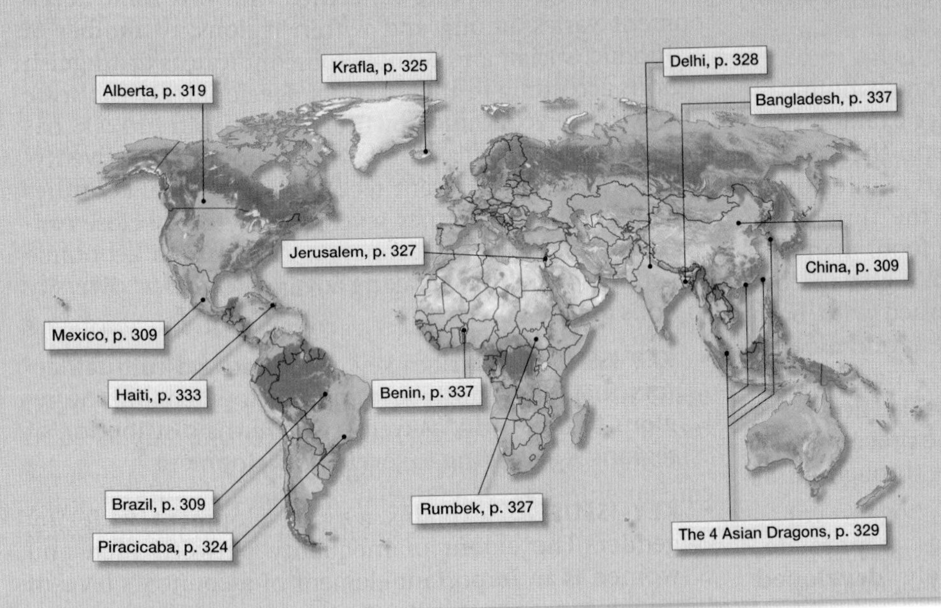

KEY ISSUE 3

Why Are Energy Resources Important for Development?

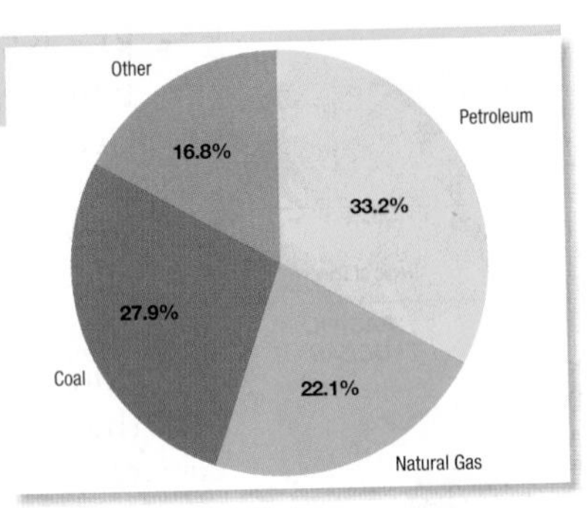

It Takes a Lot of Energy p. 314

Development needs abundant energy, but some sources are being depleted.

KEY ISSUE 4

Why Do Countries Face Obstacles to Development?

Trade or Stand Alone? p. 328

Countries choose paths to development and look for money to finance it.

A Decent Standard of Living

Learning Outcome 9.1.1
Identify the HDI standard of living factor.

Having enough wealth for a decent standard of living is key to development. The average individual in a developed country earns a much higher income than the average individual in a developing one. Geographers observe that people generate and spend their wealth in different ways in developed countries than in developing countries.

INCOME

The UN measures the standard of living in countries through a complex index called annual gross national income per capita at purchasing power parity:

- **Gross national income (GNI)** is the value of the output of goods and services produced in a country in a year, including money that leaves and enters the country.
- **Purchasing power parity (PPP)** is an adjustment made to the GNI to account for differences among countries in the cost of goods. For example, if a resident of country A has the same income as a resident of country B but must pay more for a Big Mac or a Starbucks latte, the resident of country B is better off.

By dividing GNI by total population, it is possible to measure the contribution made by the average individual toward generating a country's wealth in a year. For example, GNI in the United States was approximately $15 trillion in 2011, and its population was approximately 312 million, so GNI per capita was approximately $47,000. In 2011, per capita GNI was approximately $34,000 in developed countries compared to approximately $7,000 in developing countries (Figure 9-4).

Some studies refer to **gross domestic product (GDP)**, which is also the value of the output of goods and services produced in a country in a year, but it does not account for money that leaves and enters the country.

Per capita GNI—or, for that matter, any other single indicator—cannot measure perfectly the level of a country's development. Few people may be starving in a developing country with per capita GNI of a few thousand dollars. And not everyone is wealthy in a developed country with per capita GNI of $40,000. Per capita GNI measures average (mean) wealth, not the distribution of wealth. If only a few people receive much of the GNI, then the standard of living for the majority may be lower than the average figure implies. The higher the per capita GNI, the greater the potential for ensuring that all citizens can enjoy a comfortable life.

ECONOMIC STRUCTURE

Average per capita income is higher in developed countries because people typically earn their living by different means than in developing countries. Jobs fall into three categories:

- The **primary sector** includes activities that directly extract materials from Earth through agriculture and sometimes by mining, fishing, and forestry.
- The **secondary sector** includes manufacturers that process, transform, and assemble raw materials into useful products, as well as industries that fabricate manufactured goods into finished consumer goods.
- The **tertiary sector** involves the provision of goods and services to people in exchange for payment, such as retailing, banking, law, education, and government.

The contribution to GNI among primary, secondary, and tertiary sectors varies between developed and developing countries (Figure 9-5):

- The share of GNI accounted for by the primary sector has decreased in developing countries, but it remains higher than in developed countries.
- The share of GNI accounted for by the secondary sector has decreased sharply in developed countries and is now less than in developing countries.
- The share of GNI accounted for by the tertiary sector is relatively large in developed countries, and it continues to grow.

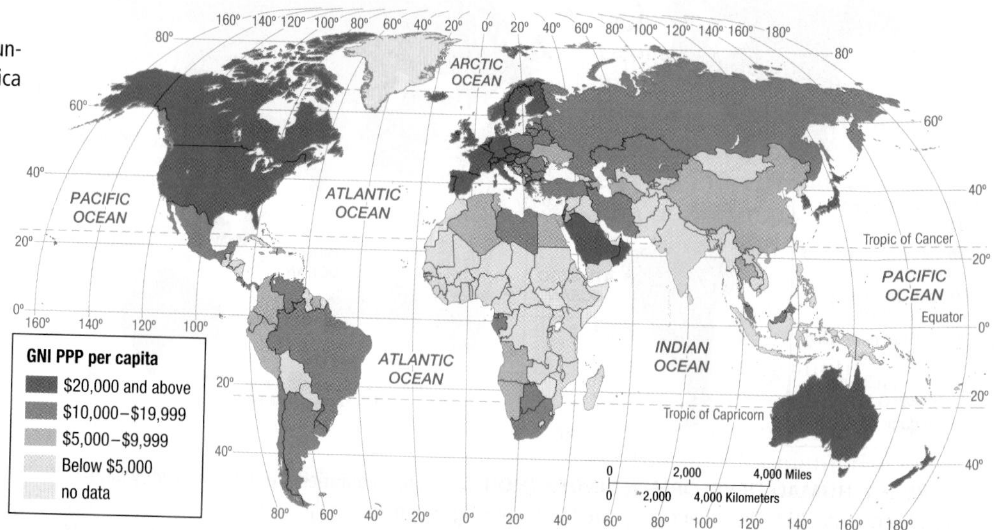

▶ **FIGURE 9-4 INCOME**
GNI per capita PPP is highest in developed countries. The lowest figures are in sub-Saharan Africa and South Asia.

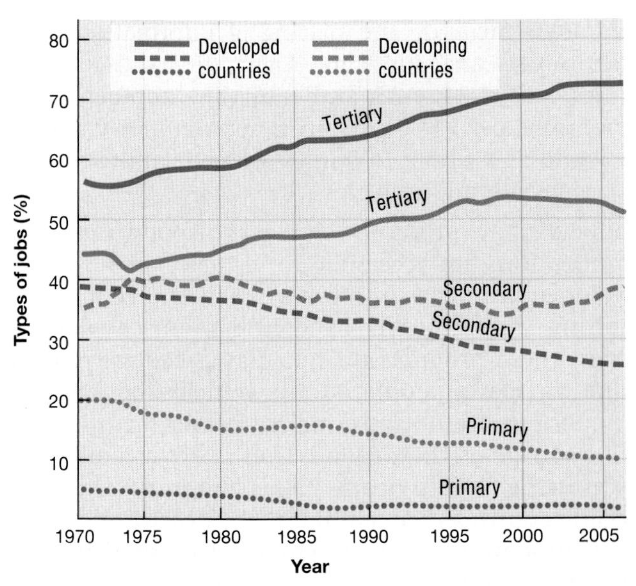

▲ **FIGURE 9-5 ECONOMIC STRUCTURE** The percentage of GNI contributed, by type of job.

The relatively low percentage of primary-sector workers in developed countries indicates that a handful of farmers produce enough food for the rest of society. Freed from the task of growing their own food, most people in a developed country can contribute to increasing the national wealth by working in the secondary and tertiary sectors.

PRODUCTIVITY

Workers in developed countries are more productive than those in developing countries. **Productivity** is the value of a particular product compared to the amount of labor needed to make it. Productivity can be measured by the value added per capita. The **value added** in manufacturing is the gross value of a product minus the costs of raw materials and energy. The value added per capita in 2010 was around $5,900 in the United States and $6,700 in Japan, compared to around $800 in China and $100 in India.

Workers in developed countries produce more with less effort because they have access to more machines, tools, and equipment to perform much of the work. On the other hand, production in developing countries relies more on human and animal power. The larger per capita GNI in developed countries in part pays for the manufacture and purchase of machinery, which in turn makes workers more productive and generates more wealth.

INEQUALITY-ADJUSTED HDI

The UN believes that every person should have access to decent standards of living, knowledge, and health. The **inequality-adjusted HDI (IHDI)** is an indicator of development that modifies the HDI to account for inequality within a country. Under perfect equality, the HDI and the IHDI are the same. If the IHDI is lower than the HDI, the country has some inequality; the greater the difference in the two measures, the greater the inequality. A country where only a few people have high incomes, college degrees, and good health care would have a lower IHDI than a country where differences in income, level of education, and access to health care are minimal.

The lowest scores (highest inequality) are in sub-Saharan Africa and South Asia. The score may be low in Southwest Asia & North Africa, but the UN lacks data from a number of the region's countries (Figure 9-6).

Pause and Reflect 9.1.1

The IHDI is 0.77 in the United States and 0.83 in Canada. Which country has greater inequality?

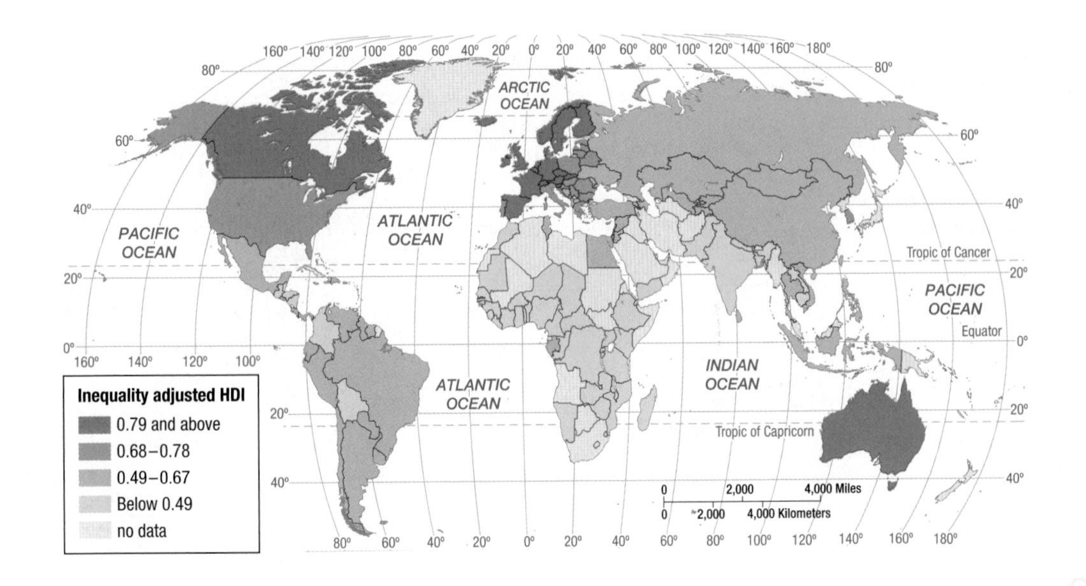

► **FIGURE 9-6 INEQUALITY-ADJUSTED HDI** The lower the score, the greater the inequality.

CONSUMER GOODS

Learning Outcome 9.1.2
Identify the HDI health factor.

Part of the wealth generated in developed countries is used to purchase goods and services. Especially important are goods and services related to transportation and communications, including motor vehicles, telephones, and computers:

- Motor vehicles provide individuals with access to jobs and services and permit businesses to distribute their products (Figure 9-7). The number of motor vehicles per 1,000 persons is approximately 170 in the world as a whole, 630 in developed countries, and 80 in developing countries.
- Telephones enhance interaction with providers of raw materials and customers for goods and services (Figure 9-8). The number of cell phones per 1,000 persons is approximately 800 in the world as a whole, 1,100 in developed countries, and 700 in developing countries.
- Computers facilitate the sharing of information with other buyers and suppliers (Figure 9-9 and refer to Figures 4-32, 4-34, 4-35, and 4-36). The number of Internet users per 1,000 persons is approximately 300 in the world as a whole, 700 in developed countries, and 200 in developing countries.

Products that promote better transportation and communications are accessible to virtually all residents in developed countries and are vital to the economy's functioning and growth. In contrast, in developing countries, these products do not play a central role in daily life for many people. Motor vehicles, computers, and telephones are not essential to people who live in the same village as their friends and relatives and work all day growing food in nearby fields. But most people in developing countries are familiar with these goods, even if they cannot afford them, and may desire them as symbols of development.

Because possession of consumer goods is not universal in developing countries, a gap can emerge between the "haves" and the "have-nots." The minority of people who have these goods may include government officials, business owners, and other elites, whereas their lack among

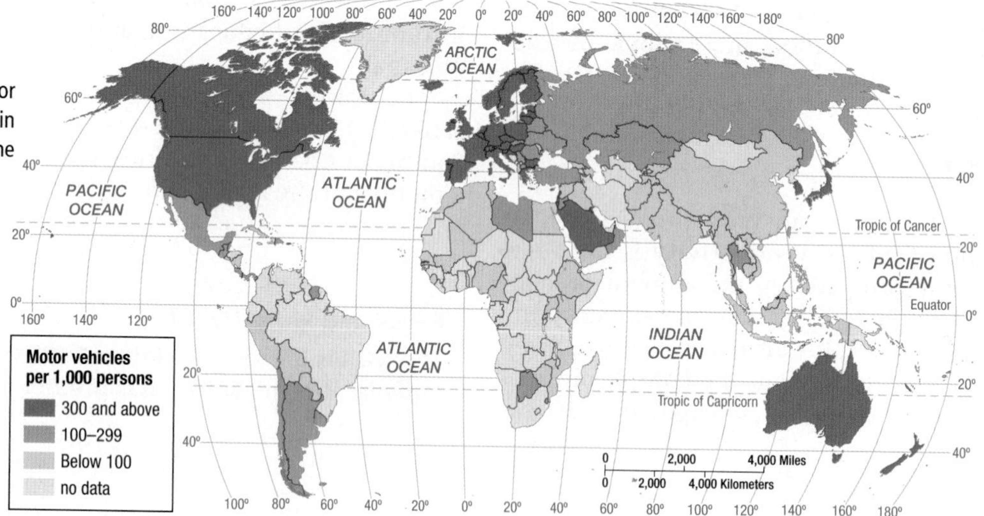

▶ **FIGURE 9-7 CONSUMER GOODS: MOTOR VEHICLES** The highest level of motor vehicle ownership is in North America, and the lowest is in South Asia.

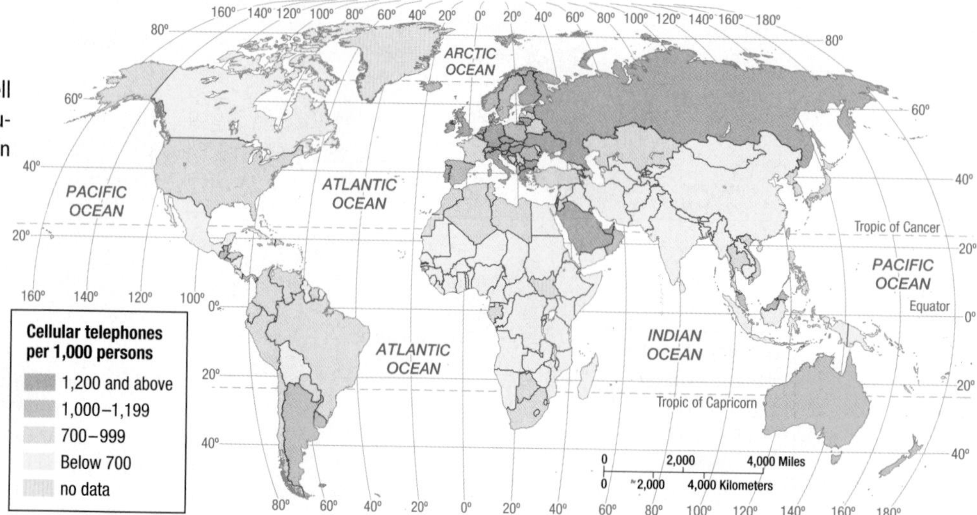

▶ **FIGURE 9-8 CONSUMER GOODS: CELL PHONES** The highest level of cell phone ownership is in Europe, and the lowest is in sub-Saharan Africa.

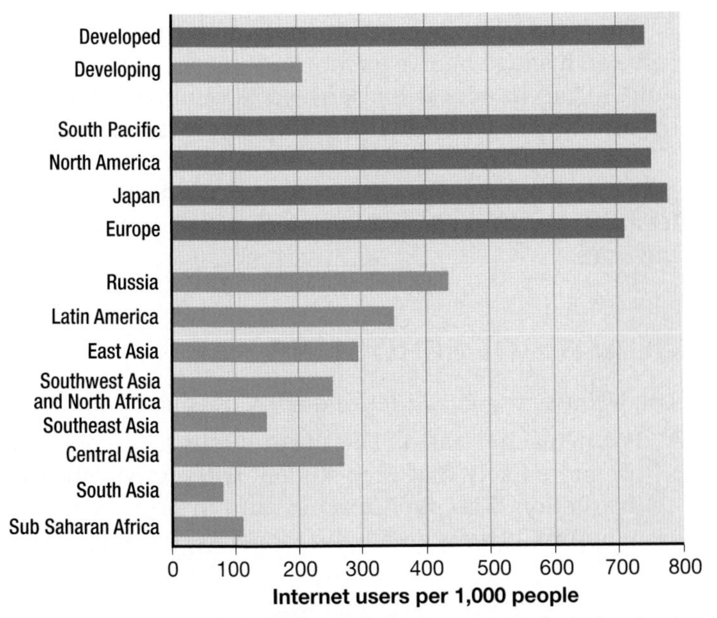

▲ **FIGURE 9-9 CONSUMER GOODS: INTERNET USERS** The highest level of Internet users is in North America, and the lowest is in South Asia.

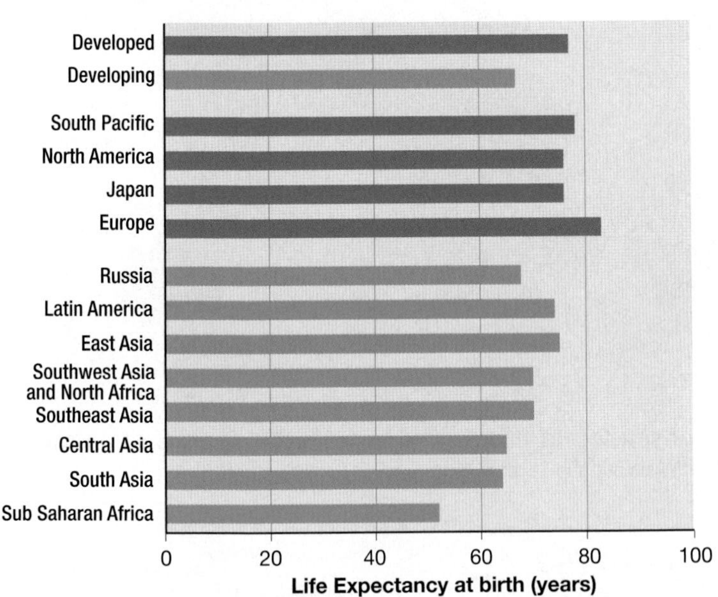

▲ **FIGURE 9-10 LIFE EXPECTANCY AT BIRTH** The highest life expectancy is in Europe, and the lowest is in sub-Saharan Africa.

the majority who are denied access may provoke political unrest. In many developing countries, those who have these products are concentrated in urban areas; those who do not live in the countryside. Technological innovations tend to diffuse from urban to rural areas. Access to these goods is more important in urban areas because of the dispersion of homes, factories, offices, and shops.

Developed countries also use some of their wealth to provide infrastructure, such as roads, bridges, airports, electricity, and water. The image of the trans-African Highway on page 299 illustrates the challenges that regions with low levels of development face in providing infrastructure that can help promote development.

Technological change is helping to reduce the gap between developed and developing countries in access to communications. Cell phone ownership, for example, is expanding rapidly in developing countries because these phones do not require the costly investment of connecting wires to each individual building, and many individuals can obtain service from a single tower or satellite.

Pause and Reflect 9.1.2
In addition to cell phones, what other electronic devices might diffuse rapidly to developing countries because of low cost of equipment and lack of need for expensive infrastructure?

A Long and Healthy Life

The UN considers good health to be an important measure of development. A goal of development is to provide the nutrition and medical services needed for people to lead long and healthy lives.

The health indicator contributing to the HDI is life expectancy at birth. On average, a baby born today is expected to live to age 70. The life expectancy is 80 in developed regions and 68 to in developing regions (Figure 9-10 and refer to Figure 2-40). Variation among developing regions is especially wide; life expectancy is 75 in Latin America, 65 in South Asia, and 55 in sub-Saharan Africa.

People are healthier in developed countries than in developing ones. When people in developed countries get sick, these countries possess the resources to care for them. Developed countries use part of their wealth to protect people who, for various reasons, are unable to work. In these countries, some public assistance is offered to those who are sick, elderly, poor, disabled, orphaned, veterans of wars, widows, unemployed, or single parents. Better health and welfare in developed countries permit people to live longer. (Refer to Key Issue 4 and Figures 2-42 through 2-46 in Chapter 2.)

With longer life expectancies, developed countries have a higher percentage of older people who have retired and receive public support and a lower percentage of children under age 15 who are too young to work and must also be supported by employed adults and government programs. The number of young people is six times higher than the number of older people in developing countries, whereas the two are nearly the same in developed countries.

Better health and welfare also permit more babies to survive infancy in developed countries. About 94 percent of infants survive and 6 percent die in developing countries, whereas in developed countries more than 99.5 percent survive and fewer than one-half of 1 percent perish (see Figure 2-40). The infant mortality rate is greater in developing countries for several reasons. Babies may die from malnutrition or lack of medicine needed to survive illness, such as dehydration from diarrhea. They may also die from poor medical practices that arise from lack of education.

Access to Knowledge

Learning Outcome 9.1.3
Identify the HDI access to knowledge factor.

Development is about more than possession of wealth. The UN believes that access to knowledge is essential for people to have the possibility of leading lives of value. In general, the higher the level of development, the greater are both the quantity and the quality of a country's education. For young people in both developed and developing countries, education is the ticket to better jobs and higher social status.

QUANTITY OF SCHOOLING

The UN considers years of schooling to be the most critical measure of the ability of an individual to gain access to knowledge needed for development. The assumption is that no matter how poor the school, the longer the pupils attend, the more likely they are to learn something. To form the access to knowledge component of HDI, the UN combines two measures of quantity of schooling:

- **Years of schooling.** This is the number of years that the average person aged 25 or older in a country has spent in school. The average pupil has attended school for approximately 7 years in the world as a whole, 11 years in developed countries, and 6 years in developing countries (Figure 9-11).
- **Expected years of schooling.** This is the number of years that an average 5-year-old child is expected to spend in school. The UN expects that today's 5-year-old will attend an average of 16 years in school in developed countries and 11 years in developing ones, as well as in the world as a whole (Figure 9-12). In other words, the average child is expected to attend college in developed countries but not finish high school in developing countries.

Thus, the UN expects children around the world to receive an average of 5 years more education in the future, but the gap in education between developed and developing regions will remain high. Otherwise stated, the UN expects that roughly half of today's 5-year-olds will graduate from college in developed countries, whereas less than half will graduate from high school in developing ones.

QUALITY OF SCHOOLING

The UN uses two measures of quality of education:

- **Pupil/teacher ratio.** The fewer pupils a teacher has, the more likely that each student will receive effective instruction. The pupil/teacher ratio in primary school is approximately 24 in the world as a whole, 14 in

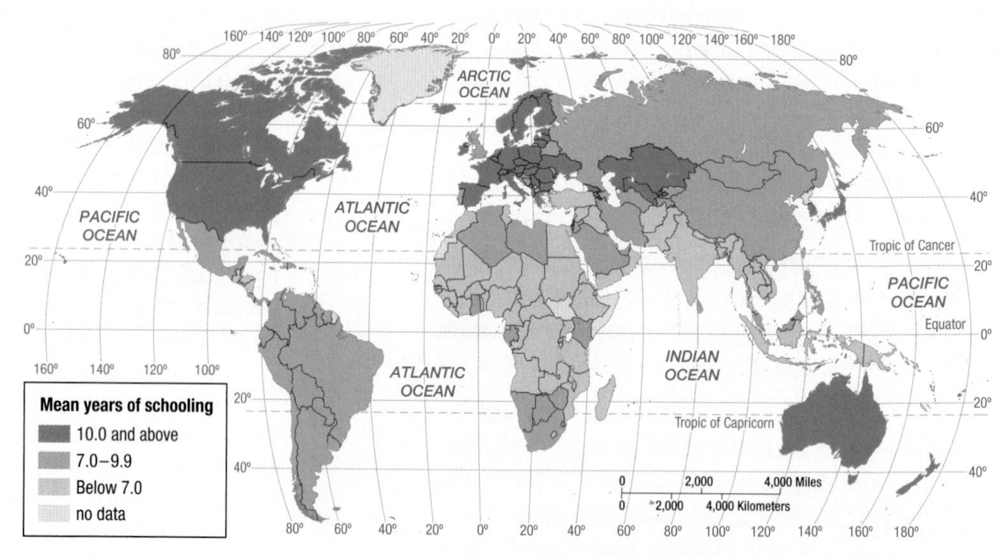

▲ **FIGURE 9-11 MEAN YEARS OF SCHOOLING** The highest number of years of schooling is in North America, and the lowest numbers are in South Asia and sub-Saharan Africa.

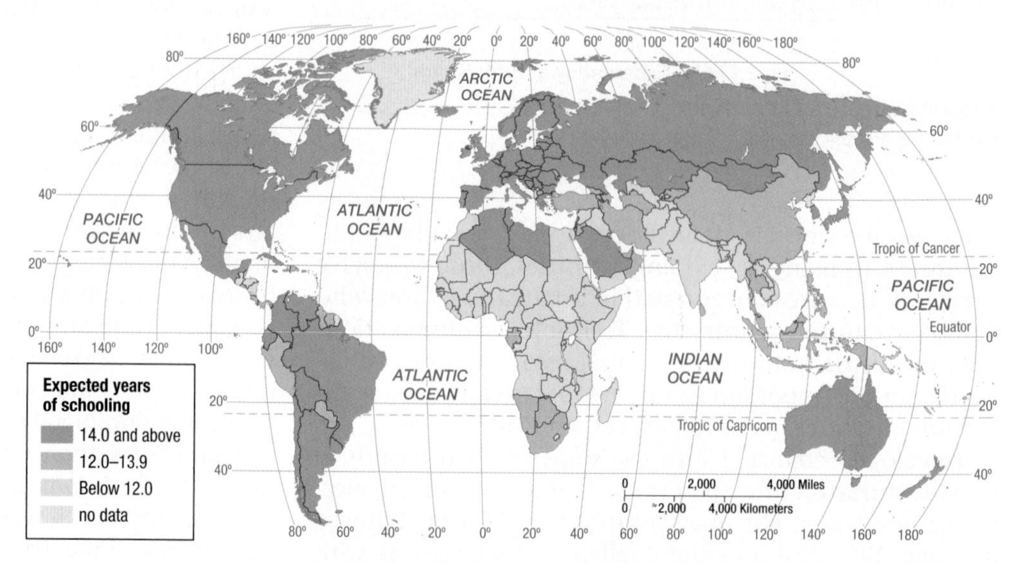

▲ **FIGURE 9-12 EXPECTED YEARS OF SCHOOLING** The highest numbers of expected years of schooling are in North America and Europe, and the lowest numbers are in sub-Saharan Africa and South Asia.

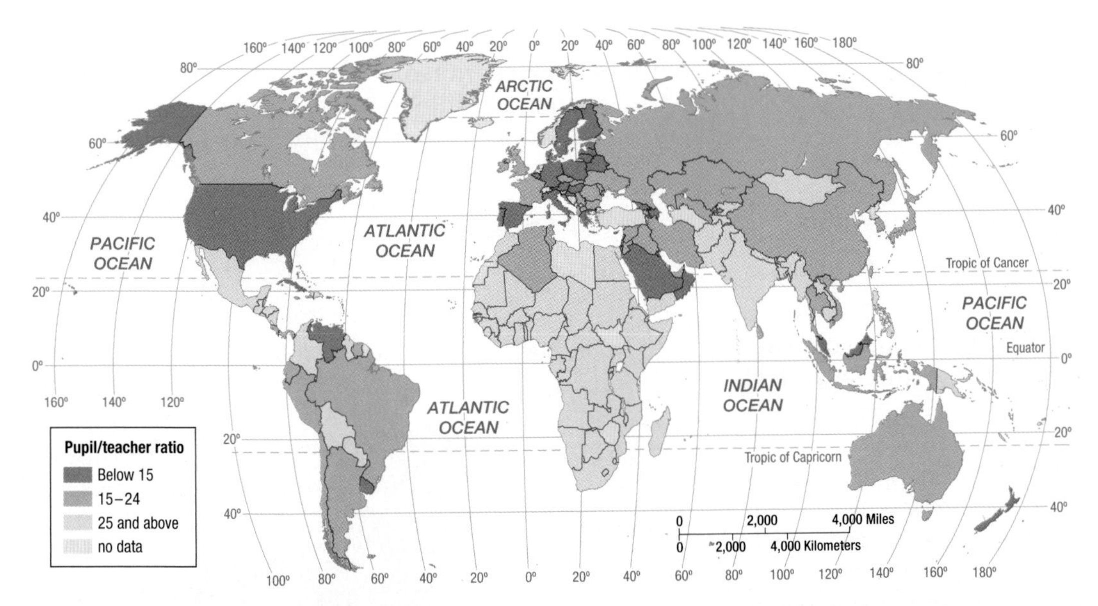

▲ **FIGURE 9-13 PUPIL/TEACHER RATIO, PRIMARY SCHOOL** The lowest pupil/teacher ratio is in North America, and the highest is in sub-Saharan Africa.

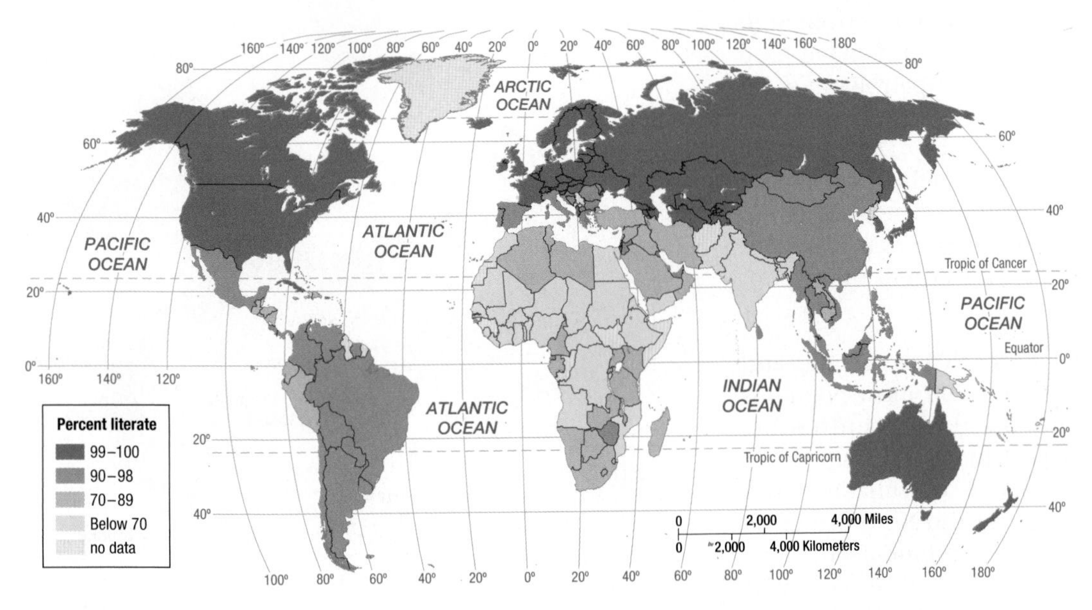

▲ **FIGURE 9-14 LITERACY RATE** Literacy is nearly 100 percent in developed countries. The lowest rates are in sub-Saharan Africa and South Asia.

developed countries, and 26 in developing countries (Figure 9-13). Thus, class size is nearly twice as large in developing countries as in developed ones.

- **Literacy rate.** A higher percentage of people in developed countries are able to attend school and as a result learn to read and write. The **literacy rate** is the percentage of a country's people who can read and write. It exceeds 99 percent in developed countries (Figure 9-14). Among developing regions, the literacy rate exceeds 90 percent in East Asia and Latin America but is less than 70 percent in sub-Saharan Africa and South Asia.

Most books, newspapers, and magazines are published in developed countries, in part because more of their citizens read and write. Developed countries dominate scientific and nonfiction publishing worldwide. (This textbook is an example.) Students in developing countries must learn technical information from books that usually are not in their native language but are printed in English, German, Russian, or French.

Improved education is a major goal of many developing countries, but funds are scarce. Education may receive a higher percentage of GNI in developing countries, but those countries' GNI is far lower to begin with, so they spend far less per pupil than do developed countries.

Pause and Reflect 9.1.3

The HDI measures the quality of schools in a country as a whole. What are ways in which differences among schools or colleges within a country might be measured?

VARIATIONS WITHIN COUNTRIES AND REGIONS

Learning Outcome 9.1.4
Describe variations in level of development within countries and regions.

Indicators of development vary widely among countries within the nine world regions, as well as within individual countries.

VARIATIONS WITHIN REGIONS. Variations in level of development are especially high in Southwest Asia & North Africa and in Central Asia. Much of Southwest Asia & North Africa is desert that can sustain only sparse concentrations of plant and animal life. This region possesses one major economic asset: a large percentage of the world's petroleum reserves. Saudi Arabia, the United Arab Emirates, and other oil-rich states in the region, most of them concentrated in states that border the Persian (Arabian) Gulf, have used the billions of dollars generated from petroleum sales to finance development. But not every country in the region has abundant petroleum reserves.. Development possibilities are limited in countries that lack significant reserves—Egypt, Jordan, Syria, and others. The large gap in per capita income between the petroleum-rich countries and those that lack resources causes tension in the region.

VARIATIONS WITHIN COUNTRIES. Brazil, China, and Mexico are among the world's largest and most populous countries. At the national scale, the three countries fall somewhere in the middle of the pack in GDP per capita and most other HDI indicators—well above sub-Saharan Africa and South Asia but well behind Europe and North America.

Hidden in nationwide statistics are substantial variations within all three countries (Figure 9-15). All three countries have GDP per capita greater than 150 percent of the national average in some provinces or states and less than 75 percent of the national average in other regions. Developed countries MDCs also have regional internal variations in GDP per capita, but they are less extreme. In the United States, for example, the GDP per capita is 122 percent of the national average in the wealthiest region (New England) and 90 percent of the national average in the poorest region (Southeast).

Regional internal variations can be traced to distinctive features of each country:

- Brazil: Wealth is highest along the Atlantic coast and lowest in the interior Amazon tropical rain forest.
- China: As in Brazil, wealth is highest along the east coast and lowest in the remote and inhospitable mountain and desert environments of the interior.
- Mexico: Wealth is relatively high in the region bordering its even wealthier neighbor to the north and in the principal tourist region on the Yucatan Peninsula.

At a local scale, wealth in these intermediate-development countries is concentrated in large urban areas, such as Rio de Janeiro and São Paulo in Brazil, Beijing and Shanghai in China, and Mexico City. These cities contain a large share

▲ **FIGURE 9-15 GDP PER CAPITA AS PERCENT OF NATIONAL AVERAGE IN THREE LARGE COUNTRIES:** (center) states of Brazil, (top) provinces of China, (bottom) states of Mexico.

of the national services and manufacturing sectors and are where many leaders of the public and private sectors live. They also contain extensive areas of poverty and slum conditions, as discussed in Chapter 13.

Pause and Reflect 9.1.4:
Russia and Canada are the world's two largest countries in land area. Which of these two countries would you expect to have regional internal variations similar to those in the United States, and which would have regional internal variations similar to those in Brazil, China, and Mexico?

CHECK-IN: KEY ISSUE 1

Why Does Development Vary among Countries?

- ✓ **The Human Development Index (HDI) measures the level of development of each country.**
- ✓ **HDI is based on three factors: a decent standard of living, a long and healthy life, and access to knowledge.**

CONTEMPORARY GEOGRAPHIC TOOLS

Collecting and Depicting Development Data

This chapter includes two dozen world maps that show a wide variety of development indicators. The concept of development involves many economic, social, and demographic dimensions.

Obtaining timely and accurate data related to development for nearly 200 countries is challenging. The data for most of the maps of world development in this chapter come from two sources:

- The United Nations Development Programme prepares the annual Human Development Report and provides much of the data contained in the report at hdr.undp.org.
- The World Bank pulls together hundreds of measures of development from a variety of sources and makes them available at data.worldbank.org.

These data can be used to depict patterns of similarities and differences among countries. For example, Figure 9-10 shows that in general, life expectancy is higher in developed countries than in developing countries. Figure 9-16 shows the same data on a graph. Each country is represented by a circle. The more populous the country, the larger the circle. The y-axis shows life expectancy, and the x-axis shows HDI level. The very high developed countries are in red, and the high, medium, and low developing countries are in yellow, green, and blue, respectively. The arc of the circles from lower left to upper right shows that countries with high HDIs have longer life expectancies.

Figure 9-16 helps to illustrate exceptions to the pattern. Circles that are way off to the bottom have life expectancies that are less than expected by their HDI. Most of the countries with lower-than-expected life expectancy are in sub-Saharan Africa. What might explain the low figures in sub-Saharan Africa? Refer to Figure 2-37, the world map of AIDS; most of the countries with the highest rates of AIDS are in sub-Saharan Africa.

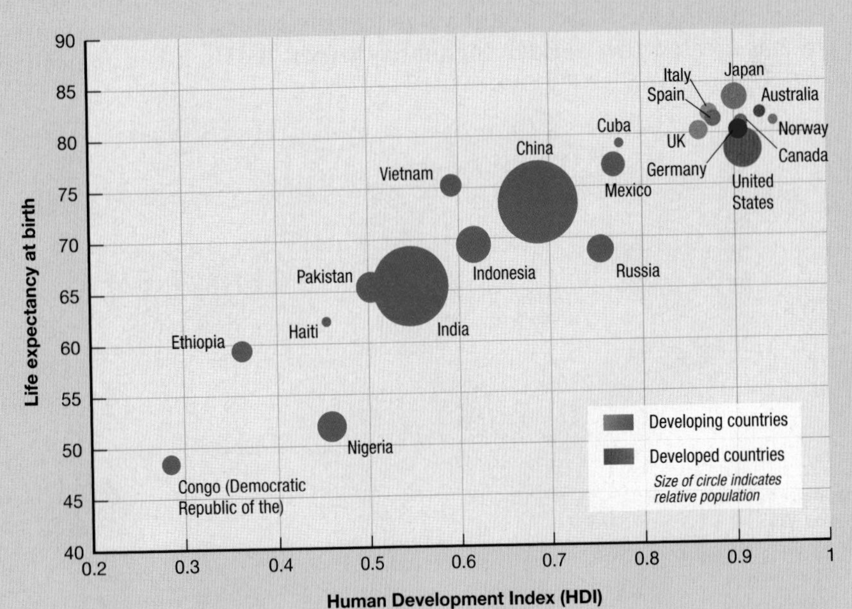

▲ **FIGURE 9-16 LIFE EXPECTANCY GRAPH** The higher the HDI, the longer the life expectancy.

KEY ISSUE 2

Why Does Development Vary by Gender?

- Gender Inequality Measures
- Gender Inequality Trends

Learning Outcome 9.2.1
Describe the UN's measures of gender inequality.

A country's overall level of development can mask inequalities in the status of men and women. The quest for an improved standard of living, access to knowledge, health, and a sustainable future are aspirations of people in all countries. Yet long-standing cultural and legal obstacles can limit women's participation in development and access to its benefits.

The UN has not found a single country in the world where the women are treated as well as the men. At best, women have achieved near-equality with men in some countries, but in other countries, the level of development for women lags far behind the level for men. The UN argues that inequality between men and women is a major factor that keeps a country from achieving a higher level of development.

Gender Inequality Measures

To measure the extent of each country's gender inequality, the UN has created the **Gender Inequality Index (GII).** As with the other indices, the GII combines multiple measures, including empowerment, labor, and reproductive health. The GII replaces other gender-related development measures formerly used by the UN, including the Gender-related Development Index and the Gender Empowerment Measure.

The higher the GII, the greater the inequality between men and women (Figure 9-17). A score of 0 would mean that men and women fare equally, and a score of 1.0 would mean that women fare as poorly as possible in all measures.

The GII is higher in developing countries than in developed ones. Sub-Saharan Africa, South Asia, Central Asia, and Southwest Asia are the developing regions with the highest levels of gender inequality. Reproductive health is the largest contributor to gender inequality in these regions. South and Southwest Asia also have relatively poor female empowerment scores. At the other extreme, 10 countries in Europe have GIIs less than 0.1, meaning that men and women are nearly equal. In general, countries with high HDIs have low GIIs and vice versa.

EMPOWERMENT

In the context of gender inequality, empowerment refers to the ability of women to achieve improvements in their own status—that is, to achieve economic and political power. The empowerment dimension of GII is measured by two indicators:

- **The percentage of seats held by women in the national legislature.** No particular gender-specific skills are required to be elected as a representative and to serve effectively. But in every country of the world, both developed and developing, fewer women than men hold positions of political power (Figure 9-18). Although more women than men vote in most places, no country has a national parliament or congress with a majority of women. The highest percentages are in Europe, where women comprise approximately one-fourth of the members of national parliaments. In the United States, one-sixth of the members of the U.S. Senate and House of Representatives are women, a figure that is below the numbers in many developing regions. The lowest rates are in Southwest Asia and North Africa.
- **The percentage of women who have completed high school.** In North America, girls are more likely than boys to complete high school, and boys are slightly ahead in Europe. In developing countries, boys are much more likely than girls to

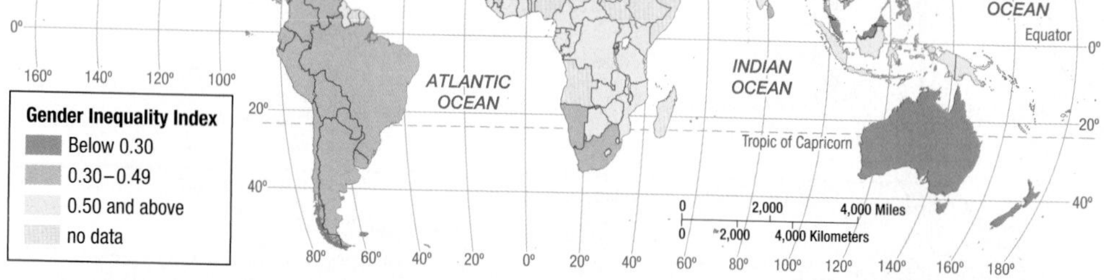

▲ **FIGURE 9-17 GENDER INEQUALITY INDEX (GII)** The lowest GII numbers and therefore the least inequality are in Europe, and the highest numbers are in sub-Saharan Africa.

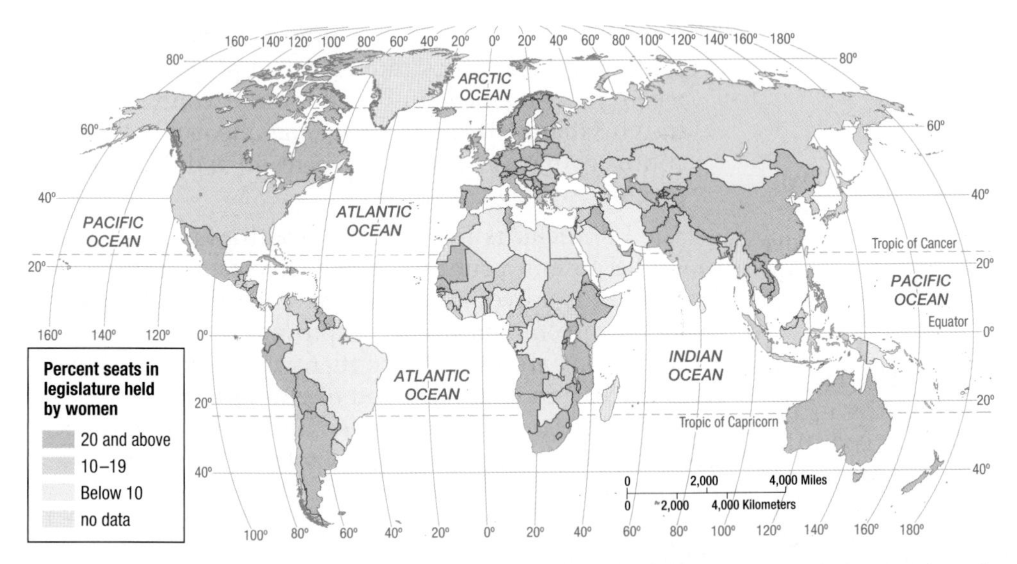

▲ **FIGURE 9-18 EMPOWERMENT: WOMEN IN THE NATIONAL LEGISLATURE** The highest numbers of women in national legislature are in Europe, and the lowest numbers are in Southwest Asia & North Africa.

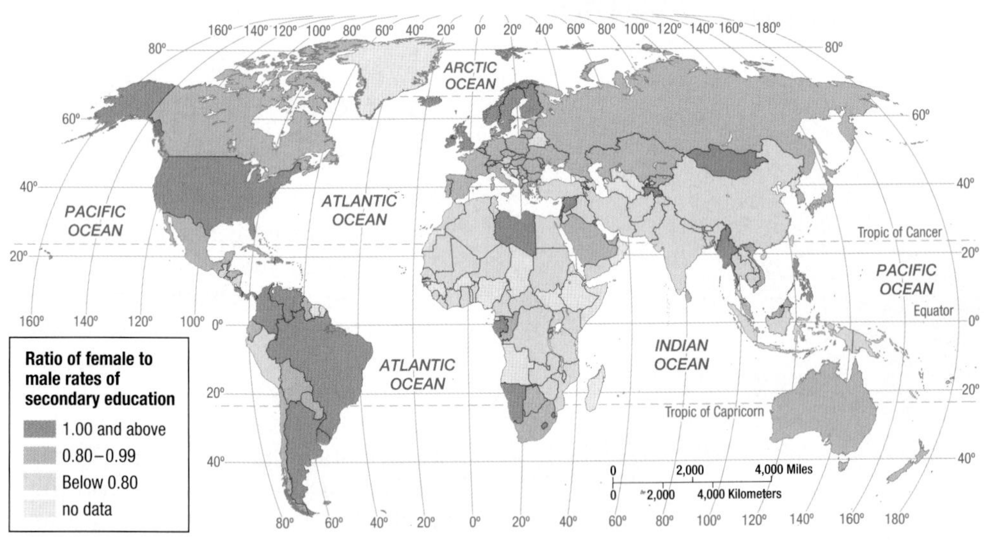

▲ **FIGURE 9-19 EMPOWERMENT: WOMEN GRADUDATING FROM HIGH SCHOOL** A figure above 1 means that more girls than boys graduate from high school.

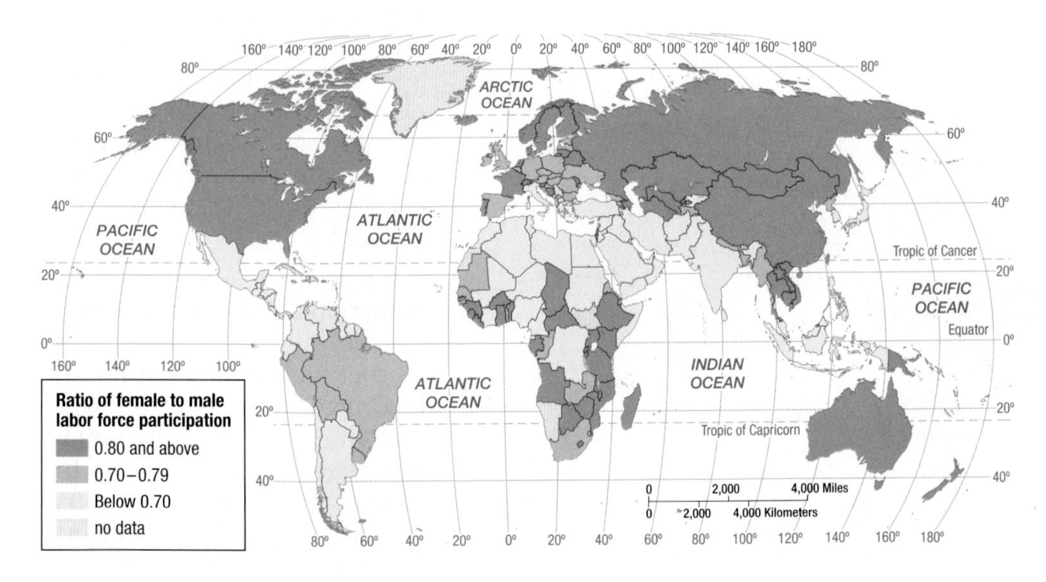

be high school graduates. For every 10 boys who graduate from high school in developing countries, only 8 girls graduate. In South Asia, for every 10 male high school graduates, there are only 5 females (Figure 9-19).

Pause and Reflect 9.2.1
Can you name a major political leader in your community or in another country who is a woman?

LABOR FORCE

The **female labor force participation rate** is the percentage of women holding full-time jobs outside the home. In general, women in developed countries are more likely than women in developing countries to hold full-time jobs outside the home (Figure 9-20). For every 100 men in the labor force, there are 75 women in the labor force in developed countries and 65 in developing countries. The lowest rates of female participation are in Southwest Asia & North Africa, where there are only 35 women for every 100 men in the labor force. However, in sub-Saharan Africa—the region with the lowest HDI—the ratio is the world's highest, with 77 women for every 100 men in the labor force. Women hold jobs in agriculture or services in sub-Saharan Africa, even while they have the world's highest fertility rates.

◀ **FIGURE 9-20 FEMALE LABOR FORCE PARTICIPATION** A lower number means that relatively few women participate in the labor force.

REPRODUCTIVE HEALTH

Learning Outcome 9.2.2
Describe changes since the 1990s in gender inequality.

Poor reproductive health is a major contributor to gender inequality around the world. The reproductive health dimension is based on two indicators:

- The **maternal mortality ratio** is the number of women who die giving birth per 100,000 births. The ratio is 15 deaths of mothers per 100,000 live births in developed countries and 140 in developing countries (Figure 9-21). The highest rates (most deaths per births) are in sub-Saharan Africa. The UN estimates that 150,000 women and 1.6 million children die each year between the onset of labor and 48 hours after birth.
- The **adolescent fertility rate** is the number of births per 1,000 women ages 15 to 19 (Figure 9-22). The rate is 20 births per 1,000 women ages 15 to 19 in developed countries and 60 in developing countries. The lowest teenage pregnancy rate is in Europe (8 per 1,000), where most couples use some form of contraception. In sub-Saharan Africa, where gender inequality is high, contraceptive use is below 10 percent, and the teenage pregnancy rate exceeds 100.

The UN includes reproductive health as a contributor to GII because in countries where effective control of reproduction is universal, women have fewer children, and maternal and child health are improved. Women in developing regions are more likely than women in developed regions to die in childbirth and to give birth as teenagers. Every country that offers women a full range of reproductive health options has a very low total fertility rate.

Pause and Reflect 9.2.2
The GII is 0.299 in the United States and 0.140 in Canada. Which country has greater gender inequality?

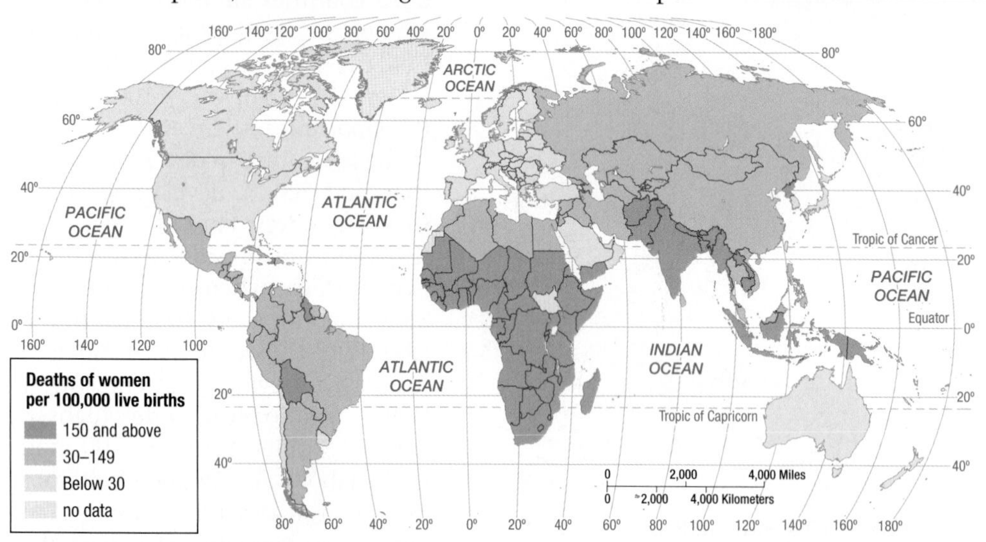

▲ **FIGURE 9-21 MATERNAL MORTALITY RATIO** The maternal mortality ratio is the number of deaths of mothers in childbirth compared to the number of live births.

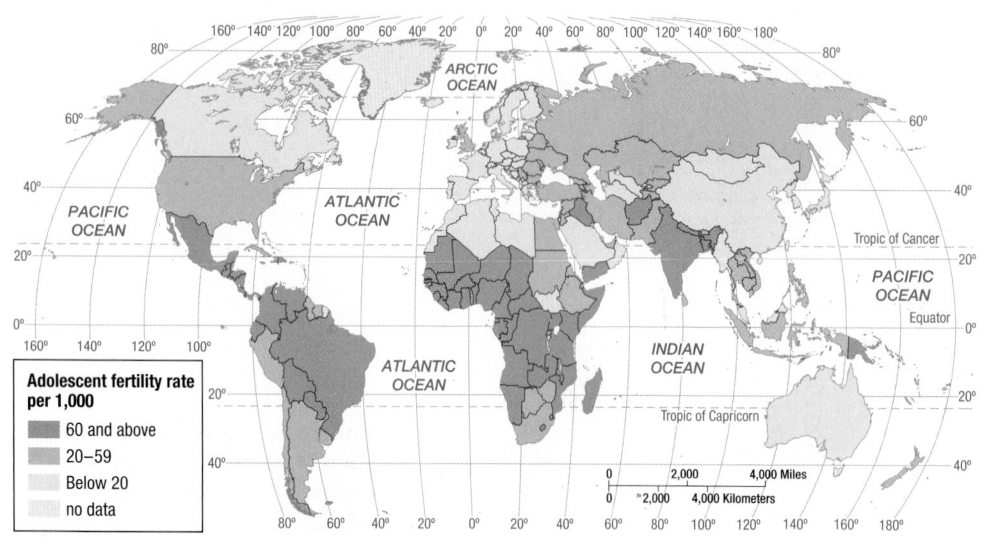

▲ **FIGURE 9-22 ADOLESCENT FERTILITY RATE** The adolescent fertility rate is the number of births per women per 1,000 women ages 15 to 19.

Gender Inequality Trends

The UN has found that in nearly every country, gender inequality has declined since the 1990s (Figure 9-24). The greatest improvements have been in Southwest Asia & North Africa. The United States is one of the few countries where the GII has increased. Furthermore, the United States has a GII rank of only 47, although it ranks fourth on the HDI. The UN points to two factors accounting for the relatively low U.S. GII ranking:

- Reproductive rights are much lower in the United States than in other very high HDI countries. For example, the maternal mortality rate is 24 in the United States, compared to 12 in Canada and less than 10 throughout Europe.
- The percentage of women in the national legislature is much lower in the United States than in other high HDI countries. In the United States, 17 of 100 senators and 74 of 435 representatives were women in 2012. In Canada, for example, 36 of 105 senators and 76 of 307 members of parliament in the House of Commons were women in 2012.

SUSTAINABILITY AND INEQUALITY IN OUR GLOBAL VILLAGE

Gender Inequality and the Environment

According to the UN, gender inequality adversely affects the environment. Countries with less gender inequality (that is, relatively high GIIs) are more likely to:

- Ratify international environmental treaties.
- Take steps to reduce carbon dioxide emissions.
- Set aside protected land areas and reduce deforestation.
- Undertake recycling and water conservation.

The reasons for variations in environmental policies extend beyond gender inequality, but the UN concludes that if women are more likely to be elected, highly educated, and in possession of reproductive rights, they are more likely to support and carry out environmental protection initiatives (Figure 9-23).

The attitudes of men and women toward the environment differ little in the world as a whole, according to a Gallup Poll. However, responses of men and women vary somewhat between the richest and poorest countries. In countries with the highest HDIs (and lowest GIIs), women are more likely than men to express concern for environmental issues, such as climate change and water and air quality, whereas men are more likely to express environmental concerns in countries with the lowest HDIs (and highest GIIs).

▲ **FIGURE 9-23 WOMEN AND ENVIRONMENTAL AWARENESS** A woman in the United Kingdom recycles bottles.

CHECK-IN: KEY ISSUE 2

Why Does Development Vary by Gender?

✓ **The Gender Inequality Index (GII) measures the extent of inequality between men and women in a country.**

✓ **GII is based on three factors: empowerment, labor force participation, and reproductive health.**

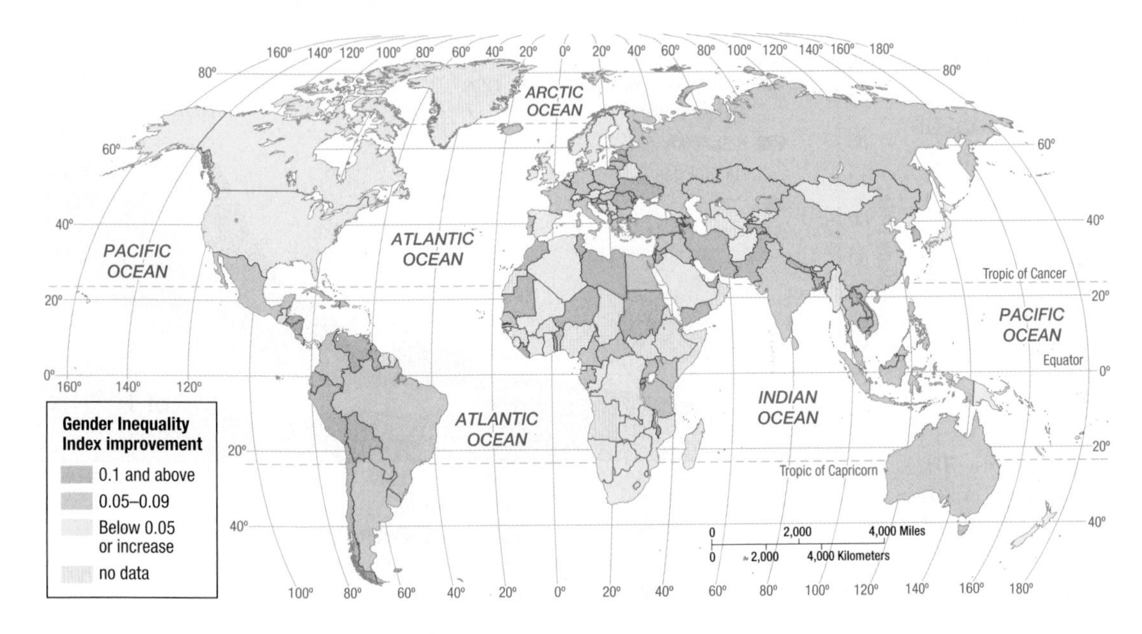

◄ **FIGURE 9-24 TRENDS IN GENDER INEQUALITY** The map shows the change in GII from the late 1990s to approximately 2010.

KEY ISSUE 3

Why Are Energy Resources Important for Development?

Learning Outcome 9.3.1
Explain the principal sources of demand for fossil fuels.

- **Energy Supply and Demand**
- **Alternative Energy Sources**

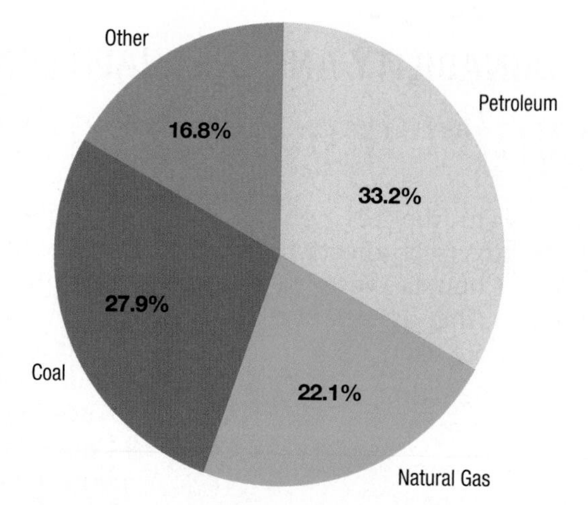

▲ **FIGURE 9-25 WORLD ENERGY DEMAND** Petroleum, coal, and natural gas account for most of the world's energy consumption.

Development is based on availability of abundant low-cost energy. Developed countries use large quantities of energy to produce food, run factories, keep homes comfortable, and transport people and goods. Developing countries expect to use more energy to improve the lives of their citizens.

In Chapter 1, we distinguished between renewable resources (those produced in nature more rapidly than consumed by humans) and nonrenewable resources (those produced in nature more slowly than consumed by humans). Most of the energy resources used by humans are nonrenewable. In the long run, sustainable development will necessitate increased reliance on renewable energy.

Energy Demand and Supply

Supply is the quantity of something that producers have available for sale. **Demand** is the quantity that consumers are willing and able to buy. Five-sixths of the world's energy needs are supplied by three of Earth's substances (Figure 9-25):

- **Coal.** Coal supplanted wood as the leading energy source in North America and Europe in the late 1800s, as these regions developed rapidly.
- **Petroleum.** Petroleum was first pumped in 1859 but did not become an important source of energy until the diffusion of motor vehicles in the twentieth century.
- **Natural gas.** Natural gas was originally burned off as a waste product of petroleum drilling, but it is now used to heat homes and to produce electricity.

In a developed country like the United States, dependency on these three sources of energy increased rapidly during the twentieth century (Figure 9-26).

Petroleum, natural gas, and coal are known as fossil fuels. A **fossil fuel** is an energy source formed from the residue of plants and animals buried millions of years ago. As sediment accumulated over these remains, intense pressure and chemical reactions slowly converted them into the fossil fuels that are currently used. When these substances are burned, energy that was stored in plants and animals millions of years ago is released.

Geographers observe two important inequalities in the global distribution of fossil fuels:

- **Demand.** The heaviest consumers of fossil fuel are in developed countries, whereas most of the reserves are in developing countries.
- **Supply.** Some developing regions have abundant reserves, whereas others have little.

Given the centrality of fossil fuels in contemporary economy and culture, unequal consumption and reserves of fossil fuels have been major sources of instability between developed and developing countries.

Pause and Reflect 9.3.1
Which energy source increased most rapidly in the United States during the twentieth century?

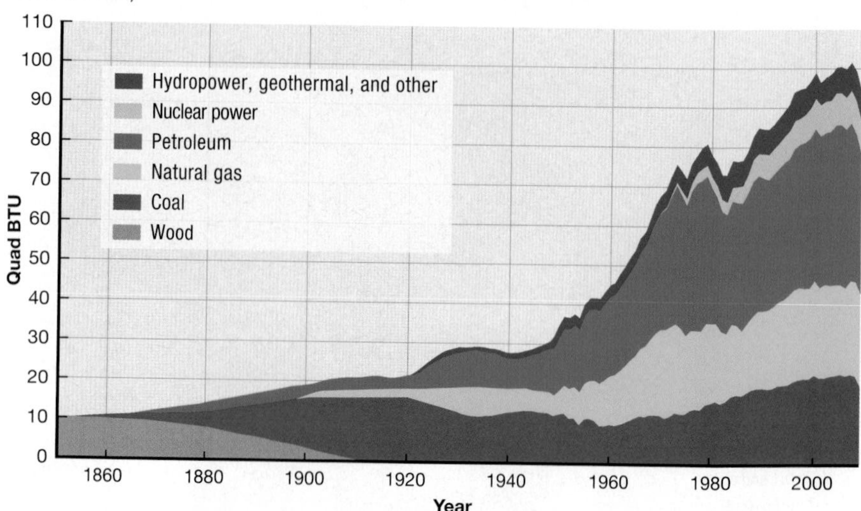

▲ **FIGURE 9-26 CHANGING U.S. ENERGY DEMAND** Coal was the principal energy source in the nineteenth century. Petroleum and natural gas became important in the twentieth century.

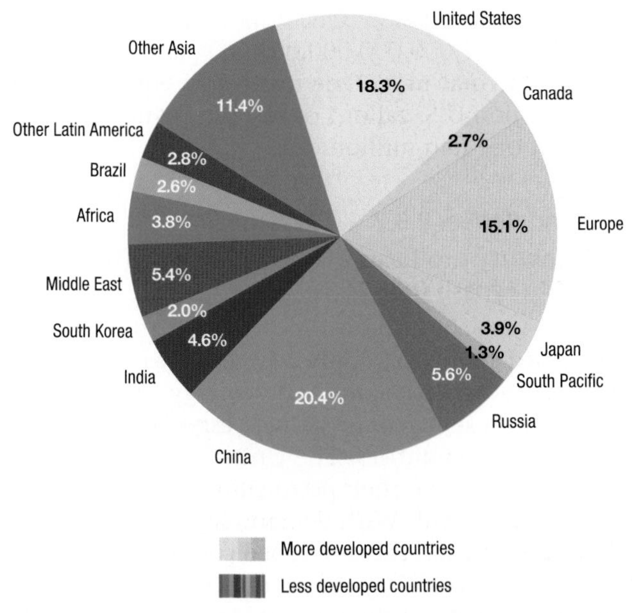

▲ **FIGURE 9-27 SHARE OF WORLD ENERGY DEMAND** Developed and developing countries each consume around one-half of the world's energy.

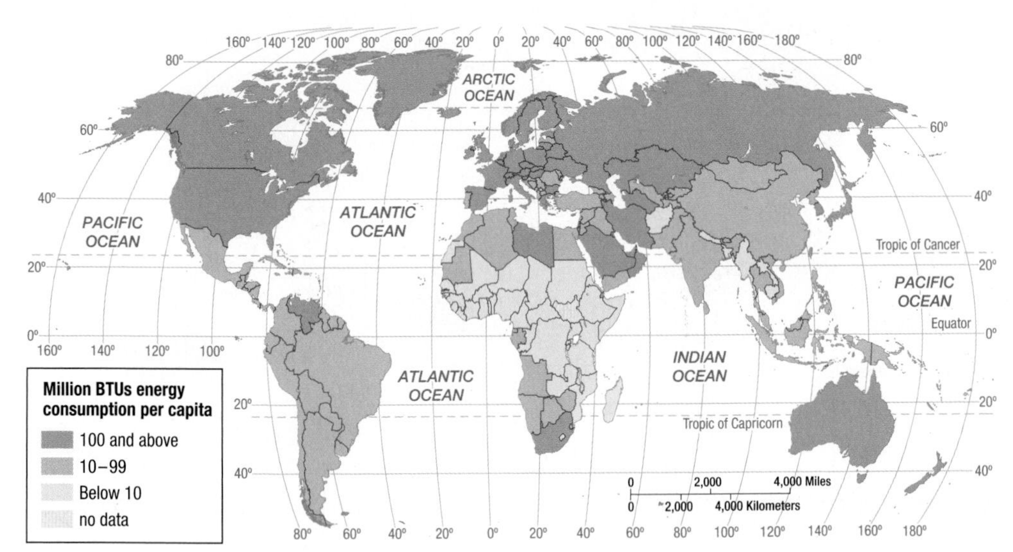

▲ **FIGURE 9-28 ENERGY DEMAND PER CAPITA** The highest per capita consumption is in North America, and the lowest is in sub-Saharan Africa.

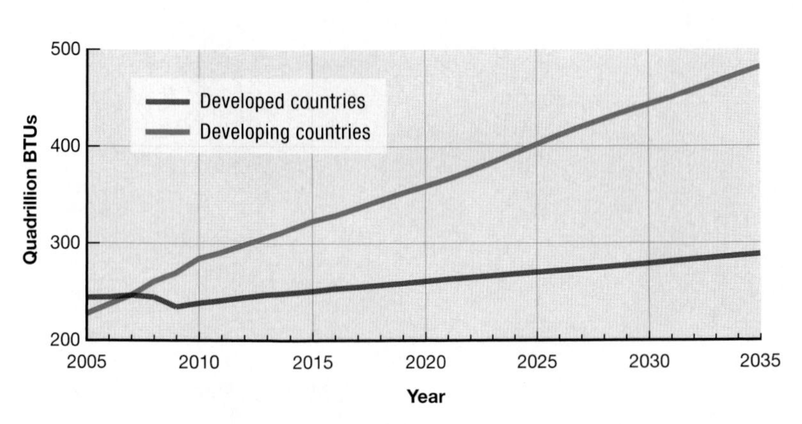

▲ **FIGURE 9-29 FUTURE ENERGY DEMAND** Developing countries are expected to consume 62 percent of the world's energy in 2035.

DEMAND FOR ENERGY

Around one-half of the world's energy is consumed in developed countries and one-half in developing countries (Figure 9-27). The United States had long been the leading consumer of energy, but China now consumes 20 percent of the world's energy, followed by the United States, at 18 percent. The highest per capita consumption of energy is in North America; the region contains one-twentieth of the world's people but consumes one-fourth of the world's energy (Figure 9-28). Developed countries contain only around one-third of the population of developing countries, so per capita consumption of energy is thus around three times higher in developed countries than in developing countries.

Demand for energy comes from three principal types of consumption in the United States:

- **Businesses.** The main energy demand is for coal, followed by natural gas and petroleum. Some businesses directly burn coal in their own furnaces. Others rely on electricity, mostly generated at coal-burning power plants.
- **Homes.** Energy is demanded primarily for the heating of living spaces and water. Natural gas is the most common source, followed by petroleum (heating oil and kerosene).
- **Transportation.** Almost all transportation systems demand petroleum products, including cars, trucks, buses, airplanes, and most railroads. Only subways, streetcars, and some trains run on coal-generated electricity.

In 2007, demand for fossil fuel consumption in developing countries surpassed that of developed countries for the first time (Figure 9-29). The gap in demand between developing and developed countries is expected to widen considerably in the years ahead because consumption of fossil fuels has been increasing at a much faster rate in developing countries—around 3 percent per year, compared to 1 percent per year in developed countries. Increasing reliance on fossil fuels also undermines the goals of sustainable development.

ENERGY SUPPLY

Learning Outcome 9.3.2
Describe the distribution of production of the three fossil fuels.

Energy is required for development, but Earth's energy resources are not distributed evenly. Why do some regions have an abundant supply of reserves of one or more fossil fuels, but other regions have little? This partly reflects how fossil fuels form:

- **Coal.** Coal formed in tropical locations, in lush, swampy areas rich in plants. Thanks to the slow movement of Earth's drifting continents, the tropical swamps of 250 million years ago have relocated to the mid-latitudes. As a result, today's main reserves of coal are in mid-latitude countries rather than in the tropics. China is responsible for supplying nearly one-half of the world's coal, other developing countries one-fourth, and developed countries (primarily the United States) the remaining one-fourth (Figure 9-30).
- **Petroleum.** Petroleum formed millions of years ago from residue deposited on the seafloor. Some still lies beneath such seas as the Persian Gulf and the North Sea, but other reserves are located beneath land that was under water millions of years ago. Russia and Saudi Arabia together supply one-fourth of the world's petroleum, other developing countries (primarily in Southwest and Central Asia) one-half, and developed countries (primarily the United States) the remaining one-fourth (Figure 9-31).
- **Natural gas.** Natural gas, like petroleum, formed millions of years ago from sediment deposited on the seafloor. One-third of natural gas production is supplied by Russia and Southwest Asia, one-third by other developing regions, and one-third by developed countries (primarily the United States) (Figure 9-32). Within the United States, the principal natural gas fields are in Texas, Oklahoma, and the Appalachian Mountains (Figure 9-33).

Figures 9-30, 9-31, and 9-32 use the same units (quad BTU), as well as the same classes. "Quad" is short for quadrillion (1 quadrillion = 1,000,000,000,000,000), and BTU is short for British thermal unit. One quad BTU equals approximately 8 million U.S. gallons of gasoline, which would fill the tanks of one-half million cars.

Pause and Reflect 9.3.2
Which country produces at least 20 quad BTUs of all three of the fossil fuels?

Developed countries supply a large share of the world's fossil fuels, but they demand more energy than they produce, so they must import fossil fuels, especially petroleum, from developing countries. The United States and Europe import more than half their petroleum, and Japan imports more than 90 percent. With demand increasing rapidly in developing countries, the developed countries face greater competition in obtaining the world's remaining supplies of fossil fuels. Many of the developing countries with low

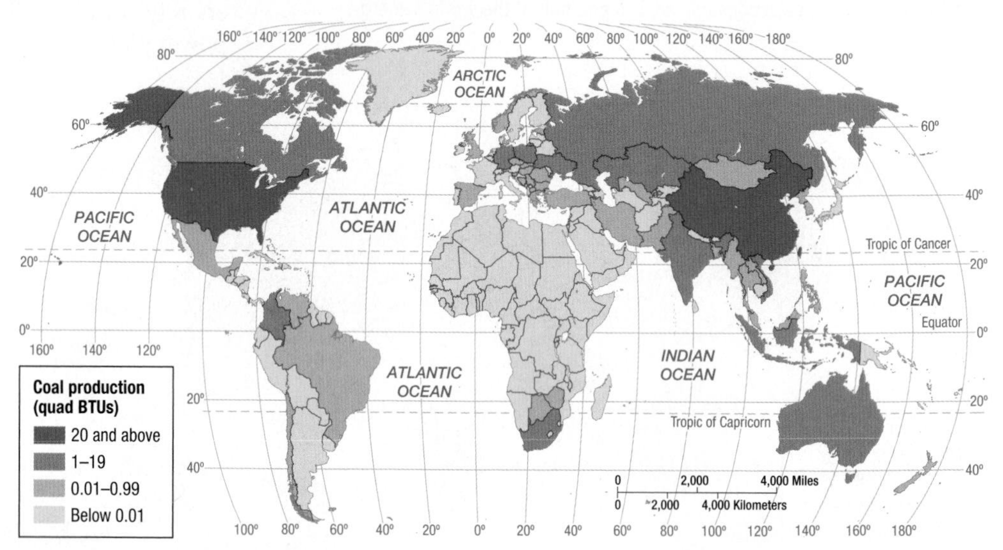

▲ **FIGURE 9-30 COAL PRODUCTION** China is the world's leading producer of coal, followed by the United states.

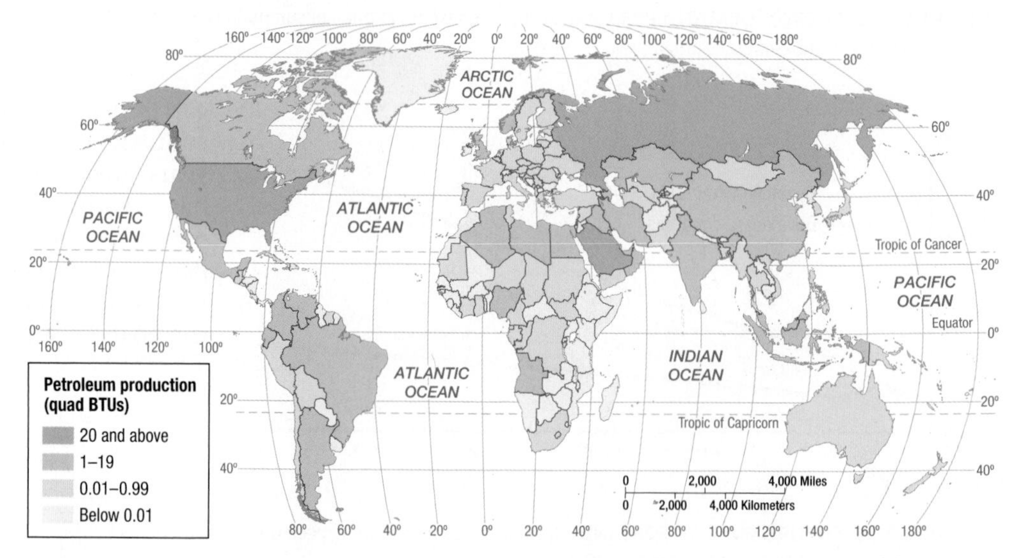

▲ **FIGURE 9-31 PETROLEUM PRODUCTION** Russia, Saudi Arabia, and the United States are the leading producers of petroleum.

HDIs also lack energy resources, and they lack the funds to pay for importing them.

Compounding future energy challenges, Earth's energy resources are divided between those that are renewable and those that are not:

- **Renewable energy** has an essentially unlimited supply and is not depleted when used by people. Examples include hydroelectric, geothermal, fusion, wind, biomass, and solar energy.
- **Nonrenewable energy** forms so slowly that for practical purposes, it cannot be renewed. Examples are the three fossil fuels that currently supply most of the world's energy needs.

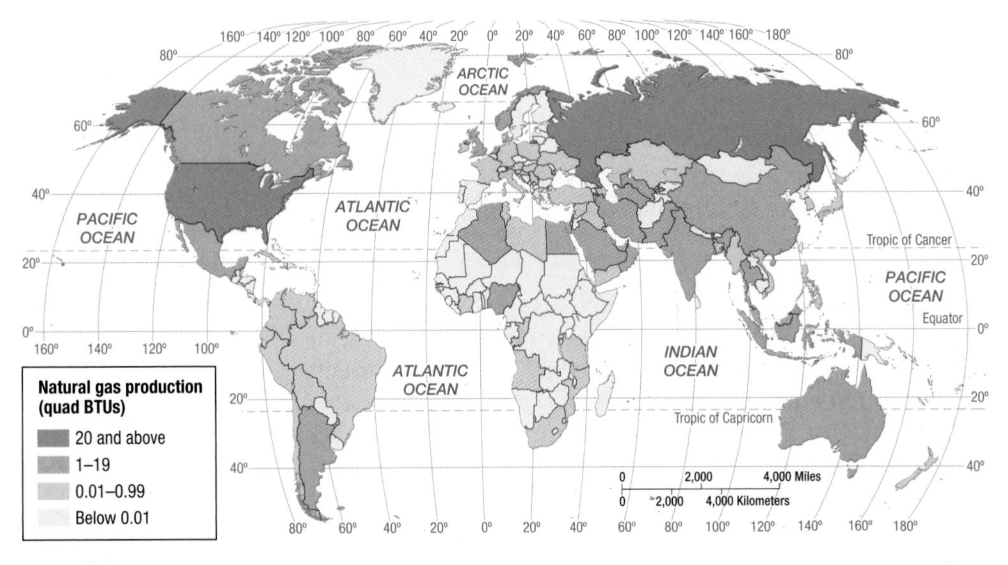

▲ **FIGURE 9-32 NATURAL GAS PRODUCTION** The United States and Russia are the leading producers of natural gas.

Because of dwindling supplies of fossil fuels, most of the buildings in which people live, work, and study will have to be heated another way. Cars, trucks, and buses will have to operate on some other energy source. Because plastic is made from petroleum, objects made of plastic will have to be made from other materials. Other resources can be used for heat, fuel, and manufacturing, but they are likely to be more expensive and less convenient to use than fossil fuels. And converting from fossil fuels will likely disrupt daily lives and cause hardship. On the other hand, the search for alternatives to fossil fuels may also create development opportunities.

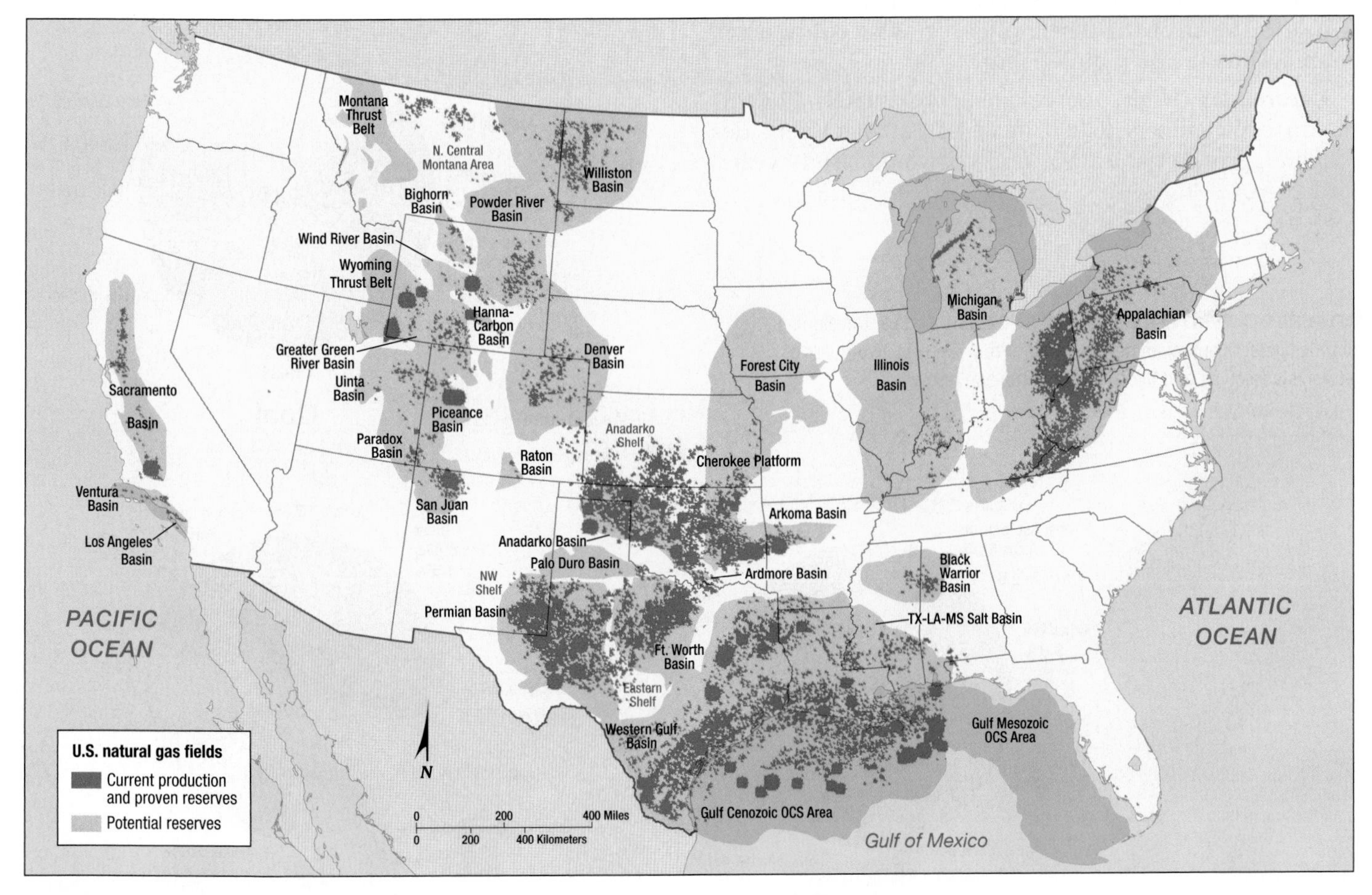

▲ **FIGURE 9-33 NATURAL GAS FIELDS IN THE UNITED STATES** The principal natural gas fields are in Oklahoma, Texas, and the Appalachians.

ENERGY RESERVES

Learning Outcome 9.3.3
Analyze the distribution of reserves of fossil fuels and differentiate between proven and potential reserves.

The world faces an energy challenge because of rapid depletion of the remaining supply of the three fossil fuels that current meet most of the world's energy needs. How much fossil fuel remains? Despite the critical importance of this question for the future, no one can answer it precisely. Because petroleum, natural gas, and coal are deposited beneath Earth's surface, considerable technology and skill are required to locate these substances and estimate their volume.

PROVEN RESERVES. The supply of energy remaining in deposits that have been discovered is called a **proven reserve**. Proven reserves can be measured with reasonable accuracy:

- **Coal.** World reserves are approximately 1 quadrillion metric tons (23 million quad BTUs). At current demand, proven coal reserves would last 131 years. Developed and developing regions each have about one-half of the supply of proven reserves. The United States has approximately one-fourth of the proven reserves, and other developed countries have one-fourth. Most of the developing regions' coal reserves are in Russia and China (Figure 9-34).
- **Natural gas.** World reserves are approximately 175 trillion cubic meters (6,000 quad BTUs). At current demand, proven natural gas reserves would last 49 years. Less than 10 percent of natural gas reserves are in developed countries, primarily the United States. The dark red areas in Figure 9-33 show proven reserve fields in the United States, as well as areas of current production. Russia, Iran, and Qatar together have nearly 60 percent of the world's proven natural gas reserves, and five other developing countries have most of the remainder.
- **Petroleum.** World reserves are approximately 1.3 trillion barrels (5,000 quad BTUs). At current demand, proven petroleum reserves would last 43 years. Developing countries possess 85 percent of the proven petroleum reserves, most of which is in Southwest Asia and North Africa and Central Asia. Saudi Arabia, Canada, and Iran together have more than 40 percent of the world's proven petroleum reserves.

Developed countries have historically possessed a disproportionately high supply of the world's proven fossil-fuel reserves. Europe's nineteenth-century industrial development depended on its abundant coal fields, and extensive coal and petroleum supplies helped the United States become the leading industrial power of the twentieth century.

But this dominance is ending in the twenty-first century. Many of Europe's coal mines have closed because either the coal has been exhausted or extracting the remaining supply would be too expensive, and the region's

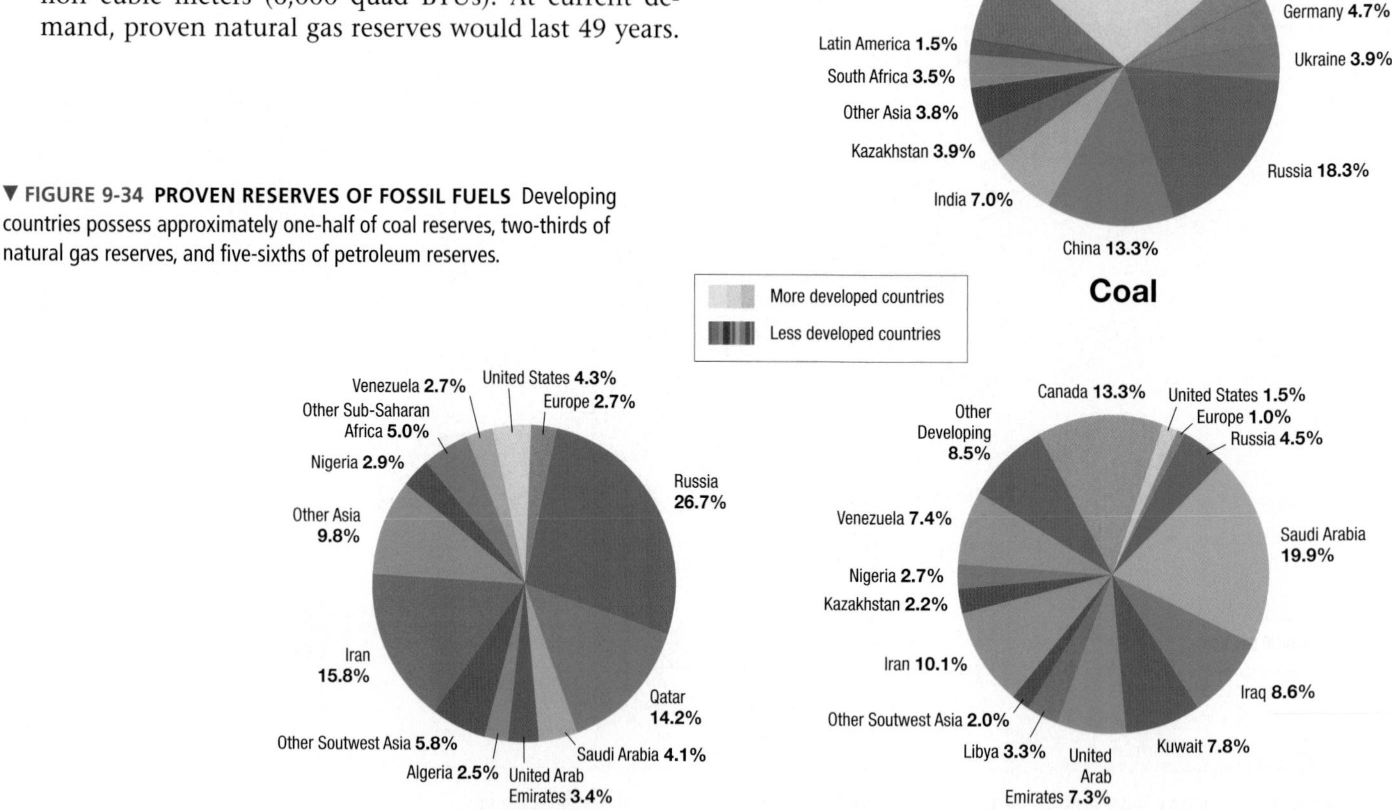

▼ **FIGURE 9-34 PROVEN RESERVES OF FOSSIL FUELS** Developing countries possess approximately one-half of coal reserves, two-thirds of natural gas reserves, and five-sixths of petroleum reserves.

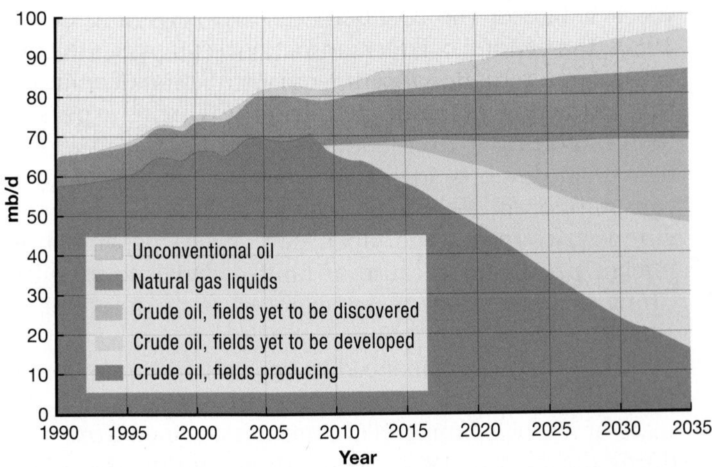

▲ **FIGURE 9-35 PETROLEUM PRODUCTION OUTLOOK** The International Energy Agency forecasts that potential reserves will be converted to proven reserves through discovery and development of new fields at about the same rate as already proven reserves are depleted.

petroleum and natural gas (in the North Sea) account for only small percentages of worldwide supplies. Japan has never had significant fossil fuel reserves. The United States still has extensive coal reserves, but its petroleum and natural gas reserves are being depleted rapidly.

Pause and Reflect 9.3.3

No country ranks among the leaders in proven reserves in all three fossil fuels. Which two countries possess at least 10 percent of the proven reserves of two of the three fossil fuels?

POTENTIAL RESERVES. Some fossil fuel deposits have not yet been discovered. The supply in deposits that are undiscovered but thought to exist is a **potential reserve**. When a potential reserve is actually discovered, it is reclassified as a proven reserve (Figure 9-35). Potential reserves can be converted to proven reserves in several ways:

- **Undiscovered fields.** The largest, most accessible deposits of petroleum, natural gas, and coal have already been exploited. Newly discovered reserves are generally smaller and more remote, such as beneath the seafloor, and extraction is costly. Exploration costs have increased because methods are more elaborate and the probability of finding new reserves is less. But as energy prices climb, exploration costs may be justified.
- **Enhanced recovery from already discovered fields.** When it was first exploited, petroleum "gushed" from wells drilled into rock layers saturated with it. Coal was quarried in open pits. But now extraction is more difficult. Sometimes pumping is not sufficient to remove petroleum, but water or carbon dioxide may be forced into wells to push out the remaining resource. The problem of removing the last supplies from a proven field is comparable to wringing out a soaked towel. It is easy to quickly remove the main volume of water, but the last few percent require more time and patience and special technology.
- **Unconventional sources.** Some sources are called unconventional because methods currently used to extract resources won't work. Also, we do not currently have economically feasible, environmentally sound technology with which to extract these sources.

An important example of an unconventional source is oil sands, which are saturated with a thick petroleum commonly called tar because of its dark color and strong odor. Native Americans used the tar to caulk canoes in the eighteenth century. The oil must be extracted from the sands through mining, which can be environmentally damaging, and current technology makes processing expensive. Abundant oil sands are found in Alberta, Canada, as well as in Venezuela and Russia. As demand has increased for petroleum, and as prices have risen, the mining of Alberta oil sands has become profitable, and extensive deposits of oil in Alberta oil sands have been reclassified from potential to proven reserves in recent years (Figure 9-36). As a result, Canada is now thought to have 13 percent of world's petroleum proven reserves, second behind Saudi Arabia, although overall oil sands are still classified as unconventional sources.

Another important unconventional source that has been increasingly exploited in recent years is extraction of natural gas through hydraulic fracturing, commonly called **fracking**. Rocks break apart naturally, and gas can fill the space between the rocks. Fracking involves pumping water at high pressure to further break apart rocks and thereby release more gas that can be extracted. Opponents of fracking fear environmental damage from pumping high-pressure water beneath Earth's surface. Safety precautions can minimize the environmental threat, but fracking does require the use of a large supply of water, and water is in high demand for other important uses, such as human consumption and agriculture.

▲ **FIGURE 9-36 CANADA'S OIL SANDS** Canada has the world's second-largest proven reserves of petroleum, which must be extracted from oil sands in Alberta.

CONTROLLING PETROLEUM RESERVES

Learning Outcome 9.3.4
Describe the role of OPEC and changes in the price and availability of petroleum.

Developed countries import most of their petroleum from Southwest Asia & North Africa and Central Asia, where most of the world's proven reserves are concentrated. These regions are the center of ethnic and political conflicts, as discussed in Chapters 7 and 8.

OPEC. Several developing countries possessing substantial petroleum reserves created the Organization of the Petroleum Exporting Countries (OPEC) in 1960. Arab OPEC members in Southwest Asia & North Africa include Algeria, Iraq, Kuwait, Libya, Qatar, Saudi Arabia, and the United Arab Emirates. OPEC members in other regions include Angola, Ecuador, Iran, Nigeria, and Venezuela.

OPEC was originally formed to enable oil-rich developing countries to gain more control over their resource. U.S. and European transnational companies, which had originally explored and exploited the oil fields, were selling the petroleum at low prices to consumers in developed countries and keeping most of the profits. Countries possessing the oil reserves nationalized or more tightly controlled the fields, and prices were set by governments rather than by petroleum companies. Under OPEC control, world oil prices have increased sharply on several occasions, especially during the 1970s and 1980s and in the early twenty-first century (Figure 9-37).

CHANGING U.S. PETROLEUM SOURCES. The United States produced more petroleum than it consumed during the first half of the twentieth century. Beginning in the 1950s, the handful of large transnational companies then in control of international petroleum distribution determined that extracting domestic petroleum was more expensive than importing it from Southwest and Central Asia. U.S. petroleum imports increased from 14 percent of total consumption in 1954 to 58 percent in 2009 (Figure 9-38). European countries and Japan have always depended on foreign petroleum because of limited domestic supplies. China changed from a net exporter to an importer of petroleum during the 1990s.

The United States reduced its dependency on imported oil in the immediate wake of the 1970s shocks, and the share of imports from OPEC countries declined from two-thirds during the 1970s to one-third during the 1980s (Figure 9-39). Conservation measures also dampened demand for petroleum in most developed countries during the late twentieth century. The average vehicle driven in the United States, for example, got 14 miles per gallon in 1975, compared to 22 miles per gallon in 1985.

The price of petroleum plummeted during the 1980s and settled during the 1990s at the lowest level in modern history, adjusting for inflation (Figure 9-40). With petroleum prices remaining low into the twenty-first century, consumption increased. Americans bought more gas-guzzling trucks and sport-utility vehicles and drove longer distances. Developed countries entered the twenty-first century optimistic that oil prices would remain low for some time. But in 2008, prices hit a record high, in both real terms and accounting for inflation. The 2008 oil shock contributed to the severe global recession that began then.

The world will not literally "run out" of petroleum during the twenty-first century. However, at some point, extracting the remaining petroleum reserves will prove so expensive and environmentally damaging that use of

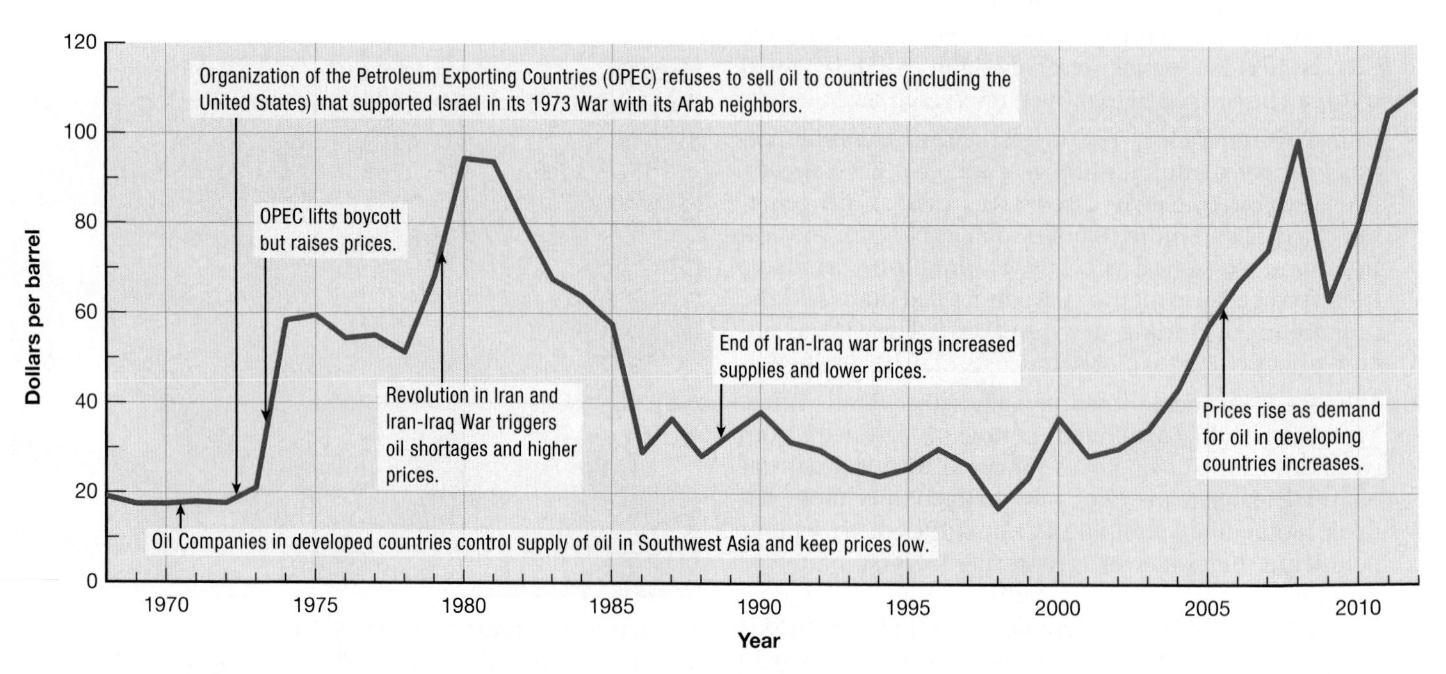

▲ **FIGURE 9-37 OIL PRICE HISTORY** Oil prices have increased sharply on several occasions.

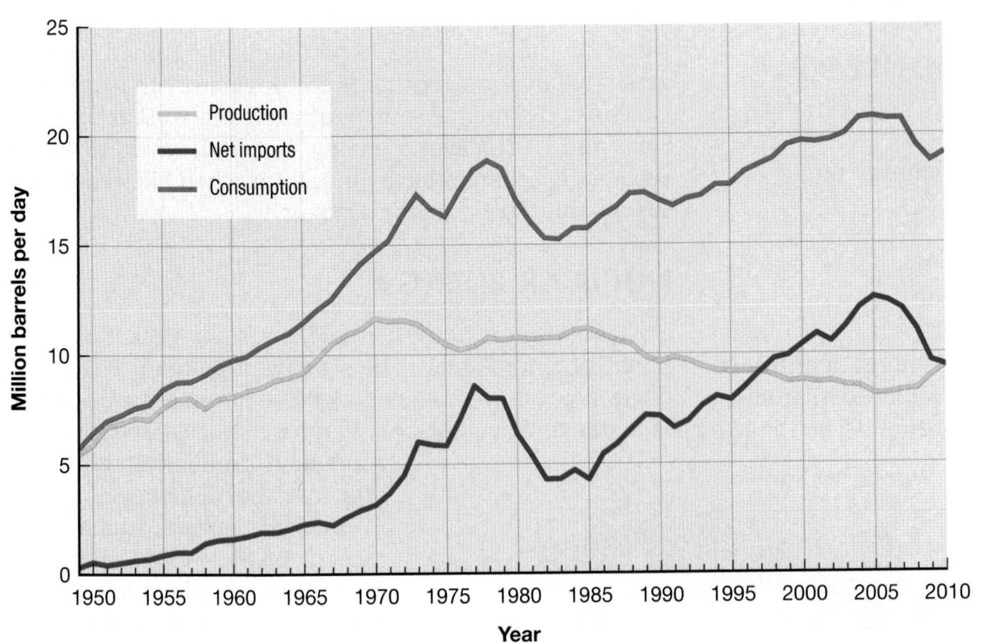

◀ **FIGURE 9-38 U.S. PETROLEUM CONSUMPTION, PRODUCTION, AND IMPORTS** U.S. production has remained relatively constant since the 1960s. Increasing consumption has been served by increasing imports.

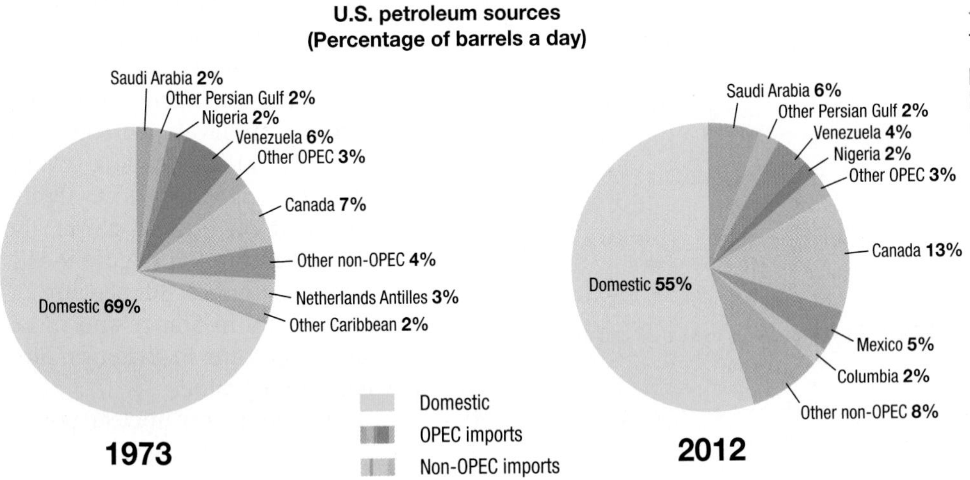

◀ **FIGURE 9-39 U.S. PETROLEUM SOURCES** The United States imports a higher percentage of petroleum now than in the 1970s. The increase has come primarily from elsewhere in the Western Hemisphere.

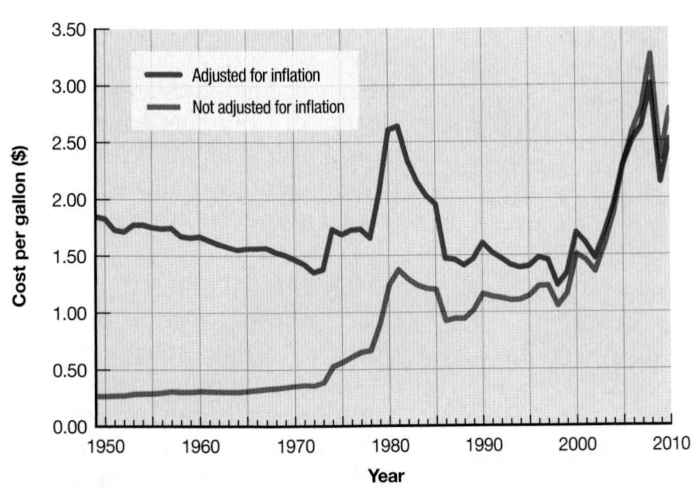

▲ **FIGURE 9-40 U.S. GASOLINE PRICES** The line adjusted for inflation is in 2005 dollars.

alternative energy sources will accelerate, and dependency on petroleum will diminish. The issues for the world are whether dwindling petroleum reserves are handled wisely and other energy sources are substituted peacefully. Given the massive growth in petroleum consumption expected in developing countries such as China and India, the United States and other developed countries may have little influence over when prices rise and supplies decline. In this challenging environment, all countries will need to pursue sustainable development strategies based on increased reliance on renewable energy sources.

Pause and Reflect 9.3.4
What country exports the most petroleum to the United States?

Alternative Energy Sources

Learning Outcome 9.3.5

Describe the distribution of nuclear energy and challenges in using it.

An especially strong challenge in the quest for sustainable development is substituting renewable energy resources for nonrenewable ones. Although renewable resources can be harnessed for energy, continued reliance on the three main nonrenewable fossil fuels—petroleum, natural gas, and coal—continues to be the less expensive alternative. About 20 percent of energy consumed in the world, and 15 percent in the United States, is generated by sources other than the three main fossil fuels.

The two principal sources other than fossil fuels are nuclear and hydroelectric energy. But both of these widely used energy sources have limitations when viiewed from a sustainable development perspective.

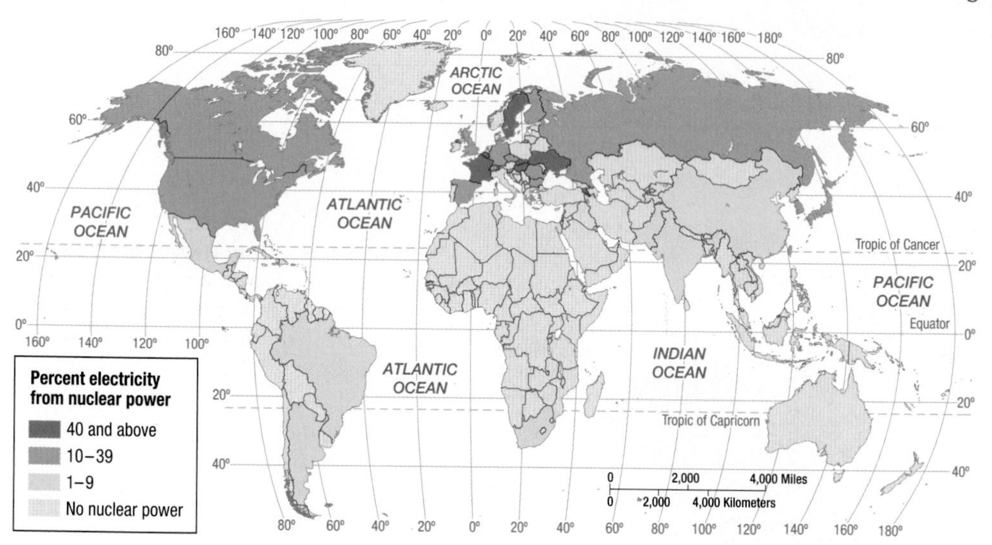

▲ **FIGURE 9-41 ELECTRICITY FROM NUCLEAR POWER** Nuclear power is used in 37 countries, primarily in Europe and North America.

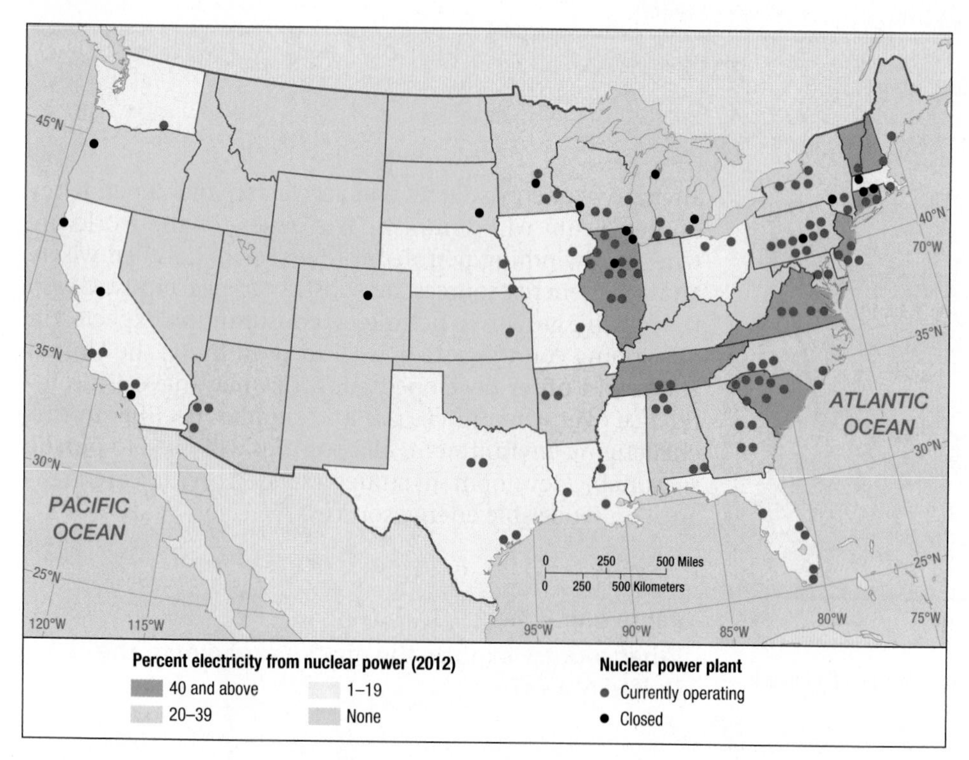

▲ **FIGURE 9-42 NUCLEAR POWER BY U.S. STATE** One-third of electricity is generated from nuclear power in the Northeast, compared to less than one-tenth in the West.

NUCLEAR ENERGY

Nuclear power is not renewable, but some view it as an alternative to fossil fuels. The big advantage of nuclear power is the large amount of energy released from a small amount of material. One kilogram of enriched nuclear fuel contains more than 2 million times the energy in 1 kilogram of coal.

Nuclear power supplies 14 percent of the world's electricity. Two-thirds of the world's nuclear power is generated in developed countries, with Europe and North America responsible for generating one-third each. Only 30 of the world's nearly 200 countries make some use of nuclear power, including 19 developed countries and only 11 developing countries. The countries most highly dependent on nuclear power are clustered in Europe (Figure 9-41), where it supplies 80 percent of all electricity in France and more than 50 percent in Belgium, Slovakia, and Ukraine.

Dependency on nuclear power varies widely among U.S. states (Figure 9-42). Nuclear power accounts for more than 70 percent of electricity in Vermont and more than one-half in Connecticut, New Jersey, and South Carolina. At the other extreme, 20 states and the District of Columbia have no nuclear power plants. Nuclear power presents serious challenges, as described in the following sections.

POTENTIAL ACCIDENTS. A nuclear power plant produces electricity from energy released by splitting uranium atoms in a controlled environment, a process called **fission**. One product of all nuclear reactions is **radioactive waste**, certain types of which are lethal to people exposed to it. Elaborate safety precautions are taken to prevent the leaking of nuclear fuel from a power plant.

Nuclear power plants cannot explode, like a nuclear bomb, because the quantities of uranium are too small and cannot be brought together fast enough. However, it is possible to have a runaway reaction, which overheats the reactor, causing a meltdown, possible steam explosions, and scattering of radioactive material into the atmosphere. This happened in 1986 at Chernobyl, then in the Soviet Union and now in the north of Ukraine, near the Belarus border. The accident caused 56 deaths due to exposure to high radiation doses and an estimated 4,000 cancer-related deaths to people who lived near the plant.

Following an earthquake and tsunami in 2011, three of the six reactors at Japan's Fukushima Daiichi nuclear power plant experienced full meltdown, resulting in release of radioactive materials. Three workers died; the death toll among nearby residents exposed to high levels of radioactivity won't be known for years.

RADIOACTIVE WASTE. The waste from nuclear fission is highly radioactive and lethal, and it remains so for many years. Plutonium for making nuclear weapons can be harvested from this waste. Pipes, concrete, and water near the fissioning fuel also become "hot" with radioactivity. No one has yet devised permanent storage for radioactive waste. The waste cannot be burned or chemically treated, and it must be isolated for several thousand years until it loses its radioactivity. Spent fuel in the United States is stored "temporarily" in cooling tanks at nuclear power plants, but these tanks are nearly full. The United States is Earth's third-largest country in land area, yet it has failed to find a suitable underground storage site because of worry about groundwater contamination. In 2002, the U.S. Department of Energy approved a plan to store the waste in Nevada's Yucca Mountains. But soon after taking office in 2009, the Obama administration reversed the decision and halted construction on the nearly complete repository.

BOMB MATERIAL. Nuclear power has been used in warfare twice, in August 1945, when the United States dropped atomic bombs on Hiroshima and Nagasaki, Japan, ending World War II. No government has dared to use these bombs in a war since then because leaders recognize that a full-scale nuclear conflict could terminate human civilization.

The United States and Russia (previously the Soviet Union) each have several thousand nuclear weapons. China, France, and the United Kingdom have several hundred nuclear weapons each, India and Pakistan several dozen each, and North Korea a handful. Israel is suspected of possessing nuclear weapons but has not admitted to it, and Iran has been developing the capability. Other countries have initiated nuclear programs over the years but have not advanced to the weapons stage. The diffusion of nuclear programs to countries sympathetic to terrorists has been particularly worrying to the rest of the world and has been a major factor in long-time tensions between Iran and other countries that do not want Iran to gain the capability of building a nuclear weapon.

Pause and Reflect 9.3.5
Iran has claimed that it is interested in nuclear power for peaceful uses. Review the maps and charts of fossil fuel production and proven reserves on the previous two spreads. Does Iran appear to have other resources for generating electricity?

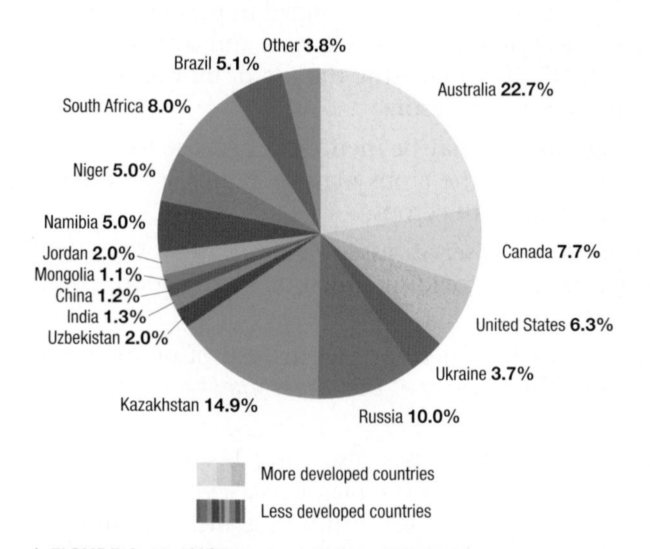

▲ **FIGURE 9-43 WORLD URANIUM RESERVES**
Australia, Kazakhstan, and Russia have the most uranium reserves.

LIMITED URANIUM RESERVES. Like fossil fuels, uranium is a nonrenewable resource. Proven uranium reserves will last about 124 years at current rates of use. And they are not distributed uniformly around the world: Australia has 23 percent of the world's proven uranium reserves, Kazakhstan 15 percent, and Russia 10 percent (Figure 9-43). The chemical composition of natural uranium further aggravates the scarcity problem. Uranium ore naturally contains only 0.7 percent U-235; a greater concentration is needed for power generation.

A **breeder reactor** turns uranium into a renewable resource by generating plutonium, also a nuclear fuel. However, plutonium is more lethal than uranium and could cause more deaths and injuries in an accident. It is also easier to fashion into a bomb. Because of these risks, few breeder reactors have been built, and none are in the United States.

HIGH COST. Nuclear power plants cost several billion dollars to build, primarily because of the elaborate safety measures required. Without double and triple backup systems at nuclear power plants, nuclear energy would be too dangerous to use. Uranium is mined in one place, refined in another, and used in still another. As with coal, mining uranium can pollute land and water and damage miners' health. The complexities of safe transportation add to the cost. As a result, generating electricity from nuclear plants is much more expensive than from coal-burning plants. The future of nuclear power has been seriously hurt by the high costs associated with reducing risks.

RENEWABLE ENERGY

Learning Outcome 9.3.6
Identify challenges to increasing the use of alternative energy sources.

By a wide margin, hydroelectric power is currently the leading source of renewable energy for sustainable development in both developed and developing regions. Biomass and wind power have some usages, and geothermal and solar trail even further in current usage.

HYDROELECTRIC POWER. Generating electricity from the movement of water is called **hydroelectric power**. Water has been a source of mechanical power since before recorded history. It was used to turn water wheels, and the rotational motion was used to grind grain, saw timber, pump water, and operate machines. Hydroelectric is now the world's second-most-popular source of electricity, after coal. Worldwide generation of hydroelectric power is approximately 30 quad BTU, compared to 150 quad BTU for coal.

Two-thirds of the world's hydroelectric power is generated in developing countries and one-third in developed countries. A number of developing countries depend on hydroelectric power for most of their electricity (Figure 9-44). The most populous country to depend primarily on hydroelectric power is Brazil. Overall, Brazil has made considerable progress towards sustainable development by generating approximately 85 percent of its electricity from renewable energy sources. Among developed countries, Canada gets two-thirds of its electricity from hydroelectric power; although the United States is the fourth-leading producer of hydroelectric power, it obtains only 8 percent of its electricity from that source. And this percentage may decline because few acceptable sites to build new dams remain.

▼ **FIGURE 9-44 ELECTRICITY FROM HYDROELECTRIC POWER** Hydroelectricity provides a large percentage of electricity in a number of developing countries, especially in Latin America and sub-Saharan Africa. The Itaipú hydroelectric dam is on the Paraná River in Brazil.

▲ **FIGURE 9-45 BIOMASS FUEL IN BRAZIL** Ethanol is produced from sugarcane in Brazil. This ethanol-producing plant is in Piracicaba.

BIOMASS. **Biomass fuel** is fuel derived from plant material and animal waste. Biomass energy sources include wood and crops. When carefully harvested in forests, wood is a renewable resource that can be used to generate electricity and heat. The waste from processing wood, such as for building construction and demolition, is also available. And crops such as sugarcane, corn, and soybeans can be processed into motor-vehicle fuels. Worldwide production of biomass fuel is approximately 3 quad BTUs, including one-third each in North America, Europe, and developing regions (Figure 9-45). Brazil in particular makes extensive use of biomass to fuel its cars and trucks.

The potential for increasing the use of biomass for fuel is limited, for several reasons:

- Burning biomass may be inefficient because the energy used to produce the crops may be as much as the energy supplied by the crops.
- Biomass already serves essential purposes other than energy, such as providing much of Earth's food, clothing, and shelter.
- When wood is burned for fuel instead of being left in the forest, the fertility of the forest may be reduced.

WIND POWER. Wind has also long been a source of energy, the most obvious examples of its uses being sailboats for travel and windmills for grinding grain. Like moving water turning a water wheel, moving air can turn a turbine.

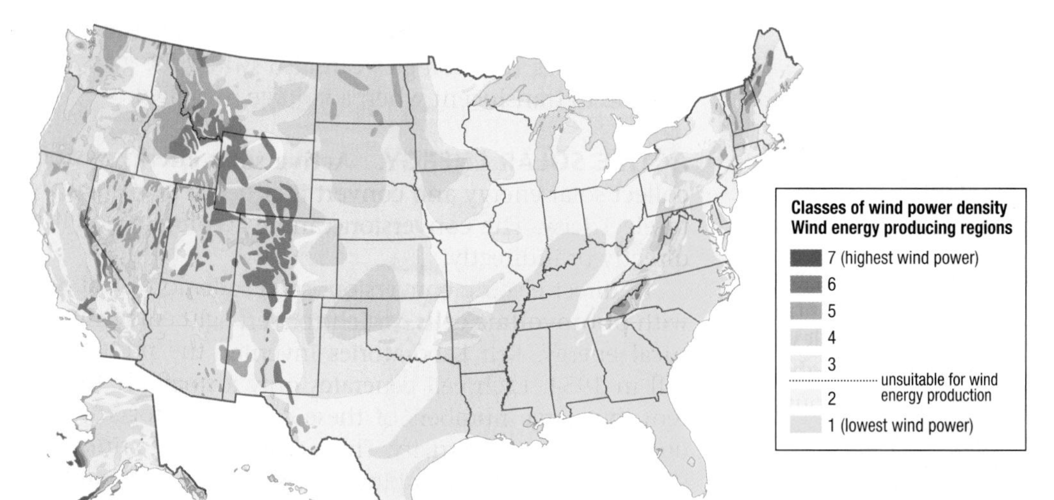

◀ **FIGURE 9-46 WIND POWER** Winds are especially strong enough to support generation of power in the U.S. Plains states.

The benefits of wind-generated power seem irresistible. Construction of a windmill modifies the environment much less severely than construction of a dam across a river. And wind power has greater potential for increased use because only a small portion of the potential resource has been harnessed. However, wind power has divided the environmental community. Some oppose construction of windmills because they can be noisy and lethal for birds and bats. They can also constitute a visual blight when constructed on mountaintops or offshore in places of outstanding beauty.

Wind usage is similar to the pattern for biomass: Worldwide production is 3 quad BTUs, divided one-third each among North America, Europe, and developing regions. Hundreds of wind "farms" consisting of dozens of windmills each have been constructed across the United States; one-third of the country is considered windy enough to make wind power economically feasible (Figure 9-46), especially North Dakota, Texas, Kansas, South Dakota, and Montana. Twenty percent of Denmark's electricity is being generated through wind power. Wind power has been used only to a limited extent in developing countries. A significant obstacle is the cost of constructing the wind turbines.

Pause and Reflect 9.3.6
Chicago is nicknamed "the Windy City." Based on Figure 9-46, does the Chicago area appear to be a good location for wind power?

GEOTHERMAL ENERGY. Natural nuclear reactions make Earth's interior hot. Toward the surface, in volcanic areas, this heat is especially pronounced. The hot rocks can encounter groundwater, producing heated water or steam that can be tapped by wells. Energy from this hot water or steam is called **geothermal energy**.

Harnessing geothermal energy is most feasible at sites along Earth's surface where crustal plates meet, which are also the sites of many earthquakes and volcanoes. Geothermal energy is being tapped in several locations, including California, Italy, New Zealand, and Japan, and other plate boundary sites are being explored. Iceland and Indonesia make extensive use of geothermal energy. Ironically, in Iceland, an island named for its glaciers, nearly all homes and businesses in the capital of Reykjavik are heated with geothermal steam (Figure 9-47). Worldwide production is less than 1 quad BTU, divided about evenly between developed and developing regions.

NUCLEAR FUSION. Some nuclear power issues could be addressed through nuclear **fusion**, which is the fusing of hydrogen atoms to form helium. Fusion releases spectacular amounts of energy: A gnat-sized amount of hydrogen releases the energy of thousands of tons of coal. But fusion can occur only at very high temperatures (millions of degrees). Such high temperatures have been briefly achieved in hydrogen bomb tests but not on a sustained basis in a power-plant reactor, given present technology. Sources such as fusion are not yet practical, so do not appear in statistics of current energy production.

▼ **FIGURE 9-47 GEOTHERMAL** Geothermal plant near Krafla, Iceland.

KEY ISSUE 4

Why Do Countries Face Obstacles to Development?

- **Two Paths to Development**
- **Financing Development**
- **Making Progress in Development**

Learning Outcome 9.4.1
Summarize the two paths to development.

The gap between rich and poor countries is substantial. Poorer countries lack much of what people in richer countries take for granted, such as access to electricity, safe drinking water, and paved roads. To reduce disparities between rich and poor countries, developing countries must develop more rapidly. This means increasing per capita GNI more rapidly and using the additional funds to make more rapid improvements in social and economic conditions. Developing countries face two fundamental obstacles in trying to encourage more rapid development:

- Adopting policies that successfully promote development
- Finding funds to pay for development

Two Paths to Development

To promote development, developing countries choose one of two models:

- **Self-sufficiency.** In the self-sufficiency model, countries encourage domestic production of goods, discourage foreign ownership of businesses and resources, and protect their businesses from international competition.
- **International trade.** In the international trade model, countries open themselves to foreign investment and international markets.

Each has important advantages and faces serious challenges.

For most of the twentieth century, self-sufficiency, or balanced growth, was the more popular of the development alternatives. International trade became more popular beginning in the late twentieth century. However, the global economic slowdown since 2008 has caused some countries to question the international trade approach.

SELF-SUFFICIENCY PATH

Key elements of the self-sufficiency path to development include the following:

- Barriers limit the import of goods from other places. Three widely used barriers include setting high taxes (tariffs) on imported goods to make them more expensive than domestic goods, fixing quotas to limit the quantity of imported goods, and requiring licenses in order to restrict the number of legal importers.
- Fledgling businesses are nursed to success by being isolated from competition with large international corporations. Such insulation from the potentially adverse impacts of decisions made by businesses and governments in developed countries encourages a country's fragile businesses to achieve independence.
- Investment is spread as equally as possible across all sectors of a country's economy and in all regions.
- Incomes in the country side keep pace with those in the city, and reducing poverty takes precedence over encouraging a few people to become wealthy consumers. The pace of development may be modest, but the system is fair because residents and enterprises throughout the country share the benefits of development.

CASE STUDY: INDIA'S QUEST FOR DEVELOPMENT. For several decades after it gained independence from Britain in 1947, India was a leading example of the self-sufficiency strategy. India made effective use of many barriers to trade:

- To import goods into India, most foreign companies had to secure a license, which was a long and cumbersome process because several dozen government agencies had to approve the request (Figure 9-50).
- Once a company received an import license, the government severely restricted the quantity of goods it could sell in India.
- The government imposed heavy taxes on imported goods, which doubled or even tripled the prices to consumers.

▼ **FIGURE 9-50 SELF-SUFFICIENCY: INDIA** Clerks work on the street in Delhi, India.

- Indian businesses were discouraged from producing goods for export.
- Indian money could not be converted to other currencies.

Effectively cut off from the world economy, businesses were supposed to produce goods for consumption inside India:

- A business needed government permission to sell a new product, modernize a factory, expand production, set prices, hire or fire workers, and change the job classification of existing workers.
- If private companies were unable to make a profit selling goods only inside India, the government provided subsidies, such as cheap electricity, or wiped out debts.
- The government owned not just communications, transportation, and power companies, which is common around the world, but it also owned businesses such as insurance companies and automakers, which are left to the private sector in most countries.

By following the self-sufficiency path, India achieved only modest development.

INTERNATIONAL TRADE PATH

The international trade model of development calls for a country to identify its distinctive or unique economic assets. What animal, vegetable, or mineral resources does the country have in abundance that other countries are willing to buy? What product can the country manufacture and distribute at a higher quality and a lower cost than other countries? According to the international trade approach, a country can develop economically by concentrating scarce resources on expansion of its distinctive local industries. The sale of these products in the world market brings funds into the country that can be used to finance other development.

ROSTOW MODEL. A pioneering advocate of the international trade approach was W. W. Rostow, who in the 1950s proposed a five-stage model of development. Several countries adopted this approach during the 1960s, although most continued to follow the self-sufficiency approach. The five stages were as follows:

1. **Traditional society.** A traditional society has not yet started a process of development. It contains a very high percentage of people engaged in agriculture and a high percentage of national wealth allocated to what Rostow called "nonproductive" activities, such as the military and religion.
2. **Preconditions for takeoff.** An elite group initiates innovative economic activities. Under the influence of these well-educated leaders, the country starts to invest in new technology and infrastructure, such as water supplies and transportation systems. Support from international funding sources often emphasizes the importance of constructing new infrastructure. These projects will ultimately stimulate an increase in productivity.
3. **Takeoff.** Rapid growth is generated in a limited number of economic activities, such as textiles or food products. These few takeoff industries achieve technical advances and become productive, whereas other sectors of the economy remain dominated by traditional practices.
4. **Drive to maturity.** Modern technology, previously confined to a few takeoff industries, diffuses to a wide variety of industries, which then experience rapid growth comparable to the growth of the take off industries. Workers become more skilled and specialized.
5. **Age of mass consumption.** The economy shifts from production of heavy industry, such as steel and energy, to consumer goods, such as motor vehicles and refrigerators.

According to the international trade model, each country is in one of these five stages of development:

INTERNATIONAL TRADE EXAMPLES. When most developing countries were following the self-sufficiency approach during the twentieth century, two groups of countries chose the international trade approach:

- **The Four Asian Dragons.** Among the first places to adopt the international trade path were South Korea, Singapore, Taiwan, and Hong Kong known as the "four dragons," Singapore and Hong Kong, British colonies until 1965 and 1997, respectively, were large cities surrounded by very small amounts of rural land and had virtually no natural resources. Lacking many natural resources, the four dragons promoted development by concentrating on producing a handful of manufactured goods, especially clothing and electronics. Low labor costs enabled these countries to sell products inexpensively in developed countries.
- **Petroleum-rich Arabian Peninsula states.** The Arabian Peninsula includes Saudi Arabia, the region's largest and most populous country, as well as Kuwait, Bahrain, Oman, and the United Arab Emirates. Once among the world's least developed countries, they were transformed overnight into some of the wealthiest countries, thanks to escalating petroleum prices beginning in the 1970s. Arabian Peninsula countries used petroleum revenues to finance large-scale projects, such as housing, highways, hospitals, airports, universities, and telecommunications networks. Their steel, aluminum, and petrochemical factories competed on world markets with the help of government subsidies. The landscape of these countries has been further changed by the diffusion of consumer goods, such as motor vehicles and electronics. Supermarkets in Arabian Peninsula countries are stocked with food imported from Europe and North America.

Pause and Reflect 9.4.1

Many countries that have adopted the international trade model are relatively small states (see Chapter 8). Why might a nation's size be a factor in the early adoption of the international trade path?

SHORTCOMINGS OF THE TWO DEVELOPMENT PATHS

Learning Outcome 9.4.2

Analyze shortcomings of the two development paths and give reasons why international trade has triumphed.

Shortcomings have been identified with both the self-sufficiency and international trade paths to development.

SELF-SUFFICIENCY CHALLENGES. The experience of India and other developing countries with self-sufficiency revealed two major difficulties:

- **Protection of inefficient businesses.** Businesses could sell all they made, at high government-controlled prices, to customers culled from long waiting lists, so they had little incentive to improve quality, lower production costs, reduce prices, or increase production. Companies protected from international competition were not pressured to keep abreast of rapid technological changes or give high priority to sustainable development and environmental protection.
- **Need for large bureaucracy.** The complex administrative system needed to administer the controls encouraged inefficiency, abuse, and corruption. A large number of people were employed in countries such as India to fill out documents that other countries considered unnecessary intrusions into the prerogatives of private businesses. Potential entrepreneurs found that struggling to produce goods or offer services was less rewarding financially than advising others how to get around the complex government regulations. Other potential entrepreneurs earned more money by illegally importing goods and selling them at inflated prices on the black market.

INTERNATIONAL TRADE CHALLENGES. Three difficulties have hindered countries outside the four Asian dragons and the Arabian Peninsula from developing through the international trade approach:

- **Uneven resource distribution.** Arabian Peninsula countries achieved successful development by means of rising petroleum prices. Other countries, however, have found that the prices of their commodities have not increased and in some cases have actually decreased. Developing countries that have depended on the sale of one product have suffered if the price of their leading commodity did not rise as rapidly as the cost of the products they needed to buy. For example, Zambia has extensive copper reserves, but it has been unable to use this asset to promote development because of declining world prices for copper.
- **Increased dependence on developed countries.** Building up a handful of take off industries that sell to people in developed countries may force developing countries to cut back on production of food, clothing, and other necessities for their own people. Rather than finance sustainable development that is environmentally sensitive, these countries may need to use funds generated from the sale of products to other countries to buy these necessities from developed countries for the employees of the take off industries.
- **Market decline.** Countries that depend on selling low-cost manufactured goods find that the world market for many products has declined sharply in recent years. Even before the recent severe recession, developed countries had limited growth in population and market size.

INTERNATIONAL TRADE APPROACH TRIUMPHS

Most countries embraced the international trade approach as the preferred alternative for stimulating development in the late twentieth century. During the late twentieth and early twenty-first centuries trade increased more rapidly than wealth (as measured by GDP), a measure of the growing importance of the international trade approach, especially in developing countries (Figure 9-51).

Optimism about the benefits of the international trade development model was based on three observations:

- Developed countries in Europe and North America were joined by others in Southern and Eastern Europe and Japan during the second half of the twentieth century. If they could become more developed by following this model, why couldn't other countries?
- Developing countries contained an abundant supply of many raw materials sought by manufacturers and producers in developed countries. In the past, European colonial powers extracted many of these resources without paying compensation to the colonies. In a global economy, the sale of these raw materials could generate funds for developing countries with which they could promote development.

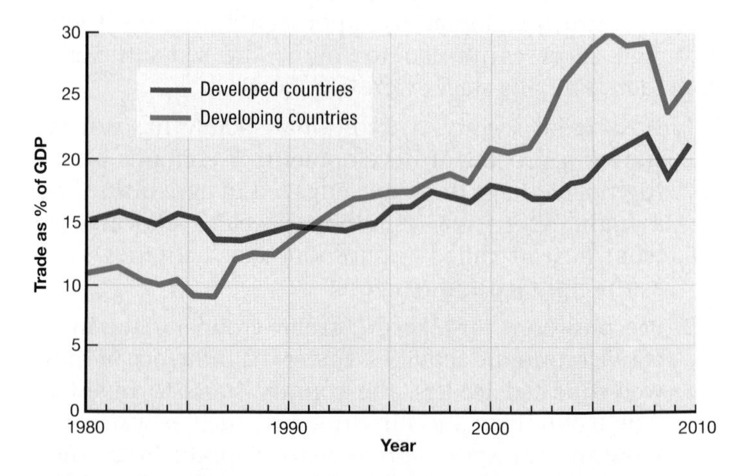

▲ **FIGURE 9-51 WORLD TRADE AS A PERCENTAGE OF INCOME** Trade as a percentage of GDP increased rapidly in developing countries, beginning in the 1990s. The severe recession that began in 2008 caused a sharp decline in trade.

- A country that concentrates on international trade benefits from exposure to the demands, needs, and preferences of consumers in other countries. To remain competitive, the take off industries constantly evaluate changes in international consumer preferences, marketing strategies, production engineering, and design technologies. Concern for international competitiveness in the exporting take off industries can filter through other sectors of the economy.

Longtime advocates of the self-sufficiency approach converted to international trade during the 1990s. India, for example, dismantled its formidable collection of barriers to international trade:

- Foreign companies were allowed to set up factories and sell in India.
- Tariffs and restrictions on the import and export of goods were reduced or eliminated.
- Monopolies in communications, insurance, and other industries were eliminated.
- With increased competition, Indian companies have improved the quality of their products.

During the self-sufficiency era, India's auto industry was dominated by Maruti-Udyog Ltd., which was controlled by the Indian government. Nursed by import duties that rose from 15 percent in 1984 to 66 percent in 1991, Maruti captured more than 80 percent of the Indian market by selling cars that would be considered out-of-date in other countries. In the international trade era, the government sold control of Maruti to the Japanese company Suzuki, which now holds only 45 percent of India's market.

Countries like India converted from self-sufficiency to international trade during the 1990s because of overwhelming evidence at the time that international trade better promoted development (Figure 9-52). After converting to international trade, India's GNI per capita increased on average 6.5 percent per year, compared to 1.8 percent per year under self-sufficiency. Worldwide, GNI increased more than 4 percent annually in countries strongly oriented toward international trade compared with less than 1 percent in countries strongly oriented toward self-sufficiency.

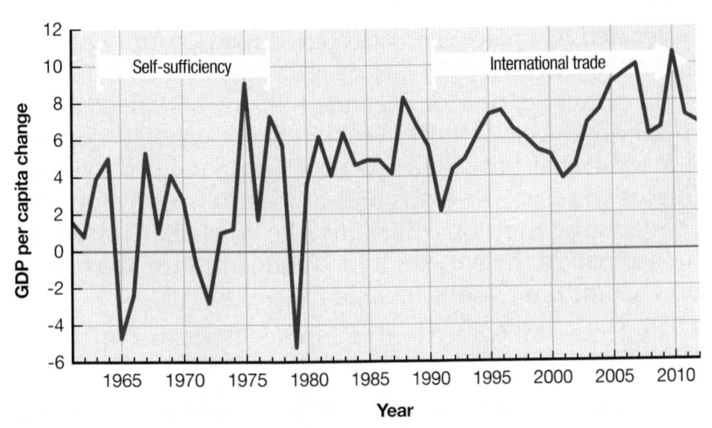

▲ **FIGURE 9-52 GDP PER CAPITA CHANGE IN INDIA** India's per capita GDP has grown much more rapidly since the country converted from the self-sufficiency model to the international trade model.

WORLD TRADE ORGANIZATION

To promote the international trade development model, countries representing 97 percent of world trade established the World Trade Organization (WTO) in 1995. The WTO works to reduce barriers to international trade in two principal ways. First, through the WTO, countries negotiate reduction or elimination of international trade restrictions on manufactured goods, such as government subsidies for exports, quotas for imports, and tariffs on both imports and exports. Also reduced or eliminated are restrictions on the international movement of money by banks, corporations, and wealthy individuals.

The WTO also promotes international trade by enforcing agreements. One country can bring to the WTO an accusation that another country has violated a WTO agreement. The WTO is authorized to rule on the validity of the charge and order remedies. The WTO also protects intellectual property in the age of the Internet. An individual or a corporation can also bring charges to the WTO that someone in another country has violated a copyright or patent, and the WTO can order illegal actions to stop.

Critics have sharply attacked the WTO. Protesters routinely gather in the streets outside high-level meetings of the WTO (Figure 9-53). Progressive critics charge that the WTO is antidemocratic because decisions made behind closed doors promote the interests of large corporations rather than poor people. Conservatives charge that the WTO compromises the power and sovereignty of individual countries because it can order changes in taxes and laws that it considers unfair trading practices.

Pause and Reflect 9.4.2

Top WTO officials meet every two years in a so-called ministerial conference. Where was the most recent conference held? Google "WTO ministerial conference" to find out and to see if there were protests at the conference.

▼ **FIGURE 9-53 WORLD TRADE ORGANIZATION PROTEST** South Korean farmers march in protest during 2005 WTO meetings in Hong Kong.

Financing Development

Learning Outcome 9.4.3
Identify the main sources of financing development.

Developing countries lack money to fund development, so they obtain financial support from developed countries. Finance comes from two primary sources: direct investment by transnational corporations and loans from banks and international organizations.

FOREIGN DIRECT INVESTMENT

International trade requires corporations based in a particular country to invest in other countries. Investment made by a foreign company in the economy of another country is known as **foreign direct investment (FDI)**.

Foreign direct investment grew rapidly during the 1990s, from $130 billion in 1990 to $1.5 trillion in 2000 and 2010. FDI does not flow equally around the world (Figure 9-54). Only two-fifths of foreign investment in 2010 went from a developed country to a developing country; the other three-fifths went from one developed country to another. And FDI is not evenly distributed among developing countries. In 2010, nearly 40 percent of all FDI destined for developing countries went to China, and 20 percent went to Brazil, Russia, and Singapore.

The major sources of FDI are transnational corporations that invest and operate in countries other than the one in which the company headquarters are located. Of the 500 largest transnational corporations in 2011, 384 had headquarters in developed countries, including 133 in the United States and 164 in Europe. China was the location of 61 of the 116 with headquarters in developing countries.

Pause and Reflect 9.4.3
Fortune magazine names the 500 largest transnational corporations every year. What is the world's largest transnational corporation?

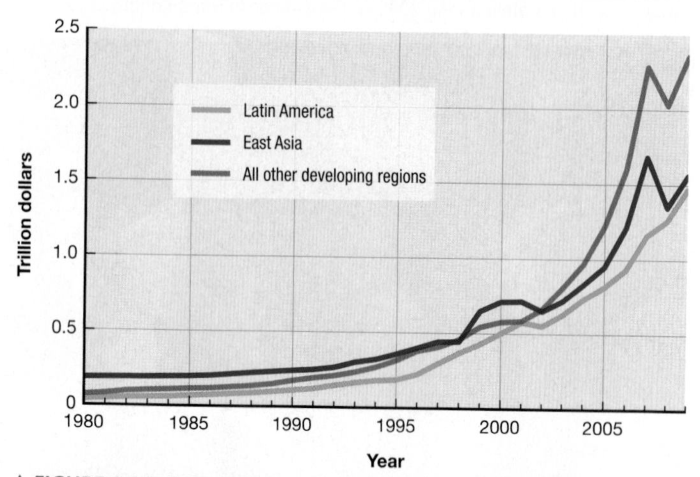

▲ **FIGURE 9-54 GROWTH IN FOREIGN DIRECT INVESTMENT** East Asia and Latin America have received the most FDI.

LOANS

The two major lenders to developing countries are the World Bank and the International Monetary Fund (IMF):

- **World Bank.** The World Bank includes the International Bank for Reconstruction and Development (IBRD) and the International Development Association (IDA). The IBRD provides loans to countries to reform public administration and legal institutions, develop and strengthen financial institutions, and implement transportation and social service projects (Figure 9-55). The IDA provides support to poor countries considered too risky to qualify for IBRD loans. The IBRD has loaned about $400 billion since 1945, primarily in Europe and Latin America (Figure 9-56), and the IDA has loaned about $150 billion since 1960, primarily in Asia and Africa. The IBRD lends money raised from sales of bonds to private investors; the IDA lends money from government contributions.
- **International Monetary Fund (IMF).** The IMF provides loans to countries experiencing balance-of-payments problems that threaten expansion of international trade. IMF assistance is designed to help a country rebuild international reserves, stabilize currency exchange rates, and pay for imports without the imposition of harsh trade restrictions or capital controls that could hamper the growth of world trade. Unlike development banks, the IMF does not lend for specific projects. Funding of the IMF is based on each member country's relative size in the world economy.

The World Bank and IMF were conceived at a 1944 United Nations Monetary and Financial Conference in Bretton Woods, New Hampshire, to promote economic development and stability after the devastation of World War II and to avoid a repetition of the disastrous economic policies contributing to the Great Depression of the 1930s. The IMF and World Bank became specialized agencies of the UN when it was established in 1945.

Developing countries borrow money to build new infrastructure, such as hydroelectric dams, electric transmission lines, flood-protection systems, water supplies, roads, and hotels. The theory is that new infrastructure will make conditions more favorable for domestic and foreign businesses to open or expand. After all, no business wants to be located in a place that lacks paved roads, running water, and electricity.

In principle, new or expanded businesses are attracted to an area because improved infrastructure will contribute additional taxes that the developing country will use in part to repay the loans and in part to improve its citizens' living conditions. In reality, the World Bank itself has judged half of the projects it has funded in Africa to be failures. Common reasons include the following:

- Projects don't function as intended because of faulty engineering.
- Recipient nations squander or spend aid on armaments, or the aid is stolen.
- New infrastructure does not attract other investment.

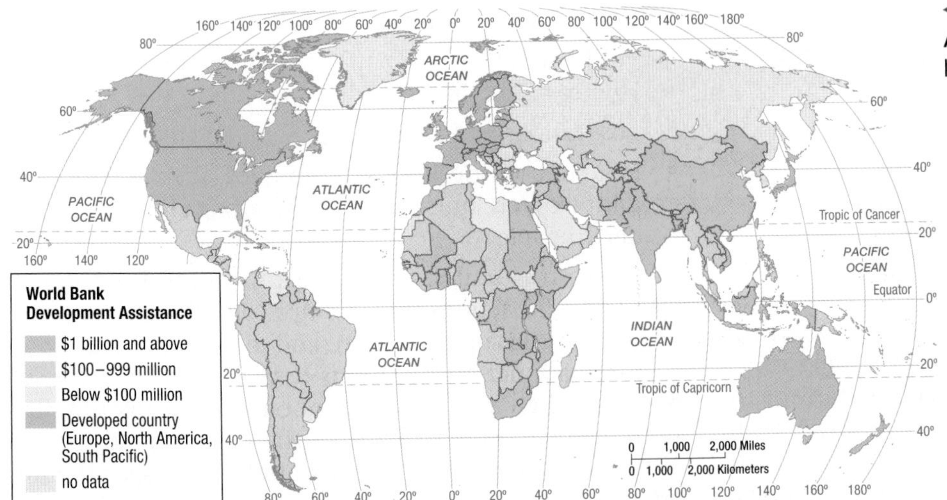

◀ **FIGURE 9-55 WORLD BANK DEVELOPMENT ASSISTANCE** Iraq and Afghanistan have been the leading recipients of aid.

▲ **FIGURE 9-56 WORLD BANK INVESTMENT PROJECT** The World Bank has assisted in the reconstruction of Haiti after the devastating earthquake in 2010.

Many developing countries have been unable to repay the interest on their loans, let alone the principal (Figure 9-57). Debt actually exceeds annual income in a number of countries. When these countries cannot repay their debts, financial institutions in developed countries refuse to make further loans, so construction of needed infrastructure stops. The inability of many developing countries to repay loans also damages the financial stability of banks in developed countries.

The economic downturn that started in 2008 also revealed that many developed countries also have extremely high debts. Among developed countries, especially high debts have been incurred by European countries, including Ireland, Italy, Greece, Portugal, and Spain.

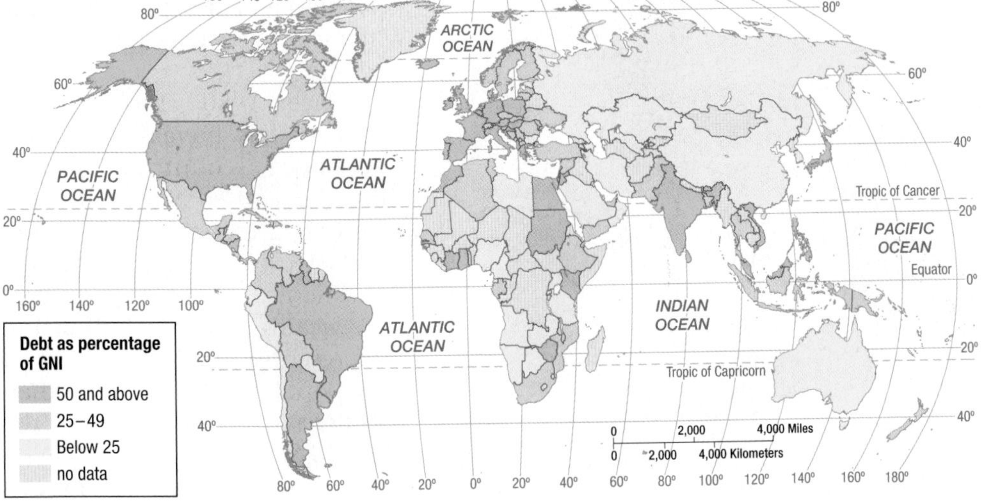

◀ **FIGURE 9-57 DEBT AS A PERCENTAGE OF GNI** Developed countries have joined developing countries in accumulating substantial debts.

FINANCING CHALLENGES IN DEVELOPING COUNTRIES

Learning Outcome 9.4.4
Explain problems with financing development in developing and developed countries.

The IMF, World Bank, and developed countries fear that granting, canceling, or refinancing debts without strings attached will perpetuate bad habits in developing countries. Therefore, to apply for debt relief, a developing country is required to prepare a Policy Framework Paper (PFP) outlining a **structural adjustment program**, which includes economic goals, strategies for achieving the objectives, and external financing requirements.

A structural adjustment program includes economic "reforms" or "adjustments." Requirements placed on a developing country typically include:

- Spending only what it can afford
- Directing benefits to the poor, not just the elite
- Diverting investment from military to health and education spending
- Investing scarce resources where they will have the most impact
- Encouraging a more productive private sector
- Reforming the government, including making the civil service more efficient, increasing accountability in accountable fiscal management, implementing more predictable rules and regulations, and disseminating more information to the public

Critics charge that poverty worsens under structural adjustment programs. By placing priority on reducing government spending and inflation, structural adjustment programs may result in the following:

- Cuts in health, education, and social services that benefit the poor
- Higher unemployment
- Loss of jobs in state enterprises and civil service
- Less support for those most in need, such as poor pregnant women, nursing mothers, young children, and elderly people

In short, structural reforms allegedly punish Earth's poorest people for actions they did not commit, such as waste, corruption, misappropriation, and military buildup.

International organizations respond that the poor suffer more when a country does not undertake reforms. Economic growth is what benefits the poor the most in the long run. Nevertheless, in response to criticisms, the IMF and the World Bank now encourage innovative programs to reduce poverty and corruption and consult more with average citizens. A safety net must be included to ease short-term pain experienced by poor people.

FINANCING CHALLENGES IN DEVELOPED COUNTRIES

Developed countries were especially hard hit by the severe economic downturn that began in 2008. GDP per capita declined between 2008 and 2009 in nearly all developed countries (Figure 9-58). The economic difficulties in developed regions spilled over into developing regions that were especially dependent on international trade, especially Latin America with North America and Southwest Asia with Europe. Citizens and political leaders in many developed countries questioned the benefits of orienting a country's economy to facilitate international trade, especially in Europe.

WIDENING INEQUALITY. Through most of the twentieth century, the gap between rich and poor narrowed in developed countries. Inequality was reduced because developed countries used some of their wealth to extend health care and education to more people, and to provide some financial assistance to poorer people.

Since 1980, however, inequality has increased in most developed countries, including the United States and the United Kingdom (Figure 9-59). In 2010, the richest 1 percent of Americans held 20 percent of the wealth, and 421 billionaires (representing 0.0001 percent of the population) held more than 10 percent of the wealth in the United States.

The severe recession exacerbated an inequality trend that

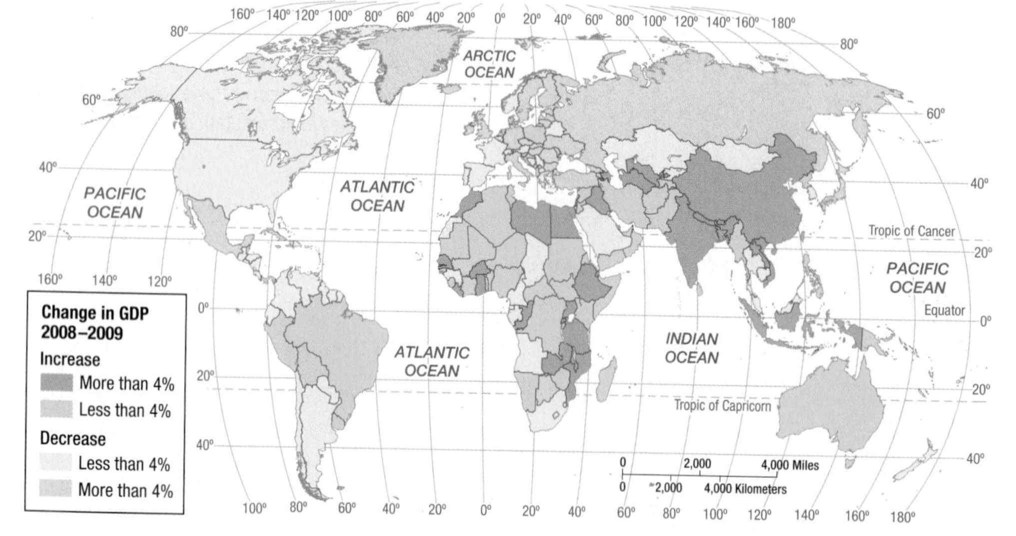

▲ **FIGURE 9-58 GDP PER CAPITA CHANGE, 2008–2009** GNI per capita declined in nearly all developed countries. East and South Asia were the principal regions with increases.

had begun a quarter-century earlier. Many Americans perceived that it was unfair for very large banks to be rescued by the government and to quickly resume making substantial profits, at a time when the income of most Americans was stagnant or declining.

Pause and Reflect 9.4.4
What government policies have helped to increase the share of wealth held by the top 1 percent? What policies have tried to reduce that share?

STIMULUS OR AUSTERITY? Political leaders and independent analysts have been sharply divided on the optimal strategy for fighting the severe economic downturn:

- **Stimulus strategy.** Proponents of stimulus argue that during a downturn, governments should spend more money than they collect in taxes. Governments should stimulate the economy by putting people to work building bridges and other needed infrastructure projects. Once the economy recovers, they say, people and businesses will be in position to pay more taxes to pay off the debt.
- **Austerity strategy.** Proponents of austerity argue that government should sharply reduce taxes so that people and businesses can revive the economy by spending their tax savings. Spending on government programs should be sharply cut as well in order to keep the debt from swelling and hampering the economy in the future.

In the United States, the stimulus strategy was initially employed by Presidents Bush and Obama. After the success of Tea Party candidates in 2010, more attention was paid to the austerity strategy. European countries divided between supporting stimulus and austerity. The lack of agreement has led to serious difficulties in Europe and may possibly result in the demise of the euro currency.

EUROPE'S SOVEREIGN DEBT CRISIS. Europe has faced an especially difficult challenge in responding to the sharp economic slowdown of the early twenty-first century.

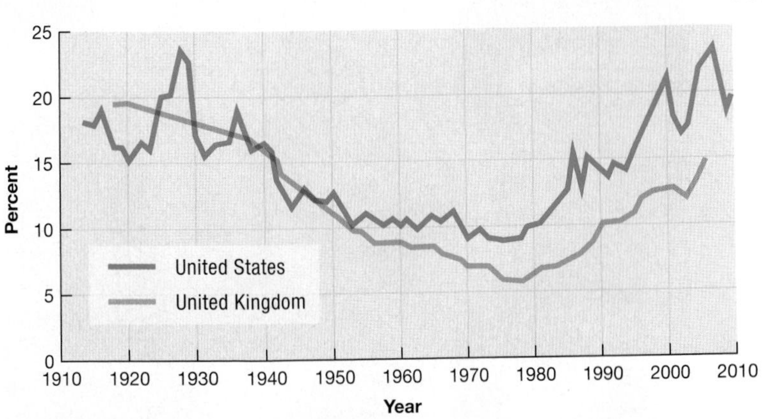

▲ **FIGURE 9-59 TOP 1% INCOME SHARE** The percent of national wealth held by the richest 1 percent of people in the United States and the United Kingdom declined during most of the twentieth century, but has increased since 1980.

Economic difficulties call into question the region's ability to continue supporting the international trade development path.

Most European countries had adopted the euro as their common currency in 1999. Europeans believed that if every country in the region operated with the same currency, trade within the region would be enhanced. In reality, once the severe economic downturn hit, having each country saddled with the same currency proved to be a burden for the countries in Europe that had weaker economies.

Consider Germany and Italy. Germany has a strong economy, with businesses producing cars, electronics, and other goods at higher quality and lower cost than can be done in Italy. If Germany and Italy had two different currencies, as in the past, Italy could lower the value of its currency so that German goods cost more and Italian goods cost less. But with both countries using the same currency, the euro, Italy no longer has that option.

The Northern European countries argue that the Southern European countries with weaker economies need to adopt austerity programs, similar to those imposed on developing countries through structural adjustment programs. The Southern European countries argue that Northern European countries with stronger economies should fund stimulus programs that would in the long run lead to more prosperity through Europe as a whole.

HOUSING BUBBLE

The heart of the global economic crisis was the poor condition of many banks and other financial institutions in developed countries. A number of financial institutions closed or were rescued by governments in North America and Europe. The shaky status of many financial institutions resulted from making loans to businesses and individuals that could not be repaid, especially after the bursting of the housing bubble beginning in 2007.

As discussed in Chapter 1, a housing bubble is a rapid increase in the value of houses followed by a sharp decline in their value. In 1637, the world's first recorded bubble occurred in the Netherlands, when tulip bulbs rapidly increased greatly in price and just as suddenly decreased.

Refer ahead on the next page to Figure 9-60, which shows the housing bubble that occurred in the United States during the first decade of the twenty-first century. The price of an average house in the United States increased rapidly from 1998 to 2006 and then decreased rapidly between 2006 and 2009 down to the level in 2002. Most developed countries and some developing ones experienced housing bubbles during the first decade of the twenty-first century.

FAIR TRADE

Learning Outcome 9.4.5
Explain the principles of fair trade.

Fair trade has been proposed as a variation of the international trade model of development that promotes sustainability. **Fair trade** is commerce in which products are made and traded according to standards that protect workers and small businesses in developing countries.

In North America, fair trade products have been primarily craft products such as decorative home accessories, jewelry, textiles, and ceramics. Ten Thousand Villages is the largest fair trade organization in North America, specializing in handicrafts. In Europe, most fair trade sales are in food, including coffee, tea, banana, chocolate, cocoa, juice, sugar, and honey products.

Two sets of standards distinguish fair trade: One set applies to workers on farms and in factories and the other applies to producers. Standards for fair trade are set internationally by Fairtrade Labelling Organizations International (FLO). A nonprofit organization, TransFair USA, certifies the products sold in the United States that are fair trade.

FAIR TRADE PRODUCER STANDARDS. Critics of international trade charge that only a tiny percentage of the price a consumer pays for a good reaches the individual in the developing country who is responsible for making or growing it. A Haitian sewing clothing for the U.S. market, for example, earns less than 1 percent of the retail price of the garment, according to the National Labor Committee. In contrast, fair trade returns on average one-third of the price to the producer in the developing country. The rest goes to the wholesaler who imports the item and for the retailer's rent, wages, and other expenses.

Fair trade advocates work with small businesses, especially worker-owned and democratically run cooperatives. Small-scale farmers and artisans in developing countries are unable to borrow from banks the money they need to invest in their businesses. By banding together, they can get credit, reduce their raw material costs, and maintain higher and fairer prices for their products. Cooperatives thus benefit the local farmers and artisans who are members rather than benefit absentee corporate owners interested only in maximizing profits. Because cooperatives are managed democratically, farmers and artisans learn leadership and organizational skills. The people who grew or made the products thereby have a say in how local resources are utilized and sold. Safe and healthy working conditions can be protected.

Consumers pay higher prices for fair trade coffee than for grocery store brands, but prices are comparable to those charged for gourmet brands. However, fair trade coffee producers receive a significantly higher price per pound than traditional coffee producers. North American consumers pay $4 to $11 a pound for coffee that is bought from growers for about 80 cents a pound. Growers who sell to fair trade organizations earn $1.12 to $1.26 a pound. Because fair trade organizations bypass distributors and work directly with producers, they can cut costs and return a greater percentage of the retail price to the producers. In some cases, the quality is higher because fair traders factor in the environmental cost of production (Figure 9-61). For instance, in the

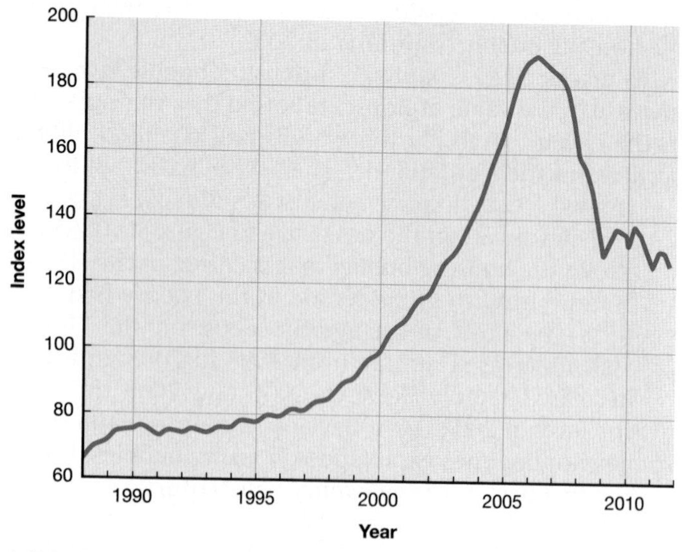

▲ **FIGURE 9-60 HOUSING BUBBLE** House prices doubled in the United States between 1998 and 2006 and declined by one-third between 2006 and 2009. The graph displays price as an index set at 100 in 2000. For example, a house that sold for $100,000 in 2000 would have been sold for $80,000 in 1995, $190,000 in 2006, and $125,000 in 2012.

▼ **FIGURE 9-61 FAIR TRADE** Fair trade coffee is widely available.

case of coffee, fair trade coffee is usually organic and shade grown, which results in higher-quality coffee.

Pause and Reflect 9.4.5
Do you have any fair trade products?

FAIR TRADE WORKER STANDARDS. Protection of workers' rights is not a high priority in the international trade development approach, according to its critics. With minimal oversight by governments and international lending agencies, workers in developing countries allegedly work long hours in poor conditions for low pay. The workforce may include children or forced labor. Health problems may result from poor sanitation and injuries from inadequate safety precautions. Injured, ill, or laid-off workers are not compensated.

In contrast, fair trade requires employers to pay workers fair wages, permit union organizing, and comply with minimum environmental and safety standards. Under fair trade, workers are paid at least the country's minimum wage. Approximately two-thirds of the artisans providing fair trade hand-crafted products are women. Often these women are mothers and the sole wage earners in the home. Because the minimum wage is often not enough for basic survival, whenever feasible, workers are paid enough to cover food, shelter, education, health care, and other basic needs. Cooperatives are encouraged to reinvest profits back into the community, such as by providing health clinics, child care, and training.

Paying fair wages does not necessarily mean that products cost the consumer more. Because fair trade organizations bypass exploitative intermediaries and work directly with producers, they are able to cut costs and return a greater percentage of the retail price to the producers. The cost remains the same as for traditionally traded goods, but the distribution of the cost of the product is different because the large percentage taken by intermediaries is removed from the equation.

DEVELOPMENT THROUGH MICROFINANCE. Many would-be business owners in developing countries are too poor to qualify for regular bank loans. An alternative source of loans is **microfinance**, which is provision of small loans and other financial services to individuals and small businesses in developing countries that are unable to obtain loans from commercial banks (Figure 9-62).

A prominent example of microfinance is the Grameen Bank, which was established in 1977 (Figure 9-63). Based in Bangladesh, Grameen specializes in making loans to women, who make up three-fourths of the borrowers. Women have borrowed money to buy cows, make perfume, bind books, and sell matches, mirrors, and bananas. For founding the bank, Muhammad Yunus was awarded the Nobel Peace Prize in 2006.

The Grameen Bank has made several hundred thousand loans to women in Bangladesh and neighboring South Asian countries, and only 1 percent of the borrowers have failed to make their weekly loan repayments, an extraordinarily low percentage for a bank. Several million loans have also been provided to women by the Bangladesh Rural Advancement Committee. The average loan is about $60. The smallest loan the bank has made was $1, to a woman who wanted to sell plastic bangles door to door.

▲ **FIGURE 9-62 MICROFINANCE** Microfinance helped these women open a tailor shop in north Benin.

▲ **FIGURE 9-63 GRAMEEN BANK**
A representative of the Grameen Bank collects loan payments from women in Bangladesh.

Making Progress in Development

Learning Outcome 9.4.6
Describe ways in which differences in development have narrowed or stayed wide.

Since the UN began measuring HDI in 1980, both developed and developing regions have made progress (Figure 9-64). The overall HDI score has increased by about the same level in developed countries and in developing countries with high HDI scores (primarily Russia and countries in Latin America) and low HDI scores (primarily countries in sub-Saharan Africa). The HDI score for developing countries with medium HDI scores, which includes most of East and South Asia, has increased more rapidly than for the other regions.

Progress in reducing the gap in level of development between developed and developing countries varies depending on the variable. Consider differences among these three prominent variables:

- **Infant mortality rate.** The gap between developed and developing countries has narrowed considerably since 1980. The infant mortality rate has decreased from 17 to 6 (per 1,000) in developed countries and from 107 to 44 in developing countries with medium HDI, which includes most of East and South Asia (Figure 9-65).
- **Life expectancy.** The number of years a baby is expected to live has increased by 8 years in developing countries (Figure 9-66). However, life expectancy at birth has increased by 7 years in developed countries. So the gap between developed and developing countries has not narrowed.
- **GNI per capita.** The gap in wealth between developed and developing countries has widened (Figure 9-67). Since 1980, GNI per capita has increased from $20,000 to $33,000 in developed countries and from $1,000 to $5,000 in developing countries with medium HDI. Progress in improving GNI per capita has been modest in developing countries with high HDI and developing countries with low HDI.

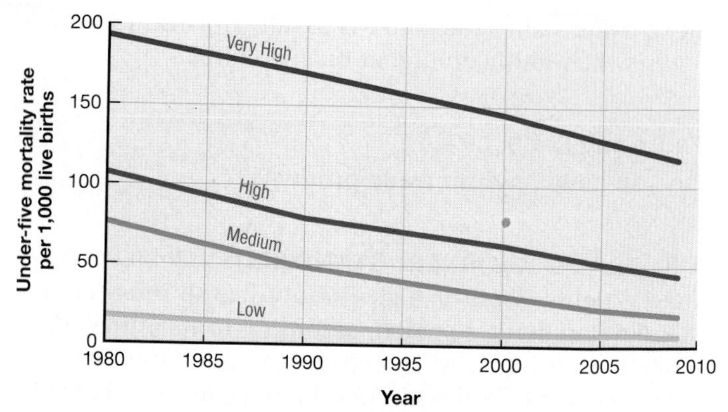

▲ **FIGURE 9-65 INFANT MORTALITY RATE CHANGE BY HDI LEVEL, 1980–2011**
Developing regions have closed the gap in infant mortality rates.

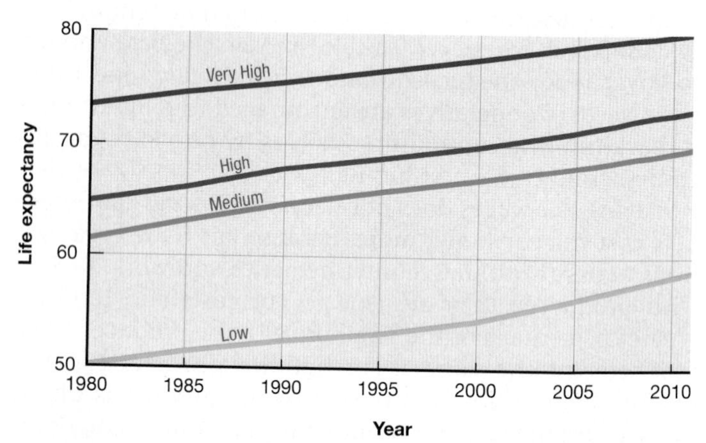

▲ **FIGURE 9-66 LIFE EXPECTANCY CHANGE BY HDI LEVEL, 1980–2011**
All regions have seen substantial progress in increasing life expectancy.

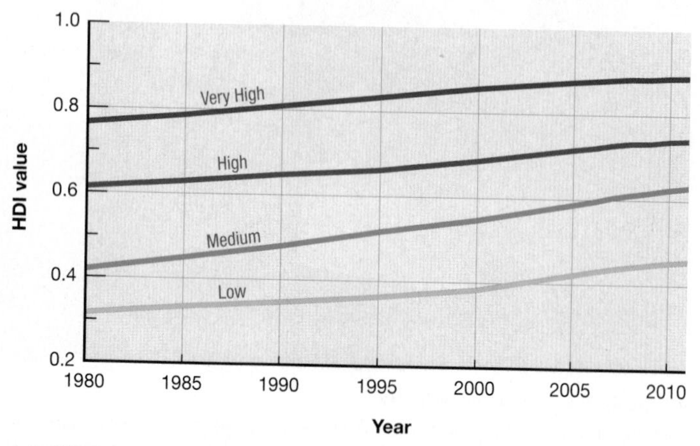

▲ **FIGURE 9-64 HDI CHANGE BY HDI LEVEL, 1980–2011** The HDI has improved relatively rapidly in developing countries with medium HDI scores.

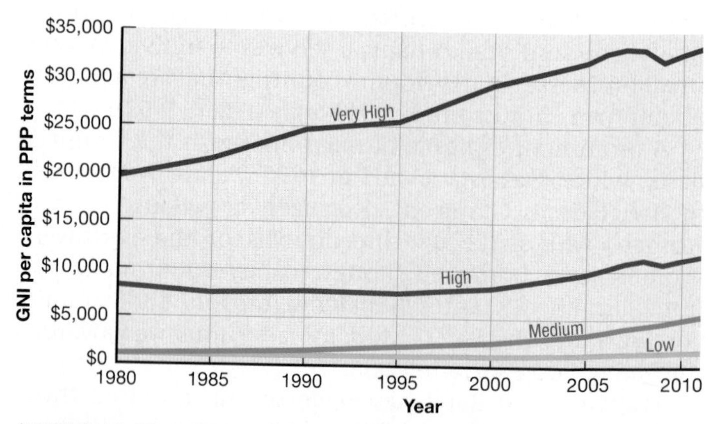

▲ **FIGURE 9-67 GNI PER CAPITA CHANGE BY HDI LEVEL, 1980–2011**
The gap in wealth between developed and developing regions has increased.

MILLENNIUM DEVELOPMENT GOALS

To reduce disparities between developed countries and developing countries the UN has set eight so-called **Millennium Development Goals**, which all UN members have agreed to achieve by 2015. Table 9-1 displays the goals and the progress that has actually been made, according to the UN:

Pause and Reflect 9.4.6

Based on Table 9-1, which Millennium Development Goal appears to be making the most limited progress?

TABLE 9–1 MILLENNIUM DEVELOPMENT GOALS AND PROGRESS TOWARDS ACHIEVEMENT

Goal 1: Eradicate extreme poverty and hunger
• Prior to the severe recession, the depth of poverty had diminished in almost every region. The global economic crisis has slowed progress, as more workers found themselves and their families living in extreme poverty. • One of the consequences of the severe recession was an increase in hunger, and progress to end hunger has been stymied. One in four children in developing countries is underweight.
Goal 2: Achieve universal primary education
• Sub-Saharan Africa and Southern Asia are home to the vast majority of children out of school, as shown in Figure 9-11. Although progress has been made, the UN concludes that the goal is not being achieved.
Goal 3: Promote gender equality and empower women
• The UN concludes that major barriers to gender equality remain, as discussed in Key Issue 2 of this chapter. The UN cites higher levels of poverty, fewer education opportunities, and lack of political representation. Top-level jobs still go to men, whereas women are relegated to jobs with low pay, limited benefits, and little security.
Goal 4: Reduce child mortality
• The UN concludes that child deaths are falling, but not quickly enough to reach the target. Infant mortality rates remain especially high in sub-Saharan Africa, as shown in Figure 9-65.
Goal 5: Improve maternal health
• Progress has been made in reducing maternal mortality, but as shown in Key Issue 2, giving birth is especially risky in Southern Asia and sub-Saharan Africa, where most women deliver without skilled care. • Progress has stalled in reducing the number of teenage pregnancies, putting more young mothers at risk. • Progress in expanding the use of contraceptives by women has slowed. Use of contraception is lowest among the poorest women and those with no education. • Inadequate funding for family planning is a major failure in fulfilling commitments to improving women's reproductive health
Goal 6: Combat HIV/AIDS, malaria, and other diseases
• The spread of HIV appears to have stabilized in most regions, as discussed in Chapter 2. Many young people still lack the knowledge to protect themselves against HIV, especially in sub-Saharan Africa. However, the rate of new HIV infections continues to outstrip the expansion of treatment. • More drugs to fight malaria are being distributed. Expanded use of insecticide-treated bed nets is protecting communities from malaria, especially in sub-Saharan Africa. • Tuberculosis prevalence is falling in most regions, but TB remains the second leading killer after HIV.
Goal 7: Ensure environmental sustainability
• The rate of deforestation is decreasing, but is still alarmingly high, and a decisive response to climate change is urgently needed, according to the UN. • Key habitats for threatened species are not being adequately protected, and the number of species facing extinction is growing by the day, especially in developing countries. • The world is on track to meet the drinking water target, though safe water supply remains a challenge in many parts of the world, especially in rural areas. • With half the population of developing regions without sanitation, the 2015 target is out of reach, according to the UN. • Slum improvements, though considerable, are failing to keep pace with the growing ranks of the urban poor, as discussed in Chapter 13.
Goal 8: Develop a global partnership for development
• Aid from developed countries continues to rise, but relatively little of it is reaching sub-Saharan Africa.

CORE AND PERIPHERY

The relationship between developed countries and developing countries is often described as a north-south split, because most of the developed countries are north of the equator, whereas many developing countries are south. Immanuel Wallerstein, a U.S. social scientist, depicted the relationship between developed and developing countries as one of "core" and "periphery." According to Wallerstein's world-systems analysis, in an increasingly unified world economy, developed countries form an inner core area, whereas developing countries occupy peripheral locations. As a result, global development patterns are sometimes referred to as **uneven development**, with countries at the core benefiting at the expense of countries on the periphery.

The unorthodox north polar map projection in Figure 9-68 emphasizes the central role that developed countries play in the world economy. North America, Europe, and Japan account for a high percentage of the world's economic activity and wealth. Developing countries in the periphery have less access to the world centers of consumption, communications, wealth, and power, which are clustered in the core.

The unorthodox projection in Figure 9-60 also shows connections between particular core and periphery regions. The development prospects of Latin America are tied to governments and businesses primarily in North America, those of Africa and Eastern Europe to Western Europe, and those of Asia to Japan and to a lesser extent Europe and North America. As countries like China, India, and Brazil develop, relationships between core and periphery are changing, and the line between core and periphery may need to be redrawn.

CHECK-IN: KEY ISSUE 4

Why Do Countries Face Obstacles to Development?

- ✓ **Two paths to development are self-sufficiency and international trade; international trade has become more important in recent decades.**
- ✓ **Development is financed through foreign direct investment by corporations and loans by governments and international organizations.**
- ✓ **The severe recession of the early twenty-first century has posed challenges to developing countries and developed countries to continue development policies.**
- ✓ **Progress has been made in achieving development in most regions.**

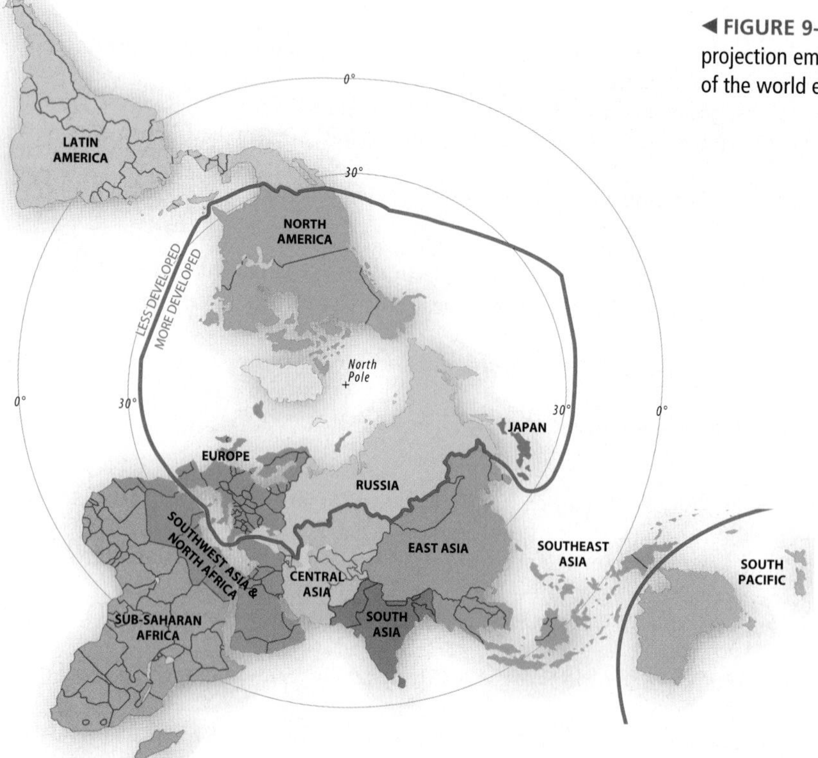

◀ **FIGURE 9-68 CORE AND PERIPHERY** This unorthodox world map projection emphasizes the central role that developed countries play at the core of the world economy.

Summary and Review

KEY ISSUE 1

Why Does Development Vary among Countries?

Development is the process by which the material conditions of a country's people are improved. The world is divided into developed countries and developing ones. Developed and developing countries can be compared according to a number of indicators.

LEARNING OUTCOME 9.1.1: Identify the HDI standard of living factor.

- The HDI, which measures the level of development of each country, is calculated by combining three measures.
- Standard of living is measured through gross national income per capita at purchasing power parity.

LEARNING OUTCOME 9.1.2: Identify the HDI health factor.

- The HDI health factor is life expectancy at birth.

LEARNING OUTCOME 9.1.3: Identify the HDI access to knowledge factor.

- The HDI knowledge factors are years of schooling and expected years of schooling.

LEARNING OUTCOME 9.1.4: Describe variations in level of development within countries and regions.

- Some developing countries, especially larger ones, have large variations among regions in level of development.

THINKING GEOGRAPHICALLY 9.1: In what ways would you expect the severe recession of the early twenty-first century to change some of the development indicators?

GOOGLE EARTH 9.1: Vehicle ownership rates are extremely low in Kenya, yet if you zoom into the center of Nairobi, what is the volume of traffic on the roads?

KEY ISSUE 2

Why Does Development Vary by Gender?

The UN has not found a single country in the world where the women are treated as well as the men.

LEARNING OUTCOME 9.2.1: Describe the UN's measures of gender inequality.

- The GII measures the extent of gender inequality.
- The GII combines measures of empowerment, labor force participation, and reproductive rights.

LEARNING OUTCOME 9.2.2: Describe changes since the 1990s in gender inequality.

- Gender inequality has declined in most countries since 1990, although not in the United States.

THINKING GEOGRAPHICALLY 9.2: Review the major economic, social, and demographic characteristics that contribute to a country's level of development. Which indicators can vary significantly by gender within countries and between countries at various levels of development? Why?

GOOGLE EARTH 9.2: Women comprise nearly one-half of Sweden's Parliament. Fly to Parliament of Sweden, Stockholm. What is the distinctive physical site on which the Parliament is located?

Key Terms

Active solar energy systems (p. 326) Solar energy systems that collects energy through the use of mechanical devices such as photovoltaic cells or flat-plate collectors.

Adolescent fertility rate (p. 312) The number of births per 1,000 women ages 15 to 19.

Biomass fuel (p. 324) Fuel that derives from plant material and animal waste.

Breeder reactor (p. 323) A nuclear power plant that creates its own fuel from plutonium.

Demand (p. 314) The quantity of something that consumers are willing and able to buy.

Developed country (more developed country [MDC] or relatively developed country) (p. 300) A country that has progressed relatively far along a continuum of development.

Developing country (less developed country [LDC]) (p. 300) A country that is at a relatively early stage in the process of economic development.

Development (p. 300) A process of improvement in the material conditions of people through diffusion of knowledge and technology.

Fair trade (p. 336) An alternative to international trade that emphasizes small businesses and worker-owned and democratically run

KEY ISSUE 3

Why Are Energy Resources Important for Development?

Development depends on abundant low-cost energy.

LEARNING OUTCOME 9.3.1: Explain the principal sources of demand for fossil fuels.

- Most energy is supplied by three fossil fuels: coal, petroleum, and natural gas.
- Developed countries and developing countries each consume approximately half of the world's energy.

LEARNING OUTCOME 9.3.2: Describe the distribution of production of the three fossil fuels.

- Fossil fuels are not distributed uniformly around the world, and they are nonrenewable sources of energy.

LEARNING OUTCOME 9.3.3: Analyze the distribution of reserves of fossil fuels and differentiate between proven and potential reserves.

- Reserves are divided into proven (fields already discovered) and potential (fields thought to exist).
- Proven reserves are not distributed uniformly.

LEARNING OUTCOME 9.3.4: Describe the role of OPEC and changes in the price and availability of petroleum.

- Much of the world's petroleum reserves are located in countries that belong to OPEC.
- The United States has increased its dependence on petroleum imported from neighbors in the Western Hemisphere.

LEARNING OUTCOME 9.3.5: Describe the distribution of nuclear energy and challenges in using it.

- Nuclear is the principal source of energy other than the three fossil fuels in the United States and a couple dozen other countries.
- Numerous problems limit the use of nuclear power, including threat of accidents, disposal of waste, use in making weapons, limited reserves, and high costs.

LEARNING OUTCOME 9.3.6: Identify challenges to increasing the use of alternative energy sources.

- Leading renewable energy sources include biomass, hydroelectric, geothermal, wind, and solar.

LEARNING OUTCOME 9.3.7: Compare and contrast between passive and active solar energy.

- Active solar energy captures energy with special devices, such as photovoltaic cells, whereas passive solar energy does not.

THINKING GEOGRAPHICALLY 9.3: The average American consumes approximately 500 gallons of gas a year in his or her car. Does your family use more or less than the average? To answer this, you need to know how many miles your or your family's vehicles are driven and the vehicles' fuel efficiency (average miles per gallon). The fuel efficiency can be found by Googling "fuel efficiency" plus the vehicle model and year.

GOOGLE EARTH 9.3: If you fly to 1301 W 120 St., Chicago, what type of energy is being supplied by the large rectangular feature?

cooperatives and requires employers to pay workers fair wages, permit union organization, and comply with minimum environmental and safety standards.

Female labor force participation rate (p. 311) The percentage of women holding full-time jobs outside the home.

Fission (p. 322) The splitting of an atomic nucleus to release energy.

Foreign direct investment (FDI) (p. 332) Investment made by a foreign company in the economy of another country.

Fossil fuel (p. 314) An energy source formed from the residue of plants and animals buried millions of years ago.

Fracking (hydraulic fracturing) (p. 319) The pumping of water at high pressure to break apart rocks in order to release natural gas.

Fusion (p. 325) Creation of energy by joining the nuclei of two hydrogen atoms to form helium.

Gender Inequality Index (GII) (p. 310) A measure of the extent of each country's gender inequality.

Geothermal energy (p. 325) Energy from steam or hot water produced from hot or molten underground rocks.

Gross domestic product (GDP) (p. 302) The value of the total output of goods and services produced in a country in a given time period (normally one year).

Gross national income (GNI) (p. 302) The value of the output of goods and services produced in a country in a year, including money that leaves and enters the country.

Housing bubble (p. 335)A rapid increase in the value of houses followed by a sharp decline in their value.

Human Development Index (HDI) (p. 301) An indicator of the level of development for each country, constructed by the United Nations, that is based on income, literacy, education, and life expectancy.

Hydroelectric power (p. 324) Power generated from moving water.

Inequality-adjusted HDI (IHDI) (p. 303) Modification of the HDI to account for inequality within a country.

Literacy rate (p. 307) The percentage of a country's people who can read and write.

Maternal mortality ratio (p. 312) The number of women who die giving birth per 100,000 births.

Microfinance (p. 337) Provision of small loans and other financial services to individuals and small businesses in developing countries.

Millennium Development Goals (p. 339) Eight international development goals that all members of the United Nations have agreed to achieve by 2015.

KEY ISSUE 4

Why Do Countries Face Obstacles to Development?

To develop more rapidly, developing countries must adopt policies that successfully promote development and find funds to pay for it.

LEARNING OUTCOME 9.4.1: Summarize the two paths to development.

- To promote development, developing countries choose either the self-sufficiency path or the international trade path.

LEARNING OUTCOME 9.4.2: Analyze shortcomings of the two development paths and give reasons international trade has triumphed.

- Self-sufficiency has protected inefficient businesses.
- International trade has increased dependency on declining resources and developed countries.
- Most countries have adopted international trade because of evidence that it promotes more rapid development.

LEARNING OUTCOME 9.4.3: Identify the main sources of financing development.

- Finance comes from direct investment by transnational corporations and loans from banks and international organizations.

LEARNING OUTCOME 9.4.4: Explain problems with financing development in developing and developed countries.

- Developing countries have been required to adopt structural adjustment programs.
- Developed countries have had to choose between policies that promote short-term growth and those that promote austerity.

LEARNING OUTCOME 9.4.5: Explain the principles of fair trade.

- Fair trade attempts to protect workers and small businesses in developing countries.
- Fair trade involves a combination of producer and worker standards.

LEARNING OUTCOME 9.4.6: Describe ways in which differences in development have narrowed or stayed wide.

- Developing countries have closed the gap with developed countries in some respects, such as health, but not in other respects, such as income.

THINKING GEOGRAPHICALLY 9.4: Some developing countries claim that the requirements placed on them by lending organizations such as the World Bank impede rather than promote development. Should developing countries be given a greater role in deciding how much the international organizations should spend and how such funds should be spent? Why or why not?

GOOGLE EARTH 9.4: A portion of the Trans-African Highway can be seen in the center of Voi, Kenya, running east-west in a curving arc immediately south of the center. Drag to street view, exit street view, and rotate so that north is to the right. For approximately what distance is the highway divided?

Nonrenewable energy (p. 317) A source of energy that has a finite supply capable of being exhausted.

Passive solar energy systems (p. 326) Solar energy systems that collect energy without the use of mechanical devices.

Photovoltaic cell (p. 326) A solar energy cell, usually made from silicon, that collects solar rays to generate electricity.

Potential reserve (p. 319) The amount of a resource in deposits not yet identified but thought to exist.

Primary sector (p. 302) The portion of the economy concerned with the direct extraction of materials from Earth's surface, generally through agriculture, although sometimes by mining, fishing, and forestry.

Productivity (p. 303) The value of a particular product compared to the amount of labor needed to make it.

Proven reserve (p. 318) The amount of a resource remaining in discovered deposits.

Purchasing power parity (PPP) (p. 302) The amount of money needed in one country to purchase the same goods and services in another country; PPP adjusts income figures to account for differences among countries in the cost of goods.

Radioactive waste (p. 322) Materials from a nuclear reaction that emit radiation; contact with such particles may be harmful or lethal to people; therefore, the materials must be safely stored for thousands of years.

Renewable energy (p. 317) A resource that has a theoretically unlimited supply and is not depleted when used by humans.

Secondary sector (p. 302) The portion of the economy concerned with manufacturing useful products through processing, transforming, and assembling raw materials.

Structural adjustment program (p. 334) Economic policies imposed on less developed countries by international agencies to create conditions encouraging international trade, such as raising taxes, reducing government spending, controlling inflation, selling publicly owned utilities to private corporations, and charging citizens more for services.

Supply (p. 314) The quantity of something that producers have available for sale.

Tertiary sector (p. 302) The portion of the economy concerned with transportation, communications, and utilities, sometimes extended to the provision of all goods and services to people, in exchange for payment.

Uneven development (p. 340) Development of core regions at the expense of those on the periphery.

Value added (p. 303) The gross value of a product minus the costs of raw materials and energy.

Chapter 10 Food and Agriculture

What food is this African girl carrying? Page 353

Why is this field deliberately flooded? Page 363

KEY ISSUE 1

Where Did Agriculture Originate?

Inventing Agriculture p. 347

Agriculture was invented around 10,000 years ago in multiple hearths.

KEY ISSUE 2

Why Do People Consume Different Foods?

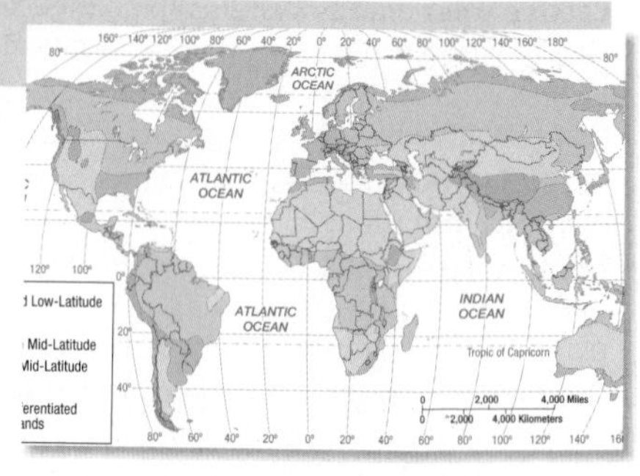

The Foods We Consume p. 352

Humans consume most of their calories through grains.

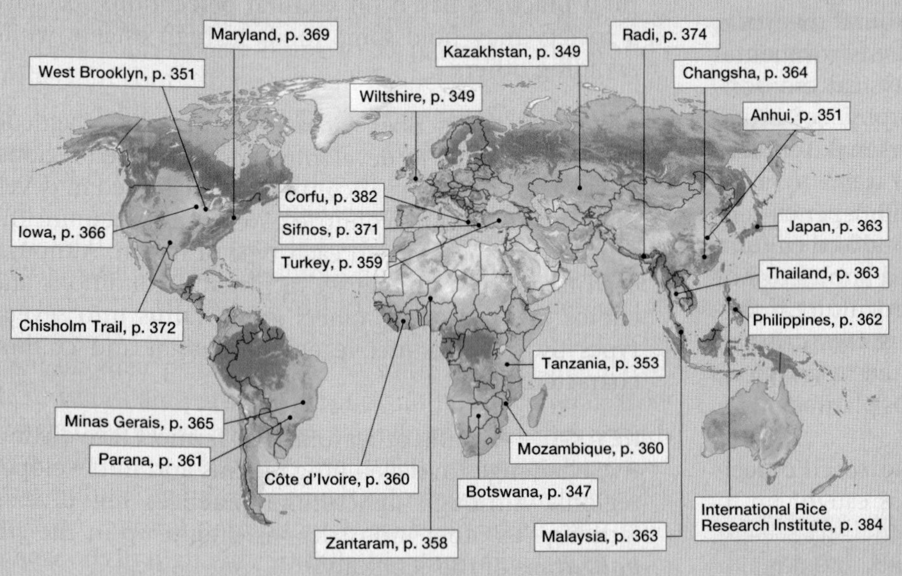

▲ These farmers in northwestern India are harvesting wheat seeds by beating the stalks by hand. The farm produces about 1,500 kilograms (3,300 pounds) of wheat per year—enough to feed a family. In contrast, the average farmer in Kansas produces 175,000 kilograms (400,000 pounds) of wheat a year. In a developed country, the work of separating the seeds from the stalks is done by a machine, whereas in developing countries, most farmers can't afford machinery, so they must do the work by hand.

KEY ISSUE 3

Where Is Agriculture Distributed?

Agricultural Regions p. 360

Eleven major agricultural regions approach agriculture differently.

KEY ISSUE 4

Why Do Farmers Face Economic Difficulties?

Challenges for Farmers p. 374

Farmers have trouble making ends meet in both developing and developed regions.

AGRICULTURAL REVOLUTION

Learning Outcome 10.1.1
Identify the major crop and livestock hearths.

The **agricultural revolution** was the time when human beings first domesticated plants and animals and no longer relied entirely on hunting and gathering. When did the agricultural revolution occur? About the year 8000 B.C., the world's population began to grow at a more rapid rate than it had in the past. Geographers and other scientists believe that the reason for the sudden population increase was the agricultural revolution. By growing plants and raising animals, human beings created larger and more stable sources of food, so more people could survive.

Scientists do not agree on whether the agricultural revolution originated primarily because of environmental factors or cultural factors. Probably a combination of both factors contributed:

- **Environmental factors.** Those favoring environmental reasons point to the coinciding of the first domestication of crops and animals with climate change around 10,000 years ago. This marked the end of the last ice age, when permanent ice cover receded from Earth's mid-latitudes to polar regions, resulting in a massive redistribution of humans, other animals, and plants at that time.
- **Cultural factors.** Human behavior may be primarily responsible for the origin of agriculture. A preference for living in a fixed place rather than as nomads may have led hunters and gatherers to build permanent settlements and to store surplus vegetation there. In gathering wild vegetation, people inevitably cut plants and dropped berries, fruits, and seeds. These hunters probably observed that, over time, damaged or discarded food produced new plants. They may have deliberately cut plants or dropped berries on the ground to see if they would produce new plants. Subsequent generations learned to pour water over the site and to introduce manure and other soil improvements. Over thousands of years, plant cultivation apparently evolved from a combination of accident and deliberate experiment.

CROP HEARTHS. Scientists also do not agree on how agriculture diffused or why most nomadic groups convert from hunting, gathering, and fishing to agriculture. They do agree that agriculture originated in multiple hearths around the world:

- **Southwest Asia.** The earliest crops domesticated in Southwest Asia are thought to have been barley and wheat, around 10,000 years ago (Figure 10-3). Lentil and olive were also early domestications in Southwest Asia. From this hearth, cultivation diffused west to Europe and east to Central Asia.
- **East Asia.** Rice is now thought to have been domesticated in East Asia more than 10,000 years ago, along the Yangtze River in eastern China. Millet was cultivated at an early date along the Yellow River.
- **Sub-Saharan Africa.** Sorghum was domesticated in central Africa around 8,000 years ago. Yams may have

▼ **FIGURE 10-3 CROP HEARTHS** Agriculture originated in multiple hearths. Domestication of some crops can be dated back more than 10,000 years.

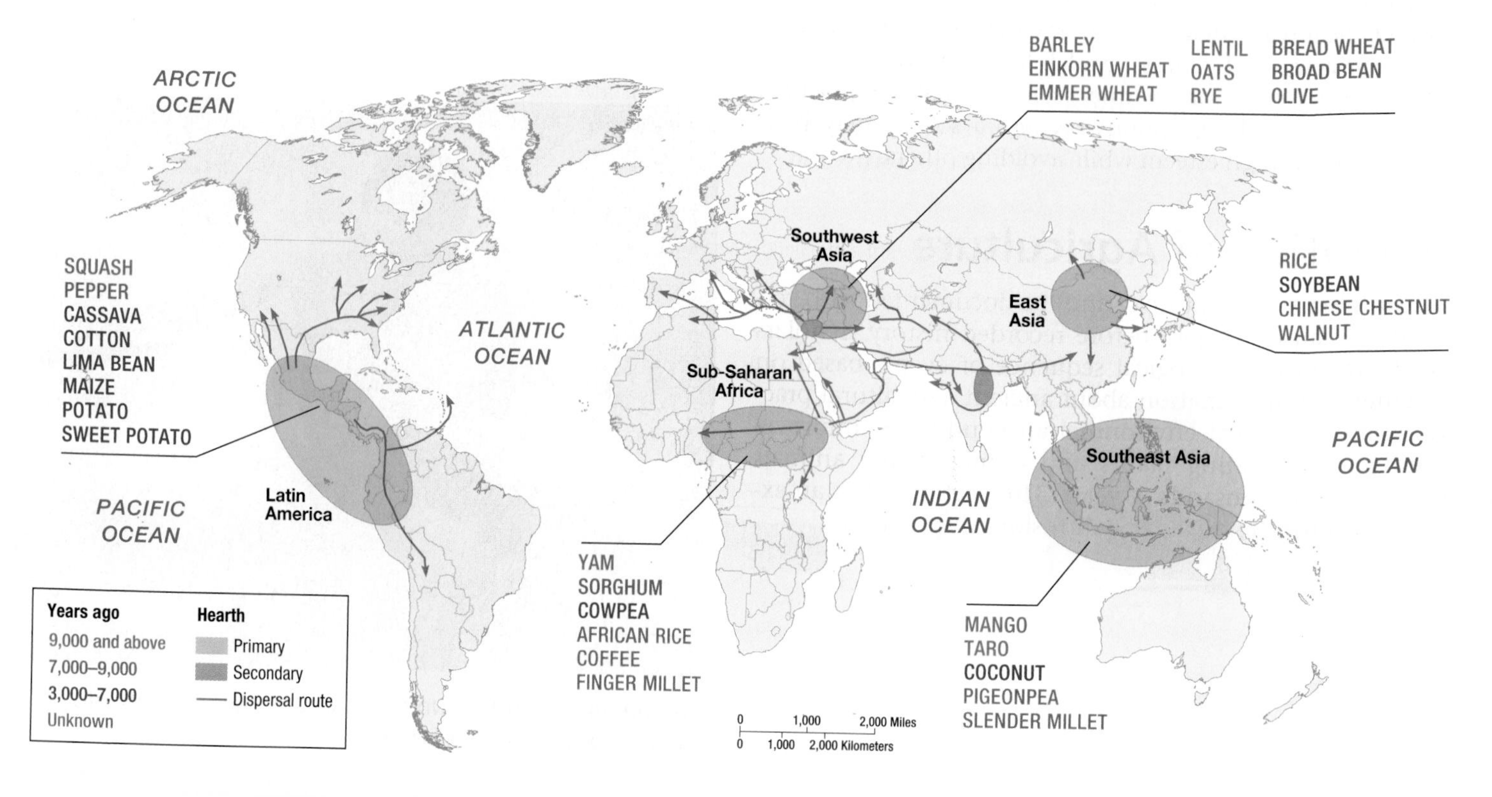

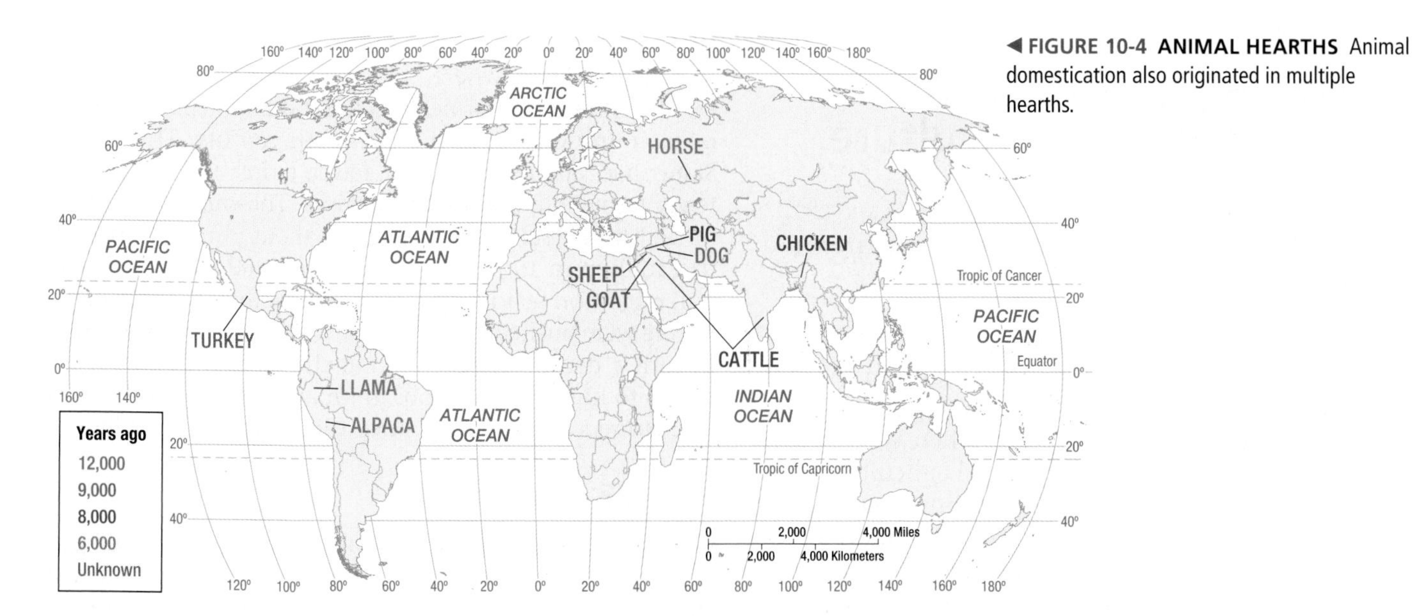

◀ **FIGURE 10-4 ANIMAL HEARTHS** Animal domestication also originated in multiple hearths.

been domesticated even earlier. Millet and rice may have been domesticated in sub-Saharan Africa independently of the hearth in East Asia. From central Africa, domestication of crops probably diffused further south in Africa.

- **Latin America.** Two important hearths of crop domestication are thought to have emerged in Mexico and Peru around 4,000 to 5,000 years ago. Mexico is considered a hearth for beans and cotton, and Peru for potato. The most important contribution of the Americas to crop domestication, maize (corn), may have emerged in the two hearths independently around the same time. From these two hearths, cultivation of maize and other crops diffused northward into North America and southward into tropical South America. Some researchers place the origin of squash in the southeastern present-day United States.

Pause and Reflect 10.1.1
Which crops appear to have reached the present-day United States first, according to Figure 10-3?

ANIMAL HEARTHS. Animals were also domesticated in multiple hearths at various dates. Southwest Asia is thought to have been the hearth for the domestication of the largest number of animals that would prove to be most important for agriculture, including cattle, goats, pigs, and sheep, between 8,000 and 9,000 years ago (Figure 10-4). Domestication of the dog is thought to date from around 12,000 years ago or possibly earlier in Southwest Asia, East Asia, and/or Europe. The horse is considered to have been domesticated in Central Asia; diffusion of the domesticated horse is thought to be associated with the diffusion of the Indo-European language, as discussed in Chapter 5 (Figure 10-5).

Inhabitants of Southwest Asia may have been the first to integrate cultivation of crops with domestication of herd animals such as cattle, sheep, and goats. These animals were used to prepare the land before planting seeds and, in turn, were fed part of the harvested crop. Other animal products, such as milk, meat, and skins, may have been exploited at a later date. This integration of plants and animals is a fundamental element of modern agriculture.

That agriculture had multiple origins means that, from earliest times, people have produced food in distinctive ways in different regions. This diversity derives from a unique legacy of wild plants, climatic conditions, and cultural preferences in each region. Improved communications in recent centuries have encouraged the diffusion of some plants to varied locations around the world. Many plants and animals thrive across a wide portion of Earth's surface, not just in their place of original domestication. Only after 1500, for example, were wheat, oats, and barley introduced to the Western Hemisphere and maize to the Eastern Hemisphere.

▼ **FIGURE 10-5 HORSES IN KAZAKHSTAN** The horse is thought to have been domesticated in this region roughly 6,000 years ago.

Comparing Subsistence and Commercial Agriculture

Learning Outcome 10.1.2

Describe the major differences between subsistence and commercial agriculture.

The most fundamental differences in agricultural practices are between those in developing countries and those in developed countries. Farmers in developing countries generally practice subsistence agriculture, whereas farmers in developed countries practice commercial agriculture. **Subsistence agriculture**, found in developing countries, is the production of food primarily for consumption by the farmer's family. **Commercial agriculture**, found in developed countries, is the production of food primarily for sale off the farm. The main features that distinguish commercial agriculture from subsistence agriculture include the percentage of farmers in the labor force, the use of machinery, and farm size.

PERCENTAGE OF FARMERS IN THE LABOR FORCE

A priority for all people is to secure the food they need to survive. In developing countries most people are subsistence farmers who work in agriculture to produce the food they and their families require. In developed countries the relatively few people engaged in farming are commercial farmers, and most people buy food with money earned by working in factories or offices or by performing other services.

In developed countries, around 5 percent of workers are engaged directly in farming, compared to around 44 percent in developing countries (Figure 10-6). The percentage of farmers is even lower in North America—only around 2 percent. Yet the small percentage of farmers in the United States and Canada produces not only enough food for themselves and the rest of the region but also a surplus to feed people elsewhere.

The number of farmers declined dramatically in developed countries during the twentieth century. The United States had about 60 percent fewer farms and 85 percent fewer farmers in 2000 than in 1900. The number of farms in the United States declined from about 6 million in 1940 to 4 million in 1960 and 2 million in 1980. Both push and pull migration factors have been responsible for the decline: People were pushed away from farms by lack of opportunity to earn a decent income, and at the same time they were pulled to higher-paying jobs in urban areas. The number of U.S. farmers has stabilized since 1980 at around 2 million.

USE OF MACHINERY

In developed countries, a small number of commercial farmers can feed many people because they rely on machinery to perform work rather than on people or animals (Figure 10-7). In developing countries, subsistence farmers do much of the work with hand tools and animal power.

Traditionally, the farmer or local craftspeople made equipment from wood, but beginning in the late eighteenth century, factories produced farm machinery. The first all-iron plow was made in the 1770s and was followed in the nineteenth and twentieth centuries by inventions that made farming less dependent on human or animal power. Tractors, combines, corn pickers, planters, and other factory-made farm machines have replaced or supplemented manual labor.

Transportation improvements have also aided commercial farmers. The building of railroads in the nineteenth century and highways and trucks in the twentieth century have enabled farmers to transport crops and livestock farther and faster. Cattle arrive at market heavier and in better condition when transported by truck or train than when driven on hoof. Crops reach markets without spoiling.

Commercial farmers use scientific advances to increase productivity. Experiments conducted in university laboratories, industry, and research organizations generate new fertilizers, herbicides, hybrid plants, animal breeds, and farming practices, which produce higher crop yields and healthier animals. Access to other scientific information has enabled farmers to make more intelligent decisions concerning proper agricultural practices. Some farmers conduct their own on-farm research.

Electronics also help commercial farmers. Farmers use Global Positioning System (GPS) devices to determine the precise coordinates for spreading different types and amounts of fertilizers. On large ranches, they also use GPS devices to monitor the location

▼ **FIGURE 10-6 AGRICULTURAL WORKERS** The percentage of the workforce engaged in agriculture is higher in developing countries than in developed countries.

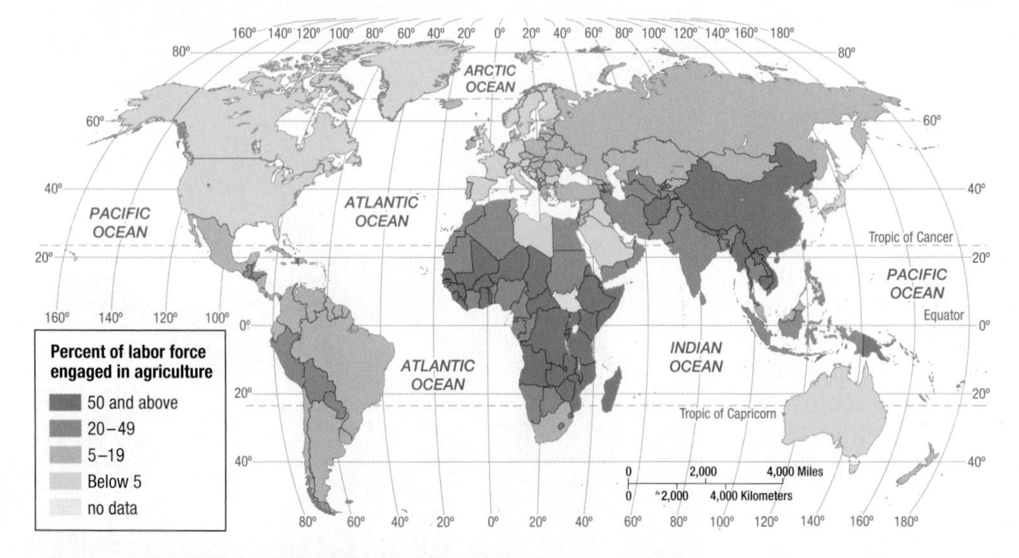

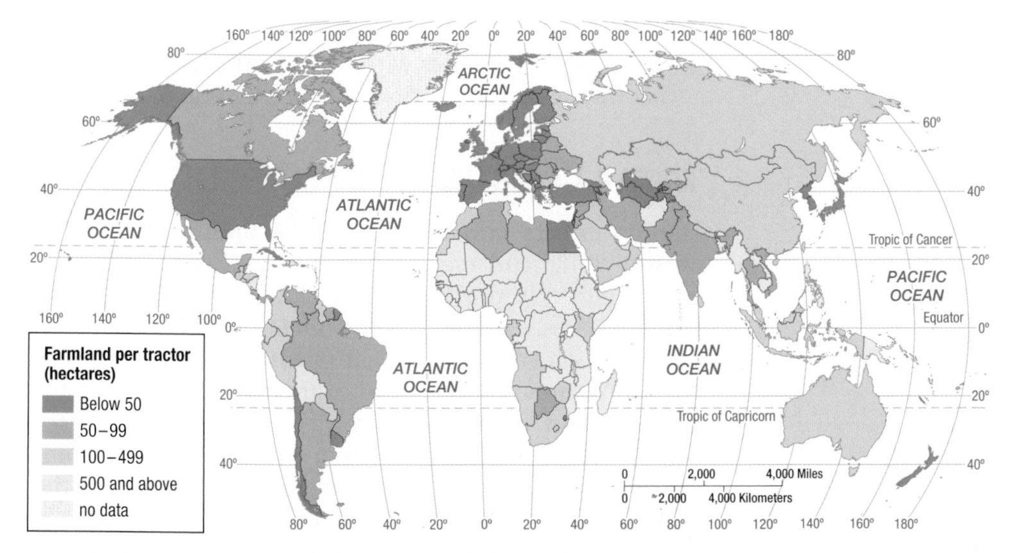

▲ **FIGURE 10-7 AREA OF FARMLAND PER TRACTOR** Farmers in developing countries have more hectares or acres of land per tractor than do farmers in developed countries. The machinery makes it possible for commercial farmers to farm extensive areas, a practice necessary to pay for the expensive machinery.

of cattle. They ue satellite imagery to measure crop progress and yield monitors attached to combines to determine the precise number of bushels being harvested.

Pause and Reflect 10.1.2
What other electronics, in addition to GPS devices, might help a farmer on a very large farm?

FARM SIZE

The average farm is relatively large in commercial agriculture. Farms average 161 hectares (418 acres) in the United States, compared to about 1 hectare (2.5 acres) in China (Figure 10-8). Large size partly depends on mechanization. Combines, pickers, and other machinery perform most efficiently at very large scales, and their considerable expense cannot be justified on a small farm. As a result of the large size and the high level of mechanization, commercial agriculture is an expensive business. Farmers spend hundreds of thousands of dollars to buy or rent land and machinery before beginning operations. This money is frequently borrowed from a bank and repaid after output is sold.

Commercial agriculture is increasingly dominated by a handful of large farms. In the United States, the largest 5 percent of farms produce 75 percent of the country's total agriculture. Despite their size, most commercial farms in developed countries—90 percent in the United States—are family owned and operated. Commercial farmers frequently expand their holdings by renting nearby fields.

Although the United States had fewer farms and farmers in 2000 than in 1900, the amount of land devoted to agriculture increased by 13 percent, primarily due to irrigation and reclamation. However, in the twenty-first century, the United States has been losing 1.2 million hectares (3 million acres) per year of its 400 million hectares (1 billion acres) of farmland, primarily because of the expansion of urban areas.

CHECK-IN: KEY ISSUE 1

Where Did Agriculture Originate?

- ✓ **Before the invention of agriculture, most humans were hunters and gatherers.**
- ✓ **Agriculture was invented in multiple hearths beginning approximately 10,000 years ago.**
- ✓ **Modern agriculture is divided between subsistence agriculture in developing countries and commercial agriculture in developed countries. They differ according to the percentage of farmers, use of machinery, and farm size.**

▼ **FIGURE 10-8 FARM SIZE** The average size of a family farm in China is much smaller than in the United States. (left) Family farm in Anhui Province, China. (right) Family farm in West Brooklyn, Illinois.

(a)

(b)

KEY ISSUE 2

Why Do People Consume Different Foods?

- Diet
- Nutrition and Hunger

Learning Outcome 10.2.1
Explain differences between developed and developing countries in food consumption.

When you buy food in a supermarket, are you reminded of a farm? Not likely. The meat is carved into pieces that no longer resemble an animal and is wrapped in paper or plastic film. Often the vegetables are canned or frozen. The milk and eggs are in cartons.

The food industry in the United States and Canada is vast, but only a few people are full-time farmers, and they may be more familiar with the operation of computers and advanced machinery than the typical factory or office worker. The mechanized, highly productive American or Canadian farm contrasts with the subsistence farm found in much of the world. The most "typical" human—if there is such a person—is an Asian farmer who grows enough food to survive, with little surplus. This sharp contrast in agricultural practices constitutes one of the most fundamental differences between the more developed and less developed countries of the world.

Diet

Everyone needs food to survive. Consumption of food varies around the world, both in total amount and source of nutrients. The variation results from a combination of:

- **Level of development.** People in developed countries tend to consume more food and from different sources than do people in developing countries.
- **Physical conditions.** Climate is important in influencing what can be most easily grown and therefore consumed in developing countries. In developed countries, though, food is shipped long distances to locations with different climates.
- **Cultural preferences.** Some food preferences and avoidances are expressed without regard for physical and economic factors, as discussed in Chapter 4.

TOTAL CONSUMPTION OF FOOD

Dietary energy consumption is the amount of food that an individual consumes. The unit of measurement of dietary energy is the kilocalorie (kcal), or Calorie in the United States. One gram (or ounce) of each food source delivers a kilocalorie level that nutritionists can measure.

Most humans derive most of their kilocalories through consumption of **cereal grain**, or simply **cereal**, which is a grass that yields grain for food. **Grain** is the seed from a cereal grass. The three leading cereal grains—wheat, rice, and maize (corn in North America)—together account for nearly 90 percent of all grain production and more than 40 percent of all dietary energy consumed worldwide:

- **Wheat.** The principal cereal grain consumed in the developed regions of Europe and North America is wheat, which is consumed in bread, pasta, cake, and many other forms. It is also the most consumed grain in the developing regions of Central and Southwest Asia, where relatively dry conditions are more suitable for growing wheat than other grains (Figure 10-9).
- **Rice.** The principal cereal grain consumed in the developing regions of East, South, and Southeast Asia is rice. It is the most suitable crop for production in tropical climates.
- **Maize.** The leading crop in the world is maize (called corn in North America), though much of it is grown for purposes other than direct human consumption, especially as animal feed. It is the leading crop in some countries of sub-Saharan Africa.
- **Other crops.** A handful of countries obtain the largest share of dietary energy from other crops, especially in sub-Saharan Africa (Figure 10-10). These include cassava, sorghum, millet, plantains, sweet potatoes, and

▼ **FIGURE 10-9 DIETARY ENERGY BY SOURCE** Wheat, rice, and maize are the three main sources of kilocalories.

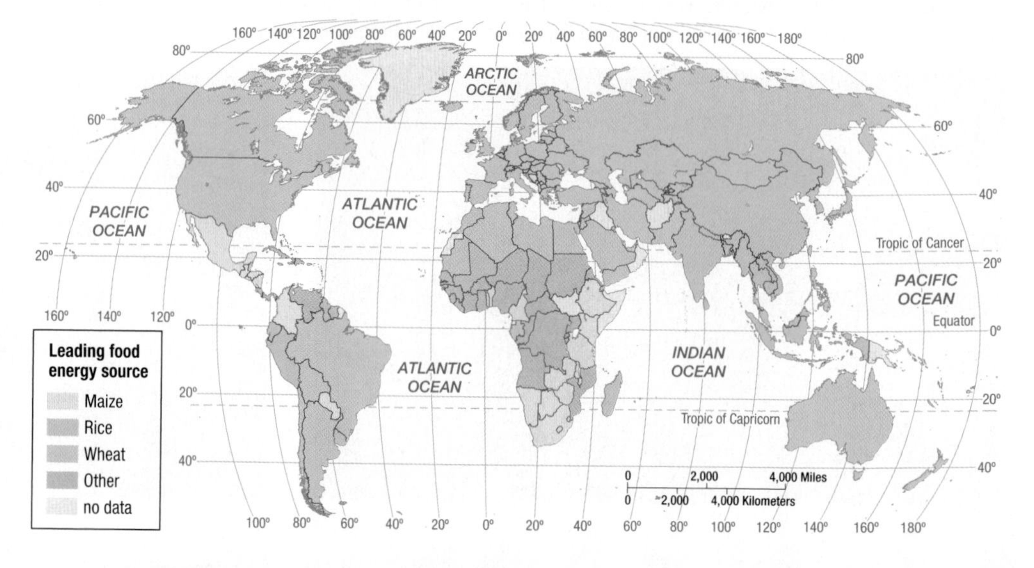

▲ **FIGURE 10-10 AFRICAN FOOD** The girl is carrying cassavas in Tanzania. At home, these roots will be pounded to break up the fibrous texture and cooked into a porridge.

yams. Sugar is the leading source of dietary energy in Venezuela.

Pause and Reflect 10.2.1
Which of the three main cereal grains is most prevalent in your diet?

SOURCE OF NUTRIENTS

Protein is a nutrient needed for growth and maintenance of the human body. Many food sources provide protein of varying quantity and quality. One of the most fundamental differences between developed and developing regions is the primary source of protein (Figure 10-11).

In developed countries, the leading source of protein is meat products, including beef, pork, and poultry (Figure 10-12). Meat accounts for approximately one-third of all protein intake in developed countries, compared to approximately one-tenth in developing ones. In most developing countries, cereal grains provide the largest share of protein.

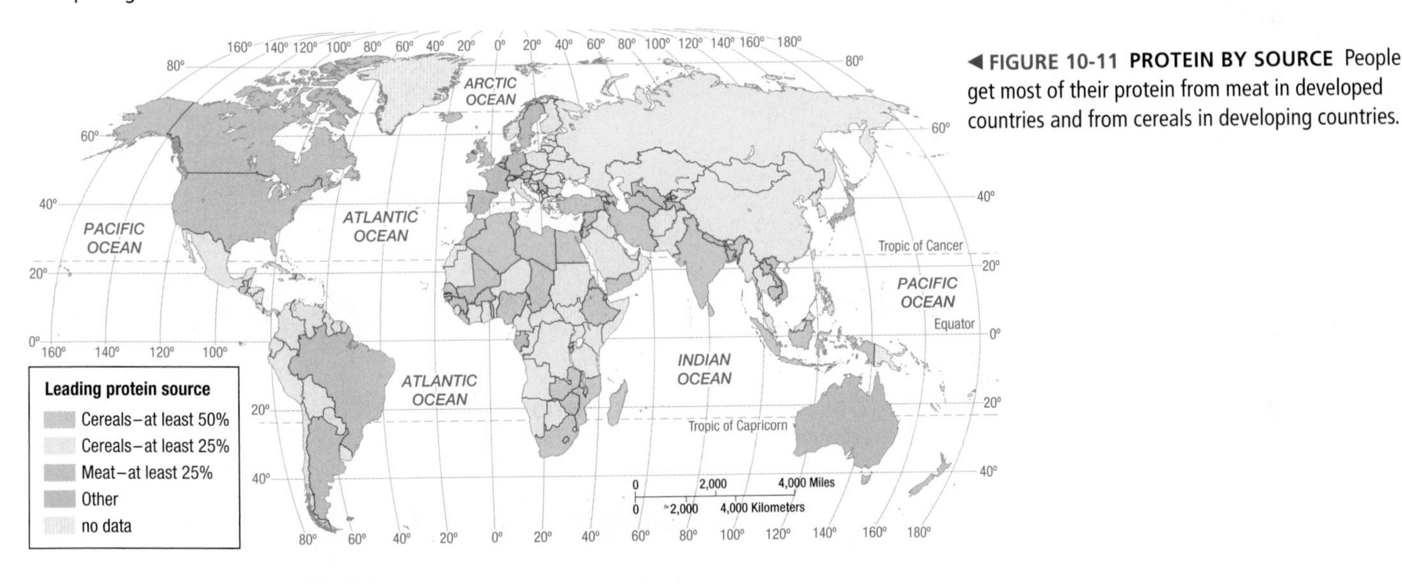

◀ **FIGURE 10-11 PROTEIN BY SOURCE** People get most of their protein from meat in developed countries and from cereals in developing countries.

◀ **FIGURE 10-12 PROTEIN FROM MEAT** The percentage of protein from meat is much higher for people in developed countries than for those in developing countries.

Nutrition and Hunger

Learning Outcome 10.2.2
Explain the global distribution of undernourishment.

The United Nations defines **food security** as physical, social, and economic access at all times to safe and nutritious food sufficient to meet dietary needs and food preferences for an active and healthy life. By this definition, roughly one-eighth of the world's inhabitants do not have food security.

DIETARY ENERGY NEEDS

To maintain a moderate level of physical activity, an average individual needs to consume at least 1,800 kcal per day, according to the UN Food and Agricultural Organization. The figure must be adjusted for age, sex, and region of the world.

Average consumption worldwide is approximately 2,800 kcal per day, or roughly 50 percent more than the recommended minimum. Thus, most people are getting enough food to eat. People in developed countries are consuming on average twice the recommended minimum, approximately 3,600 kcal per day (Figure 10-13). Austria and the United States have the world's highest consumption, approximately 3,800 kcal per day per person. The consumption of so much food is one reason that obesity is more prevalent than hunger in developed countries.

In developing regions, average daily consumption is approximately 2,600 kcal, still above the recommended minimum. However, the average in sub-Saharan Africa is only 2,400 kcal, an indication that a large percentage of Africans are not getting enough to eat. Diets are more likely to be deficient in countries where people have to spend a high percentage of their income to obtain food (Figure 10-14).

Pause and Reflect 10.2.2
How many kilocalories are in a Big Mac? You can use Google to find the answer. How does one Big Mac compare to the daily caloric intake of the average African?

UNDERNOURISHMENT

Undernourishment is dietary energy consumption that is continuously below the minimum requirement for maintaining a healthy life and carrying out light physical activity. The UN estimates that 870 million people in the world are undernourished; 99 percent of the world's undernourished people are in developing countries. India has by far the largest the number of undernourished people, 225 million, followed by China, with 130 million (Figure 10-15). One-fourth of the population in sub-Saharan Africa, one-fifth in South Asia, and one-sixth in all developing countries are classified as undernourished (Figure 10-16).

Worldwide, the total number of undernourished people has not changed much in several decades (Figure 10-17). With population growth, though, the percentage of undernourished people has decreased. Among developing regions, East Asia, led by China, has had by far the largest decrease in number undernourished, and South Asia and

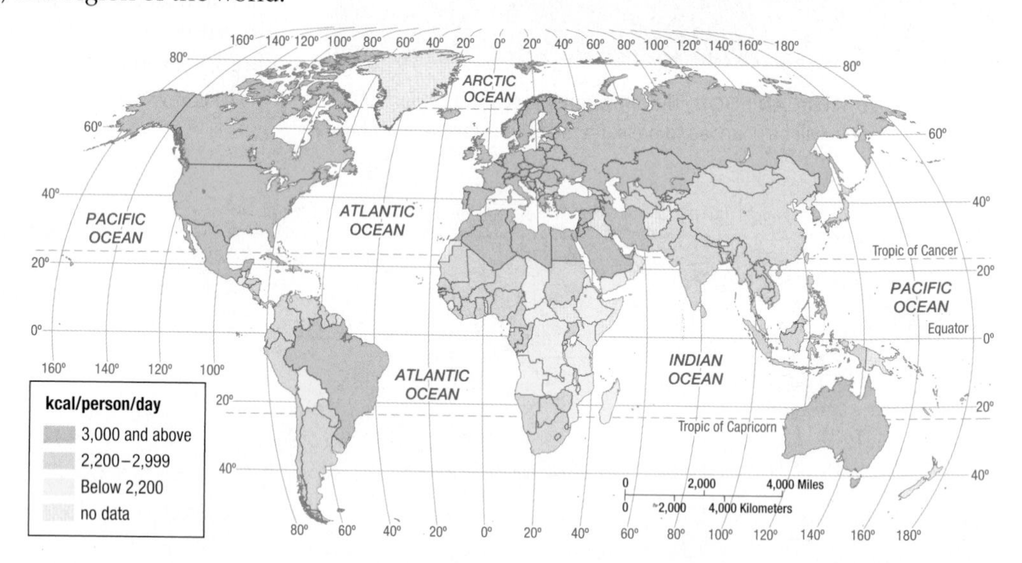

▲ **FIGURE 10-13 DIETARY ENERGY CONSUMPTION** Per capita caloric intake is approximately 3,600 kcal per day in developed countries and 2,600 in developing countries.

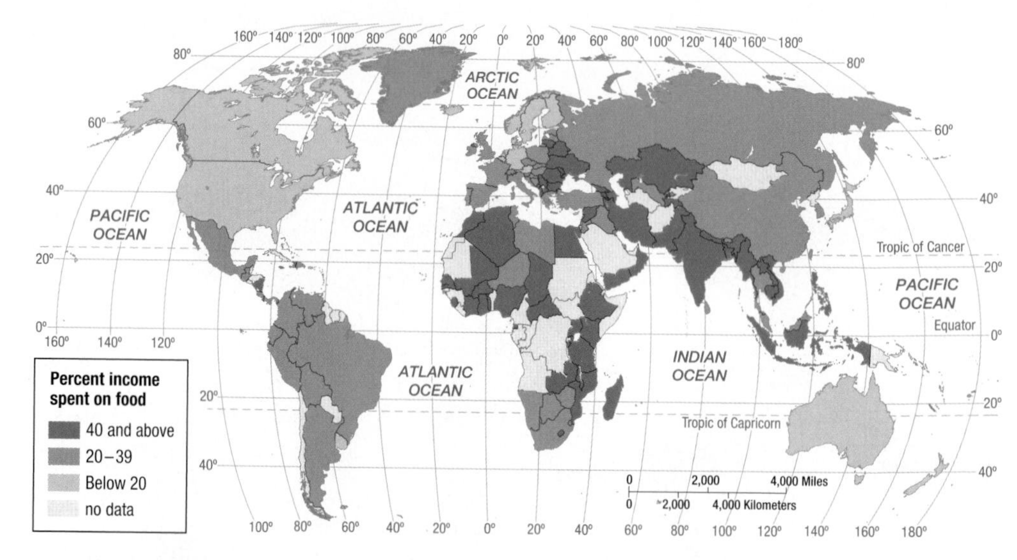

▲ **FIGURE 10-14 INCOME SPENT ON FOOD**
People spend on average less than 20 percent of income for food in developed countries compared to more than 40 percent in developing countries.

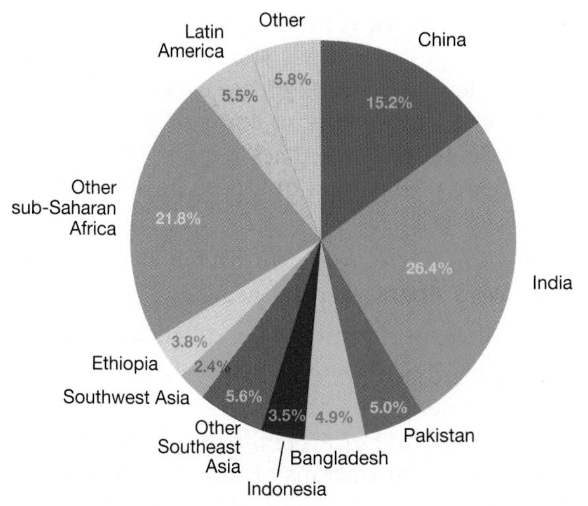

▲ **FIGURE 10-15 DISTRIBUTION OF UNDERNOURISHMENT** More than half of the world's undernourished people are in South Asia and East Asia.

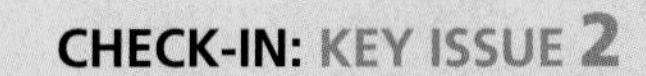

CHECK-IN: KEY ISSUE 2

Why Do People Consume Different Foods?

- ✓ **Most food is consumed in the form of cereal grains, especially wheat, rice, and maize.**
- ✓ **People in developed countries consume more total calories and a higher percentage through animal products.**
- ✓ **Most humans consume more than the recommended minimum calories, but undernourishment is widespread in Asia and sub-Saharan Africa.**

sub-Saharan Africa have had the largest increases. Southeast Asia, led by Myanmar and Vietnam, has also had a large decrease.

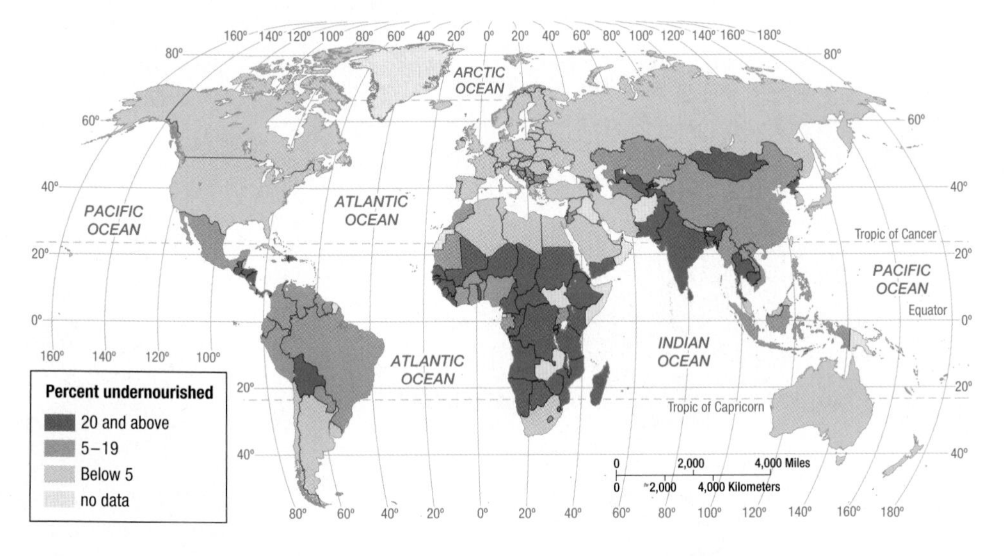

◀ **FIGURE 10-16 EXTENT OF UNDERNOURISHMENT** Less than 5 percent of the population is undernourished in developed countries compared to 15 percent in developing countries.

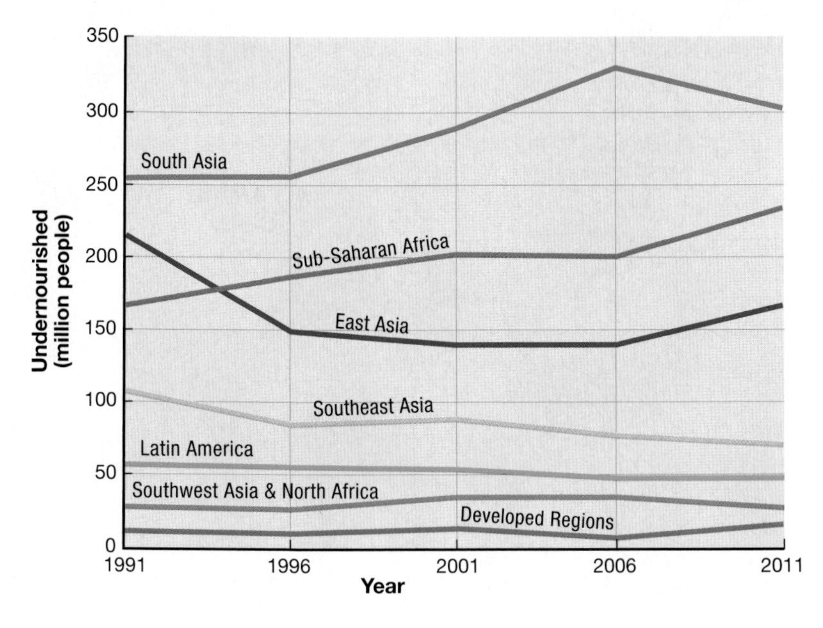

◀ **FIGURE 10-17 CHANGE IN UNDERNOURISHMENT** South Asia has seen the largest increase in number of undernourished people.

KEY ISSUE 3

Where Is Agriculture Distributed?

- **Agriculture in Developing Regions**
- **Agriculture in Developed Regions**

Learning Outcome 10.3.1
Identify the 11 major agricultural regions.

People have been able to practice agriculture in a wide variety of places. The most widely used map of world agricultural regions is based on work done by geographer Derwent Whittlesey in 1936. Whittlesey identified 11 main agricultural regions, plus an area where agriculture was nonexistent. Whittlesey's 11 regions are divided between 5 that are important in developing countries and 6 that are important in developed countries (Figure 10-18). The 5 agricultural regions that predominate in developing countries are:

- *Pastoral nomadism*—primarily the drylands of Southwest Asia & North Africa, Central Asia, and East Asia
- *Shifting cultivation*—primarily the tropical regions of Latin America, sub-Saharan Africa, and Southeast Asia
- *Intensive subsistence, wet rice dominant*—primarily the large population concentrations of East Asia and South Asia
- *Intensive subsistence, crops other than rice dominant*—primarily the large population concentrations of East Asia and South Asia, where growing rice is difficult

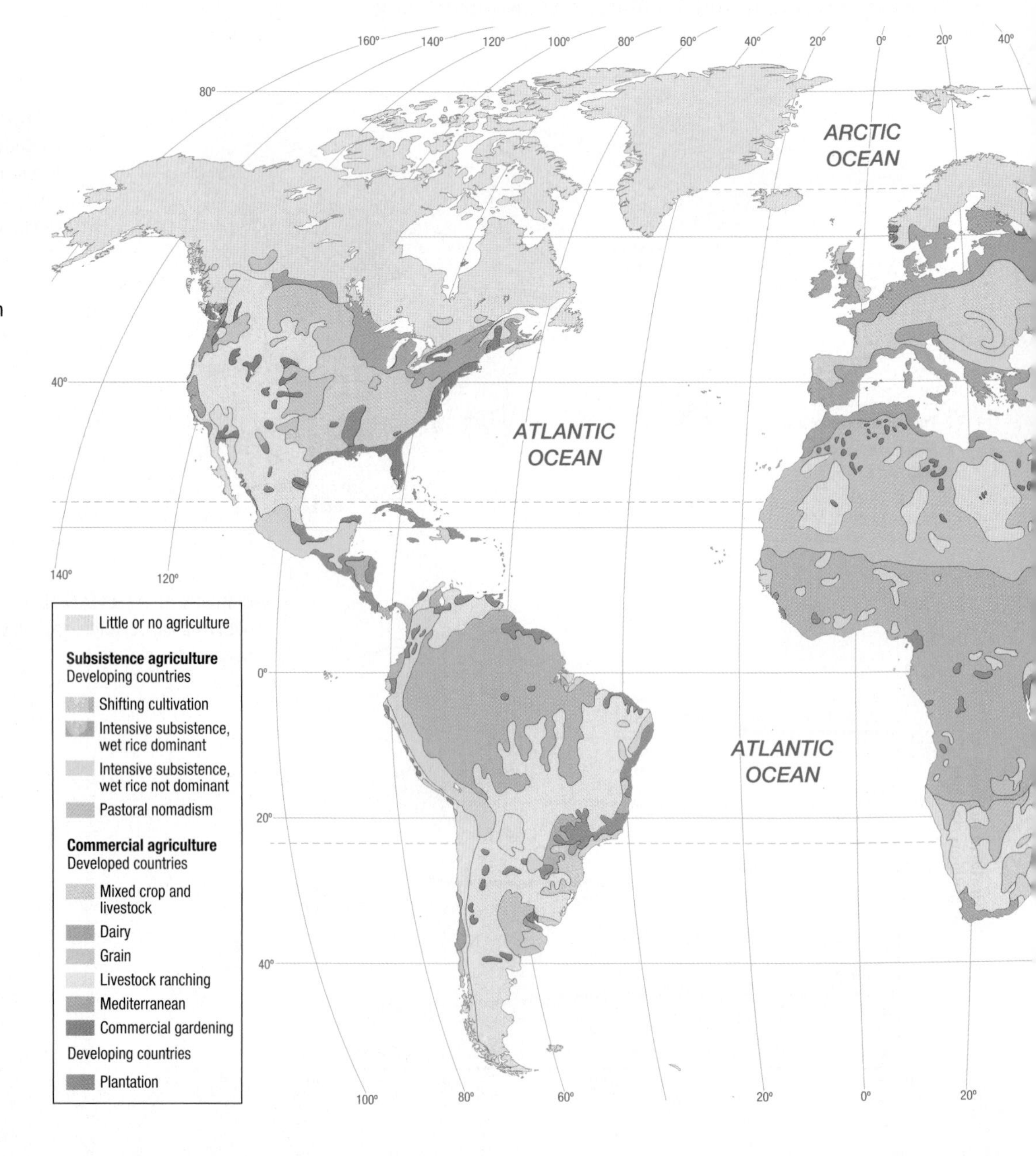

▶ **FIGURE 10-18 AGRICULTURE AND CLIMATE REGIONS**
(right) The major agricultural practices of the world can be divided into those that are prevalent in developing countries and those that are prevalent in developed countries (upper right). Climate plays a large role in the practice of agriculture. Figure 1-40 is a more detailed version of the climate map shown here.

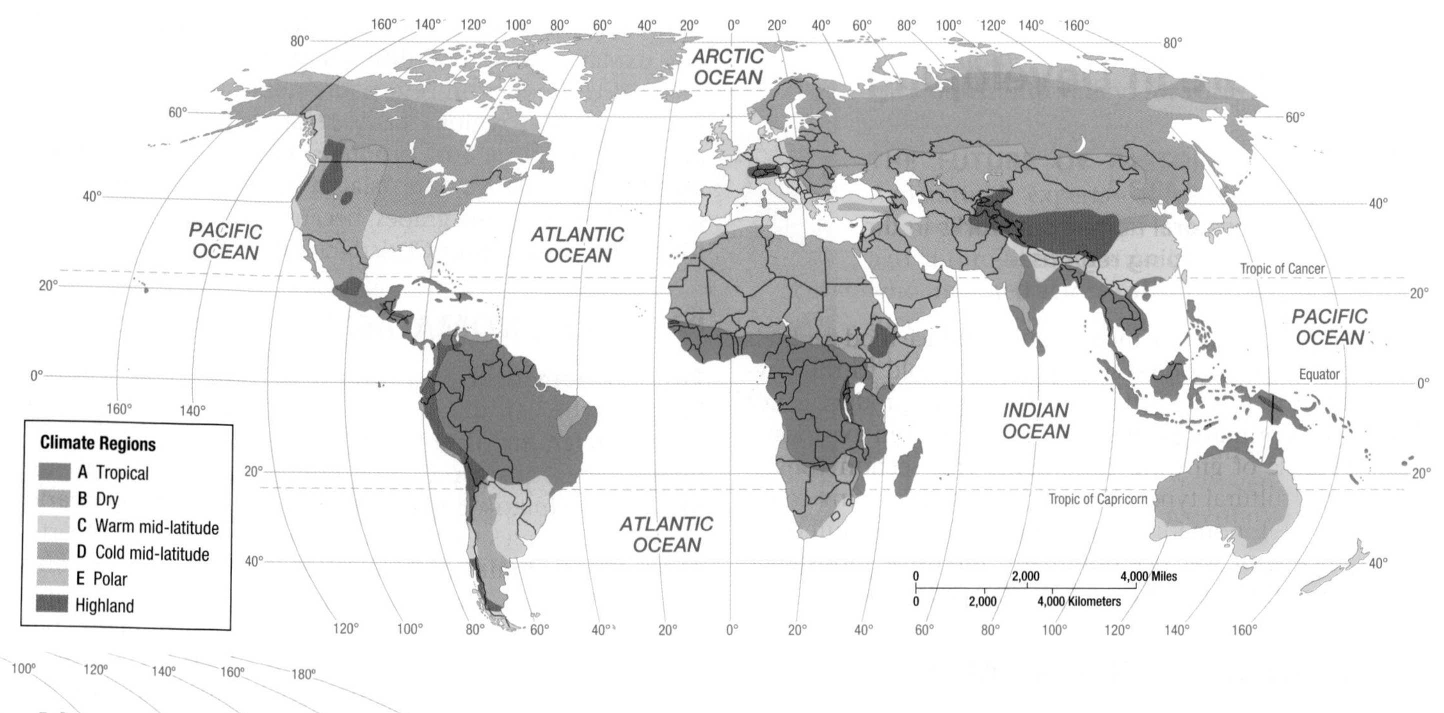

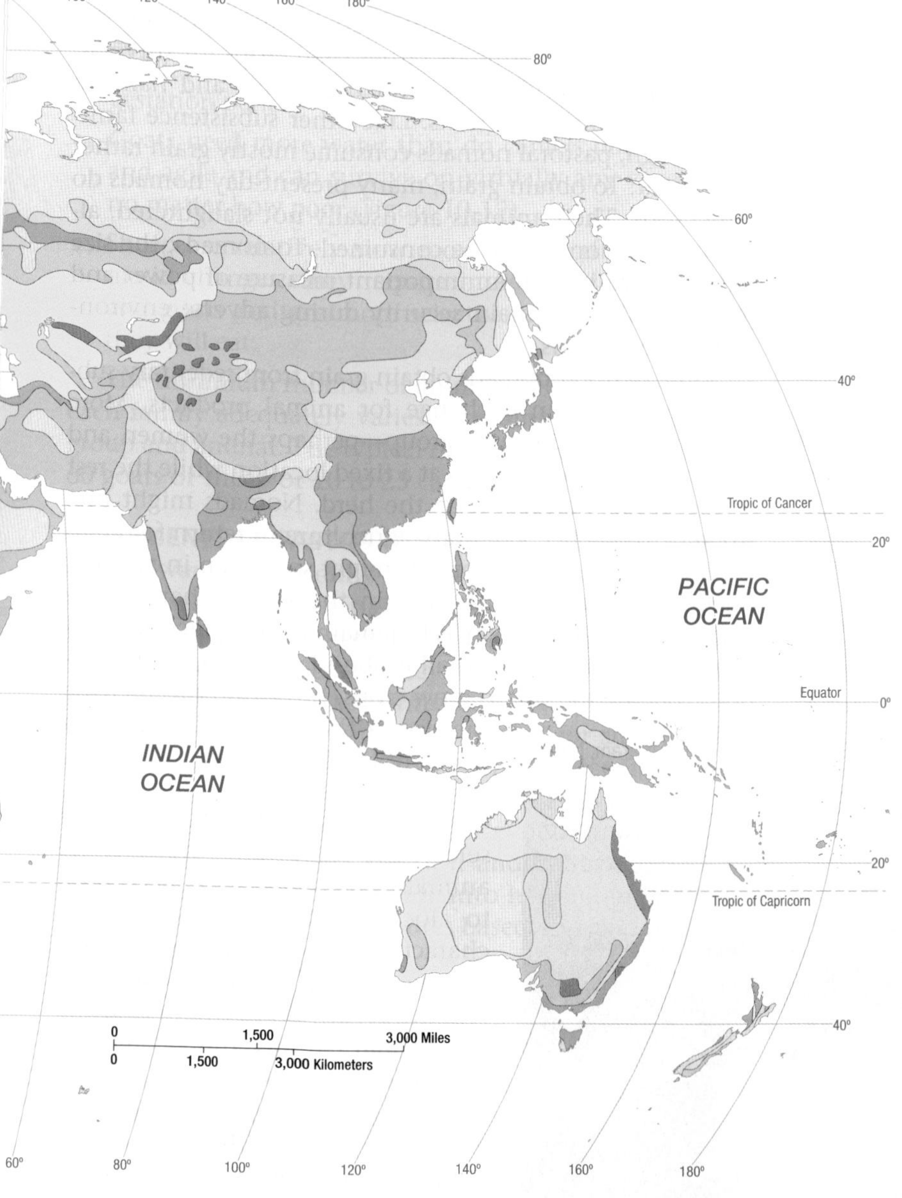

- *Plantation*—primarily the tropical and subtropical regions of Latin America, sub-Saharan Africa, South Asia, and Southeast Asia

The six agricultural regions that predominate in developed countries are:

- *Mixed crop and livestock*—primarily the U.S. Midwest and central Europe
- *Dairying*—primarily near population clusters in the northeastern United States, southeastern Canada, and northwestern Europe
- *Grain*—primarily the north-central United States, south-central Canada, and Eastern Europe
- *Ranching*—primarily the drylands of western North America, southeastern Latin America, Central Asia, sub-Saharan Africa, and the South Pacific
- *Mediterranean*—primarily lands surrounding the Mediterranean Sea, the western United States, the southern tip of Africa, and Chile
- *Commercial gardening*—primarily the southeastern United States and southeastern Australia

Pause and Reflect 10.3.1
In which agricultural region do you live?

INTENSIVE SUBSISTENCE WITH WET RICE NOT DOMINANT

Learning Outcome 10.3.5
Describe reasons for growing crops other than wet rice in intensive subsistence regions.

Climate prevents farmers from growing wet rice in portions of Asia, especially where summer precipitation levels are too low and winters are too harsh (refer to Figure 10-18). Agriculture in much of the interior of India and northeastern China is devoted to crops other than wet rice. Wheat is the most important crop, followed by barley (Figure 10-29). Various other grains and legumes are grown for household consumption, including millet, oats, corn, sorghum, and soybeans. In addition, some crops are grown in order to be sold for cash, such as cotton, flax, hemp, and tobacco.

Aside from what is grown, this region shares most of the characteristics of intensive subsistence agriculture with the wet-rice region. Land is used intensively and worked primarily by human power, with the assistance of some hand implements and animals. In milder parts of the region where wet rice does not dominate, more than one harvest can be obtained some years through skilled use of **crop rotation**, which is the practice of rotating use of different fields from crop to crop each year to avoid exhausting the soil. In colder climates, wheat or another crop is planted in the spring and harvested in the fall, but no crops can be sown through the winter.

Since the Communist Revolution in 1949, private individuals have owned little agricultural land in China. Instead, the Communist government organized agricultural producer communes, which typically consisted of several villages of several hundred people each. By combining several small fields into a single large unit, China's government hoped to promote agricultural efficiency; scarce equipment and animals and larger improvement projects, such as flood control, water storage, and terracing, could be shared (Figure 10-30). In reality, productivity did not increase as much as the government had expected because people worked less efficiently for the commune than when working for themselves.

▼ FIGURE 10-29 **CHINA BARLEY FIELDS** Hillsides are terraced to create flat fields.

▲ FIGURE 10-30 **CHINA COMMUNE** A commune in Changsha, China.

China has therefore dismantled the agricultural communes. The communes still hold legal title to agricultural land, but villagers sign contracts that entitle them to farm portions of the land as private individuals. Chinese farmers may sell to others the right to use the land and to pass on the right to their children. Reorganization has been difficult because irrigation systems, equipment, and other infrastructure were developed to serve large communal farms rather than small individually managed ones, which cannot afford to operate and maintain the machinery. But production has increased greatly.

PLANTATION FARMING

The types of agriculture in developing countries discussed so far are considered subsistence agriculture because the principal purpose is production of food for consumption by the farmer's family. Plantation farming is a form of commercial agriculture found in developing countries. A **plantation** is a large commercial farm in a developing country that specializes in one or two crops.

Most plantations are located in the tropics and subtropics, especially in Latin America, Africa, and Asia (Figure 10-31). Although generally situated in developing countries, plantations are often owned or operated by Europeans or North Americans, and they grow crops for sale primarily to developed countries. Crops are normally processed at the plantation before being shipped because processed goods are less bulky and are therefore cheaper to ship the long distances to the North American and European markets.

Among the most important crops grown on plantations are cotton, sugarcane, coffee, rubber, and tobacco (Figure 10-32). Also produced in large quantities are cocoa, jute, bananas, tea, coconuts, and palm oil. Latin American

▲ **FIGURE 10-31 COFFEE PLANTATION** This plantation is in Minas Gerais, Brazil.

plantations are most likely to grow coffee, sugarcane, and bananas, whereas Asian plantations may provide rubber and palm oil. Crops such as tobacco, cotton, and sugarcane, which can be planted only once a year, are less likely to be grown on large plantations today than in the past.

Because plantations are usually situated in sparsely settled locations, they must import workers and provide them with food, housing, and social services (Figure 10-33).

▲ **FIGURE 10-33 PLANTATION WORKERS** Temporary laborers are transported to a coffee plantation in Minas Gerais, Brazil.

Plantation managers try to spread the work as evenly as possible throughout the year to make full use of the large labor force. Where the climate permits, more than one crop is planted and harvested annually. Rubber tree plantations try to spread the task of tapping the trees throughout the year.

Until the Civil War, plantations were important in the U.S. South, where the principal crop was cotton, followed by tobacco and sugarcane. Demand for cotton increased dramatically after the establishment of textile factories in England at the start of the Industrial Revolution in the late eighteenth century. Cotton production was stimulated by the improvement of the cotton gin by Eli Whitney in 1793 and the development of new varieties of cotton that were hardier and easier to pick. Slaves brought from Africa performed most of the labor until the abolition of slavery and the defeat of the South in the Civil War. Thereafter, plantations declined in the United States; they were subdivided and either sold to individual farmers or worked by tenant farmers.

Pause and Reflect 10.3.5
What foods do you consume that are grown on plantations?

▼ **FIGURE 10-32 COFFEE BEAN PRODUCTION** One-third of the world's coffee beans are grown in Brazil.

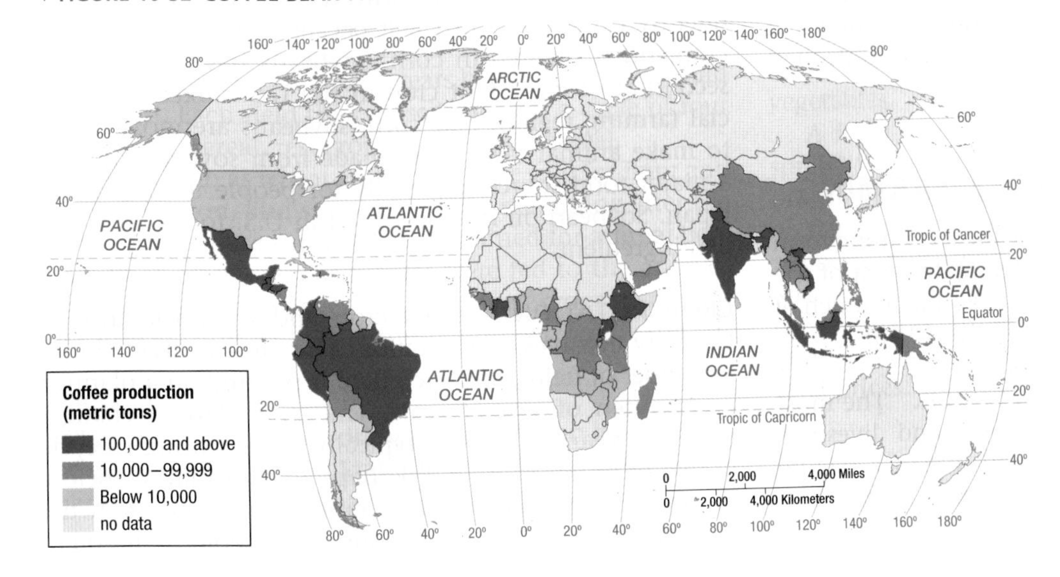

DAIRY FARMING

Learning Outcome 10.3.7
Describe how dairy farming and commercial gardening work.

Dairy farming is the most important commercial agriculture practiced on farms near the large urban areas of the northeastern United States, southeastern Canada, and northwestern Europe (Figure 10-36). Dairying has also become an important type of farming in South and East Asia. Traditionally, fresh milk was rarely consumed except directly on the farm or in nearby villages. With the rapid growth of cities in developed countries during the nineteenth century, demand for the sale of milk to urban residents increased. Rising incomes permitted urban residents to buy milk products, which were once considered luxuries.

REGIONAL DISTRIBUTION OF DAIRYING. For most of the twentieth century, the world's milk production was clustered in a handful of developed countries. However, the share of the world's dairy farming conducted in developing countries has risen dramatically, from 26 percent in 1980 to 53 percent in 2010 (Figure 10-37). In the twenty-first century, India has become the world's largest milk producer, ahead of the United States, the traditional leader, and China and Pakistan are now third and fourth largest (Figure 10-38).

In developed countries, dairying is the most important type of commercial agriculture in the first ring outside large cities because of transportation factors. Dairy farms must be closer to their market than other types of farms because their products are highly perishable. The ring surrounding a city from which milk can be supplied without spoiling is known as the **milkshed**. Improvements in transportation have permitted dairying to be undertaken farther from the market. Until the 1840s, when railroads were first used for transporting dairy products, milksheds rarely had a radius beyond 50 kilometers (30 miles). Today, refrigerated railcars and trucks enable farmers to ship milk more than 500 kilometers (300 miles). As a result, nearly every farm in the northeastern United States and northwestern Europe is within the milkshed of at least one urban area.

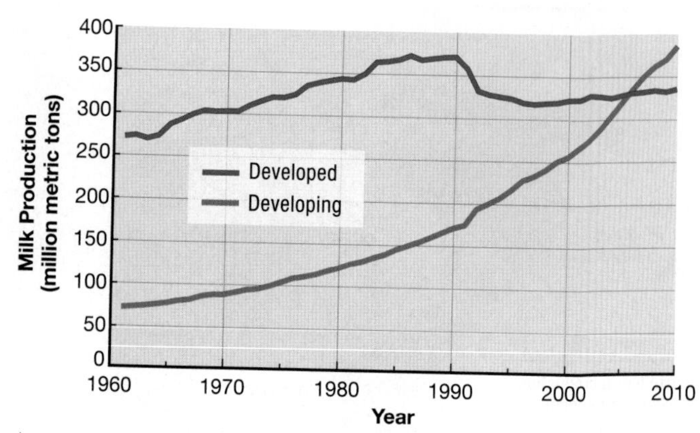

▲ **FIGURE 10-37 CHANGING MILK PRODUCTION**
Developing countries now produce more milk than developed countries.

Dairy farmers, like other commercial farmers, usually do not sell their products directly to consumers. Instead, they generally sell milk to wholesalers, who distribute it in turn to retailers. Retailers then sell milk to consumers in shops or at home. Farmers also sell milk to butter and cheese manufacturers.

In general, the farther the farm is from large urban concentrations, the smaller is the percentage of output devoted to fresh milk. Farms located farther from consumers are more likely to sell their output to processors that make butter, cheese, or dried, evaporated, and condensed milk. The reason is that these products keep fresh longer than milk does and therefore can be safely shipped from remote farms.

Countries likewise tend to specialize in certain products. New Zealand, the world's largest per capita producer of dairy products, devotes about 5 percent to liquid milk, compared to more than 50 percent in the United Kingdom. New Zealand farmers do not sell much liquid milk because the country is too far from North America and northwestern Europe, the two largest relatively wealthy population concentrations.

▼ **FIGURE 10-36 DAIRY FARM** Many cows are milked simultaneously at this dairy farm in Wiltshire, England.

CHALLENGES FOR DAIRY FARMERS. Like other commercial farmers, dairy farmers face economic difficulties because of declining revenues and rising costs. Dairy farmers who have quit farming most often cite lack of profitability and excessive workload as reasons for getting out of the business. Distinctive features of dairy farming have exacerbated the economic difficulties:

- **Labor intensive.** Cows must be milked twice a day, every day; although the actual milking can be done by machines, dairy farming nonetheless requires constant attention throughout the year.

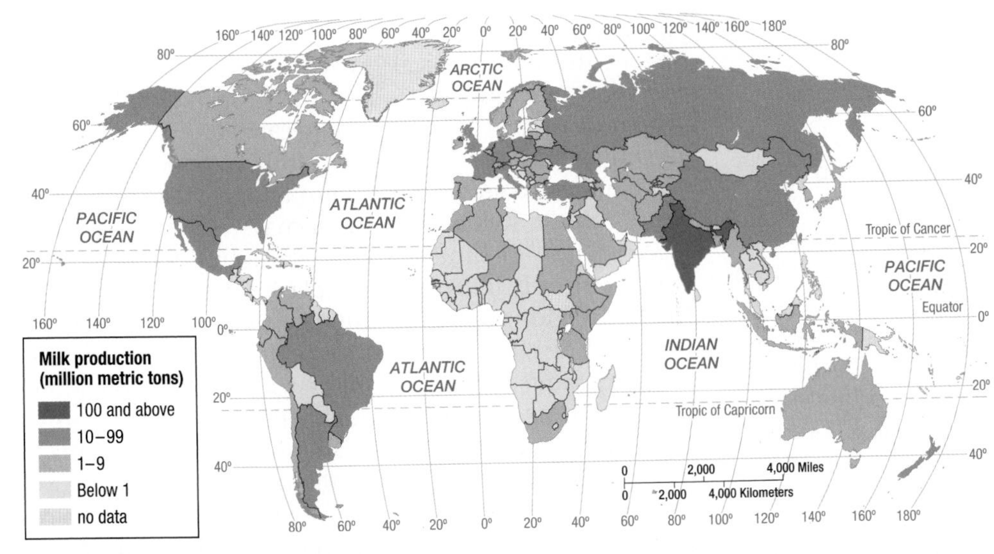

◀ **FIGURE 10-38 MILK PRODUCTION**
India has replaced the United States as the world's leading milk producer.

- **Winter feed.** Dairy farmers face the expense of feeding the cows in the winter, when they may be unable to graze on grass. In northwestern Europe and in the northeastern United States, farmers generally purchase hay or grain for winter feed. In the western part of the U.S. dairy region, crops are more likely to be grown in the summer and stored for winter feed on the same farm.

Pause and Reflect 10.3.7
Look on the label of your milk carton. How far away from you is the dairy?

CONTEMPORARY GEOGRAPHIC TOOLS
Protecting Farmland

Loss of farmland to urban growth is especially severe at the edge of the string of large metropolitan areas along the East Coast of the United States. Some of the most threatened agricultural land lies in Maryland, a small state where two major cities—Washington and Baltimore—have coalesced into a continuous built-up area (see Chapter 13). In Maryland, a geographic information system (GIS) was used to identify which farms should be preserved.

Maps generated through GIS were essential in identifying agricultural land to protect because the most appropriate farms to preserve were not necessarily those with the highest-quality soil. Why should the state and nonprofit organizations spend scarce funds to preserve "prime" farmland that is nowhere near the path of urban sprawl? Conversely, why purchase an expensive, isolated farm already totally surrounded by residential developments, when the same amount of money could buy several large contiguous farms that effectively blocked urban sprawl elsewhere?

To identify the "best" lands to protect, GIS consultants produced a series of soil quality, environmental, and economic maps that were combined into a single composite map (Figure 10-39). The map shows that 4 percent of the state's farmland had prime soils, significant environmental features, and high projected population growth, and 25 percent had two of the three factors.

Maryland officials are making use of the results of the GIS as part of an overall strategy to minimize sprawl. For example, state highway money is allocated to improving roads in existing built-up areas rather than extending new roads through important conservation areas.

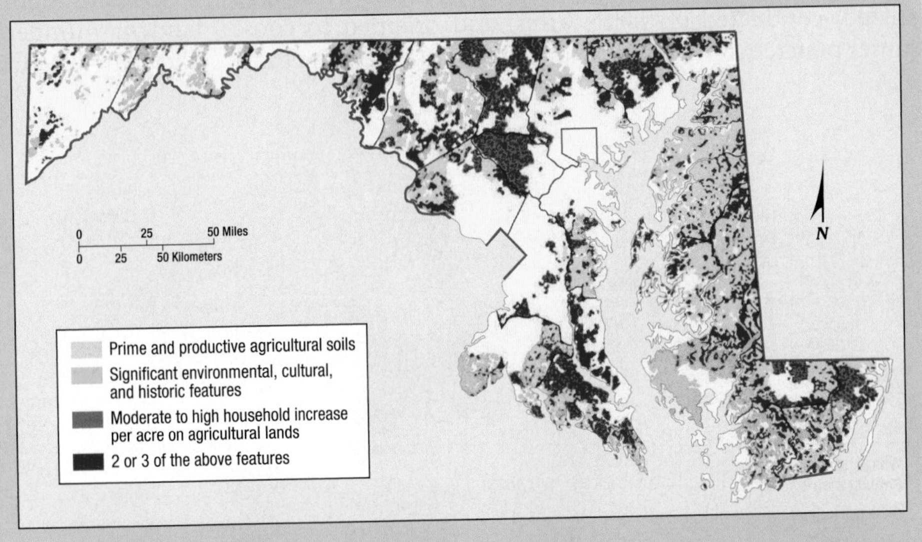

▲ **FIGURE 10-39 PROTECTING FARMLAND IN MARYLAND** Prime farmland is typically flat and well drained. Significant environmental features included water quality, flood control, species habitats, historic sites, and especially attractive scenery.

GRAIN FARMING

Learning Outcome 10.3.8
Describe how grain and Mediterranean farming work.

Some form of grain is the major crop on most farms. Grain is the seed from various grasses, such as wheat, corn, oats, barley, rice, millet, and others. Commercial grain agriculture is distinguished from mixed crop and livestock farming because crops on a grain farm are grown primarily for consumption by humans rather than by livestock. Farms in developing countries also grow crops for human consumption, but the output is directly consumed by the farmers. Commercial grain farms sell their output to manufacturers of food products, such as breakfast cereals and breads.

The most important crop grown is wheat, used to make bread flour. Wheat generally can be sold for a higher price than other grains, such as rye, oats, and barley, and it has more uses as human food. It can be stored relatively easily without spoiling and can be transported a long distance. Because wheat has a relatively high value per unit weight, it can be shipped profitably from remote farms to markets.

As was the case with milk production, the share of world production of wheat in developing countries has increased rapidly. Much of this increased production results from growth in large-scale commercial agriculture. Developing countries accounted for more than one-half of world wheat production in 2010, compared to only one-fourth in 1960. The United States is by far the largest producer of wheat among developed countries, but it now ranks third among all countries, behind China and India (Figure 10-40). China has been the world leader since 1983, and India has been second since 1999.

Large-scale grain production, like other commercial farming ventures in developed countries, is heavily mechanized, conducted on large farms, and oriented to consumer preferences. The McCormick **reaper** (a machine that cuts grain standing in the field), invented in the 1830s, first permitted large-scale wheat production. Today the **combine** machine performs in one operation the three tasks of reaping, threshing, and cleaning.

Unlike work on a mixed crop and livestock farm, the effort required to grow wheat is not uniform throughout the year. Some individuals or firms may therefore have two sets of fields—one in the spring wheat belt and one in the winter wheat belt. Because the planting and harvesting in the two regions occur at different times of the year, the workload can be distributed throughout the year. In addition, the same machinery can be used in the two regions, thus spreading the cost of the expensive equipment. Combine harvesting contractors start working in Oklahoma in early summer and work their way northward.

Commercial grain farms are generally located in regions that are too dry for mixed crop and livestock agriculture. Within North America, large-scale grain production is concentrated in three areas:

- **The winter wheat belt through Kansas, Colorado, and Oklahoma.** The **winter wheat** crop is planted in the autumn and develops a strong root system before growth stops for the winter. The wheat survives the winter, especially if it is insulated beneath a snow blanket, and is ripe by the beginning of summer.
- **The spring wheat belt through the Dakotas, Montana, and southern Saskatchewan in Canada.** Winters are usually too severe for winter wheat in this region, so **spring wheat** is planted in the spring and harvested in the late summer.
- **The Palouse region of Washington State.** Wheat comprises a smaller percentage of agricultural output than in the other two wheat-growing regions. The Palouse is also an important source of legumes; for example, 80 percent of U.S. lentils are grown in the region.

Wheat's significance extends beyond the amount of land or number of people involved in growing it. Unlike other agricultural products, wheat is grown to a

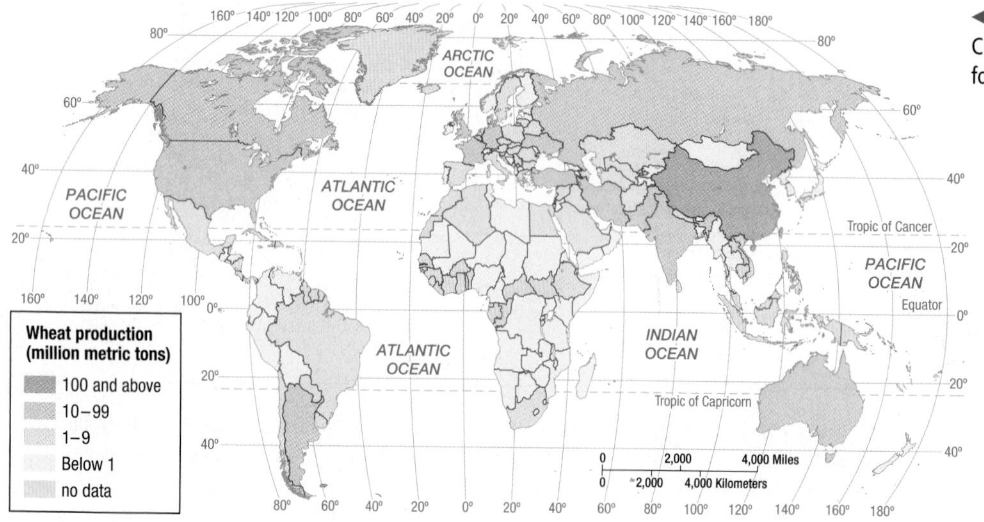

◄ **FIGURE 10-40 WHEAT PRODUCTION** China and India are the leading wheat producers, followed by the United States.

▲ **FIGURE 10-41 MEDITERRANEAN AGRICULTURE** Nearly all olives are produced in countries that border the Mediterranean Sea or have similar climates, including Sifnos, Greece.

considerable extent for international trade, and it is the world's leading export crop. The United States and Canada account for about half of the world's wheat exports; consequently, the North American prairies are accurately called the world's "breadbasket." The ability to provide food for many people elsewhere in the world is a major source of economic and political strength for these two countries.

MEDITERRANEAN AGRICULTURE

Mediterranean agriculture exists primarily on the lands that border the Mediterranean Sea in Southern Europe, North Africa, and western Asia (Figure 10-41). Farmers in California, central Chile, the southwestern part of South Africa, and southwestern Australia practice Mediterranean agriculture as well.

These Mediterranean areas share a similar physical environment (refer to Figures 10-18 and 10-40). Every Mediterranean area borders a sea, and most are on west coasts of continents (except for some lands surrounding the Mediterranean Sea). Prevailing sea winds provide moisture and moderate the winter temperatures. Summers are hot and dry, but sea breezes provide some relief. The land is very hilly, and mountains frequently plunge directly to the sea, leaving very narrow strips of flat land along the coast.

Farmers derive a smaller percentage of income from animal products in the Mediterranean region than in the mixed crop and livestock region. Livestock production is hindered during the summer by the lack of water and good grazing land. Some farmers living along the Mediterranean Sea traditionally used transhumance to raise animals, although the practice is now less common. Under transhumance, animals—primarily sheep and goats—are kept on the coastal plains in the winter and transferred to the hills in the summer.

Most crops in Mediterranean lands are grown for human consumption rather than for animal feed. **Horticulture**—which is the growing of fruits, vegetables, and flowers—and tree crops form the commercial base of Mediterranean farming. A combination of local physical and cultural characteristics determines which crops are grown in each area. The hilly landscape encourages farmers to plant a variety of crops within one farming area.

In the lands bordering the Mediterranean Sea, the two most important cash crops are olives and grapes. Two-thirds of the world's wine is produced in countries that border the Mediterranean, especially Italy, France, and Spain. Mediterranean agricultural regions elsewhere in the world produce most of the remaining one-third (refer to Figure 4-22). The lands near the Mediterranean Sea are also responsible for a large percentage of the world's supply of olives, an important source of cooking oil. Despite the importance of olives and grapes to commercial farms bordering the Mediterranean Sea, approximately half of the land is devoted to growing cereals, especially wheat for pasta and bread. As in the U.S. winter wheat belt, the seeds are sown in the fall and harvested in early summer. After cultivation, cash crops are planted on some of the land, and the remainder of the land is left fallow for a year or two to conserve moisture in the soil.

Cereals occupy a much lower percentage of the cultivated land in California than in other Mediterranean climates. Instead, a large portion of California farmland is devoted to fruit and vegetable horticulture, which supplies a large portion of the citrus fruits, tree nuts, and deciduous fruits consumed in the United States. Horticulture is practiced in other Mediterranean climates, but not to the extent found in California. The rapid growth of urban areas in California, especially Los Angeles, has converted high-quality agricultural land into housing developments. Thus far, the loss of farmland has been offset by the expansion of agriculture into arid lands. However, farming in drylands requires massive irrigation to provide water. In the future, California agriculture may face stiffer competition for the Southwest's increasingly scarce water supply.

Pause and Reflect 10.3.8

At least 1 million metric tons of wine are produced in eight countries (Argentina, Australia, China, France, Italy, South Africa, Spain, and the United States). Referring to Figures 4-22 and 10-18, which one of the eight countries does not appear to have Mediterranean agriculture?

LIVESTOCK RANCHING

Learning Outcome 10.3.9
Describe how livestock ranching works.

Ranching is the commercial grazing of livestock over an extensive area (Figure 10-42). This form of agriculture is adapted to semiarid or arid land and is practiced in developed countries where the vegetation is too sparse and the soil too poor to support crops.

CATTLE RANCHING IN THE UNITED STATES. The importance of ranching in the United States extends beyond the people who choose this form of commercial farming. Its prominence in popular culture, especially in Hollywood films and television, has not only helped to draw attention to this form of commercial farming but has also served to illustrate, albeit in sometimes romanticized ways, the crucial role that ranching played in the history and settlement of areas of the United States. Cattle ranching in Texas, as glamorized in popular culture, did actually dominate commercial agriculture, but only for a short period—from 1867 to 1885.

Cattle ranching expanded in the United States during the 1860s because of the demand for beef in East Coast cities. If they could get their cattle to Chicago, ranchers were paid $30 to $40 per head, compared to only $3 or $4 per head in Texas. Once in Chicago, the cattle could be slaughtered and processed by meat-packing companies and shipped in packages to consumers in the East. To reach Chicago, cattle were driven on hoof by cowboys over trails from Texas to the nearest railhead. There the cattle were driven into cattle cars for the rest of their journey. The western terminus of the rail line reached Abilene, Kansas, in 1867. Wichita, Caldwell, Dodge City, and other towns in Kansas took their turns as the main destination for cattle driven north on trails from Texas. The most famous route from Texas northward to the rail line was the Chisholm Trail, which began near Brownsville at the Mexican border and extended northward through Texas (Figure 10-43).

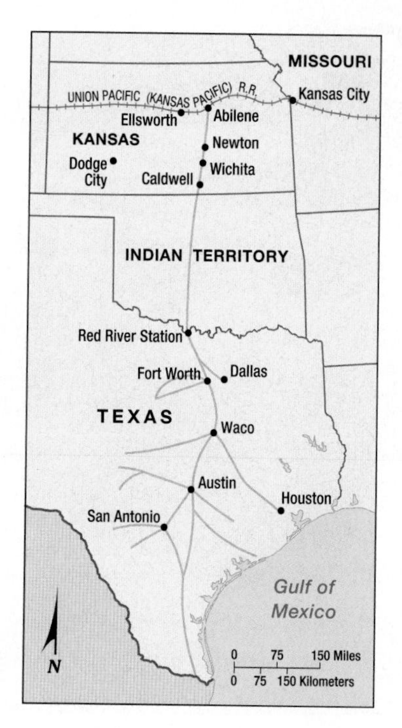

▲ **FIGURE 10-43 CHISHOLM TRAIL** The Chisholm Trail was used to move cattle from Texas to railroad stations in Kansas during the 1860s and 1870s.

Cattle ranching declined in importance during the 1880s, after it came into conflict with sedentary agriculture. Most early U.S. ranchers adhered to "the Code of the West," although the system had no official legal status. Under the code, ranchers had range rights—that is, their cattle could graze on any open land and had access to scarce water sources and grasslands. The early cattle ranchers in the West owned little land, only cattle. The U.S. government, which owned most of the land used for open grazing, began to sell it to farmers to grow crops, leaving cattle ranchers with no legal claim to it. For a few years the ranchers tried to drive out the farmers by cutting fences and then illegally erecting their own fences on public land, and "range wars" flared. The farmers' most potent weapon proved to be barbed wire, first commercially produced in 1873. The farmers eventually won the battle, and ranchers were compelled to buy or lease land to accommodate their cattle. Large cattle ranches were established, primarily on land that was too dry to support crops. Ironically, 60 percent of cattle grazing today takes place on land leased from the U.S. government.

▼ **FIGURE 10-42 RANCHING** Cattle on a west Texas ranch are rounded up for shipping.

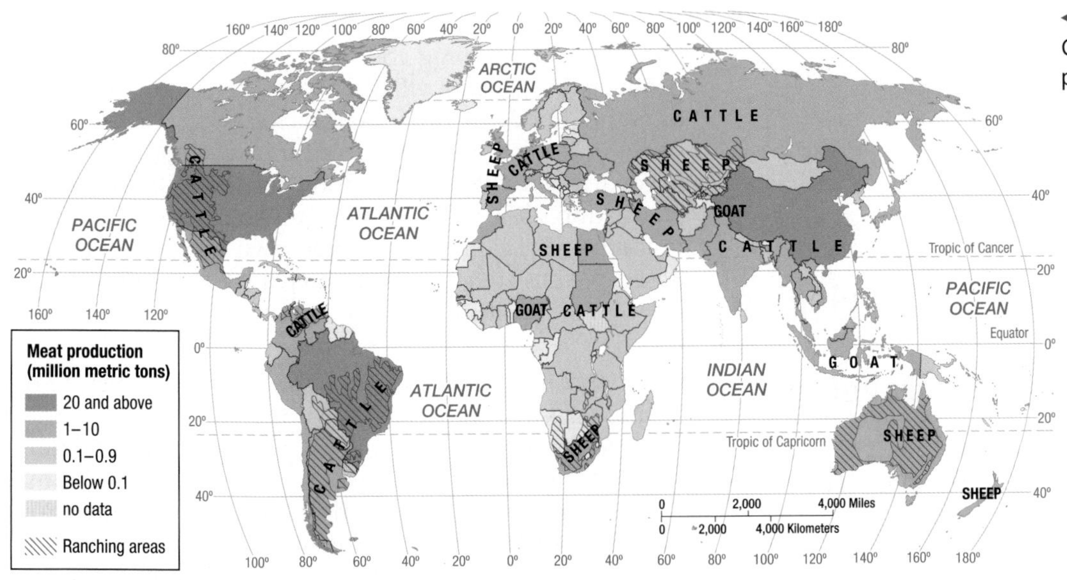

◀ **FIGURE 10-44 MEAT PRODUCTION** China is now the world's largest meat producer.

With the spread of irrigation techniques and hardier crops, land in the United States has been converted from ranching to crop growing. Ranching generates lower income per area of land, although it has lower operating costs. Cattle are still raised on ranches but are frequently sent for fattening to farms or to local feed lots along major railroad and highway routes rather than directly to meat processors.

COMMERCIAL RANCHING IN OTHER REGIONS. Commercial ranching is conducted in several developed countries besides the United States and, increasingly, in developing countries. The interior of Australia was opened for grazing in the nineteenth century, although sheep are more common there than cattle. Ranching is rare in Europe, except in Spain and Portugal. In South America, a large portion of the pampas of Argentina, southern Brazil, and Uruguay is devoted to grazing cattle and sheep. The cattle industry grew rapidly in Argentina in part because the land devoted to ranching was relatively accessible to the ocean, making it possible for meat to be transported to overseas markets.

As with other forms of commercial agriculture, the growth in ranching has been in developing countries. China is the leading producer of meat, ahead of the United States, and Brazil is third (Figure 10-44). China passed the United States as the world's leading meat producer in 1990 and now produces twice as much. Developed countries were responsible for only one-third of world meat production in 2010, compared to two-thirds in 1980.

Ranching has followed similar stages around the world. First was the herding of animals over open ranges, in a seminomadic style. Then ranching was transformed into fixed farming by dividing the open land into ranches. When many of the farms converted to growing crops, ranching was confined to the drier lands. To survive, the remaining ranches experimented with new methods of breeding and sources of water and feed. Ranching has become part of the meat-processing industry rather than an economic activity carried out on isolated farms. In this way, commercial ranching differs from pastoral nomadism, the form of animal herding practiced in less developed regions.

Pause and Reflect 10.3.9
What are the two most important ranched animals, according to Figure 10-45?

CHECK-IN: KEY ISSUE 3

Where Is Agriculture Distributed?

- ✓ **Agriculture can be divided into 11 major regions, including 5 in developing regions and 6 in developed regions.**
- ✓ **In developing regions, pastoral nomadism is prevalent in drylands, shifting cultivation in tropical forests, and intensive subsistence in regions with high population concentrations.**
- ✓ **In developed regions, mixed crop and livestock is the most common form of agriculture. Dairy, commercial gardening, grain, Mediterranean, and livestock ranching are also important.**

KEY ISSUE 4

Why Do Farmers Face Economic Difficulties?

- Challenges for Farmers in Developing Countries
- Challenges for Farmers in Developed Countries
- Strategies to Increase the World's Food Supply
- Sustainable Agriculture

Learning Outcome 10.4.1
Describe the impact of population growth and trade on farming in developing countries.

Commercial farmers in developed countries and subsistence farmers in developing countries face comparable challenges. Farmers in both developing and developed countries have difficulty generating enough income to continue farming. The underlying reasons, though, are different. Commercial farmers can produce a surplus of food, whereas many subsistence farmers are barely able to produce enough food to survive.

Challenges for Farmers in Developing Countries

Two issues discussed in earlier chapters influence the choice of crops planted by subsistence farmers in developing countries:

- Subsistence farmers must feed an increasing number of people because of rapid population growth in developing countries (discussed in Chapter 2).
- Farmers who have traditionally practiced subsistence farming are pressured to grow food for export instead of for direct consumption due to the adoption of the international trade approach to development (discussed in Chapter 9).

SUBSISTENCE FARMING AND POPULATION GROWTH

Population growth influences the distribution of types of subsistence farming, according to economist Ester Boserup. It compels subsistence farmers to consider new farming approaches that produce enough food to take care of the additional people.

For hundreds if not thousands of years, subsistence farming in developing countries yielded enough food for people living in rural villages to survive, assuming that no drought, flood, or other natural disaster occurred. Suddenly in the late twentieth century, developing countries needed to provide enough food for a rapidly increasing population as well as for the growing number of urban residents who cannot grow their own food. According to Boserup, subsistence farmers increase the supply of food through intensification of production, achieved in two ways:

- **New farming methods are adopted.** Plows replace axes and sticks. More weeding is done, more manure is applied, more terraces are carved out of hillsides, and more irrigation ditches are dug (Figure 10-45). The additional labor needed to perform these operations comes from the population growth. The farmland yields more food per area of land, but with the growing population, output per person remains about the same.
- **Land is left fallow for shorter periods.** This expands the amount of land area devoted to growing crops at any given time. Boserup identified five basic stages in the intensification of farmland:
    - **Forest fallow.** Fields are cleared and utilized for up to 2 years and left fallow for more than 20 years, long enough for the forest to grow back.
    - **Bush fallow.** Fields are cleared and utilized for up to 8 years and left fallow for up to 10 years, long enough for small trees and bushes to grow back.

▼ **FIGURE 10-45 INTENSIVE FARMING METHODS** Hillsides in Radi, Bhutan, are terraced into fields for intensive planting of rice.

- **Short fallow.** Fields are cleared and utilized for perhaps 2 years (Boserup was uncertain) and left fallow for up to 2 years, long enough for wild grasses to grow back.
- **Annual cropping.** Fields are used every year and rotated between legumes and roots.
- **Multi-cropping.** Fields are used several times a year and never left fallow.

Contrast shifting cultivation, practiced in regions of low population density, such as sub-Saharan Africa, with intensive subsistence agriculture, practiced in regions of high population density, such as East Asia. Under shifting cultivation, cleared fields are utilized for a couple years and then left fallow for 20 years or more. This type of agriculture supports a small population living at low density. As the number of people living in an area increases (that is, as the population density increases) and more food must be grown, fields will be left fallow for shorter periods of time. Eventually, farmers achieve the very intensive use of farmland characteristic of areas of high population density.

SUBSISTENCE FARMING AND INTERNATIONAL TRADE

To expand production, subsistence farmers need higher-yield seeds, fertilizer, pesticides, and machinery. Some needed supplies can be secured by trading food with urban dwellers. For many African and Asian countries, though, the main way to obtain agricultural supplies is to import them from other countries. However, subsistence farmers lack the money to buy agricultural equipment and materials from developed countries.

To generate the funds they need to buy agricultural supplies, developing countries must produce something they can sell in developed countries. The developing countries sell some manufactured goods (see Chapter 11), but most raise funds through the sale of crops in developed countries. Consumers in developed countries are willing to pay high prices for fruits and vegetables that would otherwise be out of season or for crops such as coffee and tea that cannot be grown in developed countries because of the climate.

In a developing country such as Kenya, families may divide by gender between traditional subsistence agriculture and contributing to international trade. Women practice most of the subsistence agriculture—that is, growing food for their families to consume—in addition to the tasks of cooking, cleaning, and carrying water from wells. Men may work for wages, either growing crops for export or at jobs in distant cities. Because men in Kenya frequently do not share the wages with their families, many women try to generate income for the household by making clothes, jewelry, baked goods, and other objects for sale in local markets.

The sale of export crops brings a developing country foreign currency, a portion of which can be used to buy agricultural supplies. But governments in developing countries face a dilemma: The more land that is devoted to growing export crops, the less that is available to grow crops for domestic consumption. Rather than help to increase productivity, the funds generated through the sale of export crops may be needed to feed the people who switched from subsistence farming to growing export crops.

Pause and Reflect 10.4.1
What is an example of a product available in supermarkets in the United States that was exported from a developing country?

AFRICA'S FOOD-SUPPLY STRUGGLE

Sub-Saharan Africa is struggling to keep food production ahead of population growth. Since 1961, food production has increased substantially in sub-Saharan Africa, but so has population (Figure 10-46). As a result, food production per capita has changed little in a half-century.

The threat of famine is particularly severe in the Horn of Africa and the Sahel. Traditionally, this region supported limited agriculture. With rapid population growth, farmers overplanted, and herd size increased beyond the capacity of the land to support the animals. Animals overgrazed the limited vegetation and clustered at scarce water sources.

Government policies have aggravated the food-shortage crisis. To make food affordable for urban residents, governments keep agricultural prices low. Constrained by price controls, farmers are unable to sell their commodities at a profit and therefore have little incentive to increase production.

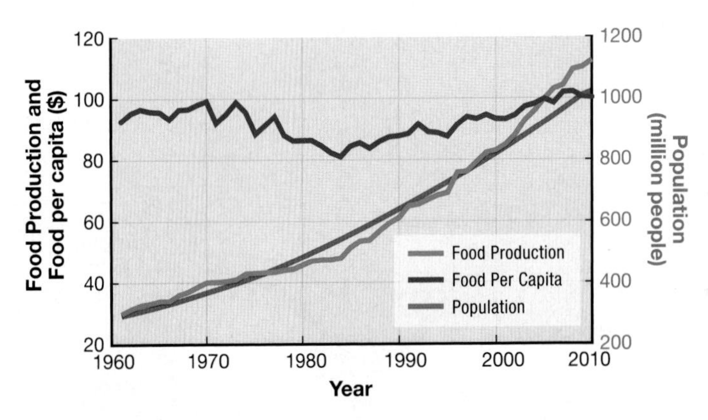

▲ **FIGURE 10-46 POPULATION AND FOOD IN AFRICA** Food production is increasing at about the same rate as population in Africa. As a result, food production per capita is staying about the same.

DRUG CROPS

Learning Outcome 10.4.2
Understand distinctive challenges for developing countries to increase food supply.

The export crops grown in some developing countries, especially in Latin America and Asia, are those that can be converted to drugs. Cocaine and heroin, the two leading, especially dangerous drugs, are abused by 16 to 17 million people each, and marijuana, the most popular drug, is estimated to be used by 140 million worldwide:

- Cocaine is derived from coca leaf, most of which is grown in Colombia or the neighboring countries Peru and Bolivia. Most consumers are located in developed countries, especially in North America. The principal shipping route is from Colombia by sea to Mexico or other Central American countries and then by land through Mexico to the United States (Figure 10-47).
- Heroin is derived from raw opium gum, which is produced by the opium poppy plant. Afghanistan is the source of nearly 90 percent of the world's opium; most of the remainder is grown in Myanmar (Burma) and Laos. Most traffic flows from Afghanistan through Iran, Turkey, and the Balkans to Western Europe, where the largest numbers of the world's users live. A second route goes through Central Asia to Russia (Figure 10-48).
- Marijuana, produced from the *Cannabis sativa* plant, is cultivated widely around the world. The overwhelming majority of the marijuana that reaches the United States is grown in Mexico. Cultivation of *C. sativa* is not thought to be expanding worldwide, whereas cultivation of opium poppies and coca leaf are.

Pause and Reflect 10.4.2
Why does most consumption of cocaine and heroin occur in developed countries?

▲ **FIGURE 10-47 POPPY FIELD** Afghanistan is the leading producer of poppies, which are cultivated for opium production.

FOOD PRICES

The greatest challenge to world food supply in the twenty-first century has been food prices rather than food supply. Food prices more than doubled between 2006 and 2008, and they have remained at record high levels since then (Figure 10-49). The UN attributes the record high food prices to four factors:

- Poor weather, especially in major crop-growing regions of the South Pacific and North America
- Higher demand, especially in China and India
- Smaller growth in productivity, especially without major new "miracle" breakthroughs
- Use of crops as biofuels instead of food, especially in Latin America

On the other side of the coin, record high food prices have stimulated record high prices for prime agricultural land. Adjusting for inflation, the price of farmland in Iowa doubled from around $2,500 per acre in 2000 to $5,000 in 2010.

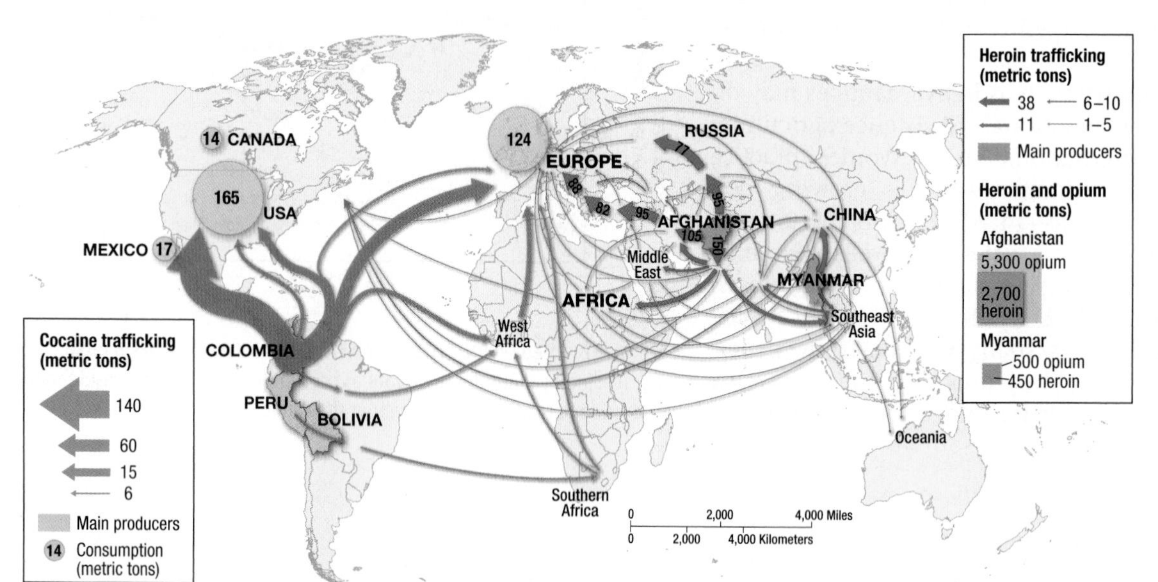

◀ **FIGURE 10-48 INTERNATIONAL DRUG TRAFFICKING** The main routes for heroin are from Afghanistan through Southwest Asia to Europe and through Central Asia to Russia. The main routes for cocaine are from Colombia to North America through Mexico and to Europe by sea.

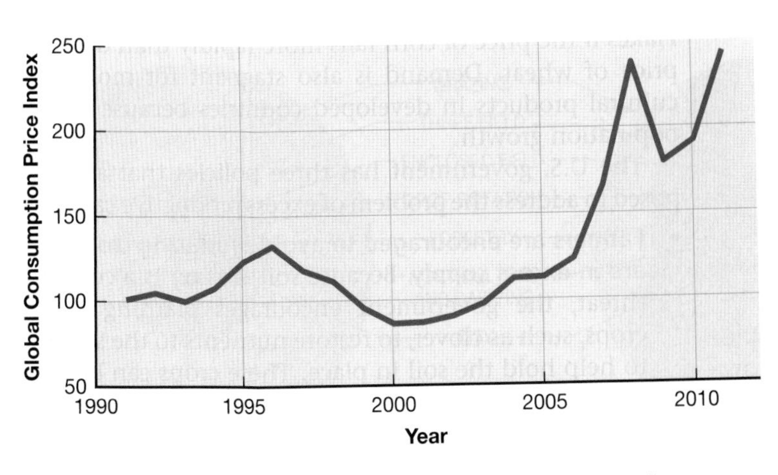

▲ **FIGURE 10-49 FOOD PRICE INDEX** Worldwide food prices rose rapidly between 2006 and 2008 and have remained high since then.

SUSTAINABILITY AND INEQUALITY IN OUR GLOBAL VILLAGE

Asian Carp and Chicago's Economy

The growth of aquaculture has led to the farming of nonnative species. One example is the Asian carp, which were imported to the United States in the 1970s to stock a fish farm in Arkansas. Flooding allowed the carp to escape the farm and enter U.S. waterways. Fast-growing and voracious eaters, Asian carp can grow to over 45 kilograms (100 pounds) (Figure 10-50). Once in the waterways, the extremely aggressive Asian carp have competed successfully with native fish for food and habitat, and they have even attacked people fishing in small boats. Asian carp have traveled up the Mississippi and Illinois rivers, and they now constitute 97 percent of the fish in these rivers. Now the Asian carp threaten to reach the Great Lakes.

The most likely point of entry into the Great Lakes for the Asian carp is through Chicago-area waterways. To connect Lake Michigan and the rest of the Great Lakes with the inland waterways of the United States, canals were constructed during the nineteenth century. The U.S. Army Corps of Engineers has installed electric barriers to try to keep the Asian carp from traveling through the canals to Lake Michigan. However, in the long run, the only effective way to keep the carp out of the Great Lakes is to shut the canals. However, the canals play a major role in the economy of the Chicago area and the United States as a whole. Barges carry petroleum, coal, and other important raw materials from domestic and international sources to factories. Shutting the canals could devastate the region's economy; estimates of the impact on Chicago's economy range from $70 million to $235 million per year.

▲ **FIGURE 10-50 ASIAN CARP** Asian carp are in the Illinois River and threaten to reach the Great Lakes through Chicago-area canals.

Strategies to Increase the World's Food Supply

Learning Outcome 10.4.4
Explain the contribution of expanding exports and farmland to world food supply.

Whereas developed countries often produce more food than they need, many developing countries struggle to produce enough to feed their rapidly growing populations. Four strategies are being employed to distribute food to everyone in the world:

- Increasing exports from countries with surpluses
- Expanding the land area used for agriculture
- Expanding fishing
- Increasing the productivity of land now used for agriculture

Challenges underlie each of these strategies.

INCREASING EXPORTS FROM COUNTRIES WITH SURPLUSES

Trade in food has increased rapidly, especially since 2000, exceeding $1 billion for the first time in 2008 (Figure 10-53). On a global scale, agricultural products are moving primarily from the Western Hemisphere to the Eastern Hemisphere. Latin America, led by Brazil and Argentina, is the by far the leading region for export of agricultural products; North America, Southeast Asia, and the South Pacific are the other major exporting regions (Figure 10-54).

Prior to the 1980s, the only major food importing regions were Europe, East Asia, and the former Soviet Union. Historically, European countries used their colonies as suppliers of food; after they became independent countries, the former colonies sold food to Europe. Joining East

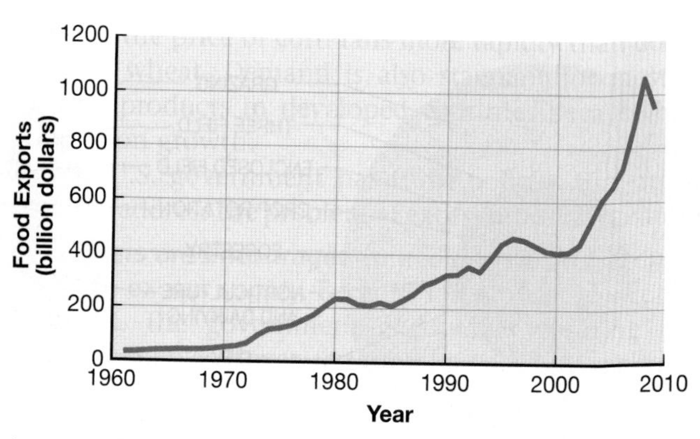

▲ **FIGURE 10-53 GROWTH IN AGRICULTURAL EXPORTS**
Agricultural trade increased from $400 billion in 2000 to $1 trillion in 2010.

Asia as net food importers were Southwest Asia and North Africa during the 1970s, South Asia and sub-Saharan Africa during the 1980s, and Central Asia in 2008. Food production was unable to keep up with rapid population growth in these regions, and as they embraced the international trade path of development, agriculture was increasingly devoted to growing export crops for sale in developed countries. Japan is by far the leading importer of food, followed by the United Kingdom, China, and Russia.

In response to the increasing global demand for food imports, the United States passed Public Law 480, the Agricultural, Trade, and Assistance Act of 1954 (referred to as P.L.-480). Title I of the act provided for the sale of grain at low interest rates, and Title II gave grants to needy groups of people. The United States remains the world's leading exporter of grain, including nearly one-half of the world's maize exports. But the overall share of exports accounted for by the United States has declined rapidly, from 18 to 19 percent of the world total in the 1970s to 10 to 11 percent in the twenty-first century. Agricultural exports from the United States have continued to increase rapidly, but developing regions—especially Latin America and Southeast Asia—have had more rapid increases.

▼ **FIGURE 10-54 TRADE IN AGRICULTURAL PRODUCTS** The principal flow of agriculture in the world is from the Western Hemisphere to Europe and Asia.

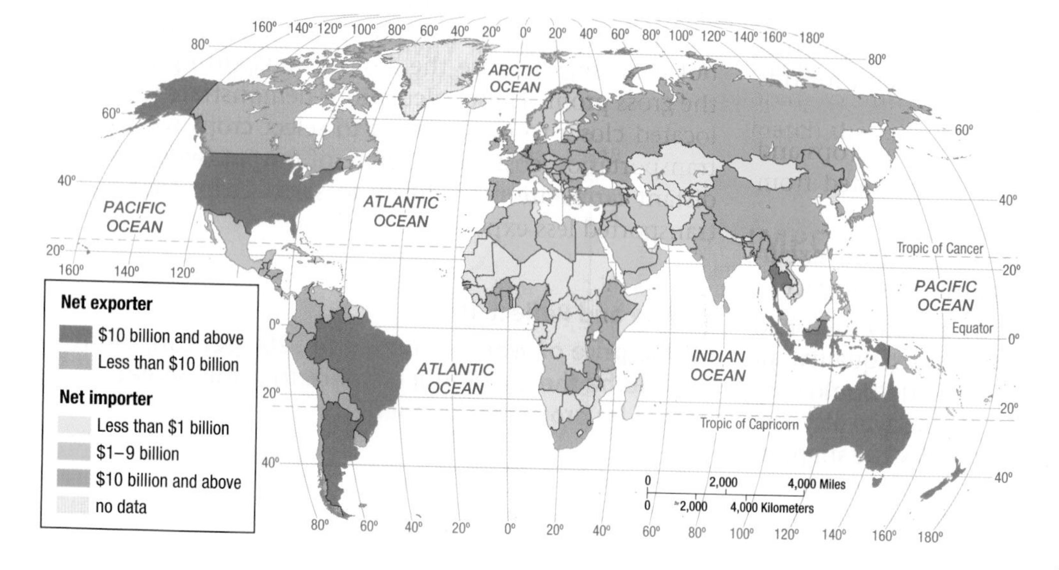

EXPANDING AGRICULTURAL LAND

Historically, world food production has increased primarily by expanding the amount of land devoted to agriculture. When the world's population began to increase more rapidly in the late eighteenth and early nineteenth centuries, during the Industrial Revolution, pioneers could migrate to uninhabited territory and cultivate the land. Sparsely inhabited land suitable for agriculture was available in western North America, central Russia, and Argentina's pampas.

Two centuries ago, people believed that good agricultural land would always be available for willing pioneers. Today few scientists believe that further expansion of agricultural land can feed the growing world population. At first glance, new agricultural land appears to be available because only 11 percent of the world's land area is currently cultivated. However, in recent decades, population has increased much more rapidly than agricultural land (Figure 10-55).

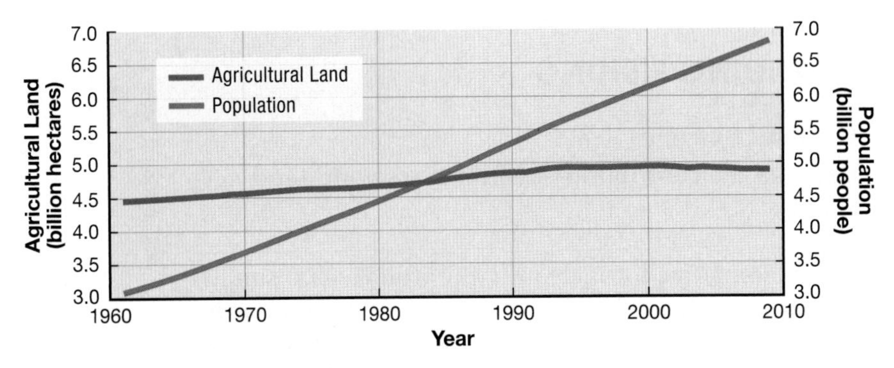

▲ **FIGURE 10-55 AGRICULTURAL LAND AND POPULATION GROWTH** Land devoted to agriculture has remained virtually unchanged since 1990, whereas population has increased by more than 50 percent.

In some regions, farmland is abandoned for lack of water. Especially in semiarid regions, human actions are causing land to deteriorate to a desertlike condition, a process called **desertification** (or, more precisely, semiarid land degradation). Semiarid lands that can support only a handful of pastoral nomads are overused because of rapid population growth. Excessive crop planting, animal grazing, and tree cutting exhaust the soil's nutrients and preclude agriculture. The Earth Policy Institute estimates that 2 billion hectares (5 million acres) of land have been degraded around the world (Figure 10-56). Overgrazing is thought to be responsible for 34 percent of the total, deforestation for 30 percent, and agricultural use for 28 percent. The UN estimates that desertification removes 27 million hectares (70 million acres) of land from agricultural production each year, an area roughly equivalent to Colorado.

Excessive water threatens other agricultural areas, especially drier lands that receive water from human-built irrigation systems. If the irrigated land has inadequate drainage, the underground water level rises to the point where roots become waterlogged. The UN estimates that 10 percent of all irrigated land is waterlogged, mostly in Asia and South America. If the water is salty, it can damage plants. The ancient civilization of Mesopotamia may have collapsed in part because of waterlogging and excessive salinity in its agricultural lands near the Tigris and Euphrates rivers.

Urbanization can also contribute to reducing agricultural land. As urban areas grow in population and land area, farms on the periphery are replaced by homes, roads, shops, and other urban land uses. In North America, farms outside urban areas are left idle until the speculators who own them can sell them at a profit to builders and developers, who convert the land to urban uses. A serious problem in the United States has been the loss of 200,000 hectares (500,000 acres) of the most productive farmland, known as **prime agricultural land**, as urban areas sprawl into the surrounding countryside (see the Contemporary Geographic Tools feature).

Pause and Reflect 10.4.4

By itself, GIS can't rank the relative importance of the various factors in protecting farmland. Policymakers and the public must make these value judgments. Do you think that prime soils, significant environmental features, and high population growth should be valued the same or differently in deciding which farmland to protect?

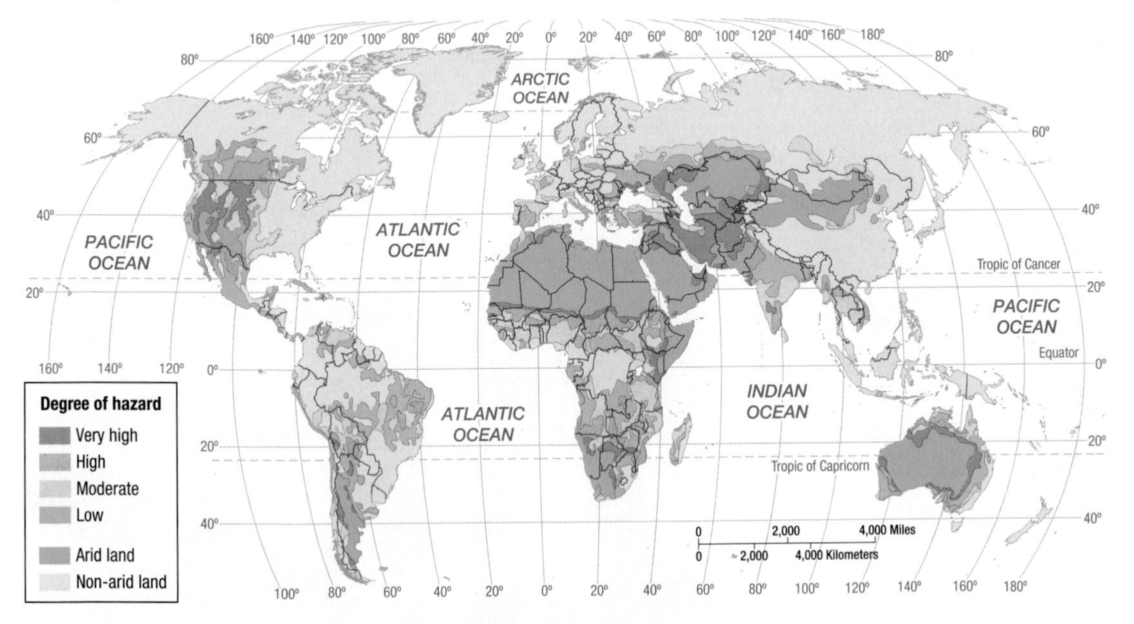

◀ **FIGURE 10-56 DESERTIFICATION (SEMIARID LAND DEGRADATION)** The most severe problems are in northern Africa, central Australia, and the southwestern parts of Africa, Asia, North America, and South America.

EXPANDING FISHING

Learning Outcome 10.4.5
Describe the contribution of fishing to world food supply.

A third alternative for increasing the world's food supply is to expand fishing. The agriculture discussed thus far in this chapter is land based. At first glance, increased use of food from the sea is attractive. Oceans are vast, covering nearly three-fourths of Earth's surface and lying near most population concentrations. Historically the sea has provided only a small percentage of the world food supply.

Food acquired from Earth's waters includes fish, crustaceans (such as shrimp and crabs), mollusks (such as clams and oysters), and aquatic plants (such as watercress). Water-based food is acquired in two ways:

- Fishing, which is the capture of wild fish and other seafood living in the waters.
- **Aquaculture**, or **aquafarming**, which is the cultivation of seafood under controlled conditions. (See the Sustainability and Inequality in Our Global Village feature.)

FISH CONSUMPTION. Human consumption of fish and seafood has increased from 27 million metric tons in 1960 to 110 million metric tons in 2010 (Figure 10-57). Developing countries are responsible for five-sixths of the increase. Fish consumption has increased more rapidly than population growth. During the past half-century, per capita consumption of fish has nearly doubled in both developed and developing countries, from 17 kcal per person per day in 1960 to 30 kcal per person per day in 2010. Still, fish and seafood account for only 1 percent of all calories consumed by humans (refer to Figure 10-13).

▼ FIGURE 10-57 **GROWTH IN HUMAN CONSUMPTION OF FISH** Human consumption of fish has increased in both developed and developing regions.

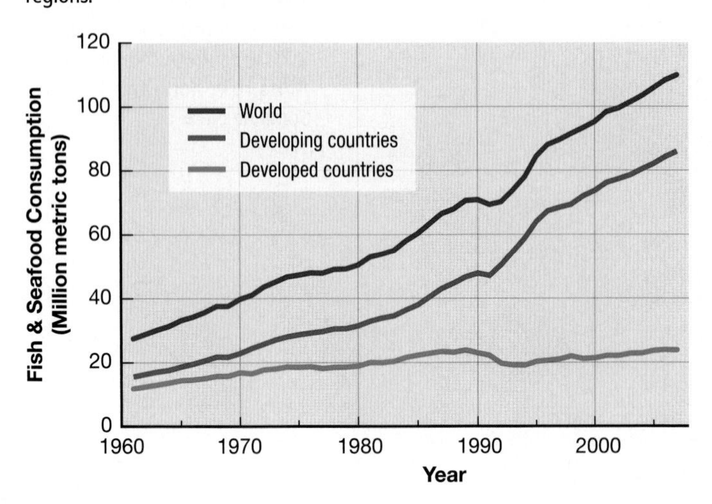

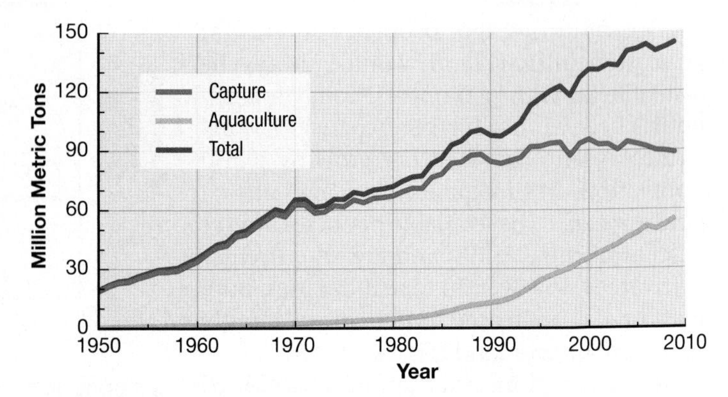

▲ FIGURE 10-58 **GROWTH IN FISH PRODUCTION** Increased fish production has come primarily from aquaculture rather than wild capture of fish.

FISH PRODUCTION. During the past half-century, global fish production has increased from approximately 36 to 145 million metric tons (Figure 10-58). The growth results entirely from expansion of aquaculture (Figure 10-59). The capture of wild fish in the oceans and lakes has stagnated since the 1990s, despite population growth and increased demand to consume fish. The reason that production is higher than human consumption is that a large portion of the fish that is caught is converted to fish meal and fed to poultry and hogs. Only two-thirds of the fish caught from the ocean is consumed directly by humans.

The world's oceans are divided into 18 major fishing regions, including seven each in the Atlantic and Pacific oceans, three in the Indian Ocean, and the Mediterranean (Figure 10-60). Fishing is also conducted in inland waterways, such as lakes and rivers. The areas with the largest yields are the Pacific Northwest and Asia's inland waterways. China is responsible for one-third of the world's yield of fish (Figure 10-61). The other leading countries are naturally those with extensive ocean boundaries, such as Chile, Indonesia, and Peru.

▼ FIGURE 10-59 **AQUACULTURE** Fish are raised inside the containers at this fish farm in Corfu, Greece.

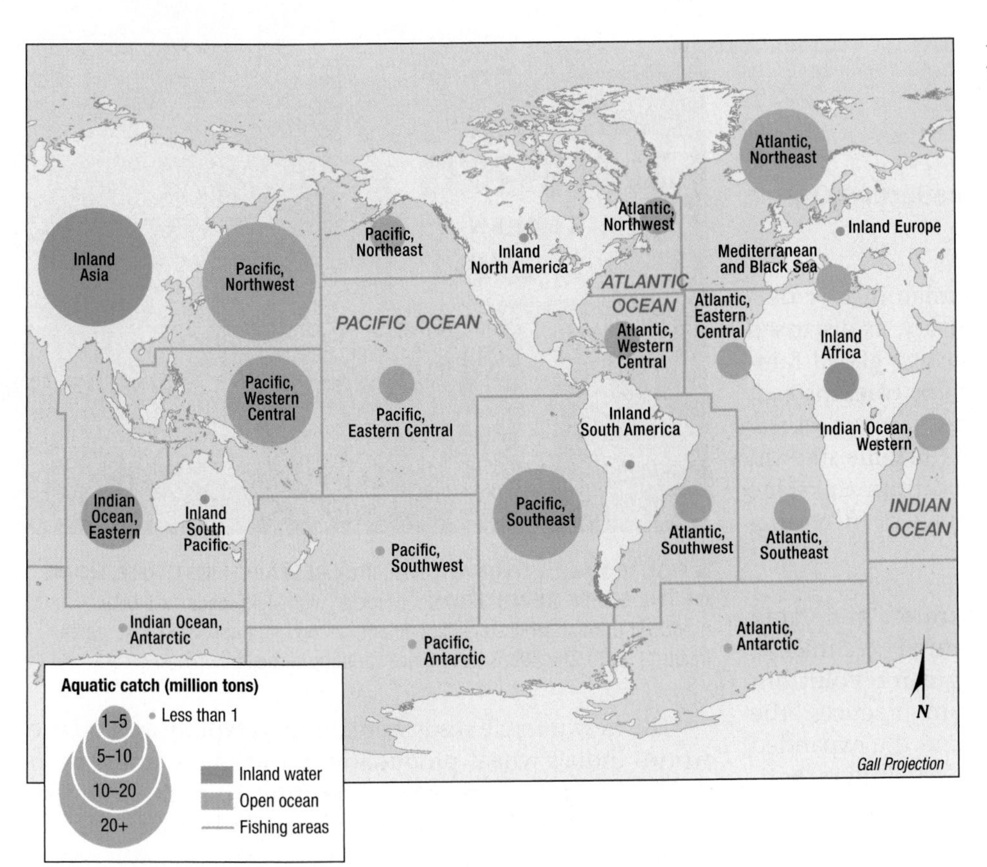

◀ **FIGURE 10-60 MAJOR FISHING REGIONS** The largest yields are in the Pacific and Asia.

OVERFISHING. Hope grew during the mid-twentieth century that increased fish consumption could meet the needs of a rapidly growing global population. However, the population of some fish species declined because they were harvested faster than they could reproduce. Overfishing has been particularly acute in the North Atlantic and Pacific oceans. Because of overfishing, the population of large predatory fish, such as tuna and swordfish, has declined by 90 percent in the past half-century. The UN estimates that one-quarter of fish stocks have been overfished and one-half fully exploited, leaving only one-fourth underfished. Consequently, the total world fish catch has remained relatively constant since the 1980s, despite population growth.

Pause and Reflect 10.4.5

Should Chicago's canals be shut to protect the Great Lakes from Asian carp? Why or why not?

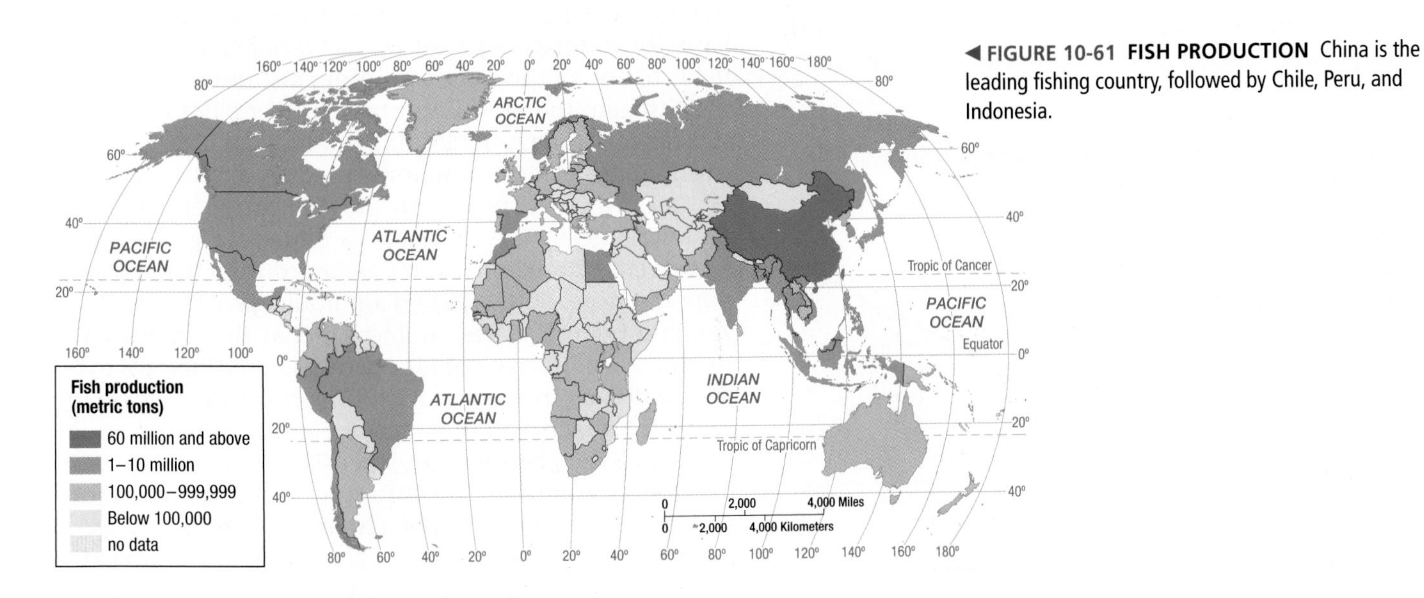

◀ **FIGURE 10-61 FISH PRODUCTION** China is the leading fishing country, followed by Chile, Peru, and Indonesia.

INCREASING PRODUCTIVITY

Learning Outcome 10.4.6
Describe the contribution of higher productivity to world food supply.

Population grew at the fastest rate in human history during the second half of the twentieth century, as discussed in Chapter 2. Many experts forecast massive global famine, but these dire predictions did not come true. Instead, new agricultural practices have permitted farmers worldwide to achieve much greater yields from the same amount of land. Worldwide, obtaining more food from the same amount of land has been the leading source of increasing the food supply.

THE GREEN REVOLUTION. The invention and rapid diffusion of more productive agricultural techniques during the 1970s and 1980s is called the **green revolution.** The green revolution involves two main practices: the introduction of new higher-yield seeds and the expanded use of fertilizers. Because of the green revolution, agricultural productivity at a global scale has increased faster than population growth (Figure 10-62).

Scientists began an intensive series of experiments during the 1950s to develop a higher-yield form of wheat. A decade later, the "miracle wheat seed" was ready. Shorter and stiffer than traditional breeds, the new wheat was less sensitive to variation in day length, responded better to fertilizers, and matured faster. The Rockefeller and Ford foundations sponsored many of the studies, and the program's director, Dr. Norman Borlaug, won the Nobel Peace Prize in 1970. The International Rice Research Institute, established in the Philippines by the Rockefeller and Ford foundations, worked to create a miracle rice seed (Figure 10-63). During the 1960s, their scientists introduced a hybrid of Indonesian rice and Taiwan dwarf rice that was hardier and that increased yields. More recently, scientists have developed new high-yield maize (corn).

▼ **FIGURE 10-62 POPULATION AND FOOD PRODUCTION** World population has increased less rapidly than food production.

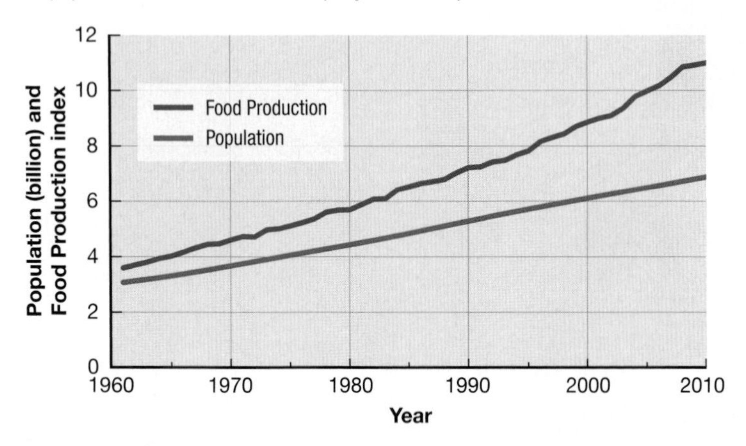

▲ **FIGURE 10-63 INTERNATIONAL RICE RESEARCH INSTITUTE, HOME OF THE GREEN REVOLUTION** "Miracle" high-yield seeds have been produced through laboratory experiments at the International Rice Research Institute (IRRI). The IRRI is testing rice varieties in the Philippines.

The new miracle seeds were diffused rapidly around the world. India's wheat production, for example, more than doubled in five years. After importing 10 million tons of wheat annually in the mid-1960s, India had a surplus of several million tons by 1971. Other Asian and Latin American countries recorded similar productivity increases. The green revolution was largely responsible for preventing a food crisis in these regions during the 1970s and 1980s. But will these scientific breakthroughs continue in the twenty-first century?

To take full advantage of the new miracle seeds, farmers must use more fertilizer and machinery. Farmers have known for thousands of years that application of manure, bones, and ashes somehow increases, or at least maintains, the fertility of the land. Not until the nineteenth century did scientists identify nitrogen, phosphorus, and potassium (potash) as the critical elements in these substances that improve fertility. Today these three elements form the basis for fertilizers—products that farmers apply to their fields to enrich the soil by restoring lost nutrients.

Nitrogen, the most important fertilizer, is a ubiquitous substance. China is the leading producer of nitrogen fertilizer. Europeans most commonly produce a fertilizer known as urea, which contains 46 percent nitrogen. In North America, nitrogen is available as ammonia gas, which is 82 percent nitrogen but more awkward than urea to transport and store. Both urea and ammonia gas combine nitrogen and hydrogen. The problem is that the cheapest way to produce both types of nitrogen-based fertilizers is to obtain hydrogen from natural gas or petroleum. As fossil fuel prices increase, so do the prices for nitrogen-based fertilizers, which then become too expensive for many farmers in developing countries. In contrast to nitrogen, phosphorus and potash reserves are not distributed uniformly across Earth's surface. Phosphate rock reserves are clustered in China, Morocco, and the United States. Proven potash reserves are concentrated in Canada, Russia, and Ukraine.

Farmers need tractors, irrigation pumps, and other machinery to make the most effective use of the new miracle seeds. In developing countries, farmers cannot afford such equipment and cannot, in view of high energy costs, buy fuel to operate the equipment. To maintain the green revolution, governments in developing countries must allocate scarce funds to subsidize the cost of seeds, fertilizers, and machinery.

GENETICALLY MODIFIED FOODS. Farmers have been manipulating crops and livestock for thousands of years. The very nature of agriculture is to deliberately manipulate nature. Humans control selective reproduction of plants and animals in order to produce a larger number of stronger, hardier survivors. Beginning in the nineteenth century, the science of genetics expanded understanding of how to manipulate plants and animals to secure dominance of the most favorable traits. However, genetic modification (GM), which became widespread in the late twentieth century, marks a sharp break with the agricultural practices of the past several thousand years. Under GM, the genetic composition of an organism is not merely studied, it is actually altered; GM involves mixing genetic material of two or more species that would not otherwise mix in nature.

Worldwide, 160 million hectares—10 percent of all farmland—were devoted to genetically modified crops in 2010; 77 percent of the world's soybeans, 49 percent of cotton, and 26 percent of maize were genetically modified in 2010. GM is especially widespread in the United States: 94 percent of soybeans, 90 percent of cotton, and 88 percent of maize; usage increased rapidly during the first decade of the twenty-first century (Figure 10-64). Three-fourths of the processed food that Americans consume has at least one GM ingredient. North America was responsible for one-half of the world's genetically modified foods, and developing countries—especially in Latin America—were responsible for the other one-half.

The United States has urged sub-Saharan African countries to increase their food supply in part through increased use of GM of crops and livestock. Africans are divided on whether to accept genetically modified organisms. The positives of GM are higher yields, increased nutrition, and more resistance to pests. Genetically modified foods are also better tasting, at least to some palates. Despite these benefits, opposition to GM is strong in Africa for several reasons:

- **Health problems.** Consuming large quantities of genetically modified foods may reduce the effectiveness of antibiotics and could destroy long-standing ecological balances in local agriculture.
- **Export problems.** European countries, the main markets for Africa's agricultural exports, require genetically modified foods to be labeled. Europeans are especially strongly opposed to GM because they believe genetically modified food is not as nutritious as food from traditionally bred crops and livestock. Because European consumers shun genetically modified food, African farmers fear that if they are no longer able to certify their exports as being not genetically modified, European customers will stop buying them (Figure 10-65).
- **Increased dependence on the United States.** U.S.-based transnational corporations, such as Monsanto, manufacture most of the GM seeds. Africans fear that the biotech companies could—and would—introduce a so-called "terminator" gene in the GM seeds to prevent farmers from replanting them after harvest and require them to continue to purchase seeds year after year from the transnational corporations.

"We don't want to create a habit of using genetically modified maize that the country cannot maintain," explained Mozambique's prime minister. If agriculture is regarded as a way of life, not just a food production business, GM represents for many Africans an unhealthy level of dependency on developed countries.

Pause and Reflect 10.4.6
What are the benefits and drawbacks for sub-Saharan Africa to plant more genetically modified crops?

▼ **FIGURE 10-64 GENETICALLY MODIFIED CROPS IN THE UNITED STATES** Approximately 90 percent of major crops in the United States are genetically modified.

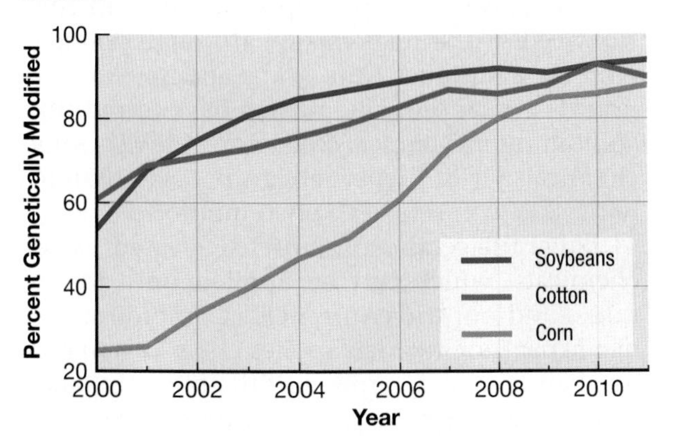

▼ **FIGURE 10-65 GENETICALLY MODIFIED FOOD** Genetically modified food is widespread in the United States but shunned by most consumers in Europe.

Sustainable Agriculture

Learning Outcome 10.4.7
Describe the role of sustainable agriculture in world food supply.

Some commercial farmers are converting their operations to **sustainable agriculture**, agricultural practices that preserve and enhance environmental quality. Farmers practicing sustainable agriculture typically generate lower revenues than do conventional farmers, but they also have lower costs.

An increasingly popular form of sustainable agriculture is organic farming. Worldwide, the UN classified 37 million hectares (75 million acres), or 0.6 percent of farmland, as organic in 2009. Australia was the leader, with 12 million of the hectares, or 32 percent of the worldwide total (Figure 10-66). Argentina accounted for 12 percent of the worldwide total, and the United States, China, and Brazil for 5 percent each. Three principal practices distinguish sustainable agriculture (and, at its best, organic farming) from conventional agriculture:

- Sensitive land management
- Limited use of chemicals
- Better integration of crops and livestock

SENSITIVE LAND MANAGEMENT

Sustainable agriculture protects soil in part through **ridge tillage**, which is a system of planting crops on ridge tops. Crops are planted on 10- to 20-centimeter (4- to 8-inch) ridges that are formed during cultivation or after harvest. A crop is planted on the same ridges, in the same rows, year after year. Ridge tillage is attractive for two main reasons: lower production costs and greater soil conservation.

Production costs are lower with ridge tillage in part because it requires less investment in tractors and other machinery than conventional planting. An area that would be prepared for planting under conventional farming with three to five tractors can be prepared for ridge tillage with only one or two tractors. The primary tillage tool is a row-crop cultivator that can form ridges. There is no need for a plow, or a field cultivator, or a 300-horsepower four-wheel-drive tractor. With ridge tillage, the space between rows needs to match the distance between wheels of the machinery. If 75 centimeters (30 inches) are left between rows, tractor tires will typically be on 150-centimeter (60-inch) centers and combine wheels on 300-centimeter (120-inch) centers. Wheel spacers are available from most manufacturers to fit the required spacing.

Ridge tillage features a minimum of soil disturbance from harvest to the next planting. A compaction-free zone is created under each ridge and in some row middles. Keeping the trafficked area separate from the crop-growing area improves soil properties. Over several years, the soil will tend to have increased organic matter, greater water-holding capacity, and more earthworms. The channels left by earthworms and decaying roots enhance drainage.

Ridge tillage compares favorably with conventional farming for yields while lowering the cost of production. Although more labor intensive than other systems, it is profitable on a per-acre basis. In Iowa, for example, ridge tillage has gained favor for production of organic and herbicide-free soybeans, which sell for more than regular soybeans.

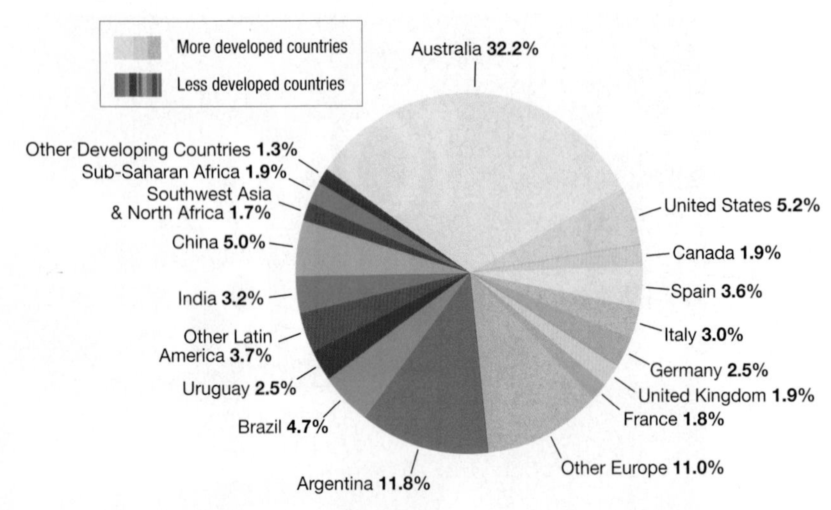

▼ **FIGURE 10-66 DISTRIBUTION OF ORGANIC FARMING** Australia accounts for nearly one-third of the world's organic farming.

LIMITED USE OF CHEMICALS

In conventional agriculture, seeds are often genetically modified to survive when herbicides and insecticides are sprayed on fields to kill weeds and insects. These are known as "Roundup Ready" seeds because their creator, Monsanto, sells its weed killers under the brand name Roundup. Roundup Ready seeds were planted in 90 percent of all soybean fields and 70 percent of all cotton and maize (corn) fields in the United States in 2010. In addition to the adverse impacts of herbicides on soil and water quality, widespread use of Roundup Ready seeds is causing some weeds to become resistant to herbicides.

Sustainable agriculture, on the other hand, involves application of limited if any herbicides to control weeds. In principle, farmers can control weeds without chemicals, although doing so requires additional time and expense that few farmers can afford. Researchers have found that combining mechanical weed control with some chemicals yields higher returns per acre than relying solely on one of the two methods.

Ridge tillage also promotes decreased use of chemicals, which can be applied only to the ridges and not the entire field. Combining herbicide banding—which applies chemicals in narrow bands over crop rows—with cultivating may be the best option for many farmers.

INTEGRATED CROP AND LIVESTOCK

Mixed crop and livestock is a common form of farming in the United States, as discussed earlier in the chapter. But many farmers in the mixed crop and livestock region actually choose to only grow crops or raise more animals than the crops they grow can feed. They sell their crops off the farm or purchase feed for their animals from outside suppliers. Sustainable agriculture attempts to integrate the growing of crops and the raising of livestock as much as possible at the level of the individual farm. Animals consume crops grown on the farm and are not confined to small pens.

Integration of crops and livestock reflects a return to the historical practice of mixed crop and livestock farming, in which growing crops and raising animals were regarded as complementary activities on the farm. This was the common practice for centuries, until the mid-1900s, when technology, government policy, and economics encouraged farmers to become more specialized.

Sustainable agriculture is sensitive to the complexities of biological and economic interdependencies between crops and livestock:

- **Number of livestock.** The correct number, as well as the distribution, of livestock for an area is determined based on the landscape and forage sources. Prolonged concentration of livestock in a specific location can result in permanent loss of vegetative cover, so a farmer needs to move the animals to reduce overuse in some areas. Growing row crops on the more level land while confining pastures to steeper slopes will reduce soil erosion, so it may be necessary to tolerate some loss of vegetation in specific locations.
- **Animal confinement.** The moral and ethical debate over animal welfare is particularly intense regarding confined livestock production systems (Figure 10-67). Confining livestock leads to surface and ground water pollution, particularly where the density of animals is high. Expensive waste management facilities are a necessary cost of confined production systems. If animals are not confined, manure can contribute to soil fertility. However, quality of life in nearby communities may be adversely affected by the smell.
- **Management of extreme weather conditions.** Herd size may need to be reduced during periods of short- and long-term drought. On the other hand, livestock can buffer the negative impacts of low rainfall periods by consuming crops that in conventional farming would be left as failures. Especially in Mediterranean climates such as California's, properly managed grazing significantly reduces fire hazards by reducing fuel buildup in grasslands and brushlands.
- **Flexible feeding and marketing.** Flexibility in feeding livestock and sending livestock to market can help cushion farmers against trade and price fluctuations and, in conjunction with cropping operations, make more efficient use of farm labor. Feed costs are the largest single variable cost in any livestock operation. Most of the feed may come from other enterprises on a ranch, though some is usually purchased off the farm. Feed costs can be kept to a minimum by monitoring animal condition and performance and understanding seasonal variations in feed and forage quality on the farm.

▲ **FIGURE 10-67 (TOP) CONVENTIONAL VERSUS (BOTTOM) ORGANIC FARMING** Chickens are not penned up in cages on an organic farm.

Pause and Reflect 10.4.7

Are you willing to pay more for food that is organically produced? Why or why not?

CHECK-IN: KEY ISSUE 4

Why Do Farmers Face Economic Difficulties?

- ✓ **Farmers in developing countries face challenges of meeting the needs of rapid population growth and growing food for export.**
- ✓ **Farmers in developed countries face challenges of overproduction and access to markets.**
- ✓ **Four strategies for increasing the world's food supplies include increasing exports, expanding agricultural land, expanding fishing, and increasing productivity of land.**
- ✓ **Sustainable agriculture involves sensitive land management, limited use of chemicals, and better integration of crops and livestock.**

Summary

KEY ISSUE 1

Where Did Agriculture Originate?

Prior to the development of agriculture, people survived by hunting animals, gathering wild vegetation, and fishing. Current agricultural practices vary between developed and developing countries.

LEARNING OUTCOME 10.1.1: Identify the major crop and livestock hearths.

- Agriculture was invented approximately 10,000 years ago in multiple hearths of crops and livestock.

LEARNING OUTCOME 10.1.2: Describe the major differences between subsistence and commercial agriculture.

- Subsistence agriculture, practiced in developing countries, is characterized by a high percentage of farmers in the labor force, limited use of machinery, and small average farm size.
- Commercial agriculture, practiced in developed countries, is characterized by a small percentage of farmers in the labor force, heavy use of machinery, and large average farm size.

THINKING GEOGRAPHICALLY 10.1: Compare agricultural hearths with the origin of Indo-European (Figures 5-18 and 5-19). What similarities appear between the diffusion of language and of agriculture?

GOOGLE EARTH 10.1: Little Andaman Island is home to approximately 100 Onge people, who traditionally live by hunting and gathering. More than 90 percent of the land area of the island appears to be dense forests. Why is this type of land cover especially suitable habitat for animals being hunted?

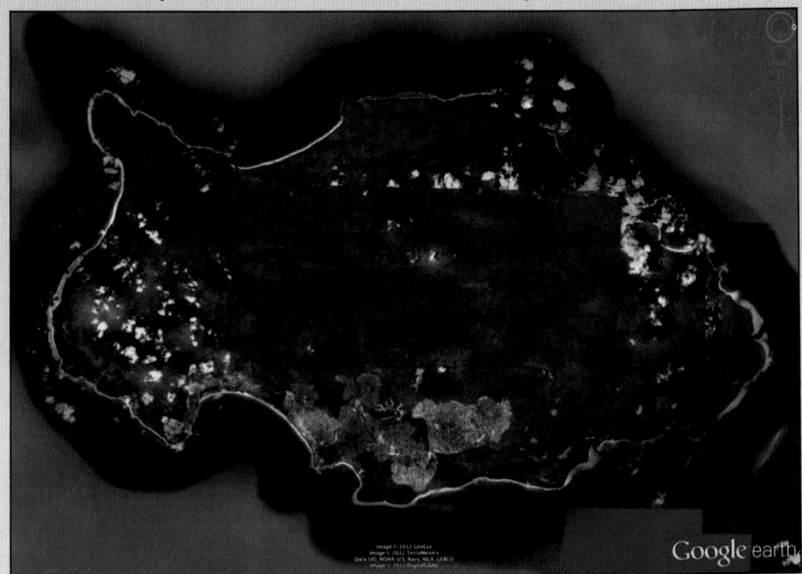

Key Terms

Agribusiness (p. 366) Commercial agriculture characterized by the integration of different steps in the food-processing industry, usually through ownership by large corporations.

Agricultural revolution (p. 348) The time when human beings first domesticated plants and animals and no longer relied entirely on hunting and gathering.

Agriculture (p. 347) The deliberate effort to modify a portion of Earth's surface through the cultivation of crops and the raising of livestock for sustenance or economic gain.

Aquaculture (or aquafarming) (p. 382) The cultivation of seafood under controlled conditions.

Cereal grain (or cereal) (p. 352) A grass that yields grain for food.

Chaff (p. 363) Husks of grain separated from the seed by threshing.

Combine (p. 370) A machine that reaps, threshes, and cleans grain while moving over a field.

Commercial agriculture (p. 350) Agriculture undertaken primarily to generate products for sale off the farm.

Crop (p. 347) Any plant gathered from a field as a harvest during a particular season.

Crop rotation (p. 364) The practice of rotating use of different fields from crop to crop each year to avoid exhausting the soil.

Desertification (p. 381) Degradation of land, especially in semiarid areas, primarily because of human actions such as excessive crop planting, animal grazing, and tree cutting. Also known as semiarid land degradation.

Dietary energy consumption (p. 352) The amount of food that an individual consumes, measured in kilocalories (Calories in the United States).

Double cropping (p. 363) Harvesting twice a year from the same field.

Food security (p. 354) Physical, social, and economic access at all times to safe and nutritious food sufficient to meet dietary needs and food preferences for an active and healthy life.

Grain (p. 352) Seed of a cereal grass.

Green revolution (p. 384) Rapid diffusion of new agricultural technology, especially new high-yield seeds and fertilizers.

Horticulture (p. 371) The growing of fruits, vegetables, and flowers.

Hull (p. 363) The outer covering of a seed.

Intensive subsistence agriculture (p. 362) A form of subsistence agriculture in which farmers must expend a relatively large amount of effort to produce the maximum feasible yield from a parcel of land.

Milkshed (p. 368) The area surrounding a city from which milk is supplied.

Paddy (p. 363) The Malay word for wet rice, commonly but incorrectly used to describe a sawah.

Pastoral nomadism (p. 358) A form of subsistence agriculture based on herding domesticated animals.

KEY ISSUE 2

Why Do People Consume Different Foods?

Everyone needs food to survive. The amount of food and the dietary composition of the food vary between developed and developing countries.

LEARNING OUTCOME 10.2.1: Explain differences between developed and developing countries in food consumption.

- Most humans derive most of their dietary energy through cereal grains, especially wheat, rice, and maize.
- The primary source of protein is meat products in developed countries and grain in developing countries.

LEARNING OUTCOME 10.2.2: Explain the global distribution of undernourishment.

- The average individual consumes 50 percent more calories than the recommended minimum, but many in sub-Saharan Africa are getting less than the recommended minimum.
- Worldwide, an estimated 850 million people are undernourished, nearly all of them in developing countries.

THINKING GEOGRAPHICALLY 10.2: Compare world distributions of wheat, rice, and maize production. To what extent do differences derive from environmental conditions and to what extent from food preferences and other social customs?

GOOGLE EARTH 10.2: Fly to Jungle Jim's in Fairfield, Ohio, at 30,000 square meters, possibly the largest supermarket in the United States. Under Find Businesses, type Kroger. Move to the nearest Kroger to the west of Jungle Jim's. How many square meters is it?

Pasture (p. 359) Grass or other plants grown for feeding grazing animals, as well as land used for grazing.

Plantation (p. 364) A large farm in tropical and subtropical climates that specializes in the production of one or two crops for sale, usually to a more developed country.

Prime agricultural land (p. 381) The most productive farmland.

Ranching (p. 372) A form of commercial agriculture in which livestock graze over an extensive area.

Reaper (p. 370) A machine that cuts cereal grain standing in a field.

Ridge tillage (p. 386) A system of planting crops on ridge tops in order to reduce farm production costs and promote greater soil conservation.

Sawah (p. 363) A flooded field for growing rice.

Shifting cultivation (p. 360) A form of subsistence agriculture in which people shift activity from one field to another; each field is used for crops for a relatively few years and left fallow for a relatively long period.

Slash-and-burn agriculture (p. 360) Another name for shifting cultivation, so named because fields are cleared by slashing the vegetation and burning the debris.

Spring wheat (p. 370) Wheat planted in the spring and harvested in the late summer.

Subsistence agriculture (p. 350) Agriculture designed primarily to provide food for direct consumption by the farmer and the farmer's family.

Sustainable agriculture (p. 386) Farming methods that preserve long-term productivity of land and minimize pollution, typically by rotating soil-restoring crops with cash crops and reducing inputs of fertilizer and pesticides.

Swidden (p. 360) A patch of land cleared for planting through slashing and burning.

Thresh (p. 363) To beat out grain from stalks.

Transhumance (p. 359) The seasonal migration of livestock between mountains and lowland pastures.

Truck farming (p. 367) Commercial gardening and fruit farming, so named because *truck* was a Middle English word meaning "bartering" or "exchange of commodities."

Undernourishment (p. 354) Dietary energy consumption that is continuously below the minimum requirement for maintaining a healthy life and carrying out light physical activity.

Wet rice (p. 362) Rice planted on dry land in a nursery and then moved to a deliberately flooded field to promote growth.

Winnow (p. 363) To remove chaff by allowing it to be blown away by the wind.

Winter wheat (p. 370) Wheat planted in the autumn and harvested in the early summer.

KEY ISSUE 3

Where Is Agriculture Distributed?

Most people in developing countries are subsistence farmers, growing crops primarily to feed themselves. Important types of subsistence agriculture include shifting cultivation, pastoral nomadism, and intensive farming. The most common type of farm in developed countries is mixed crop and livestock. Where mixed crop and livestock farming is not suitable, commercial farmers practice other types of agriculture, including dairy farming, commercial gardening, grain, Mediterranean, and ranching.

LEARNING OUTCOME 10.3.1: Identify the 11 major agricultural regions.

- The most widely used map of agriculture divides the world into 11 major regions, including 5 in developing countries and 6 in developed countries.

LEARNING OUTCOME 10.3.2: Explain how pastoral nomadism works in the dry lands of developing regions.

- Pastoral nomadism, which is the herding of animals, is the principal form of agriculture adapted to the dry lands of developing countries.

LEARNING OUTCOME 10.3.3: Explain how shifting cultivation works in the tropics of developing regions.

- Distinctive features of shifting cultivation include the clearing of land through slashing and burning and the use of fields for only a few years.

LEARNING OUTCOME 10.3.4: Explain how intensive subsistence farming works in the high population concentrations of developing regions.

- The principal crop in the intensive subsistence region is wet rice.
- Growing rice is an intensive operation that depends primarily on abundant labor.

LEARNING OUTCOME 10.3.5: Describe reasons for growing crops other than wet rice in intensive subsistence regions.

- In intensive subsistence areas where the climate is unsuitable for rice, hardier crops are grown, such as wheat and barley.
- Plantation farming is a form of commercial agriculture conducted in developing regions. Plantations grow crops primarily for export to developed countries.

LEARNING OUTCOME 10.3.6: Describe how mixed crop and livestock farming works.

- Mixed crop and livestock is the most common form of agriculture in the center of the United States.
- Crops, especially maize and soybeans, are grown primarily to feed animals.

LEARNING OUTCOME 10.3.7: Describe how dairy farming and commercial gardening work.

- Dairy farming is especially important near major population concentrations in developed countries.
- Commercial gardening is the predominant form of agriculture in the southeastern United States. These farms specialize in fruits and vegetables preferred by relatively wealthy consumers in developed countries.

LEARNING OUTCOME 10.3.8: Describe how grain and Mediterranean farming work.

- Grain, especially wheat, is grown in areas that are too dry for mixed crop and livestock farming.
- Mediterranean agriculture specializes in crops such as grapes and olives.

LEARNING OUTCOME 10.3.9: Describe how livestock ranching works.

- Livestock is raised on land that is too dry for growing crops.

THINKING GEOGRAPHICALLY 10.3: Review the concept of overpopulation (the number of people in an area exceeding the capacity of the environment to support life at a decent standard of living). What agricultural regions have relatively limited capacities to support intensive food production? Which of these regions face rapid population growth?

GOOGLE EARTH 10.3: Terraces for planting rice are carved into the hillsides surrounding the village of Banaue, Philippines. What step in growing rice, as described in Learning Outcome 10.3.4, makes it necessary to terrace the hillsides?

KEY ISSUE 4

Why Do Farmers Face Economic Difficulties?

Agriculture in developing countries faces distinctive economic problems resulting from rapid population growth and pressure to adopt international trade strategies to promote development. Agriculture in developed nations faces problems resulting from access to markets and overproduction.

LEARNING OUTCOME 10.4.1: Describe the impact of population growth and trade on farming in developing countries.

- Due to rapid population growth, subsistence farmers must feed more people.
- Pressure to contribute to international trade means that subsistence farmers increasingly grow crops to export rather than to consume at home.

LEARNING OUTCOME 10.4.2 Understand distinctive challenges for developing countries to increase food supply.

- Africa faces the greatest challenge in providing enough food for a growing population.
- Export crops such as drugs are increasingly been grown in some developing countries.

LEARNING OUTCOME 10.4.3: Explain the impact of overproduction and market access on farming in developed countries.

- Because of their efficiency, commercial farmers produce more food than can be consumed in developed countries.

LEARNING OUTCOME 10.4.4: Explain the contribution of expanding exports and farmland to world food supply.

- Export of food has increased rapidly, although only a handful of countries produce enough to be major exporters.
- Historically, agricultural output was increased by expanding the amount of land that is farmed, but expansion of farmland has slowed in recent decades.

LEARNING OUTCOME 10.4.5: Describe the contribution of fishing to world food supply.

- Fish consumption is increasing but accounts for a small percentage of the average human's diet.
- Fish production has increased primarily through aquaculture rather than catching of wild fish.

LEARNING OUTCOME 10.4.6: Describe the contribution of higher productivity to world food supply.

- Agricultural productivity has increased sharply, especially through the invention of higher-yield seeds and expanded use of fertilizers.
- Despite advances, food prices in the early twenty-first century have been at a record high.

LEARNING OUTCOME 10.4.7: Describe the role of sustainable agriculture in world food supply.

- Sustainable agriculture involves sensitive land management, limited use of chemicals, and better integration of crops and livestock.
- Sustainable agriculture accounts for a small but increasing share of world agriculture.

THINKING GEOGRAPHICALLY 10.4: New Zealand once sold nearly all its dairy products to the British, but since the United Kingdom joined the European Union in 1973, New Zealand has been forced to find other markets. What are some other examples of countries that have restructured their agricultural production in the face of increased global interdependence and regional cooperation?

GOOGLE EARTH 10.4: The eastern end of the Chicago Sanitary and Ship Canal joins with the Chicago River near the center of Chicago. The canal was constructed to provide the only water link between the Great Lakes and the Mississippi River. If Asian carp now migrating up the Mississippi River are to be prevented from reaching Lake Michigan, the canal will have to be blocked. What is the approximate distance between the end of the canal at the Chicago River and Lake Michigan?

Chapter

11 Industry and Manufacturing

Why are most potato chips manufactured near their consumers? Page 401

Why are most fabrics made in Asia? Page 411

KEY ISSUE 1

Where Is Industry Distributed?

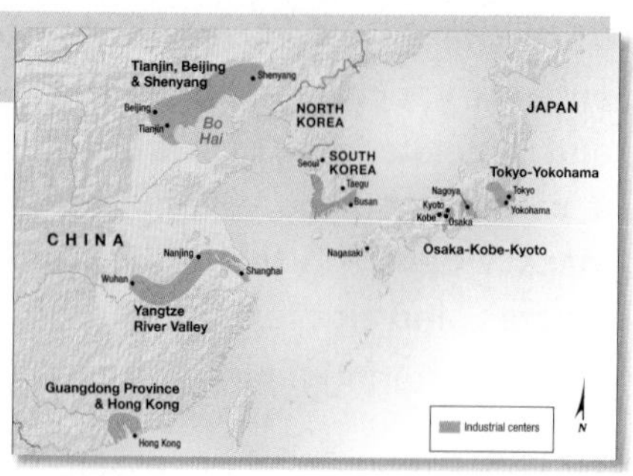

Factories Past and Present p. 395

Much of the world's industry is clustered in three regions.

KEY ISSUE 2

Why Are Situation and Site Factors Important?

Factors of Production p. 398

Geographers can explain reasons for the location of factories.

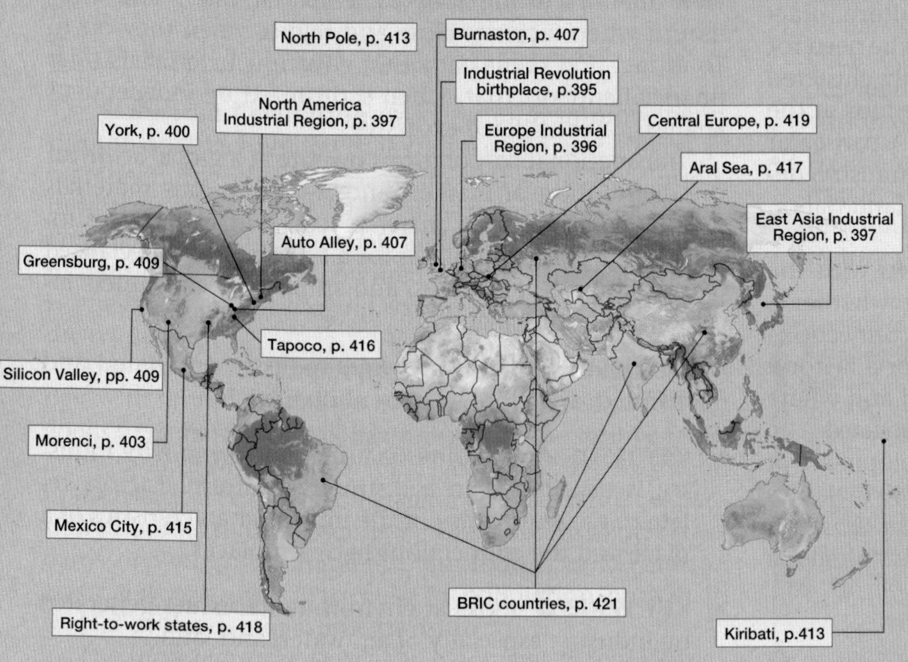

▲ Foxconn may not be a familiar brand name, but it is the world's largest manufacturer of electronic components. Owned by Hon Hai Precision Industry Co., Foxconn is the largest exporter of products from China. Its largest main factory in Shenzhen, China, employs several hundred thousand people. Foxconn has become the world's dominant electronics manufacturer because it does the actual manufacturing for several well-known products, including the iPad, iPhone, Kindle, PlayStation 3, and Xbox 360. Geographers study why a company like Apple, which is based in the United States, chooses to have its products made by another company in another country.

KEY ISSUE 3

Where Does Industry Cause Pollution?

Factories Clean and Dirty p. 412

Some factories pollute our air, land, and water.

KEY ISSUE 4

Why Are Situation and Site Factors Changing?

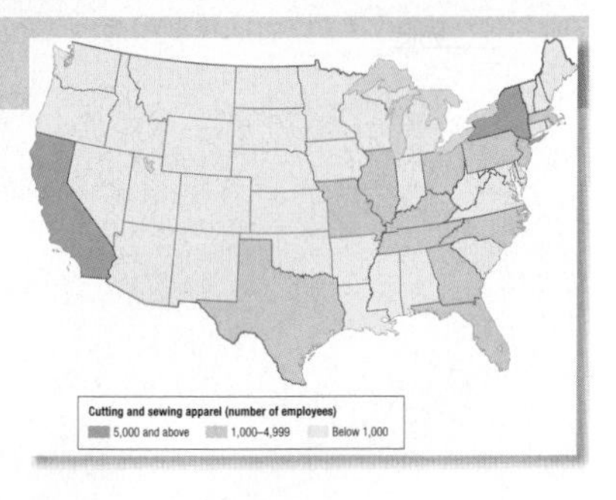

Industry on the Move p. 418

Manufacturing is expanding into new regions.

Industrial Regions

Learning Outcome 11.1.1
Describe the locations of the principal industrial regions.

Industry is concentrated in three of the nine world regions discussed in Chapter 9 regions of the world: Europe (Figure 11-3), North America (Figure 11-4), and East Asia (Figure 11-5). Each of the three regions accounts for roughly one-fourth of the world's total industrial output. Outside these three regions, the leading industrial producers are Brazil and India.

EUROPE'S INDUSTRIAL AREAS

Major industrial areas in Europe include:

- The **United Kingdom** dominated world production of steel and textiles during the nineteenth century. These industries have declined, but the country has attracted international investment through new high-tech industries that serve the European market.
- The **Rhine-Ruhr Valley** has a concentration of iron and steel manufacturing because of proximity to large coalfields. Rotterdam, the world's largest port, lies at the mouth of several branches of the Rhine River as it flows into the North Sea.
- The **Mid-Rhine** is Europe's most centrally located industrial area. Frankfurt is a financial and commercial center and the hub of Germany's transport network. Stuttgart specializes in high-value goods that require skilled labor. Mannheim, an inland port along the Rhine, has a large chemical industry that manufactures synthetic fibers, dyes, and pharmaceuticals.
- The **Po Basin** has attracted textiles and other industries because of two key assets, compared to Europe's other industrial regions: numerous workers willing to accept lower wages and inexpensive hydroelectricity from the nearby Alps.
- **Northeastern Spain** was Europe's fastest-growing manufacturing area during the late twentieth century. Spain's leading industrial area, Catalonia, centered on the city of Barcelona, is the center of Spain's textile industry and the country's largest motor-vehicle plant.
- **Moscow** is Russia's oldest industrial region, centered around the country's capital and largest city.
- **St. Petersburg**, Russia's second-largest city, specializes in shipbuilding and other industries serving Russia's navy and ports in the Baltic Sea.
- The **Urals**, contain the world's most varied collection of minerals. Proximity to these minerals has attracted iron and steel, chemicals, machinery, and metal fabricating plants.
- **Volga** is the region containing Russia's largest petroleum and natural gas fields. To the northeast, the Ural mountain range contains more than 1,000 types of minerals, the most varied collection found in any mining region in the world.
- **Kuznetsk** is Russia's most important manufacturing district east of the Ural Mountains, with the country's largest reserves of coal and an abundant supply of iron ore.
- **Donetsk**, in Eastern Ukraine, has one of the world's largest coal reserves.
- **Silesia**, Europe's most rapidly growing industrial area, takes advantage of a skilled but low-paid workforce and proximity to wealthy markets in Western Europe.

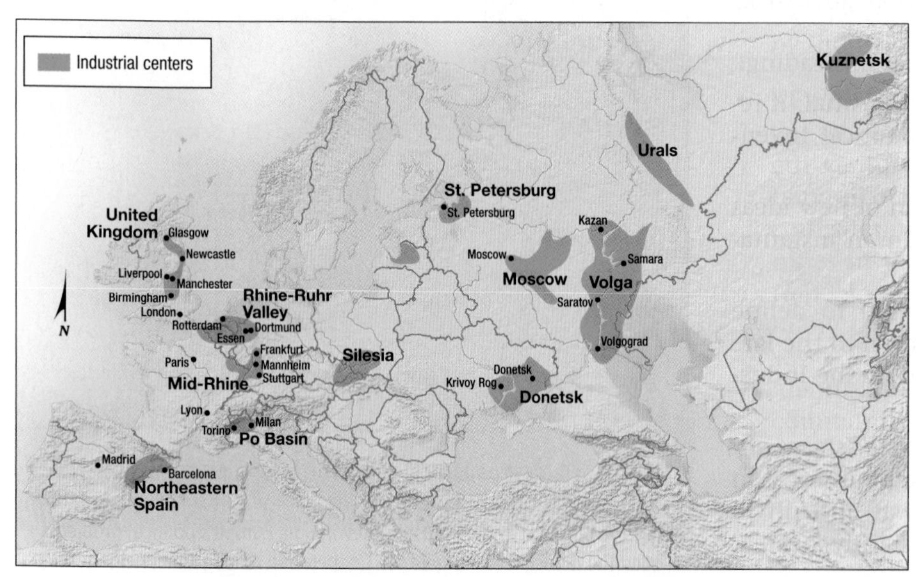

▲ **FIGURE 11-3 EUROPE'S INDUSTRIAL AREAS** Europe was the first region to industrialize during the nineteenth century. Numerous industrial centers emerged in Europe as countries competed with each other for supremacy.

NORTH AMERICA'S INDUSTRIAL AREAS

Major industrial areas in North America include:

- **New England** was a cotton textile center in the early nineteenth century. Cotton was imported from southern states, and finished cotton products were shipped to Europe.
- The **Middle Atlantic** is the largest U.S. market, so the region attracts industries that need proximity to a large number of consumers and depend on foreign trade through one of this region's large ports.
- The **Mohawk Valley**, a linear industrial belt in upper New York State, takes advantage of inexpensive electricity generated at nearby Niagara Falls.
- **Pittsburgh–Lake Erie** was the leading steel-producing area in the nineteenth century because of its proximity to Appalachian coal and iron ore.

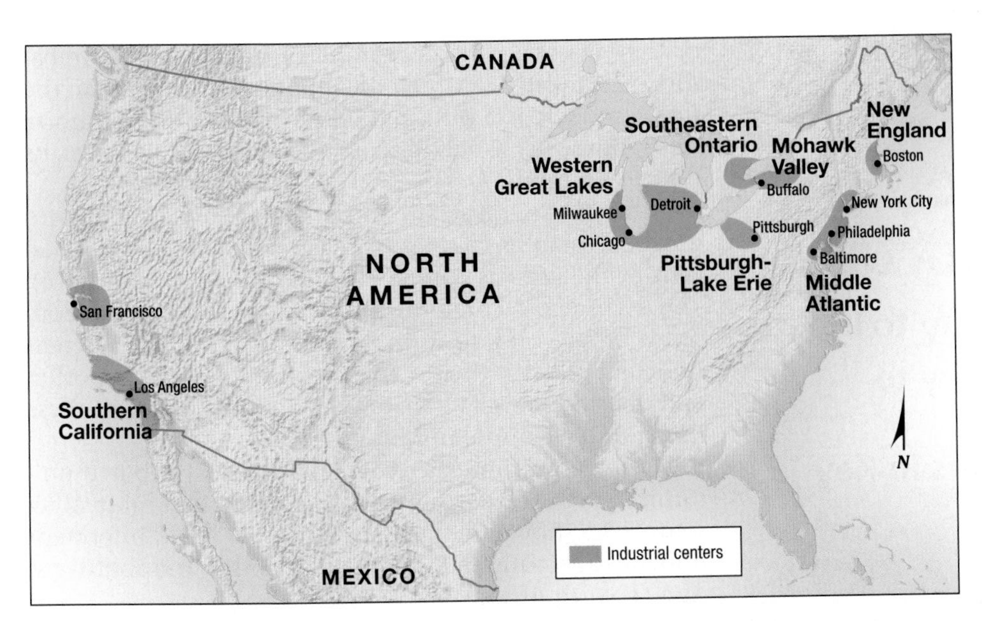

◄ **FIGURE 11-4 NORTH AMERICA'S INDUSTRIAL AREAS** Industry arrived a bit later in North America than in Europe, but it grew much faster in the nineteenth century. North America's manufacturing was traditionally highly concentrated in the northeastern United States and southeastern Canada. In recent years, manufacturing has relocated to the South, lured by lower wages and legislation that has made it difficult for unions to organize factory workers.

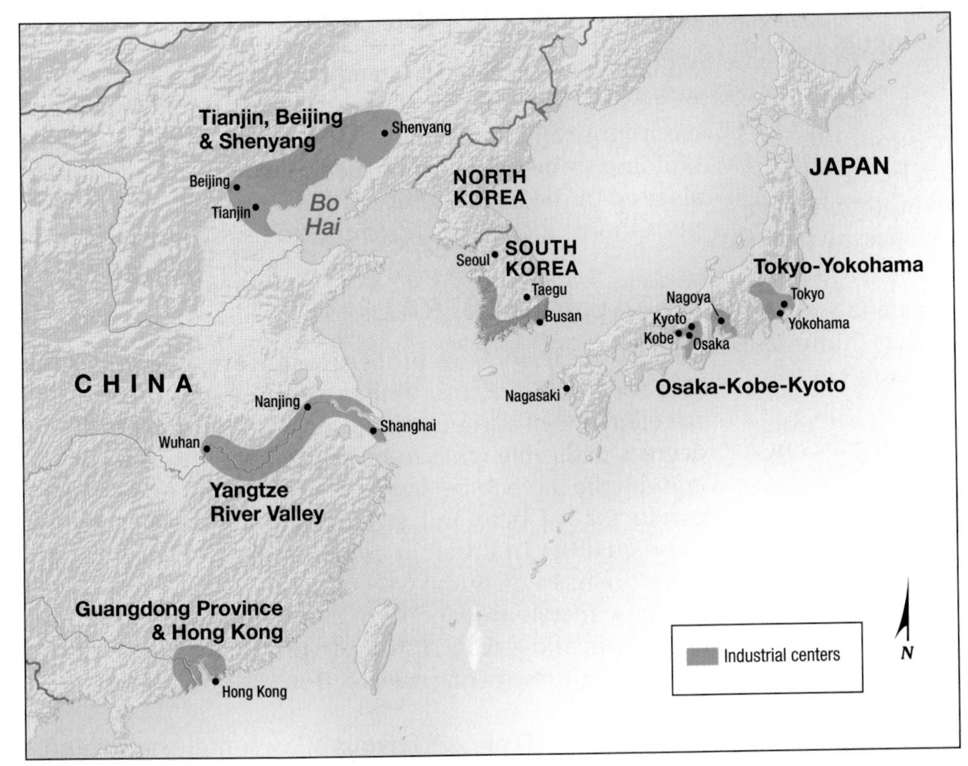

▲ **FIGURE 11-5 EAST ASIA'S INDUSTRIAL AREAS** East Asia became an important industrial region in the second half of the twentieth century, beginning with Japan. Into the twenty-first century, China has emerged as the world's leading manufacturing country by most measures.

- The **Western Great Lakes**, centered on Chicago, is the hub of the nation's transportation network and is now the center of steel production.
- **Southern California** is now the country's largest area of clothing and textile production, the second-largest furniture producer, and a major food-processing center.
- **Southeastern Ontario**, Canada's most important industrial area, is central to the Canadian and U.S. markets and near the Great Lakes and Niagara Falls.

ASIA'S INDUSTRIAL AREAS

Major industrial areas in Asia include:

- **Japan** became an industrial power in the 1950s and 1960s, initially by producing goods that could be sold in large quantity at cut-rate prices to consumers in other countries. Manufacturing is concentrated in the central region, between Tokyo and Nagasaki.
- **China** has the world's largest supply of low-cost labor and is the world's largest market for many consumer products. Manufacturers cluster in three areas along the east coast: near Guangdong and Hong Kong, in the Yangtze River valley between Shanghai and Wuhan, and along the Gulf of Bo Hai, from Tianjin and Beijing to Shenyang.
- **South Korea** followed Japan's lead in focusing on export-oriented manufacturers. The country is a leading producer of ocean-going ships. Manufacturing is centered along the rim of the country between the capital and largest city Seoul and Busan, the largest port.

CHECK-IN: KEY ISSUE 1

Where Is Industry Distributed?

✓ **The Industrial Revolution was a series of improvements that transformed manufacturing. Most of the improvements occurred first in the United Kingdom.**

✓ **The world's three principal industrial regions are Europe, North America, and East Asia.**

KEY ISSUE 2

Why Are Situation and Site Factors Important?

- Situation Factors: Proximity to Inputs
- Situation Factors: Proximity to Markets
- Changing Situation Factors in Key Industries
- Site Factors

Learning Outcome 11.2.1
Identify the two types of situation factors and explain why some industries locate near inputs.

Having looked at the "where" question for industrial location, we can next consider the "why" question: Why are industries located where they are? Geographers try to explain why one location may prove more profitable for a factory than others. A company ordinarily faces two geographic costs—situation and site:

- **Situation factors** involve transporting materials to and from a factory. A firm seeks a location that minimizes the cost of transporting inputs to the factory and finished goods to consumers.
- **Site factors** result from the unique characteristics of a location.

Situation Factors: Proximity to Inputs

Manufacturers buy from companies and individuals who supply inputs, such as minerals, materials, energy, machinery, and supporting services. They sell to companies and individuals who purchase the product. The farther something is transported, the higher the cost, so a manufacturer tries to locate its factory as close as possible to its inputs and markets:

- **Proximity to inputs.** The optimal plant location is as close as possible to inputs if the cost of transporting raw materials to the factory is *greater than* the cost of transporting the product to consumers.
- **Proximity to markets.** The optimal plant location is as close as possible to the customer if the cost of transporting raw materials to the factory is *less than* the cost of transporting the product to consumers.

Every industry uses some inputs. The inputs may be resources from the physical environment, such as minerals, or they may be parts or materials made by other companies. An industry in which the inputs weigh more than the final products is a **bulk-reducing industry**. To minimize transport costs, a bulk-reducing industry locates near its sources of inputs.

Minerals are especially important inputs for many industries. Earth has 92 natural elements, but about 99 percent of the crust is composed of 8 of them (Figure 11-6). The eight most common elements combine with thousands of rare ones to form approximately 3,000 different minerals, all with their own properties of hardness, color, and density, as well as spatial distribution. Many of these minerals have important industrial uses.

Like energy, mineral resources are not distributed uniformly across Earth. Countries with important mineral resources are shown in orange in Figure 11-7. Few important minerals are found in Europe, Central Asia, and Southwest Asia & North Africa.

NONMETALLIC MINERALS

Minerals are either nonmetallic or metallic. In weight, more than 90 percent of the minerals that humans use are nonmetallic. Important nonmetallic minerals include building stones, gemstones such as diamonds, and minerals used in the manufacture of fertilizers such as nitrogen, phosphorus, potassium, calcium, and sulfur.

METALLIC MINERALS

Metallic minerals have properties that are especially valuable for fashioning machinery, vehicles, and other essential elements of contemporary society. They are to varying degrees malleable (able to be hammered into thin plates) and ductile (able to be drawn into fine wire) and are good conductors of heat and electricity. Each metal possesses these qualities in different combinations and degrees and therefore has a distinctive set of uses.

Many metals are capable of combining with other metals to form alloys with distinctive properties important for industry. Alloys are known as ferrous or nonferrous.

FERROUS ALLOYS. A **ferrous** alloy contains iron, and a **nonferrous** one does not. The word *ferrous* comes from the

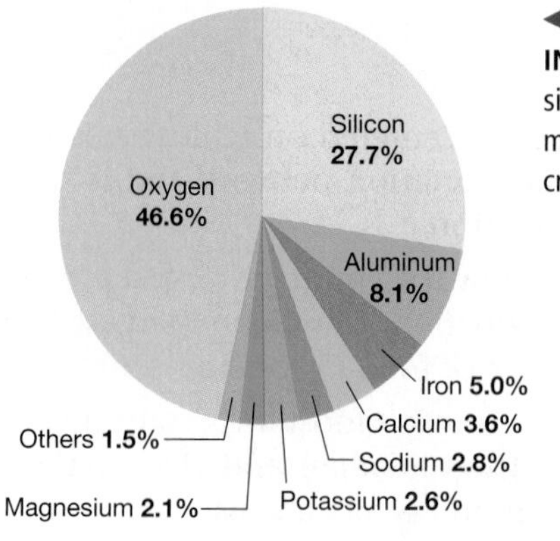

FIGURE 11-6 ELEMENTS IN EARTH'S CRUST Oxygen, silicon, and aluminum are the most common elements in Earth's crust.

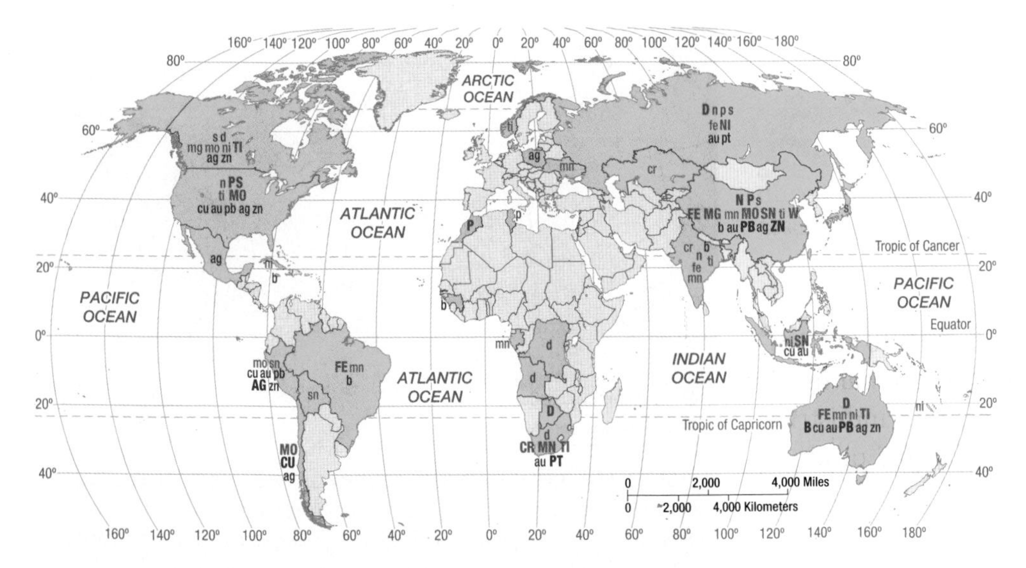

▲ **FIGURE 11-7 DISTRIBUTION OF MINERALS**
Australia and China are especially well endowed with minerals that are important for industry.

Nonmetallic minerals

D	d	Diamonds	P	p	Phosphorus
N	n	Nitrogen	S	s	Sulfur

Ferrous metals

FE	fe	Iron ore	NI	ni	Nickel
CR	cr	Chromium	SN	sn	Tin
MG	mg	Magnesium	TI	ti	Titanium
MN	mn	Manganese	W	w	Tungsten
MO	mo	Molybdenum			

Nonferrous metals

B	b	Bauxite	PT	pt	Platinum
CU	cu	Copper	AG	ag	Silver
AU	au	Gold	ZN	zn	Zinc
PB	pb	Lead			

XX represents 15% and above of world production
xx represents 5–15% of world production

Latin for "iron." Iron is extracted from iron ore, by far the world's most widely used ore. Humans began fashioning tools and weapons from iron 4,000 years ago. Important metals used to make ferrous alloys include:

- **Chromium** is a principal component of stainless steel, extracted from chromite ore, one-half of which is mined in South Africa.
- **Manganese** imparts toughness and carries off undesirable sulfur and oxygen during the smelting process. Brazil, Gabon, and South Africa are the leading producers.
- **Molybdenum** imparts toughness and resilience to steel. The United States is the leading producer.
- **Nickel** is used primarily for stainless steel and high-temperature and electrical alloys. Russia, Australia, and Canada are the leading producers.
- **Tin** is valued for its corrosion-resistant properties and is used for plating iron and steel. China is the leading producer.
- **Titanium** is used as white pigment in paint. It is extracted primarily from the mineral ilmenite, and Australia is the leading producer.
- **Tungsten** is used to manufacture tungsten carbide for cutting tools. China is responsible for 90 percent of world production.

NONFERROUS METALS. Important metals utilized to manufacture products that don't contain iron and steel include:

- **Aluminum** is the most abundant nonferrous metal. Lighter, stronger, and more resistant to corrosion than iron and steel, aluminum is obtained primarily through extraction from bauxite ore. Australia is the leading producer.
- **Copper** is valued for its high ductility, malleability, thermal and electrical conductivity, and resistance to corrosion. It is used primarily in electronics and constructing buildings. Chile is the leading producer.
- **Lead** is has been used for thousands of years, first in building materials and pipes; then in ammunition, brass, glass, and crystal; and now primarily in motor-vehicle batteries. Australia and China are the leading producers.
- **Lithium** is used in batteries for a wide variety of devices such as cell phones, laptop computers, and hybrid and electric-powered vehicles. Chile and Australia each produce about one-third of global output.
- **Magnesium** is relatively light yet strong, so it is used to produce lightweight, corrosion-resistant alloys, especially with aluminum to make beverage cans. China supplies three-fourths of the world's magnesium.
- **Zinc** is primarily used as a coating to protect iron and steel from corrosion, and it is also used as an alloy to make bronze and brass. China is the leading producer.
- **Precious metals** include silver, gold, and the platinum group. Silver and gold have been prized since ancient times for their beauty and durability. Platinum is used in motor vehicles for catalytic converters and fuel cells.
- **Rare earth metals** comprise 17 elements, 15 of which are lanthanides, such as cerium. They are called "rare" because only a few deposits in the world are economically profitable to mine, nearly all of them in China. Rare earth metals are used in electronics and motors.

Pause and Reflect 11.2.1
North America is a leading source of which minerals?

SHIP, RAIL, TRUCK, OR AIR?

Learning Outcome 11.2.3
Explain why industries use different types of transportation.

Inputs and products are transported in one of four ways: via ship, rail, truck, or air. Firms seek the lowest-cost mode of transport, but which of the four alternatives is cheapest changes with the distance that goods are being sent.

The farther something is transported, the lower is the cost per kilometer (or mile). Longer-distance transportation is cheaper per kilometer in part because firms must pay workers to load goods on and off vehicles, whether the material travels 10 kilometers or 10,000. The cost per kilometer decreases at different rates for each of the four modes because the loading and unloading expenses differ for each mode:

- **Trucks** are most often used for short-distance delivery, because they can be loaded and unloaded quickly and cheaply. Truck delivery is especially advantageous if the driver can reach the destination within one day, before having to stop for an extended rest.
- **Trains** are often used to ship to destinations that take longer than one day to reach, such as between the East and West coasts of the United States. Trains take longer than trucks to load, but once under way, they aren't required to make daily rest stops like trucks.
- **Ships** are attractive for transport over very long distances because the cost per kilometer is very low. Ships are slower than land-based transportation, but unlike trains or trucks, they can cross oceans, such as to North America from Europe or Asia (Figure 11-12).
- **Air** is most expensive for all distances so is usually reserved for speedy delivery of small-bulk, high-value packages.

Modes of delivery are often mixed. For example, air-freight companies pick up packages in the afternoon and transport them by truck to the nearest airport. Late at night, planes filled with packages are flown to a central hub airport in the interior of the country, such as Memphis, Tennessee, or Louisville, Kentucky. The packages are transferred to other planes, flown to airports nearest their destination, transferred to trucks, and delivered the next morning.

Containerization has facilitated transfer of packages between modes. Containers may be packed into a rail car, transferred quickly to a container ship to cross the ocean, and unloaded onto trucks at the other end. Large ships have been specially built to accommodate large numbers of rectangular box-like containers.

Regardless of transportation mode, cost rises each time inputs or products are transferred from one mode to another. For example, workers must unload goods from a truck and then reload them onto a plane. The company may need to build or rent a warehouse to store goods temporarily after unloading from one mode and before loading to another mode. Some companies may calculate that the cost of one mode is lower for some inputs and products, whereas another mode may be cheaper for other goods. Many companies that use multiple transport modes locate at a **break-of-bulk point**, which is a location where transfer among transportation modes is possible. Important break-of-bulk points include seaports and airports. For example, a steel mill near the port of Baltimore receives iron ore by ship from South America and coal by train from Appalachia.

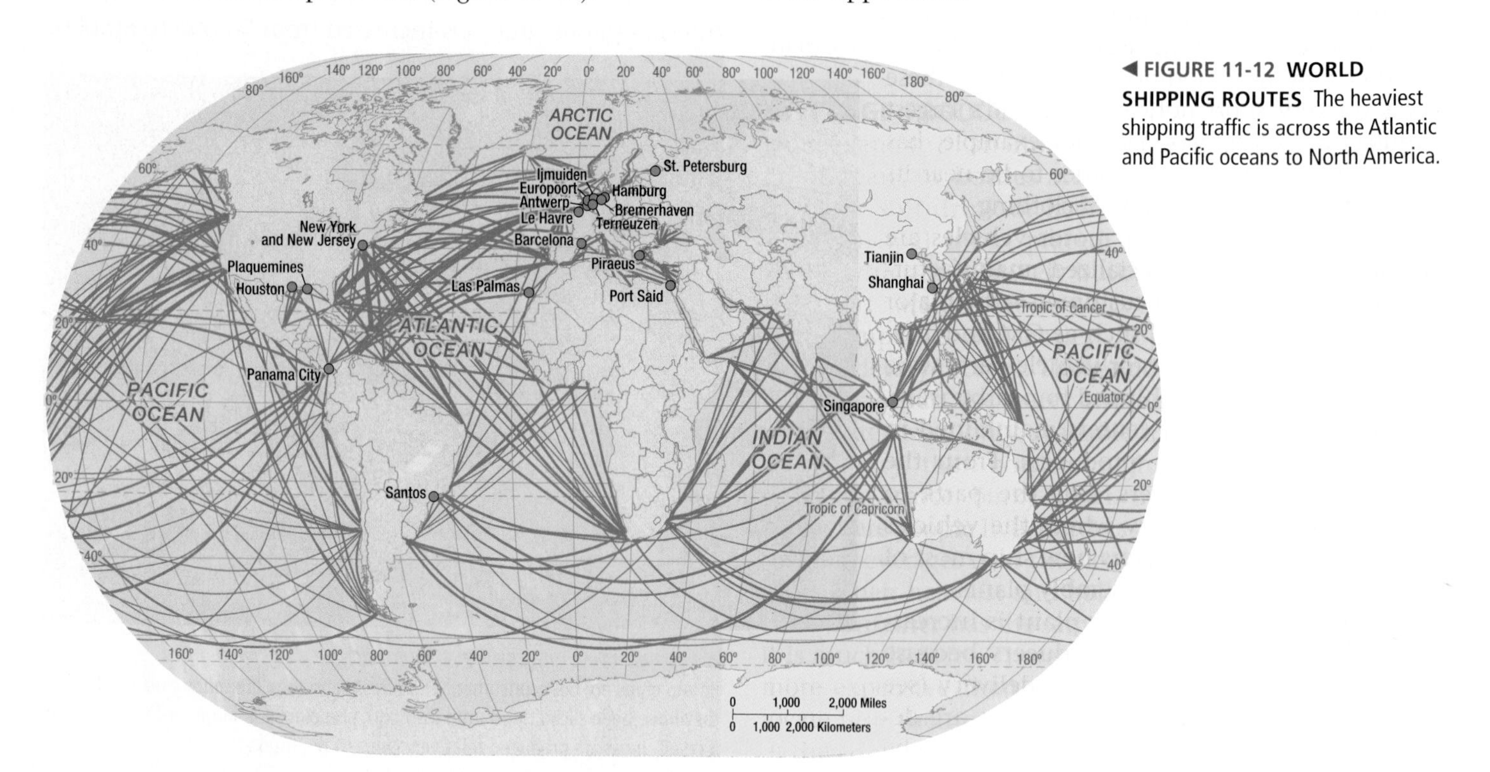

◀ **FIGURE 11-12 WORLD SHIPPING ROUTES** The heaviest shipping traffic is across the Atlantic and Pacific oceans to North America.

Changing Situation Factors in Key Industries

Each step in the production process can result in a different combination of situation factors. As a result, the optimal locations for the different steps can vary. In other cases, the relative importance of various situation factors can change over time, or their costs can change. If the mix of situation factors changes, the optimal location for an individual factory, or for an entire industry, can change.

COPPER: PROXIMITY TO INPUTS OR MARKETS?

Copper production involves several steps. The first three steps are good examples of bulk-reducing activities that need to be located near their sources of inputs (Figure 11-13). The fourth step is not bulk reducing, so does not need to be near inputs:

1. **Mining.** The first step in copper production is mining the copper ore. Mining in general is bulk reducing because the heavy, bulky ore extracted from mines is mostly waste, known as *gangue*. Copper ore mined in North America is especially low grade, less than 0.7 percent copper.
2. **Concentration.** Concentration mills crush and grind the ore into fine particles, mix them with water and chemicals, and filter and dry them. Copper concentrate is about 25 percent copper. Concentration mills are always near the mines because concentration transforms the heavy, bulky copper ore into a product of much higher value per weight (Figure 11-14).
3. **Smelting.** The concentrated copper becomes the input for smelters, which remove more impurities. Smelters produce copper matte (about 60 percent copper), blister copper (about 97 percent copper), and anode copper (about 99 percent copper). As another bulk-reducing industry, smelters are built near their main inputs—the concentration mills—again to minimize transportation cost.
4. **Refining.** The purified copper produced by smelters is treated at refineries to produce copper cathodes, about 99.99 percent pure copper. Most refineries are located near smelters.

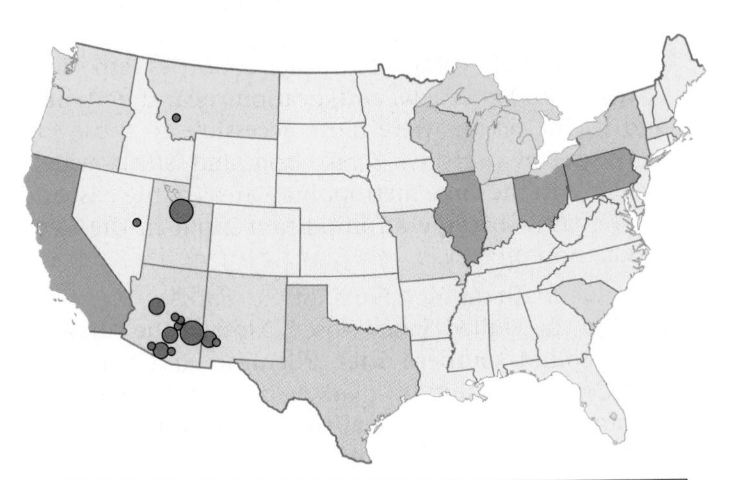

▲ **FIGURE 11-13 U.S. COPPER INDUSTRY** Copper mining, concentrating, and smelting are examples of bulk-reducing industries. In the United States, most plants that concentrate, smelt, and refine copper are in or near Arizona, where most copper mines are located. In contrast, most foundries, where copper products are manufactured, are located near markets in the East and West coasts.

▲ **FIGURE 11-14 COPPER MINING AND CONCENTRATION** Morenci Mine, Arizona, is the largest copper mine in the United States. Nearby are other bulk-reducing facilities, including the concentrator shown here.

Another important locational consideration is the source of energy to power these energy-demanding operations. In general, metal processors such as the copper industry try to locate near economical electrical sources and to negotiate favorable rates from power companies.

Figure 11-13 shows the distribution of the U.S. copper industry. Two-thirds of U.S. copper is mined in Arizona, so the state also has most of the concentration mills and smelters. Most foundries, where copper is manufactured, are located near markets on the East and West coasts.

Pause and Reflect 11.2.3
What is an example of a product purchased by consumers that is made of copper?

STEEL: CHANGING INPUTS

Learning Outcome 11.2.4
Describe how the optimal location for steel production has changed.

Steel is an alloy of iron that is manufactured by removing impurities in iron, such as silicon, phosphorus, sulfur, and oxygen, and adding desirable elements, such as manganese and chromium. Steel was a luxury item until Henry Bessemer (1813–1898) patented an efficient process for casting steel in 1855. The Bessemer process remained the most common method of manufacturing steel until the mid-twentieth century.

Steelmaking is an example of a bulk-reducing industry that traditionally located its facilities because of situation factors. Two changes in situation factors have influenced changes in the distribution of steel mills within the United States and worldwide:

- Changes in the relative importance of the main inputs.
- Increasing importance of proximity to markets rather than proximity to inputs.

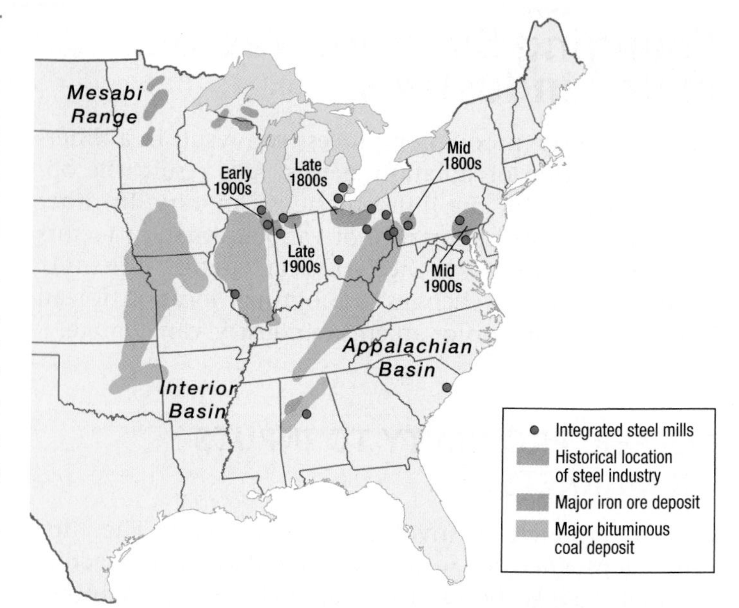

▲ **FIGURE 11-15 INTEGRATED STEEL MILLS IN THE UNITED STATES** Integrated steel mills are highly clustered near the southern Great Lakes, especially Lake Erie and Lake Michigan. Historically, the most critical factor in situating a steel mill was to minimize transportation cost for raw materials, especially heavy, bulky iron ore and coal. In recent years, many integrated steel mills have closed. Most surviving mills are in the Midwest to maximize access to consumers.

CHANGING DISTRIBUTION OF THE U.S. STEEL INDUSTRY. The two principal inputs in steel production are iron ore and coal. Because of the need for large quantities of bulky, heavy iron ore and coal, steelmaking traditionally clustered near sources of the two key raw materials. Within the United States, the distribution of steel production changed several times because of changing inputs (Figure 11-15):

- **Mid-nineteenth century: Southwestern Pennsylvania.** The U.S. steel industry concentrated around Pittsburgh in southwestern Pennsylvania because iron ore and coal were both mined there. The area no longer has steel mills, but it remains the center for research and administration.
- **Late nineteenth century: Lake Erie.** Steel mills were built around Lake Erie, in the Ohio cities of Cleveland, Youngstown, and Toledo, and near Detroit. The locational shift was largely influenced by the discovery of rich iron ore in the Mesabi Range, a series of low mountains in northern Minnesota. This area soon became the source for virtually all iron ore used in the U.S. steel industry. The ore was transported by way of Lake Superior, Lake Huron, and Lake Erie. Coal was shipped from Appalachia by train.
- **Early twentieth century: Southern Lake Michigan.** Most new steel mills were located near the southern end of Lake Michigan—in Gary, Indiana, Chicago, and other communities. The main raw materials continued to be iron ore and coal, but changes in steelmaking required more iron ore in proportion to coal. Thus, new steel mills were built closer to the Mesabi Range to minimize transportation cost. Coal was available from nearby southern Illinois, as well as from Appalachia.
- **Mid-twentieth century: East and West coasts.** Most new U.S. steel mills were located in communities near the East and West coasts, including Baltimore, Los Angeles, and Trenton, New Jersey. These coastal locations partly reflected further changes in transportation cost. Iron ore increasingly came from other countries, especially Canada and Venezuela, and locations near the Atlantic and Pacific oceans were more accessible to those foreign sources. Further, scrap iron and steel—widely available in the large metropolitan areas of the East and West coasts—became an important input in the steel-production process.
- **Late twentieth century: Proximity to markets.** Most steel mills in the United States closed. Most of the survivors were around southern Lake Michigan and along the East Coast. Proximity to markets has become more important than the traditional situation factor of proximity to inputs. Coastal plants provide steel to large East Coast population centers, and southern Lake Michigan plants are centrally located to distribute their products countrywide.

The increasing importance of proximity to markets is also demonstrated by the recent growth of steel minimills, which have captured one-fourth of the U.S. steel market (Figure 11-16). Rather than iron ore and coal, the main input into minimill production is scrap metal. In the past, most steel was produced at large integrated mill complexes. They processed iron ore, converted coal into coke, converted the iron into steel, and formed the steel into sheets, beams, rods, or other shapes. Minimills, generally limited to one step in the process—steel production—are

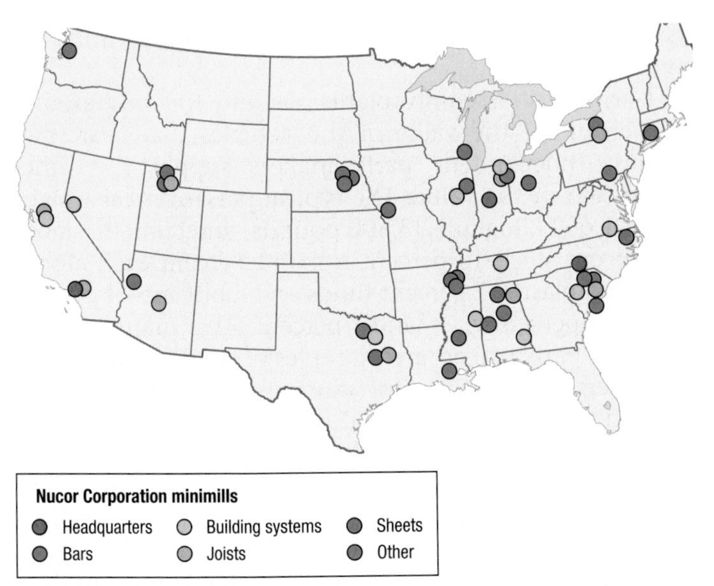

▲ **FIGURE 11-16 MINIMILLS** Minimills, which produce steel from scrap metal, are more numerous than integrated steel mills, and they are distributed around the country near local markets. Shown are the plants of Nucor, the largest minimill operator in the United States.

less expensive than integrated mills to build and operate, and they can locate near their markets because their main input—scrap metal—is widely available.

CHANGING DISTRIBUTION OF THE WORLD STEEL INDUSTRY. The shift of world manufacturing to new industrial regions can be seen clearly in steel production. In 1980, 80 percent of world steel was produced in developed countries and 20 percent in developing countries (Figure 11-17, top). Between 1980 and 2010, the share of world steel production declined to 37 percent in developed countries and increased to 68 percent in developing countries (Figure 11-17, bottom).

World steel production doubled between 1980 and 2010, from around 700 million to around 1,400 million metric tons. China was responsible for 600 million of the 700 million metric ton increase, and other developing countries (primarily India and South Korea) for the other 100 million (Figure 11-18). Production in developed countries remained unchanged, at approximately 100 million metric tons.

China's steel industry has grown in part because of access to the primary inputs iron ore and coal. However, the principal factor in recent years has been increased demand by growing industries in China that use a lot of steel, such as motor vehicles.

Pause and Reflect 11.2.4
Although Pittsburgh's football team is named "Steelers," based on Figure 11-15, what city's team might be more appropriately given this nickname?

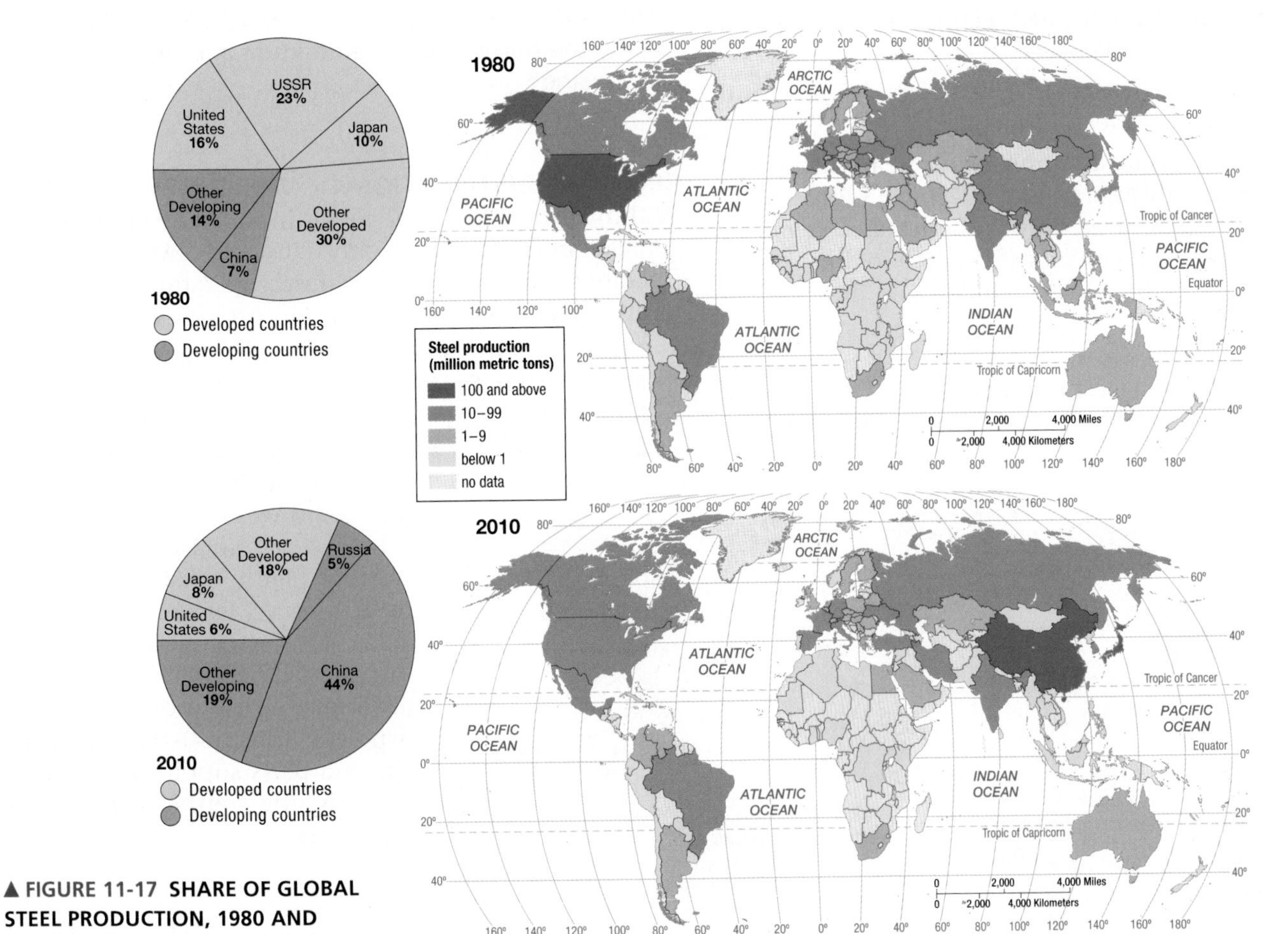

▲ **FIGURE 11-17 SHARE OF GLOBAL STEEL PRODUCTION, 1980 AND 2010** The share of world steel produced in developing countries increased from 21 percent in 1980 to 68 percent in 2010.

▲ **FIGURE 11-18 WORLD STEEL PRODUCTION, 1980 AND 2010** The leading steel producer in 1980 was the United States, and in 2010 it was China.

MOTOR VEHICLES: CHANGING MARKETS

Learning Outcome 11.2.5
Explain the distribution of motor vehicle production.

The motor vehicle is a prominent example of a fabricated metal product, described earlier as one of the main types of bulk-gaining industries. Motor vehicles are therefore built near their markets. As the markets for new cars change, the distribution of factories changes.

GLOBAL DISTRIBUTION OF VEHICLE PRODUCTION. Carmakers manufacture vehicles at final assembly plants, using thousands of parts supplied by independent companies. The world's three major industrial regions house 80 percent of the world's final assembly production, including 40 percent in East Asia, 25 percent in Europe, and 15 percent in North America (Figure 11-19). Most assembly plants are clustered in these three regions because most of the world's car buyers are there.

Ten carmakers control 85 percent of the world's sales:

- Two based in North America: Ford and GM.
- Four based in Europe: Germany's Volkswagen, Italy's Fiat (which controls Chrysler), France's Renault (which controls Nissan) and Peugeot.
- Four based in East Asia: Japan's Toyota, Honda, and Suzuki and South Korea's Hyundai.

These carmakers operate assembly plants in at least two of the three major industrial regions (Figure 11-20). Three-fourths of vehicles sold in North America are assembled in North America. Similarly, most vehicles sold in Europe are assembled in Europe, most vehicles sold in Japan are assembled in Japan, and most vehicles sold in China are assembled in China.

Carmakers' assembly plants account for only around 30 percent of the value of the vehicles that bear their names. Independent parts makers supply the other 70 percent of the value. The typical passenger car weighs about 1,600 kilograms (3,500 pounds) and contains about 45 percent steel, 13 percent iron, 11 percent each aluminum and plastic, 7 percent fluids and lubricants, 4 percent rubber, 2 percent glass, and 7 percent other materials.

Many parts makers are examples of single-market manufacturers because they ship most of their products to one or perhaps a handful of final assembly plants. As single-market manufacturers, parts makers cluster near the final assembly plants. Motor vehicle seats, for example, are invariably manufactured within an hour of the final assembly plant. A seat is an especially large and bulky object, and carmakers do not want to waste valuable space in their assembly plants by piling up an inventory of them.

On the other hand, some parts do not need to be manufactured close to the customer. For them, changing site factors are more important, discussed beginning on the next page. Some locate in countries that have relatively low labor costs, such as Mexico, China, and Czech Republic.

Pause and Reflect 11.2.5
Why is the percentage of steel in vehicles declining, while the percentage of aluminum and plastic is increasing?

REGIONAL DISTRIBUTION OF VEHICLE PRODUCTION. Within each of the three major industrial regions, motor vehicle production is highly clustered. Because a final assembly plant is a bulk-gaining operation, its critical location factor is minimizing transportation to the market:

- **North America.** Most of the assembly and parts plants are located in the interior of the United States, between Michigan and Alabama, centered in a corridor known as "auto alley," formed by north–south interstate highways 65 and 75, with an extension into southwestern Ontario (Figure 11-21). The principal cluster of assembly plants outside auto alley is in central Mexico. Within auto alley, U.S.-owned carmakers and suppliers have clustered in Michigan and nearby northern states, whereas foreign-owned carmakers and parts suppliers have clustered in the southern portion of auto alley.

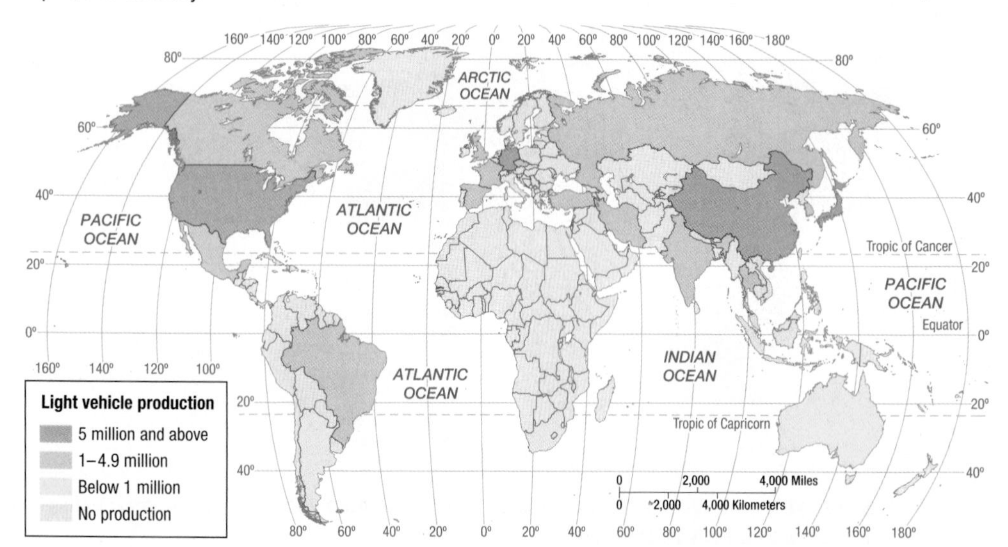

▼ **FIGURE 11-19 MOTOR VEHICLE PRODUCTION**
China is the world's leading producer of cars, followed by the United States, Japan, and Germany.

◀ **FIGURE 11-20 ASSEMBLY PLANT IN EUROPE** Toyota's factory near Burnaston, in the United Kingdom, is surrounded by farmland.

- **Europe.** Most plants are clustered in an east–west corridor between the United Kingdom and Russia (Figure 11-22). Germany is the leading producer of vehicles in Europe. Since the end of communism in Eastern Europe in the early 1990s, that region has had most of the growth in vehicle production. The large carmakers have modernized inefficient Communist-era factories or built entirely new ones in Eastern Europe. Labor costs are lower there than in Western Europe, and demand for vehicles has increased with the end of Communist restrictions on the ability of private individuals to buy consumer goods such as cars.
- **East Asia.** China's assembly plants are clustered in the east in order to be near the major population centers (Figure 11-23). Most car buyers in China are located in the large cities, such as Shanghai and Beijing.

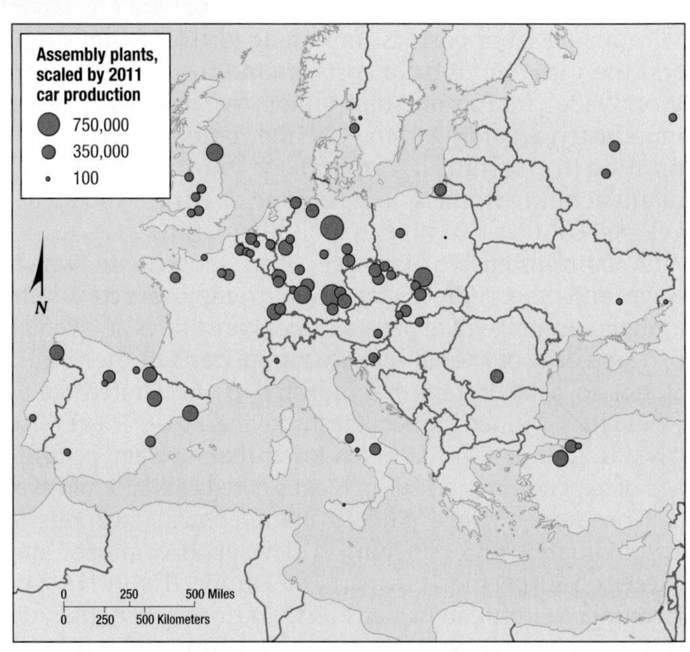

▲ **FIGURE 11-22 MOTOR VEHICLE PRODUCTION IN EUROPE** Within Europe, most vehicles are produced in an east–west corridor centered on Germany.

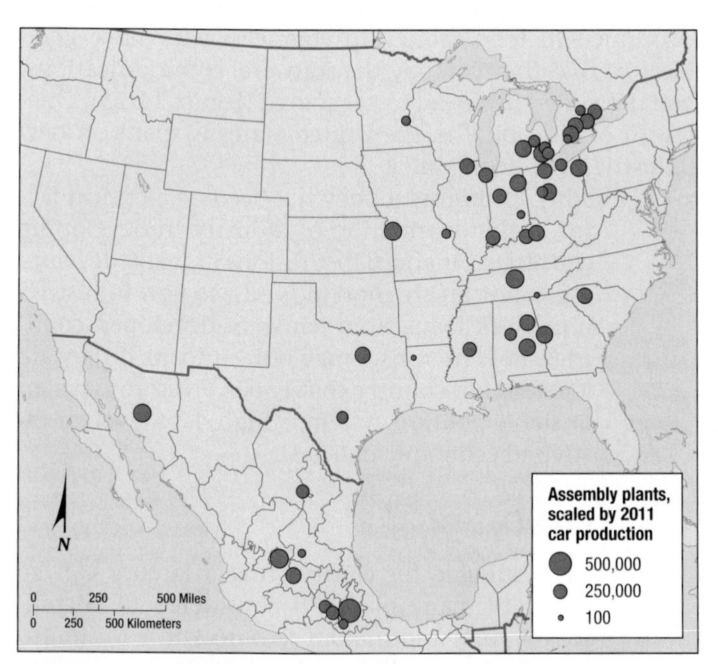

▲ **FIGURE 11-21 MOTOR VEHICLE PRODUCTION IN NORTH AMERICA** Most vehicles are produced in auto alley. Most U.S.-owned companies are clustered in the north, and most foreign-owned ones in the south.

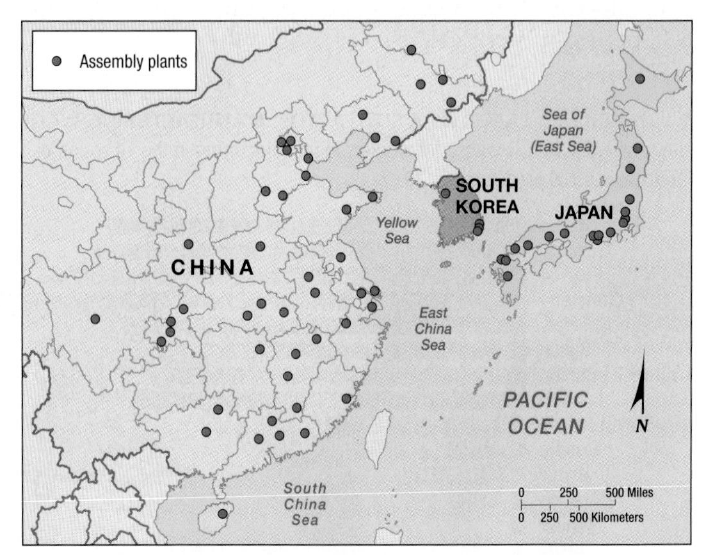

▲ **FIGURE 11-23 MOTOR VEHICLE PRODUCTION IN EAST ASIA** Most vehicles are produced near major metropolitan areas, especially in western China.

Site Factors

Learning Outcome 11.2.6
List the three types of site factors.

Firms take into consideration site factors as well as situation factors (see the Contemporary Geography Tools feature). Labor, capital, and land are the three traditional production factors that may vary among locations.

LABOR

The most important site factor on a global scale is labor. Minimizing labor costs is important for some industries, and the variation of labor costs around the world is large. Worldwide, around one-half billion workers are engaged in industry, according to the UN International Labor Organization (ILO). China has around one-fourth of the world's manufacturing workers, India around one-fifth, and all developed countries combined around one-fifth.

A **labor-intensive industry** is an industry in which wages and other compensation paid to employees constitute a high percentage of expenses. Labor constitutes an average of 11 percent of overall manufacturing costs in the United States, so a labor-intensive industry in the United States would have a much higher percentage than that. The reverse case, an industry with a much lower-than-average percentage of expenditures on labor, is considered capital intensive.

The average wage paid to manufacturing workers is approximately $35 per hour in developed countries and exceeds $40 per hour in parts of Europe (Figure 11-24). Health-care, retirement pensions, and other benefits add substantially to the compensation. In China and India, average wages are approximately $1 per hour and include limited additional benefits. For some manufacturers—but not all—the difference between paying workers $1 and $35 per hour is critical.

▼ **FIGURE 11-24 LABOR AS A SITE FACTOR: MANUFACTURING WAGES** The chart shows average hourly wages for workers in manufacturing in the 14 countries with the largest industrial production in 2010.

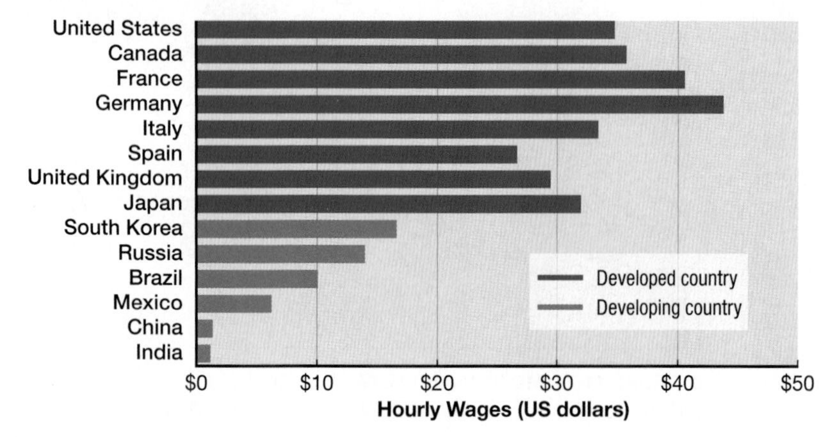

A labor-intensive industry is not the same as a high-wage industry. "Labor-intensive" is measured as a percentage, whereas "high-wage" is measured in dollars or other currencies. For example, motor-vehicle workers are paid much higher hourly wages than textile workers, yet the textile industry is labor intensive, and the auto industry is not. Although auto workers earn relatively high wages, most of the value of a car is accounted for by the parts and the machinery needed to put together the parts. On the other hand, labor accounts for a large percentage of the cost of producing a towel or shirt compared with materials and machinery.

Pause and Reflect 11.2.6
Labor accounts for around 5 percent of the cost of manufacturing a car. Does this mean that motor vehicle manufacturing is a labor-intensive industry? Explain.

CAPITAL

Manufacturers typically borrow capital—the funds to establish new factories or expand existing ones. The U.S. motor-vehicle industry concentrated in Michigan early in the twentieth century largely because that region's financial institutions were more willing than eastern banks to lend money to the industry's pioneers. The most important factor in the clustering of high-tech industries in California's Silicon Valley—even more important than proximity to skilled labor—was the availability of capital. Banks in Silicon Valley have long been willing to provide money for new software and communications firms, even though lenders elsewhere have hesitated. High-tech industries have been risky propositions—roughly two-thirds of them fail—but Silicon Valley financial institutions have continued to lend money to engineers who have good ideas so that they can buy the software, communications, and networks they need to get started (Figure 11-25). One-fourth of all capital in the United States is spent on new industries in Silicon Valley.

The ability to borrow money has become a critical factor in the distribution of industry in developing countries. Financial institutions in many developing countries are short of funds, so new industries must seek loans from banks in developed countries. But enterprises may not get loans if they are located in a country that is perceived to have an unstable political system, a high debt level, or ill-advised economic policies.

LAND

Land suitable for constructing a factory can be found in many places. If considered to encompass natural and human resources in addition to terra firma, "land" is a critical site factor.

Early factories located inside cities due to a combination of situation and site factors. A city

▲ **FIGURE 11-25 CAPITAL AS A SITE FACTOR: SILICON VALLEY**
A Google employee bicycles to work past the Green Android statue at Googleplex, Google's world headquarters in Mountain View, California, in the heart of Silicon Valley.

offered an attractive situation—proximity to a large local market and convenience in shipping to a national market by rail. A city also offered an attractive site—proximity to a large supply of labor as well as to sources of capital. The site factor that cities have always lacked is abundant land. To get the necessary space in cities, early factories were typically multistory buildings. Raw materials were hoisted to the upper floors to make smaller parts, which were then sent downstairs on chutes and pulleys for final assembly and shipment. Water was stored in tanks on the roof.

Contemporary factories operate most efficiently when laid out in one-story buildings (see for example,Figure 11-20). Raw materials are typically delivered at one end and moved through the factory on conveyors or forklift trucks. Products are assembled in logical order and shipped out at the other end. The land needed to build one-story factories is now more likely to be available in suburban and rural locations. Also, land is much cheaper in suburban and rural locations than near the center of a city.

In addition to providing enough space for one-story buildings, locations outside cities are also attractive because they facilitate delivery of inputs and shipment of products. In the past, when most material moved in and out of a factory by rail, a central location was attractive because rail lines converged there. With trucks now responsible for transporting most inputs and products, proximity to major highways is more important for a factory. Especially attractive is the proximity to the junction of a long-distance route and the beltway, or ring road, that encircles most cities. Thus, factories cluster in industrial parks located near suburban highway junctions.

CONTEMPORARY GEOGRAPHIC TOOLS

Honda Selects a Factory Location

When Honda decided that it needed another assembly plant in the United States, it applied situation and site factors to select a location for the factory:

- **Situation factors were considered first:**
  - **Proximity to markets.** To minimize the cost of shipping vehicles, Honda looked for locations within auto alley (Figure 11-26).
  - **Proximity to inputs.** Honda's most important inputs, the engine and transmission, were to come from existing factories in western Ohio. That guided Honda to the portion of auto alley encompassing Illinois, Indiana, and Ohio.
- **Site factors helped Honda find specific locations within auto alley:**
  - **Land.** Honda wanted a large tract of land near at least one interstate highway and a rail line.
  - **Labor.** Honda needed a large labor supply within a one-hour commuting range, but it didn't want to compete for workers with existing assembly plants. That could lead to a shortage of skilled workers and push up wages. So Honda looked for areas outside the one-hour commuting range around existing assembly plants.

Honda's short list of locations included Decatur in eastern Illinois, Greensburg in southwestern Indiana, and unnamed communities in west-central Ohio. Honda considered Indiana the safest choice, because the governors of the other two states at the time were involved in financial scandals.

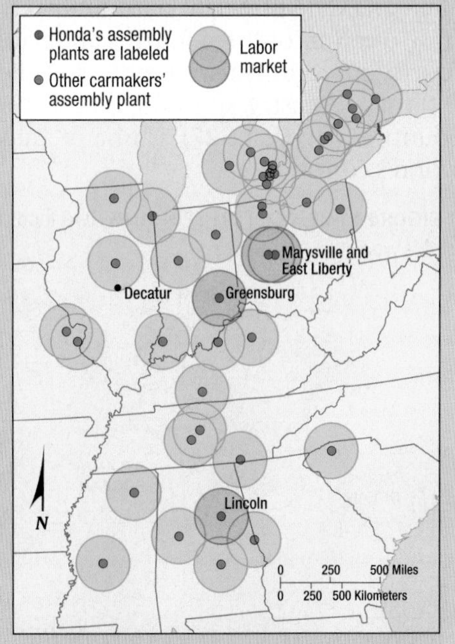

▲ **FIGURE 11-26 HONDA PICKS AN ASSEMBLY PLANT SITE** An assembly plant draws its workforce from within a radius of roughly one hour. New plants have been located outside the labor market areas of existing plants to minimize competition for workers.

TEXTILES AND APPAREL: CHANGING INPUTS

Learning Outcome 11.2.7
Explain the distribution of textile and apparel production.

Production of **textiles** (woven fabrics) and **apparel** (clothing) is a prominent example of an industry that generally requires less-skilled, low-cost workers. The textile and apparel industry accounts for 6 percent of the dollar value of world manufacturing but a much higher 14 percent of world manufacturing employment, an indicator that it is a labor-intensive industry. The percentage of the world's women employed in this type of manufacturing is even higher.

Textile and apparel production involves three principal steps:

- Spinning of fibers and other preparatory work to make yarn from natural or human-made materials
- Weaving or knitting of yarn into fabric (as well as finishing of fabric by bleaching or dyeing)
- Cutting and sewing of fabric for assembling into clothing and other products

Spinning, weaving, and sewing are all labor intensive compared to other industries, but the importance of labor varies somewhat among them. As a result, their global distributions are not identical because the three steps are not equally labor intensive.

SPINNING. Fibers can be spun from natural or synthetic elements. The principal natural fiber is cotton. Synthetics now account for three-fourths and natural fibers only one-fourth of world thread production. Because it is a labor-intensive industry, spinning is done primarily in low-wage countries (Figure 11-27). China produces two-thirds of the world's cotton thread.

TEXTILE AND APPAREL WEAVING. For thousands of years, fabric has been woven or laced together by hand on a loom, which is a frame on which two sets of threads are placed at right angles to each other. One set of threads, called the warp, is strung lengthwise. A second set of threads, called the weft, is carried in a shuttle that is inserted over and under the warp. Because the process of weaving by hand is physically hard work, weavers were traditionally men.

For mechanized weaving, labor constitutes a high percentage of the total production cost. Consequently, weaving is highly clustered in low-wage countries (Figure 11-28). Despite their remoteness from European and North American markets, China and India have become the dominant fabric producers because their lower labor costs offset the expense of shipping inputs and products long distances. China accounts for nearly 60 percent of the world's woven cotton fabric production and India another 30 percent.

TEXTILE AND APPAREL ASSEMBLY. Sewing is probably an even older human activity than spinning and weaving. Needles made from animal horns or bones date back tens of thousands of years, and iron needles date from the fourteenth century.

The first functional sewing machine was invented by French tailor Barthelemy Thimonnier in 1830. In 1841, Thimonnier installed 80 sewing machines in a factory in St.-Etienne, France, to sew uniforms for the French army. However, Parisian tailors, fearing that the machines would put them out of work, stormed the factory and destroyed the machines. Isaac Singer manufactured the first commercially successful sewing machine in the United States during the 1850s, but he was convicted of infringing a patent filed by Elias Howe in 1846.

Textiles are assembled into four main types of products: garments, carpets, home products such as bed linens and curtains, and industrial items such as headliners for inside motor vehicles. Developed countries play a larger role in

▼ **FIGURE 11-27 COTTON SPINNING** Two-thirds of world cotton yarn is produced in China, including by this woman.

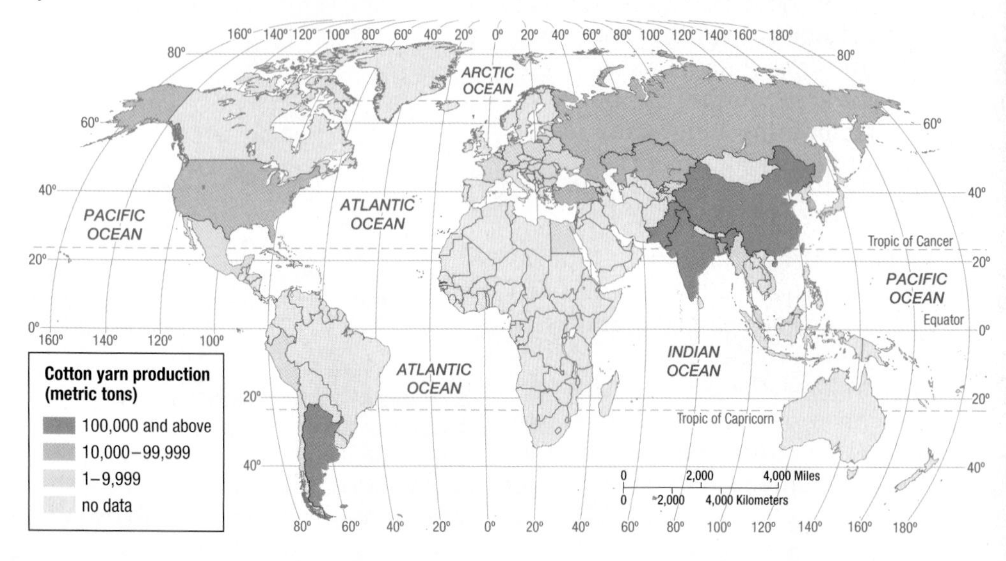

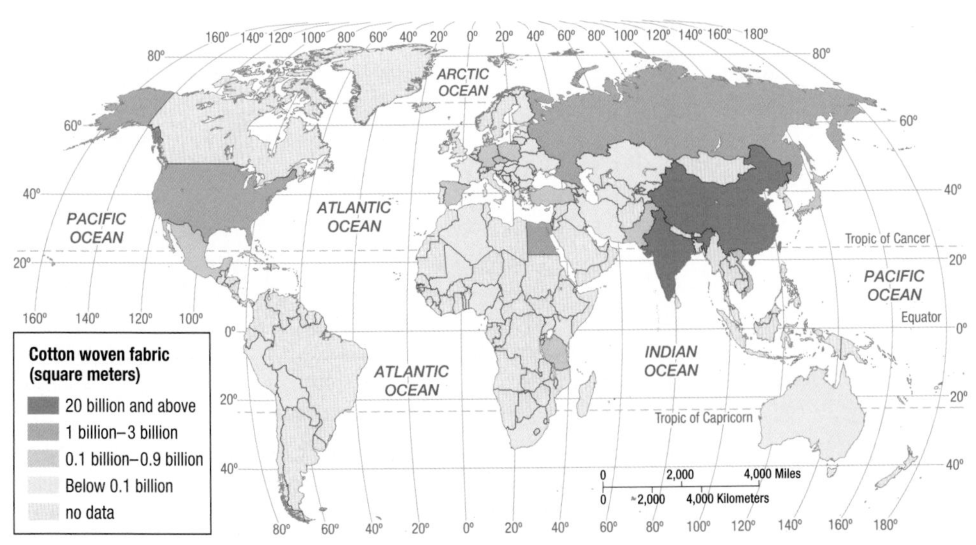

▲ **FIGURE 11-28 COTTON WEAVING** China and India together account for nearly 90 percent of the world's woven cotton production. In the image, cotton is being woven in China.

assembly than in spinning and weaving because most of the consumers of assembled products are located in developed countries (Figure 11-29). For example, two-thirds of the women's blouses sold worldwide in a year are sewn in developed countries.

Pause and Reflect 11.2.7
Check the labels on the clothes you are wearing. Where were they made?

CHECK-IN: KEY ISSUE 2

Why Are Situation and Site Factors Important?

- ✓ **Situation factors involve transporting materials to and from a factory.**
- ✓ **Bulk-reducing industries are located near their sources of inputs.**
- ✓ **Bulk-gaining, single-market, and perishable industries locate near their markets.**
- ✓ **Site factors derive from distinctive features of a particular place, including labor, capital, and land.**

▼ **FIGURE 11-29 DISTRIBUTION OF WOMEN'S BLOUSE PRODUCTION** The United States is the leading producer of women's blouses. These women are sewing blouses in China, which is the leading producer among developing countries.

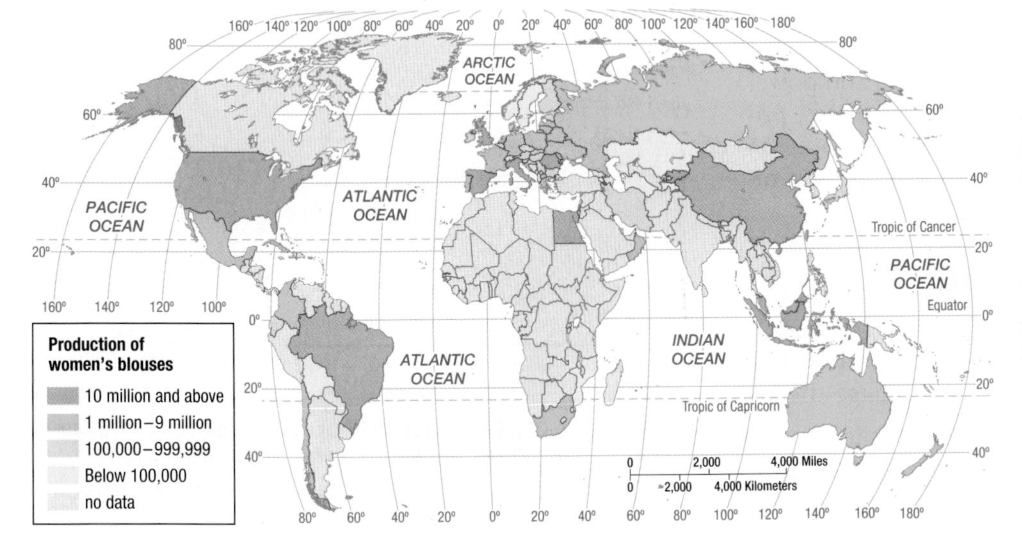

KEY ISSUE 3

Where Does Industry Cause Pollution?

- Air Pollution
- Solid Waste Pollution
- Water Pollution

Learning Outcome 11.3.1
Describe causes and effects of global warming and damage to the ozone layer.

Industry is a major polluter of air, water, and land. People rely on air, water, and land to remove and disperse waste from factories as well as from other human activities. Pollution occurs when more waste is added than air, water, and land resources can handle.

As a country's per capita income increases, its per capita carbon dioxide emissions also increase. Some of the wealthiest countries, located primarily in Europe, with gross national income (GNI) per capita between $30,000 and $50,000, show declines in pollution. However, the world's richest countries, including the United States and several countries in Southwest Asia, display the highest pollution levels (Figure 11-30).

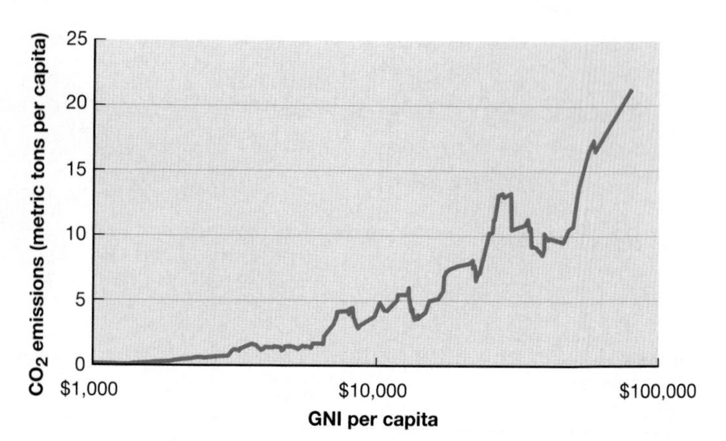

▲ **FIGURE 11-30 GNI AND POLLUTION** Carbon dioxide emissions generally increase with rising income. The principal exception is in Europe, where some relatively wealthy countries have curbed emissions.

Air Pollution

At ground level, Earth's average atmosphere is made up of about 78 percent nitrogen, 21 percent oxygen, and less than 1 percent argon. The remaining 0.04 percent includes several trace gases, some of which are critical. **Air pollution** is concentration of trace substances at a greater level than occurs in average air. Concentrations of these trace gases in the air can damage property and adversely affect the health of people, other animals, and plants.

Most air pollution is generated from factories and power plants, as well as from motor vehicles. Factories and power plants produce sulfur dioxides and solid particulates, primarily from burning coal. Burning petroleum in motor vehicles produces carbon monoxide, hydrocarbons, and nitrogen oxides.

GLOBAL-SCALE AIR POLLUTION

Air pollution concerns geographers at three scales—global, regional, and local. At the global scale, air pollution may contribute to global warming. It may also damage the atmosphere's ozone layer.

GLOBAL WARMING. The average temperature of Earth's surface has increased by 1°C (2°F) since 1880 (Figure 11-31). Human actions, especially the burning of fossil fuels in factories and vehicles, may have caused this.

Earth is warmed by sunlight that passes through the atmosphere, strikes the surface, and is converted to heat. When the heat tries to pass back through the atmosphere to space, some gets through and some is trapped. This process keeps Earth's temperatures moderate and allows life to flourish on the planet. A concentration of trace gases in the atmosphere can block or delay the return of some of the heat leaving the surface heading for space, thereby raising Earth's temperatures. When fossil fuels are burned, one of the trace gases, carbon dioxide, is discharged into the atmosphere. Plants and oceans absorb much of the discharges, but increased fossil fuel burning during the past 200 years, as shown in Figure 11-30, has caused the level of carbon dioxide in the atmosphere to rise by more than one-fourth, according to the UN Intergovernmental Panel on Climate Change.

The anticipated increase in Earth's temperature, caused by carbon dioxide and other greenhouse gases trapping some of the radiation emitted by the surface, is called the **greenhouse effect**. The term is somewhat misleading because a greenhouse does not work in the same way as do trace gases in the atmosphere. In a real greenhouse, the interior gets very warm when the windows remain closed on a sunny day. The Sun's light energy passes through the glass

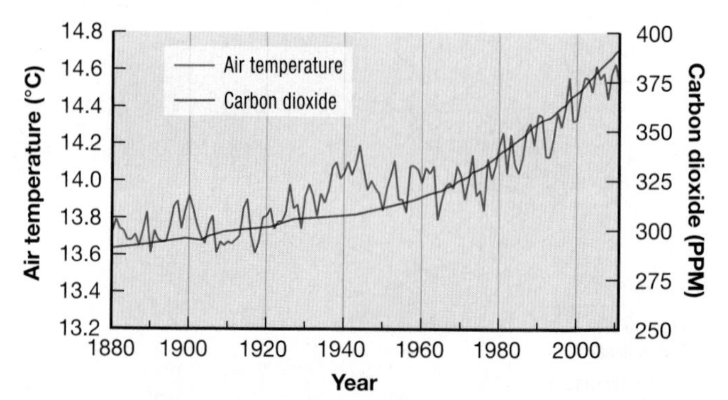

▲ **FIGURE 11-31 GLOBAL-SCALE AIR POLLUTION: GLOBAL WARMING AND CARBON DIOXIDE CONCENTRATIONS, 1880–2010** Since 1880, carbon dioxide concentration has increased by more than one-third, and Earth has warmed by about 1°C (2°F).

into the greenhouse and is converted to heat, and the heat trapped inside the building is unable to escape out through the glass. Although this is an imprecise analogy, "greenhouse effect" is a term that has been widely adopted to describe the anticipated warming of Earth's surface when trace gases block some of the heat trying to escape into space.

Regardless of what it is called, global warming of only a few degrees could melt the polar ice sheets and raise the level of the oceans many meters (Figure 11-32). Coastal cities such as New York, Los Angeles, Rio de Janeiro, and Hong Kong would flood (see the Sustainability and Inequality in Our Global Village feature). Global patterns of precipitation could shift: Some deserts could receive more rainfall, and currently productive agricultural regions, such as the U.S. Midwest, could become too dry for farming. Humans can adapt to a warmer planet, but the shifts in coastlines and precipitation patterns could require massive migration and could be accompanied by political disputes.

GLOBAL-SCALE OZONE DAMAGE. Earth's atmosphere has zones with distinct characteristics. The stratosphere—the zone 15 to 50 kilometers (9 to 30 miles) above Earth's surface—contains a concentration of **ozone** gas. The ozone layer absorbs dangerous ultraviolet (UV) rays from the Sun. Were it not for the ozone in the stratosphere, UV rays would damage plants, cause skin cancer, and disrupt food chains.

◀ **FIGURE 11-32 RECEDING NORTH POLAR ICE SHEET** These images taken by NASA show that between 1979 (top) and 2005 (bottom), the north polar ice sheet melted visibly.

Earth's protective ozone layer is threatened by pollutants called **chlorofluorocarbons (CFCs)**. CFCs such as Freon were once widely used as coolants in refrigerators and air conditioners. When they leak from these appliances, the CFCs are carried into the stratosphere, where they break down Earth's protective layer of ozone gas. In 2007, virtually all countries of the world agreed to cease using CFCs, by 2020 in developed countries and by 2030 in developing countries.

Pause and Reflect 11.3.1

What gas is now most commonly used as a coolant instead of CFC? Google "what replaced CFCs?"

SUSTAINABILITY AND INEQUALITY IN OUR GLOBAL VILLAGE

Climate Change in the South Pacific

One consequence of global warming is a rise in the level of the oceans. The large percentage of the world's population—including one-half of Americans—who live near the sea face increased threat of flooding. The threat is especially severe for island countries in the Pacific Ocean; they could be wiped off the map entirely.

Kiribati is a collection of approximately 32 small islands, one of the world's most isolated countries (Figure 11-33). Despite its extreme isolation, global forces threaten Kiribati's existence. Rising sea levels due to global warming threaten Kiribati because the entire country is within a few meters of sea level. Two of Kiribati's islands—Tebua Tarawa and Abanuea—have already disappeared.

Kiribati and other Pacific island microstates are atolls—that is, islands made of coral reefs. A coral is a small sedentary marine animal that has a horny or calcareous skeleton. Corals form colonies, and the skeletons build up to form coral reefs. Coral is very fragile. Humans are attracted to coral for its beauty and the diversity of species it supports, but handling coral can kill it. The threat of global warming to coral is especially severe: Coral stays alive in only a narrow range of ocean temperatures, between 23°C and 25°C (between 73°F and 77°F), so global warming threatens the ecology of Kiribati, even if it remains above sea level.

Kiribati has an emergency response to rising sea levels. The government has negotiated with Fiji to purchase 2,000 hectares (5,000 acres) of land on the island of Vanua Levu to relocate people from Kiribati someday.

◀ **FIGURE 11-33 KIRIBATI** Global warming may cause the oceans to rise, submerging small island countries such as Kiribati.

REGIONAL-SCALE AIR POLLUTION

Learning Outcome 11.3.2
Describe causes and effects of regional and local-scale air pollution and solid waste pollution.

At the regional scale, air pollution may damage a region's vegetation and water supply through acid deposition. The world's three principal industrial regions are especially affected by acid deposition.

Sulfur oxides and nitrogen oxides, emitted by burning fossil fuels, enter the atmosphere, where they combine with oxygen and water. Tiny droplets of sulfuric acid and nitric acid form and return to Earth's surface as **acid deposition**. When dissolved in water, the acids may fall as **acid precipitation**—rain, snow, or fog. The acids can also be deposited in dust. Before they reach the surface, these acidic droplets might be carried hundreds of kilometers.

Acid precipitation damages lakes, killing fish and plants. On land, concentrations of acid in the soil can injure plants by depriving them of nutrients and can harm worms and insects. Buildings and monuments made of marble and limestone have suffered corrosion from acid rain.

Geographers are particularly interested in the effects of acid precipitation because the worst damage is not experienced at the same location as the emission of the pollutants. Within the United States the major generators of acid deposition are in Ohio and other industrial states along the southern Great Lakes. However, the severest effects of acid rain are felt in several areas farther east. The United States reduced sulfur dioxide emissions significantly during the late twentieth century (Figure 11-34).

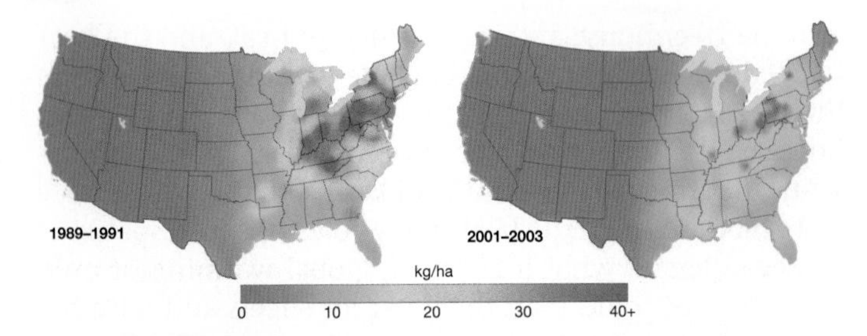

▲ **FIGURE 11-34 REGIONAL-SCALE AIR POLLUTION: ACID DEPOSITION IN THE UNITED STATES** As a result of emissions controls, the rate of acid deposition has declined.

LOCAL-SCALE AIR POLLUTION

At the local scale, air pollution is especially severe in places where emission sources are concentrated, such as in urban areas. The air above urban areas may be polluted because a large number of factories, motor vehicles, and other polluters emit residuals in a concentrated area. Urban air pollution has three basic components:

- **Carbon monoxide.** Breathing carbon monoxide reduces the oxygen level in blood, impairs vision and alertness, and threatens those with breathing problems.
- **Hydrocarbons.** In the presence of sunlight, hydrocarbons, as well as nitrogen oxides, form **photochemical smog**, which causes respiratory problems, stinging in the eyes, and an ugly haze over cities.
- **Particulates.** They include dust and smoke particles. The dark plume of smoke from a factory stack and the exhaust of a diesel truck are examples of particulate emission.

The worst urban air pollution occurs when winds are slight, skies are clear, and a temperature inversion exists. When the wind blows, it disperses pollutants; when it is calm, pollutants build. Sunlight provides the energy for the formation of smog. Air is normally cooler at higher elevations, but during temperature inversions—in which air is warmer at higher elevations—pollutants are trapped near the ground.

According to the American Lung Association, the worst area in the United States for concentrations of particulates is in southern California, including Los Angeles and nearby communities. Worldwide, according to the World Health Organization, the 10 most polluted cities are all in developing regions, including 4 each in Iran and South Asia. Mexico City is an example of a city in a developing country that has improved its air quality since the 1990s (Figure 11-35).

Pause and Reflect 11.3.2
What environmental features can be seen in Mexico City on a clear day but not during smog periods? What is their role in the city's air pollution problem?

Progress in controlling urban air pollution is mixed. In developed countries, air has improved where strict clean-air regulations are enforced. Limited emission controls in developing countries are contributing to severe urban air pollution. Changes in manufacturing processes, motor vehicle engines, and electric generation have all helped. For example, since the 1970s, when the U.S. government began to require catalytic converters on motor vehicles, carbon monoxide emissions have been reduced by more than three-fourths, and nitrogen oxide and hydrocarbon emissions have been reduced by more than 95 percent. But more people are driving, offsetting gains made by emission controls.

Solid Waste Pollution

About 2 kilograms (4 pounds) of solid waste per person is generated daily in the United States, about 60 percent from residences and 40 percent from businesses. Paper products, such as corrugated cardboard and newspapers, account for the largest percentage of solid waste in the United States, especially among residences and retailers. Manufacturers discard large quantities of metals as well as paper.

SANITARY LANDFILL

Using a **sanitary landfill** is by far the most common strategy for disposal of solid waste in the United States: More than one-half of the country's waste is trucked to landfills

▲ **FIGURE 11-35 LOCAL-SCALE AIR POLLUTION: MEXICO CITY SMOG** Downtown Mexico City without smog (left) and with smog (right).

and buried under soil. But the number of landfills in the United States has declined by three-fourths since 1990.

Given the shortage of space in landfills, alternatives have been sought to disposal of solid waste. A rapidly growing alternative is incineration. Burning trash reduces its bulk by about three-fourths, and the remaining ash demands less landfill space. Incineration also provides energy: The incinerator's heat can boil water to produce steam heat or operate a turbine that generates electricity.

HAZARDOUS WASTE

Disposing of hazardous waste is especially difficult. Hazardous wastes include heavy metals (including mercury, cadmium, and zinc), PCB oils from electrical equipment, cyanides, strong solvents, acids, and caustics. These may be unwanted by-products generated in manufacturing or waste to be discarded after usage.

According to the toxic waste inventory published by the U.S. Environmental Protection Agency (EPA), 1.78 billion kilograms (3.93 billion pounds) of toxic chemicals were released into the environment in 2010. Mining operations were the largest polluters. Ohio had 10 of the 100 largest polluting firms (Figure 11-36).

If poisonous industrial residuals are not carefully placed in protective containers, the chemicals may leach into the soil and contaminate groundwater or escape into the atmosphere. Breathing air or consuming water contaminated with toxic wastes can cause cancer, mutations, chronic ailments, and even immediate death.

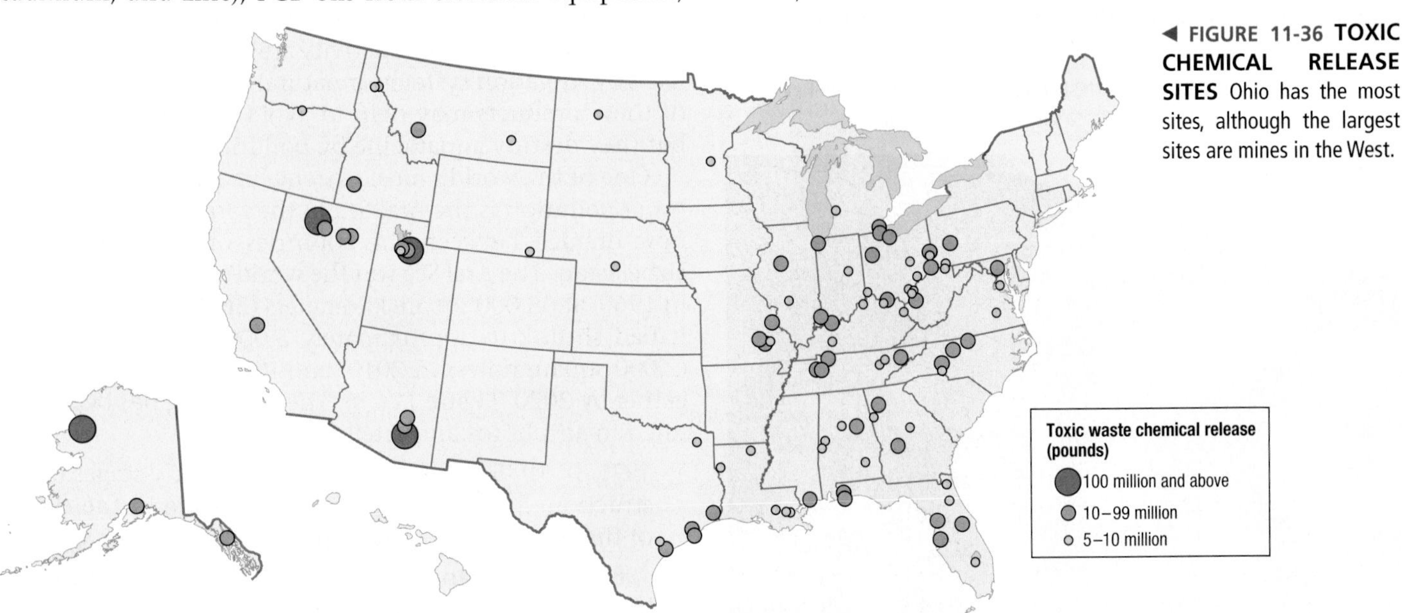

◀ **FIGURE 11-36 TOXIC CHEMICAL RELEASE SITES** Ohio has the most sites, although the largest sites are mines in the West.

Water Pollution

Learning Outcome 11.3.3
Compare and contrast point and nonpoint sources of water pollution.

Some manufacturers are heavy users of water. One example is the aluminum industry. Aluminum producers locate near dams to take advantage of cheap hydroelectric power. A large amount of electricity is needed to separate pure aluminum from bauxite ore (Figure 11-37). Alcoa, the world's largest aluminum producer, even owns dams in North Carolina and Tennessee.

Water also serves many human purposes:

- It must be drunk to survive.
- It is used for cooking.
- It is used for bathing.
- It provides a location for boating, swimming, fishing, and other recreation activities.
- It is home to fish and other edible aquatic life.

When all these uses are totaled, the average American consumes 5,300 liters (1,400 gallons) of water per day, including 680 liters (180 gallons) for drinking, cooking, and bathing. These uses require fresh, clean, unpolluted water.

But clean water is not always available because people and industries also use water for purposes that pollute it. Pollution is widespread because it is easy to dump waste into a river and let the water carry it downstream, where it becomes someone else's problem. By polluting water, humans harm the health of aquatic life and the health of land-based life (including humans themselves).

WATER POLLUTION SOURCES

The sources of pollution can be divided into point sources and nonpoint sources. **Point-source pollution** enters a body of water at a specific location, whereas **nonpoint-source pollution** comes from a large, diffuse area.

▼ **FIGURE 11-37 HYDROELECTRIC POWER** The Cheoah Dam in Tapoco, Tennessee, provides electricity for Alcoa's nearby aluminum factory.

POINT SOURCES. Point-source pollutants are usually smaller in quantity and much easier to control than nonpoint-source pollutants. Point-source water pollution originates from a specific point, such as a pipe from a wastewater treatment plant. The two main point sources of pollution are manufacturers and municipal sewage systems:

- **Water-using manufacturers.** Steel, chemicals, paper products, and food processing are major industrial polluters of water. Each requires a large amount of water in the manufacturing process and generates a lot of wastewater. Food processors, for example, wash pesticides and chemicals from fruit and vegetables. They also use water to remove skins, stems, and other parts. Water can also be polluted by industrial accidents, such as petroleum spills from ocean tankers and leaks from underground tanks at gasoline stations.
- **Municipal sewage.** In developed countries, sewers carry wastewater from sinks, bathtubs, and toilets to a municipal treatment plant, where most—but not all—of the pollutants are removed. The treated wastewater is then typically dumped back into a river or lake. Since passage of the U.S. Clean Water Act and equivalent laws in other developed countries, most treatment plants meet high water-quality standards. In developing countries, sewer systems are rare, and wastewater usually drains, untreated, into rivers and lakes. The drinking water, usually removed from the same rivers, may be inadequately treated as well. The combination of untreated water and poor sanitation makes drinking water deadly in developing countries. Waterborne diseases such as cholera, typhoid, and dysentery are major causes of death.

NONPOINT SOURCES. Nonpoint sources usually pollute in greater quantities and are much harder to control than point sources of pollution. The principal nonpoint source is agriculture. Fertilizers and pesticides spread on fields to increase agricultural productivity are carried into rivers and lakes by irrigation systems or natural runoff. Expanded use of these products may help to avoid a global food crisis, but they destroy aquatic life by polluting rivers and lakes.

One of the world's most extreme instances of nonpoint water pollution is the Aral Sea in the former Soviet Union, now divided between the countries of Kazakhstan and Uzbekistan. The Aral Sea was the world's fourth-largest lake in 1960, at 68,000 square kilometers (26,000 square miles). It had shrunk to approximately 5,000 square kilometers (2,000 square miles) in 2010, and it could disappear altogether by 2020 (Figure 11-38). The shrinking has been captured in air photos and satellite imagery:

- **1975.** In 1975, the Aral Sea was in the early stages of destruction. Small islands are barely visible in the center of the sea (Figure 11-38, upper left).
- **1989.** A large island had formed in the middle of the sea by 1989 (Figure 11-38, upper right).

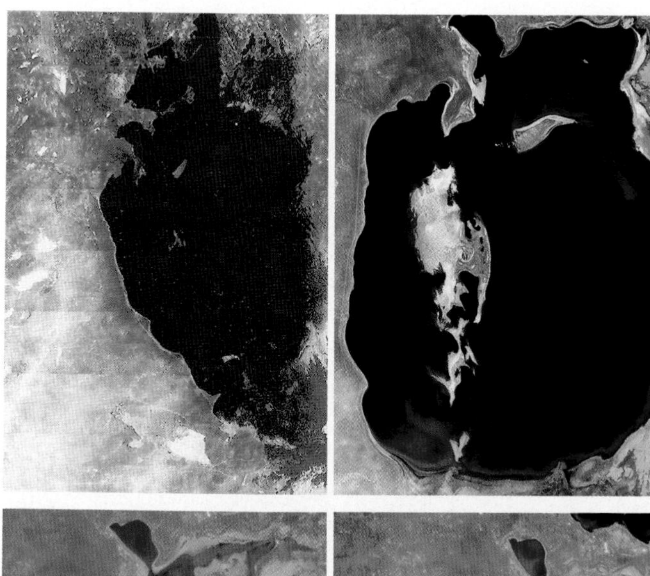

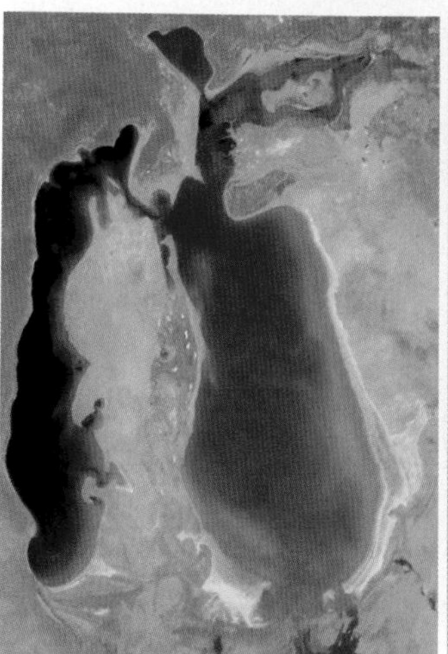

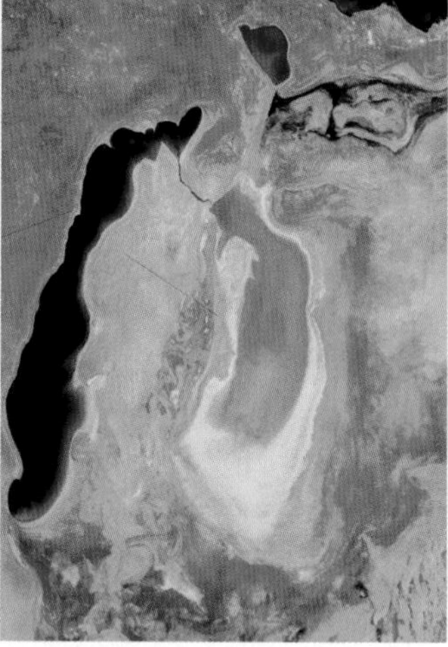

▲ **FIGURE 11-38 THE DISAPPEARING ARAL SEA** In 1975 (upper left), 1989 (upper right), 2003 (lower left), and 2009 (lower right).

- **2003.** By 2003, the sea was divided into two portions, western and eastern (Figure 11-38, lower left).
- **2009.** In 2009, the western portion had not changed much, but the eastern portion had dried up into a wasteland of salt. A small northern lake also remained (Figure 11-38, lower right).

The Aral Sea died because beginning in 1954, the Soviet Union diverted its tributary rivers, the Amu Dar'ya and the Syr Dar'ya, to irrigate cotton fields. Ironically, the cotton now is withering because winds pick up salt from the exposed lakebed and deposit it on the cotton fields. Carp, sturgeon, and other fish species have disappeared; the last fish died in 1983. Large ships lie aground in salt flats that were once the lakebed, outside abandoned fishing villages that now lay tens of kilometers from the rapidly receding shore.

Pause and Reflect 11.3.3

How might sustainable agriculture practices , as discussed in Chapter 10, help to improve water quality?

IMPACT OF WATER POLLUTION ON AQUATIC LIFE

Polluted water can harm aquatic life. Aquatic plants and animals consume oxygen, and so does the decomposing organic waste that humans dump in the water. The oxygen consumed by the decomposing organic waste constitutes the **biochemical oxygen demand (BOD)**. If too much waste is discharged into water, the water becomes oxygen starved and fish die.

This condition is typical when water becomes loaded with municipal sewage or industrial waste. The sewage and industrial pollutants consume so much oxygen that the water can become unlivable for normal plants and animals, creating a "dead" stream or lake. Similarly, when runoff carries fertilizer from farm fields into streams or lakes, the fertilizer nourishes excessive aquatic plant production—a "pond scum" of algae—that consumes too much oxygen. Either type of pollution reduces the normal oxygen level, threatening aquatic plants and animals. Some of the residuals may become concentrated in the fish, making them unsafe for human consumption. For example, salmon from the Great Lakes became unfit to eat because of high concentrations of the pesticide DDT, which washed into streams from farm fields.

Many factories and power plants use water for cooling and then discharge the warm water back into the river or lake. The warm water may not be polluted with chemicals, but it raises the temperature of the body of water it enters. Fish adapted to cold water, such as salmon and trout, might not be able to survive in the warmer water.

CHECK IN: KEY ISSUE 3

Where Does Industry Cause Pollution?

- ✓ **Industry is a major polluter of air, land, and water.**
- ✓ **Air pollution can occur at global, regional, and local scales.**
- ✓ **Solid waste that is not recycled is either transported to landfills or incinerated; some of it is hazardous.**
- ✓ **Water pollution can have point or nonpoint sources.**

KEY ISSUE 4

Why Are Situation and Site Factors Changing?

- Changes within Developed Regions
- Emerging Industrial Regions
- Renewed Attraction of Traditional Industrial Regions

Learning Outcome 11.4.1
Explain reasons for changing distribution of industry within the United States.

Industry is on the move around the world. Changing site factors have been especially important in stimulating industrial growth in new regions internationally and within developed countries. At the same time, some industries remain in the traditional regions, primarily because of changing situation factors.

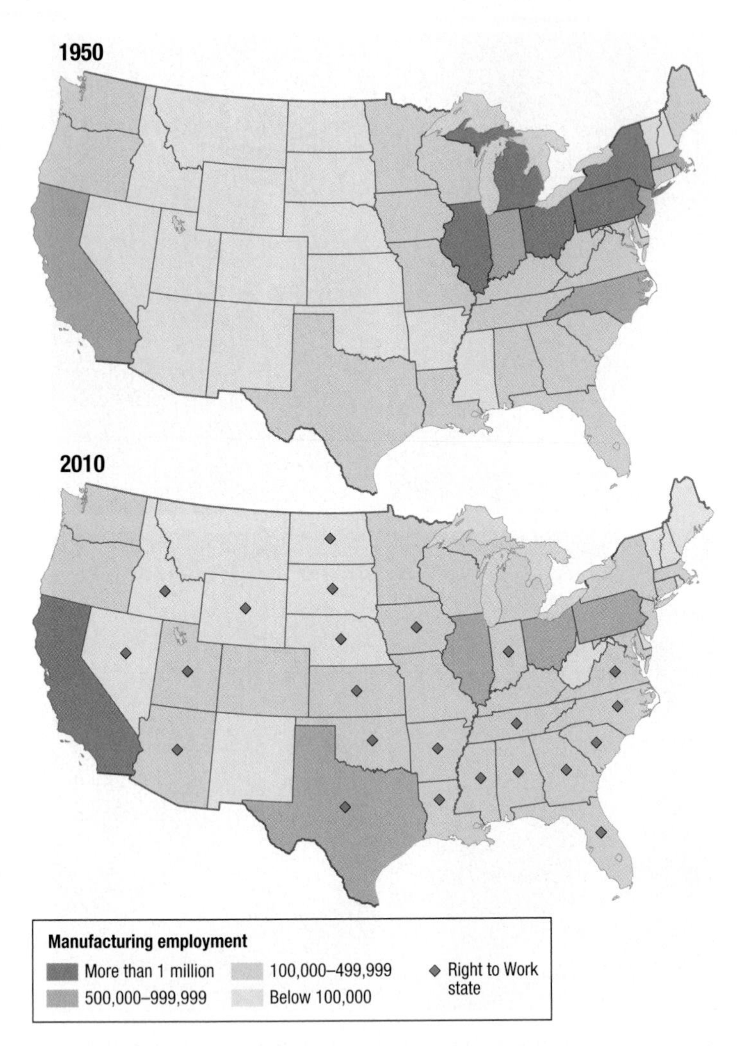

▲ **FIGURE 11-39 CHANGING U.S. MANUFACTURING** Manufacturing has decreased in the Northeast.

Changes within Developed Regions

Within developed countries, industry is shifting away from the traditional industrial areas of northwestern Europe and the northeastern United States. In the United States, industry has shifted from the Northeast toward the South and West. In Europe, government policies have encouraged relocation toward economically distressed peripheral areas.

SHIFTS WITHIN THE UNITED STATES

The northeastern United States lost 6 million jobs in manufacturing between 1950 and 2010 (Figure 11-39). Especially large declines were recorded by New York State and Pennsylvania, states that once served as centers for clothing, textile, steel, and fabricated metal manufacturing. Meanwhile, 2 million manufacturing jobs were added in the South and West between 1950 and 2009. California and Texas had the largest increases.

Industrialization during the late nineteenth and early twentieth centuries largely bypassed the South, which had not recovered from losing the Civil War. The South lacked the infrastructure needed for industrial development: Road and rail networks were less intensively developed in the South, and electricity was less common than in the North. As a result, the South was the poorest region of the United States. Industrial growth in the South since the 1930s has been stimulated in part by government policies to reduce historical disparities. The Tennessee Valley Authority brought electricity to much of the rural South, and roads were constructed in previously inaccessible sections of the Appalachians, the Piedmont, and the Ozarks. Air-conditioning made living and working in the South more tolerable during the summer.

Steel, textiles, tobacco products, and furniture industries have become dispersed through smaller communities in the South, many in search of a labor force willing to work for less pay than in the North and forgo joining a union. The Gulf Coast has become an important industrial area because of its access to oil and natural gas. Along the Gulf Coast are oil refining, petrochemical manufacturing, food processing, and aerospace product manufacturing.

RIGHT-TO-WORK LAWS. The principal lure for many manufacturers has been right-to-work laws. A **right-to-work law** requires a factory to maintain a so-called "open shop" and prohibits a "closed shop." In a "closed shop," a company and a union agree that everyone must join the union to work in the factory. In an "open shop," a union and a company may not negotiate a contract that requires workers to join a union as a condition of employment.

Twenty-three U.S. states (refer to Figure 11-39) have right-to-work laws that make it much more difficult for unions to organize factory workers, collect dues, and bargain with employers from a position of strength. Right-to-work laws send a powerful signal that antiunion attitudes will be tolerated and perhaps even actively supported. As a result, the percentage of workers who are members of a union is much lower in the South than elsewhere in the United States. More importantly, the region has been especially attractive for companies working hard to keep out unions altogether.

Pause and Reflect 11.4.1

Laws to curb unions have been enacted or proposed in several U.S. states in the past few years. What are the arguments in favor of and against restricting unions?

TEXTILE PRODUCTION. The textile and apparel industry has been especially prominent in opening production in lower-wage locations while shutting down production in higher-wage locations. The U.S. textile and apparel industry was heavily concentrated in the Northeast during the early twentieth century, and then it shifted to the South and West.

Most textile and apparel production in the United States moved from the Northeast to the Southeast during the mid-twentieth century. Favored sites were small towns in the Appalachian, Piedmont, and Ozark mountains, especially western North and South Carolina and northern Georgia and Alabama. The area is home to 99 percent of U.S. hosiery and sock producers, half of them in North Carolina.

In the mid-twentieth century, prevailing wage rates were much lower in the Southeast than elsewhere in the United States. Even more important for manufacturers, workers in the Southeast showed little interest in joining the unions established by Northeastern textile and apparel workers to bargain for higher wages and safer working conditions.

INTERREGIONAL SHIFTS IN EUROPE

Manufacturing has diffused from traditional industrial centers in northwestern Europe toward Southern and Eastern Europe. In contrast to the United States, European government policies have explicitly encouraged this industrial relocation (Figure 11-40). The European Union Structural Funds provide assistance to what it calls convergence regions and competitive and employment regions:

- Convergence regions are primarily in Eastern and Southern Europe, where incomes lag behind Europe's average.
- Competitive and employment regions are primarily Western Europe's traditional core industrial areas, which have experienced substantial manufacturing job losses in recent years.

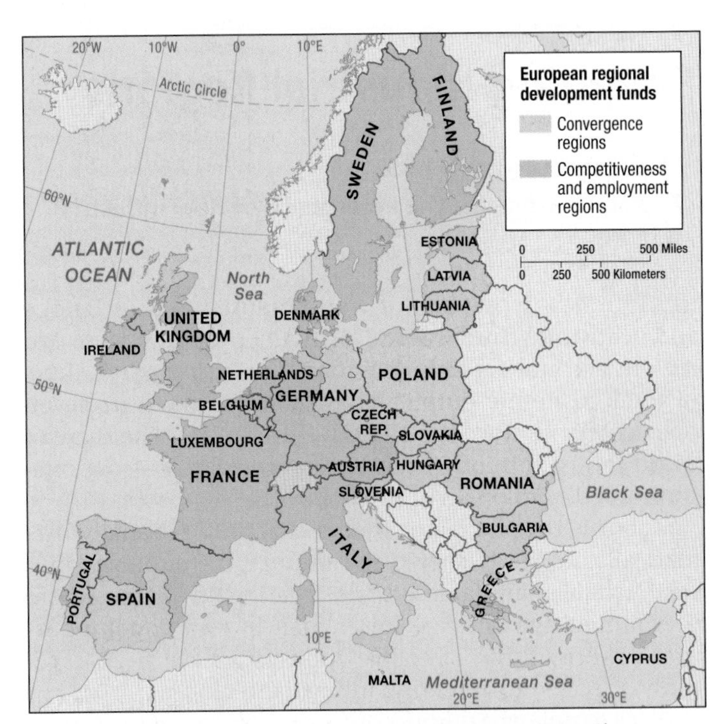

▲ **FIGURE 11-40 EUROPEAN UNION STRUCTURAL FUNDS** The European Union provides subsidies to regions with economic difficulties because of declining industries, as well as to regions that have lower-than-average incomes.

The Western European country with the most rapid manufacturing growth during the late twentieth century was Spain, especially after its admission to the European Union in 1986. Until then, Spain's manufacturing growth had been impeded by physical and political isolation. Spain's motor-vehicle industry has grown into the second largest in Europe, behind only Germany's, although it is entirely foreign owned. Spain's leading industrial area is Catalonia, in the northeast, centered on the city of Barcelona. The region has the country's largest motor-vehicle plant and is the center of Spain's textile industry as well. Spain's industry, though, has been especially hard hit by the severe recession of the early twenty-first century.

Several European countries situated east of Germany and west of Russia have become major centers of industrial investment since the fall of communism in the early 1990s. Poland, Czech Republic, and Hungary have had the most industrial development, though other countries in the region have shared in the growth. The region prefers to be called *Central Europe*, reverting to a common pre–Cold War term, to signify its more central location in Europe's changing economy. Central Europe offers manufacturers an attractive combination of two important site and situation factors: labor and market proximity. Central Europe's workers offer manufacturers good value for money; they are less skilled but much cheaper than in Western Europe, and they are more expensive but much more skilled than in Asia and Latin America. At the same time, the region offers closer proximity to the wealthy markets of Western Europe than other emerging industrial centers.

Emerging Industrial Regions

Learning Outcome 11.4.2
Explain reasons for the emergence of new industrial regions.

In 1970, nearly one-half of world industry was in Europe and nearly one-third was in North America; now these two regions account for only one-fourth each. Industry's share of total economic output has steadily declined in developed countries since the 1970s (Figure 11-41). The share of world industry in other regions has increased—from one-sixth in 1970 to one-half in 2010.

Labor is the site factor that is changing especially dramatically in the twenty-first century. To minimize labor costs, some manufacturers are locating in places where prevailing wage rates are lower than in traditional industrial regions. Labor-intensive industries have been especially attracted to emerging industrial regions.

For example, the number of apparel workers in the United States declined from 900,000 in 1990 to 500,000 in 2000 and to 150,000 in 2010. During this period, most apparel sold in the United States switched from being domestically made to being foreign made (Figure 11-42). As apparel from other countries has become less expensive and less complicated to import into the United States, mills in the Southeast paying wages of $10 to $15 per hour have been unable to compete with manufacturers in countries paying less than $1 per hour. European countries have been even harder hit by international competition. Compensation for manufacturing employees exceeds $30 per hour in much of Europe.

OUTSOURCING

Transnational corporations have been especially aggressive in using low-cost labor in developing countries. To remain competitive in the global economy, they carefully review their production processes to identify steps that can be performed by low-paid, low-skilled workers in developing countries. Despite the greater transportation cost, transnational corporations can profitably transfer some work

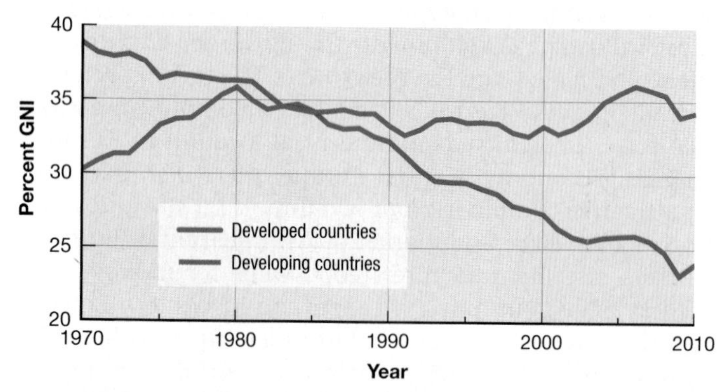

▲ **FIGURE 11-41 MANUFACTURING VALUE AS A PERCENTAGE OF GNI** Manufacturing has accounted for a much higher share of GNI in developing countries than in developed countries since the 1990s.

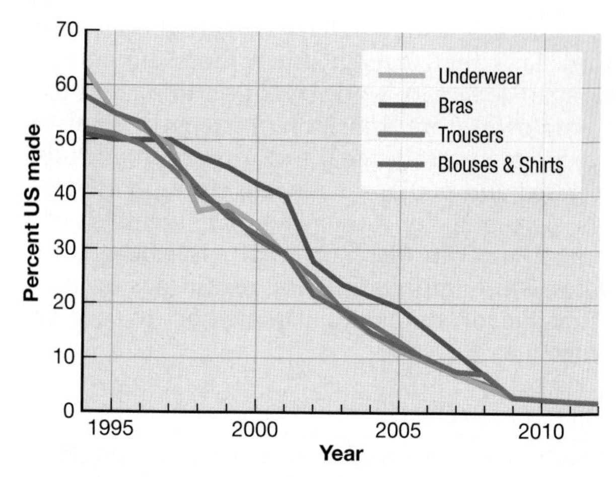

▲ **FIGURE 11-42 U.S. CLOTHING** The percentage of clothing made in the United States declined from around 50 percent in the 1990s to around 2 percent today.

to developing countries, given their substantially lower wages compared to those in developed countries. At the same time, operations that require highly skilled workers remain in factories in developed countries. This selective transfer of some jobs to developing countries is known as the **new international division of labor**.

Transnational corporations allocate production to low-wage countries through **outsourcing**, which is turning over much of the responsibility for production to independent suppliers. Outsourcing contrasts with the approach typical of traditional mass production, called **vertical integration**, in which a company controls all phases of a highly complex production process. Vertical integration was traditionally regarded as a source of strength for manufacturers because it gave them the ability to do and control everything. Carmakers once made nearly all their own parts, for example, but now most of this operation is outsourced to other companies that are able to make the parts cheaper and better. As another example, the parts in an iPhone are made by independent companies.

Outsourcing has had a major impact on the distribution of manufacturing because each step in the production process is now scrutinized closely in order to determine the optimal location. For example, most of the cost of an iPhone is in the parts, which are made by relatively skilled workers in Japan, Germany, and South Korea. Most of the profits go to the United States, where Apple is based. But one step in the production process is especially labor intensive—snapping all the parts together at an assembly plant—and this step is done in China, by relatively low-wage, low-skilled workers (Figure 11-43).

MEXICO AND NAFTA

Manufacturing has been increasing in Mexico. The North American Free Trade Agreement (NAFTA), effective in 1994, eliminated most barriers to moving goods among Mexico, the United States, and Canada. Because it is the nearest low-wage country to the United States, Mexico attracts labor-intensive industries that also need proximity to the U.S. market.

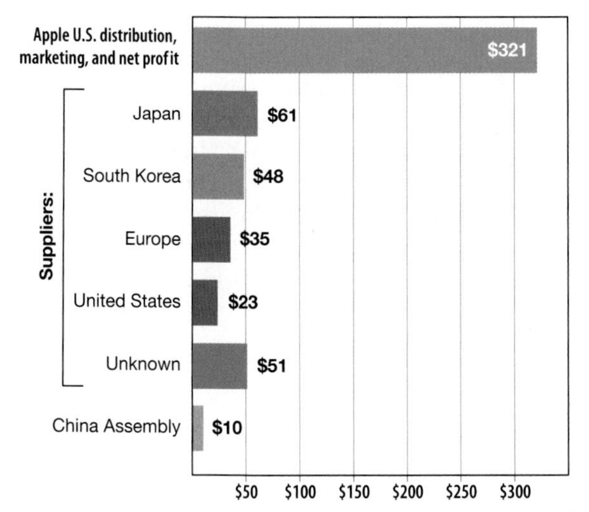

▲ **FIGURE 11-43 iPHONE PRODUCTION** iPhones are assembled in China from parts made in the United States, Europe, and East Asia.

Plants in Mexico near the U.S. border are known as **maquiladoras**. The term originally applied to a tax when Mexico was a Spanish colony. Under U.S. and Mexican laws, companies receive tax breaks if they ship materials from the United States, assemble components at a *maquiladora* plant in Mexico, and export the finished product back to the United States. More than 1 million Mexicans are employed at over 3,000 *maquiladoras*.

Integration of North American industry has generated fear in the United States and Canada:

- Labor leaders fear that more manufacturers relocate production to Mexico to take advantage of lower wage rates. Labor-intensive industries such as food processing and textile manufacturing are especially attracted to regions where prevailing wage rates are lower.
- Environmentalists fear that NAFTA encourages firms to move production to Mexico because laws governing air- and water-quality standards are less stringent than in the United States and Canada. Mexico has adopted regulations to reduce air pollution in Mexico City; catalytic converters have been required on Mexican automobiles since 1991. But environmentalists charge that environmental protection laws are still not strictly enforced in Mexico.

Mexico faces its own challenges: It lost a quarter-million *maquiladora* jobs during the first decade of the twenty-first century. Electronics firms were especially likely to pull out of Mexico. The reason: Although much lower than in the United States, Mexican wages at $6 an hour were higher than $1 wages in China and India. Despite the higher site costs, however, Mexico still competes effectively with China because of situation factors. Because of its proximity, Mexico has much lower shipping costs to the United States than does China.

Pause and Reflect 11.4.2
Can you identify any products in your house that were made in Mexico?

BRIC AND BRICS

Much of the world's future growth in manufacturing is expected to locate outside the principal industrial regions described earlier. The investment banking firm Goldman Sachs coined the acronym BRIC to indicate the countries it expects to dominate global manufacturing during the twenty-first century: Brazil, Russia, India, and China. The foreign ministers of these four countries started meeting in 2006. The four BRIC countries together currently control one-fourth of the world's land area and contain 3 billion of the world's 7 billion inhabitants, but the four countries combined account for only one-sixth of world GDP (Figure 11-44). Their economies rank second (China), seventh (Brazil), ninth (Russia), and eleventh (India) in the world.

China is expected to pass the United States as the world's largest economy around 2020, and India is expected to become second around 2035. In 2050, Brazil and Russia are expected to rank sixth and seventh. Two other developing countries, Indonesia and Nigeria, are expected to be fourth and fifth. Thus, in 2050 the United States would be the only developed country to rank among the world's seven largest economies.

China and India have the two largest labor forces, whereas Russia and Brazil are especially rich in inputs critical for industry. As an industrial region, BRIC has the obvious drawback of Brazil's being on the other side of the planet from the other three. China, India, and Russia could form a contiguous region, but long-standing animosity among them has limited their economic interaction so far. Still, the BRIC concept is that if the four giants work together, they can be the world's dominant industrial bloc in the twenty-first century.

In 2010, South Africa was invited to join a meeting with the other four emerging countries, and the group adopted the acronym BRICS. Although South Africa has the largest economy, population, and land area in the southern portion of sub-Saharan Africa, it is much smaller by all of these measures than the four original BRIC members.

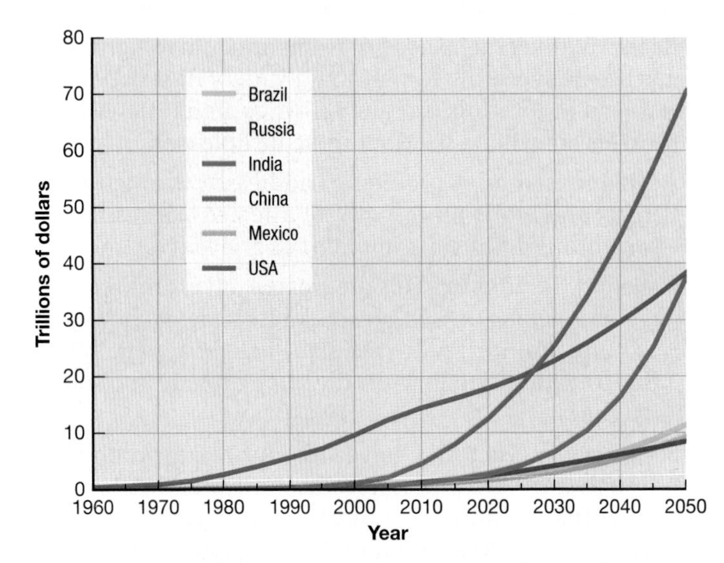

▲ **FIGURE 11-44 GDP FOR BRIC COUNTRIES** The BRIC countries are expected to increase GDP relatively rapidly during the twenty-first century.

Renewed Attraction of Traditional Industrial Regions

Learning Outcome 11.4.3
Explain reasons for renewed attraction of traditional industrial regions.

Given the strong lure of low-cost labor in new industrial regions, why would any industry locate in one of the traditional regions, especially in the northeastern United States or northwestern Europe? Two location factors influence industries to remain in these traditional regions: availability of skilled labor and rapid delivery to market.

PROXIMITY TO SKILLED LABOR

Henry Ford boasted that he could take people off the street and put them to work with only a few minutes of training. That has changed for some industries, which now want skilled workers instead. The search for skilled labor has important geographic implications because it is an asset found principally in the traditional industrial regions.

Traditionally, factories assigned each worker one specific task to perform repeatedly. Some geographers call this approach **Fordist production**, or mass production, because the Ford Motor Company was one of the first companies to organize its production this way early in the twentieth century. At its peak, Ford's factory complex along the River Rouge in Dearborn, Michigan, near Detroit, employed more than 100,000. Most of these workers did not need education or skills to do their jobs, and many were immigrants from Europe or the southern United States.

Many industries now follow a lean, or flexible, production approach. The term **post-Fordist production** is sometimes used to describe lean production, in contrast with Fordist production. Another carmaker is best known for pioneering lean production—in this case, Toyota. Four types of work rules distinguish post-Fordist lean production:

- **Teams.** Workers are placed in teams and told to figure out for themselves how to perform a variety of tasks. Companies are locating production in communities where workers are willing to adopt more flexible work rules.
- **Problem solving.** A problem is addressed through consensus after consulting with all affected parties rather than through filing a complaint or grievance.
- **Leveling.** Factory workers are treated alike, and managers and veterans do not get special treatment; they wear the same uniform, eat in the same cafeteria, park in the same lot, and participate in the same athletic and social activities.
- **Productivity.** Factories have become more productive through introduction of new machinery and processes. Rather than requiring physical strength, these new machines and processes require skilled operators, typically with college degrees.

Computer manufacturing is an example of an industry that has concentrated in relatively high-wage, high-skilled communities of the United States (Figure 11-45). Even the clothing industry has not completely abandoned the Northeast. Dresses, woolens, and other "high-end" clothing products are still made in the region. They require more skill in cutting and assembling the material, and skilled textile workers are more plentiful in the Northeast and California than in the South (Figure 11-46).

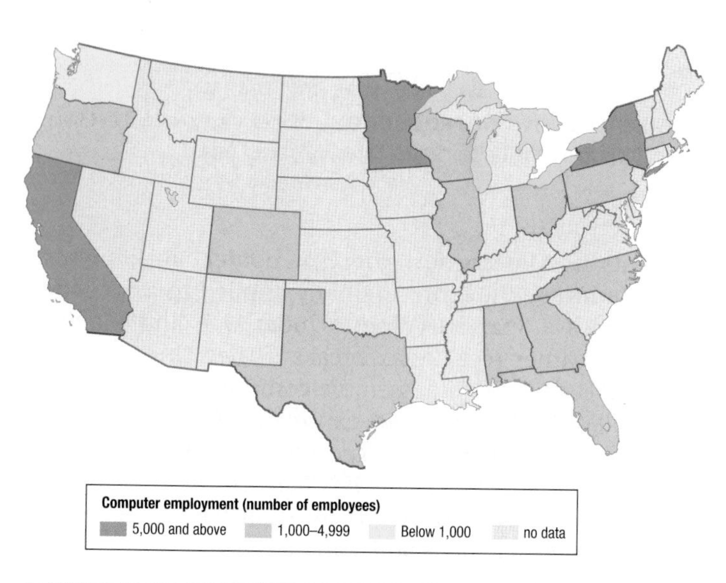

▲ **FIGURE 11-45 COMPUTER AND PERIPHERAL EQUIPMENT MANUFACTURING** Manufacturers of computing equipment seek access to skilled workers to perform precision tasks. The assembly work that requires lower-skilled workers is done abroad, mostly in Asia, as shown in the case of the iPhone (Figure 11-46).

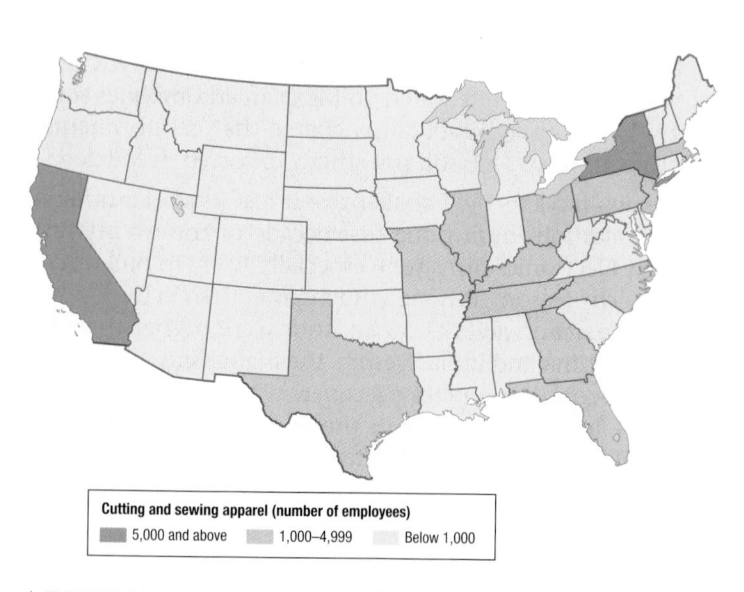

▲ **FIGURE 11-46 THE APPAREL INDUSTRY** What's left of the U.S. apparel industry is concentrated in California and the Northeast.

JUST-IN-TIME DELIVERY

Proximity to market has long been important for many types of manufacturers, as discussed earlier in this chapter. This factor has become even more important in recent years because of the rise of **just-in-time delivery**. As the name implies, just-in-time is shipment of parts and materials to arrive at a factory moments before they are needed. Just-in-time delivery is especially important for delivery of inputs, such as parts and raw materials, to manufacturers of fabricated products, such as cars and computers.

Under just-in-time, parts and materials arrive at a factory frequently, in many cases daily or even hourly. Suppliers of the parts and materials are told a few days in advance how much will be needed over the next week or two, and first thing each morning, they are told exactly what will be needed at precisely what time that day. To meet a tight timetable, a supplier of parts and materials must locate factories near its customers. If given only an hour or two of notice, a supplier has no choice but to locate a factory within 50 miles or so of the customer.

Just-in-time delivery reduces the money that a manufacturer must tie up in wasteful inventory. In fact, the percentage of the U.S. economy tied up in inventory has been cut in half during the past three decades. Manufacturers also save money through just-in-time delivery by reducing the size of the factory because space does not have to be wasted on piling up a mountain of inventory. Leading computer manufacturers have eliminated inventory altogether. They build computers only in response to customer orders placed primarily over the Internet or by telephone. In some cases, just-in-time delivery merely shifts the burden of maintaining inventory to suppliers. Wal-Mart, for example, holds low inventories but tells its suppliers to hold high inventories "just in case" a sudden surge in demand requires restocking on short notice.

Just-in-time delivery means that producers have less inventory to cushion against disruptions in the arrival of needed parts. Three kinds of disruptions can result from reliance on just-in-time delivery:

- **Labor unrest.** A strike at one supplier plant can shut down the entire production within a couple of days. A strike in the logistics industry, such as a strike by truckers or dockworkers, could also disrupt deliveries.
- **Traffic.** Deliveries may be delayed when traffic is slowed by accident, construction, or unusually heavy volume. Trucks and trains are both subject to these types of delays, especially crossing international borders.
- **Natural hazards.** Poor weather conditions can afflict deliveries anywhere in the world. Blizzards and floods can close highways and rail lines. The 2011 earthquake and tsunami in Japan put many factories and transportation lines out of service for months. Carmakers around the world had to curtail production because key parts had been made at the damaged factories. Superstorm Sandy, which hit the East Coast of the United States in 2012, severely disrupted transportation and delivery of goods and energy in the most densely population region of the country (Figure 11-47).

▶ **FIGURE 11-47** NATURAL HAZARDS: SUPERSTORM SANDY Superstorm Sandy, which hit the East Coast of the United States in 2012, disrupted travel for several days. In New York City, subways and tunnels were closed because of flooding. People walked across the Brooklyn Bridge to get to work, while private cars, taxis, and delivery trucks sat bumper-to-bumper on the bridge.

A GLOBAL INDUSTRY: WHAT IS AN AMERICAN CAR?

Distinctions between "American" and "foreign" motor vehicles have been blurred for the past three decades. Popular media have delighted in showcasing examples of "American" vehicles produced by the Detroit 3 (Chrysler, Ford, and General Motors) that have lower U.S. content than those produced by "Japanese" carmakers such as Honda and Toyota. The U.S. government distinguishes between domestic and foreign vehicles in three ways:

- For measuring fuel efficiency, the U.S. Environmental Protection Agency considers a vehicle domestic if at least 75 percent of its content comes from North America, originally defined as the United States and Canada, and, after enactment of the North American Free Trade Agreement (NAFTA), including Mexico.
- For setting import tariffs, the U.S. Department of Treasury Customs Service considers as domestic a vehicle having at least 50 percent U.S. and Canadian content.
- For informing consumers, the American Automobile Labeling Act of 1992 considers a vehicle domestic if at least 85 percent of the parts originate in the United States and Canada; a part is counted as domestic if at least 70 percent of its overall content comes from the United States and Canada.

According to data derived from Labeling Act reports, vehicles built by foreign-owned carmakers at assembly plants located in the United States have around 60 percent domestic content. Domestic content for the Detroit 3 is 76 percent. The lower domestic content for foreign carmakers masks differences among individual companies. Honda and Toyota have a level of U.S. content comparable to that of the Detroit 3. German-owned carmakers such as BMW and Daimler-Benz have much lower percentages.

After opening assembly plants in the United States during the 1980s, Japanese-owned carmakers convinced many of their Japanese-owned suppliers to build factories in the United States. The gap in domestic content has also narrowed because the Detroit 3 bought more foreign parts. More than one-fourth of all new vehicle parts are imported. Mexico has become the leading source of imported parts, and China has been increasing its share rapidly.

Figure 11-48 shows the extent to which several popular vehicles are "American." The x axis shows the percentage of these vehicles sold in the United States that were assembled in the United States in 2011. The y axis shows the percentage of U.S.-made parts in these vehicles.

- GM's Chevrolet Malibu was assembled entirely in the United States with all but a handful of U.S.-made parts.
- Toyota's Prius was assembled in Japan with Japanese-made parts.
- Ford's Fusion was assembled in Mexico with about one-half U.S. parts.
- BMW's X3 was assembled in the United States with parts mostly imported from Germany.
- Honda Civics were assembled either in the United States with mostly U.S.-made parts, assembled in Canada with mostly U.S.-made parts, or imported from Japan with mostly Japanese-made parts.

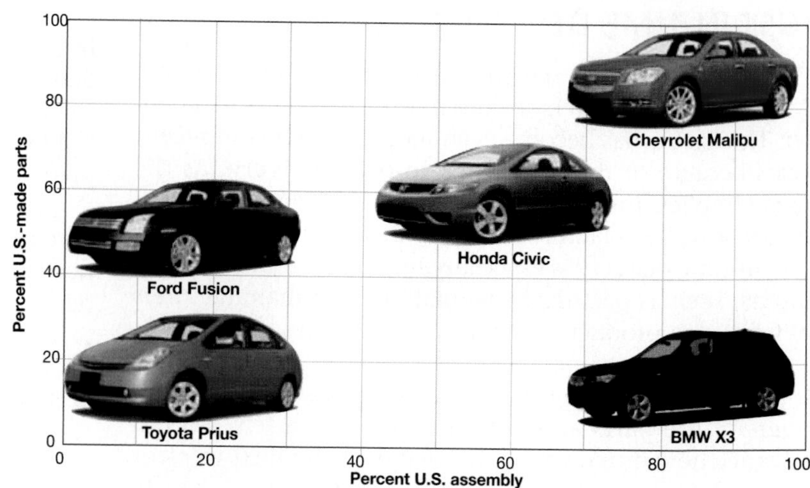

▲ **FIGURE 11-48** "AMERICAN" AND "FOREIGN" CARS The x axis shows the percentage of these vehicles sold in the United States that were assembled in the United States in 2011. The y axis shows the percentage of U.S.-made parts in these vehicles.

Pause and Reflect 11.4.3

Why might weather conditions encourage companies to locate factories in the U.S. South rather than the North?

CHECK-IN: KEY ISSUE 4

Why Are Situation and Site Factors Changing?

- ✓ **Industry is moving from the North to the South within the United States; in many cases, lower-cost nonunion labor is the principal factor.**
- ✓ **Low-cost labor is also inducing firms to locate in countries that are not part of the traditional industrial regions.**
- ✓ **On the other hand, some industry is attracted to traditional industrial regions because of the need for skilled labor or rapid delivery to consumers.**

Summary

KEY ISSUE 1

Where Is Industry Distributed?

The concept of manufacturing goods in a factory originated with the Industrial Revolution in the United Kingdom.

LEARNING OUTCOME 11.1.1: Describe the locations of the principal industrial regions.

- Most of the world's industry is clustered in the three regions: Europe, North America, and East Asia.

THINKING GEOGRAPHICALLY 11.1: What are the principal manufacturers in your community or area? How have they been affected by increasing global competition?

GOOGLE EARTH 11.1: Coalbrookdale, England, is considered the birthplace of the Industrial Revolution, because a factory here was the first to produce high-quality iron using coal. What structure, visible in 3D, was the first in the world to be made of cast iron?

Key Terms

Acid deposition (p. 414) Sulfur oxides and nitrogen oxides, emitted by burning fossil fuels, that enter the atmosphere—where they combine with oxygen and water to form sulfuric acid and nitric acid—and return to Earth's surface.

Acid precipitation (p. 414) Conversion of sulfur oxides and nitrogen oxides to acids that return to Earth as rain, snow, or fog.

Air pollution (p. 412) Concentration of trace substances, such as carbon monoxide, sulfur dioxide, nitrogen oxides, hydrocarbons, and solid particulates, at a greater level than occurs in average air.

Apparel (p. 410) An article of clothing.

Biochemical oxygen demand (BOD) (p. 417) The amount of oxygen required by aquatic bacteria to decompose a given load of organic waste; a measure of water pollution.

Break-of-bulk point (p. 402) A location where transfer is possible from one mode of transportation to another.

Bulk-gaining industry (p. 400) An industry in which the final product weighs more or comprises a greater volume than the inputs.

Bulk-reducing industry (p. 398) An industry in which the final product weighs less or comprises a lower volume than the inputs.

Chlorofluorocarbon (CFC) (p. 413) A gas used as a solvent, a propellant in aerosols, a refrigerant, and in plastic foams and fire extinguishers.

Cottage industry (p. 395) Manufacturing based in homes rather than in factories, commonly found prior to the Industrial Revolution.

Ferrous (p. 398) Metals, including iron, that are utilized in the production of iron and steel

Fordist production (p. 422) A form of mass production in which each worker is assigned one specific task to perform repeatedly.

Greenhouse effect (p. 412) The anticipated increase in Earth's temperature caused by carbon dioxide (emitted by burning fossil fuels) trapping some of the radiation emitted by the surface.

Industrial Revolution (p. 395) A series of improvements in industrial technology that transformed the process of manufacturing goods.

Just-in-time delivery (p. 423) Shipment of parts and materials to arrive at a factory moments before they are needed.

Labor-intensive industry (p. 408) An industry for which labor costs comprise a high percentage of total expenses.

KEY ISSUE 2

Why Are Situation and Site Factors Important?

Manufacturers select locations for factories based on assessing a combination of situation and site factors.

LEARNING OUTCOME 11.2.1: Identify the two types of situation factors and explain why some industries locate near inputs.

- Situation factors involve minimizing the cost of shipping from sources of inputs or to markets.
- A location near sources of inputs is optimal for bulk-reducing industries.
- Industries that extract a large amount of minerals tend to be bulk-reducing industries.

LEARNING OUTCOME 11.2.2: Explain why some industries locate near markets.

- Bulk-gaining industries, single-market manufacturers, and perishable products companies tend to locate near markets.

LEARNING OUTCOME 11.2.3: Explain why industries use different types of transportation.

- Trucks are most often used for short-distance delivery, trains for longer trips within a region, ships for ocean crossings, and planes for very high-value packages.
- Some firms locate near break-of-bulk points, where goods are transferred between modes of transportation.

LEARNING OUTCOME 11.2.4:

Describe how the optimal location for steel production has changed.

- Steel production has traditionally been located near inputs, but the relative importance of the two main inputs—coal and iron ore—has changed.
- Some steel production, especially minimills, is now located near the markets.
- Industries that extract a large amount of minerals tend to be bulk-reducing industries.

LEARNING OUTCOME 11.2.5:

Explain the distribution of motor vehicle production.

- Because they are bulk-gaining products, most motor vehicles are assembled near their markets.
- The distribution of motor vehicle production has changed because the distribution of buyers has changed.

LEARNING OUTCOME 11.2.6: List the three types of site factors.

- The three site factors are labor, capital, and land.
- A labor-intensive industry has a high percentage of labor in the production process.

LEARNING OUTCOME 11.2.7: Explain the distribution of textile and apparel production.

- The clothing industry is a labor-intensive industry.
- Three steps in production are spinning, weaving, and sewing. Most spinning and weaving occur in low-wage countries, but some sewing occurs in developed countries near consumers.

THINKING GEOGRAPHICALLY 11.2: To induce Kia to build its U.S. production facility in Georgia, the state spent $36 million to buy the site; $25 million to prepare the site, including grading; $30 million to provide road improvements, including an interchange off I-85; $6 million to build a rail spur; $20 million to construct a training center; $6 million to operate the center for five years; $6 million to develop a training course; $76 million in tax credits; $14 million in sales tax exemptions; and $41 million in training equipment. Did Georgia overpay to win the Kia factory? Explain.

GOOGLE EARTH 11.2: The largest steel works in the United States, the US Steel complex at Gary, Indiana, sits at the south end of Lake Michigan. How many modes of transport delivering raw materials to the plant can you see?

Maquiladora (p. 421) A factory built by a U.S. company in Mexico near the U.S. border, to take advantage of the much lower labor costs in Mexico.

New international division of labor (p. 420) Transfer of some types of jobs, especially those requiring low-paid, less-skilled workers, from more developed to less developed countries.

Nonferrous (p. 398) Metals utilized to make products other than iron and steel.

Nonpoint-source pollution (p. 416) Pollution that originates from a large, diffuse area.

Outsourcing (p. 420) A decision by a corporation to turn over much of the responsibility for production to independent suppliers.

Ozone (p. 413) A gas that absorbs ultraviolet solar radiation, found in the stratosphere, a zone 15 to 50 kilometers (9 to 30 miles) above Earth's surface.

Photochemical smog (p. 414) An atmospheric condition formed through a combination of weather conditions and pollution, especially from motor vehicle emissions.

Point-source pollution (p. 416) Pollution that enters a body of water from a specific source.

Post-Fordist production (p. 422) Adoption by companies of flexible work rules, such as the allocation of workers to teams that perform a variety of tasks.

Right-to-work law (p. 418) A U.S. law that prevents a union and a company from negotiating a contract that requires workers to join the union as a condition of employment.

Sanitary landfill (p. 414) A place to deposit solid waste, where a layer of earth is bulldozed over garbage each day to reduce emissions of gases and odors from the decaying trash, to minimize fires, and to discourage vermin.

KEY ISSUE 3

Why Does Industry Cause Pollution?

Industry is a major polluter of air, land, and water.

LEARNING OUTCOME 11.3.1: Describe the causes and effects of global warming and damage to the ozone layer.

- Air pollution occurs at global, regional, and local scales.
- At the global scale, the principal pollution is global warming, caused primarily by burning of fossil fuels in factories and vehicles.

LEARNING OUTCOME 11.3.2: Describe the causes and effects of regional and local-scale air pollution and solid waste pollution.

- Acid deposition is a major form of regional-scale air pollution. Sulfuric acid and nitric acid generated by burning of fossil fuels fall into bodies of water.
- Carbon monoxide, hydrocarbons, and particulates are the major forms of local-scale air pollution.
- Solid waste is typically placed in landfills or incinerated.

LEARNING OUTCOME 11.3.3: Compare and contrast point and nonpoint sources of water pollution.

- Point-source pollution originates from a specific place, such as a pipe, generated principally by factories and sewage disposal.
- Nonpoint sources are generated primarily by agricultural runoff.

THINKING GEOGRAPHICALLY 11.3: What are the major polluters in or near your community?

GOOGLE EARTH 11.3: The world's largest electronics manufacturer, FoxConn, has a large complex in Longhua, Shenzhen, China. How many different FoxConn buildings are labeled in Longhua?

KEY ISSUE 4

Why Are Situation and Site Factors Changing?

Industry is on the move within developed countries, as well as to emerging developing countries.

LEARNING OUTCOME 11.4.1: Explain reasons for changing distribution of industry within the United States.

- Industry is moving from the North to the South within the United States.
- Lower labor costs and absence of unions are major factors in the migration.

LEARNING OUTCOME 11.4.2: Explain reasons for the emergence of new industrial regions.

- Some jobs have been transferred to low-wage countries as part of the new international division of labor.
- The BRIC countries (Brazil, Russia, India, and China) are expected to be the top industrial powers by the middle of the twenty-first century.

LEARNING OUTCOME 11.4.3: Explain reasons for renewed attraction of traditional industrial regions.

- Traditional industrial regions attract and retain industries that need skilled labor.
- Just-in-time delivery has increased the attraction of locating near consumers.

THINKING GEOGRAPHICALLY 11.4: What have been the benefits and costs to Canada, Mexico, and the United States as a result of NAFTA?

GOOGLE EARTH 11.4: If you fly to Ciudad Acuna, Mexico, several *maquiladora* plants can be seen on the northern edge of the city, near the U.S. border, along the Rio Grande River (Rio Bravo in Spanish). What is the distance from the *maquiladora* complex to the nearest border crossing?

Site factors (p. 398) Location factors related to the costs of factors of production inside a plant, such as land, labor, and capital.

Situation factors (p. 398) Location factors related to the transportation of materials into and from a factory.

Textile (p. 410) A fabric made by weaving, used in making clothing.

Vertical integration (p. 420) An approach typical of traditional mass production in which a company controls all phases of a highly complex production process.

Chapter

12 Services and Settlements

Why is this man carrying raw pig meat on his back? Page 440

Why are these farm fields long and narrow rather than square? Page 449

KEY ISSUE 1

Where Are Services Distributed?

More and More Services p. 431

Most jobs—and most of the growth in jobs—is in services.

KEY ISSUE 2

Where Are Consumer Services Distributed?

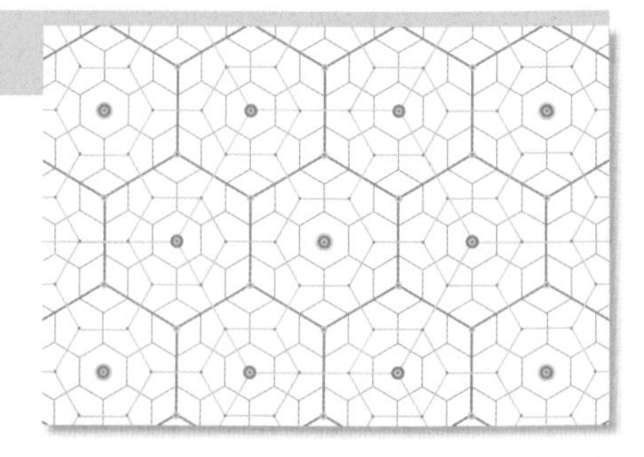

Services for People p. 434

Services for people are located where the people are.

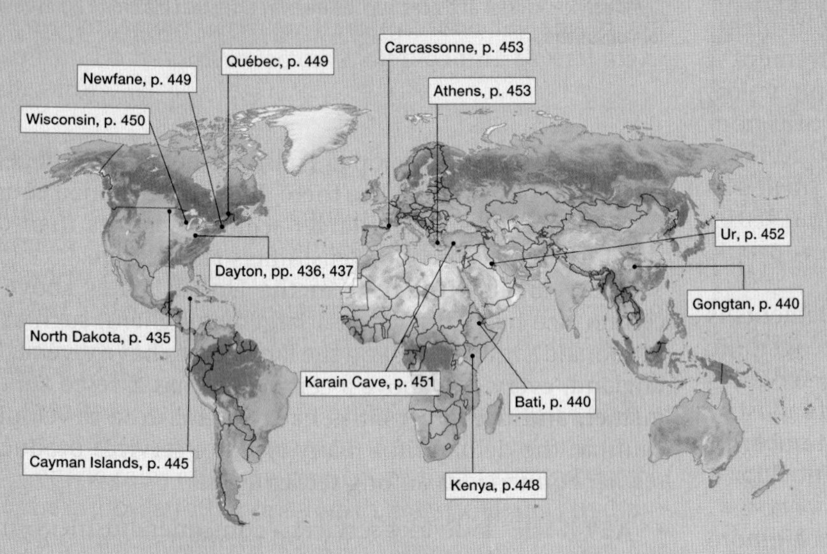

▲ Need to have your computer fixed? Correct a mistake on your credit card bill? Change your plane reservation? The company whose name is on the computer, credit card, or airplane may not actually employ the person who answered your call. Instead, the call-answering job may have been contracted out to another company known as a call center. Call centers are one of the fastest-growing services in the global economy. Many of them are located in India, including this one in Kolkata.

KEY ISSUE 3

Where Are Business Services Distributed?

Services for Businesses p. 442

Most business services are in very large settlements.

KEY ISSUE 4

Why Do Services Cluster in Settlements?

A World of Urban Services p. 448

Settlements can be rural or urban; the urban ones are growing.

BUSINESS SERVICES

Learning Outcome 12.1.1
Describe the three types of services and changing numbers of types of jobs.

The principal purpose of **business services** is to facilitate the activities of other businesses. One-fourth of all jobs in the United States are in business services. Professional services, financial services, and transportation services are the three main types of business services (Figure 12-4):

- *Professional services* comprise about 11 percent of all U.S. jobs. Technical services, including law, management, accounting, architecture, engineering, design, and consulting, comprise 60 percent of professional services jobs. Support services, such as clerical, secretarial, and custodial work, account for the other 40 percent.
- *Financial services* comprise about 7 percent of all U.S. jobs. This sector is often called "FIRE," an acronym for finance, insurance, and real estate. One-half of the financial services jobs are in banks and other financial institutions, one-third in insurance companies, and the remainder in real estate.
- *Transportation and information services* comprise about 7 percent of all U.S. jobs. Transportation, primarily trucking and warehousing, account for 60 percent of these jobs. The other 40 percent are in information services such as publishing and broadcasting, as well as utilities such as water and electricity.

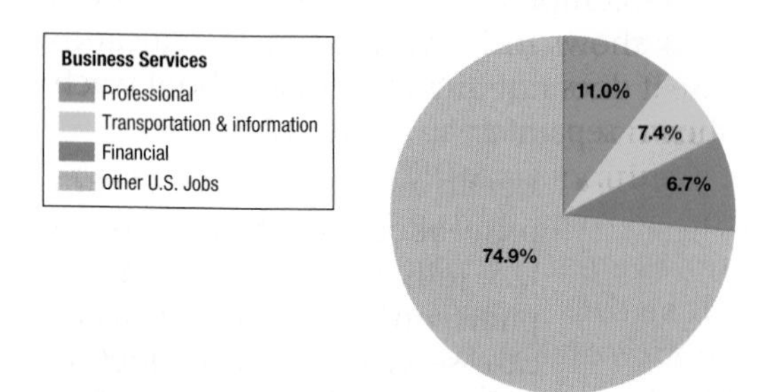

▲ FIGURE 12-4 **U.S. BUSINESS SERVICES** Most business service jobs are in professional services.

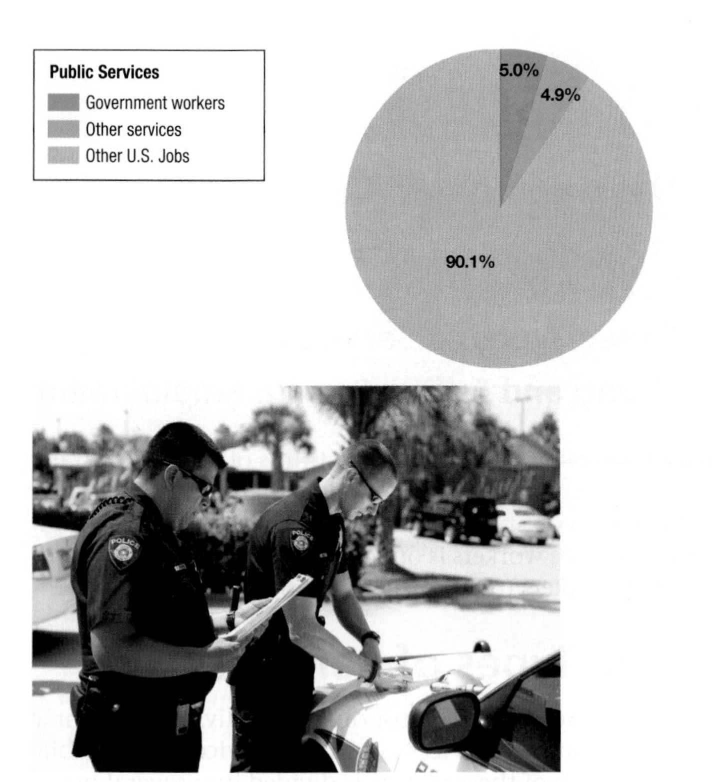

▲ FIGURE 12-5 **U.S. PUBLIC SERVICES** Most public service jobs are in local government.

PUBLIC SERVICES

The purpose of **public services** is to provide security and protection for citizens and businesses. About 10 percent of all U.S. jobs are in the public sector (Figure 12-5). Excluding educators, one-sixth of public-sector employees work for the federal government, one-fourth for one of the 50 state governments, and three-fifths for one of the tens of thousands of local governments (Figure 12-5). The census classifies another 5 percent of jobs as "other services" because they don't fall logically under the categories of consumer, business, and public services.

Pause and Reflect 12.2.1
In which sectors of the economy do you or members of your family work? If in the service sector, in which types of services are these jobs?

Rising and Falling Service Employment

The service sector of the economy has seen nearly all the growth in employment worldwide. It is also the sector that has been impacted the most by the severe recession that began in 2008.

CHANGES IN NUMBER OF EMPLOYEES

Figure 12-6 shows changes in employment in the United States between 1972 and 2010. All the growth in employment in the United States has been in services, whereas

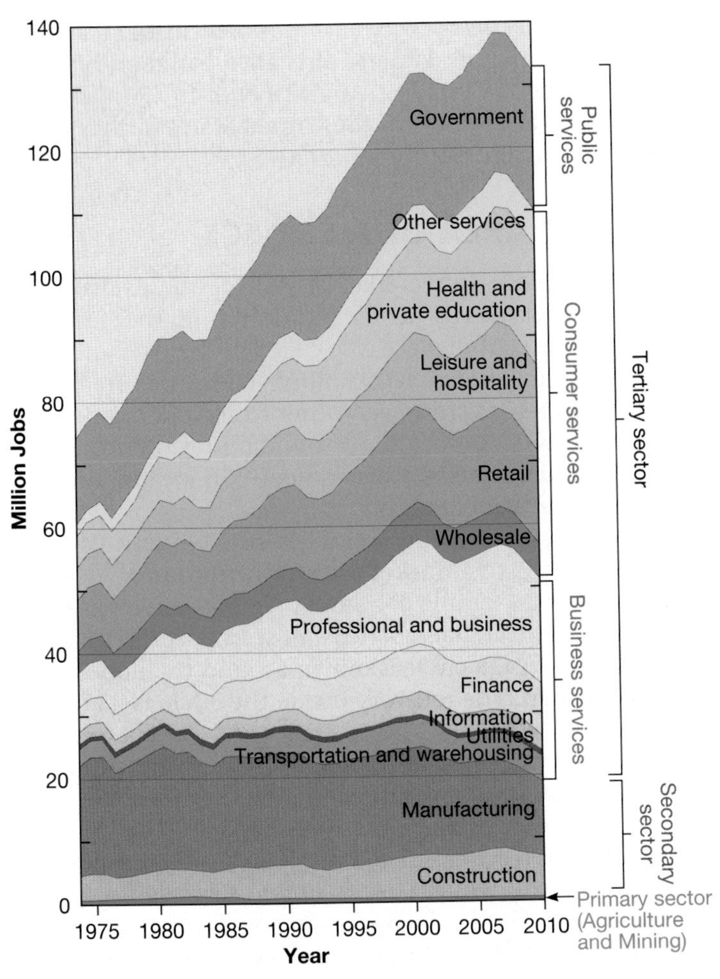

▲ **FIGURE 12-6 CHANGES IN U.S. EMPLOYMENT** Jobs have increased in the service sector.

employment in primary- and secondary-sector activities has declined.

Within business services, jobs expanded most rapidly in professional services (such as engineering, management, and law), data processing, advertising, and temporary employment agencies. Jobs grew more slowly in finance and transportation services because of improved efficiency—fewer workers are needed to run trains and answer phones, for example.

On the consumer services side, the most rapid increase has been in the provision of health care, including hospital staff, clinics, nursing homes, and home health-care programs. Other large increases have been recorded in education, entertainment, and recreation. The share of jobs in retailing has not increased; more stores are opening all the time, but they don't need as many employees as in the past.

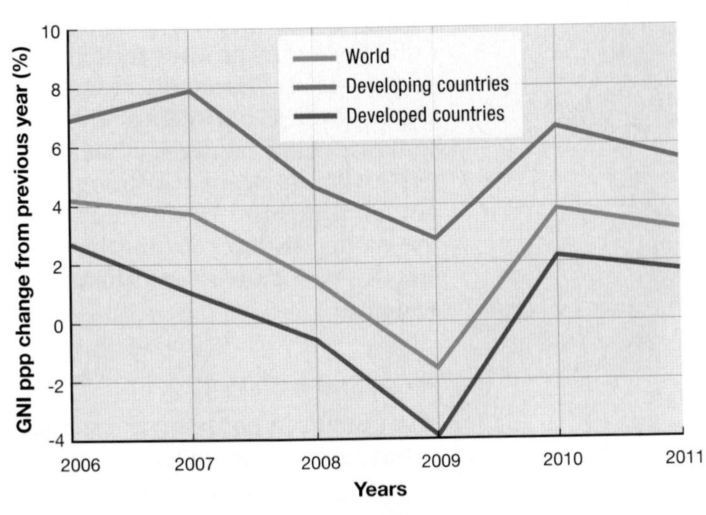

▲ **FIGURE 12-7 GNI CHANGE** GNI per capita declined during the severe recession that began in 2008.

SERVICES IN THE RECESSION

The service sector of the economy has been the engine of growth in the economy of developed countries, even as industry and agriculture have declined. But it was the service sector that triggered the severe economic recession that began in 2008. Principal contributors to the recession were some of the practices involved in financial services and real estate services, including:

- A rapid rise in real estate prices, encouraging speculators to acquire properties for the purpose of reselling them quickly at even higher prices.
- Poor judgment in lending by financial institutions, especially by offering "subprime" mortgages to individuals whose poor credit history made the loans highly risky.
- Invention of new financial services practices, such as derivatives, in which investors bought and sold risky assets, with the expectation that the value of the assets would continually rise.
- Decisions by government agencies to reduce or eliminate regulation of the practices of financial institutions.
- Unwillingness of financial institutions to make loans once the recession started.

The early twenty-first century recession was also distinctive because it rapidly affected every other region of the world. At the same time, the impact of the global recession varied by region and locality.

The early twenty-first century recession resulted in an absolute decline in world GNI for the first time since the 1930s (Figure 12-7). GNI grew by an annual average of 3.7 percent between 1960 and the start of the recession in 2008. Only twice in that time did GNI grow at a rate of less than 1 percent per year.

CHECK-IN: KEY ISSUE 1

Where Are Services Distributed?

✓ **Three types of services are consumer, business, and public.**

✓ **The fastest-growing consumer service is health care, and the fastest-growing business service is professional.**

KEY ISSUE 2

Where Are Consumer Services Distributed?

- Central Place Theory
- Hierarchy of Consumer Services
- Market Area Analysis

Learning Outcome 12.2.1
Explain the concepts of market area, range, and threshold.

Consumer services and business services do not have the same distributions. Consumer services generally follow a regular pattern based on size of settlements, with larger settlements offering more consumer services than smaller ones. The next Key Issue will describe how business services cluster in specific settlements, creating a specialized pattern.

Central Place Theory

Selecting the right location for a new shop is probably the single most important factor in the profitability of a consumer service. **Central place theory** helps to explain how the most profitable location can be identified.

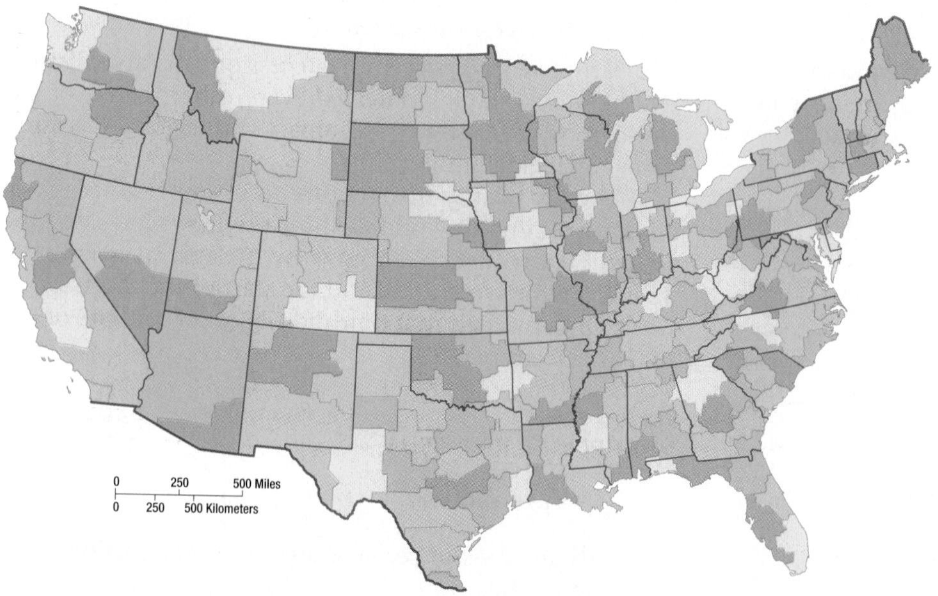

Central place theory was first proposed in the 1930s by German geographer Walter Christaller, based on his studies of southern Germany. August Lösch in Germany and Brian Berry and others in the United States further developed the concept during the 1950s.

MARKET AREA OF A SERVICE

A **central place** is a market center for the exchange of goods and services by people attracted from the surrounding area. The central place is so called because it is centrally located to maximize accessibility. Businesses in central places compete against each other to serve as markets for goods and services for the surrounding region. According to central place theory, this competition creates a regular pattern of settlements.

The area surrounding a service from which customers are attracted is the **market area**, or **hinterland**. A market area is a good example of a nodal region—a region with a core where the characteristic is most intense. To establish the market area, a circle is drawn around the node of service on a map. The territory inside the circle is its market area.

Because most people prefer to get services from the nearest location, consumers near the center of the circle obtain services from local establishments. The closer to the periphery of the circle, the greater the percentage of consumers who will choose to obtain services from other nodes. People on the circumference of the market-area circle are equally likely to use the service or go elsewhere. The United States can be divided into market areas based on the hinterlands surrounding the largest urban settlements (Figure 12-8). Studies conducted by C. A. Doxiadis, Brian Berry, and the U.S. Department of Commerce allocated the 48 contiguous states to 171 functional regions centered around commuting hubs, which they called "daily urban systems."

To represent market areas in central place theory, geographers draw hexagons around settlements (Figure 12-9). Hexagons represent a compromise between circles and squares. Like squares, hexagons nest without gaps. Although all points along the hexagon are not the same distance from the center, the variation is less than with a square.

0 250 500 Miles
0 250 500 Kilometers

▲ FIGURE 12-8 DAILY URBAN SYSTEMS The U.S. Department of Commerce divided the 48 contiguous states into "daily urban systems," delineated by functional ties, especially commuting to the nearest metropolitan area. This division of the country into daily urban systems demonstrates that everyone in the United States has access to services in at least one large settlement. Compare this information to the information on TV market areas in Figure 1-18.

Pause and Reflect 12.2.1
What occurs in nature in the shape of hexagons? Google "naturally occurring hexagons." Infer why human economic activities also create a hexagonal pattern.

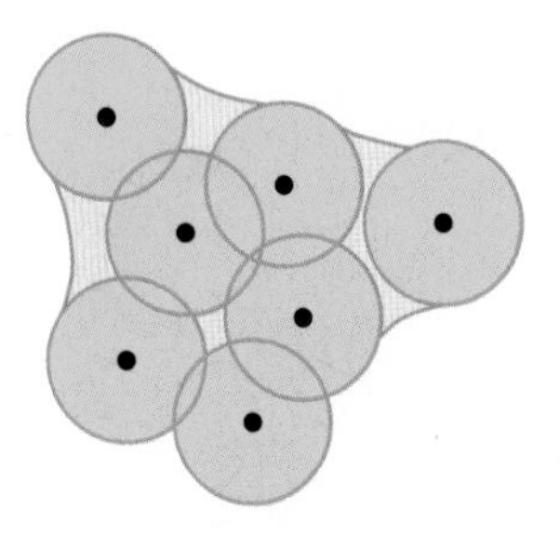

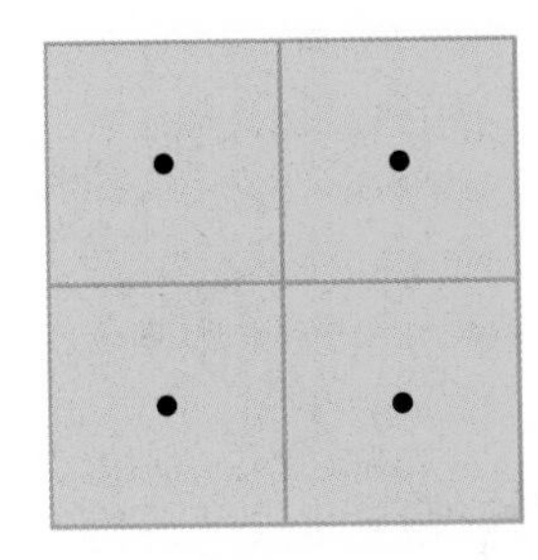

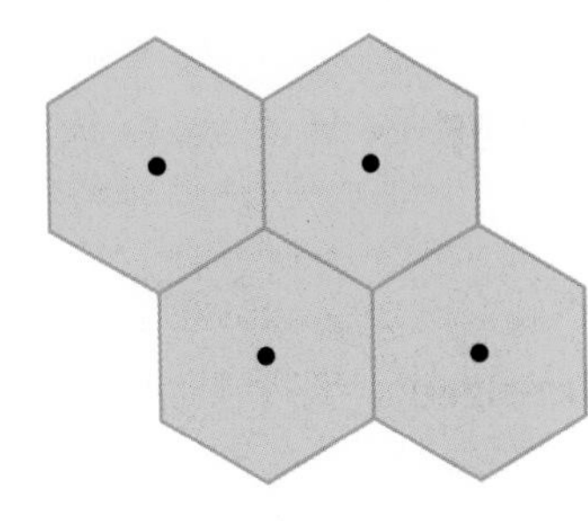

▲ **FIGURE 12-9 WHY GEOGRAPHERS USE HEXAGONS TO DELINEATE MARKET AREAS** (left) **The problem with circles.** Circles are equidistant from center to edge, but they overlap or leave gaps. An arrangement of circles that leaves gaps indicates that people living in the gaps are outside the market area of any service, which is obviously not true. Overlapping circles are also unsatisfactory, for one service or another will be closer, and people will tend to patronize it. (center) **The problem with squares.** Squares nest together without gaps, but their sides are not equidistant from the center. If the market area is a circle, the radius—the distance from the center to the edge—can be measured because every point around a circle is the same distance from the center. But in a square, the distance from the center varies among points along a square. (right) **The hexagon compromise.** Geographers use hexagons to depict the market area of a good or service because hexagons offer a compromise between the geometric properties of circles and squares.

RANGE AND THRESHOLD OF A MARKET AREA

The market area of every service varies. To determine the extent of a market area, geographers need two pieces of information about a service: its range and its threshold (Figure 12-10).

RANGE OF A SERVICE. How far are you willing to drive for a pizza? To see a doctor for a serious problem? To watch a ball game? The **range** is the maximum distance people are willing to travel to use a service. The range is the radius of the circle (or hexagon) drawn to delineate a service's market area.

People are willing to go only a short distance for everyday consumer services, such as groceries and pharmacies. But they will travel longer distances for other services, such as a concert or professional ball game. Thus a convenience store has a small range, whereas a stadium has a large range. In a large urban settlement, for example, the range of a fast-food franchise such as McDonald's is roughly 5 kilometers (3 miles); the range of a casual dining chain such as Steak 'n Shake is roughly 8 kilometers (5 miles), and the range of a stadium is 100 kilometers (60 miles) or more.

As a rule, people tend to go to the nearest available service: Someone in the mood for a McDonald's hamburger is likely to go to the nearest McDonald's. Therefore, the range of a service must be determined from the radius of a circle that is irregularly shaped rather than perfectly round. The irregularly shaped circle takes in the territory for which the proposed site is closer than competitors' sites.

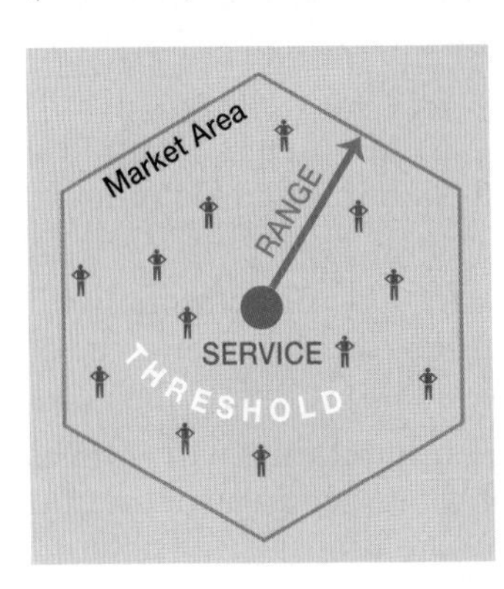

◀ **FIGURE 12-10 MARKET AREA, RANGE, AND THRESHOLD** The market area is the area of the hexagon, the range is the radius, and the threshold is a sufficient number of people inside the area to support the service.

The range must be modified further because most people think of distance in terms of time rather than in terms of a linear measure such as kilometers or miles. If you ask people how far they are willing to travel to a restaurant or a baseball game, they are more likely to answer in minutes or hours than in distance. If the range of a good or service is expressed in travel time, then the irregularly shaped circle must be drawn to acknowledge that travel time varies with road conditions. "One hour" may translate into traveling 90 kilometers (60 miles) while driving on an expressway but only 50 kilometers (30 miles) while driving congested city streets.

THRESHOLD OF A SERVICE. The second piece of geographic information needed to compute a market area is the **threshold**, which is the minimum number of people needed to support the service. Every enterprise has a minimum number of customers required to generate enough sales to make a profit. So once the range has been determined, a service provider must determine whether a location is suitable by counting the potential customers inside the irregularly shaped circle. Census data help to estimate the potential population within the circle.

How expected consumers inside the range are counted depends on the product. Convenience stores and fast-food restaurants appeal to nearly everyone, whereas other goods and services appeal primarily to certain consumer groups:

- Movie theaters attract younger people; chiropractors attract older folks.
- Poorer people are drawn to thrift stores; wealthier ones might frequent upscale department stores.
- Amusement parks attract families with children; nightclubs appeal to singles.

Developers of shopping malls, department stores, and large supermarkets may count only higher-income people, perhaps those whose annual incomes exceed $50,000. Even though the stores may attract individuals of all incomes, higher-income people are likely to spend more and purchase items that carry higher profit margins for the retailer.

Hierarchy of Consumer Services

Learning Outcome 12.2.2
Explain the distribution of different-sized settlements.

Only consumer services that have small thresholds, short ranges, and small market areas are found in small settlements because too few people live in small settlements to support many services. A large department store or specialty store cannot survive in a small settlement because the threshold (the minimum number of people needed) exceeds the population within range of the settlement.

Larger settlements provide consumer services that have larger thresholds, ranges, and market areas. Neighborhoods within large settlements provide services that have small thresholds and ranges. Services patronized by a small number of locals ("mom-and-pop stores") can coexist in a neighborhood with services that attract many from throughout the settlement. This difference is vividly demonstrated by comparing an on-line business directory for a small settlement with one for a major city. The major city's directory is much more extensive, with more services and diverse headings showing widely varied services that are unavailable in small settlements.

We spend as little time and effort as possible in obtaining consumer services and thus go to the nearest place that fulfills our needs. There is no point in traveling to a distant department store if the same merchandise is available at a nearby one. We travel greater distances only if the price is much lower or if the item is unavailable locally.

NESTING OF SERVICES AND SETTLEMENTS

According to central place theory, market areas across a developed country would be a series of hexagons of various sizes, unless interrupted by physical features such as mountains and bodies of water. Developed countries have numerous small settlements with small thresholds and ranges and far fewer large settlements with large thresholds and ranges.

The nesting pattern can be illustrated with overlapping hexagons of different sizes. Four different levels of market area—hamlet, village, town, and city—are shown in Figure 12-11. Hamlets with very small market areas are represented by the smallest contiguous hexagons. Larger hexagons represent the market areas of larger settlements and are overlaid on the smaller hexagons because consumers from smaller settlements shop for some goods and services in larger settlements.

In his original study, Walter Christaller showed that the distances between settlements in southern Germany followed a regular pattern. He identified seven sizes of settlements (market hamlet, township center, county seat, district city, small state capital, provincial head capital, and regional capital city). In southern Germany, the smallest settlement (market hamlet) had an average population of 800 and a market area of 45 square kilometers (17 square miles). The average distance between market hamlets was 7 kilometers (4.4 miles). The figures were higher for the average settlement at each increasing level in the hierarchy. Brian Berry has documented a similar hierarchy of settlements in parts of the U.S. Midwest.

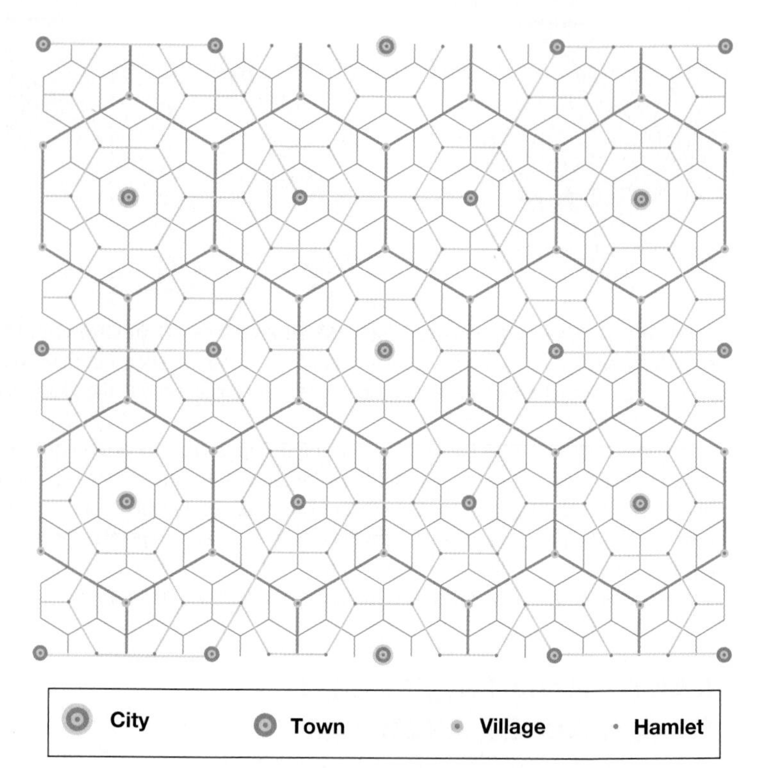

▲ **FIGURE 12-11 CENTRAL PLACE THEORY** According to central place theory, market areas are arranged in a regular pattern. Larger market areas, based in larger settlements, are fewer in number and farther apart from each other than smaller market areas and settlements. However, larger settlements also provide goods and services with smaller market areas; consequently, larger settlements have both larger and smaller market areas drawn around them.

Across much of the interior of the United States, a regular pattern of settlements can be observed, even if not precisely the same as the generalized model shown in Figure 12-11. North-central North Dakota is an example (Figure 12-12). Minot—the largest city in the area, with 41,000 inhabitants—is surrounded by:

- 7 small towns of between 1,000 and 5,000 inhabitants, with average ranges of 30 kilometers (20 miles) and market areas of around 2,800 square kilometers (1,200 square miles)
- 15 villages of between 100 and 999 inhabitants, with ranges of 20 kilometers (12 miles) and market areas of around 1,200 square kilometers (500 square miles)
- 19 hamlets of fewer than 100 inhabitants, with ranges of 15 kilometers (10 miles) and market areas of around 800 square kilometers (300 square miles)

◀ **FIGURE 12-12 CENTRAL PLACE THEORY IN NORTH DAKOTA** Central place theory helps explain the distribution of settlements of varying sizes in North Dakota. Larger settlements are fewer and farther apart, whereas smaller settlements are more numerous and closer together.

RANK-SIZE DISTRIBUTION OF SETTLEMENTS

In many developed countries, geographers observe that ranking settlements from largest to smallest (population) produces a regular pattern. This is the **rank-size rule**, in which the country's *n*th-largest settlement is $1/n$ the population of the largest settlement. In other words, the second-largest city is one-half the size of the largest, the fourth-largest city is one-fourth the size of the largest, and so on. When plotted on logarithmic paper, the rank-size distribution forms a fairly straight line. In the United States and a handful of other countries (Figure 12-13), the distribution of settlements closely follows the rank-size rule.

If the settlement hierarchy does not graph as a straight line, then the country does not follow the rank-size rule. Instead, it may follow the **primate city rule**, in which the largest settlement has more than twice as many people as the second-ranking settlement. In this distribution, the country's largest city is called the **primate city**. Mexico is an example of a country that follows the primate city distribution. Its largest city, Mexico City, is five times larger than its second-largest city, Guadalajara.

The existence of a rank-size distribution of settlements is not merely a mathematical curiosity. It has a real impact on the quality of life for a country's inhabitants. A regular hierarchy—as in the United States—indicates that the society is sufficiently wealthy to justify the provision of goods and services to consumers throughout the country. Conversely, the absence of the rank-size distribution in a developing country indicates that there is not enough wealth in the society to pay for a full variety of services. The absence of a rank-size distribution constitutes a hardship for people who must travel long distances to reach an urban settlement with shops and such services as hospitals. Because most people in developing countries do not have cars, buses must be provided to reach larger towns. A trip to a shop or a doctor that takes a few minutes in the United States could take several hours in a developing country.

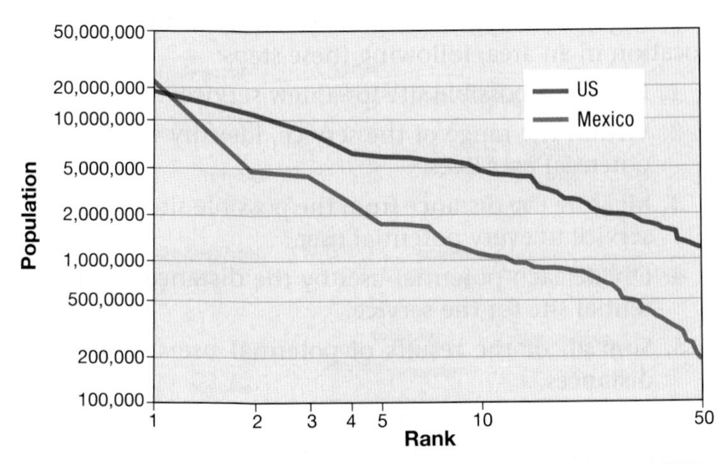

▲ **FIGURE 12-13 RANK-SIZE DISTRIBUTION OF SETTLEMENTS IN THE UNITED STATES AND MEXICO** The size of settlements follows the rank-size rule in the United States and the primate city rule in Mexico.

Pause and Reflect 12.2.2

According to the rank-size rule, the second-largest city in a country should have one-half the population of the largest city, and the tenth-largest city should have one-tenth the population of the largest city. Does Peru follow the rank-size rule or the primate city rule? Google "most populous cities in Peru."

PERIODIC MARKETS

Learning Outcome 12.2.4
Understand the role of periodic markets in the provision of services in developing countries.

Services at the lower end of the central place hierarchy may be provided at a periodic market, which is a collection of individual vendors who come together to offer goods and services in a location on specified days. A periodic market typically is set up in a street or other public space early in the morning, taken down at the end of the day, and set up in another location the next day (Figure 12-16).

A periodic market provides goods to residents of developing countries, as well as rural areas in developed countries, where sparse populations and low incomes produce purchasing power too low to support full-time retailing. A periodic market makes services available in more villages than would otherwise be possible, at least on a part-time basis. In urban areas, periodic markets offer residents fresh food brought in that morning from the countryside (Figure 12-17).

Many of the vendors in periodic markets are mobile, driving their trucks from farm to market, back to the farm to restock, then to another market. Other vendors, especially local residents who cannot or prefer not to travel to other villages, operate on a part-time basis, perhaps only a few times a year. Other part-time vendors are individuals who are capable of producing only a small quantity of food or handicrafts.

The frequency of periodic markets varies by culture:

- **Muslim countries.** Muslim countries typically conform to the weekly calendar—once a week in each of six cities and no market on Friday, the Muslim day of rest.

▼ **FIGURE 12-16 PERIODIC MARKET** The weekly market at Bati is considered the largest in Ethiopia.

▲ **FIGURE 12-17 BRINGING FOOD TO THE PERIODIC MARKET** Meat is carried to the periodic market at Gongtan, China.

- **Rural China.** According to G. William Skinner, rural China has a three-city, 10-day cycle of periodic markets. The market operates in a central market on days 1, 4, and 7; in a second location on days 2, 5, and 8; in a third location on days 3, 6, and 9; and no market on the tenth day. Three 10-day cycles fit in a lunar month.
- **Korea.** Korea has two 15-day market cycles in a lunar month.
- **Africa.** In Africa, the markets occur every 3 to 7 days. Variations in the cycle stem from ethnic differences.

Pause and Reflect 12.2.4
Identify an example of a periodic market in developed countries.

CHECK-IN: KEY ISSUE 2

Where Are Consumer Services Distributed?

- ✓ **Central place theory helps determine the most profitable location for a consumer service.**
- ✓ **A central place is surrounded by a market area that has a range and a threshold.**
- ✓ **Market areas of varying sizes nest and overlap.**
- ✓ **Regular patterns of settlements that provide consumer services can be observed, especially in developed countries.**

SUSTAINABILITY AND INEQUALITY IN OUR GLOBAL VILLAGE

Unequal Spatial Impacts of the Severe Recession

The severe global recession that began in 2008 hit some communities harder than others. As Figure 12-7 shows, developed countries were more severely impacted by the global recession. GNI declined more sharply in developed countries than in developing countries. The countries least affected by the global recession were the poorest countries of sub-Saharan Africa. Those countries are the most peripheral to the global economy.

Within the United States, the recession hit some communities harder than others (Figure 12-18). Some of the hardest-hit communities were industrial centers in the Midwest, where bankrupt carmakers Chrysler and GM were based. But most of the hardest-hit communities were in the South and West, regions that had been the most prosperous. Those communities were especially affected by declines in services, especially real estate and finance (Figure 12-19).

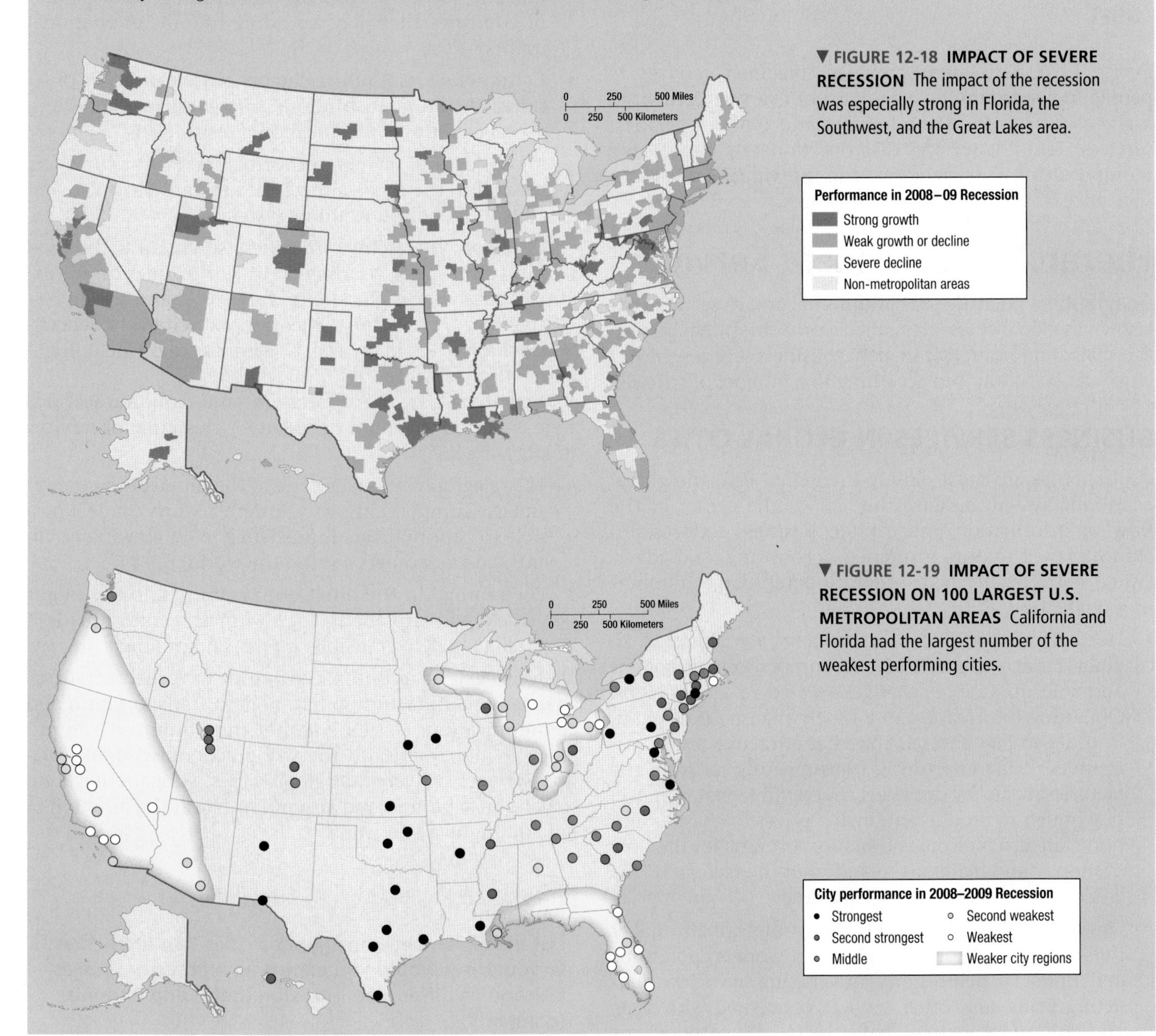

▼ **FIGURE 12-18 IMPACT OF SEVERE RECESSION** The impact of the recession was especially strong in Florida, the Southwest, and the Great Lakes area.

▼ **FIGURE 12-19 IMPACT OF SEVERE RECESSION ON 100 LARGEST U.S. METROPOLITAN AREAS** California and Florida had the largest number of the weakest performing cities.

KEY ISSUE 3

Where Are Business Services Distributed?

- Hierarchy of Business Services
- Business Services in Developing Countries
- Economic Base of Settlements

Learning Outcome 12.3.1
Describe the factors that are used to identify global cities.

Every urban settlement provides consumer services to people in a surrounding area, but not every settlement of a given size has the same number and types of business services. Business services disproportionately cluster in a handful of urban settlements, and individual settlements specialize in particular business services.

Hierarchy of Business Services

Geographers identify a handful of urban settlements known as global cities (also called world cities) that play an especially important role in global business services. Global cities can be subdivided according to a number of criteria.

BUSINESS SERVICES IN GLOBAL CITIES

Global cities are most closely integrated into the global economic system because they are at the center of the flow of information and capital. Business services, including law, banking, insurance, accounting, and advertising, concentrate in disproportionately large numbers in global cities:

- Headquarters of large corporations are clustered in global cities, and shares of these corporations are bought and sold on stock exchanges located in global cities. Obtaining information in a timely manner is essential in order to buy and sell shares at attractive prices. Executives of manufacturing firms meeting far from the factories make key decisions concerning what to make, how much to produce, and what prices to charge. Support staff also far from the factory accounts for the flow of money and materials to and from the factories. This work is done in offices in global cities.
- Lawyers, accountants, and other professionals cluster in global cities to provide advice to major corporations and financial institutions. Advertising agencies, marketing firms, and other services concerned with style and fashion locate in global cities to help corporations anticipate changes in taste and to help shape those changes.
- As centers for finance, global cities attract the headquarters of the major banks, insurance companies, and specialized financial institutions where corporations obtain and store funds for expansion of production.

Global cities are divided into three levels: alpha, beta, and gamma. These three levels in turn are further subdivided (Figure 12-20). A combination of economic, political, cultural, and infrastructure factors are used to identify global cities and to distinguish among the various ranks:

- **Economic factors.** Economic factors include number of headquarters for multinational corporations, financial institutions, and law firms that influence the global economy.
- **Political factors.** Political factors include hosting headquarters for international organizations and capitals of countries that play a leading role in international events.
- **Cultural factors.** Cultural factors include presence of renowned cultural institutions, influential media outlets, sports facilities, and educational institutions.
- **Infrastructural factors.** Infrastructural factors include a major international airport, health-care facilities, and advanced communications systems.

The same hierarchy of business services can be used within countries or continents. In North America, for example, below the alpha++ city (New York) and the alpha+ city (Chicago) are 4 alpha cities, 5 alpha– cities, 11 beta cities (including + and –), and 17 gamma cities (including + and –) (Figure 12-21).

New forms of transportation and communications were expected to reduce the need for clustering of services in large cities:

- The telegraph and telephone in the nineteenth century and the computer in the twentieth century made it possible to communicate immediately with coworkers, clients, and customers around the world.
- The railroad in the nineteenth century and the motor vehicle and airplane in the twentieth century made it possible to deliver people, inputs, and products quickly.

To some extent, economic activities have decentralized, especially manufacturing, but modern transportation and communications reinforce rather than diminish the primacy of global cities in the world economy. Transportation services converge on global cities. Global cities tend to have busy harbors and airports and lie at the junction of rail and highway networks.

Pause and Reflect 12.3.1
List the alpha, beta, and gamma cities that are nearest to you. How would you expect an alpha city such as Chicago to differ from Houston (beta) and Phoenix (gamma)?

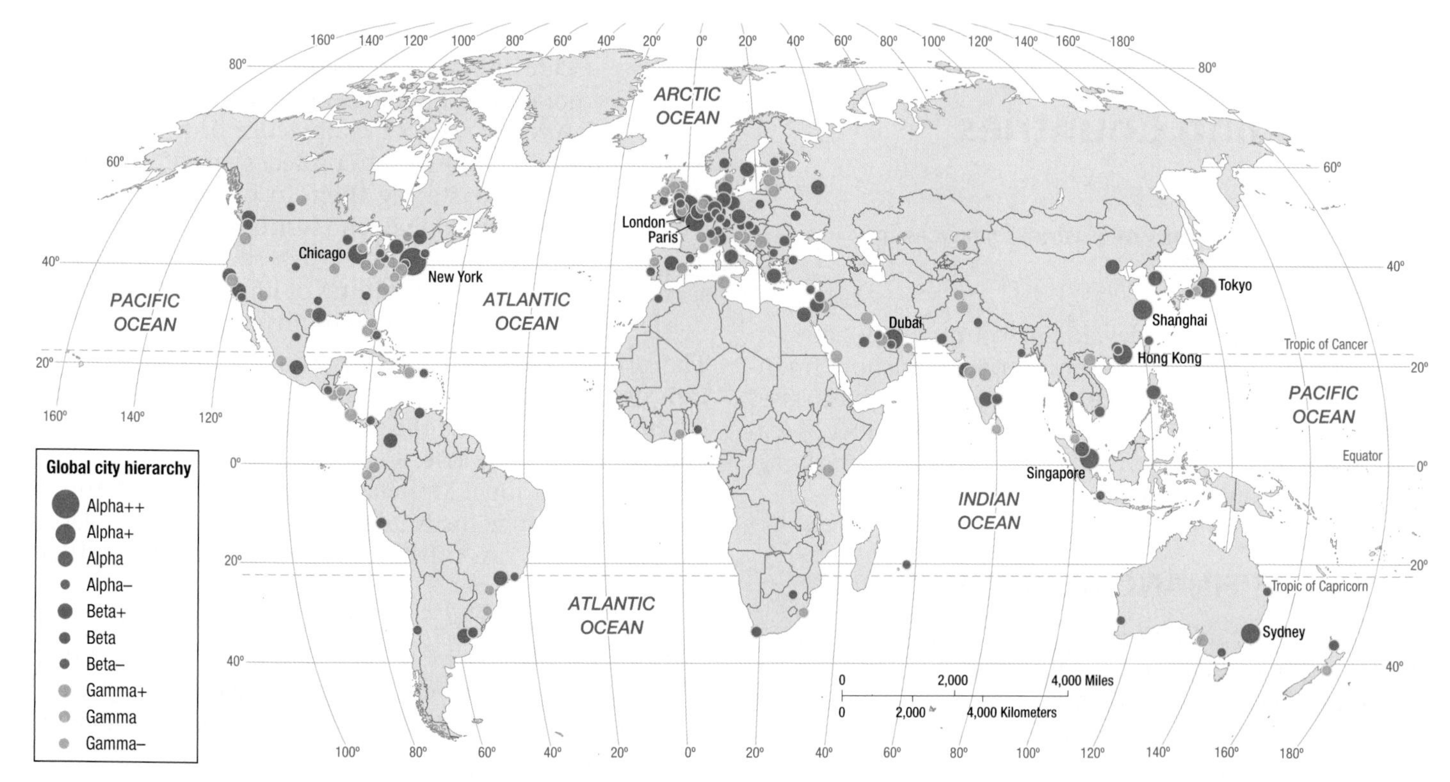

▲ **FIGURE 12-20 GLOBAL CITIES** Global cities are centers for the provision of services in the global economy. London and New York, the two dominant global cities, are ranked as alpha++. Other alpha, beta, and gamma global cities play somewhat less central roles in the provision of services than the two dominant global cities. Cities ranked alpha++ and alpha+ are labeled on the map.

CONSUMER AND PUBLIC SERVICES IN GLOBAL CITIES

Because of their large size, global cities have retail services with extensive market areas, but they may have even more retailers than large size alone would predict. A disproportionately large number of wealthy people live in global cities, so luxury and highly specialized products are especially likely to be sold there. Global cities typically offer the most plays, concerts, operas, night clubs, restaurants, bars, and professional sporting events. They contain the largest libraries, museums, and theaters. London presents more plays than the rest of the United Kingdom combined, and New York nearly has more theaters than the rest of the United States combined. Leisure services of national significance are especially likely to cluster in global cities, in part because they require large thresholds and large ranges and in part because of the presence of wealthy patrons.

Global cities may be centers of national or international political power. Most are national capitals, and they contain mansions or palaces for the head of state, imposing structures for the national legislature and courts, and offices for the government agencies. Also clustered in global cities are offices for groups having business with the government, such as representatives of foreign countries, trade associations, labor unions, and professional organizations. Unlike other global cities, New York is not a national capital. But as the home of the world's major international organization, the United Nations, it attracts thousands of diplomats and bureaucrats, as well as employees of organizations with business at the United Nations. Brussels is a global city because it is the most important center for European Union activities.

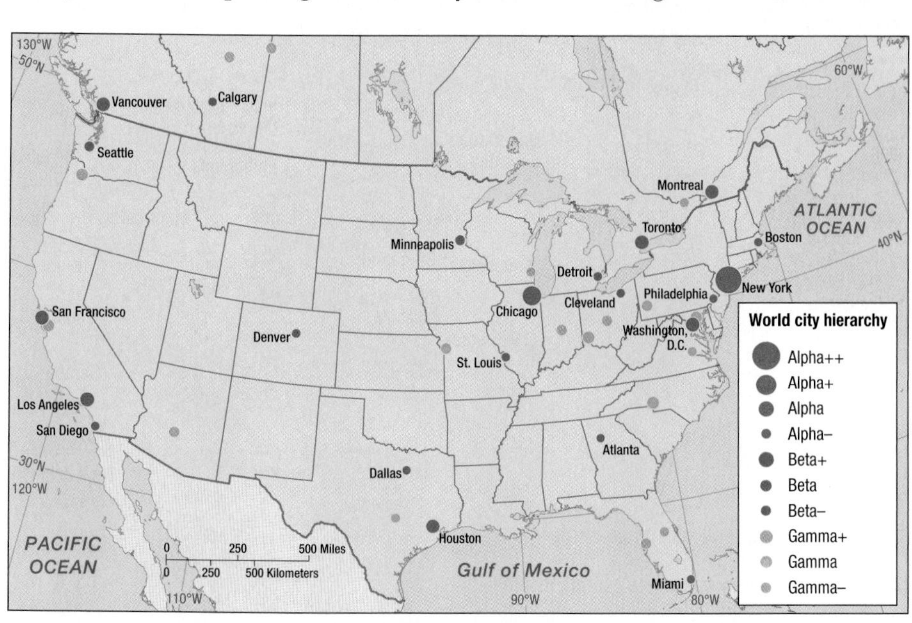

◀ **FIGURE 12-21 GLOBAL CITIES IN NORTH AMERICA** Atop the hierarchy of business services are New York and Chicago.

Business Services in Developing Countries

Learning Outcome 12.3.2
Explain the two types of business services in developing countries.

In the global economy, developing countries specialize in two distinctive types of business services: offshore financial services and back-office functions. These businesses tend to locate in developing countries for a number of reasons, including the presence of supportive laws, weak regulations, and low-wage workers.

OFFSHORE FINANCIAL SERVICES

Small countries, usually islands and microstates, exploit niches in the circulation of global capital by offering offshore financial services. Offshore centers provide two important functions in the global circulation of capital:

- **Taxes.** Taxes on income, profits, and capital gains are typically low or nonexistent. Companies incorporated in an offshore center also have tax-free status, regardless of the nationality of the owners. The United States loses an estimated $70 billion in tax revenue each year because companies operating in the country conceal their assets in offshore tax havens.
- **Privacy.** Bank secrecy laws can help individuals and businesses evade disclosure in their home countries. People and corporations in litigious professions, such as a doctor or lawyer accused of malpractice or the developer of a collapsed building, can protect some of their assets from lawsuits by storing them in offshore centers, as can a wealthy individual who wants to protect assets in a divorce. Creditors cannot reach such assets in bankruptcy hearings. Short statutes of limitation protect offshore accounts from long-term investigation.

The privacy laws and low tax rates in offshore centers can also provide havens to tax dodges and other illegal schemes. By definition, the extent of illegal activities is unknown and unknowable.

The International Monetary Fund, the United Nations, and the Tax Justice Institute identify the following places, among others, as offshore financial services centers (Figure 12-22):

- **Dependencies of the United Kingdom,** such as Anguilla, Cayman Islands, Montserrat, and the British Virgin Islands in the Caribbean; Guernsey/Sark/Alderney, Isle of Man, and Jersey in the English Channel; and Gibraltar, off Spain.
- **Dependencies of other countries,** such as Cook Island and Niue, controlled by New Zealand; Aruba, Curaçao, and Sint Maarten, controlled by the Netherlands; and Hong Kong and Macau, controlled by China.
- **Independent island countries,** such as The Bahamas, Barbados, Dominica, Grenada, St. Kitts & Nevis, St. Lucia,

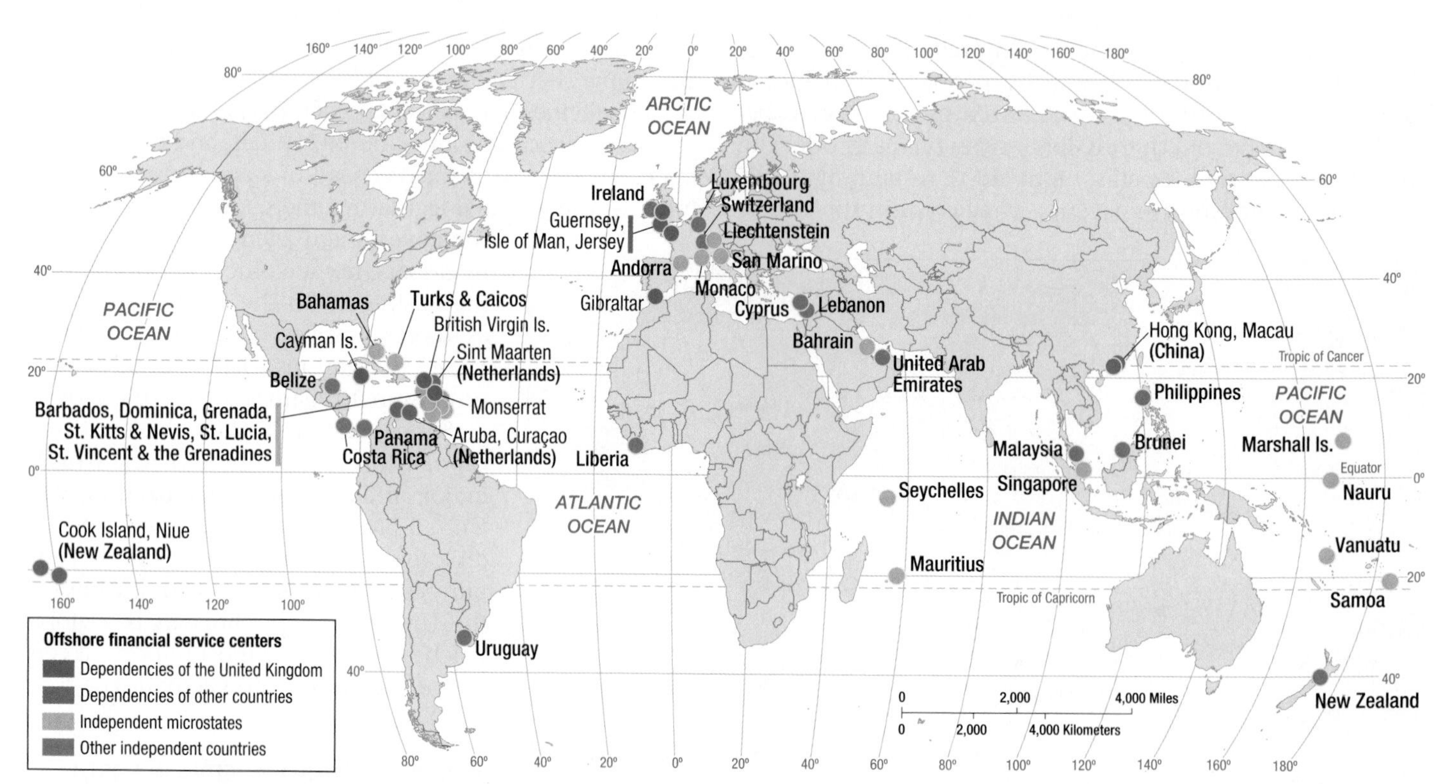

▲ **FIGURE 12-22 OFFSHORE FINANCIAL SERVICE CENTERS** Offshore financial service centers include microstates and dependencies of other countries.

St. Vincent & the Grenadines, and Turks & Caicos in the Caribbean; the Marshall Islands, Nauru, Samoa, and Vanuatu in the Pacific Ocean; and Mauritius and Seychelles in the Indian Ocean.

- **Other independent countries,** such as Andorra, Cyprus, Ireland, Liechtenstein, Luxembourg, Monaco, San Marino, and Switzerland in Europe; Belize, Costa Rica, Panama, and Uruguay in Latin America; Liberia in Africa; and Bahrain, Brunei, Lebanon, Malaysia, New Zealand, the Philippines, Singapore, and the United Arab Emirates in Asia.

A prominent example is the Cayman Islands, a British Crown Colony in the Caribbean near Cuba. The Caymans comprise three main islands and several smaller ones totaling around 260 square kilometers (100 square miles), with 40,000 inhabitants. Several hundred banks with assets of more than $1 trillion are legally based in the Caymans. Most of these banks have only a handful of people, if any, actually working in the Caymans.

In the Caymans, it is a crime to discuss confidential business—defined as matters learned on the job—in public. Assets placed in an offshore center by an individual or a corporation in a trust are not covered by lawsuits originating in the United States, Britain, or other service centers. To get at those assets, additional lawsuits would have to be filed in the offshore centers, where privacy laws would shield the individual or corporation from undesired disclosures.

BUSINESS-PROCESS OUTSOURCING

The second distinctive type of business service found in peripheral regions is back-office functions, also known as business-process outsourcing (BPO). Typical back-office functions include insurance claims processing, payroll management, transcription work, and other routine clerical activities (Figure 12-23). Back-office work also includes centers for responding to billing inquiries related to credit cards, shipments, and claims, or technical inquiries related to installation, operation, and repair.

▲ **FIGURE 12-23 CALL CENTER** Young Indians are recruited to work in call centers.

Traditionally, companies housed their back-office staff in the same office building downtown as their management staff, or at least in nearby buildings. A large percentage of the employees in a downtown bank building, for example, would be responsible for sorting paper checks and deposit slips. Proximity was considered important to assure close supervision of routine office workers and rapid turnaround of information.

Rising rents downtown have induced many business services to move routine work to lower-rent buildings elsewhere. In most cases, sufficiently low rents can be obtained in buildings in the suburbs or nearby small towns. However, for many business services, improved telecommunications have eliminated the need for spatial proximity.

Selected developing countries have attracted back offices for two reasons related to labor:

- **Low wages.** Most back-office workers earn a few thousand dollars per year—higher than wages paid in most other sectors of the economy, but only one-tenth the wages paid for workers performing similar jobs in developed countries. As a result, what is regarded as menial and dead-end work in developed countries may be considered relatively high-status work in developing countries and therefore able to attract better-educated, more-motivated employees in developing countries than would be possible in developed countries.
- **Ability to speak English.** Many developing countries offer lower wages than developed countries, but only a handful of developing countries possess a large labor force fluent in English. In Asia, countries such as India, Malaysia, and the Philippines have substantial numbers of workers with English-language skills, a legacy of British and American colonial rule. Major multinational companies such as American Express and General Electric have extensive back-office facilities in those countries.

The ability to communicate in English over the telephone is a strategic advantage in competing for back offices with neighboring countries, such as Indonesia and Thailand, where English is less commonly used. Familiarity with English is an advantage not only for literally answering the telephone but also for gaining a better understanding of the preferences of American consumers through exposure to English-language music, movies, and television.

Workers in back offices are often forced to work late at night, when it's daytime in the United States, peak demand for inquiries. Many employees must arrive at work early and stay late because they lack their own transportation, so they depend on public transportation, which typically does not operate late at night. Sleeping and entertainment rooms are provided at work to fill the extra hours.

Pause and Reflect 12.3.2

When it is 3 P.M. on a Tuesday where you live, what time and day is it at a call center in India? Refer to Figure 1-11.

Economic Base of Settlements

Learning Outcome 12.3.3
Explain the concept of economic base.

A settlement's distinctive economic structure derives from its **basic industries**, which export primarily to consumers outside the settlement. **Nonbasic industries** are enterprises whose customers live in the same community—essentially, consumer services. A community's unique collection of basic industries defines its **economic base**.

A settlement's economic base is important because exporting by the basic industries brings money into the local economy, thus stimulating the provision of more nonbasic consumer services for the settlement. New basic industries attract new workers to a settlement, and these workers bring their families with them. The settlement then attracts additional consumer services to meet the needs of the new workers and their families. Thus a new basic industry stimulates establishment of new supermarkets, laundromats, restaurants, and other consumer services. But a new nonbasic service, such as a supermarket, will not induce construction of new basic industries.

A community's basic industries can be identified by computing the percentage of the community's workers employed in different types of businesses. The percentage of workers employed in a particular industry in a settlement is then compared to the percentage of all workers in the country employed in that industry. If the percentage is much higher in the local community, then that type of business is a basic economic activity.

SPECIALIZATION OF CITIES IN DIFFERENT SERVICES

Settlements in the United States can be classified by their type of basic activity (Figure 12-24). Each type of basic activity has a different spatial distribution. The concept of basic industries originally referred to manufacturing. Some communities specialize in durable manufactured goods, such as steel and automobiles, others in nondurable manufactured goods, such as textiles, apparel, food, chemicals, and paper. Most communities that have an economic base of manufacturing durable goods are clustered between northern Ohio and southeastern Wisconsin, near the southern Great Lakes. Nondurable manufacturing industries, such as textiles, are clustered in the Southeast, especially in the Carolinas.

In a postindustrial society, such as the United States, increasingly the basic economic activities are in business, consumer, or public services. Geographers Ó hUallacháin and Reid have documented examples of settlements that specialize in particular types of services. Examples of settlements specializing in business services include:

- **General business:** Large metropolitan areas, especially Chicago, Los Angeles, New York, and San Francisco.
- **Computing and data processing services:** Boston and San Jose.
- **High-tech industries support services:** Austin, Orlando, and Raleigh-Durham.
- **Military activity support services:** Albuquerque, Colorado Springs, Huntsville, Knoxville, and Norfolk.
- **Management-consulting services:** Washington, D.C.

Examples of settlements specializing in consumer services include:

- **Entertainment and recreation:** Atlantic City, Las Vegas, and Reno.
- **Medical services:** Rochester, Minnesota.

Examples of settlements specializing in public services include:

- **State capitals:** Sacramento and Tallahassee.
- **Large universities:** Tuscaloosa.
- **Military bases:** Arlington.

Although the populations of cities in the South and West have grown more rapidly in recent years, Ó hUallacháin and Reid found that cities in the North and East have expanded their provision of business services more rapidly. Northern and eastern cities that were once major manufacturing centers have been transformed

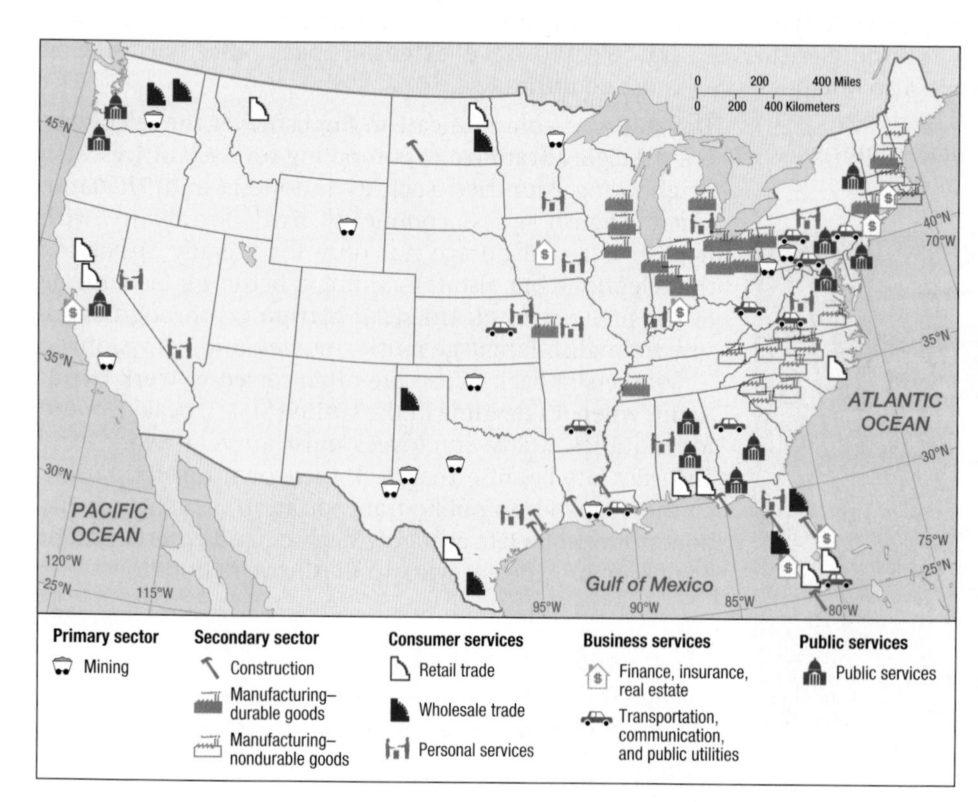

◀ **FIGURE 12-24 ECONOMIC BASE OF U.S. CITIES** Cities specialize in different economic activities.

into business service centers. These cities have moved more aggressively to restructure their economic bases to offset sharp declines in manufacturing jobs.

Steel was once the most important basic industry of Cleveland and Pittsburgh, but now health services such as hospitals and clinics and medical high-technology research are more important. Baltimore once depended for its economic base on manufacturers of fabricated steel products, such as Bethlehem Steel, General Motors, and Westinghouse. The city's principal economic asset was its port, through which raw materials and fabricated products passed. As these manufacturers declined, the city's economic base turned increasingly to services, taking advantage of its clustering of research-oriented universities, especially in medicine. The city is trying to become a center for the provision of services in biotechnology.

DISTRIBUTION OF TALENT

Individuals possessing special talents are not distributed uniformly among cities. Some cities have a higher percentage of talented individuals than others (Figure 12-25). To some extent, talented individuals are attracted to the cities with the most job opportunities and financial incentives. But the principal enticement for talented individuals to cluster in some cities more than others is cultural rather than economic, according to research conducted by Richard Florida. Individuals with special talents gravitate toward cities that offer more cultural diversity.

Florida measured talent as a combination of the percentage of people in the city with college degrees, the percentage employed as scientists or engineers, and the percentage employed as professionals or technicians. He used three measures of cultural diversity: the number of cultural facilities per capita, the percentage of gay men, and a "coolness" index. The "coolness" index, developed by *POV Magazine*, combined the percentage of population in their 20s, the number of bars and other nightlife places per capita, and the number of art galleries per capita (Figure 12-26). A city's gay population was based on census figures for the percentage of households consisting of two adult men. Two adult men who share a house may not be gay, but Florida assumed that the percentage of adult men living together who were gay did not vary from one city to another.

Florida found a significant positive relationship between the distribution of talent and the distribution of diversity in the largest U.S. cities. In other words, cities with high cultural diversity tended to have relatively high percentages of talented individuals. Washington, San Francisco, Boston, and Seattle ranked among the top in both talent and diversity, whereas Las Vegas was near the bottom in both. Attracting talented individuals is important for a city because these individuals are responsible for promoting economic innovation. They are likely to start new businesses and infuse the local economy with fresh ideas.

CHECK-IN: KEY ISSUE 3

Where Are Business Services Distributed?

- ✓ **Business services cluster in global cities.**
- ✓ **Developing countries provide offshore financial services and business-process outsourcing.**
- ✓ **Communities specialize in the provision of particular services; the specialized services constitute a community's economic base.**

▲ **FIGURE 12-25 GEOGRAPHY OF TALENT** Some cities have concentrations of scientists and professionals.

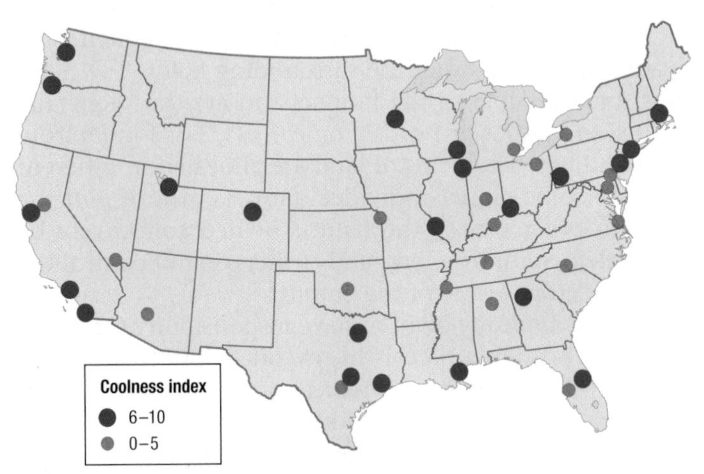

▲ **FIGURE 12-26 GEOGRAPHY OF CULTURAL DIVERSITY** The map is based on a "coolness" index developed by *POV Magazine*.

KEY ISSUE 4

Why Do Services Cluster in Settlements?

- Services in Rural Settlements
- Urbanization

Learning Outcome 12.4.1
Describe the difference between clustered and dispersed rural settlements.

Services are clustered in settlements. Rural settlements are centers for agriculture and provide a small number of services. Urban settlements are centers for consumer and business services. One-half of the people in the world live in rural settlements and the other half in urban settlements.

Services in Rural Settlements

Rural settlements are either clustered or dispersed. A **clustered rural settlement** is an agricultural-based community in which a number of families live in close proximity to each other, with fields surrounding the collection of houses and farm buildings. A **dispersed rural settlement**, typical of the North American rural landscape, is characterized by farmers living on individual farms isolated from neighbors rather than alongside other farmers in settlements.

CLUSTERED RURAL SETTLEMENTS

A clustered rural settlement typically includes homes, barns, tool sheds, and other farm structures, plus consumer services, such as religious structures, schools, and shops. A handful of public and business services may also be present in a clustered rural settlement. In common language, such a settlement is called a *hamlet* or *village*.

Each person living in a clustered rural settlement is allocated strips of land in the surrounding fields. The fields must be accessible to the farmers and are thus generally limited to a radius of 1 or 2 kilometers (1/2 or 1 mile) from the buildings. The strips of land are allocated in different ways. In some places, individual farmers own or rent the land. In other places, the land is owned collectively by the settlement or by a lord, and farmers do not control the choice of crops or use of the output.

Farmers typically own, or have responsibility for, a collection of scattered parcels in several fields. This pattern of controlling several fragmented parcels of land has encouraged living in a clustered rural settlement to minimize travel time to the various fields. Traditionally, when the population of a settlement grew too large for the capacity of the surrounding fields, new settlements were established nearby. This was possible because not all land was under cultivation.

Homes, public buildings, and fields in a clustered rural settlement are arranged according to local cultural and physical characteristics. Clustered rural settlements are often arranged in one of two types of patterns: circular or linear.

CIRCULAR RURAL SETTLEMENTS. Circular rural settlements comprise a central open space surrounded by structures. The following are examples:

- Kraal villages in sub-Saharan Africa were built by the Maasi people, who are pastoral nomads. Women have the principal responsibility for constructing them. The kraal villages have enclosures for livestock in the center, surrounded by a ring of houses. Compare *kraal* to the English word *corral* (Figure 12-27).
- Gewandorf settlements were once found in rural Germany. von Thünen observed this circular rural pattern in his landmark agricultural studies during the nineteenth century (refer to Figure 10-52). Gewandorf settlements consisted of a core of houses, barns, and churches, encircled by different types of agricultural activities. Small garden plots were located in the first ring surrounding the village, with cultivated land, pastures, and woodlands in successive rings.

LINEAR RURAL SETTLEMENTS. Linear rural settlements comprise buildings clustered along a road, river, or dike to facilitate communications. The fields extend behind the buildings in long, narrow strips. Long-lot farms can be seen today along the St. Lawrence River in Québec (Figure 12-28).

In the French long-lot system, houses were erected along a river, which was the principal water source and means of communication. Narrow lots from 5 to 100 kilometers (3 to 60 miles) deep were established perpendicular

▼ **FIGURE 12-27 CIRCULAR RURAL SETTLEMENT** A kraal village, Kenya.

▲ FIGURE 12-28 **CLUSTERED LINEAR RURAL SETTLEMENT** Québec long lots.

to the river, so that each original settler had river access. This created a linear settlement along the river. These long, narrow lots were eventually subdivided. French law required that each son inherit an equal portion of an estate, so the heirs established separate farms in each division. Roads were constructed inland parallel to the river for access to inland farms. In this way, a new linear settlement emerged along each road, parallel to the original riverfront settlement.

CLUSTERED SETTLEMENTS IN COLONIAL AMERICA. New England colonists built clustered settlements centered on an open area called a common (Figure 12-29). Settlers grouped their homes and public buildings, such as the church and school, around the common. In addition to their houses, each settler had a home lot of ½ to 2 hectares (1 to 5 acres), which contained a barn, a garden, and enclosures for feeding livestock. New England colonists favored clustered settlements for several reasons:

- They typically traveled to the New World in a group. The English government granted an area of land, in New England perhaps 4 to 10 square miles (10 to 25 square kilometers). Members of the group then traveled to America to settle the land and usually built the settlement near the center of the land grant.
- The colonists wanted to live close together to reinforce common cultural and religious values. Most came from the same English village and belonged to the same church. Many of them left England in the 1600s to gain religious freedom. The settlement's leader was often an official of the Puritan Church, and the church played a central role in daily activities.
- They clustered their settlements for defense against attacks by Native Americans.

Each villager owned several discontinuous parcels on the periphery of the settlement to provide the variety of land types needed for different crops. Beyond the fields, the town held pastures and woodland for the common use of all residents. Outsiders could obtain land in the settlement only by gaining permission from the town's residents. Land was not sold but rather was awarded to an individual when the town's residents felt confident that the recipient would work hard. Settlements accommodated a growing population by establishing new settlements nearby. As in the older settlements, the newer ones contained central commons surrounded by houses and public buildings, home lots, and outer fields.

The contemporary New England landscape contains remnants of the old clustered rural settlement pattern. Many New England towns still have a central common surrounded by the church, school, and various houses. However, quaint New England towns are little more than picturesque shells of clustered rural settlements because today's residents work in shops and offices rather than on farms.

▼ FIGURE 12-29 **CLUSTERED COLONIAL AMERICAN SETTLEMENT** Newfane, Vermont, includes a courthouse and church buildings clustered around a central common.

Pause and Reflect 12.4.1
How might the presence of clustered rural settlements in New England have contributed to the region's distinctive dialect of English noted in Chapter 5?

DISPERSED RURAL SETTLEMENTS

Learning Outcome 12.4.2
Explain the types of services in early settlements.

Dispersed rural settlements were more common in the American colonies outside New England. Meanwhile, in New England and in the United Kingdom, clustered rural settlements were converted to a dispersed pattern.

DISPERSED RURAL SETTLEMENTS IN THE UNITED STATES. The Middle Atlantic colonies were settled by more heterogeneous groups than those in New England. Colonists came from Germany, the Netherlands, Ireland, Scotland, and Sweden, as well as from England. Most arrived in Middle Atlantic colonies individually rather than as members of a cohesive religious or cultural group. Some bought tracts of land from speculators. Others acquired land directly from individuals who had been given large land grants by the English government, including William Penn (Pennsylvania), Lord Baltimore (Maryland), and Sir George Carteret (the Carolinas).

Dispersed settlement patterns dominated in the American Midwest in part because the early settlers came primarily from the Middle Atlantic colonies. The pioneers crossed the Appalachian Mountains and established dispersed farms on the frontier (Figure 12-30). Land was plentiful and cheap, and people bought as much as they could manage. In New England, a dispersed distribution began to replace clustered settlements in the eighteenth century. Eventually people bought, sold, and exchanged land to create large, continuous holdings instead of several isolated pieces.

The clustered rural settlement pattern worked when the population was low, but settlements had no spare land to meet the needs of a population that was growing through natural increase and net in-migration. A shortage of land eventually forced immigrants and children to strike out alone and claim farmland on the frontier. In addition, the cultural bonds that had created clustered rural settlements were weakened. Descendants of the original settlers were less interested in the religious and cultural values that had unified the original immigrants.

DISPERSED RURAL SETTLEMENTS IN THE UNITED KINGDOM. To improve agricultural production, a number of European countries converted their rural landscapes from clustered settlements to dispersed patterns. Dispersed settlements were considered more efficient for agriculture than clustered settlements. A prominent example was the **enclosure movement** in Great Britain, between 1750 and 1850. The British government transformed the rural landscape by consolidating individually owned strips of land surrounding a village into a single large farm owned by an individual. When necessary, the government forced people to give up their former holdings.

Owning several discontinuous fields around a clustered rural settlement had several disadvantages: Farmers lost time moving between fields, villagers had to build more roads to connect the small lots, and farmers were restricted in what they could plant. With the introduction of farm machinery, farms operated more efficiently at a larger scale.

The enclosure movement brought greater agricultural efficiency, but it destroyed the self-contained world of village life. Village populations declined drastically as displaced farmers moved to urban settlements. Because the enclosure movement coincided with the Industrial Revolution, villagers who were displaced from farming moved to urban settlements and became workers in factories and services. Some villages became the centers of the new, larger farms, but villages that were not centrally located to a new farm's extensive land holdings were abandoned and replaced with entirely new farmsteads at more strategic locations. As a result, the isolated, dispersed farmstead, unknown in medieval England, is now a common feature of that country's rural landscape.

▼ **FIGURE 12-30 DISPERSED RURAL SETTLEMENT** Wisconsin.

SERVICES IN EARLY SETTLEMENTS

Before the establishment of permanent settlements as service centers, people lived as nomads, migrating in small groups across the landscape in search of food and water. They gathered wild berries and roots or killed wild animals for food (see Chapter 10). At some point, groups decided to build permanent settlements. Several families clustered together in a rural location and obtained food in the surrounding area. What services would these nomads require? Why would they establish permanent settlements to provide these services?

▲ **FIGURE 12-31 EARLY SETTLEMENT** Karain Cave, Turkey. Evidence of human settlement has been found in the cave dating back 150,000–200,000 years.

No one knows the precise sequence of events through which settlements were established to provide services. Based on archaeological research, settlements probably originated to provide consumer and public services. Business services came later.

EARLY CONSUMER SERVICES. The earliest permanent settlements may have been established to offer consumer services, specifically places to bury the dead (Figure 12-31). Perhaps nomadic groups had rituals honoring the deceased, including ceremonies commemorating the anniversary of a death. Having established a permanent resting place for the dead, the group might then install priests at the site to perform the service of saying prayers for the deceased. This would have encouraged the building of structures—places for ceremonies and dwellings. By the time recorded history began about 5,000 years ago, many settlements existed, and some featured temples. In fact, until the invention of skyscrapers in the late nineteenth century, religious buildings were often the tallest structures in a community.

Settlements also may have been places to house families, permitting unburdened males to travel farther and faster in their search for food. Women kept "home and hearth," making household objects, such as pots, tools, and clothing, and educating the children. These household-based services evolved over thousands of years into schools, libraries, theaters, museums, and other institutions that create and store a group's values and heritage and transmit them from one generation to the next.

People also needed tools, clothing, shelter, containers, fuel, and other material goods. Settlements therefore became manufacturing centers. Men gathered the materials needed to make a variety of objects, including stones for tools and weapons, grass for containers and matting, animal hair for clothing, and wood for shelter and heat. Women used these materials to manufacture household objects and maintain their dwellings. The variety of consumer services expanded as people began to specialize. One person could be skilled at repairing tools, another at training horses. People could trade such services with one another. Settlements took on a retail-service function.

EARLY PUBLIC SERVICES. Public services probably followed religious activities into the early permanent settlements. A group's political leaders also chose to live permanently in the settlement, which may have been located for strategic reasons, to protect the group's land claims.

Everyone in a settlement was vulnerable to attack from other groups, so for protection, some members became soldiers, stationed in the settlement. The settlement likely was a good base from which the group could defend nearby food sources against competitors. For defense, the group might surround the settlement with a wall. Defenders were stationed at small openings or atop the wall, giving them a great advantage over attackers. Thus settlements became citadels—centers of military power. Walls proved an extremely effective defense for thousands of years, until warfare was revolutionized by the introduction of gunpowder in Europe in the fourteenth century.

EARLY BUSINESS SERVICES. Everyone in settlements needed food, which was supplied by the group through hunting or gathering. At some point, someone probably wondered: Why not bring in extra food for hard times, such as drought or conflict? This perhaps was the origin of transportation services.

Not every group had access to the same resources because of the varied distribution of vegetation, animals, fuel wood, and mineral resources across the landscape. People brought objects and materials they collected or produced into the settlement and exchanged them for items brought by others. Settlements became warehousing centers to store the extra food. The settlement served as neutral ground where several groups could safely come together to trade goods and services. To facilitate this trade, officials in the settlement provided producer services, such as regulating the terms of transactions, setting fair prices, keeping records, and creating a currency system.

Through centuries of experiments and accidents, residents of early settlements realized that some of the wild vegetation they had gathered could generate food if deliberately placed in the ground and nursed to maturity—in other words, agriculture, as described in Chapter 10. Over time, settlements became surrounded by fields, where people produced most of their food by planting seeds and raising animals rather than by hunting and gathering.

Pause and Reflect 12.4.2
Infer what functions caves might have served for early humans, in addition to burying the dead.

Urbanization

Learning Outcome 12.4.3
Identify important prehistoric, ancient, and medieval urban settlements.

Settlements existed prior to the beginning of recorded history around 5,000 years ago. With a few exceptions, these were rural settlements. As recently as 1800, only 3 percent of Earth's population lived in urban settlements. Two centuries later, one-half of the world's people live in urban settlements.

▲ FIGURE 12-33 **PREHISTORIC URBAN SETTLEMENT: UR** The remains of Ur, in present-day Iraq, provide evidence of early urban civilization. Ancient Ur was compact, perhaps covering 100 hectares (250 acres), and was surrounded by a wall. The most prominent building, the stepped temple, called a *ziggurat*, was originally constructed around 4,000 years ago. The ziggurat was originally a three-story structure with a base that was 64 by 46 meters (210 by 150 feet) and the upper stories stepped back. Four more stories were added in the sixth century B.C. Surrounding the ziggurat was a dense network of small residences built around courtyards and opening onto narrow passageways. The excavation site was damaged during the two wars in Iraq.

EARLIEST URBAN SETTLEMENTS

Settlements may have originated in Mesopotamia, part of the Fertile Crescent of Southwest Asia (see Figure 8-10), and diffused at an early date west to Egypt and east to China and to South Asia's Indus Valley. Or settlements may have originated independently in each of the four hearths. In any case, from these four hearths, settlements diffused to the rest of the world.

PREHISTORIC URBAN SETTLEMENTS. The earliest urban settlements were probably in the Fertile Crescent of Southwest Asia and North Africa (Figure 12-32). Among the oldest well-documented urban settlements is Ur in Mesopotamia (present-day Iraq). Ur, which means "fire," was where Abraham lived prior to his journey to Canaan in approximately 1900 B.C., according to the Bible. Archaeologists have unearthed ruins in Ur that date from approximately 3000 B.C. (Figure 12-33).

ANCIENT URBAN SETTLEMENTS. Settlements were first established in the eastern Mediterranean about 2500 B.C. The oldest settlements include Knossos on the island of Crete, Troy in Asia Minor (Turkey), and Mycenae in Greece. These settlements were trading centers for the thousands of islands dotting the Aegean Sea and the eastern Mediterranean and provided the government, military protection, and other public services for their surrounding hinterlands. They were organized into **city-states**—independent self-governing communities that included the settlement and nearby countryside.

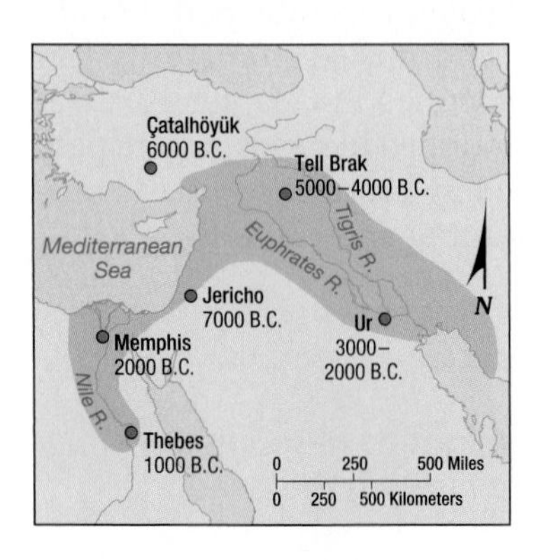

◀ FIGURE 12-32 **LARGEST URBAN SETTLEMENTS IN PREHISTORIC TIMES** The earliest known large urban settlements were in the Fertile Crescent of Southwest Asia and Egypt.

Athens, the largest city-state in ancient Greece (Figure 12-34), made substantial contributions to the development of culture, philosophy, and other elements of Western civilization, an example of the traditional distinction between urban settlements and rural. The urban settlements provided not only public services but also a concentration of consumer services, notably cultural activities, not found in smaller settlements.

The rise of the Roman Empire encouraged urban settlement. With much of Europe, North Africa, and Southwest Asia under Roman rule, settlements were established as centers of administrative, military, and other public services, as well as retail and other consumer services. Trade was encouraged through transportation and utility services, notably construction of many roads and aqueducts, and the security the Roman legions provided.

The city of Rome—the empire's center for administration, commerce, culture, and all other services—grew to at least 250,000 inhabitants, although some claim that the population may have reached 1 million. The city's centrality in the empire's communications network was reflected in the old saying "All roads lead to Rome."

With the fall of the Roman Empire in the fifth century, urban settlements declined. The empire's prosperity had rested on trading in the secure environment of imperial Rome. But with the empire fragmented under hundreds of rulers, trade diminished. Large urban settlements shrank or were abandoned. For several hundred years, Europe's cultural heritage was preserved largely in monasteries and isolated rural areas.

MEDIEVAL URBAN SETTLEMENTS. Urban life began to revive in Europe in the eleventh century, as feudal

▲ **FIGURE 12-34 ANCIENT URBAN SETTLEMENT: ATHENS** Dominating the skyline of modern Athens is the ancient hilltop site of the city, the Acropolis. Ancient Greeks selected this high place because it was defensible, and they chose it as a place to erect shrines to their gods. The most prominent structure on the Acropolis is the Parthenon, built in the fifth century B.C. to honor the goddess Athena. The structure in the foreground is the Herodes Atticus Odeon, a theater built in 161 A.D. Behind the Odeon is the Propylaea, which was the entrance gate to the Acropolis. To the right of the Parthenon, in the background, is the Chapel of St. George, built in the nineteenth century atop Mount Lycabettus, the highest point in Athens.

▲ **FIGURE 12-35 MEDIEVAL URBAN SETTLEMENT: CARCASSONNE** Medieval European cities, such as Carcassonne in southwestern France, were often surrounded by walls for protection. The walls have been demolished in most places, but they still stand around the medieval center of Carcassonne.

lords established new urban settlements. The lords gave residents charters of rights with which to establish independent cities in exchange for their military service. Both the lord and the urban residents benefited from this arrangement. The lord obtained people to defend his territory at less cost than maintaining a standing army. For their part, urban residents preferred periodic military service to the burden faced by rural serfs, who farmed the lord's land and could keep only a small portion of their own agricultural output.

With their newly won freedom from the relentless burden of rural serfdom, the urban dwellers set about expanding trade. Surplus from the countryside was brought into the city for sale or exchange, and markets were expanded through trade with other free cities. The trade among different urban settlements was enhanced by new roads and greater use of rivers. By the fourteenth century, Europe was covered by a dense network of small market towns serving the needs of particular lords.

The largest medieval European urban settlements served as power centers for the lords and church leaders, as well as major market centers. The most important public services occupied palaces, churches, and other prominent buildings arranged around a central market square. The tallest and most elaborate structures were usually churches, many of which still dominate the landscape of smaller European towns. In medieval times, European urban settlements were usually surrounded by walls even though by then cannonballs could destroy them (Figure 12-35). Dense and compact within the walls, medieval urban settlements lacked space for construction, so ordinary shops and houses nestled into the side of the walls and the large buildings. Most of these modest medieval shops and homes, as well as the walls, have been demolished in modern times, with only the

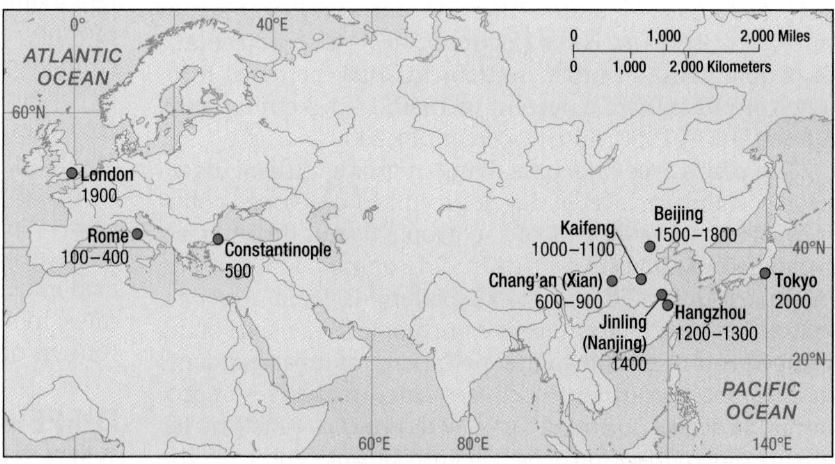

▲ **FIGURE 12-36 LARGEST SETTLEMENTS SINCE 1 A.D.** The largest cities have been in China for most of the past 2,000 years.

massive churches and palaces surviving. Modern tourists can appreciate the architectural beauty of these medieval churches and palaces, but they do not receive an accurate image of a densely built medieval town.

After the collapse of the Roman Empire, most of the world's largest urban settlements were clustered in China (Figure 12-36). Several cities in China are estimated to have exceeded 1 million inhabitants between 700 and 1800 A.D., including Chang'an (now Xian), Kaifeng, Hangzhou, Jinling (now Nanjing), and Beijing. London grabbed the title of world's largest urban settlement during the nineteenth century, as part of the Industrial Revolution. New York held the title briefly during the mid-twentieth century, and Tokyo is now considered to be the world's largest urban settlement.

Pause and Reflect 12.4.3

Medieval walled cities were constructed near political boundaries. How far is the medieval walled city of Carcassonne, France, from an international boundary?

RAPID GROWTH OF URBAN SETTLEMENTS

Learning Outcome 12.4.4
Explain the two dimensions of urbanization.

The process by which the population of urban settlements grows, known as **urbanization**, has two dimensions: an increase in the *number* of people living in urban settlements and an increase in the *percentage* of people living in urban settlements. The distinction between these two factors is important because they occur for different reasons and have different global distributions.

INCREASING PERCENTAGE OF PEOPLE IN URBAN SETTLEMENTS. The population of urban settlements exceeded that of rural settlements for the first time in human history in 2008 (Figure 12-37). The percentage of people living in urban settlements had increased from 3 percent in 1800 to 6 percent in 1850, 14 percent in 1900, 30 percent in 1950, and 47 percent in 2000.

The percentage of people living in urban settlements reflects a country's level of development. In developed countries, about three-fourths of the people live in urban areas, compared to about two-fifths in developing countries. The major exception to the global pattern is Latin America, where the urban percentage is comparable to the level of developed countries. The higher percentage of urban residents in developed countries is a consequence of changes in economic structure during the past two centuries—first the Industrial Revolution in the nineteenth century and then the growth of services in the twentieth. The world map of urban percentages looks very much like the world map of percentage of GDP derived from services (see Figure 12-2).

The percentage of urban dwellers is high in developed countries because over the past 200 years, rural residents have migrated from the countryside to work in the factories and services that are concentrated in cities. The need for fewer farm workers has pushed people out of rural areas, and rising employment opportunities in manufacturing and services have lured them into urban areas. Because everyone resides either in an urban settlement or a rural settlement, an increase in the percentage living in urban areas has produced a corresponding decrease in the percentage living in rural areas.

INCREASING NUMBER OF PEOPLE IN URBAN SETTLEMENTS. Developed countries have a higher percentage of urban residents, but developing countries have more of the very large urban settlements (Figure 12-38). Eight of the 10 most populous cities are currently in developing countries: Cairo, Delhi, Jakarta, Manila, Mexico City, São Paulo, Seoul, and Shanghai. New York and Tokyo are the two large cities in developed countries. In addition, 44 of the 50 largest urban settlements are in developing countries. That developing countries dominate the list of largest urban settlements is remarkable because urbanization was once associated with economic development. In 1800, 7 of the world's 10 largest cities were in Asia. In 1900, after diffusion of the Industrial Revolution from the United Kingdom to today's developed countries, all 10 of the world's largest cities were in Europe and North America.

In developing countries, migration from the countryside is fueling half of the increase in population in urban settlements, even though job opportunities may not be available. The other half results from high natural increase rates; in Africa, the natural increase rate accounts for three-fourths of urban growth.

DIFFERENCES BETWEEN URBAN AND RURAL SETTLEMENTS

A century ago, social scientists observed striking differences between urban and rural residents. Louis Wirth argued during the 1930s that an urban dweller follows a different way of life than does a rural dweller. Thus Wirth defined a city as a permanent settlement that has three characteristics: large size, high population density, and socially heterogeneous people. These characteristics produced differences in the social behavior of urban and rural residents.

LARGE SIZE. If you live in a rural settlement, you know most of the other inhabitants and may even be related to many of them. The people with whom you relax are probably the same ones you see in local shops and at church.

In contrast, if you live in an urban settlement, you can know only a small percentage of the other residents. You meet most of them in specific roles—your supervisor, your lawyer, your supermarket cashier, your electrician.

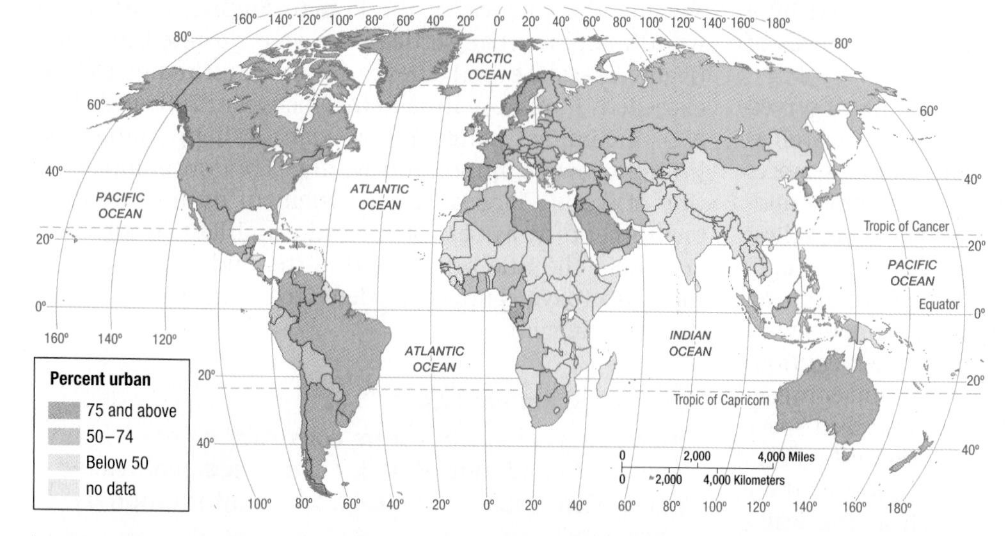

▲ **FIGURE 12-37 PERCENTAGE LIVING IN URBAN SETTLEMENTS** Developed countries have higher percentages of urban residents than do developing countries.

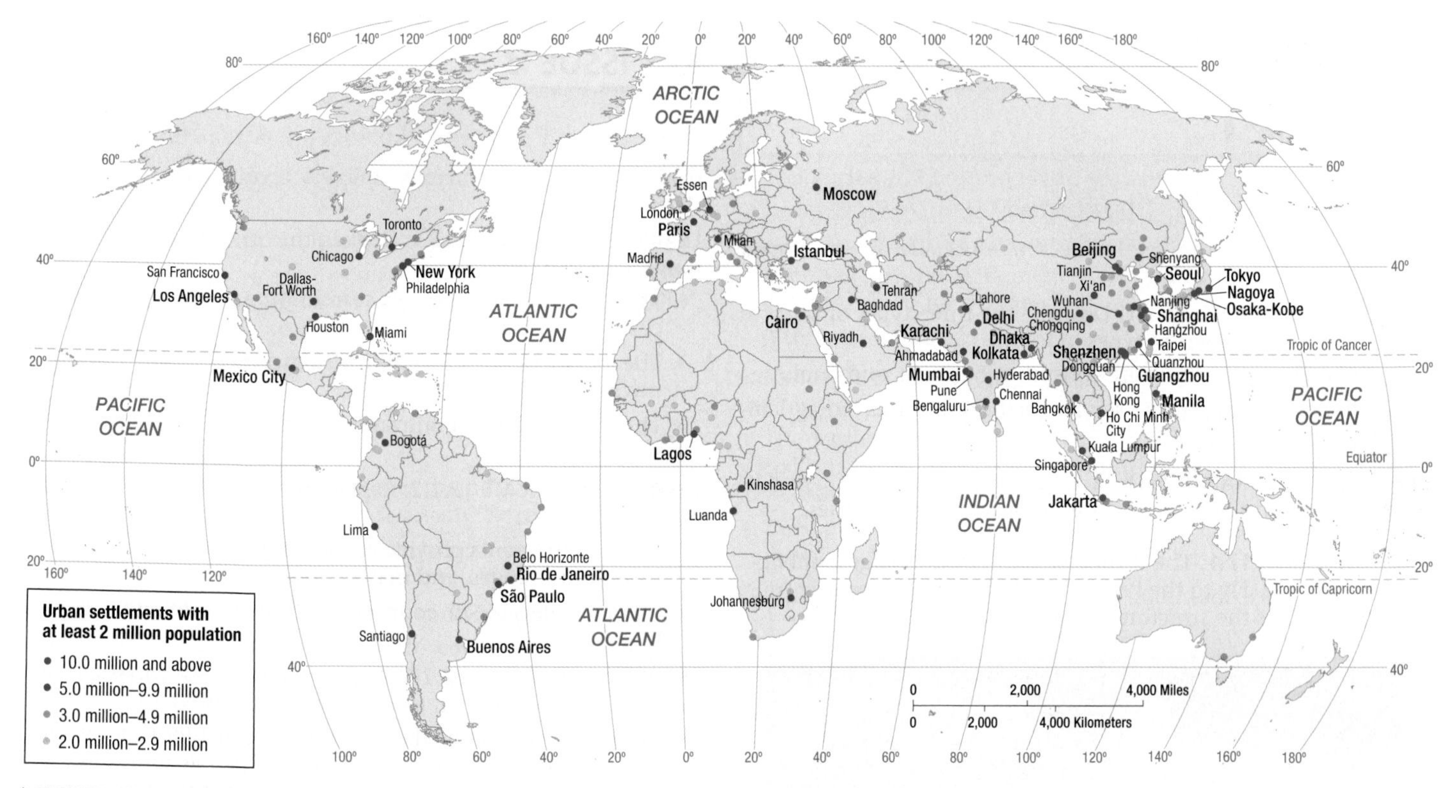

▲ **FIGURE 12-38 URBAN SETTLEMENTS WITH AT LEAST 2 MILLION INHABITANTS**
Most of the world's largest urban settlements are in developing countries, especially in East Asia, South Asia, and Latin America.

Most of these relationships are contractual: You are paid wages according to a contract, and you pay others for goods and services. Consequently, the large size of an urban settlement produces different social relationships than those formed in rural settlements.

HIGH DENSITY. High density also produces social consequences for urban residents, according to Wirth. The only way that a large number of people can be supported in a small area is through specialization. Each person in an urban settlement plays a special role or performs a specific task to allow the complex urban system to function smoothly. At the same time, high density also encourages social groups to compete to occupy the same territory.

SOCIAL HETEROGENEITY. The larger the settlement, the greater the variety of people. A person has greater freedom in an urban settlement than in a rural settlement to pursue an unusual profession, sexual orientation, or cultural interest. In a rural settlement, unusual actions might be noticed and scorned, but urban residents are more tolerant of diverse social behavior. Regardless of values and preferences, in a large urban settlement, individuals can find people with similar interests. But despite the freedom and independence of an urban settlement, people may also feel lonely and isolated. Residents of a crowded urban settlement often feel that they are surrounded by people who are indifferent and reserved.

Wirth's three-part distinction between urban and rural settlements may still apply in developing countries. But in developed countries, social distinctions between urban and rural life have blurred. According to Wirth's definition, nearly everyone in a developed country now is urban. All but 1 percent of workers in developed societies hold "urban" types of jobs. Nearly universal ownership of automobiles, telephones, televisions, and other modern communications and transportation has also reduced the differences between urban and rural lifestyles in developed countries. Almost regardless of where you live in a developed country, you have access to urban jobs, services, culture, and recreation.

CHECK-IN: KEY ISSUE 4

Why Do Services Cluster in Settlements?

- ✓ **Settlements are either rural or urban; rural settlements, which specialize in agricultural services, may be clustered or dispersed.**
- ✓ **Few humans lived in urban settlements until the nineteenth century.**
- ✓ **Developed countries have higher percentages of urban residents, but developing countries have most of the very large cities.**

Chapter 13 Urban Patterns

Why do British suburbs look different from American suburbs? Page 481

Why were these buildings blown up? Page 491

KEY ISSUE 1

Why Do Services Cluster Downtown?

The Center of it All 461

Downtown is the most distinctive area of most cities.

KEY ISSUE 2

Where Are People Distributed Within Urban Areas?

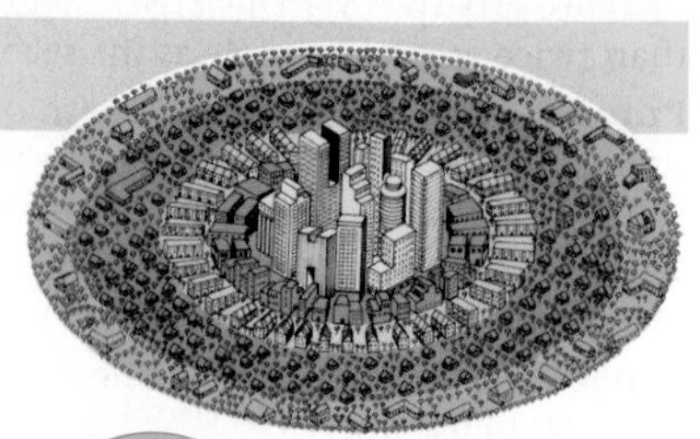

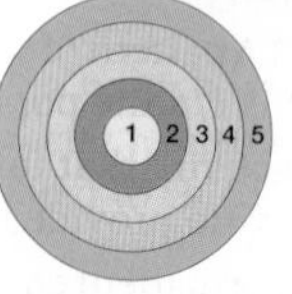

Rings, Wedges, and Nodes 466

Three models help to explain where different groups live within a city.

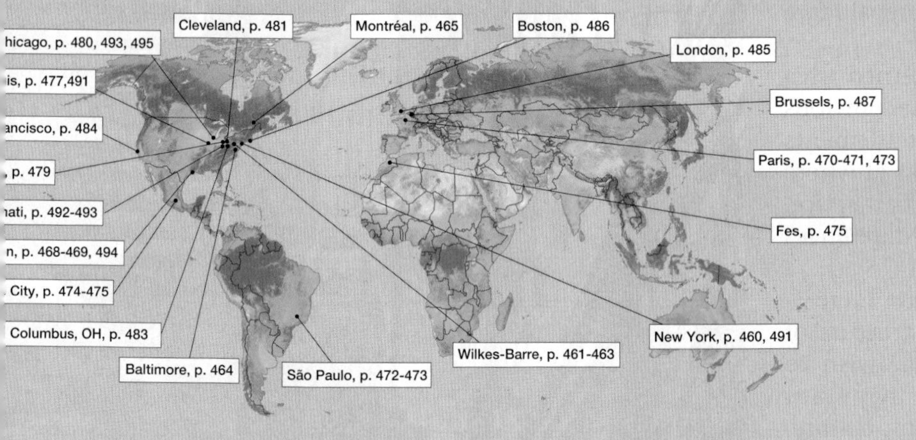

▲ These two passengers on a New York City bus represent the contrasts and diversity of a large city. The well-dressed woman in front uses her smart phone while the simply dressed woman behind her clutches her walker. When you are in a city, you are more likely than when you are in a small town to sit next to people who are different from you, but do the other passengers smile at you and chat, or do they mind their own business?

KEY ISSUE 3

Why Are Urban Areas Expanding?

Sprawling into Suburbs 476

Cities have spread out far into the countryside, along highway corridors.

KEY ISSUE 4

Why Do Cities Face Challenges?

Decline and Renewal 490

Cities display physical, social, and economic challenges and accomplishments.

PUBLIC SERVICES IN CBDs

Learning Outcome 13.1.1
Describe the three types of services found in a CBD.

Public services typically located in a CBD include city hall, courts, county and state agencies, and libraries (Figure 13-4). These facilities historically clustered downtown, in many cases in substantial structures. Today, many remain in the CBD to facilitate access for people living in all parts of town. Similarly, semipublic services such as places of worship and social service agencies also cluster downtown in handsome historic structures.

Sports facilities and convention centers have been constructed or expanded downtown in many cities. These structures attract a large number of people, including many suburbanites and out-of-towners. Cities place these facilities in the CBD because they hope to stimulate more business for downtown restaurants, bars, and hotels. Cities such as Wilkes-Barre have reclaimed their waterfronts as public park space.

BUSINESS SERVICES IN CBDs

Offices cluster in a CBD for accessibility (Figure 13-5). People in business services such as advertising, banking, finance, journalism, and law particularly depend on proximity to professional colleagues. Lawyers, for example, choose locations near government offices and courts. Services such as temporary secretarial agencies and instant printers locate downtown to be near lawyers, forming a chain of interdependency that continues to draw offices to the center city.

▲ **FIGURE 13-5 BUSINESS SERVICES IN WILKES-BARRE'S CBD** Downtown buildings house offices for financial and professional services.

Even with the diffusion of modern telecommunications, many professionals still exchange information with colleagues primarily through face-to-face contact. Financial analysts discuss attractive stocks or impending corporate takeovers. Lawyers meet to settle disputes out of court. Offices are centrally located to facilitate rapid communication of fast-breaking news through spatial proximity. Face-to-face contact also helps establish a relationship of trust based on shared professional values.

A central location also helps businesses that employ workers from a variety of neighborhoods. Top executives may live in one neighborhood, junior executives in another, secretaries in another, and custodians in still another. Only a central location is readily accessible to all groups. Firms that need highly specialized employees are more likely to find them in the central area, perhaps currently working for another company downtown.

▼ **FIGURE 13-4 PUBLIC SERVICES IN WILKES-BARRE'S CBD Much of Wilkes-Barre's CBD is devoted to public services, such as the Luzerne County Courthouse.**

CONSUMER SERVICES IN CBDs

In the past, three types of retail services clustered in a CBD because they required accessibility to everyone in the region: retailers with a high threshold, those with a high range, and those that served people who worked in the CBD. Changing shopping habits and residential patterns have reduced the importance of retail services in the CBD. Some downtowns have actively encouraged leisure services, such as theaters. In Wilkes-Barre, for example, an abandoned movie theater built in 1938 was converted into the F. M. Kirby Center for the Performing Arts in 1986 (Figure 13-6).

RETAILERS WITH A HIGH THRESHOLD. Retailers with high thresholds, such as department stores, traditionally preferred a CBD location in order to be accessible to many people. Large department stores in the CBD would cluster near one intersection, which was known as the "100 percent corner." Rents were highest there because that location had the highest accessibility for the most customers.

▲ **FIGURE 13-6 CONSUMER SERVICES IN WILKES-BARRE'S CBD** Kirby Center for the Performing Arts is in downtown Wilkes-Barre.

Most high-threshold shops such as large department stores have closed their downtown branches. CBDs that once boasted three or four stores now have none, or perhaps one struggling survivor. The customers for downtown department stores now consist of downtown office workers, inner-city residents, and tourists. Department stores with high thresholds are now more likely to be in suburban malls.

RETAILERS WITH A HIGH RANGE. High-range retailers are often specialists, with customers who patronize them infrequently. These retailers once preferred CBD locations because their customers were scattered over a wide area. For example, a jewelry or clothing store attracted shoppers from all over the urban area, but each customer visited infrequently. Like those with high thresholds, high-range retailers have moved with department stores to suburban locations.

RETAILERS SERVING DOWNTOWN WORKERS. A third type of retail activity in the CBD serves the many people who work in the CBD and shop during lunch or working hours. These retailers sell office supplies, computers, and clothing or offer shoe repair, rapid photocopying, dry cleaning, and so on. In contrast to the other two types of retailers, shops that appeal to nearby office workers are expanding in the CBD, in part because the number of downtown office workers has increased and in part because downtown offices require more services.

Patrons of downtown shops tend increasingly to be downtown employees who shop during the lunch hour. Thus, although the total volume of sales in downtown areas has been stable, the pattern of demand has changed. Large department stores have difficulty attracting their old customers, whereas smaller shops that cater to the special needs of the downtown labor force are expanding.

ACTIVITIES EXCLUDED FROM CBDs

High rents and land shortage discourage two principal activities in the CBD—industrial and residential.

LACK OF MANUFACTURING IN CBDs. Modern factories require large parcels of land to spread operations among one-story buildings. Suitable land is generally available in suburbs. In the past, inner-city factories and retail establishments relied on waterfront CBDs that were once lined with piers for cargo ships to load and unload and warehouses to store the goods. Today's large oceangoing vessels are unable to maneuver in the tight, shallow waters of the old CBD harbors. Consequently, port activities have moved to more modern facilities downstream.

Port cities have transformed their waterfronts from industry to commercial and recreational activities. Derelict warehouses and rotting piers have been replaced with new offices, shops, parks, and museums. As a result, CBD waterfronts have become major tourist attractions in a number of North American cities, including Boston, Toronto, Baltimore, and San Francisco, as well as in European cities such as Barcelona and London. The cities took the lead in clearing the sites and constructing new parks, docks, walkways, museums, and parking lots. They have also built large convention centers to house professional meetings and trade shows. Private developers have added hotels, restaurants, boutiques, and entertainment centers to accommodate tourists and conventioneers.

LACK OF RESIDENTS IN CBDs. Many people used to live in or near the CBD. Poorer people jammed into tiny, overcrowded apartments, and richer people built mansions downtown. In the twentieth century, most residents abandoned downtown living because of a combination of pull and push factors. They were pulled to suburbs that offered larger homes with private yards and modern schools. And they were pushed from CBDs by high rents that business and retail services were willing to pay and by the dirt, crime, congestion, and poverty that they experienced by living downtown.

In the twenty-first century, however, the population of many U.S. CBDs has increased. New apartment buildings and townhouses have been constructed, and abandoned warehouses and outdated office buildings have been converted into residential lofts. Downtown living is especially attractive to people without school-age children, either "empty nesters" whose children have left home or young professionals who have not yet had children. These two groups are attracted by the entertainment, restaurants, museums, and nightlife that are clustered downtown, and they are not worried about the quality of neighborhood schools.

Despite the growth in population in the center of some U.S. cities, some consumer services, such as grocery stores, may still be lacking (see Sustainability and Inequality feature on the next page).

Pause and Reflect 13.1.1

Do you ever spend time in a CBD? If so, for what reasons?

COMPETITION FOR LAND IN THE CBD

Learning Outcome 13.1.2
Explain the three-dimensional nature of a CBD.

A CBD's accessibility produces extreme competition for the limited sites available. As a result, land values are very high in the CBD, and it is too expensive for some activities. In a rural area a hectare of land might cost several thousand dollars. In a suburb it might run tens of thousands of dollars. In the CBD of a global city like London, if a hectare of land were even available, it would cost more than two hundred million dollars. If this page were a parcel of land in the CBD of London, it would sell for $1,000.

The intensive demand for space has given the CBD a three-dimensional character, pushing it vertically. Compared to other parts of a city, the CBD uses more space below and above ground level.

SUSTAINABILITY AND INEQUALITY IN OUR GLOBAL VILLAGE

Identifying Food Deserts

A **food desert** is an area in a developed country where healthy food is difficult to obtain. Food deserts are especially common in low-income inner-city areas.

In Baltimore, the Baltimore Food Policy Initiative is a joint venture of the Johns Hopkins University Center for a Livable Future and several local government agencies. The initiative prepared a food environment map and found that approximately 20 percent of Baltimore's residents lived in a food desert; the percentages were highest for children and for African Americans (Figure 13-7). An area was determined to be a food desert if it met all four of these criteria:

- The distance to the nearest supermarket was more than ¼ mile. This distance was chosen as the maximum convenient distance for walking with grocery bags.
- The median household income was at or below 185 percent of the federal poverty level.
- At least 40 percent of the area's households did not have any motor vehicles.
- The average Healthy Food Availability Index score was low for nearby supermarkets and convenience stores. This index was calculated by sending researchers into each market and assessing the availability of fresh and healthy food in the store, using a survey form called the Nutrition Environment Measures Survey developed at the University of Pennsylvania.

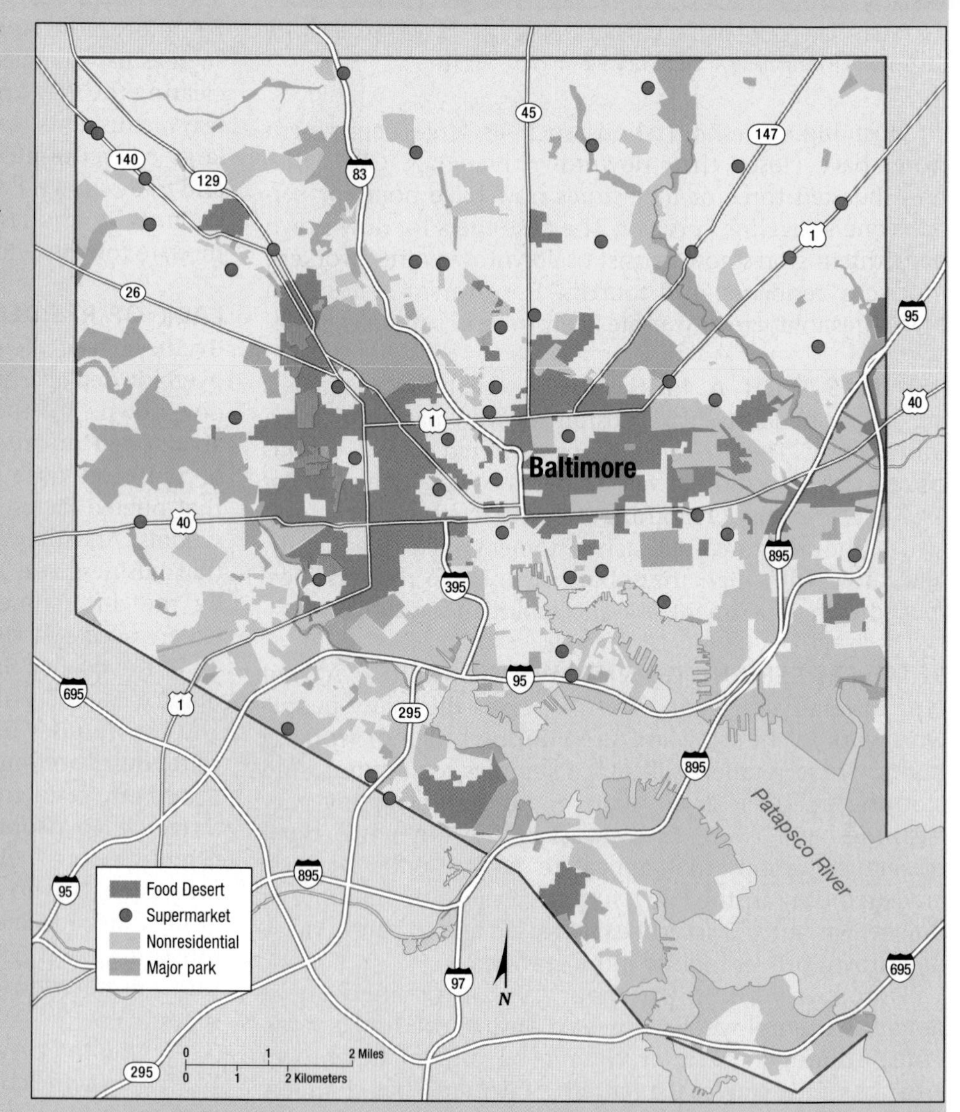

▲ FIGURE 13-7 **FOOD DESERTS IN BALTIMORE** Baltimore's food deserts are clustered in predominantly low-income African American inner-city neighborhoods.

◀ **FIGURE 13-8 UNDERGROUND CBD** Montreal's CBD has an extensive network of underground walkways lined with retail services.

THE UNDERGROUND CBD. A vast underground network exists beneath most CBDs. The typical "underground city" includes garages, loading docks for deliveries to offices and shops, and pipes for water and sewer service. Telephone, electric, TV, and broadband cables run beneath the surface as well because not enough space is available in the CBD for the large number of overhead poles that would be needed for such a dense network, and the wires would be unsightly and hazardous. Subway trains run beneath the streets of large CBDs. And cities in cold-weather climates, such as Minneapolis, Montreal, and Toronto, have built extensive underground pedestrian passages and shops. These underground areas segregate pedestrians from motor vehicles and shield them from harsh winter weather (Figure 13-8).

SKYSCRAPERS. Demand for space in CBDs has also made high-rise structures economically feasible. Downtown skyscrapers give a city one of its most distinctive images and unifying symbols. Suburban houses, shopping malls, and factories look much the same from one city to another, but each city has a unique downtown skyline, resulting from the particular arrangement and architectural styles of its high-rise buildings.

The first skyscrapers were built in Chicago in the 1880s, made possible by several inventions, including the elevator, steel girders, and glass structures because they blocked light and air movement. Artificial lighting, ventilation, central heating, and air-conditioning have helped solve these problems. Most North American and European cities enacted **zoning ordinances** early in the twentieth century in part to control the location and height of skyscrapers.

Skyscrapers are an interesting example of "vertical geography." The nature of an activity influences which floor it occupies in a typical high-rise:

- Retail services pay high rents for street-level space to entice customers.
- Business services, less dependent on walk-in trade, occupy offices on the middle levels at lower rents.
- Apartments on the upper floors take advantage of lower noise levels and panoramic views.

The one large U.S. CBD without skyscrapers is Washington, D.C., where no building is allowed to be higher than the U.S. Capitol dome. Consequently, offices in downtown Washington rise no more than 13 stories. As a result, the typical Washington office building uses more horizontal space—land area—than in other cities. Thus the city's CBD spreads over a much wider area than those in comparable cities.

Pause and Reflect 13.1.2

The Capitol is the tallest building in the CBD of Washington, D.C. Is Washington's CBD typical of American cities? Why or why not?

CHECK-IN: KEY ISSUE 1

Why Do Services Cluster Downtown?

- ✓ **Business, public, and some consumer services cluster in the CBD.**
- ✓ **The CBD has relatively few manufacturers and residents.**
- ✓ **North American CBDs are characterized by high-rise office buildings, as well as extensive underground services.**
- ✓ **Historic European CBDs have fewer high-rises and more residents and consumer services.**

KEY ISSUE 2

Where Are People Distributed within Urban Areas?

- **Models of Urban Structure**
- **Geographic Application of the Models**
- **Applying the Models Outside North America**

Learning Outcome 13.2.1
Describe the concentric zone, sector, and multiple nuclei models.

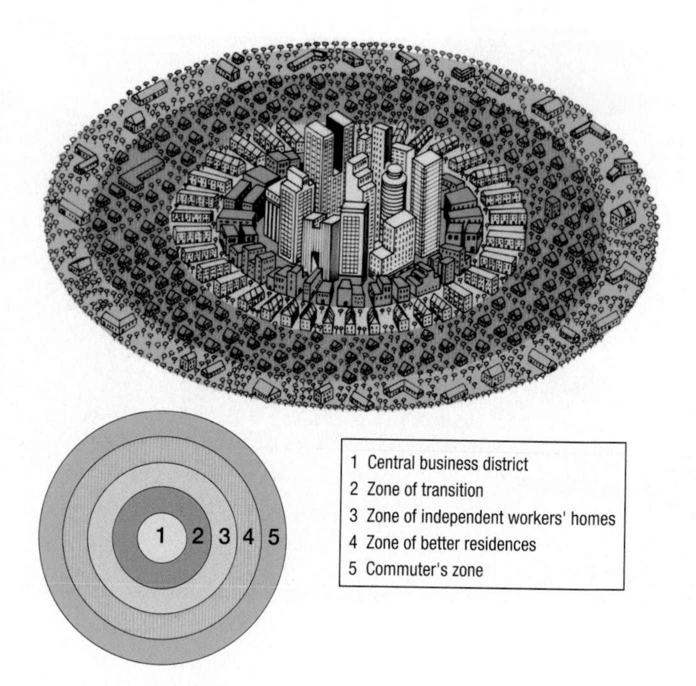

▲ FIGURE 13-9 **CONCENTRIC ZONE MODEL** According to this model, a city grows in a series of rings that surround the central business district.

People are not distributed randomly within an urban area. They concentrate in particular neighborhoods, depending on their social characteristics. Geographers describe where people with particular characteristics are likely to live within an urban area, and they offer explanations for why these patterns occur.

Models of Urban Structure

Sociologists, economists, and geographers have developed three models to help explain where different types of people tend to live in an urban area—the concentric zone, sector, and multiple nuclei models. The three models describing the internal social structure of cities were developed in Chicago, a city on a prairie. The three models were later applied to cities elsewhere in the United States and in other countries. Chicago includes a CBD known as the Loop because transportation lines (originally cable cars, now El trains) loop around it. Surrounding the Loop are residential suburbs to the south, west, and north. Except for Lake Michigan to the east, few physical features have interrupted Chicago's growth.

CONCENTRIC ZONE MODEL

The **concentric zone model** was the first model to explain the distribution of different social groups within urban areas (Figure 13-9). It was created in 1923 by sociologist E. W. Burgess. According to the concentric zone model, a city grows outward from a central area in a series of concentric rings, like the growth rings of a tree. The precise size and width of the rings vary from one city to another, but the same basic types of rings appear in all cities in the same order. Back in the 1920s, Burgess identified five rings:

1. **CBD**, the innermost ring, where nonresidential activities are concentrated.
2. **A zone in transition**, which contains industry and poorer-quality housing. Immigrants to the city first live in this zone in small dwelling units, frequently created by subdividing larger houses into apartments. The zone also contains rooming houses for single individuals.
3. **A zone of working-class homes**, which contains modest older houses occupied by stable, working-class families.
4. **A zone of better residences**, which contains newer and more spacious houses for middle-class families.
5. **A commuters' zone**, beyond the continuous built-up area of the city. Some people who work in the CBD nonetheless choose to live in small villages that have become dormitory towns for commuters.

Pause and Reflect 13.2.1
If you cut down a large tree, the cross-section will appear to be a circle with concentric rings. Which rings of the tree are the newest? Are tree rings a good analogy to the concentric zone model? Why or why not?

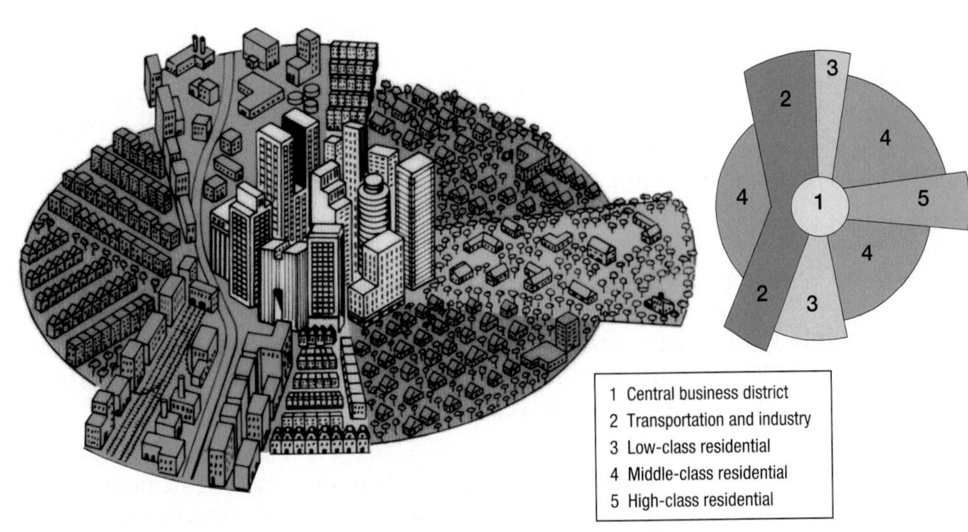

▲ **FIGURE 13-10 SECTOR MODEL** According to this model, a city grows in a series of wedges or corridors, which extend out from the central business district.

SECTOR MODEL

A second theory of urban structure, the **sector model**, was developed in 1939 by land economist Homer Hoyt (Figure 13-10). According to Hoyt, a city develops in a series of sectors, not rings. Certain areas of the city are more attractive for various activities, originally because of an environmental factor or even by mere chance. As a city grows, activities expand outward in a wedge, or sector, from the center.

Once a district with high-class housing is established, the most expensive new housing is built on the outer edge of that district, farther out from the center. The best housing is therefore found in a corridor extending from downtown to the outer edge of the city. Industrial and retailing activities develop in other sectors, usually along good transportation lines.

To some extent the sector model is a refinement of the concentric zone model rather than a radical restatement. Hoyt mapped the highest-rent areas for a number of U.S. cities at different times and showed that the highest social-class district usually remained in the same sector, although it moved farther out along that sector over time.

Hoyt and Burgess both claimed that social patterns in Chicago supported their model. According to Burgess, Chicago's CBD was surrounded by a series of rings, broken only by Lake Michigan on the east. Hoyt argued that the best housing in Chicago developed north from the CBD along Lake Michigan, whereas industry located along major rail lines and roads to the south, southwest, and northwest.

MULTIPLE NUCLEI MODEL

Geographers C. D. Harris and E. L. Ullman developed the **multiple nuclei model** in 1945. According to the multiple nuclei model, a city is a complex structure that includes more than one center around which activities revolve (Figure 13-11). Examples of these nodes include a port, a neighborhood business center, a university, an airport, and a park.

The multiple nuclei theory states that some activities are attracted to particular nodes, whereas others try to avoid them. For example, a university node may attract well-educated residents, pizzerias, and bookstores, whereas an airport may attract hotels and warehouses. On the other hand, incompatible land-use activities avoid clustering in the same locations. Heavy industry and high-class housing, for example, rarely exist in the same neighborhood.

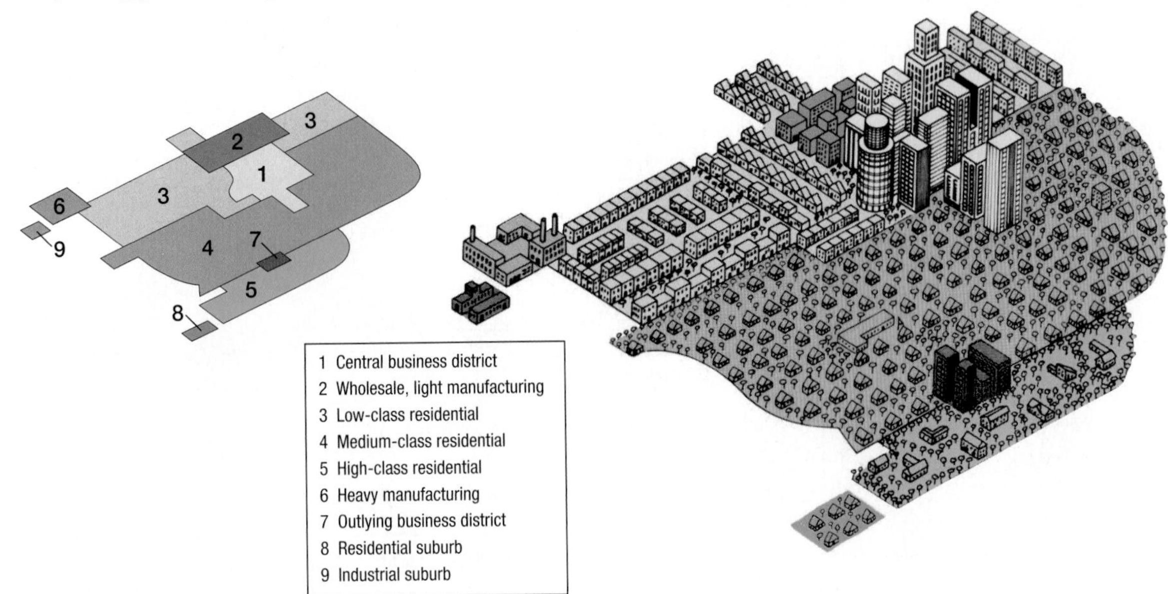

◀ **FIGURE 13-11 MULTIPLE NUCLEI MODEL** According to this model, a city consists of a collection of individual nodes, or centers, around which different types of people and activities cluster.

Geographic Applications of the Models

Learning Outcome 13.2.2:

Analyze how the three models help to explain where people live in an urban area.

The three models help us understand where people with different social characteristics tend to live within an urban area. They can also help explain why certain types of people tend to live in particular places. Effective use of the models depends on the availability of data at the scale of individual neighborhoods. In the United States and many other countries, that information comes from the census.

Urban areas in the United States are divided into **census tracts** that each contain approximately 5,000 residents and correspond, where possible, to neighborhood boundaries. Every decade the U.S. Bureau of the Census publishes data summarizing the characteristics of the residents and the housing in each tract. Estimates are also issued annually through the American Fact Finder service of the census's American Community Survey program. Examples of information the census provides include the number of nonwhites, the median income of all families, and the percentage of adults who finished high school. The spatial distribution of any of these social characteristics can be plotted on a map of the community's census tracts. Computers have become invaluable in this task because they permit rapid creation of maps and storage of voluminous data about each census tract. Social scientists can compare the distributions of characteristics and create an overall picture of where various types of people tend to live. This kind of study is known as **social area analysis**.

None of the three models taken individually completely explains why different types of people live in distinctive parts of a city. Critics point out that the models are too simple and fail to consider the variety of reasons that lead people to select particular residential locations. Because the three models are all based on conditions that existed in U.S. cities between the two world wars, critics also question their relevance to contemporary urban patterns in the United States or in other countries.

But if the models are combined rather than considered independently, they help geographers explain where different types of people live in a city. People tend to reside in certain locations, depending on their particular personal characteristics. This does not mean that everyone with the same characteristics must live in the same neighborhood, but the models say that most people live near others who have similar characteristics:

- **Applying the concentric zone model.** Consider two families with the same income and ethnic background. One family lives in a newly constructed home, whereas the other lives in an older one. The family in the newer house is much more likely to live in an outer ring and the family in the older house in an inner ring (Figure 13-12).

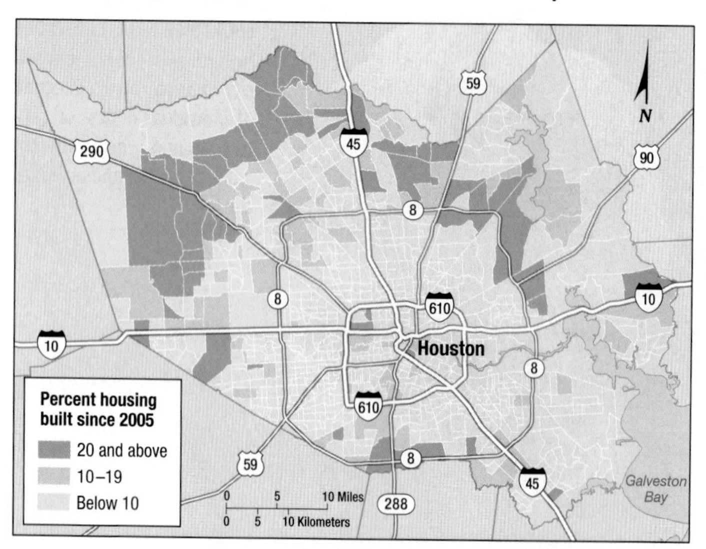

▲ **FIGURE 13-12 CONCENTRIC ZONES IN HOUSTON** Age of housing. Housing is newer in the outer rings of the city than in the inner rings.

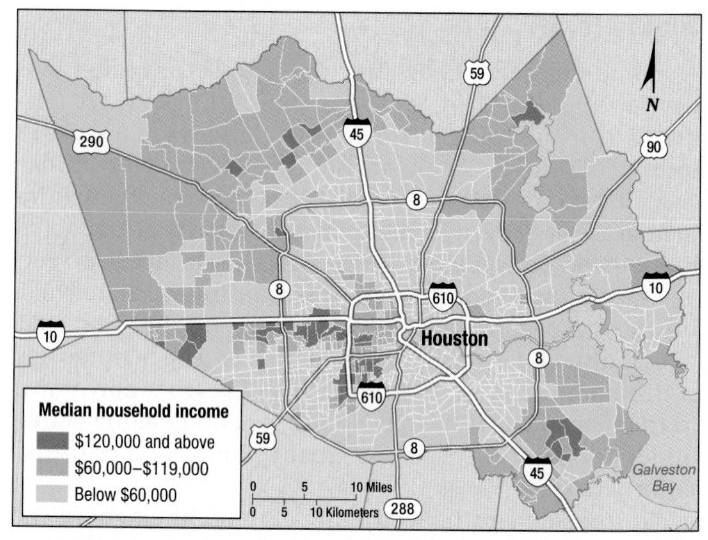

▲ **FIGURE 13-13 SECTOR MODEL IN HOUSTON** Distribution of high-income households. The median household income is the highest in a sector to the west.

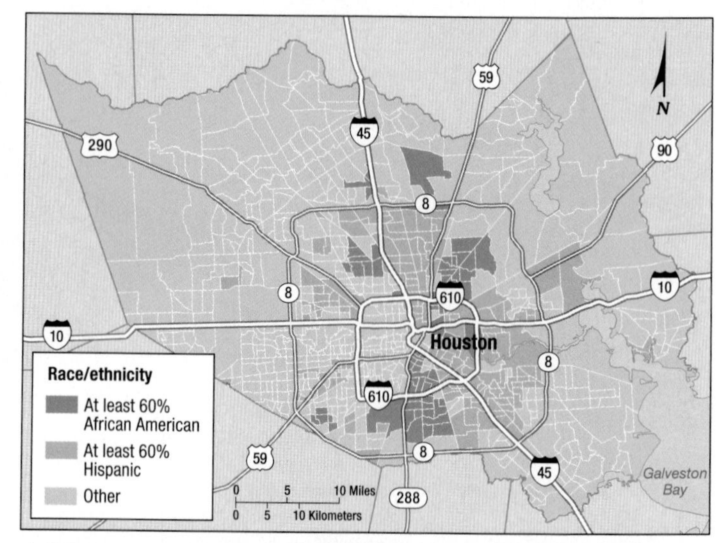

▲ **FIGURE 13-14 MULTIPLE NUCLEI MODEL IN HOUSTON** Distribution of minorities. Hispanics occupy nodes to the north and southeast of downtown, and African Americans occupy nodes to the south and northeast.

- **Applying the sector model.** Given two families who own their homes, the family with the higher income will not live in the same sector of the city as the family with the lower income (Figure 13-13).
- **Applying the multiple nuclei model.** People with the same ethnic or racial background are likely to live near each other (Figure 13-14).

Putting the three models together, we can identify, for example, the neighborhood in which a high-income, Asian American owner-occupant is most likely to live (see the Contemporary Geographic Tools feature).

Pause and Reflect 13.2.2

What are the five most important PRIZM clusters for your zip code? Google Nielsen Claritas PRIZM or go to www.claritas.com/MyBestSegments/Default.jsp?ID=20.

CONTEMPORARY GEOGRAPHIC TOOLS

Market Segmentation: You Are Where You Live

Marketing geographers identify sectors, rings, and nodes that come closest to matching customers preferred by a retailer. Companies use this information to understand, locate, and reach their customers better and to determine where to put new stores and where advertising should appear.

Segmentation is the process of partitioning markets into groups of potential customers with similar needs and characteristics who are likely to exhibit similar purchasing behavior. A prominent example of geographic segmentation is the Potential Rating Index by Zip Market (PRIZM) clusters created by Nielsen Claritas. As Nielsen Claritas states, "birds of a feather flock together"—in other words, a person is likely to live near people who are similar.

Nielsen Claritas combines two types of geographic information: distribution of the social and economic characteristics of people obtained from the census and the addresses of purchasers of various products obtained from service providers. The variables are organized into 66 clusters that are given picturesque names. For each zip code in the United States, Nielsen Claritas determines the five clusters that are most prevalent. Nielsen Claritas calls this analysis "you are where you live."

We can compare PRIZM clusters for two zip codes in the Houston area (Figure 13-15). Refer to Figures 13-12, 13-13, and 13-14 to see the close relationship between the Nielsen Claritas PRIZM clusters and the models of urban structure. Zip code 77004 is south of downtown Houston. The five most common clusters (in alphabetical order) are as follows:

- City Roots: older low-income ethnic minorities, living in older homes and apartments.
- Low-Rise Living: The lowest income of any PRIZM cluster; many are single parents who rent their homes and travel by bus rather than personal car.
- Multi-Culti Ethnic: Hispanics with modest incomes.
- Urban Achievers: young Hispanics.
- Urban Elders: elderly Hispanics.

Compare the above to the five most common PRIZM clusters in zip code 77079 in the western suburbs of Houston:

- Beltway Boomers: college-education, affluent, home-owning baby boomers.
- Executive Suites: upper-middle class couples with professional jobs.
- Gray Power: older couples living in quiet comfort.
- Pools & Patios: high-income older couples, with back-yard pools.
- Upper Crust: very high income couples, especially those with grown children, living an opulent lifestyle.

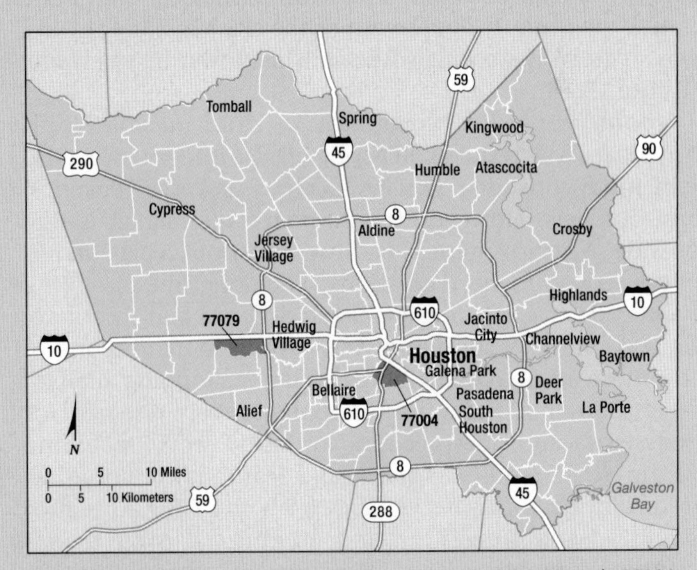

▲ **FIGURE 13-15 NIELSEN CLARITAS PRIZM CLUSTERS** Zip code 77004, immediately southeast of downtown Houston, is home to a large percentage of older homes occupied by Hispanics with modest incomes. Zip code 77079, in Houston's western suburbs, contains a large number of high-income, older college-educated couples.

Applying the Models Outside North America

Learning Outcome 13.2.3
Describe how the three models explain patterns in European cities.

The three models may describe the spatial distribution of social classes in the United States, but American urban areas differ from those elsewhere in the world. These differences do not invalidate the models, but they do point out that social groups in other countries may not have the same reasons for selecting particular neighborhoods within their cities.

CBDs IN EUROPE

Compared to CBDs in the United States, those outside North America are less dominated by skyscrapers for business services. The most prominent structures may be public and semipublic services, such as churches and former royal palaces, situated on the most important public squares, at road junctions, or on hilltops. Parks in the center of European cities often were first laid out as private gardens for aristocratic families and later were opened to the public.

European cities display a legacy of low-rise structures and narrow streets, built as long ago as medieval times. Some European cities try to preserve their historic CBDs by limiting high-rise buildings and the number of cars. Several high-rise offices were built for business services in Paris during the 1970s, including Europe's tallest office building (the 210-meter [688-foot] Tour Montparnasse). The public outcry over this disfigurement of the city's historic skyline was so great that officials reestablished lower height limits (Figure 13-16).

More people live downtown in cities outside North America. As a result, CBDs outside North America contain more consumer services, such as groceries, bakeries, and butchers. However, the 24-hour supermarket is rare outside North America because of shopkeeper preferences, government regulations, and longtime shopping habits. Many CBDs outside North America ban motor vehicles from busy shopping streets, thus emulating one of the most attractive attributes of large shopping malls—pedestrian-only walkways. Shopping streets reserved for pedestrians are widespread in Northern Europe, including in the Netherlands, Germany, and Scandinavia. Rome periodically bans private vehicles from the CBD to reduce pollution and congestion and minimize damage to ancient monuments.

Although constructing large new buildings is difficult, many shops and offices still wish to be in the center of European cities. The alternative to new construction is renovation of older buildings. However, renovation is more expensive and does not always produce enough space to meet the demand. As a result, rents are much higher in the center of European cities than in U.S. cities of comparable size.

▲ **FIGURE 13-16 SKYSCRAPER IN PARIS** The Tour Montparnasse, Europe's tallest building, dominates the skyline of Paris. The image illustrates the allocation of much of the land in a European CBD to public services, including the Ecole Military (Military Academy), which takes up the entire foreground, and behind it France's Finance Ministry (left) and the United Nations offices (right).

APPLYING THE MODELS IN EUROPE

To some extent, the models look similar in Europe to the way they look in the United States. In Paris, as in U.S. cities, newer housing is in the outer ring, and higher-income people cluster in a sector. These similarities mask important differences.

SECTORS IN EUROPEAN CITIES. In contrast to most U.S. cities, in Europe, wealthy people still live in the inner portions of the upper-class sector, not just in the suburbs. A central location provides proximity to the region's best shops, restaurants, cafés, and cultural facilities. Wealthy people are also attracted by the opportunity to occupy elegant residences in carefully restored, beautiful old buildings (Figure 13-17).

In Paris, for example, the wealthy lived near the royal palace (the Louvre) beginning in the twelfth century and the Palace of Versailles from the sixteenth century until the French Revolution in 1789. The preference of Paris's wealthy to cluster in a southwest sector was reinforced in the nineteenth century during the Industrial Revolution. Factories were built to the south, east, and north, along the Seine and Marne River valleys, and relatively few were built on the southwestern hills. Similar upper-class sectors emerged in the inner areas other European cities, typically on higher elevations and near royal palaces.

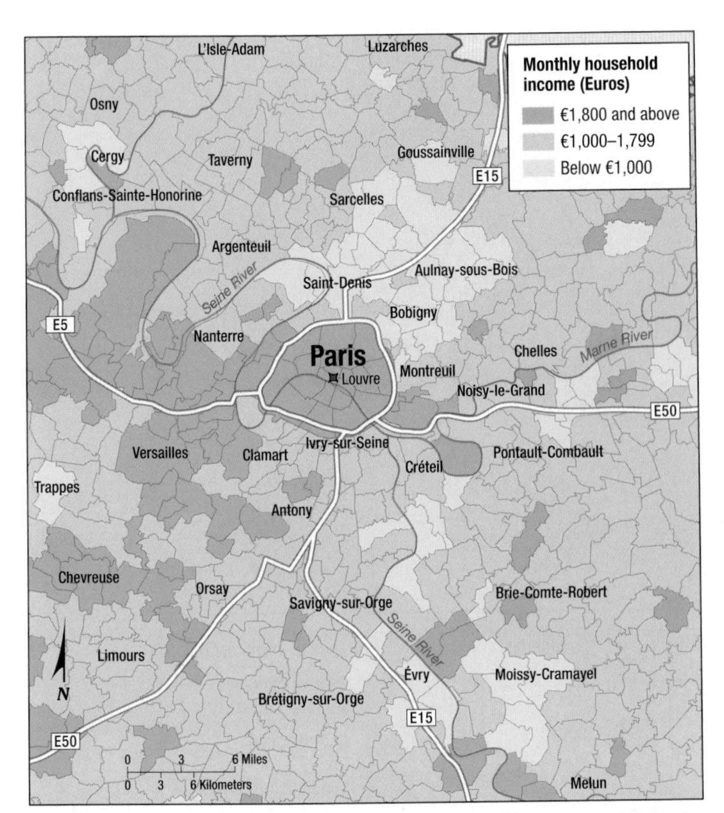

▲ **FIGURE 13-17 SECTORS IN PARIS, FRANCE** Wealthier people live in the center and to the southwest sector, often above sidewalk cafés and other consumer services.

CONCENTRIC ZONES IN EUROPEAN CITIES. Unlike in U.S. cities, in European cities such as Paris, most of the newer housing built in the suburbs is high-rise apartment buildings for low-income people and persons of color who have immigrated from Africa or Asia. (see ahead to Figure 13-21)

European officials encouraged the construction of high-density suburbs to help preserve the countryside from development and to avoid the inefficient sprawl that characterizes American suburbs, as discussed in the last section of this chapter. And tourists are attracted to the historic, lively centers of European cities. But these policies have resulted in the clustering of people with social and economic problems in remote suburbs rarely seen by wealthier individuals.

European suburban residents face the prospect of long commutes by public transportation to reach jobs and other downtown amenities. Shops, schools, and other services are worse in the suburbs than in inner neighborhoods; the suburbs are centers for crime, violence, and drug dealing; and people lack the American suburban amenity of large private yards. Many residents of these dreary suburbs are persons of color or recent immigrants from Africa or Asia who face discrimination and prejudice from "native" Europeans.

In the past, low-income people also lived in the center of European cities. Before the invention of electricity in the nineteenth century, social segregation was vertical: Wealthier people lived on the first or second floors, whereas poorer people occupied the dark, dank basements or climbed many flights of stairs to reach the attics. As the city expanded during the Industrial Revolution, housing for low-income people was constructed in sectors near the factories and away from the wealthy. Today, low-income people are less likely to live in European inner-city neighborhoods. Poor-quality housing has been renovated for wealthy people or demolished and replaced by offices or luxury apartment buildings. Building and zoning codes prohibit anyone from living in basements, and upper floors are attractive to wealthy individuals once elevators are installed.

Pause and Reflect 13.2.3

European cities contain many famous tourist sites, such as the Parthenon in Athens (Figure 12-34) and the Church of the Holy Sepulchre in Jerusalem (Figures 6-14 and 9-48). Are these tourist sites in an inner ring or an outer ring? What do you think explains their location?

APPLYING THE MODELS IN DEVELOPING COUNTRIES

Learning Outcome 13.2.4
Describe how the three models explain patterns in cities in developing countries.

In developing countries, as in Europe, the poor are accommodated in the suburbs, whereas the wealthy live near the center of a city, as well as in a sector extending from the center.

SECTORS IN CITIES OF DEVELOPING COUNTRIES. Geographers Ernest Griffin and Larry Ford show that in Latin American cities, wealthy people push out from the center in a well-defined elite residential sector. The elite sector forms on either side of a narrow spine that contains offices, shops, and amenities attractive to wealthy people, such as restaurants, theaters, parks, and zoos (Figure 13-18). The wealthy are also attracted to the center and spine because services such as water and electricity are more readily available and reliable there than elsewhere (Figure 13-19). Wealthy and middle-class residents avoid living near sectors of "disamenity," which are land uses that may be noisy or polluting or that cater to low-income residents.

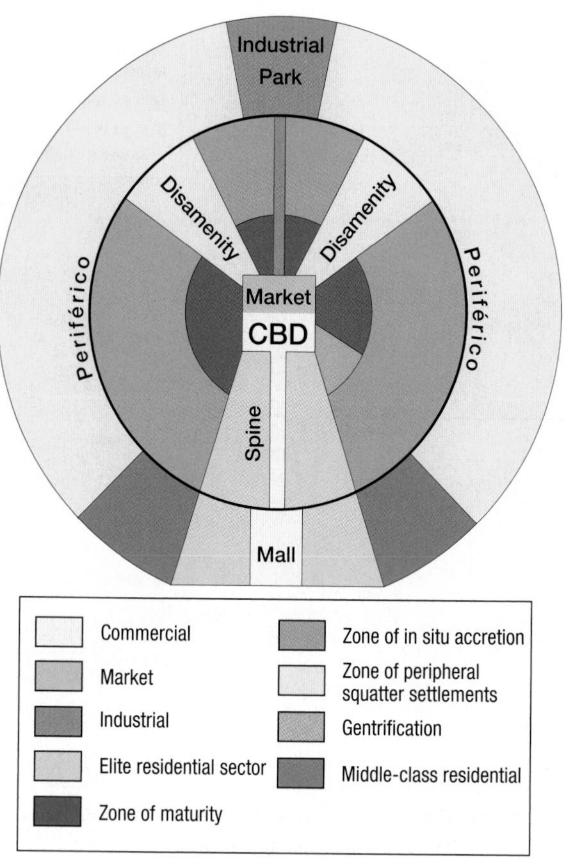

▲ **FIGURE 13-18 MODEL OF A LATIN AMERICAN CITY** Wealthy people live in the inner city and a sector extending along a commercial spine. (Adapted from Larry R. Ford, "A New and Improved Model of Latin American City Structure," *Geographical Review* 86 (1996): 438. Used by permission of the publisher.)

CONCENTRIC ZONES IN CITIES OF DEVELOPING COUNTRIES. Cities in developing countries have zones of the most intensive land uses and highest land values toward the center or along the commercial spine. Surrounding these zones is a ring of less-developed, lower-value land. Within this framework, cities in developing countries are unable to house the rapidly growing number of poor people. Their cities are growing because of overall population increase and migration from rural areas for job opportunities. Because of the housing shortage, a large percentage of poor immigrants to urban areas in developing countries live in squatter settlements.

Squatter settlements are known by a variety of names, including *barriadas* and *favelas* in Latin America, *bidonvilles* in North Africa, *bastees* in India, *gecekondu* in Turkey, *kampongs* in Malaysia, and *barong-barong* in the Philippines. Estimates of the number of people living in squatter settlements vary widely, between 175 million and 1 billion. Squatter settlements have few services because neither the city nor the residents can afford them. The settlements generally lack schools, paved roads, telephones, and sewers. Latrines are usually designated by the settlement's leaders, and

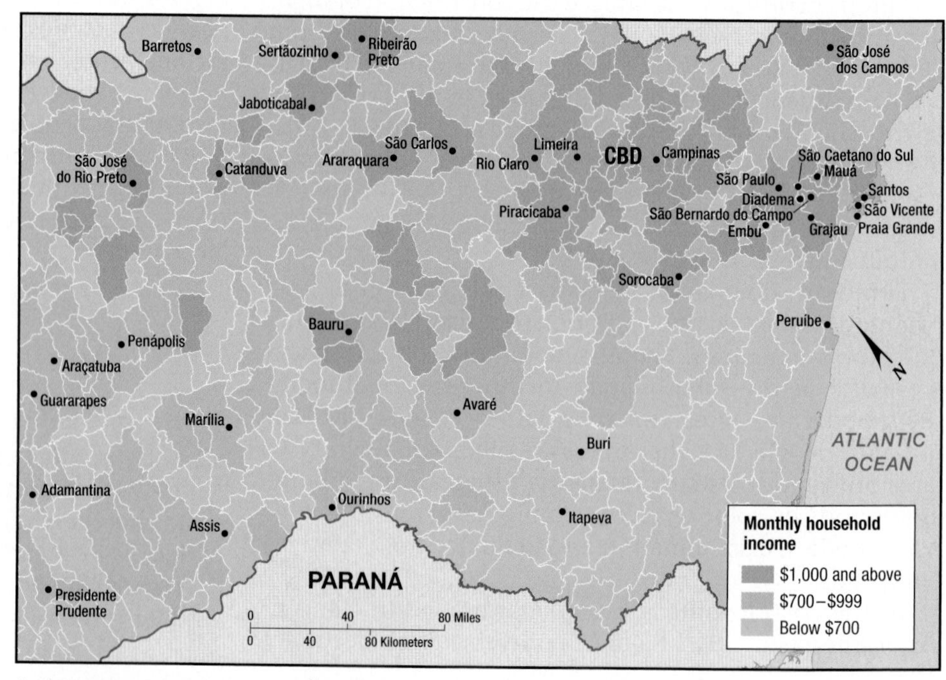

▲ **FIGURE 13-19 SECTORS IN SÃO PAULO** The high-income sector extends south from the CBD to the Atlantic Ocean.

water is carried from a central well or dispensed from a truck. Electricity service may be stolen by running a wire from the nearest power line. In the absence of bus service or available private cars, a resident may have to walk two hours to reach a place of employment (Figure 13-20).

Pause and Reflect 13.2.4
In Google Earth, go to Rua Oscar Freire, São Paulo, Brazil. How does this street in São Paulo's high-income sector compare to the suburban neighborhood in Figure 13-20?

At first, squatters do little more than camp on the land or sleep in the street. In severe weather, they may take shelter in markets and warehouses. Families then erect primitive shelters with scavenged cardboard, wood boxes, sackcloth, and crushed beverage cans. As they find new bits of material, they add them to their shacks. After a few years they may build a tin roof and partition the space into rooms, and the structure acquires a more permanent appearance.

▲ **FIGURE 13-20 SQUATTER SETTLEMENT IN SÃO PAULO** The *favela* in the foreground is in a suburb, and the high-rises in the background are close to downtown.

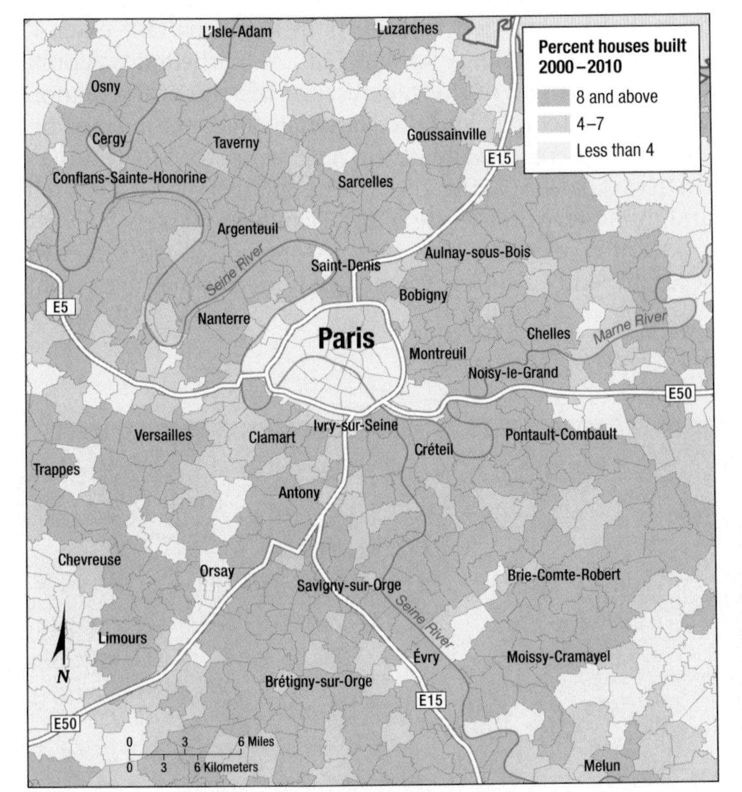

▲ **FIGURE 13-21 CONCENTRIC ZONES IN PARIS, FRANCE** Newer housing is in the outer rings, but much of it is high-rise apartments for poorer people and immigrants.

STAGES OF CITIES IN DEVELOPING COUNTRIES

Learning Outcome 13.2.5

Describe the history of development of cities in developing countries.

The similarity between cities in Europe and in developing countries is not a coincidence: European colonial policies left a heavy mark on cities in developing countries, many of which have passed through three stages of development—pre-European colonization, the European colonial period, and postcolonial independence. Mexico City provides a good example of these three stages.

PRECOLONIAL CITIES. Few cities existed in Africa, Asia, and Latin America before the Europeans established colonies. Most people lived in rural settlements. In Latin America, some cities were located in interior Mexico and the Andean highlands of northwestern South America. In Africa, cities could be found along the western coast, in Egypt's Nile River valley, and in Islamic empires in the north and east (as well as in Southwest Asia). Cities were also built in South and East Asia, especially in India, China, and Japan.

In Mexico, the Aztecs founded Mexico City—which they called Tenochtitlan—on a hill known as Chapultepec ("the hill of the grasshopper"). When forced by other people to leave the hill, they migrated a few kilometers south, near the present-day site of the University of Mexico, and then in 1325 to a marshy 10-square-kilometer (4-square-mile) island in Lake Texcoco (Figure 13-22).

The node of religious life was the Great Temple. Three causeways with drawbridges linked Tenochtitlan to the mainland and also helped control flooding. An aqueduct brought fresh water from Chapultepec. Most food, merchandise, and building materials crossed from the mainland to the island by canoe, barge, or other type of boat, and the island was laced with canals to facilitate pickup and delivery of people and goods. Over the next two centuries, the Aztecs conquered the neighboring peoples and extended their control through much of present-day Mexico. As their wealth and power grew, Tenochtitlan grew to a population of a half-million.

COLONIAL CITIES. When Europeans gained control of Africa, Asia, and Latin America, they sometimes expanded existing cities to provide colonial services, such as administration, military command, and international trade, as well as housing for European colonists. Sometimes, Existing native towns were either left to one side or demolished because they were totally at variance with European ideas.

Colonial cities followed standardized plans. All Spanish cities in Latin America, for example, were built according to the Laws of the Indies, drafted in 1573. The laws explicitly outlined how colonial cities were to be constructed—a gridiron street plan centered on a church and central plaza, walls around individual houses, and neighborhoods built around central, smaller plazas with parish churches or monasteries. Compared to the existing cities, these European districts typically contain wider streets and public squares, larger houses surrounded by gardens, and much lower density. In contrast, the old quarters have narrow, winding streets, little open space, and cramped residences.

After the Spanish conquered Tenochtitlan in 1521, after a two-year siege, they destroyed the city and dispersed or killed most of the inhabitants. The city, renamed Mexico City, was rebuilt around a main square, called the Zócalo, in the center of the island, on the site of the Aztecs' sacred precinct. The Spanish reconstructed the streets in a grid pattern extending from the Zócalo. A Roman Catholic cathedral was built on the north side of the square, near the site of the demolished Great Temple, and the National Palace was erected on the east side, on the site of the Aztec emperor Moctezuma's destroyed palace (Figure 13-23). The Spanish placed a church and monastery on the site of the Tlatelolco market.

Another example, Fes (Fez), Morocco, consists of two separate and distinct towns—the precolonial city that

▼ **FIGURE 13-22 PRECOLONIAL MEXICO CITY** (left) The Aztec city of Tenochtitlan was built on an island in Lake Texcoco. (right) The center of the city was dominated by the Templo Mayor. The twin shrines on the top of the temple were dedicated to the Aztec god of rain and agriculture (in blue) and to the Aztec god of war (in red).

▲ **FIGURE 13-23 COLONIAL MEXICO CITY** The main square in downtown Mexico City, the Zócalo, was laid out by the Spanish. The Metropolitan Cathedral is at the near end of the square. The National Palace is to the left, and City Hall is facing the square. Excavations at the site of the Templo Mayor are in the lower left.

existed before the French gained control and one built by the French colonialists (Figure 13-24). The precolonial Muslim city was laid out surrounding a mosque. The center also had a bazaar, or marketplace, which served as the commercial core. Government buildings and the homes of wealthy families surrounded the mosque and bazaar. Narrow, winding streets led from the core to other quarters. Families with less wealth and lower status located farther from the core, and recent migrants to the city lived on the edge.

Pause and Reflect 13.2.5

In Google Earth, go to Fes, Morocco, and zoom in on the buildings to see the colonial city. Then go to Fes el Bali, Morocco, and zoom in on the buildings to see the precolonial city. How do the buildings differ? Which has the higher density? Which has more trees and green space?

▼ **FIGURE 13-24 FES, MOROCCO** The precolonial part of Fes, in the foreground, is characterized by narrow, winding streets and high density. The tower in the foreground is the Karaouine Mosque. The colonial city laid out by the French is in the background, separate and distinct from the precolonial city.

CITIES SINCE INDEPENDENCE. Following independence, cities have become the focal points of change in developing countries. Millions of people have migrated to the cities in search of work.

In Mexico City, Emperor Maximilian (1864–1867) designed a 14-lane, tree-lined boulevard patterned after the Champs-Elysées in Paris. The boulevard (now known as the Paseo de la Reforma) extended 3 kilometers southwest from the center to Chapultepec (Figure 13-25). The Reforma between downtown and Chapultepec became the spine of an elite sector. During the late nineteenth century, the wealthy built pretentious palacios (palaces) along it. Physical factors also influenced the movement of wealthy people toward the west, along the Reforma. Because elevation was higher than elsewhere in the city, sewage flowed eastward and northward, away from Chapultepec. In 1903, most of Lake Texcoco was drained by a gigantic canal and tunnel project, allowing the city to expand to the north and east. The dried-up lakebed was a less desirable residential location than the west side because prevailing winds from the northeast stirred up dust storms. As Mexico City's population grew rapidly during the twentieth century, the social patterns inherited from the nineteenth century were reinforced.

CHECK-IN: KEY ISSUE 2

Where Are People Distributed Within Urban Areas?

- ✓ **According to the concentric zone model, a city grows by adding rings. The outer rings contain the newer housing.**
- ✓ **According to the sector model, a city grows along corridors. Some sectors contain higher-income households than others.**
- ✓ **According to the multiple nuclei model, a city grows through a series of nodes. Different ethnicities cluster around individual nodes.**
- ✓ **The three models show some similarities and some differences in the patterns within cities of North America and other regions.**
- ✓ **Cities in developing countries are further influenced by colonial history.**

▼ **FIGURE 13-25 INDEPENDENT MEXICO CITY** The Paseo de la Reforma, in the heart of the high-income sector, is traffic-free on Sunday mornings.

KEY ISSUE 3

Why Are Urban Areas Expanding?

- Suburban Expansion
- Suburban Segregation
- Urban Transportation

Learning Outcome 13.3.1
State three definitions of urban settlements.

In 1950, only 20 percent of Americans lived in suburbs compared to 40 percent in cities and 40 percent in small towns and rural areas. In 2000, after a half-century of rapid suburban growth, 50 percent of Americans lived in suburbs compared to only 30 percent in cities and 20 percent in small towns and rural areas.

Suburban Expansion

Until recently in the United States, as cities grew, they expanded by adding peripheral land. Now cities are surrounded by a collection of suburban jurisdictions whose residents prefer to remain legally independent of the large city.

THE PERIPHERAL MODEL

North American urban areas follow what Chauncey Harris (one of the creators of the multiple nuclei model) called the **peripheral model**. According to the peripheral model, an urban area consists of an inner city surrounded by large suburban residential and business areas tied together by a beltway or ring road (Figure 13-26). Peripheral areas lack the severe physical, social, and economic problems of inner-city neighborhoods. But the peripheral model points to problems of sprawl and segregation that characterize many suburbs.

Around the beltway are nodes of consumer and business services called *edge cities*. Edge cities originated as suburban residences for people who worked in the central city, and then shopping malls were built to be near the residents. Now edge cities contain manufacturing centers spread out over a single story for more efficient operations and office parks where producer services cluster. Specialized nodes emerge in the edge cities—for example, a collection of hotels and warehouses around an airport, a large theme park, a distribution center near the junction of the beltway, and a major long-distance interstate highway.

▼ **FIGURE 13-26 PERIPHERAL MODEL OF URBAN AREAS** The central city is surrounded by a beltway or ring road. Around the beltway are suburban residential areas and nodes, or edge cities, where consumer and business services and manufacturing cluster. (Adapted from Chauncy D. Harris, "The Nature of Cities and Urban Geography in the Last Half Century." Reprinted with permission from *Urban Geography*, vol. 18, no. 1 (1997), p. 17. © V. H. Winston & Son, Inc., 360 South Ocean Blvd., Palm Beach, FL 33480. All rights reserved.)

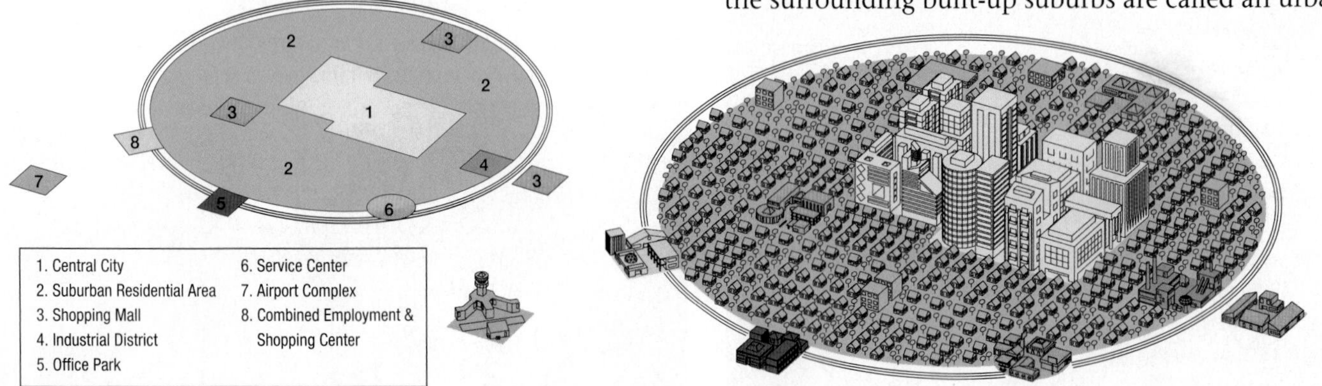

DEFINING URBAN SETTLEMENTS

Several definitions have been created to characterize cities and their suburbs:

- A city is a legal entity.
- An urban area is a continuously built-up area.
- A metropolitan area is a functional area.

LEGAL DEFINITION OF CITY. The term **city** defines an urban settlement that has been legally incorporated into an independent, self-governing unit (Figure 13-27). In the United States, a city surrounded by suburbs is sometimes called a central city.

Virtually all countries have a local government system that recognizes cities as legal entities with fixed boundaries. A city has locally elected officials, the ability to raise taxes, and responsibility for providing essential services. The boundaries of the city define the geographic area within which the local government has legal authority.

Population has declined since 1950 by more than one-half in the central cities of Buffalo, Cleveland, Detroit, Pittsburgh, and St. Louis, and by at least one-third in more than a dozen other cities. The number of tax-paying middle-class families and industries has invariably declined by much higher percentages in these cities.

URBAN AREA. In the United States, the central city and the surrounding built-up suburbs are called an urban area.

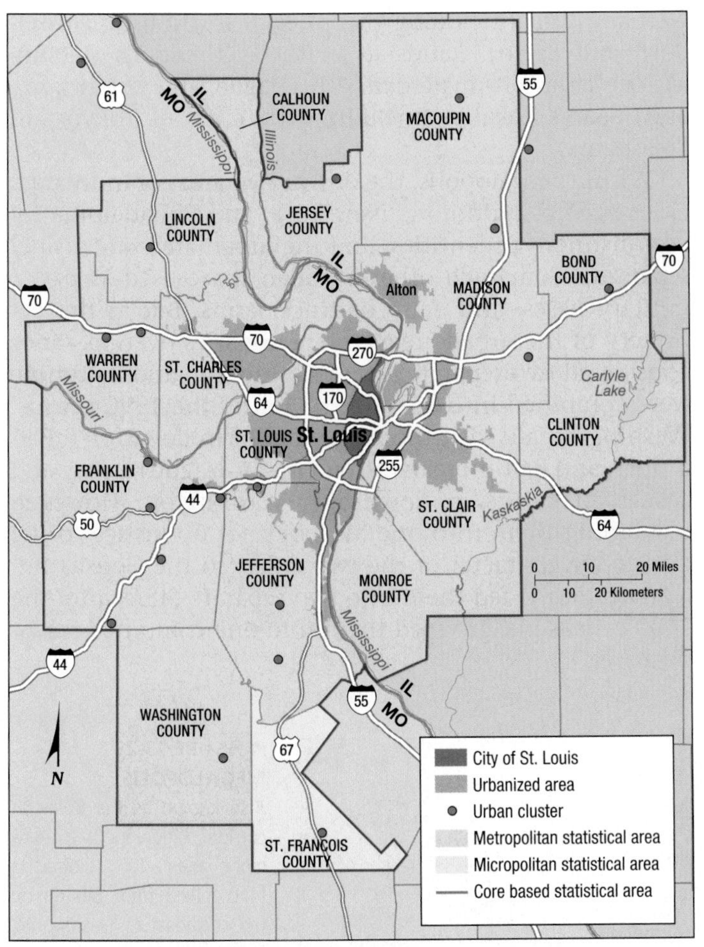

▲ **FIGURE 13-27 DEFINITIONS OF ST LOUIS** The City of St. Louis comprises only 6 percent of the land area and 11 percent of the population of the MSA.

An **urban area** consists of a dense core of census tracts, densely settled suburbs, and low-density land that links the dense suburbs with the core. The census recognizes two types of urban areas:

- An **urbanized area** is an urban area with at least 50,000 inhabitants.
- An **urban cluster** is an urban area with between 2,500 and 50,000 inhabitants.

The census identified 486 urbanized areas and 3,087 urban clusters in the United States in 2010. Approximately 70 percent of the U.S. population lived in one of the 486 urbanized areas, including about 30 percent in central cities and 40 percent in surrounding jurisdictions. Approximately 10 percent of the U.S. population lived in one of the 3,087 urban clusters. The census does not have a precise definition of suburbs, but they can be considered roughly equivalent to the urban clusters and the urbanized areas outside the central cities.

Working with urbanized areas is difficult because few statistics are available about them. Most data in the United States and other countries are collected for cities, counties, and other local government units, but urbanized areas do not correspond to government boundaries. The term *urban area* also has limited applicability because it does not accurately reflect the full influence that an urban settlement has in contemporary society.

METROPOLITAN STATISTICAL AREA. The area of influence of a city extends beyond legal boundaries and adjacent built-up jurisdictions. For example, commuters may travel a long distance to work and shop in the city or built-up suburbs. People in a wide area watch the city's television stations, read the city's newspapers, and support the city's sports teams. Therefore, we need another definition of urban settlement to account for its more extensive zone of influence.

The U.S. Bureau of the Census has created a method of measuring the functional area of a city, known as the **metropolitan statistical area (MSA)**. An MSA includes the following:

- An urbanized area with a population of at least 50,000
- The county within which the city is located
- Adjacent counties with a high population density and a large percentage of residents working in the central city's county (specifically, a county with a density of 25 persons per square mile and at least 50 percent working in the central city's county)

Studies of metropolitan areas in the United States are usually based on information about MSAs. MSAs are widely used because many statistics are published for counties, the basic MSA building block.

The Census Bureau had designated 366 MSAs as of 2012, encompassing 84 percent of the U.S. population. Older studies may refer to SMSAs, or standard metropolitan statistical areas, which the census used before 1983 to designate metropolitan areas in a manner similar to MSAs. An MSA is not a perfect tool for measuring the functional area of a city. One problem is that some MSAs include extensive land area that is not urban. For example, Great Smoky Mountains National Park is partly in the Knoxville, Tennessee, MSA; Sequoia National Park is in the Visalia-Porterville, California, MSA. MSAs comprise some 20 percent of total U.S. land area, compared to only 2 percent for urbanized areas. The urbanized area typically occupies only 10 percent of an MSA land area but contains nearly 90 percent of its population.

The census has also designated smaller urban areas as **micropolitan statistical areas (µSAs)**. A µSA includes an urbanized area of between 10,000 and 50,000 inhabitants, the county in which it is found, and adjacent counties tied to the city. The United States had 576 micropolitan statistical areas as of 2012, for the most part found around southern and western communities previously considered rural in character. About 10 percent of Americans live in micropolitan statistical areas. The 366 MSAs and 576 µSAs together are known as **core based statistical areas (CBSAs)**.

Recognizing that many MSAs and µSAs have close ties, the census had combined some of them into 128 **combined statistical areas (CSAs)** as of 2012. A CSA is defined as two or more contiguous CBSAs tied together by commuting patterns. The 125 CSAs plus the remaining 187 MSAs and 406 µSAs not combined into CSAs together are known as **primary census statistical areas (PCSAs)**.

Pause and Reflect 13.3.1

In what metropolitan or micropolitan statistical area do you live? Google [your city and state] statistical area.

OVERLAPPING METROPOLITAN AREAS

Learning Outcome 13.3.2:
Describe how metropolitan areas contain many local governments and overlap with each other.

A county between two central cities may send a large number of commuters to jobs in each. In the northeastern United States, large metropolitan areas are so close together that they now form one continuous urban complex, extending from north of Boston to south of Washington, D.C. Geographer Jean Gottmann named this region **Megalopolis**, a Greek word meaning "great city"; others have called it the Boswash corridor (Figure 13-28).

Other continuous urban complexes exist in the United States—the southern Great Lakes between Chicago and Milwaukee on the west and Pittsburgh on the east, and southern California, from Los Angeles to Tijuana. Among important examples in other developed countries are the German Ruhr (including the cities of Dortmund, Düsseldorf, and Essen), Randstad in the Netherlands (including the cities of Amsterdam, The Hague, and Rotterdam), and Japan's Tokaido (including the cities of Tokyo and Yokohama).

Within Megalopolis, the downtown areas of individual cities such as Baltimore, New York, and Philadelphia retain distinctive identities, and the urban areas are visibly separated from each other by open space used as parks, military bases, and dairy or truck farms. But at the periphery of the urban areas, the boundaries overlap. Once considered two separate areas, Washington and Baltimore were combined into a single MSA after the 1990 census. Washingtonians visit the Inner Harbor in downtown Baltimore, and Baltimoreans attend major-league hockey and basketball games in downtown Washington. However, combining them into one MSA did not do justice to the distinctive character of the two cities, so the Census Bureau again divided them into two separate MSAs after the 2000 census but grouped them into one combined statistical area.

◀ **FIGURE 13-28**
MEGALOPOLIS
Also known as the Boswash corridor, Megalopolis extends more than 700 kilometers (440 miles) from Boston on the northeast to Washington, D.C., on the southwest. Megalopolis contains one-fourth of the U.S. population on 2 percent of the country's total land area.

LOCAL GOVERNMENT FRAGMENTATION

The fragmentation of local government in the United States makes it difficult to solve regional problems of traffic management, solid-waste disposal, and the building of affordable housing. According to the 2002 census, the United States had 87,525 local governments, including 3,034 counties, 19,429 cities, 16,504 townships, 13,506 school districts, and 35,052 special-purpose districts, such as police and fire. The larger metropolitan areas have thousands of local governments, with widely varying levels of wealth (Figure 13-29).

The large number of local government units has led to calls for a metropolitan government that could coordinate—if not replace—the numerous local governments in an urban area. Most U.S. metropolitan areas have a **council of government**, which is a cooperative agency consisting of representatives of the various local governments in the region. The council of government may be empowered to do some overall planning for the area that local governments cannot logically do. Strong metropolitan-wide governments have been established in a few places in North America. Two kinds exist:

- **Consolidations of city and county governments.** Examples of consolidations of city and county governments include Indianapolis and Miami. The boundaries of Indianapolis were changed to match those of Marion County, Indiana. Government functions that were handled separately by the city and the county now are combined into a joint operation in the same office building. In Florida, the city of Miami and surrounding Dade County have combined some services, but the city boundaries have not been changed to match those of the county.
- **Federations.** Examples of federations include Toronto and other large Canadian cities. Toronto's metropolitan government was created in 1953, through a federation of 13 municipalities. A two-tier system of government existed until 1998, when the municipalities were amalgamated into a single government.

Pause and Reflect 13.3.2

Canada has a method of delineating urban and metropolitan areas of various sizes. If the Canadian side of Lake Ontario were colored in Figure 13-28, most of it would also be urban. What is the largest city and metropolitan area on the Canadian side of Lake Ontario?

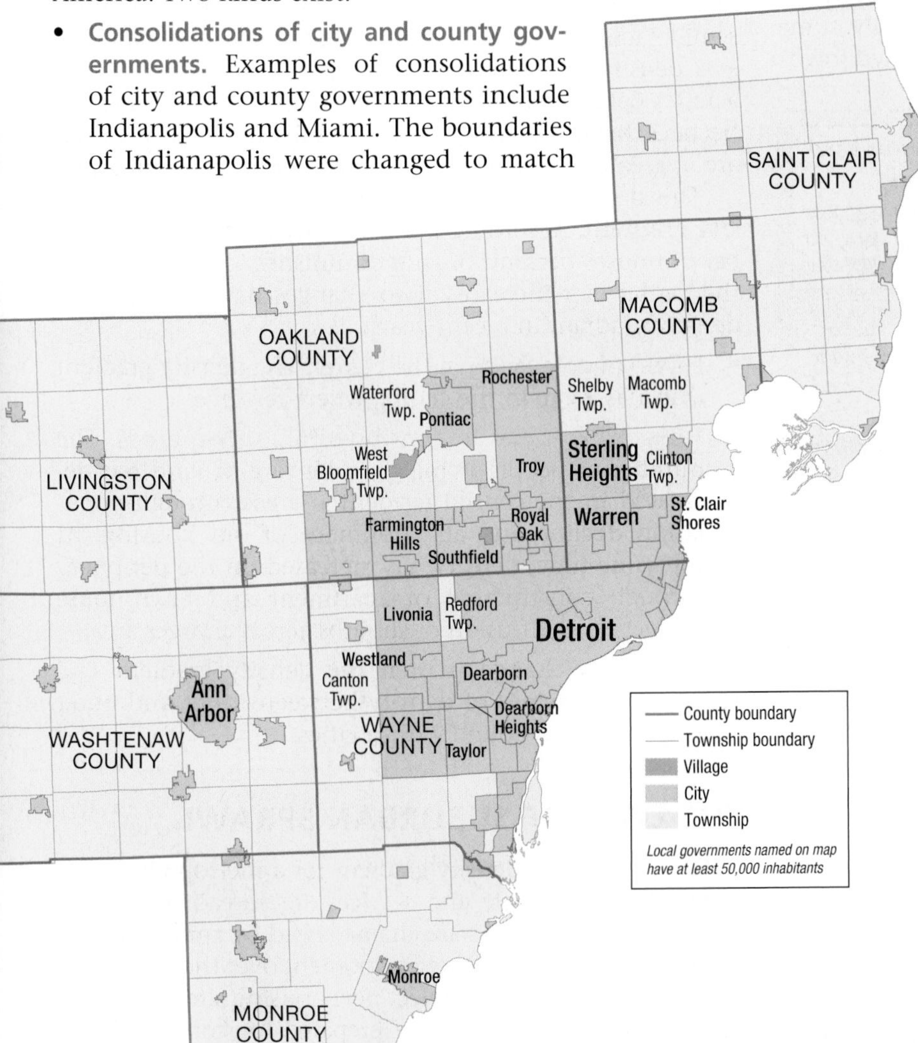

▲ **FIGURE 13-29 LOCAL GOVERNMENTS IN THE DETROIT METROPOLITAN AREA** The map does not include numerous local governments directly across the Detroit River in Canada. The city of Detroit, which has a large number of vacant and abandoned houses (top) is surrounded by wealthy suburbs (bottom).

ANNEXATION

Learning Outcome 13.3.3
Understand historical and contemporary patterns of suburban expansion.

The process of legally adding land area to a city is **annexation**. Rules concerning annexation vary among states. Normally, land can be annexed to a city only if a majority of residents in the affected area vote in favor of the annexation.

Peripheral residents generally desired annexation in the nineteenth century because the city offered better services, such as water supply, sewage disposal, trash pickup, paved streets, public transportation, and police and fire protection. Thus, as U.S. cities grew rapidly in the nineteenth century, the legal boundaries frequently changed to accommodate newly developed areas. For example, the city of Chicago expanded from 26 square kilometers (10 square miles) in 1837 to 492 square kilometers (190 square miles) in 1900 (Figure 13-30).

Today, however, cities are less likely to annex peripheral land because the residents prefer to organize their own services rather than pay city taxes for them. Originally, some of these peripheral jurisdictions were small, isolated towns that had a tradition of independent local government before being swallowed up by urban growth. Others are newly created communities whose residents wish to live close to the large city but not be legally part of it.

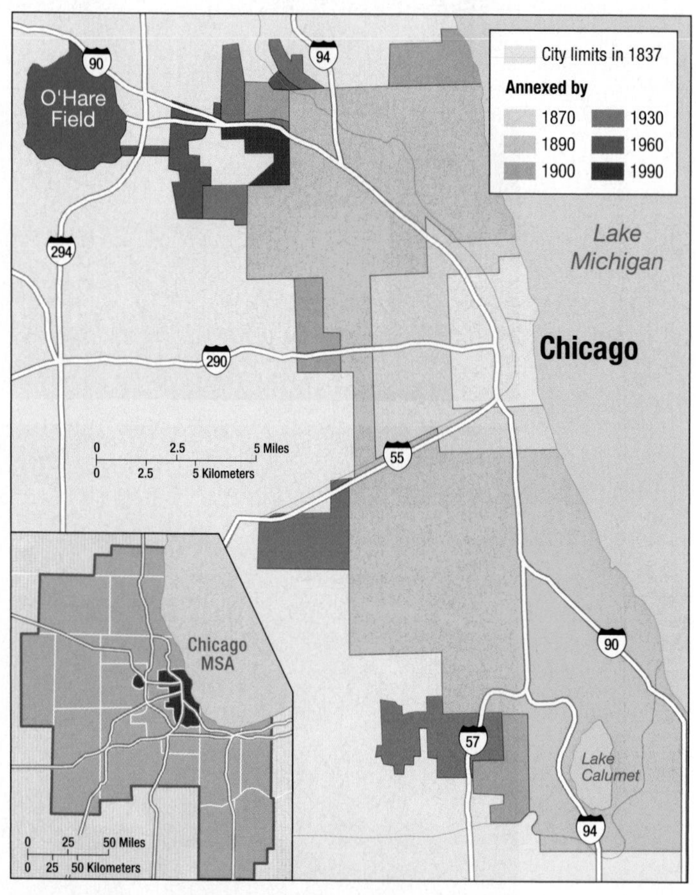

▲ FIGURE 13-30 **ANNEXATION IN CHICAGO** During the nineteenth century, the city of Chicago grew rapidly through annexation of peripheral land. Relatively little land was annexed during the twentieth century; the major annexation was on the northwest side, for O'Hare Airport. The inset shows that the city of Chicago covers only a small portion of the Chicago metropolitan statistical area.

Pause and Reflect 13.3.3
The three largest cities in Ohio are Cincinnati, Cleveland, and Columbus. In 1950, Cincinnati's land area was 72 square miles, Cleveland's was 75 square miles, and Columbus's was 40 square miles. Which of the three cities has increased its land area substantially since 1950? Refer to each city's Wikipedia site to find the current land areas. What might account for the large increase?

DENSITY GRADIENT

As you travel outward from the center of a city, you can watch the decline in the density at which people live (Figure 13-31). Inner-city apartments or row houses may pack as many as 250 dwellings on a hectare of land (100 dwellings per acre). Older suburbs have larger row houses, semidetached houses, and individual houses on small lots, at a density of about 10 houses per hectare (4 houses per acre). A detached house typically sits on a lot of 0.25 to 0.5 hectares (0.6 to 1.2 acres) in new suburbs and a lot of 1 hectare or greater (2.5 acres) on the fringe of the built-up area.

This density change in an urban area is called the **density gradient**. According to the density gradient, the number of houses per unit of land diminishes as distance from the center city increases. Two changes have affected the density gradient in recent years:

- **Fewer people living in the center.** The density gradient thus has a gap in the center, where few live.
- **Fewer differences in density within urban areas.** The number of people living on a hectare of land has decreased in the central residential areas through population decline and abandonment of old housing. At the same time, density has increased on the periphery through construction of apartment and town-house projects and diffusion of suburbs across a larger area.

These two changes flatten the density gradient and reduce the extremes of density between inner and outer areas traditionally found within cities.

THE COST OF SUBURBAN SPRAWL

A flattening of the density gradient for a metropolitan area means that its people and services are spread out over a larger area. U.S. suburbs are characterized by **sprawl**, which is the progressive spread of development over the landscape. When private developers select new housing sites, they seek cheap land that can easily be prepared for construction—land often not contiguous to the existing built-up area (Figure 13-32). Sprawl is also fostered by the desire of many families to own large tracts of land.

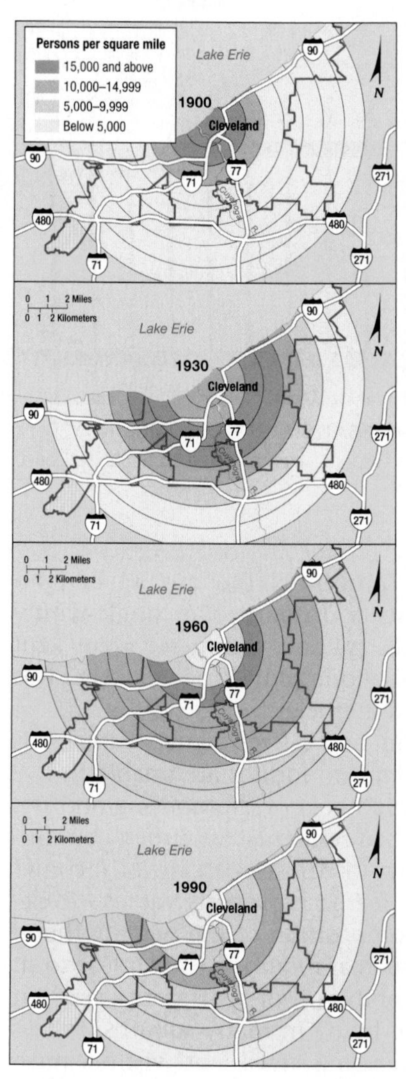

◀ **FIGURE 13-31 DENSITY GRADIENT IN CLEVELAND** In 1900, the population was highly clustered in and near the central business district (CBD). By 1930 and 1960, the population was spreading, leaving the original core less dense. By 1990, the population was distributed over a much larger area, the variation in the density among different rings was much less, and the area's lowest densities existed in the rings near the CBD. The current boundary of the city of Cleveland is shown. (First three maps adapted from Avery M. Guest, "Population Suburbanization in American Metropolitan Areas, 1940–1970," *Geographical Analysis* 7 (1975): 267–283, table 4. Used by permission of the publisher.)

As long as demand for single-family detached houses remains high, land on the fringe of urban areas will be converted from open space to residential land use. Land is not transformed immediately from farms to housing developments. Instead, developers buy farms for future construction of houses by individual builders. Developers frequently reject land adjacent to built-up areas in favor of detached isolated sites, depending on the price and physical attributes of the alternatives. The peripheries of U.S. cities therefore look like Swiss cheese, with pockets of development and gaps of open space.

Urban sprawl has some undesirable traits. Roads and utilities must be extended to connect isolated new developments to nearby built-up areas. The cost of these new roads and utilities may be funded by taxes, or the developer may install the services and pass on the cost to new residents through higher home prices. Sprawl also wastes land. Some prime agricultural land may be lost through construction of isolated housing developments. In the interim, other sites lie fallow while speculators await the most profitable time to build homes on them. In reality, sprawl has little impact on the total farmland in the United States, but it does reduce the ability of city dwellers to get to the country for recreation, and it can affect the supply of local dairy products and vegetables. Low-density suburbs also waste more energy, especially because motor vehicles are required for most trips.

The supply of land for the construction of new housing is more severely restricted in European urban areas than in the United States. Officials attack sprawl by designating areas of mandatory open space. London, Birmingham, and several other British cities are surrounded by **greenbelts**, or rings of open space. New housing is built either in older suburbs inside the greenbelts or in planned extensions to small towns and new towns beyond the greenbelts. However, restriction of the supply of land on the urban periphery has driven up house prices in Europe.

Several U.S. states have taken strong steps in the past few years to curb sprawl, reduce traffic congestion, and reverse inner-city decline. The goal is to produce a pattern of compact and contiguous development and protect rural land for agriculture, recreation, and wildlife. Legislation and regulations to limit suburban sprawl and preserve farmland has been called **smart growth**. Oregon and Tennessee have defined growth boundaries within which new development must occur. Cities can annex only lands that have been included in the urban growth areas. New Jersey, Rhode Island, and Washington were also early leaders in enacting strong state-level smart-growth initiatives. Maryland's smart-growth law discourages the state from funding new highways and other projects that would extend suburban sprawl and destroy farmland. State money must be spent to "fill in" already urbanized areas.

▼ **FIGURE 13-32 SUBURBAN DEVELOPMENT PATTERNS IN THE UNITED KINGDOM AND THE UNITED STATES** The United States has much more sprawl than the United Kingdom. In the United Kingdom, new housing is more likely to be concentrated in new towns or planned extensions of existing small towns (right), whereas in the United States, growth occurs in discontinuous developments.

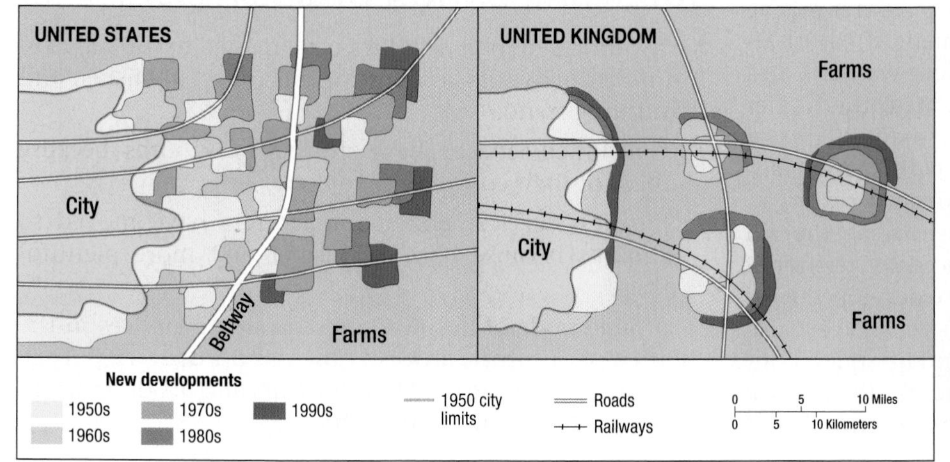

Segregation in the Suburbs

Learning Outcome 13.3.4
Explain two ways in which suburbs are segregated.

Public opinion polls in the United States show people's strong desire for suburban living. In most polls, more than 90 percent of respondents prefer the suburbs to the inner city. It is no surprise then that the suburban population has grown much faster than the overall population in the United States.

Suburbs offer varied attractions—a detached single-family dwelling rather than a row house or an apartment, private land surrounding the house, space to park cars, and a greater opportunity for home ownership. A suburban house provides space and privacy, a daily retreat from the stress of urban living. Families with children are especially attracted to suburbs, which offer more space for play and protection from the high crime rates and heavy traffic that characterize inner-city life. As incomes rose in the twentieth century, first in the United States and more recently in other developed countries, more families were able to afford to buy suburban homes.

The modern residential suburb is segregated in two ways:

- **Segregated social classes.** Housing in a given suburban community is usually built for people of a single social class, with others excluded by virtue of the cost, size, or location of the housing. Segregation by race and ethnicity also persists in some suburbs (see Chapter 7).
- **Segregated land uses.** Residents are separated from commercial and manufacturing activities that are confined to compact, distinct areas.

RESIDENTIAL SEGREGATION

The homogeneous suburb was a twentieth-century phenomenon. Before then, activities and classes in a city were more likely to be separated vertically rather than horizontally. In a typical urban building, shops were on the street level, with the shop owner or another well-to-do family living on one or two floors above the shop. Poorer people lived on the higher levels or in the basement, the least attractive parts of the building. The basement was dark and damp, and before the elevator was invented, the higher levels could be reached only by climbing many flights of stairs. Wealthy families lived in houses with space available in the basement or attic to accommodate servants.

Once cities spread out over much larger areas, the old pattern of vertical separation was replaced by territorial segregation. Large sections of the city were developed with houses of similar interior dimension, lot size, and cost, appealing to people with similar incomes and lifestyles. Zoning ordinances, developed in Europe and North America in the early decades of the twentieth century, encouraged spatial separation. They prevented the mixing of land uses within the same district. In particular, single-family houses, apartments, industry, and commerce were kept apart because the location of one activity near another was considered unhealthy and inefficient.

▲ **FIGURE 13-33 HOUSING SEGREGATION: GATED COMMUNITY** Dana Point, California, a Los Angeles suburb, has a gated community called Lantern Bay.

The strongest criticism of U.S. residential suburbs is that low-income people and minorities are unable to live in them because of the high cost of the housing and the unfriendliness of established residents. Suburban communities discourage the entry of those with lower incomes and minorities because of fear that property values will decline if the high-status composition of the neighborhood is altered. Legal devices, such as requiring each house to sit on a large lot and the prohibition of apartments, prevent low-income families from living in many suburbs. Fences are built around some housing areas, and visitors must check in at a gate house to enter (Figure 13-33).

Pause and Reflect 13.3.4
Are you able to walk from your home to consumer services? What do you think explains the spatial pattern of residential and commercial land uses in the area where you live?

SUBURBANIZATION OF BUSINESSES

Many nonresidential activities have moved to the suburbs. A number of factors account for this long-established and continuing trend:

- Consumer services have moved to suburbs because most of their customers live there.
- Business services and manufacturers have moved to suburbs because land is cheaper and more plentiful there.

A large node of business and consumer services in the suburbs of an urban area is known as an **edge city**. Edge cities are planned around freeway exits and are designed to be navigable only in motor vehicles.

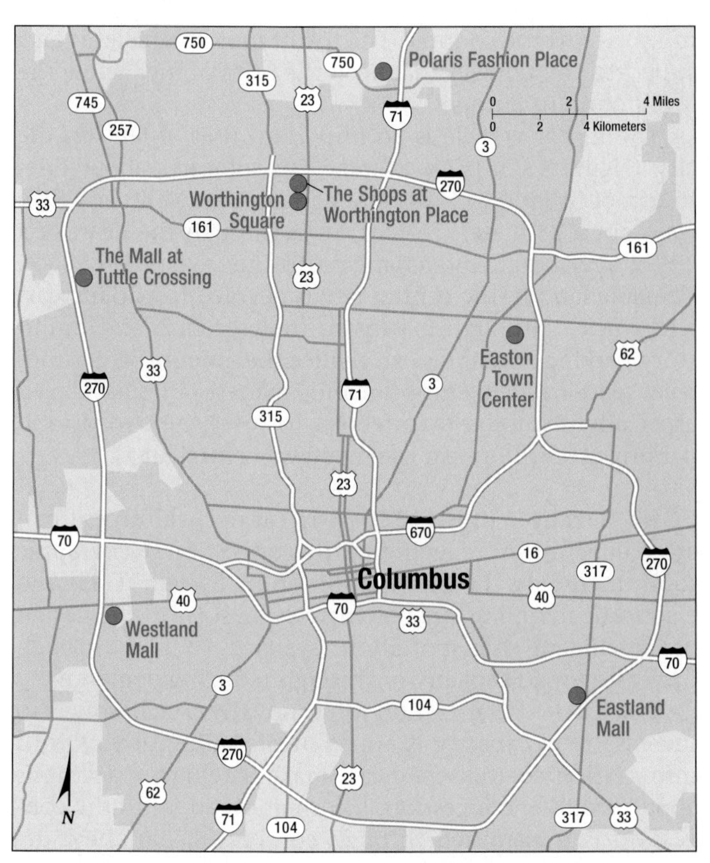

▲ **FIGURE 13-34 SHOPPING MALLS NEAR COLUMBUS, OHIO** The malls surround the city near the beltway. (right) Easton Town Center is the largest mall in the Columbus area.

SUBURBANIZATION OF CONSUMER SERVICES. Suburban residential growth has fostered change in traditional retailing patterns (Figure 13-34). Historically, urban residents bought food and other daily necessities at small neighborhood shops in the midst of housing areas and shopped in the CBD for other products. But since the end of World War II, downtown sales have not increased, whereas suburban sales have risen at an annual rate of 5 percent. Downtown sales have stagnated because suburban residents who live far from the CBD don't make the long journey there. At the same time, small corner shops do not exist in the midst of newer residential suburbs. The low density of residential construction discourages people from walking to stores, and restrictive zoning practices often exclude shops from residential areas.

Instead, retailing has been increasingly concentrated in planned suburban shopping malls of varying sizes. Corner shops have been replaced by supermarkets in small shopping centers. Larger malls contain department stores and specialty shops traditionally located only in the CBD. Generous parking lots surround the stores. A shopping mall is built by a developer, who buys the land, builds the structures, and leases space to individual merchants. Typically, a merchant's rent is a percentage of sales revenue.

Shopping malls require as many as 40 hectares (100 acres) of land and are frequently near key road junctions, such as the interchange of two interstate highways. Some shopping malls are elaborate multilevel structures exceeding 100,000 square meters (1 million square feet), with more than 100 stores arranged along covered walkways, and surrounded by an extensive parking area. The key to a successful large shopping mall is the inclusion of one or more anchors, usually large department stores. Most consumers go to a mall to shop at an anchor and, while there, patronize the smaller shops. In smaller shopping centers, the anchor is frequently a supermarket or discount store.

Malls have become centers for activities in suburban areas that lack other types of community facilities. Retired people go to malls for safe, vigorous walking exercise, or they sit on a bench to watch the passing scene. Teenagers arrive after school to meet their friends. Concerts and exhibitions are frequently set up in malls.

SUBURBANIZATION OF BUSINESS SERVICES AND FACTORIES. Offices that do not require face-to-face contact are increasingly moving to suburbs, where rents are lower than in the CBD. Executives can drive on uncongested roads to their offices from their homes in nearby suburbs and park their cars without charge. For other employees, though, suburban office locations can pose a hardship. Secretaries, custodians, and other lower-status office workers may not have cars, and public transportation may not serve the site. Other office workers might miss the stimulation and animation of a central location, particularly at lunchtime.

Factories and warehouses have migrated to suburbia for more space, cheaper land, and better truck access. Modern factories and warehouses demand more land because they spread their conveyor belts, forklift trucks, loading docks, and machinery over a single level for efficient operation. Suburban locations also facilitate truck shipments by providing good access to main highways and no central city traffic congestion, which is important because industries increasingly receive inputs and distribute products by truck.

Urban Transportation

Learning Outcome 13.3.5
Describe the impact of motor vehicles in urban areas.

People do not travel aimlessly; their trips have a precise point of origin, destination, and purpose. More than half of all trips are work related—commuting between work and home, business travel, or deliveries. Shopping or other personal business and social journeys each account for approximately one-fourth of all trips. Together, all these trips produce congestion in urban areas. Congestion imposes costs on individuals and businesses by delaying arrival at destinations, and the high concentration of slowly moving vehicles produces increased air pollution.

Historically, the growth of suburbs was constrained by poor transportation. People lived in crowded cities because they had to be within walking distance of shops and places of employment. The invention of the railroad in the nineteenth century enabled people to live in suburbs and work in the central city. Cities then built railroads at street level (called trolleys, streetcars, or trams) and underground (subways) to accommodate commuters. Many so-called streetcar suburbs built in the nineteenth century still exist and retain unique visual identities. They consist of houses and shops clustered near a station or former streetcar stop at a much higher density than is found in newer suburbs.

MOTOR VEHICLES

The suburban explosion in the twentieth century relied on motor vehicles rather than railroads, especially in the United States. Rail lines restricted nineteenth-century suburban development to narrow ribbons within walking distance of the stations. Cars and trucks permitted large-scale development of suburbs at greater distances from the center, in the gaps between the rail lines. Motor vehicle drivers have much greater flexibility in their choice of residence than was ever before possible.

Motor vehicle ownership is nearly universal among American households, with the exception of some poor families, older individuals, and people living in the centers of large cities such as New York. More than 95 percent of all trips within U.S. cities are made by car, compared to fewer than 5 percent by bus or rail. Outside the big cities, public transportation service is extremely rare or nonexistent. The U.S. government has encouraged the use of cars and trucks by paying 90 percent of the cost of limited-access, high-speed interstate highways, which stretch for 74,000 kilometers (46,000 miles) across the country. The use of motor vehicles is also supported by policies that keep the price of fuel below the level found in Europe.

The motor vehicle is an important user of land in the city (Figure 13-35). An average city allocates about one-fourth of its land to roads and parking lots. Multilane freeways cut a 23-meter (75-foot) path through the heart of a city, and elaborate interchanges consume even more space. Valuable land in the central city is devoted to parking cars and trucks, although expensive underground and multistory parking structures can reduce the amount of ground-level space needed. European and Japanese cities have been especially disrupted by attempts to insert new roads and parking areas in or near the medieval central areas.

CONTROLLING VEHICLES. The future health of urban areas depends on relieving traffic congestion. Geographic tools, including the Global Positioning System (GPS) and electronic mapping, are playing central roles in the design of intelligent transportation systems, either through increasing road capacity or through reducing demand.

The current generation of innovative techniques to increase road capacity is aimed at providing drivers with information so that they can make intelligent decisions about avoiding congestion. Information about traffic congestion is transmitted through computers, handheld devices, and vehicle monitors. Traffic hot spots are displayed on electronic maps and images, using information collected through sensors in the roadbeds and cameras placed at strategic locations. An individual wishing to know about a particular route can program an electronic device to receive a congestion alert and to suggest alternatives. Radio stations in urban areas broadcast reports to advise motorists of accidents or especially congested highways.

Demand to use congested roads is being reduced in a number of ways:

- **Congestion charges.** In London, motorists must pay a congestion charge of up to £12 ($18) to drive into

▼ **FIGURE 13-35 URBAN EXPRESSWAY** San Francisco, like most other U.S. cities, had major expressways constructed into the center of the city.

▲ **FIGURE 13-36 LONDON CONGESTION CHARGE** The sign warns motorists that they must pay a charge to drive into Central London at certain times.

the central area between 7 A.M. and 6 P.M. Monday through Friday (Figure 13-36). A similar system exists in Stockholm, where the charge varies depending on the time of day.

- **Tolls.** In Toronto and several California cities, motorists are charged higher tolls to drive on freeways during congested times than at other times. A transponder attached to a vehicle records the time of day it is on the highway. A monthly bill sent to the vehicle's owner reflects the differential tolls.
- **Permits.** In Singapore, to be permitted to drive downtown during the busiest times of the day, a motorist must buy a license and demonstrate ownership of a parking space. The government limits the number of licenses and charges high tolls to drive downtown. Several cities in China intend to require permits to drive in congested areas.

▲ **FIGURE 13-37 DRIVERLESS CAR** Google has developed technology enabling vehicles to be driven by sensors, electronic maps, and cameras.

- **Bans.** Cars have been banned from portions of the central areas of a number of European cities, including Copenhagen, Munich, Vienna, and Zurich.

Future intelligent transportation systems are likely to increase capacity through hands-free driving (Figure 13-37). A motorist will drive to a freeway entrance, where the vehicle will be subjected to a thorough diagnostic (taking a half-second) to ensure that it has enough fuel and is in good operating condition. A menu will offer a choice of predetermined destinations, such as "home" or "office," or a destination can be programmed by hand.

Pause and Reflect 13.3.5
Which methods of easing congestion appear to you to be most likely to be successful?

A release will send the vehicle accelerating automatically on the entrance ramp onto the freeway. Sensors in the bumpers and fenders, attached to radar or GPS, will alert vehicle systems to accelerate, brake, or steer, as needed. With such a system, spacing between vehicles can be as little as 2 meters. While a vehicle is automatically controlled, the "driver" will be able to swivel the seat to a workstation to make phone calls, check e-mail, or surf the Internet; read; watch television; or nap. When the vehicle nears the programmed freeway exit, a tone will warn that the driver will have to take back control. The vehicle will halted on the exit ramp until the driver firmly presses the brake to release the "autodrive" system, much as cruise control is currently disengaged.

PUBLIC TRANSIT

Learning Outcome 13.3.6
State the benefits and limitations of public transportation.

Because few people live within walking distance of their place of employment, urban areas are characterized by extensive commuting. The heaviest flow of commuters is into the CBD in the morning and out of it in the evening.

The intense concentration of people in the CBD during working hours strains transportation systems because a large number of people must reach a small area of land at the same time in the morning and disperse at the same time in the afternoon. As much as 40 percent of all trips made into or out of a CBD occur during four hours of the day—two in the morning and two in the afternoon. **Rush hour**, or peak hour, is the four consecutive 15-minute periods that have the heaviest traffic.

PUBLIC TRANSIT IN THE UNITED STATES. In the United States, public transit is used primarily for rush-hour commuting by workers into and out of the CBD. One-half of trips to work are by public transit in New York; one-third in Boston, San Francisco, and Washington; and one-fourth in Chicago and Philadelphia. But in most other cities, public transit service is minimal or nonexistent.

Despite the obvious advantages of public transportation for commuting, only 5 percent of work trips are by public transit in the United States. Overall, public transit ridership in the United States declined from 23 billion per year in the 1940s to 10 billion in 2011. The average American wastes 14 gallons of gasoline and loses 34 hours per year sitting in traffic jams, according to the Urban Mobility Report prepared by the Texas Transportation Institute. In the United States, the total cost of congestion is valued at $101 billion per year. But most Americans still prefer to commute by vehicle. Most people overlook these costs because they place higher value on the privacy and flexibility of schedule offered by a car.

Early in the twentieth century, U.S. cities had 50,000 kilometers (30,000 miles) of street railways and trolleys that carried 14 billion passengers a year, but only a few hundred kilometers of track remain. The number of U.S. and Canadian cities with trolley service declined from approximately 50 in 1950 to 8 in the 1960s. General Motors acquired many of the privately owned streetcar companies and replaced the trolleys with buses that the company made. Buses offer more flexible service than do trolleys because they are not restricted to fixed tracks. However, bus ridership in the United States declined from a peak of 11 billion riders annually in the late 1940s to 5 billion in 2011. Commuter railroad service, like trolleys and buses, has also been drastically reduced in most U.S. cities.

RAPID TRANSIT. The one exception to the downward trend in public transit in the United States is rapid transit. It is known to transportation planners as either fixed heavy rail (such as subways) or fixed light rail (such as streetcars). Cities such as Boston and Chicago have attracted new passengers through construction of new subway lines and modernization of existing service (Figure 13-38). Chicago has been a pioneer in the construction of heavy-rail rapid transit lines in the median strips of expressways. Entirely new subway systems have been built in recent years in U.S. cities, including Atlanta, Baltimore, Miami, San Francisco, and Washington.

The federal government has permitted Boston, New York, and other cities to use funds originally allocated for interstate highways to modernize rapid transit service instead. New York's subway cars, once covered with graffiti spray-painted by gang members, have been cleaned so that passengers can ride in a more hospitable environment. As a result of these improvements, subway ridership in the United States increased from 2 billion in 1995 to 3.6 billion in 2011.

The trolley—now known by the more elegant term fixed light-rail transit—was once relegated almost exclusively to a tourist attraction in New Orleans and San Francisco but is making a modest comeback in North America. New trolley lines have been built or are under construction in Baltimore, Buffalo, Calgary, Edmonton, Los Angeles, Portland

▲ **FIGURE 13-38 BOSTON PUBLIC TRANSIT** Boston's subway system, known as "the T," includes heavy rail (top) and light rail (borrom).

(Oregon), Sacramento, St. Louis, San Diego, and San Jose. Ridership in all cities combined was a half-billion in 2011.

California, the state that most symbolizes the automobile-oriented American culture, is the leader in construction of new fixed light-rail transit lines. San Diego has added more kilometers than any other city. One line that runs from the CBD south to the Mexican border has been irreverently dubbed the "Tijuana trolley" because it is heavily used by residents of nearby Tijuana, Mexico. Los Angeles—the city perhaps most associated with the motor vehicle—has planned the most extensive new light-rail system. The city had a rail network exceeding 1,600 kilometers (1,000 miles) as recently as the late 1940s, but the lines were abandoned when freeways were built to accommodate increasing automobile usage. Now Los Angeles wants to entice motorists out of their cars and trucks with new light-rail lines, but construction is very expensive, and the lines serve only a tiny percentage of the region.

The minimal level of public transit service in most U.S. cities means that low-income people may not be able to reach places of employment. Low-income people tend to live in inner-city neighborhoods, but the job opportunities, especially those requiring minimal training and skill in personal services, are in suburban areas not well served by public transportation. Inner-city neighborhoods have high unemployment rates at the same time that suburban firms have difficulty attracting workers. In some cities, governments and employers subsidize vans to carry low-income inner-city residents to suburban jobs.

Pause and Reflect 13.3.6
What strategies are being used at your college or school district to reduce dependency on private motor vehicles?

PUBLIC TRANSIT IN OTHER COUNTRIES. In dozens of major cities around the world, extensive networks of bus, tram, and subway lines have been maintained, and funds for new construction have been provided in recent years (Figure 13-39). Smaller cities have shared the construction boom. In France, new subway lines have been built since the 1970s in Lille, Lyon, and Marseille, and hundreds of kilometers of entirely new tracks have been laid between the country's major cities to operate a high-speed train known as the TGV (Trains à Grande Vitesse). Growth in the suburbs has stimulated nonresidential construction, including suburban shops, industry, and offices.

Despite modest recent successes, public transit in the United States is caught in a vicious circle because fares do not cover operating costs. As patronage declines and expenses rise, the fares are increased, which drives away passengers and leads to service reduction and still higher fares. Public expenditures to subsidize construction and operating costs have increased, but the United States does not fully recognize that public transportation is a vital utility deserving of subsidy to the degree long assumed by governments in other developed countries, as well as developing countries.

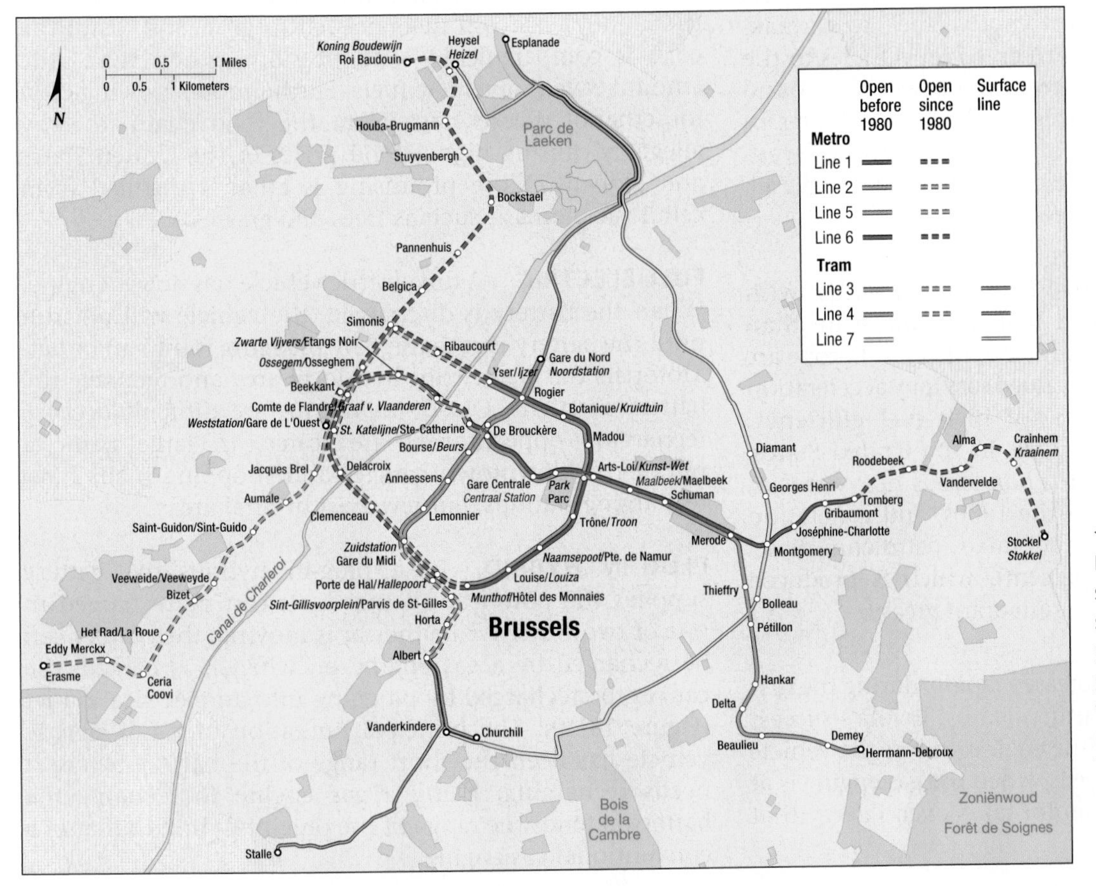

◀ **FIGURE 13-39 BRUSSELS, BELGIUM, METRO AND TRAM** European cities such as Brussels have invested substantially in improving public transportation in recent years. Brussels provides a good example of a public transport system that integrates heavy rail (Métro) with light rail (trams). Trams initially used Métro tunnels, but the tunnels were large enough to convert to heavy-rail lines as funds became available.

ADVANTAGES OF PUBLIC TRANSIT

Learning Outcome 13.3.7
Describe recent and possible future improvements in vehicles.

In larger cities, public transit is better suited than motor vehicles to moving large numbers of people because each transit traveler takes up far less space. Public transportation is cheaper, less polluting, and more energy efficient than privately operated motor vehicles. It also is particularly suited to rapidly bringing a large number of people into a small area. A bus can accommodate 30 people in the amount of space occupied by one car, whereas a double-track rapid transit line can transport the same number of people as 16 lanes of urban freeway.

Motor vehicles have costs beyond their purchase and operation—including delays imposed on others, increased need for highway maintenance, construction of new highways, and pollution. One-third of the high-priced central land is devoted to streets and parking lots, although multistory and underground garages also are constructed.

THE CAR OF THE FUTURE

Consumers in developed countries are reluctant to give up their motor vehicles, and demand for vehicles is soaring in developing countries. One of the greatest challenges to reducing pollution and conserving nonrenewable resources is reliance on petroleum as automotive fuel, so carmakers are scrambling to bring alternative-fuel vehicles to the market. The Department of Energy forecasts that around one-half of all new vehicles sold in the United States in 2020 will be powered by an alternative to the conventional gas engine. Alternative technologies include diesel, biofuel, hybrid, electric, and hydrogen.

DIESEL. Diesel engines burn fuel more efficiently, with greater compression, and at a higher temperature than conventional gas engines. Most new vehicles in Europe are diesel powered, where they are valued for zippy acceleration on crowded roads, as well as for high fuel efficiency. Diesels have made limited inroads in the United States, where they were identified with ponderous heavy trucks, poorly performing versions in the 1980s, and generation of more pollutants. Biodiesel fuel mixes petroleum diesel with biodiesel (typically 5 percent), which is produced from vegetable oils or recycled restaurant grease.

HYBRID. Sales of hybrids increased rapidly during the first decade of the twenty-first century, led by Toyota's success with the hybrid Prius. A gasoline engine powers the vehicle at high speeds, and at low speeds, when the gas engine is at its least efficient, an electric motor takes over. Energy that would otherwise be wasted in coasting and braking is also captured as electricity and stored until needed.

▲ **FIGURE 13-40 PLUG-IN HYBRID** Chevrolet Volt cars are being recharged outside the factory in Detroit where they are assembled.

ETHANOL. Ethanol is fuel made by distilling crops such as sugarcane, corn, and soybeans. Sugarcane is distilled for fuel in Brazil, where most vehicles run on ethanol. In the United States, corn has been the principal crop for ethanol, but this has proved controversial because the amount of fossil fuels needed to grow and distill the corn is comparable to—and possibly greater than—the amount saved in vehicle fuels. Furthermore, growing corn for ethanol diverts corn from the food chain, thereby allegedly causing higher food prices in the United States and globally. More promising is ethanol distilled from cellulosic biomass, such as trees and grasses.

FULL ELECTRIC. A full electric vehicle has no gas engine. When the battery is discharged, the vehicle will not run until the battery is recharged by plugging it into an outlet. Motorists can make trips in a local area and recharge the battery at night. Out-of-town trips are difficult because recharging opportunities are scarce. In large cities, a number of downtown garages and shopping malls have recharging stations, but few exist in rural areas.

PLUG-IN HYBRID. In a plug-in hybrid, the battery supplies the power at all speeds. It can be recharged in one of two ways: While the car is moving, the battery can be recharged by a gas engine or, when it is parked, the car can be recharged by plugging into an electrical outlet (Figure 13-40). The principal limitation of a full electric vehicle has been the short range of the battery before it needs recharging. Using a gas engine to recharge the battery extends the range of the plug-in hybrid to that of a conventional gas engine.

HYDROGEN FUEL CELL. Hydrogen forced through a PEM (polymer electrolyte membrane or proton exchange membrane) combines with oxygen from the air, producing an electric charge. The electricity can then be used to power an electric motor. Fuel cells are now widely used in small vehicles such as forklifts. Fuel cell vehicles are being used in a handful of large East Coast and West Coast cities, where hydrogen fueling stations have been constructed.

Pause and Reflect 13.3.7

Which alternative-fuel vehicles appear most likely to be successful at reducing dependency on fossil fuels? Which appear most successful at improving air pollution?

REGIONAL VARIATIONS IN ELECTRICITY. Electric-powered vehicles require recharging by being plugged into a source of electricity such as an outlet in the garage that ultimately comes from a power plant. Though fossil fuel is not being pumped directly into the tank of the electric-powered vehicle, fossil fuel is consumed to generate the electricity at the power plant. In fact, the United States as a whole generates around 40 percent of its electricity from coal-burning power plants and around 25 percent from natural gas. An electric vehicle does reduce consumption of an increasingly scarce and expensive resource—petroleum. But if the electricity is generated by natural gas, then plugging a vehicle into the electric grid may conserve petroleum at the expense of more rapid depletion of natural gas. If electricity is generated by coal, a plug-in may cause more air pollution.

Electricity is generated differently across the 50 U.S. states. In the Pacific Northwest, where hydroelectric is the leading source of electricity, recharging electric vehicles will have much less impact on air quality than will be the case in the Midwest (Figure 13-41). States that depend on farm production may benefit from increased use of ethanol. Thus, the "greenest" alternative varies by location.

CHECK-IN: KEY ISSUE 3

Why Are Urban Areas Expanding?

- ✓ **A city is an incorporated unit of government. An urban area includes a city and surrounding built-up suburbs. A metropolitan area includes an urban area and surrounding counties.**
- ✓ **In the northeastern United States, adjacent metropolitan areas form a continuous urban region called Megalopolis.**
- ✓ **U.S. cities once expanded by annexing surrounding land, but annexation is now less common; instead, cities are surrounded by numerous independent suburban jurisdictions.**
- ✓ **Sprawling suburbs surround U.S. cities; suburban sprawl consumes a lot of land and requires investment in a lot of new roads and utilities.**
- ✓ **Suburbs are segregated by social class and by land use activities.**
- ✓ **Suburban residents are dependent on motor vehicles to get to other places, whereas most cities offer forms of public transit.**

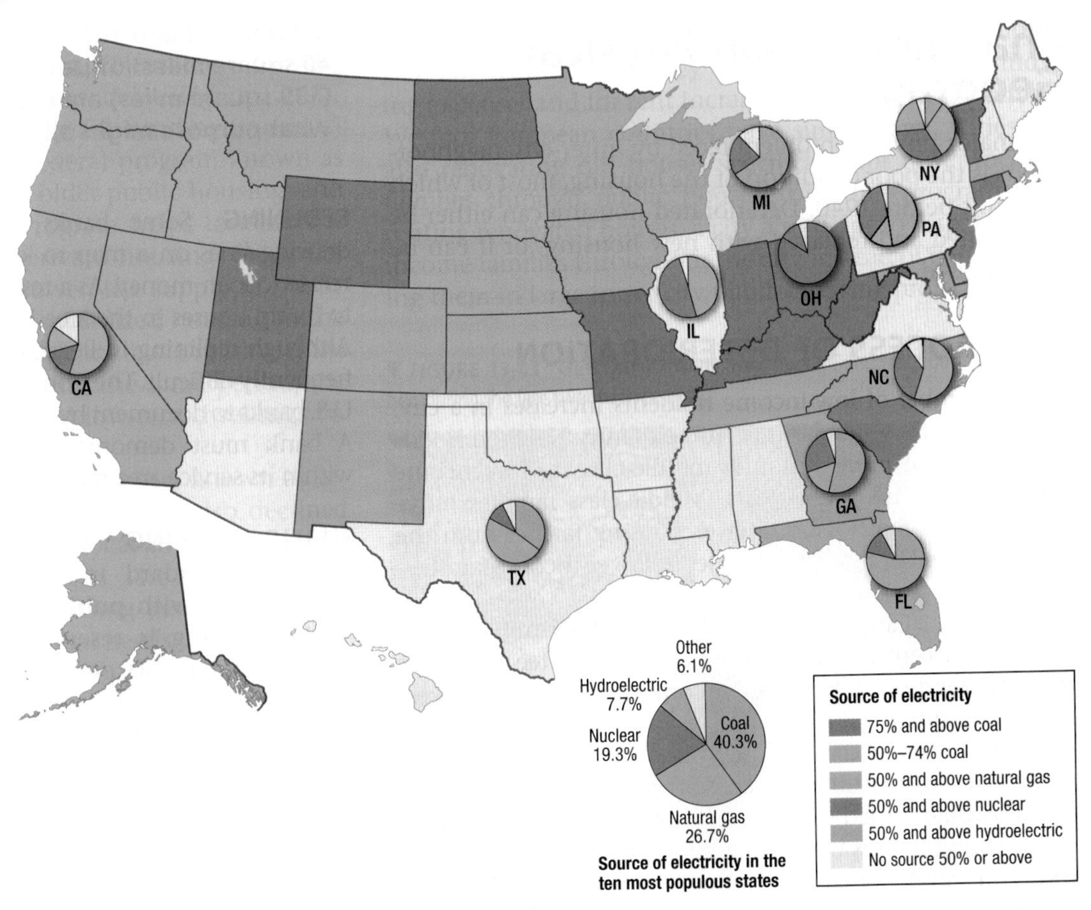

▲ **FIGURE 13-41 ELECTRICITY BY U.S. STATE** Dependency on nonrenewable and polluting fossil fuels to generate electricity varies widely among states.

Summary

KEY ISSUE 1

Why Do Services Cluster Downtown?

Services, especially public and business services, cluster in the CBD; some consumer services, especially leisure, are in the CBD.

LEARNING OUTCOME 13.1.1: Describe the three types of services found in a CBD.

- The CBD contains a large percentage of an urban area's public, business, and consumer services.
- Offices cluster in the CBD to take advantage of its accessibility.
- Retail services, as well as manufacturers and residents, are less likely than in the past to be in the CBD.

LEARNING OUTCOME 13.1.2: Explain the three-dimensional nature of a CBD.

- A CBD is characterized by an extensive underground city of services and utilities, as well as high-rise buildings.
- Outside North America, CBDs may have more consumer services and fewer high-rise offices.

THINKING GEOGRAPHICALLY 13.1: Compare the CBDs of Toronto and Detroit. What might account for the differences?

GOOGLE EARTH 13.1: The tallest structure in the CBD of Ghent, Belgium, is Saint Bavo Cathedral, built in the sixteenth century. Fly to Saint Bavo Cathedral, Bisdomplein 1-3, Ghent, Belgium, drag to enter street view, exit street view, turn on 3D, exit street view, and zoom out so that the entire cathedral and its surroundings can be seen. What other buildings are highlighted in 3D in the CBD of Ghent?

Key Terms

Annexation (p. 480) Legally adding land area to a city in the United States.

Census tract (p. 468) An area delineated by the U.S. Bureau of the Census for which statistics are published; in urban areas, census tracts correspond roughly to neighborhoods.

Central business district (CBD) (p. 461) The area of a city where retail and office activities are clustered.

City (p. 476) An urban settlement that has been legally incorporated into an independent, self-governing unit.

Combined statistical area (CSA) (p. 477) In the United States, two or more contiguous core-based statistical areas tied together by commuting patterns.

Concentric zone model (p. 466) A model of the internal structure of cities in which social groups are spatially arranged in a series of rings.

Core based statistical area (CBSA) (p. 477) In the United States, the combination of all metropolitan statistical areas and micropolitan statistical areas.

Council of government (p. 479) A cooperative agency consisting of representatives of local governments in a metropolitan area in the United States.

Density gradient (p. 480) The change in density in an urban area from the center to the periphery.

Edge city (p. 482) A large node of office and retail activities on the edge of an urban area.

KEY ISSUE 2

Where Are People Distributed Within Urban Areas?

Three models help to explain where different groups of people live within urban areas.

LEARNING OUTCOME 13.2.1: Describe the concentric zone, sector, and multiple nuclei models.

- According to the concentric zone model, a city grows outward in rings.
- According to the sector model, a city grows along transportation corridors.
- According to the multiple nuclei model, a city grows around several nodes.

LEARNING OUTCOME 13.2.2: Analyze how the three models help to explain where people live in an urban area.

- According to the concentric zone model, housing is newer in outer rings than in inner rings.
- According to the sector model, wealthier people live in different corridors than do poorer people.
- According to the multiple nuclei model, different ethnic groups cluster around various nodes.

LEARNING OUTCOME 13.2.3: Describe how the three models explain patterns in European cities.

- In other countries, wealthier people live in different sectors than poorer people, and outer rings have newer housing.
- In cities outside North America, lower-income people are more likely to live in outer rings.

LEARNING OUTCOME 13.2.4: Describe how the three models explain patterns in cities in developing countries.

LEARNING OUTCOME 13.2.5: Describe the history of development of cities in developing countries.

- Many cities in developing countries were shaped by colonial powers.
- Since gaining their independence, developing countries have seen cities grow rapidly.

THINKING GEOGRAPHICALLY 13.2: Officials of rapidly growing cities in developing countries discourage the building of houses that do not meet international standards for sanitation and construction methods. Also discouraged are privately owned transportation services because the vehicles generally lack decent tires, brakes, and other safety features. Yet the residents prefer substandard housing to no housing, and they prefer unsafe transportation to no transportation. What would be the advantages and problems for a city if health and safety standards for housing, transportation, and other services were relaxed?

GOOGLE EARTH 13.2: Sectors, nodes, and rings can be seen in a Google Earth image of Chicago. North is to the right in the image. The large white structure along the lakefront is McCormick Place convention center. Is this an example of a sector, node, or ring? The series of large buildings along the river to the top left and top right are factories and warehouses. Are these examples of sectors, nodes, or rings? The structures to the far left and far right of the image are houses, whereas the buildings closer to the CBD are apartment towers. Are these examples of sectors, nodes, or rings?

Filtering (p. 490) A process of change in the use of a house, from single-family owner occupancy to abandonment.

Food desert (p. 464) An area in a developed country where healthy food is difficult to obtain.

Gentrification (p. 491) A process of converting an urban neighborhood from a predominantly low-income, renter-occupied area to a predominantly middle-class, owner-occupied area.

Greenbelt (p. 481) A ring of land maintained as parks, agriculture, or other types of open space to limit the sprawl of an urban area.

Megalopolis (p. 478) A continuous urban complex in the northeastern United States.

Multiple nuclei model (p. 467) A model of the internal structure of cities in which social groups are arranged around a collection of nodes of activities.

KEY ISSUE 3

Why Are Urban Areas Expanding?

Urban growth has been primarily focused on suburbs that surround older cities.

LEARNING OUTCOME 13.3.1: State three definitions of urban settlements.

- A city is a legally incorporated entity that encompasses the older portion of the urban area.
- An urban area includes the city and built-up suburbs.
- A metropolitan area includes the city, built-up suburbs, and counties that are tied to the city.

LEARNING OUTCOME 13.3.2: Describe how metropolitan areas contain many local governments and overlap with each other.

- In some regions, adjacent metropolitan areas overlap with each, creating large contiguous urban complexes.
- The United States has nearly 90,000 local governments, making it difficult to address urban problems.

LEARNING OUTCOME 13.3.3: I dentify historical and contemporary patterns of suburban expansion.

- In the past, cities expanded their land area to encompass outlying areas, but now they are surrounded by independent suburban jurisdictions.
- Suburban sprawl has been documented to be costly.

LEARNING OUTCOME 13.3.4: Explain two ways in which suburbs are segregated.

- Suburbs are segregated according to social class and land uses.

LEARNING OUTCOME 13.3.5: Describe the impact of motor vehicles in urban areas.

- Motor vehicles take up a lot of space in cities, including streets, freeways, and parking areas.
- Some cities control the number of vehicles that can enter the center of the city.

LEARNING OUTCOME 13.3.6: State benefits and limitations of public transportation.

- Public transit, such as subways and buses, are more suited than private cars to move large numbers of people into and out of the CBD.
- New investment in public transit has occurred in a number of U.S. cities, though less extensively than in other countries.

LEARNING OUTCOME 13.3.7: Describe recent and possible future improvements in vehicles.

- Vehicles that are more fuel efficient and less polluting are likely to become more widely available in the future.

THINKING GEOGRAPHICALLY 13.3: Draw a sketch of your community or neighborhood. In accordance with Kevin Lynch's *The Image of the City*, place five types of information on the map: districts (homogeneous areas), edges (boundaries that separate districts), paths (lines of communication), nodes (central points of interaction), and landmarks (prominent objects on the landscape). How clear an image does your community have for you?

GOOGLE EARTH 13.3: Public transit in Brussels. A #7 tram enters a tunnel near Diamant station. Is this an example of light rail or heavy rail?

Peripheral model (p. 476) A model of North American urban areas consisting of an inner city surrounded by large suburban residential and business areas tied together by a beltway or ring road.

Primary census statistical area (PCSA) (p. 477) In the United States, all of the combined statistical areas plus all of the remaining metropolitan statistical areas and micropolitan statistical areas.

Public housing (p. 490) Housing owned by the government; in the United States, it is rented to residents with low incomes, and the rents are set at 30 percent of the families' incomes.

Redlining (p. 490) A process by which banks draw lines on a map and refuse to lend money to purchase or improve property within the boundaries.

Rush hour (p. 486) The four consecutive 15-minute periods in the morning and evening with the heaviest volumes of traffic.

Sector model (p. 467) A model of the internal structure of cities in which social groups are arranged around a series of sectors, or wedges, radiating out from the central business district.

Smart growth (p. 481) Legislation and regulations to limit suburban sprawl and preserve farmland.

Social area analysis (p. 468) Statistical analysis used to identify where people of similar living standards, ethnic background, and lifestyle live within an urban area.

Sprawl (p. 480) Development of new housing sites at relatively low density and at locations that are not contiguous to the existing built-up area.

Squatter settlement (p. 472) An area within a city in a less developed country in which people illegally establish residences on land they do not own or rent and erect homemade structures.

KEY ISSUE 4

Why Do Cities Face Challenges?

Cities face physical, social, and economic difficulties, but some improvements have also occurred.

LEARNING OUTCOME 13.4.1: Describe the processes of deterioration and gentrification in cities.

- The older housing in the inner city can deteriorate through processes of filtering and redlining.
- Massive public housing projects were once constructed for poor people, but many of them have been demolished.
- Some cities have experienced gentrification, in which higher-income people move in and renovate previously deteriorated neighborhoods.

LEARNING OUTCOME 13.4.2: Explain the problems of a permanent underclass and culture of poverty in cities.

- Inner cities have concentrations of very poor people, considered to belong to an underclass, some of whom are homeless.
- A culture of poverty traps some poor people in the inner cities.

LEARNING OUTCOME 13.4.3: Describe the difficulties that cities face in paying for services, especially in a recession.

- Cities are faced with the choice of reducing services or raising taxes to pay for needed services.
- The severe recession that started in 2008 continues to hurt the economic condition of cities.
- Some cities have seen a revival of retail services downtown.

THINKING GEOGRAPHICALLY 13.4: Jane Jacobs wrote in *Death and Life of Great American Cities* that an attractive urban environment is one that is animated with an intermingling of a variety of people and activities, such as found in many New York City neighborhoods. What are the attractions and drawbacks to living in such environments?

GOOGLE EARTH 13.4: City meets country in the United Kingdom. Harlow, a New Town built primarily during the 1950s and 1960s, shows the sharp boundary between a high-density residential suburb and the countryside. How does this landscape differ from the outer edge of a typical U.S. suburb?

Underclass (p. 492) A group in society prevented from participating in the material benefits of a more developed society because of a variety of social and economic characteristics.

Urban area (p. 477) A dense core of census tracts, densely settled suburbs, and low-density land that links the dense suburbs with the core.

Urban cluster (p. 477) In the United States, an urban area with between 2,500 and 50,000 inhabitants.

Urbanized area (p. 477) In the United States, an urban area with at least 50,000 inhabitants.

Zoning ordinance (p. 465) A law that limits the permitted uses of land and maximum density of development in a community.

AFTERWORD

CAREERS IN GEOGRAPHY

Where do you see yourself working in 5 or 10 years after graduation? You could be a retail geographer, analyzing customer behavior for a major department store. You could conduct research on land cover, vegetation structure, and snow cover in Alaska. You could be a social science analyst, evaluating redistricting plans to ensure that they do not disenfranchise voters. Or you could lead a team of experts at one of the world's largest retailers in specific dimensions of sustainable business practices.

What do all these careers have in common? In each case, enterprising individuals have found creative ways to apply the core concepts and skills of geography. An increasing number of students recognize that geographic education is practical as well as stimulating. Employment opportunities are expanding for students trained in geography, especially in geospatial technologies, teaching, government service, and business.

Geospatial Technologies

In the past 20 years, the field of geospatial technologies has been making rapid advances, thanks to developments in computer software, computer science, and geographic information systems (GIS), including remote-sensing technologies. Today jobs that make use of geospatial technologies and GIS can be found in such private and public sector areas as environmental consulting, software development, air navigation services, spatial database management for mapping companies (Figure A-1), location analysis for retail businesses and real estate, corporate transportation and logistics departments, criminology, archaeology, resource management, and infrastructure management, to name just a few.

▲ **FIGURE AF-1 GOOGLE EARTH** A Google Earth image of Paris is displayed on a wall-sized monitor in Google's Berlin, Germany, office.

Teaching

More than 100 universities in the United States and Canada offer doctorate or master's degrees in geography. A career as a geography teacher is promising because schools throughout North America are expanding their geography curriculum. Educators increasingly recognize geography's role in teaching students about global diversity (Figure AF-2). AP Human Geography is the fastest-growing AP discipline in U.S. high schools.

Some university geography departments have emphasized outstanding teaching, whereas others are concerned primarily with scholarship and research. The Association of American Geographers includes several dozen specialty groups, organized around research themes, including agricultural, industrial, medical, and transportation geography.

Government Service

Geographers contribute their knowledge of the location of activities, the patterns underlying the distribution of various activities, and the interpretation of data from maps and satellite imagery to local, state, and national governments. Employment opportunities with cities, states, provinces, and other units of local government are typically found in departments of planning, transportation, parks and recreation, economic development, housing, zoning, or other similarly titled government agencies. Geographers may be hired to conduct studies of local economic, social, and physical patterns; to prepare information through maps and reports; and to help plan the community's future.

Many federal government agencies also employ geographers:

- The Department of Agriculture hires geographers for the Forest Service and Natural Resources Conservation Service to enhance environmental quality.
- The Department of Commerce hires geographers for the Bureau of the Census to study changing population trends and for the Economic Development Administration to promote rural development.
- The Department of Defense hires geographers for the Defense Intelligence Agency and the

National Geospatial-Intelligence Agency to analyze satellite imagery.

- The Department of Energy hires geographers for the Office of Environmental Policy and Assistance to administer environmental protection programs.
- The Department of Housing and Urban Development hires geographers to help revitalize American cities.
- The Department of Interior hires geographers for the U.S. Geological Survey to study land use and create topographic maps and for the Office of Environmental Policy and Compliance to administer environmental protection programs.
- The Department of State hires geographers for foreign service.
- The Department of Transportation hires geographers to plan new transportation projects.

▲ **FIGURE AF-2 GEOGRAPHIC AWARENESS** A future geographer reflects on a U.S. electoral map at an American Presidential Experience exhibit, during the ramp-up to the 2012 Presidential election.

Business

An increasing number of American geographers are finding jobs with private companies. The list of possibilities is long, but here are some common examples:

- Developers hire geographers to find the best locations for new shopping centers.
- Real estate firms hire geographers to assess the value of properties.
- Supermarket chains, department stores, and other retailers hire geographers to determine the potential market for new stores.
- Banks hire geographers to assess the probability that a loan applicant has planned a successful development.
- Distributors and wholesalers hire geographers to find ways to minimize transportation costs.
- Transnational corporations hire geographers to predict the behavior of consumers and officials in other countries.
- Manufacturers hire geographers to identify new sources of raw materials and markets.
- Utility companies hire geographers to determine future demand at different locations for gas, electricity, and other services.

For more information on careers in geography, contact the Association of American Geographers at www.aag.org, or the National Council for Geographic Education at www.ncge.org.

The Future: Geography Still Matters

The arrival of the year 2000 sparked many forecasts of what life would be like in the new millennium. Many prophesied that geography would be irrelevant in the twenty-first century. Geography's future was thought to be grim because the diffusion of electronic communications, such as the Internet and smartphones, would make it easier for human activities to be conducted remotely. If any piece of information could be accessed from any place in the world (at least where the Internet works), why live, shop, work, or establish a business in a crowded city or a harsh climate?

To some extent this forecast has come to pass. Many of you obtained this book electronically instead of buying it in a bookstore. The United States has one-fourth fewer bookstores now than in the 1990s, and the number is declining by 2 percent per year.

Bookstores aside, into the second decade of the new millennium geography has actually become more, not less important in people's lives and the conduct of business. Here are several ways that location matters to business more now than in the past, because of—not despite—the diffusion of electronic devices:

1. Smartphones and other electronic devices match specific demand to supply in a particular locality (Figure AF-3). For example:
   - Restaurant apps match hungry people to empty seats in a locality's restaurants.
   - Real estate apps let people find housing for sale or for rent in a locality.
   - Social apps let people know where their friends in a particular locality are hanging out that night.
   - Transportation apps match a locality's empty taxis or carpools with available seats to people trying to get to specific locations.

These sorts of apps generate data on people's preferences in space, which in turn helps even more location-based

▲ **FIGURE AF-3 TIMES SQUARE, NEW YORK** A mapping application on an iPhone displays the locations of Broadway theaters located in the vicinity.

business get started and grow. No wonder that geography apps, in the form of maps (including navigation) and travel (including transportation), rank as two of the five most frequently used services on smartphones.

2. Electronic devices are essential to the smooth movement of people and goods. For example:
   - Turn-by-turn information can prevent you from getting lost or steer you back if you did get lost.
   - Traffic jams on overcrowded roads can be avoided or minimized.
   - Vehicles in the future will be driverless, so you can spend driving time working, learning stuff, or social networking.

Images from Google Earth and others that you see throughout this book will get more detailed and accurate. Mapping is expanding into indoor space such as shopping malls, or three dimensions, such as Figure AF-1.

3. The people who make all of these new location-based apps are themselves highly clustered in a handful of places in the world, especially Silicon Valley.
   - Ideas—both brilliant and far-fetched—are still easier to communicate face-to-face than across long distances.
   - Living and working in places like Silicon Valley, despite high expenses and choking traffic jams, put people next to other like-minded innovators in the electronic-based geography of the twenty-first century.

This final section has focused on business applications. But let's not forget the role of electronic devices on the changing geography of cultural diversity.

- What if you searched for an available restaurant table in a foreign language? Would you find the same places?
- What if you conducted an Internet search in a foreign country? Would you find the same information?

As more electronic-based geography diffuses through our daily lives and commerce, we are thinking in a different way. It's called geobrowsing. For example,

- Instead of looking for restaurants in the Yellow Pages, we find places to eat that are mapped on our device.
- Instead of turning on a radio to hear traffic information, we look at the red and green traffic flow patterns on an electronic map.
- Instead of waiting for a TV weather report, we look at storm patterns on our device's map.

So, as the twenty-first century unfolds, geography's perspectives on global changes and cultural diversity, together with the power of geospatial technologies to analyze and display data, will play an increasingly important role in daily life, learning, work, leisure, and the challenges of citizenship. We are all becoming participants in the world of geography. Welcome!

Appendix
Map Scale and Projections

Phillip C. Muercke

Unaided, our human senses provide a limited view of our surroundings. To overcome those limitations, humankind has developed powerful vehicles of thought and communication, such as language, mathematics, and graphics. Each of those tools is based on elaborate rules; each has an information bias, and each may distort its message, often in subtle ways. Consequently, to use those aids effectively, we must understand their rules, biases, and distortions. The same is true for the special form of graphics we call maps: we must master the logic behind the mapping process before we can use maps effectively.

A fundamental issue in cartography, the science and art of making maps, is the vast difference between the size and geometry of what is being mapped—the real world, we will call it—and that of the map itself. Scale and projection are the basic cartographic concepts that help us understand that difference and its effects.

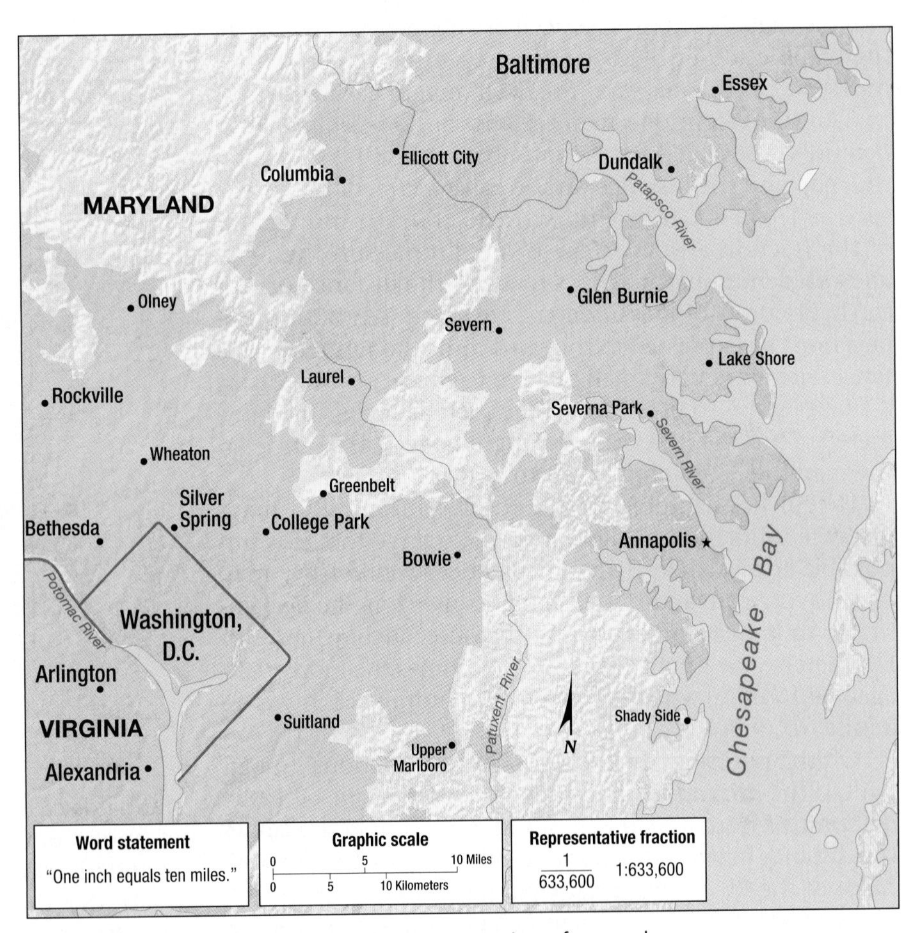

▲ **FIGURE A-1** Common expressions of map scale.

Map Scale

Our senses are dwarfed by the immensity of our planet; we can sense directly only our local surroundings. Thus, we cannot possibly look at our whole state or country at one time, even though we may be able to see the entire street where we live. Cartography helps us expand what we can see at one time by letting us view the scene from some distant vantage point. The greater the imaginary distance between that position and the object of our observation, the larger the area the map can cover, but the smaller the features will appear on the map. That reduction is defined by the *map scale*, the ratio of the distance on the map to the distance on the earth. Map users need to know about map scale for two reasons: so that they can convert measurements on a map into meaningful real-world measures and so that they can know how abstract the cartographic representation is.

REAL-WORLD MEASURES. A map can provide a useful substitute for the real world for many analytical purposes. With the scale of a map, for instance, we can compute the actual size of its features (length, area, and volume). Such calculations are helped by three expressions of a map scale: a word statement, a graphic scale, and a representative fraction.

A *word statement* of a map scale compares X units on the map to Y units on the earth, often abbreviated "X unit to Y units." For example, the expression "1 inch to 10 miles" means that 1 inch on the map represents 10 miles on the earth (Figure A-1). Because the map is always smaller than the area that has been mapped, the ground unit is always the larger number. Both units are expressed in meaningful terms, such as inches or centimeters and miles or kilometers. Word statements are not intended for precise calculations but give the map user a rough idea of size and distance.

A *graphic scale,* such as a bar graph, is concrete and therefore overcomes the need to visualize inches and miles that is associated with a word statement of scale (see Figure A-1). A graphic scale permits direct visual comparison of feature sizes and the distances between features. No ruler is required; any measuring aid will do. It needs only be compared with the scaled bar; if the length of 1 toothpick is equal to 2 miles on the ground and the map distance equals the length of 4 toothpicks, then the ground distance is 4 times 2, or 8 miles. Graphic scales are especially convenient in this age of copying machines, when we are more likely to be working with a copy than with the original map. If a map is reduced or enlarged as it is copied, the graphic scale will change in proportion to the change in the size of the map and thus will remain accurate.

The third form of a map scale is the *representative fraction* (RF). An RF defines the ratio between the distance on the map and the distance on the earth in fractional terms, such as 1/633,600 (also written 1:633,600). The numerator of the fraction always refers to the distance on the map, and the denominator always refers to the distance on the earth. No units of measurement are given, but both numbers must be expressed in the same units. Because map distances are extremely small relative to the size of the earth, it makes sense to use small units, such as inches or centimeters. Thus the RF 1:633,600 might be read as "1 inch on the map to 633,600 inches on the earth."

Herein lies a problem with the RF. Meaningful map-distance units imply a denominator so large that it is impossible to visualize. Thus, in practice, reading the map scale involves an additional step of converting the denominator to a meaningful ground measure, such as miles or kilometers. The unwieldy 633,600 becomes the more manageable 10 miles when divided by the number of inches in a mile (63,360).

On the plus side, the RF is good for calculations. In particular, the ground distance between points can be easily determined from a map with an RF. One simply multiplies the distance between the points on the map by the denominator of the RF. Thus a distance of 5 inches on a map with an RF of 1/126,720 would signify a ground distance of 5 X 126,720, which equals 633,600. Because all units are inches and there are 63,360 inches in a mile, the ground distance is 633,600/ 63,360, or 10 miles. Computation of area is equally straightforward with an RF. Computer manipulation and analysis of maps is based on the RF form of map scale.

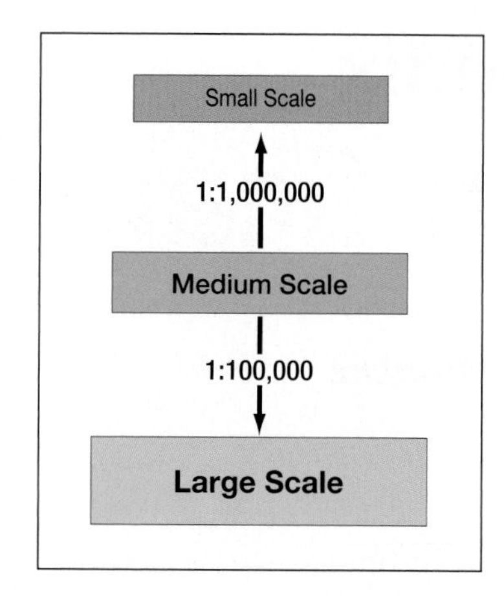

▲ **FIGURE A-2** The scale gradient can be divided into three broad categories.

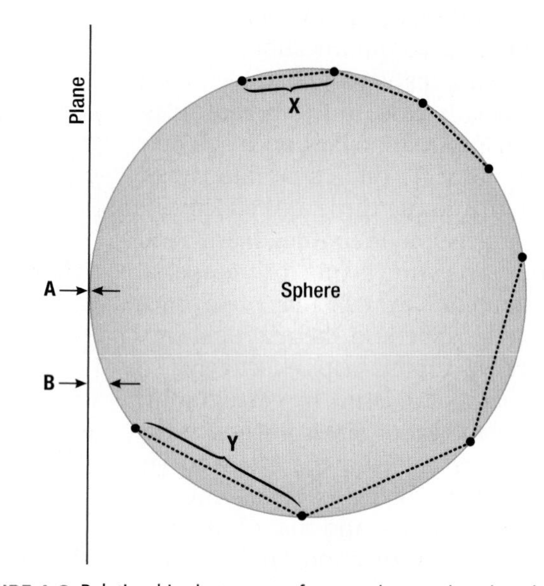

▲ **FIGURE A-3** Relationships between surfaces on the round earth and a flat map.

GUIDES TO GENERALIZATION. Scales also help map users visualize the nature of the symbolic relation between the map and the real world. It is convenient here to think of maps as falling into three broad scale categories (Figure A-2). (Do not be confused by the use of the words large AND small in this context; just remember that the larger the denominator, the smaller the scale ratio and the larger the area that is shown on the map.) Scale ratios greater than 1:100,000, such as the 1:24,000 scale of U.S. Geological Survey topographic quadrangles, are large-scale maps. Although those maps can cover only a local area, they can be drawn to rather rigid standards of accuracy. Thus they are useful for a wide range of applications that require detailed and accurate maps, including zoning, navigation, and construction.

At the other extreme are maps with scale ratios of less than 1:1,000,000, such as maps of the world that are found in atlases. Those are small-scale maps. Because they cover large areas, the symbols on them must be highly abstract. They are therefore best suited to general reference or planning, when detail is not important. Medium- or intermediate-scale maps have scales between 1:100,000 and 1:1,000,000. They are good for regional reference and planning purposes.

Another important aspect of map scale is to give us some notion of geometric accuracy; the greater the expanse of the real world shown on a map, the less accurate the geometry of that map is. Figure A-3 shows why. If a curve is represented by straight line segments, short segments (*X*) are more similar to the curve than are long segments (*Y*). Similarly, if a plane is placed in contact with a sphere, the difference between the two surfaces is slight where they touch (*A*) but grows rapidly with increasing distance from the point of contact (*B*). In view of the large diameter and slight local curvature of the earth, distances will be well represented on large-scale maps (those with small denominators) but will be increasingly poorly represented at smaller scales. This close relationship between map scale and map geometry brings us to the topic of map projections.

Map Projections

The spherical surface of the earth is shown on flat maps by means of map projections. The process of "flattening" the earth is essentially a problem in geometry that has captured the attention of the best mathematical minds for centuries. Yet no one has ever found a perfect solution; there is no known way to avoid spatial distortion of one kind or another. Many map projections have been devised, but only a few have become standard. Because a single flat map cannot preserve all aspects of the earth's surface geometry, a mapmaker must be careful to match the projection with the task at hand. To map something that involves distance, for example, a projection should be used in which distance is not distorted. In addition, a map user should be able to recognize which aspects of a map's geometry are accurate and which are distortions caused by a particular projection process. Fortunately, that objective is not too difficult to achieve.

It is helpful to think of the creation of a projection as a twostep process (Figure A-4). First, the immense earth is reduced to a small globe with a scale equal to that of the desired flat map. All spatial properties on the globe are true to those on the earth. Second, the globe is flattened. Since that cannot be done without distortion, it is accomplished in such a way that the resulting map exhibits certain desirable spatial properties.

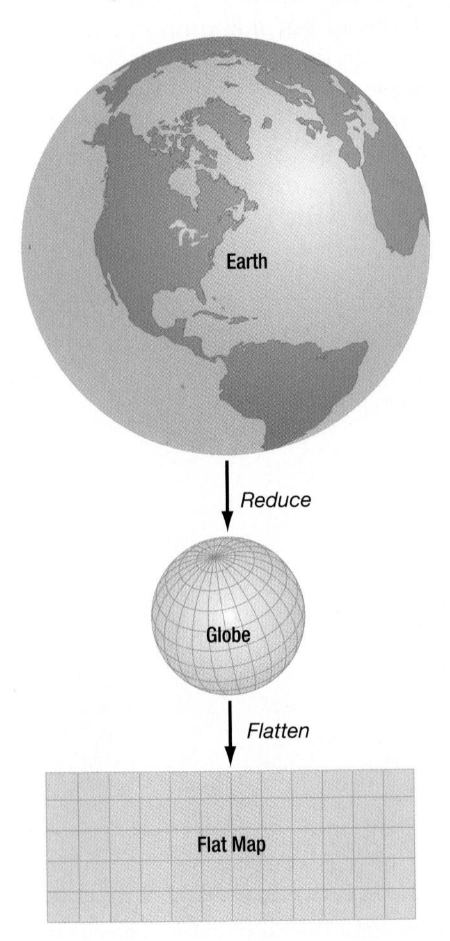

▲ **FIGURE A-4** The two-step process of creating a projection.

PERSPECTIVE MODELS. Early map projections were sometimes created with the aid of perspective methods, but that has changed. In the modern electronic age, projections are normally developed by strictly mathematical means and are plotted out or displayed on computer-driven graphics devices. The concept of perspective is still useful in visualizing what map projections do, however. Thus projection methods are often illustrated by using strategically located light sources to cast shadows on a projection surface from a latitude/longitude net inscribed on a transparent globe.

The success of the perspective approach depends on finding a projection surface that is flat or that can be flattened without distortion. The cone, cylinder, and plane possess those attributes and serve as models for three general classes of map projections: *conic, cylindrical,* and *planar* (or azimuthal). Figure A-5 shows those three classes, as well as a fourth, a false cylindrical class with an oval shape. Although the oval class is not of perspective origin, it appears to combine properties of the cylindrical and planar classes (Figure A-6).

The relationship between the projection surface and the model at the point or line of contact is critical because distortion of spatial properties on the projection is symmetrical about, and increases with distance from, that point or line. That condition is illustrated for the cylindrical and planar classes of projections in Figure A-7. If the point or line of contact is changed to some other position on the globe, the distortion pattern will be recentered on the new position but will retain the same symmetrical form. Thus centering a projection on the area of interest on the earth's surface can minimize the effects of projection distortion. And recognizing the general projection shape, associating it with a perspective model, and recalling the characteristic distortion pattern will provide the information necessary to compensate for projection distortion.

PRESERVED PROPERTIES. For a map projection to truthfully depict the geometry of the earth's surface, it

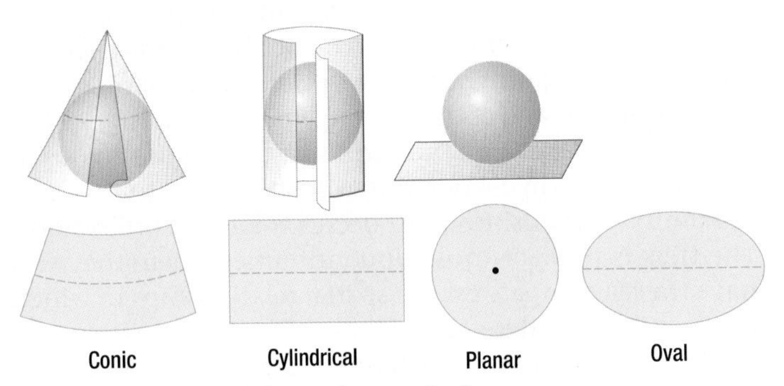

▲ **FIGURE A-5** General classes of map projections.

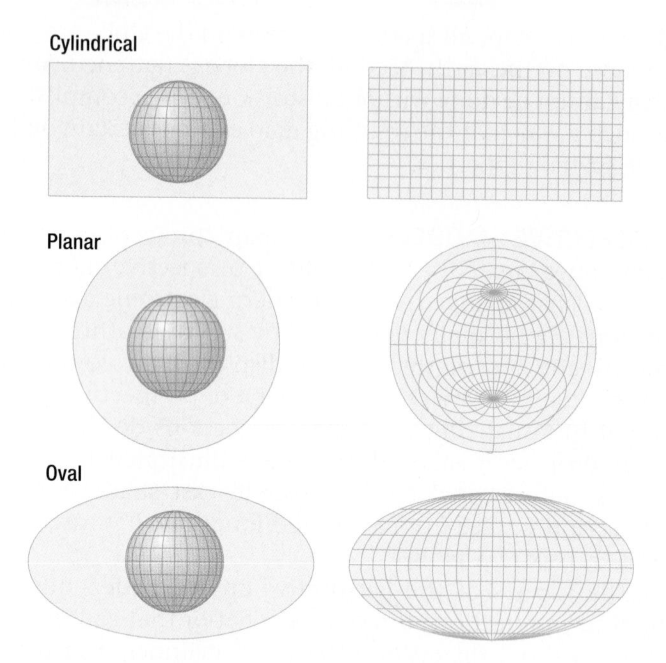

▲ **FIGURE A-6** The visual properties of cylindrical and planar projections combined in oval projections.

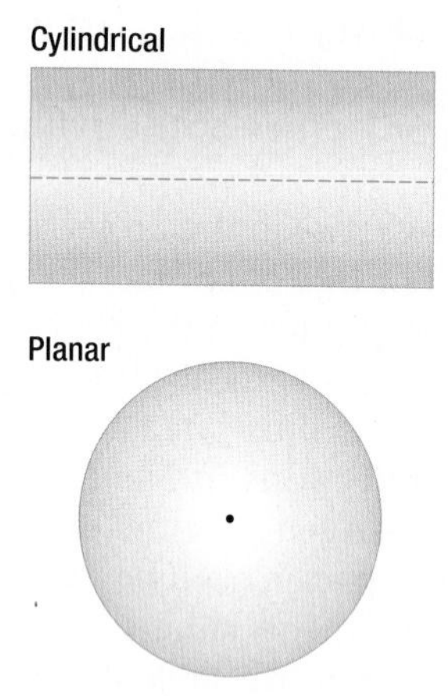

▲ **FIGURE A-7** Characteristic patterns of distortion for two projection classes. Here, darker shading implies greater distortion.

would have to preserve the spatial attributes of *distance, direction, area, shape*, and *proximity*. That task can be readily accomplished on a globe, but it is not possible on a flat map. To preserve area, for example, a mapmaker must stretch or shear shapes; thus area and shape cannot be preserved on the same map. To depict both direction and distance from a point, area must be distorted. Similarly, to preserve area as well as direction from a point, distance has to be distorted. Because the earth's surface is continuous in all directions from every point, discontinuities that violate proximity relationships must occur on all map projections. The trick is to place those discontinuities where they will have the least impact on the spatial relationships in which the map user is interested.

We must be careful when we use spatial terms, because the properties they refer to can be confusing. The geometry of the familiar plane is very different from that of a sphere; yet when we refer to a flap map, we are in fact making reference to the spherical earth that was mapped. A shape-preserving projection, for example, is truthful to local shapes—such as the rightangle crossing of latitude and longitude lines—but does not preserve shapes at continental or global levels. A distance-preserving projection can preserve that property from one point on the map in all directions or from a number of points in several directions, but distance cannot be preserved in the general sense that area can be preserved. Direction can also be generally preserved from a single point or in several directions from a number of points but not from all points simultaneously. Thus a shape-, distance-, or direction-preserving projection is truthful to those properties only in part.

Partial truths are not the only consequence of transforming a sphere into a flat surface. Some projections exploit that transformation by expressing traits that are of considerable value for specific applications. One of those is the famous shape-preserving *Mercator projection* (Figure A-8). That cylindrical projection was derived mathematically in the 1500s so that a compass bearing (called rhumb lines) between any two points on the earth would plot as straight lines on the map. That trait let navigators plan, plot, and follow courses between origin and destination, but it was achieved at the expense of extreme areal distortion toward the margins of the projection (see Antarctica in Figure A-8). Although the Mercator projection is admirably suited for its intended purpose, its widespread but inappropriate use for nonnavigational purposes has drawn a great deal of criticism.

The *gnomonic projection* is also useful for navigation. It is a planar projection with the valuable characteristic of showing the shortest (or great circle) route between any two points on the earth as straight lines. Long-distance navigators first plot the great circle course between origin

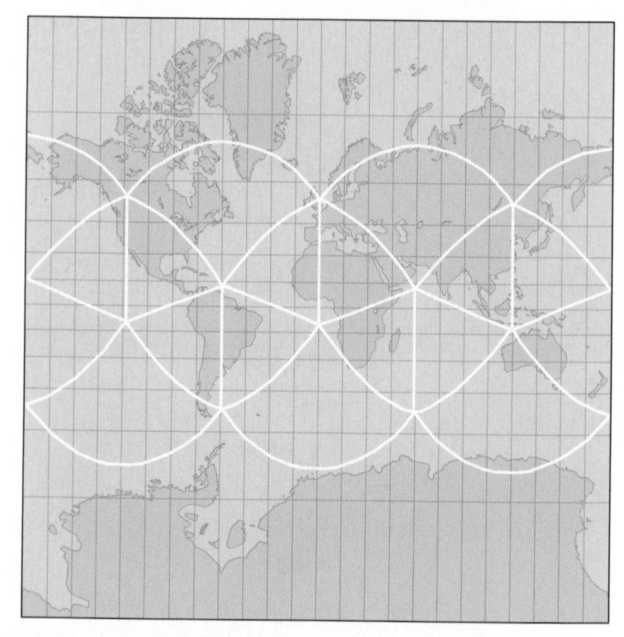

▲ **FIGURE A-8** The useful Mercator projection, showing extreme area distortion in the higher latitudes.

(A) Gnomonic

(B) Mercator

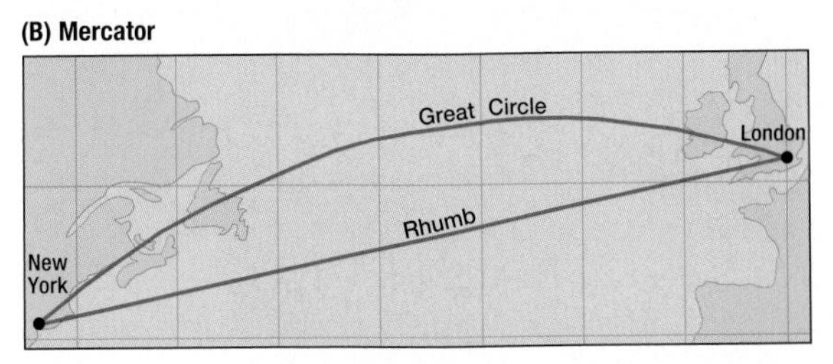

▲ **FIGURE A-9** A gnomonic projection (A) and a Mercator projection (B), both of value to long-distance navigators.

and destination on a gnomonic projection (Figure A-9, top). Next they transfer the straight line to a Mercator projection, where it normally appears as a curve (Figure A-9, bottom). Finally, using straight-line segments, they construct an approximation of that course on the Mercator projection. Navigating the shortest course between origin and destination then involves following the straight segments of the course and making directional corrections between segments. Like the Mercator projection, the specialized gnomonic projection distorts other spatial properties

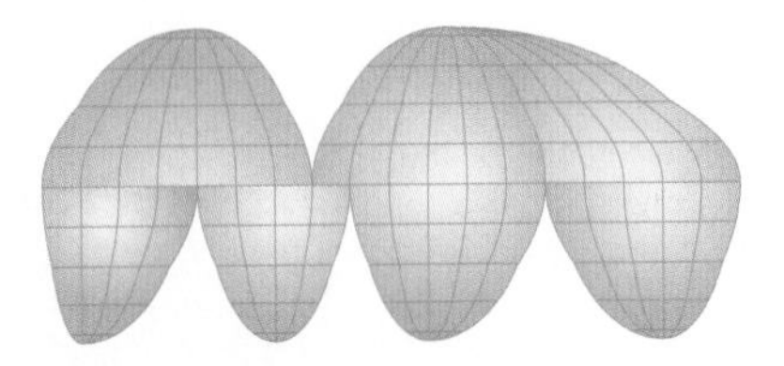

▲ **FIGURE A-11** The distortion pattern of the interrupted Goode's homolosine projection, which mimics that of cylindrical projections.

so severely that it should not be used for any purpose other than navigation or communications.

PROJECTIONS USED IN TEXTBOOKS. Although a map projection cannot be free of distortion, it can represent one or several spatial properties of the earth's surface accurately if other properties are sacrificed. The two projections used for world maps throughout this textbook illustrate that point well. *Goode's homolosine projection*, shown in Figure A-10, belongs to the oval category and shows area accurately, although it gives the impression that the earth's surface has been torn, peeled, and flattened. The interruptions in Figure A-10 have been placed in the major oceans, giving continuity to the land masses. Ocean areas could be featured instead by placing the interruptions in the continents. Obviously, that type of interrupted projection severely distorts proximity relationships. Consequently, in different locations the properties of distance, direction, and shape are also distorted to varying degrees. The distortion pattern mimics that of cylindrical projections, with the equatorial zone the most faithfully represented (Figure A-11).

An alternative to special-property projections such as the equal-area Goode's homolosine is the compromise projection. In that case no special property is achieved at the expense of others, and distortion is rather evenly distributed among the various properties, instead of being focused on one or several properties. The *Robinson projection*, which is also used in this textbook, falls into that category (Figure A-12). Its oval projection has a global feel, somewhat like that of Goode's homolosine. But the Robinson projection shows the North Pole and the South Pole as lines that are slightly more than half the length of the equator, thus exaggerating distances and areas near the poles. Areas look larger than they really are in the high

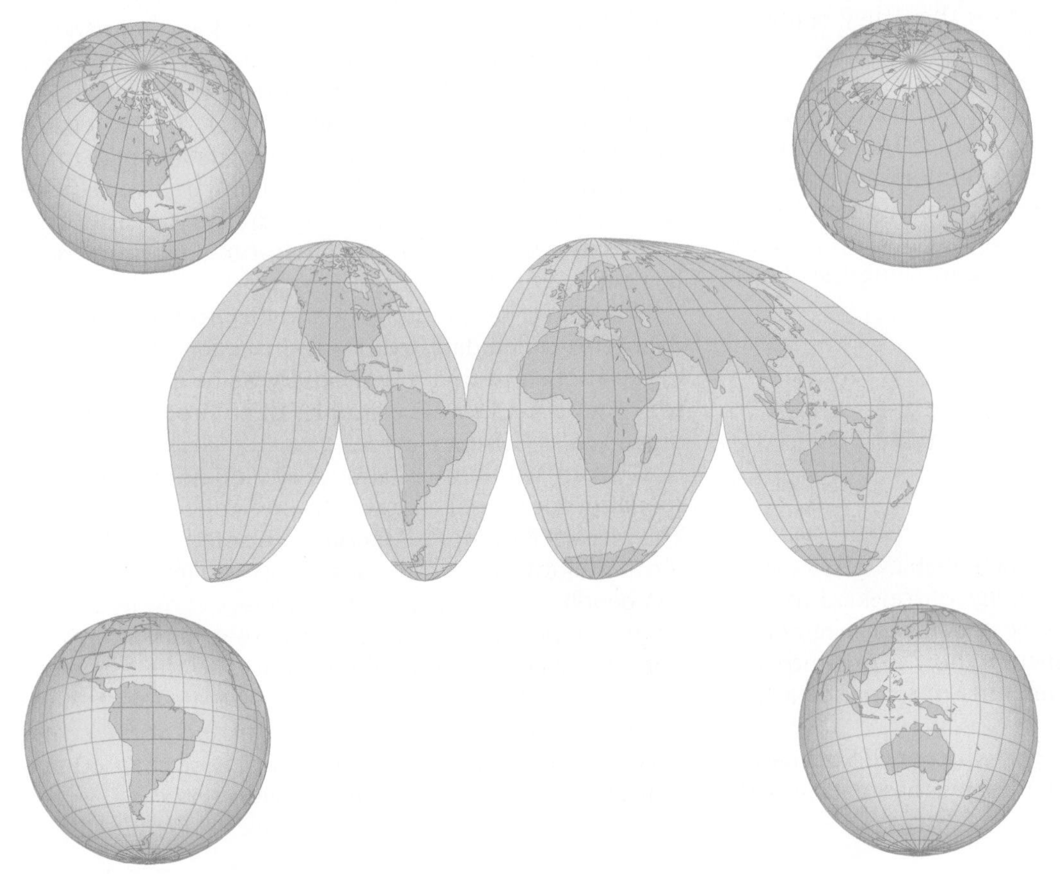

▲ **FIGURE A-10** An interrupted Goode's homolosine, an equal-area projection.

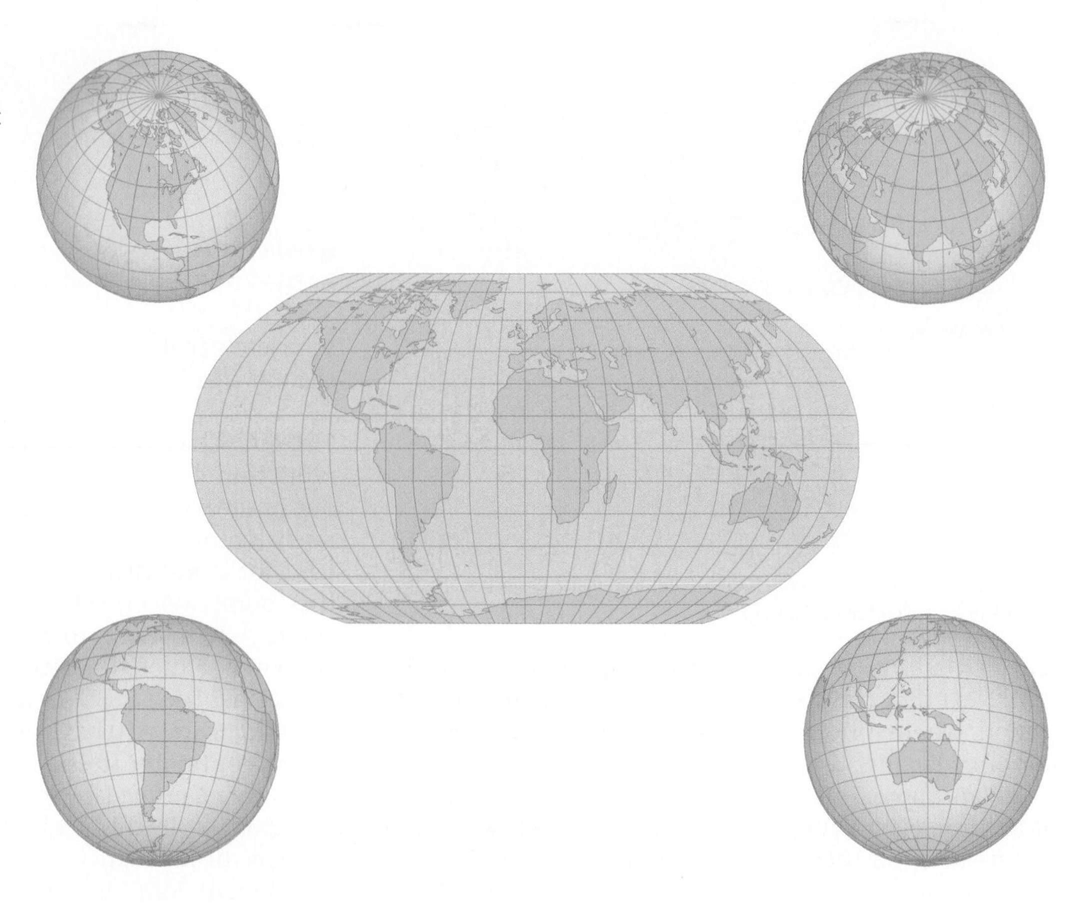

▶ **FIGURE A-12** The compromise Robinson projection, which avoids the interruptions of Goode's homolosine but preserves no special properties. (*Courtesy of ACSM*)

latitudes (near the poles) and smaller than they really are in the low latitudes (near the equator). In addition, not all latitude and longitude lines intersect at right angles, as they do on the earth, so we know that the Robinson projection does not preserve direction or shape either. However, it has fewer interruptions than the Goode's homolosine does, so it preserves proximity better. Overall, the Robinson projection does a good job of representing spatial relationships, especially in the low to middle latitudes and along the central meridian.

GIS and Geospatial Technologies

Today, user-friendly mapping software enables anyone with a computer to produce maps at a range of scales using a variety of projections. But a challenge remains to geographers and other users of spatial information: how to organize and present in map form the vast amounts of spatial data that are now available.

From data gathered by orbiting satellites or GPS devices to statistical data linked to spatial coordinates, these data provide a more detailed view of Earth's physical and human systems than has ever before been possible. To manage these data, geographers have developed a powerful tool—geographic information systems (GIS), which enable users to manipulate and display spatial data in map form. GIS maps contribute to problem solving in diverse fields such as science and engineering, industry, health care, retail sales, urban planning, environmental protection, law enforcement, and many others.

The power of GIS lies in its ability to map different data sets—called data layers—against each other, revealing relationships that might otherwise be difficult to detect. Figure A-13 shows examples of environmental data organized as GIS data layers.

GIS can help answer almost any question involving spatial or locational analysis. In one application of GIS, The city of Baltimore, Maryland, wanted to determine the best location for an emergency shelter for homeless people. Among many other factors considered, one main criterion for the shelter was that it be accessible from other facilities providing services to the homeless. As shown in the map in Figure A-14, the site selected was in a densely populated part of the city and at the center of a 1.5-mile-radius circle containing more than 60 percent of the city's providers of homeless services.

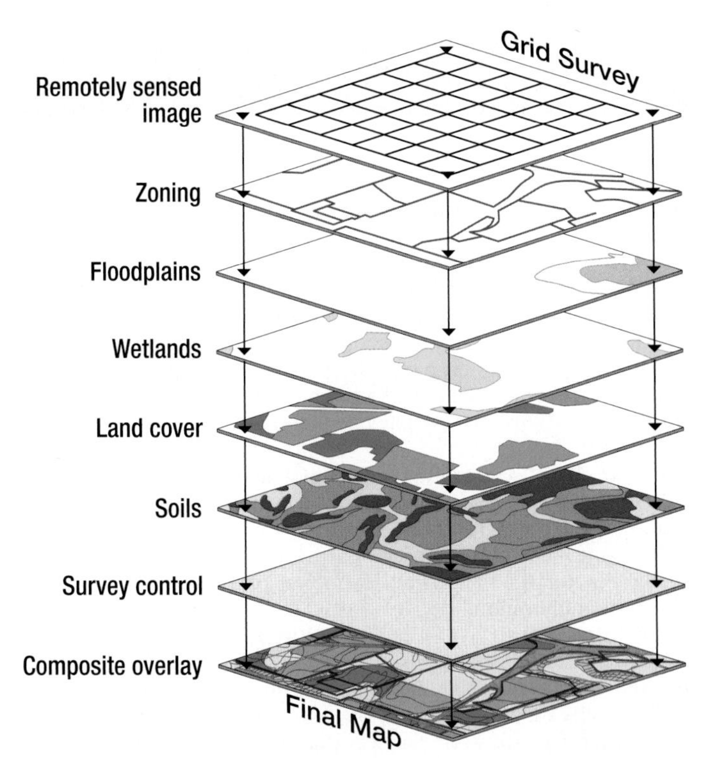

▲ **FIGURE A-13** Within a GIS, environmental data attached to a common terrestrial reference system, such as latitude/longitude, can be stacked in layers for spatial comparison and analysis.

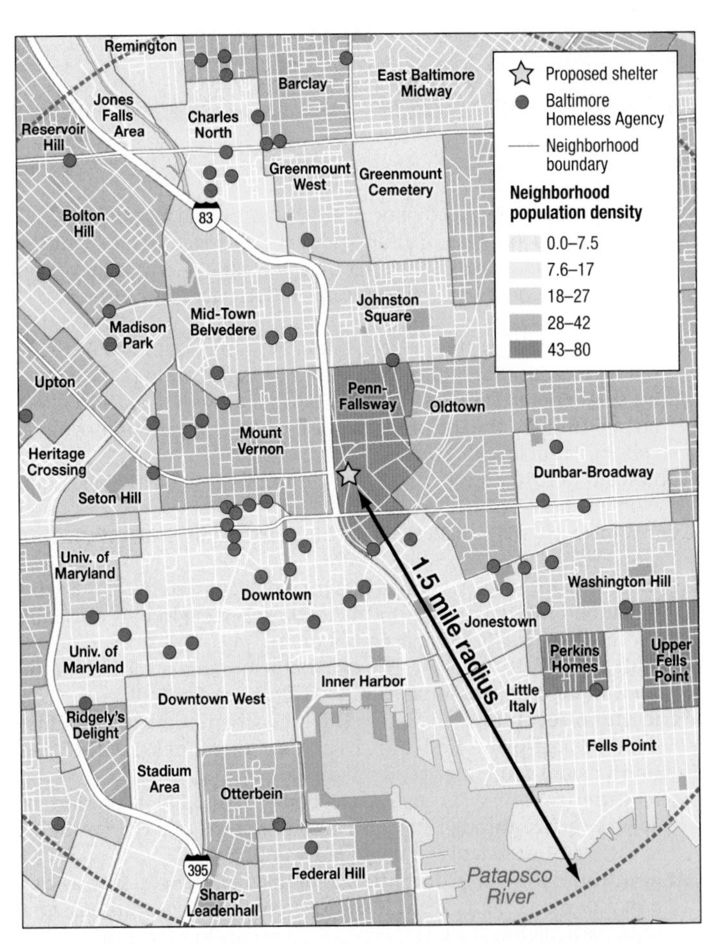

▲ **FIGURE A-14** This GIS map shows that Baltimore's proposed homeless shelter would be near a cluster of service providers. The map's data layers include: neighborhood borders, population density (by neighborhood), and locations of homeless agencies.

KEY TERMS

Abiotic (p. 32) Composed of nonliving or inorganic matter.

Acid deposition (p. 414) Sulfur oxides and nitrogen oxides, emitted by burning fossil fuels, that enter the atmosphere—where they combine with oxygen and water to form sulfuric acid and nitric acid—and return to Earth's surface.

Acid precipitation (p. 414) Conversion of sulfur oxides and nitrogen oxides to acids that return to Earth as rain, snow, or fog.

Active solar energy systems (p. 326) Solar energy systems that collect energy through the use of mechanical devices such as photovoltaic cells or flat-plate collectors.

Adolescent fertility rate (p. 312) The number of births per 1,000 women ages 15 to 19.

Agnosticism (p. 184) Belief that nothing can be known about whether God exists.

Agribusiness (p. 366) Commercial agriculture characterized by the integration of different steps in the food-processing industry, usually through ownership by large corporations.

Agricultural density (p. 49) The ratio of the number of farmers to the total amount of land suitable for agriculture.

Agricultural revolution (p. 348) The time when human beings first domesticated plants and animals and no longer relied entirely on hunting and gathering.

Agriculture (p. 347) The deliberate effort to modify a portion of Earth's surface through the cultivation of crops and the raising of livestock for sustenance or economic gain.

Air pollution (p. 412) Concentration of trace substances, such as carbon monoxide, sulfur dioxide, nitrogen oxides, hydrocarbons, and solid particulates, at a greater level than occurs in average air.

Animism (p. 191) Belief that objects, such as plants and stones, or natural events, like thunderstorms and earthquakes, have a discrete spirit and conscious life.

Annexation (p. 480) Legally adding land area to a city in the United States.

Anocracy (p. 282) A country that is not fully democratic or fully autocratic, but rather displays a mix of the two types.

Apartheid (p. 236) Laws (no longer in effect) in South Africa that physically separated different races into different geographic areas.

Apparel (p. 410) An article of clothing.

Aquaculture (or aquafarming) (p. 382) The cultivation of seafood under controlled conditions.

Arithmetic density (p. 48) The total number of people divided by the total land area.

Asylum seeker (p. 92) Someone who has migrated to another country in the hope of being recognized as a refugee.

Atheism (p. 184) Belief that God does not exist.

Atmosphere (p. 32) The thin layer of gases surrounding Earth.

Autocracy (p. 282) A country that is run according to the interests of the ruler rather than the people.

Autonomous religion (p. 211) A religion that does not have a central authority but shares ideas and cooperates informally.

Balance of power (p. 286) A condition of roughly equal strength between opposing countries or alliances of countries.

Balkanization (p. 251) A process by which a state breaks down through conflicts among its ethnicities.

Balkanized (p. 251) Descriptive of a small geographic area that could not successfully be organized into one or more stable states because it was inhabited by many ethnicities with complex, long-standing antagonisms toward each other.

Basic industries (p. 446) Industries that sell their products or services primarily to consumers outside the settlement.

Biochemical oxygen demand (BOD) (p. 417) The amount of oxygen required by aquatic bacteria to decompose a given load of organic waste; a measure of water pollution.

Biomass fuel (p. 324) Fuel that derives from plant material and animal waste.

Biosphere (p. 32) All living organisms on Earth, including plants and animals, as well as microorganisms.

Biotic (p. 32) Composed of living organisms.

Blockbusting (p. 235) A process by which real estate agents convince white property owners to sell their houses at low prices because of fear that persons of color will soon move into the neighborhood.

Boundary (p. 276) An invisible line that marks the extent of a state's territory.

Branch (p. 186) A large and fundamental division within a religion.

Brain drain (p. 96) Large-scale emigration by talented people.

Break-of-bulk point (p. 402) A location where transfer is possible from one mode of transportation to another.

Breeder reactor (p. 323) A nuclear power plant that creates its own fuel from plutonium.

Bulk-gaining industry (p. 400) An industry in which the final product weighs more or comprises a greater volume than the inputs.

Bulk-reducing industry (p. 398) An industry in which the final product weighs less or comprises a lower volume than the inputs.

Business services (p. 432) Services that primarily meet the needs of other businesses, including professional, financial, and transportation services.

Cartography (p. 5) The science of making maps.

Caste (p. 213) The class or distinct hereditary order into which a Hindu is assigned, according to religious law.

Census (p. 45) A complete enumeration of a population.

Census tract (p. 468) An area delineated by the U.S. Bureau of the Census for which statistics are published; in urban areas, census tracts correspond roughly to neighborhoods.

Central business district (CBD) (p. 461) The area of a city where retail and office activities are clustered.

Central place (p. 434) A market center for the exchange of services by people attracted from the surrounding area.

Central place theory (p. 434) A theory that explains the distribution of services based on the fact that settlements serve as centers of market areas for services; larger settlements are fewer and farther apart than smaller settlements and provide services for a larger number of people who are willing to travel farther.

Centripetal force (p. 239) An attitude that tends to unify people and enhance support for a state.

Cereal grain (or cereal) (p. 352) A grass that yields grain for food.

Chaff (p. 363) Husks of grain separated from the seed by threshing.

Chain migration (p. 97) Migration of people to a specific location because relatives or members of the same nationality previously migrated there.

Chlorofluorocarbon (CFC) (p. 413) A gas used as a solvent, a propellant in aerosols, a refrigerant, and in plastic foams and fire extinguishers.

Circulation (p. 78) Short-term, repetitive, or cyclical movements that recur on a regular basis.

City (p. 476) An urban settlement that has been legally incorporated into an independent, self-governing unit.

City-state (p. 266) A sovereign state comprising a city and its immediately surrounding countryside.

Climate (p. 32) The long-term average weather condition at a particular location.

Clustered rural settlement (p. 448) A rural settlement in which the houses and farm buildings of each family are situated close to each other, with fields surrounding the settlement.

Colonialism (p. 274) An attempt by one country to establish settlements and to impose its political, economic, and cultural principles in another territory.

Colony (p. 274) A territory that is legally tied to a sovereign state rather than completely independent.

Combine (p. 370) A machine that reaps, threshes, and cleans grain while moving over a field.

Combined statistical area (CSA) (p. 477) In the United States, two or more contiguous core based statistical areas tied together by commuting patterns.

Commercial agriculture (p. 350) Agriculture undertaken primarily to generate products for sale off the farm.

Compact state (p. 280) A state in which the distance from the center to any boundary does not vary significantly.

Concentration (p. 22) The spread of something over a given area.

Concentric zone model (p. 466) A model of the internal structure of cities in which social groups are spatially arranged in a series of rings.

Connection (p. 26) Relationships among people and objects across the barrier of space.

Conservation (p. 30) The sustainable management of a natural resource.

Consumer services (p. 431) Businesses that provide services primarily to individual consumers, including retail services and education, health, and leisure services.

Contagious diffusion (p. 26) The rapid, widespread diffusion of a feature or trend throughout a population.

Core based statistical area (CBSA) (p. 477) In the United States, the combination of all metropolitan statistical areas and micropolitan statistical areas.

Cosmogony (p. 204) A set of religious beliefs concerning the origin of the universe.

Cottage industry (p. 395) Manufacturing based in homes rather than in factories, commonly found prior to the Industrial Revolution.

Council of government (p. 479) A cooperative agency consisting of representatives of local governments in a metropolitan area in the United States.

Counterurbanization (p. 91) Net migration from urban to rural areas in more developed countries.

Creole, or creolized language (p. 163) A language that results from the mixing of a colonizer's language with the indigenous language of the people being dominated.

Crop (p. 347) Any plant gathered from a field as a harvest during a particular season.

Crop rotation (p. 364) The practice of rotating use of different fields from crop to crop each year to avoid exhausting the soil.

Crude birth rate (CBR) (p. 50) The total number of live births in a year for every 1,000 people alive in the society.

Crude death rate (CDR) (p. 50) The total number of deaths in a year for every 1,000 people alive in the society.

Cultural ecology (p. 34) A geographic approach that emphasizes human–environment relationships.

Cultural landscape (p. 16) The fashioning of a natural landscape by a cultural group.

Culture (p. 18) The body of customary beliefs, social forms, and material traits that together constitute a group's distinct tradition.

Custom (p. 109) The frequent repetition of an act, to the extent that it becomes characteristic of the group of people performing the act.

Demand (p. 314) The quantity of something that consumers are willing and able to buy.

Democracy (p. 282) A country in which citizens elect leaders and can run for office.

Demographic transition (p. 56) The process of change in a society's population from a condition of high crude birth and death rates and low rate of natural increase to a condition of low crude birth and death rates, low rate of natural increase, and higher total population.

Demography (p. 44) The scientific study of population characteristics.

Denglish (p. 175) A combination of German and English.

Denomination (p. 186) A division of a branch that unites a number of local congregations into a single legal and administrative body.

Density (p. 22) The frequency with which something exists within a given unit of area.

Density gradient (p. 480) The change in density in an urban area from the center to the periphery.

Dependency ratio (p. 54) The number of people under age 15 and over age 64 compared to the number of people active in the labor force.

Desertification (p. 381) Degradation of land, especially in semiarid areas, primarily because of human actions such as excessive crop planting, animal grazing, and tree cutting. Also known as semiarid land degradation.

Developed country (more developed country [MDC] or relatively developed country) (p. 300) A country that has progressed relatively far along a continuum of development.

Developing country (less developed country [LDC]) (p. 300) A country that is at a relatively early stage in the process of economic development.

Development (p. 300) A process of improvement in the material conditions of people through diffusion of knowledge and technology.

Dialect (p. 158) A regional variety of a language distinguished by vocabulary, spelling, and pronunciation.

Dietary energy consumption (p. 352) The amount of food that an individual consumes, measured in kilocalories (Calories in the United States).

Diffusion (p. 26) The spread of a feature or trend from one place to another over time.

Dispersed rural settlement (p. 448) A rural settlement pattern characterized by isolated farms rather than clustered villages.

Distance decay (p. 28) The diminishing in importance and eventual disappearance of a phenomenon with increasing distance from its origin.

Distribution (p. 22) The arrangement of something across Earth's surface.

Double cropping (p. 363) Harvesting twice a year from the same field.

Doubling time (p. 50) The number of years needed to double a population, assuming a constant rate of natural increase.

Ebonics (p. 174) A dialect spoken by some African Americans.

Ecology (p. 34) The scientific study of ecosystems.

Economic base (p. 446) A community's collection of basic industries.

Ecosystem (p. 34) A group of living organisms and the abiotic spheres with which they interact.

Ecumene (p. 47) The portion of Earth's surface occupied by permanent human settlement.

Edge city (p. 480) A large node of office and retail activities on the edge of an urban area.

Elderly support ratio (p. 62) The number of working-age people (ages 15–64) divided by the number of persons 65 and older.

Elongated state (p. 280) A state with a long, narrow shape.

Emigration (p. 78) Migration from a location.

Enclosure movement (p. 450) The process of consolidating small landholdings into a smaller number of larger farms in England during the eighteenth century.

Environmental determinism (p. 34) A nineteenth- and early twentieth-century approach to the study of geography which argued that the general laws sought by human geographers could be found in the physical sciences. Geography was therefore the study of how the physical environment caused human activities.

Epidemiologic transition (p. 64) Distinctive causes of death in each stage of the demographic transition.

Epidemiology (p. 64) The branch of medical science concerned with the incidence, distribution, and control of diseases that are prevalent among a population at a special time and are produced by some special causes not generally present in the affected locality.

Ethnic cleansing (p. 246) A process in which a more powerful ethnic group forcibly removes a less powerful one in order to create an ethnically homogeneous region.

Ethnic religion (p. 184) A religion with a relatively concentrated spatial distribution whose principles are likely to be based on the physical characteristics of the particular location in which its adherents are concentrated.

Ethnicity (p. 227) Identity with a group of people that share distinct physical and mental traits as a product of common heredity and cultural traditions.

Expansion diffusion (p. 26) The spread of a feature or trend among people from one area to another in an additive process.

Extinct language (p. 166) A language that was once used by people in daily activities but is no longer used.

Fair trade (p. 326) An alternative to international trade that emphasizes small businesses and worker-owned and democratically run co-operatives and requires employers to pay workers fair wages, permit union organization, and comply with minimum environmental and safety standards.

Federal state (p. 283) An internal organization of a state that allocates most powers to units of local government.

Female labor force participation rate (p. 311) The percentage of women holding full-time jobs outside the home.

Ferrous (p. 398) Metals, including iron, that are utilized in the production of iron and steel.

Filtering (p. 490) A process of change in the use of a house, from single-family owner occupancy to abandonment.

Fission (p. 322) The splitting of an atomic nucleus to release energy.

Floodplain (p. 92) The area subject to flooding during a given number of years, according to historical trends.

Folk culture (p. 108) Culture traditionally practiced by a small, homogeneous, rural group living in relative isolation from other groups.

Food desert (p. 464) An area in a developed country where healthy food is difficult to obtain.

Food security (p. 354) Physical, social, and economic access at all times to safe and nutritious food sufficient to meet dietary needs and food preferences for an active and healthy life.

Forced migration (p. 80) Permanent movement, usually compelled by cultural factors.

Fordist production (p. 422) A form of mass production in which each worker is assigned one specific task to perform repeatedly.

Foreign direct investment (FDI) (p. 332) Investment made by a foreign company in the economy of another country.

Formal region (or uniform or homogeneous region) (p. 16) An area in which everyone shares in common one or more distinctive characteristics.

Fossil fuel (p. 314) An energy source formed from the residue of plants and animals buried millions of years ago.

Fracking (hydraulic fracturing) (p. 319) The pumping of water at high pressure to break apart rocks in order to release natural gas.

Fragmented state (p. 281) A state that includes several discontinuous pieces of territory.

Franglais (p. 174) A term used by the French for English words that have entered the French language; a combination of *français* and *anglais*, the French words for *French* and *English*, respectively.

Frontier (p. 276) A zone separating two states in which neither state exercises political control.

Functional region (or nodal region) (p. 17) An area organized around a node or focal point.

Fundamentalism (p. 212) Literal interpretation and strict adherence to basic principles of a religion (or a religious branch, denomination, or sect).

Fusion (p. 325) Creation of energy by joining the nuclei of two hydrogen atoms to form helium.

Gender Inequality Index (GII) (p. 310) A measure of the extent of each country's gender inequality.

Genocide (p. 252) The mass killing of a group of people in an attempt to eliminate the entire group from existence.

Gentrification (p. 491) A process of converting an urban neighborhood from a predominantly low-income, renter-occupied area to a predominantly middle-class, owner-occupied area.

Geographic information science (GIScience) (p. 12) The development and analysis of data about Earth acquired through satellite and other electronic information technologies.

Geographic information system (GIS) (p. 12) A computer system that stores, organizes, analyzes, and displays geographic data.

Geothermal energy (p. 325) Energy from steam or hot water produced from hot or molten underground rocks.

Gerrymandering (p. 284) The process of redrawing legislative boundaries for the purpose of benefiting the party in power.

Ghetto (p. 199) During the Middle Ages, a neighborhood in a city set up by law to be inhabited only by Jews; now used to denote a section of a city in which members of any minority group live because of social, legal, or economic pressure.

Global Positioning System (GPS) (p. 12) A system that determines the precise position of something on Earth through a series of satellites, tracking stations, and receivers.

Globalization (p. 20) Actions or processes that involve the entire world and result in making something worldwide in scope.

Grain (p. 352) Seed of a cereal grass.

Gravity model (p. 438) A model which holds that the potential use of a service at a particular location is directly related to the number of people in a location and inversely related to the distance people must travel to reach the service.

Green revolution (p. 384) Rapid diffusion of new agricultural technology, especially new high-yield seeds and fertilizers.

Greenbelt (p. 481) A ring of land maintained as parks, agriculture, or other types of open space to limit the sprawl of an urban area.

Greenhouse effect (p. 412) The anticipated increase in Earth's temperature caused by carbon dioxide (emitted by burning fossil fuels) trapping some of the radiation emitted by the surface.

Greenwich Mean Time (GMT) (p. 11) The time in the zone encompassing the prime meridian, or 0° longitude.

Gross domestic product (GDP) (p. 302) The value of the total output of goods and services produced in a country in a given time period (normally one year).

Gross national income (GNI) (p. 302) The value of the output of goods and services produced in a country in a year, including money that leaves and enters the country.

Guest worker (p. 95) A term once used for a worker who migrated to the developed countries of Northern and Western Europe, usually from Southern and Eastern Europe or from North Africa, in search of a higher-paying job.

Habit (p. 109) A repetitive act performed by a particular individual.

Hearth (p. 26) The region from which innovative ideas originate.

Hierarchical diffusion (p. 26) The spread of a feature or trend from one key person or node of authority or power to other persons or places.

Hierarchical religion (p. 210) A religion in which a central authority exercises a high degree of control.

Horticulture (p. 371) The growing of fruits, vegetables, and flowers.

Housing bubble (p. 20) A rapid increase in the value of houses followed by a sharp decline in their value.

Hull (p. 363) The outer covering of a seed.

Human Development Index (HDI) (p. 301) An indicator of the level of development for each country, constructed by the United Nations, that is based on income, literacy, education, and life expectancy.

Hydroelectric power (p. 324) Power generated from moving water.

Hydrosphere (p. 32) All of the water on and near Earth's surface.

Immigration (p. 78) Migration to a new location.

Industrial Revolution (p. 56, 393) A series of improvements in industrial technology that transformed the process of manufacturing goods.

Inequality-adjusted HDI (IHDI) (p. 303) Modification of the HDI to account for inequality within a country.

Infant mortality rate (IMR) (p. 70) The total number of deaths in a year among infants under one year of age for every 1,000 live births in a society.

Intensive subsistence agriculture (p. 362) A form of subsistence agriculture in which farmers must expend a relatively large amount of effort to produce the maximum feasible yield from a parcel of land.

Internal migration (p. 80) Permanent movement within a particular country.

Internally displaced person (IDP) (p. 92) Someone who has been forced to migrate for similar political reasons as a refugee but has not migrated across an international border.

International Date Line (p. 11) An arc that for the most part follows 180° longitude, although it deviates in several places to avoid dividing land areas. When you cross the International Date Line heading east (toward America), the clock moves back 24 hours, or one entire day. When you go west (toward Asia), the calendar moves ahead one day.

International migration (p. 80) Permanent movement from one country to another.

Interregional migration (p. 80) Permanent movement from one region of a country to another.

Intervening obstacle (p. 94) An environmental or cultural feature of the landscape that hinders migration.

Intraregional migration (p. 78) Permanent movement within one region of a country.

Isogloss (p. 158) A boundary that separates regions in which different language usages predominate.

Isolated language (p. 166) A language that is unrelated to any other languages and therefore not attached to any language family.

Just-in time delivery (p. 423) Shipment of parts and materials to arrive at a factory moments before they are needed.

Labor-intensive industry (p. 408) An industry for which labor costs comprise a high percentage of total expenses.

Landlocked state (p. 281) A state that does not have a direct outlet to the sea.

Language (p. 143) A system of communication through the use of speech; a collection of sounds understood by a group of people to have the same meaning.

Language branch (p. 143) A collection of languages related through a common ancestor that existed several thousand years ago. Differences are not as extensive or as old as with language families, and archaeological evidence can confirm that the branches derived from the same family.

Language family (p. 143) A collection of languages related to each other through a common ancestor long before recorded history.

Language group (p. 143) A collection of languages within a branch that share a common origin in the relatively recent past and display relatively few differences in grammar and vocabulary.

Latitude (p. 10) The numbering system used to indicate the location of parallels drawn on a globe and measuring distance north and south of the equator (0°).

Life expectancy (p. 70) The average number of years an individual can be expected to live, given current social, economic, and medical conditions. Life expectancy at birth is the average number of years a newborn infant can expect to live.

Lingua franca (p. 172) A language mutually understood and commonly used in trade by people who have different native languages.

Literacy rate (p. 307) The percentage of a country's people who can read and write.

Literary tradition (p. 143) A language that is written as well as spoken.

Lithosphere (p. 32) Earth's crust and a portion of upper mantle directly below the crust.

Location (p. 14) The position of anything on Earth's surface.

Logogram (p. 146) A symbol that represents a word rather than a sound.

Longitude (p. 10) The numbering system used to indicate the location of meridians drawn on a globe and measuring distance east and west of the prime meridian (0°).

Map (p. 5) A two-dimensional, or flat, representation of Earth's surface or a portion of it.

Map scale (p. 8) The relationship between the size of an object on a map and the size of the actual feature on Earth's surface.

Maquiladora (p. 421) A factory built by a U.S. company in Mexico near the U.S. border to take advantage of the much lower labor costs in Mexico.

Market area (or hinterland) (p. 434) The area surrounding a central place from which people are attracted to use the place's goods and services.

Maternal mortality ratio (p. 312) The number of women who die giving birth per 100,000 births.

Medical revolution (p. 56) Medical technology invented in Europe and North America that has diffused to the poorer countries in Latin America, Asia, and Africa. Improved medical practices have eliminated many of the traditional causes of death in poorer countries and enabled more people to live longer and healthier lives.

Megalopolis (p. 478) A continuous urban complex in the northeastern United States.

Mental map (p. 17) A representation of a portion of Earth's surface based on what an individual knows about a place, containing personal impressions of what is in the place and where the place is located.

Meridian (p. 10) An arc drawn on a map between the North and South poles.

Metropolitan statistical area (MSA) (p. 477) In the United States, an urbanized area of at least 50,000 population, the county within which the city is located, and adjacent counties meeting one of several tests indicating a functional connection to the central city.

Microfinance (p. 337) Provision of small loans and other financial services to individuals and small businesses in developing countries.

Micropolitan statistical area (μSA) (p. 477) An urbanized area of between 10,000 and 50,000 inhabitants, the county in which it is found, and adjacent counties tied to the city.

Microstate (p. 261) A state that encompasses a very small land area.

Migration (p. 78) A form of relocation diffusion involving a permanent move to a new location.

Migration transition (p. 79) A change in the migration pattern in a society that results from industrialization, population growth, and other social and economic changes that also produce the demographic transition.

Milkshed (p. 368) The area surrounding a city from which milk is supplied.

Millennium Development Goals (p. 339) Eight international development goals that all members of the United Nations have agreed to achieve by 2015.

Missionary (p. 196) An individual who helps to diffuse a universalizing religion.

Mobility (p. 78) All types of movement between location.

Monotheism (p. 191) The doctrine of or belief in the existence of only one god.

Multiethnic state (p. 268) A state that contains more than one ethnicity.

Multinational state (p. 268) A state that contains two or more ethnic groups with traditions of self-determination that agree to coexist peacefully by recognizing each other as distinct nationalities.

Multiple nuclei model (p. 467) A model of the internal structure of cities in which social groups are arranged around a collection of nodes of activities.

Nation-state (p. 267) A state whose territory corresponds to that occupied by a particular ethnicity that has been transformed into a nationality.

Nationalism (p. 239) Loyalty and devotion to a particular nationality.

Nationality (p. 238) Identity with a group of people that share legal attachment and personal allegiance to a particular place as a result of being born there.

Natural increase rate (NIR) (p. 50) The percentage growth of a population in a year, computed as the crude birth rate minus the crude death rate.

Net migration (p. 78) The difference between the level of immigration and the level of emigration.

Network (p. 28) A chain of communication that connects places.

New international division of labor (p. 420) Transfer of some types of jobs, especially those requiring low-paid, less-skilled workers, from more developed to less developed countries.

Nonbasic industries (p. 446) Industries that sell their products primarily to consumers in the community.

Nonferrous (p. 398) Metals utilized to make products other than iron and steel.

Nonpoint-source pollution (p. 416) Pollution that originates from a large, diffuse area of a body of water.

Nonrenewable energy (p. 317) A source of energy that has a finite supply capable of being exhausted.

Nonrenewable resource (p. 30) Something produced in nature more slowly than it is consumed by humans.

Official language (p. 143) The language adopted for use by the government for the conduct of business and publication of documents.

Outsourcing (p. 420) A decision by a corporation to turn over much of the responsibility for production to independent suppliers.

Overpopulation (p. 44) A situation in which the number of people in an area exceeds the capacity of the environment to support life at a decent standard of living.

Ozone (p. 413) A gas that absorbs ultraviolet solar radiation, found in the stratosphere, a zone 15 to 50 kilometers (9 to 30 miles) above Earth's surface.

Paddy (p. 363) The Malay word for wet rice, commonly but incorrectly used to describe a sawah.

Pagan (p. 190) A follower of a polytheistic religion.

Pandemic (p. 64) Disease that occurs over a wide geographic area and affects a very high proportion of the population.

Parallel (p. 10) A circle drawn around the globe parallel to the equator and at right angles to the meridians.

Passive solar energy systems (p. 326) Solar energy systems that collect energy without the use of mechanical devices.

Pastoral nomadism (p. 358) A form of subsistence agriculture based on herding domesticated animals.

Pasture (p. 359) Grass or other plants grown for feeding grazing animals, as well as land used for grazing.

Pattern (p. 23) The geometric or regular arrangement of something in a study area.

Perforated state (p. 281) A state that completely surrounds another one.

Peripheral model (p. 476) A model of North American urban areas consisting of an inner city surrounded by large suburban residential and business areas tied together by a beltway or ring road.

Photochemical smog (p. 414) An atmospheric condition formed through a combination of weather conditions and pollution, especially from motor vehicle emissions.

Photovoltaic cell (p. 326) A solar energy cell, usually made from silicon, that collects solar rays to generate electricity.

Physiological density (p. 48) The number of people per unit of area of arable land, which is land suitable for agriculture.

Pidgin language (p. 173) A form of speech that adopts a simplified grammar and limited vocabulary of a lingua franca; used for communications among speakers of two different languages.

Pilgrimage (p. 202) A journey to a place considered sacred for religious purposes.

Place (p. 14) A specific point on Earth distinguished by a particular characteristic.

Plantation (p. 364) A large farm in tropical and subtropical climates that specializes in the production of one or two crops for sale.

Point-source pollution (p. 416) Pollution that enters a body of water from a specific source.

Polder (p. 36) Land created by the Dutch by draining water from an area.

Polytheism (p. 191) Belief in or worship of more than one god.

Popular culture (p. 108) Culture found in a large, heterogeneous society that shares certain habits despite differences in other personal characteristics.

Population pyramid (p. 54) A bar graph that represents the distribution of population by age and sex.

Possibilism (p. 35) The theory that the physical environment may set limits on human actions, but people have the ability to adjust to the physical environment and choose a course of action from many alternatives.

Post-Fordist production (p. 422) Adoption by companies of flexible work rules, such as the allocation of workers to teams that perform a variety of tasks.

Potential reserve (p. 319) The amount of a resource in deposits not yet identified but thought to exist.

Preservation (p. 30) The maintenance of resources in their present condition, with as little human impact as possible.

Primary census statistical area (PCSA) (p. 477) In the United States, all of the combined statistical areas plus all of the remaining metropolitan statistical areas and micropolitan statistical areas.

Primary sector (p. 302) The portion of the economy concerned with the direct extraction of materials from Earth's surface, generally through agriculture, although sometimes by mining, fishing, and forestry.

Primate city (p. 437) The largest settlement in a country, if it has more than twice as many people as the second-ranking settlement.

Primate city rule (p. 437) A pattern of settlements in a country such that the largest settlement has more than twice as many people as the second-ranking settlement.

Prime agricultural land (p. 381) The most productive farmland.

Prime meridian (p. 10) The meridian, designated as 0° longitude, that passes through the Royal Observatory at Greenwich, England.

Productivity (p. 303) The value of a particular product compared to the amount of labor needed to make it.

Projection (p. 9) A system used to transfer locations from Earth's surface to a flat map.

Prorupted state (p. 280) An otherwise compact state with a large projecting extension.

Proven reserve (p. 318) The amount of a resource remaining in discovered deposits.

Public housing (p. 490) Housing owned by the government; in the United States, it is rented to residents with low incomes, and the rents are set at 30 percent of the families' incomes.

Public services (p. 432) Services offered by the government to provide security and protection for citizens and businesses.

Pull factor (p. 92) A factor that induces people to move to a new location.

Purchasing power parity (PPP) (p. 302) The amount of money needed in one country to purchase the same goods and services in another country; PPP adjusts income figures to account for differences among countries in the cost of goods.

Push factor (p. 92) A factor that induces people to leave old residences.

Quotas (p. 96) In reference to migration, laws that place maximum limits on the number of people who can immigrate to a country each year.

Race (p. 227) Identity with a group of people descended from a biological ancestor.

Racism (p. 227) Belief that race is the primary determinant of human traits and capacities and that racial differences produce an inherent superiority of a particular race.

Racist (p. 227) A person who subscribes to the beliefs of racism.

Radioactive waste (p. 322) Materials from a nuclear reaction that emit radiation; contact with such particles may be harmful or lethal to people; therefore, the materials must be safely stored for thousands of years.

Ranching (p. 372) A form of commercial agriculture in which livestock graze over an extensive area.

Range (of a service) (p. 435) The maximum distance people are willing to travel to use a service.

Rank-size rule (p. 437) A pattern of settlements in a country such that the *n*th largest settlement is $1/n$ the population of the largest settlement.

Reaper (p. 370) A machine that cuts cereal grain standing in a field.

Received Pronunciation (RP) (p. 160) The dialect of English associated with upper-class Britons living in London and now considered standard in the United Kingdom.

Redlining (p. 490) A process by which banks draw lines on a map and refuse to lend money to purchase or improve property within the boundaries.

Refugees (p. 92) People who are forced to migrate from their home country and cannot return for fear of persecution because of their race, religion, nationality, membership in a social group, or political opinion.

Region (p. 16) An area distinguished by a unique combination of trends or features.

Regional (or cultural landscape) studies (p. 16) An approach to geography that emphasizes the relationships among social and physical phenomena in a particular study area.

Relocation diffusion (p. 26) The spread of a feature or trend through bodily movement of people from one place to another.

Remote sensing (p. 12) The acquisition of data about Earth's surface from a satellite orbiting the planet or from other long-distance methods.

Renewable energy (p. 317) A resource that has a theoretically unlimited supply and is not depleted when used by humans.

Renewable resource (p. 30) Something produced in nature more rapidly than it is consumed by humans.

Resource (p. 30) A substance in the environment that is useful to people, is economically and technologically feasible to access, and is socially acceptable to use.

Ridge tillage (p. 386) A system of planting crops on ridge tops in order to reduce farm production costs and promote greater soil conservation.

Right-to-work law (p. 418) A U.S. law that prevents a union and a company from negotiating a contract that requires workers to join the union as a condition of employment.

Rush hour (p. 486) The four consecutive 15-minute periods in the morning and evening with the heaviest volumes of traffic.

Sanitary landfill (p. 414) A place to deposit solid waste, where a layer of earth is bulldozed over garbage each day to reduce emissions of gases and odors from the decaying trash, to minimize fires, and to discourage vermin.

Sawah (p. 363) A flooded field for growing rice.

Scale (p. 20) Generally, the relationship between the portion of Earth being studied and Earth as a whole.

Secondary sector (p. 302) The portion of the economy concerned with manufacturing useful products through processing, transforming, and assembling raw materials.

Sect (p. 186) A relatively small group that has broken away from an established denomination.

Sector model (p. 467) A model of the internal structure of cities in which social groups are arranged around a series of sectors, or wedges, radiating out from the central business district.

Self-determination (p. 267) The concept that ethnicities have the right to govern themselves.

Service (p. 430) Any activity that fulfills a human want or need and returns money to those who provide it.

Settlement (p. 430) A permanent collection of buildings and inhabitants.

Sex ratio (p. 54) The number of males per 100 females in the population.

Sharecropper (p. 234) A person who works fields rented from a landowner and pays the rent and repays loans by turning over to the landowner a share of the crops.

Shifting cultivation (p. 360) A form of subsistence agriculture in which people shift activity from one field to another; each field is used for crops for a relatively few years and left fallow for a relatively long period.

Site (p. 14) The physical character of a place.

Site factors (p. 398) Location factors related to the costs of factors of production inside a plant, such as land, labor, and capital.

Situation (p. 15) The location of a place relative to another place.

Situation factors (p. 398) Location factors related to the transportation of materials into and from a factory.

Slash-and-burn agriculture (p. 360) Another name for shifting cultivation, so named because fields are cleared by slashing the vegetation and burning the debris.

Smart growth (p. 481) Legislation and regulations to limit suburban sprawl and preserve farmland.

Social area analysis (p. 468) Statistical analysis used to identify where people of similar living standards, ethnic background, and lifestyle live within an urban area.

Solstice (p. 205) An astronomical event that happens twice each year, when the tilt of Earth's axis is most inclined toward or away from the Sun, causing the Sun's apparent position in the sky to reach it most northernmost or southernmost extreme, and resulting in the shortest and longest days of the year.

Sovereignty (p. 261) Ability of a state to govern its territory free from control of its internal affairs by other states.

Space (p. 22) The physical gap or interval between two objects.

Space–time compression (p. 28) The reduction in the time it takes to diffuse something to a distant place as a result of improved communications and transportation systems.

Spanglish (p. 174) A combination of Spanish and English spoken by Hispanic Americans.

Sprawl (p. 480) Development of new housing sites at relatively low density and at locations that are not contiguous to the existing built-up area.

Spring wheat (p. 370) Wheat planted in the spring and harvested in the late summer.

Squatter settlement (p. 472) An area within a city in a less developed country in which people illegally establish residences on land they do not own or rent and erect homemade structures.

Standard language (p. 160) The form of a language used for official government business, education, and mass communications.

State (p. 261) An area organized into a political unit and ruled by an established government that has control over its internal and foreign affairs.

Stimulus diffusion (p. 27) The spread of an underlying principle even though a specific characteristic is rejected.

Structural adjustment program (p. 334) Economic policies imposed on less developed countries by international agencies to create conditions encouraging international trade, such as raising taxes, reducing government spending, controlling inflation, selling publicly owned utilities to private corporations, and charging citizens more for services.

Subsistence agriculture (p. 350) Agriculture designed primarily to provide food for direct consumption by the farmer and the farmer's family.

Supply (p. 314) The quantity of something that producers have available for sale.

Sustainability (p. 30) The use of Earth's renewable and nonrenewable natural resources in ways that do not constrain resource use in the future.

Sustainable agriculture (p. 386) Farming methods that preserve long-term productivity of land and minimize pollution, typically by rotating soil-restoring crops with cash crops and reducing inputs of fertilizer and pesticides.

Swidden (p. 360) A patch of land cleared for planting through slashing and burning.

Syncretic (p. 190) A religion that combines several traditions.

Taboo (p. 118) A restriction on behavior imposed by social custom.

Terroir (p. 118) The contribution of a location's distinctive physical features to the way food tastes.

Terrorism (p. 290) The systematic use of violence by a group in order to intimidate a population or coerce a government into granting its demands.

Tertiary sector (p. 302) The portion of the economy concerned with transportation, communications, and utilities, sometimes extended to the provision of all goods and services to people, in exchange for payment.

Textile (p. 410) A fabric made by weaving, used in making clothing.

Thresh (p. 363) To beat out grain from stalks.

Threshold (p. 435) The minimum number of people needed to support a service.

Toponym (p. 14) The name given to a portion of Earth's surface.

Total fertility rate (TFR) (p. 52) The average number of children a woman will have throughout her childbearing years.

Transhumance (p. 359) The seasonal migration of livestock between mountains and lowland pastures.

Transnational corporation (p. 20) A company that conducts research, operates factories, and sells products in many countries, not just where its headquarters or shareholders are located.

Triangular slave trade (p. 233) A practice, primarily during the eighteenth century, in which European ships transported slaves from Africa to Caribbean islands, molasses from the Caribbean to Europe, and trade goods from Europe to Africa.

Truck farming (p. 367) Commercial gardening and fruit farming, so named because *truck* was a Middle English word meaning "bartering" or "exchange of commodities."

Unauthorized immigrants (p. 98) People who enter a country without proper documents to do so.

Underclass (p. 492) A group in society prevented from participating in the material benefits of a more developed society because of a variety of social and economic characteristics.

Undernourishment (p. 354) Dietary energy consumption that is continuously below the minimum requirement for maintaining a healthy life and carrying out light physical activity.

Uneven development (p. 29) The increasing gap in economic conditions between core and peripheral regions as a result of the globalization of the economy.

Unitary state (p. 283) An internal organization of a state that places most power in the hands of central government officials.

Universalizing religion (p. 184) A religion that attempts to appeal to all people, not just those living in a particular location.

Urban area (p. 477) A dense core of census tracts, densely settled suburbs, and low-density land that links the dense suburbs with the core.

Urban cluster (p. 477) In the United States, an urban area with between 2,500 and 50,000 inhabitants.

Urbanized area (p. 477) In the United States, an urban area with at least 50,000 inhabitants.

Urbanization (p. 454) An increase in the percentage of the number of people living in urban settlements.

Value added (p. 303) The gross value of a product minus the costs of raw materials and energy.

Vertical integration (p. 420) An approach typical of traditional mass production in which a company controls all phases of a highly complex production process.

Vernacular region (or perceptual region) (p. 17) An area that people believe exists as part of their cultural identity.

Voluntary migration (p. 80) Permanent movement undertaken by choice.

Vulgar Latin (p. 155) A form of Latin used in daily conversation by ancient Romans, as opposed to the standard dialect, which was used for official documents.

Wet rice (p. 362) Rice planted on dry land in a nursery and then moved to a deliberately flooded field to promote growth.

Winnow (p. 363) To remove chaff by allowing it to be blown away by the wind.

Winter wheat (p. 370) Wheat planted in the autumn and harvested in the early summer.

Zero population growth (ZPG) (p. 57) A decline of the total fertility rate to the point where the natural increase rate equals zero.

Zoning ordinance (p. 465) A law that limits the permitted uses of land and maximum density of development in a community.

MAP INDEX

TEXT, PHOTO, AND ILLUSTRATION CREDITS

Chapter 1 Chapter openers:, (top right, Maurice Savage/Alamy; ,Picture Contact BV/Alamy; Figure 1-3, Mariner's Museum; Figure 1-4, North Wind Picture Archives/Alamy; Figure 1-5, World History Archive/ Alamy; Figure 1-6, PRISMA ARCHIVO/Alamy; Figure 1-7 (top center), Data from FEMA; Figure 1-7 (top right), Based on map "Population Decline in New Orleans" by Haeyoun Park, from THE NEW YORK TIMES website, February 3, 2011; Figure 1-7 (bottom), UPPA/Photoshot; Figure 1-8, Map data Copyright © 2012 by Google Maps; Figure 1-9, Based on American Congress on Surveying and Mapping; Figure 1-10, Based on Wikimedia Commons; Figure 1-12 and text excerpt, p. 6 "from Amerigo the Discoverer (Translation of text from a 1507 map), ZUMA/Press/Newscom; Figure 1-14, Robert Spencer/Redux Pictures; Figure 1-15, Based on "Boston: History of the Landfills," from the Boston College website; Figure 1-16, Hung Chung Chih/Shutterstock.com; Figure 1-20, Pietro Scozzari/age fotostock; Figure 1-21, Based on map "Neighborhood Health Profiles: Overview" from Baltimore City Health Department website; Figure 1-23 (left), Kevin Foy/ Alamy; Figure 1-23 (right), Christopher Papsch/Alamy; Figure 1-27 (left), Image Source/age fotostock; Figure 1-27 (right), Colin Underhill/Alamy; Figure 1-29, Based on "Le Paris gay. Eléments pour une géographie de l'homosexualite" by Stephane Leroy from ANNALES DE GEOGRAPHIE, November-December 2005, Volume 114(646); Figure 1-30, Based on maps, "Marriage and Substitutes to Marriage" and "Punishments for Male to Male Relationships" from International Lesbian, Gay, Bisexual, Trans and Intersex; Figure 1-35, "Trends in the Distribution of Household Income Between 1979 and 2007" from Congressional Budget Office, October 2011; Figure 1-36, "Data Underlying Select Figures" from Congressional Budget Office, October 2011; Figure 1-37, "Three Pillars of Sustainability," from Wikimedia Commons; Figure 1-38 (right), Die Bildagentur der Fotografen GmbH/Alamy; Figure 1-38 (left), Henry Kotowski/Alamy; Figure 1-38 (middle), Jeff Morgan 08/Alamy; Figure 1-40, Eye Ubiquitous/Glow Images; Figure 1-41, U.S. Geological Survey, sections 29 and 32; text excerpt p. 36 "God made Earth, but the Dutch made the Netherlands.", From "Impressions of Holland and France" by Booker T. Washington, as appears in *The Southern Workman*; text excerpt on p. 37 "God made the world in six days, and the Army Corps ...", From "Render Back to Nature" from The Economist, December 9, 1989; Figure 1-43 (left), Corbis Flirt/Alamy; Figure 1-43 (right), garfotos/Alamy; Figure 1-44 (right), Frans Lemmens/SuperStock; Google Earth Images end of chapter: Gray Buildings © District of Columbia (DC GIS) & CyberCity; Data SIO, NOAA, U.S. Navy, NGA, GEBCO. Image © 2012 GeoEye. Image © TerraMetrics; map data Copyright © by Google Maps; Image © 2012 Aerodata International Surveys. Data SIO, NOAA, U.S. Navy, NGA, GEBCO. Image © 2012 DigitalGlobe .

Chapter 2 Chapter openers: John Batdorff II/Alamy; Copyright © 2012 Google; Figure 2-1, Steve McCurry/Magnum Photos; Figure 2-23, Jake Lyell/Alamy; Figure 2-29, Based on Population Reference Bureau; Figure 2-31, Based on map, "Cholera, Areas Reporting Outbreaks, 2010-2011" from World Health Organization website; Figure 2-33, Based on map, "Cancer, Death Rates per 100,000 Population, Age Standardized-Males, 2008" from World Health Organization website; Figure 2-34, Based on map, "Prevalence of Obesity," ages 20+, Age Standardized-Both Sexes, 2008" from World Health Organization website; Figure 2-35, Based on map, "Estimated to Tb Incidence Rates, 2010" from World Health Organization website; Figure 2-36 (top), Based on map, "HIV Estimated Prevalence among Population Aged 15-49(%), 1990" from World Health Organization website; Figure 2-36 (bottom), Based on map, "HIV Estimated Prevalence among Population Aged 15-49%, 2007" from World Health Organization website; Figure 2-37, Mark Phillips/Photo Researchers, Inc.; Figure 2-42, Based on map, "Per Capita Total Expenditures on Health at Average Exchange Rate (US$), 2009" from World Health Organization website; Figure 2-43, Based on map, "General Government Expenditure on Health as a Percentage of Total Government Expenditure (in US$), 2009" from World Health Organization website; Figure 2-45, Based on map, "Density of physicians per 1000 population" from World Health Organization website; Figure 2-47, stringer ss/Xinhua/Photoshop/Newscom; Google Earth screen on p. 74 (left): Data SIO, NOAA, U.S. Navy, NGA, GEBCO. Copyright © 2012 Cnews/ Spot Image. Image U.S. Geological Survey; Google Earth Images end of chapter: Gray Buildings. Copyright © 2008 Sanbom; Data SIO, NOAA, U.S. Navy, NGA, GEBCO; Image NOAA.

Chapter 3, Chapter openers: Nadia Mackenzie / alamy; epa european pressphoto agency b.v./Alamy; Figure 3-1, Achive Pics/Alamy; Figure 3-2, Marmaduke St. John/Alamy; Figure 3-3, © 2012 Tele Atlas. © 2012 Google ; Figure 3-5, Gary Coronado/ZUMA Press/Newscom; Figure 3-11, Based on "Russia Population Density Map" from HowStuffWorks website; Figure 3-12, RIA Novosti/Alamy; Figure 3-13, Andre Jenny/Alamy; Figure 3-14, Based on "Tears of Trails" from Wikimedia Commons; Figure 3-15, North Wind Picture Archives/Alamy; Figure 3-17, "Interregional Migration: China," by James Rubenstein, William Renwick, and Carl Dahlman, from *Introduction to Contemporary Geography*, 1st edition, 2012. Copyright © 2012 by Pearson Education. Reprinted with permission; Figure 3-18, Reflexoes sobre os Deslocamentos Populacionais no Brasil," by Luis Antonio Pinto de Oliviera and Antonio Tadeu Ribeiro de Oliveira, 2011. Copyright © 2011 by IBGE. Reprinted with permission; Figure 3-19, © MapLink/ Tele Atlas; text excerpt on p. 92 "A refugee has been forced ...", Based on UNHCR, The United Nations Refugee Agency; Figure 3-20, Travel/Stock Collection—Homer Sykes/ Alamy; Figure 3-21, Dave Stamboulis/Alamy; Figure 3-26, NASA Archive/Alamy; Figure 3-27, Mike Goldwater/Alamy; Figure 3-29, JTB Photo Communications, Inc./Alamy; Figure 3-31, Nadia Mackenzie/Alamy; Figure 3-32, © 2012 Google. Image U.S. Geological Survey. © 2012 INEGI; Figure 3-33, © 2012 Google. Gray Buildings © 2008 Sanborn; Figure 3-36, © 2012 Google. © 2012 INEGI]; Figure 3-37, © 2012 Google. © 2012 INEGI; Figure 3-38, © 2012 Google. © 2012 INEGI; Figure 3-39, Image © 2012 GeoEye. © 2012 Google,© 2012 INEGI; Figure 3-40, Aurora Photos/Alamy; Figure 3-41, David Grossman/ Alamy; Figure 3-42, Jim West/Alamy; text excerpt p. 101 "Inclined toward violent crime ...", *Becoming American*, by Thomas J. Archdeacon, 1984; Figure 3-44 (top), Directphoto.org/Alamy; Figure 3-44 (bottom), Olivier Ogeron/EPA/Landov; Google Earth Images end of chapter: © 2012 Google; © 2012 Google. Image NOAA. Image © 2012 DigitalGlobe; © 2012 Google.

Chapter 4 Chapter opener: ZUMA Press/Newscom; Figure 4-1, Maurice Joseph/Alamy; Figure 4-2, Carlo Brambilla 2/Alamy; Figure 4-3 and on p. 106, Arterra Picture Library/Alamy; Figure 4-4, Jake Lyell/Alamy; Figure 4-5 and on p. 106, Delmi Alvarez/ZUMA Press/Newscom; Figure 4-6, Everett Collection/SuperStock; Figure 4-7, Screenshots from Interactive infographic, "How Music Travels: The Evolution of Western Dance Music," from Thomson website. Reprinted with permission; text excerpt on p. 112 "Ma chiem ba thang ...", "The Functions of Folk Songs" by Cong-Huyen-Ton-Nu Nah-Trang, from *The Performing Arts: Music and Dance*, 1979; Figure 4-8, "Christian distribution.png," from Wikimedia Commons website; Figure 4-9, Dai Kurokawa/EPA/Newscom; Figure 4-10, Screenshots of interactive map, "The Landscape of Music," by Yifan Hu from AT&T website. Copyright © 2012 by AT&T. Reprinted with permission; Figure 4-11, "Music on the Tube," by Dorian Lynskey, from Transport for London website. Copyright © 2012 by Transport for London. Reprinted by permission; Figure 4-12, "World cup appearances.png" from Wikimedia Commons website; Figure 4-13 © Nathan King/Alamy; Figure 4-14, Friedrich Stark/Alamy; Figure 4-15 Reuters/Fred Ernst; Figure 4-16, (bottom) Robert Harding Picture Library Ltd./Alamy; Figure 4-16 (top) Yoan Valat/EPA/Newscom; Figure 4-17 Jacek Kadaj/Alamy; text excerpt p. 118 "Among the connections ...", Quote from *Principles de Geographie Humaine*, by Paul Vidal de la Blache, Paris: Colin, 1922, translated as *Principles of Human Geography*, New York, Holt, 1926, by Millicent Todd Bingham; Figure 4-18 (left), "Istanbul's Bostans: A Millennium of Market Gardens," by Paul J. Kadjian, from *Geographical Review*, 94 (2004): 284-304; Figure 4-18, Ayhan Altun/Alamy; Figure 4-19, Carol Bollo/Alamy; Figure 4-21, Based on data from "Demographics of Fast Food in America" from Lexicalist.com; Figure 4-23 (left), dbimages/Alamy; Figure 4-23 (right top), LatitudeStock/Alamy; Figure 4-23 (middle), Daniel Palmer/Alamy; Figure 4-23 (right bottom), Best View Stock/Alamy; Figure 4-25, Based on "Folk Housing: Key to Diffusion" by Fred Kniffen, from the *Annuals of the Association of American Geographers*, December 1965, Volume 55(4); Figure 4-27, Sketch from *The Motel in America* by John A. Jakle, Keith A. Sculle, and Jefferson S. Rogers. Baltimore: Johns Hopkins University Press, 1996; Figure 4-28, "One Television Year in the World," by Jacques Braun, from Eurodata TV WORLDWIDE website. Copyright © 2010 by Eurodata TV Worldwide. Reprinted with permission; Figure 4-29, DIZ Muenchen GmbH, Sueddeutsche Zeitung Photo/Alamy; Figure 4-30, Yaacov Dagan/Alamy; Figure 4-33, Jeremy Sutton-Hibbert/Alamy; Figure 4-35, "Twitter Facts and Figures," Infographic Siteimpulse website, 2010. Copyright © 2010-2012 by Website-Monitoring.com – website availability monitoring – Siteimpulse®. Reprinted with permission. All rights reserved; Figure 4-36, "YouTube Facts and Figures," Infographic from Siteimpulse website, 2010. Copyright ©2010-2012 by Website-Monitoring.com – website availability monitoring; Figure 4-38, Desrus Benedicte/SIPA/Newscom; text adaptation p.. 131, "Country Profiles ...", "Country Profiles," from Opennet Initiative website. Reprinted with permission. Figure 4-40 and on p. 107, Jeff Greenberg/Alamy; Figure 4-41, ZUMA Press/Newscom; Figure 4-42, Jon Arnold Images Ltd/Alamy; Figure 4-44 (left), © 2012 Google. Image © 2012 The Geoinformation Group. Image © 2012 DigitalGlobe; Figure 4-44 (right), © 2012 Google; Google Earth Images end of chapter: Image © 2012 GeoEye. © 2012 Google . U.S. Dept. of State of Geographer. © 2012 Cnes/Spot Image; ©2009 GeoBasis-DE/BKG. Image © 2012 GeoEye; Image © 2012 GeoEye. © 2012 Google. © 2012 Mapabc.com; © 2012 Google.

Chapter 5 Chapter opener: PhotoStock-Israel/Alamy; Figure 5-1, Suzanne Long/Alamy; Figure 5-2 (left) , age fotostock/SuperStock; Figure 5-2, (right), Iain Masterton/Alamy; text excerpt on p. 144, "... languages are most complex ...," Quentin Atkinson; Figure 5-7, Mary Evans Picture Library/Alamy; Figure 5-12, Annie Watson/Alamy; Figure 5-14, Nik Wheeler/Alamy; Figure 5-16, Jean-Marc Charles/AGE Fotostock; Figure 5-17, Mauricio Lima/AFP/Getty Images/Newscom; Figure 5-23 based on map by Paul Kerswill, from "Geordie's still alreet: Some accents are becoming more distinctive and others more widespread," in The Economist, June 2, 2011; Figure 5-25 and on p. 140, Lonely Planet Images/Alamy; Figure 5-26, Design Pics Inc.–RM Content/Alamy; Figure 5-28 and on p. 141, John Elk III/Alamy; Figure 5-29 (top) and on p. 140, swissworld.org; Figure 5-30, Johnny Greig/Alamy; Figure 5-31, Javier Etxerazreta/Newscom; Figure5-32,ArticImages/Alamy;Figure5-33,Hemis/Alamy;Figure5-34, Picture Contact BV/Alamy; Figure 5-35, Keith Morris/Alamy; Figure 5-36, David Lyons/AGE Fotostock; Figure 5-37, pictureproject/Alamy; Figure 5-38, John Holmes/AGE Fotostock; Figure 5-39, Hamish Trouson/Alamy; Figure 5-40, Kevin Britland/Alamy; Figure 5-41, Guillaume Horcajuelo/EPA/Newscom; Figure 5-43 Li Qiaoqiao/Newscom; Figure 5-44, redsnapper/Alamy; Figure 5-45, Megapress/Alamy; Figure 5-46 Spencer Grant/Alamy; Figure 5-47, Imagestate Media Partners Limited-Impact Photos/Alamy; Google Earth Images end of chapter: © 2012 Google; Image © 2012 DigitalGlobe; Image © 2012 Bluesky; © 2012 Google.

Chapter 6 Chapter openers: Frans Lemmens / SuperStock; n8n photo/Alamy; Imaegs & Stories / Alamy; Figure 6-1 Bob Daemmich/Alamy; Figure 6-2 and p. 180, Hung Chung Chih/Shutterstock.com; Figure 6-8, "Muslim distribution.png" by TheGreenEditor, from Wikimedia Commons website; Figure 6-9, Map, "Buddhist distribution.png," by TheGreenEditor, from Wikimedia Commons website; Figure 6-12, "Percentage of Population Practicing Traditional Tribal Religion," by Matthew White. Copyright © 2012 by Matthew White. Reprinted with permission; Figure 6-13 and on p. 180, Robert Preston Photography/Alamy; text excerpt on p. 192 from Canadian National Household Survey "What is this person's religion ...," Statistics Canada. Reproduced and distributed on an "as is" basis with the permission of Statistics Canada; Figure 6-14, Eitan Simanor/Alamy; Figure 6-15, Tengku Mohd Yusof/Alamy; Figure 6-16, Reuters/Mohammed Ameen; Figure 6-17, Aurora Photos/Alamy; Figure 6-19, "The Spread of Christianity during 200-400 A.D., and Paul's Missionary Activities," from World Religions. Pennyslvania State University website. Copyright © 2012 by Pennsylvania State University. Reprinted with permission; Figure 6-20, Based on "Age of Caliphs" map from Wikimedia Commons website, and "Spread of Islam" map from Wall-Maps.com; Figure 6-23, age fotostock/Robert Harding; Figure 6-24, maratr/Shutterstock.com; Figure 6-25, Jens Benninghofen/Alamy; Figure 6-27, Gianni Muratore/Alamy; Figure 6-28, Leonid Plotkin/Alamy; Figure 6-29, ayazad/Shutterstock.com; Figure 6-30, DC Premiumstock/Alamy; Figure 6-31, Last Refuge/Robert Harding; Figure 6-32, David Nunuk/Glow Images; Figure 6-33, age fotostock/Robert Harding; Figure 6-34, Picture Contact BV/Alamy; Figure 6-35, EmmePi Images/Alamy; Figure 6-37, www.BibleLandPictures.com; Figure 6-38, Die Bildagentur der Fotografen GmbH/Alamy; Figure 6-39, Claudia Wiens/Alamy; Figure 6-40, "Map of the Roman Catholic Dioceses in the United States of America," from The Office of Catholic Schools, Diocense of Columbus website. Reprinted with permission; Figure 6-41, World Religions Photo Library/Alamy; Figure 6-42 (left top), Jenny Matthews/Alamy; Figure 6-42 (left bottom), Marion Kaplan/Alamy; Figure 6-43 and on p. 181, Louise Batalia Duran/Alamy; Figure 6-44, Friedrich Stark/Alamy; Figure 6-45, Walter Bibikow/age fotostock ; Figure 6-46, Robert Harding Picture Library/SuperStock; Figure 6-50, Map data Copyright © 2012 by Google maps; Figure 6-52, ZUMA Wire Service/Alamy; Figure 6-54 and on p. 180, Israel images/Alamy; Figure 6-55, David Silverman/Getty Images; Google Earth Images end of chapter:, Map data Copyright © 2012 by Google maps; © 2012 Cnes/Spot Image. Image © 2012 DigitalGlobe. Image © GeoEye; © 2012 Cnes/Spot Image. Image © 2012 GeoEye. Image © 2012 DigitalGlobe; Image © GeoEye.

Chapter 7 Chapter opener: Antony Souter / Alamy; Figure 7-1, Aurora Photos/Alamy; Figure 7-2 (top left and right), Ho New/Reuters; Figure 7-2 (bottom), Obama Press Office/Newscom; Figure 7-3 and on p. 224, Hemis/Alamy; Figure 7-4, Janice Wiedel Photolibrary/Alamy; Figure 7-5, Kim Karpeles/Alamy; Figure 7-7, "Distribution of Hispanics in the United States" map, from U.S. Census Bureau; Figure 7-8, "Distribution of African Americans in the United States" map, from the U.S. Census Bureau; Figure 7-9, "Distribution of Asian Americans in the United States" map, from the Asian Nation website; Figure 7-10, "Distribution of Ethnicities in Chicago" map, from Center for Urban Research website; Figure 7-11, "Distribution of Ethnicities in Los Angeles" map, from Center for Urban Research website; Figure 7-12, North Wind Pictures Archives/Alamy; Figure 7-14, Based on Figure 7-8 from previous (10th) edition of *The Cultural Landscape*, by James M. Rubenstein; Figure 7-17 and on p. 224, Courtesy of Everett Collection/age fotostock; Figure 7-21, Elliott Erwitt/Magnum Photos; Figure 7-22, Ian Berry/Magnum Photos; Figure 7-24 and on p. 225, Reuters/Peter Jones; Figure 7-26 Jon Arnold Images Ltd./Alamy; Figure 7-28, Everett Collection Inc./Alamy; Figure 7-30, Janine Wiedel Photolibrary/Alamy; Figure 7-33, Reuters/Azad Lashkari; Figure 7-41, United States Department of Defense; Figure 7-42 (top), Nigel Chandler/Sygma/Corbis; Figure 7-42 (middle), Otto Lang/Corbis; Figure 7-42 (bottom), Aleksandar Todorovic/Shutterstock.com; Figure 7-45 and on p. 225, Picture Contact BV/Alamy; Figure 7-47, REUTERS/Str

Old; Figure 7-48, Reuters/Ismail Taxta; Figure 7-49, Jenny Matthews/Alamy; Google Earth Images end of chapter: Map data copyright © 2012 Google maps; ©AfriGIS (Pty) Ltd. © 2012 Google; © 2012 Good Earth Image,© 2012 DigitalGlobe. © 2012 Basarsoft. US Dept of State Geographer; Image © 2012 DigitalGlobe. © 2012 Google.

Chapter 8 Chapter opener: Colouria Media/Alamy; Figure 8-1, © 2012 Google. Data SIO, NOAA, U.S. Navy, NGA, GEBCO; Figure 8-2, © 2012 Google. © 2012 IGN-France. © 2012 TerraMetrics; Figure 8-6 and on p. 258, Ulana Switucha/Alamy; Figure 8-7 and on p. 258, National Geographic Image Collection/Alamy; Figure 8-15, Based on data from "Census of Population, Households and Housing, Slovenia, 31 March 2002" from RAPID REPORTS, No. 93 (Statistical Office of the Republic of Slovenia); Figure 8-16 and on p. 258, Die Bildagentur der Fotografen GmbH/Alamy; Figure 8-18, various images GmbH & Co. KG/Alamy; Figure 8-19, "Moldova Map— Political Map of Moldova," from www.ezilon.com; Figure 8-20, "Major Ethnic Groups in Central Asia" map, from Wikimedia Commons website; Figure 8-25, Jack Barker/Alamy; Figure 8-26, David R. Frazier Photolibrary, Inc./Alamy; Figure 8-27 and on p. 259, Zute Lightfoot/Alamy; Figure 8-29 David Muenker/Alamy; Figure 8-31, Franck Fotos/Alamy; Figure 8-32, age fotostock/Robert Harding; Figure 8-42, UPI/Corbis; Figure 8-44, Ian Shaw/Alamy; Figure 8-45, Steve Stock/Alamy; Figure 8-46 and p. 259, Allan Coutts/Alamy; Figure 8-47, Sean Adair/Reuters; Figure 8-48, NYC Office of Emergency Management/Getty Images; Figure 8-49, Damon Winter KRT/Newscom; Figure 8-50 and on p. 258, Picture Contact BV/Alamy; Figure 8-52, AP Photo/Martin Cleaver; Google Earth Images end of chapter: © 2012 Tele Atlas; © 2012 GIS Innovatsia. © 2012 Basarsoft. US Dept of State Geographer. © 2012 Google; © 2012 ORION-ME. US Dept of State Geographer. © 2012 Google ©2012 Cnes/Spot Image; © 2012 Google. © 2012 Mapabc.com. © 2012 GeoEye.

Chapter 9 Chapter openers: Map data Copyright © 2012 by Google Earth; David Wall / Alamy; Michael Kemp / Alamy; Figure 9-1, Data from Human Development Report 2011, United Nations Development Programme; Figure 9-23 and on p. 298, Simon Battensby/Getty Images; Figure 9-32, "Gas Production in Conventional Fields, Lower 48 States" map from Energy Information Administration website; Figure 9-36, imago stock&people/Newscom; Figure 9-44, Sue Cunningham Photographic/Alamy; Figure 9-45, Aurora Photos/Alamy; Figure 9-47, age fotostock/Robert Harding; Figure 9-48, Rob Lacey/vividstock.net/Alamy; Figure 9-49, Joerg Boethling/Alamy; Figure 9-50 and on p. 298, Pep Roig/Alamy; Figure 9-53, Ron Yue/Alamy; Figure 9-56, Alliance Images/Alamy; Figure 9-58, Based on "GDP Real Growth Rate 2007 CIA Facebook," by Sbw01f from Wikimedia Commons website, 2009; Figure 9-61 , David Grossman/Alamy; Figure 9-62, Jake Lyell/Alamy; Figure 9-63, REUTERS/Rafiquar Rahman; Google Earth Images end of chapter: Image © 2012 GeoEye. © 2012 Google; © Google; © 2012 Google; Image © 2012 GeoEye. © 2012 Google.

Chapter 10 Chapter Opener: Dinodia Photos / Alamy; Figure 10-1, Jeanne Provost/Fotolia; Figure 10-2, Ariadne Van Zandbergen/Alamy; Figure 10-5, Arman Zhenikeyev/Alamy; Figure 10-8a and on p. 344, LMR Group/Alamy; Figure 10-8b, Russ Munn/ age fotostock; Figure 10-10 and on p. 344, Sue Cunningham Photographic/Alamy; Figure 10-16, Map "Percentage Population Undemourished World Map 2009," from Wikimedia Commons website; Figure 10-19, Robert Harding World Imagery/Alamy; Figure 10-20, Michele Burgess/Alamy; Figure 10-22 and p. 345, Charles O. Cecil/Alamy; Figure 10-23, BrazilPhotos.com/Alamy; Figure 10-25, David R. Frazier Photolibrary, Inc./Alamy; Figure 10-26 and on p. 344, Robert Harding World Imagery/Alamy; Figure 10-27, Fotosearch RM/age fotostock; Figure 10-28, Mark Chivers/Glow Images; Figure 10-29, TAO Images/age fotostock; Figure 10-30, INTERFOTO/Alamy; Figure 10-31, National Geographic Image Collection/Alamy; Figure 10-33, BrazilPhotos.com/Alamy; Figure 10-34, Peter Beck/CORBIS/Glow Images; Figure 10-36, Terry Matthews/Alamy; Figure 10-41, Kathy deWitt/Alamy; Figure 10-42, Luc Novovitch/Alamy; Figure 10-45 and p. 345, Eitan Simanor/Alamy; Figure 10-47, US Marines Photo/Alamy; Figure 10-50, Jim Weber/ZUMA Press/Newscom; Figure 10-59, EggImages/Alamy; Excerpt on p. 385, "We don't want to create ...", "Africa Bites the Bullet on Genetically Modified Food Aid" by Meron Tesfa Michael, from Worldpress.org, Sept 26, 2002; Figure 10-62, "FAO Global Consumption Price Index (1990-2010)," by Food and Agriculture Organization of the United Nations; Figure 10-63, Bloomberg/Getty Images; Figure 10-65, Charlie Newham/Alamy; Figure 10-67 (top), Simon Belcher/Alamy; Figure 10-67 (bottom), NHPA/SuperStock; Google Earth Images end of chapter: Image © 2012 DigitalGlobe. Image © 2012 GeoEye . Data SIO, NOAA, U.S. Navy, NGA, GEBCO. © 2012 Google; © 2012 Google; © 2012 Mapabc.com. © 2012 Kingway Ltd. Image © 2012 DigitalGlobe. © 2012 Google; Image NOAA. Gray Building © 2008 Sanborn. Image © 2012 TerraMetrics.

Chapter 11 Chapter opener: Reuters/Bobby Yip; NASA; Figure 11-1, Wolverhampton City Council—Arts and Heritage/Alamy; Figure 11-8, H. Mark Weidman Photography/Alamy; Figure 11-10 and unnumbered photo on p. 401, Aurora Photos; Figure 11-11 and on p. 392, brt Photo/Alamy; Figure 11-14, Robert Shantz/Alamy; Figure 11-20, Ian Shipley IND/Alamy; Figure 11-25, ZUMA Wire Service/Alamy; Figure 11-27, David South/Alamy; Figure 11-28 and on p. 392, SD-BCHINA /Alamy; Figure 11-29, Stock Connection/SuperStock; Figure 11-32, NASA; Figure 11-33, Jeremy Sutton-Hibbert/Alamy; Figure 11-35 (left), Radius Images/Alamy; Figure 11-35 (right), David R. Frazier Photolibrary, Inc./Alamy; Figure 11-37, Byron Jorjorian/Alamy; Figure 11-38 and on p. 393, NASA; Figure 11-48 (left top), Courtesy of Ford/ZUMA Press/Newscom; Figure 11-48 (left bottom), Toyota/ZUMA Press/Newscom; Figure 11-48 (right middle), Courtesy of Honda/ZUMA/Newscom; Figure 11-48 (right top), ZUMA Press/Newscom; Figure 11-48 (right bottom), eVox Productions LLC/Newscom; Google Earth Images end of chapter: © 2012 Google. Image © 2012 Getmapping pic; © 2012 Google; © 2012 Mapabc.com. Image © 2012 DigitalGlobe. © Kingway Ltd. © 2012 MapKing; © 2012 Google. Texas Orthoimagery Program. Image © GeoEye.

Chapter 12 Chapter opener: Joerg Boethling/Alamy; Figure 12-1, H. Mark Weidman Photography/Alamy; Figure 12-2, Data from the World Bank website; Figure 12-3, Superstock/Glow Images; Figure 12-4, PhotoAlto/Alamy; Figure 12-5 and on p. 428, Bob Daemmrich/Alamy; Figure 12-6, Data from the U.S. Census Bureau; Figure 12-7, Data from the Human Development Reports; Figure 12-16, Robert Harding Picture Library Ltd./Alamy; Figure 12-17 and on p. 428, John Behr/Alamy; Figure 12-20, Data from Wikipedia; Figure 12-23, Not Far From/Alamy; Figure 12-27, AfriPics.com/Alamy; Figure 12-28 and on p. 428 Yves Marcoux/age fotostock; Figure 12-29 and on p. 429, Pegez/Alamy; Figure 12-30, AgStock Images, Inc./Alamy; Figure 12-31, Marion Bull/Alamy; Figure 12-33, Nik Wheeler/Alamy; Figure 12-34, Hercules Milas/Alamy; Figure 12-35 and on p. 429, Hemis/Alamy; Figure 12-37, Data from the World Bank. Google Earth images end of chapter:

Chapter 13 Chapter opener: Terese Loeb Kreuzer / Alamy; Figure 13-1, Hemis/Alamy; Figure 13-3, Andre Jenny/Alamy; Figure 13-4, Andre Jenny/Alamy; Figure 13-5, Andre Jenny/Alamy; Figure 13-6, Andre Jenny/Alamy; Figure 13-8, Megapress/Alamy; Figure 13-12, US Census Bureau, 2006-2011 American Community Survey; Figure 13-13, US Census Bureau, 2006-2011 American Community Survey; Figure 13-14, US Census Bureau, 2006-2011 American Community Survey; Figure 13-16, guichaoua/Alamy; Figure 13-17, Ian Dagnall/Alamy; Figure 13-19, Mike Goldwater/Alamy; Figure 13-21, Jacques Demarthon/AFP/Getty Images/Newscom; Figure 13-22, National Geographic Image Collection/Alamy; Figure 13-22 (left), De Agostini/Getty Images; Figure 13-22 (right), DEA/G. Dagli Orti/Getty Images; Figure 13-24, Robert Harding World Imagery/Alamy; Figure 13-25, Dorothy Alexander/Alamy; Figure 13-29b (top), Jim West/Alamy; Figure 13-29 (bottom), Universal Images Group/DeAgostini/Alamy; Figure 13-31 and on p. 458, Photofusion Picture Library/Alamy; Figure 13-34, SG cityscapes/Alamy; Figure 13-35, Rudy Balasko/Shutterstock; Figure 13-36, Alex Segre/Alamy; Figure 13-37, Nevada DMV/

ZUMA Press/Newscom; Figure 13-38 (top), Huntstock, Inc./Alamy; Figure 13-38 (bottom), Michael Doolittle/Alamy; Figure 13-40 and on p. 459, ZUMA Wire Service/Alamy; Figure 13-41, ZUMA Press/Newscom; Figure 13-43, Alex Segre/Alamy; Figure 13-44, Tom Uhlman/Alamy; Figure 13-47 (right), Data from the Houston Police Department; Figure 13-47 (left), Data from Eric's Internet Speed Bump website; Figure 13-48, Chuck Eckert/Alamy. Google Earth images end of chapter:

Afterword Figure AF-1 epa european pressphoto agency b.v. / Alamy; Figure AF-2 AFP / Getty Images; Figure AF-3 Ian Shaw / Alamy

Appendix Adapted from map by Dr. Linda Loubert, from ESRI GIS FOR ECONOMIC DEVELOPMENT, December 2010. Copyright © 2010 by Linda Loubert. Reprinted with permission.

INDEX

RapidReturn™ by Rafter
54B365037000000000232684

pulation Distribution

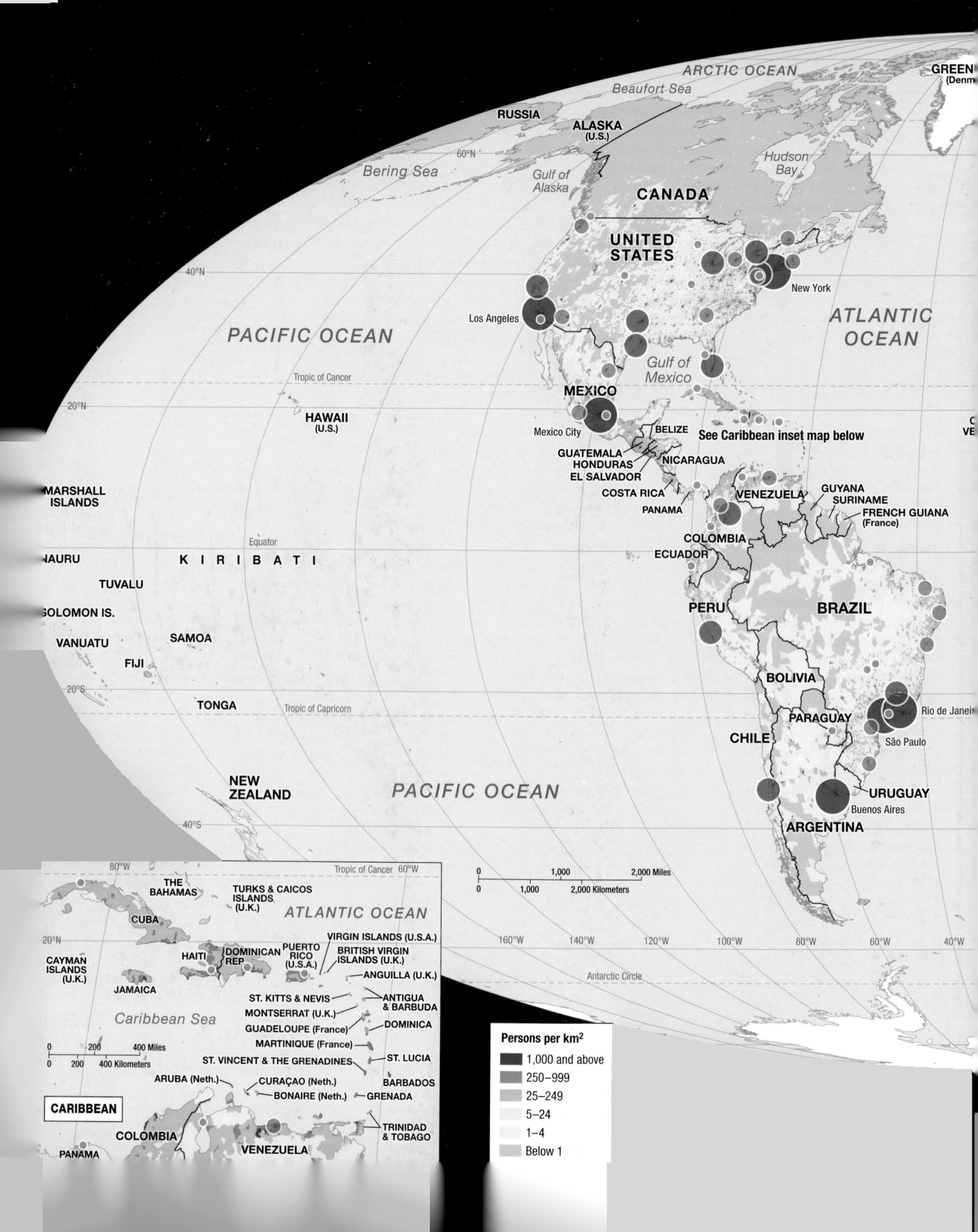